Engineering Mathematics

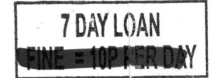

In memory of Elizabeth

Engineering Mathematics

Fifth edition

John Bird BSc(Hons), CEng, CSci, CMath, FIET, MIEE, FIIE, FIMA, FCollT

ELSEVIER

AMSTERDAM • BOSTON • HEIDELBERG • LONDON • NEW YORK • OXFORD
PARIS • SAN DIEGO • SAN FRANCISCO • SINGAPORE • SYDNEY • TOKYO

Newnes is an imprint of Elsevier

Newnes

Newnes is an imprint of Elsevier
Linacre House, Jordan Hill, Oxford OX2 8DP, UK
30 Corporate Drive, Suite 400, Burlington, MA 01803, USA

First edition 1989
Second edition 1996
Reprinted 1998 (twice), 1999
Third edition 2001
Fourth edition 2003
Reprinted 2004
Fifth edition 2007

British Library Cataloguing in Publication Data
A catalogue record for this book is available from the British Library

Library of Congress Cataloguing in Publication Data
A catalogue record for this book is available from the Library of Congress

ISBN: 978-0-75-068555-9

For information on all Newnes publications
visit our website at www.books.elsevier.com

Typeset by Charon Tec Ltd (A Macmillan Company), Chennai, India
www.charontec.com
Printed and bound in The Netherlands

7 8 9 10 11   11 10 9 8 7 6 5 4 3 2 1

Contents

Preface

Engineering Mathematics 5th Edition covers a wide range of syllabus requirements. In particular, the book is most suitable for the latest **National Certificate and Diploma courses and City & Guilds syllabuses in Engineering**.

This text will provide a foundation in mathematical principles, which will enable students to solve mathematical, scientific and associated engineering principles. In addition, the material will provide engineering applications and mathematical principles necessary for advancement onto a range of Incorporated Engineer degree profiles. It is widely recognised that a students' ability to use mathematics is a key element in determining subsequent success. First year undergraduates who need some remedial mathematics will also find this book meets their needs.

In **Engineering Mathematics 5th Edition**, new material is included on inequalities, differentiation of parametric equations, implicit and logarithmic functions and an introduction to differential equations. Because of restraints on extent, chapters on linear correlation, linear regression and sampling and estimation theories have been removed. However, **these three chapters are available to all via the internet**.

A new feature of this fifth edition is that a **free Internet download** is available of a **sample of solutions** (some 1250) of the 1750 further problems contained in the book – see below.

Another new feature is a **free Internet download (available for lecturers only) of all 500 illustrations** contained in the text – see below.

Throughout the text theory is introduced in each chapter by a simple outline of essential definitions, formulae, laws and procedures. The theory is kept to a minimum, for problem solving is extensively used to establish and exemplify the theory. It is intended that readers will gain real understanding through seeing problems solved and then through solving similar problems themselves.

For clarity, the text is divided into **eleven topic areas**, these being: number and algebra, mensuration, trigonometry, graphs, vectors, complex numbers, statistics, differential calculus, integral calculus, further number and algebra and differential equations.

This new edition covers, in particular, the following syllabuses:

(i) **Mathematics for Technicians**, the core unit for **National Certificate/Diploma** courses in Engineering, to include all or part of the following chapters:
1. **Algebraic methods:** 2, 5, 11, 13, 14, 28, 30 (1, 4, 8, 9 and 10 for revision)
2. **Trigonometric methods and areas and volumes:** 18–20, 22–25, 33, 34
3. **Statistical methods:** 37, 38
4. **Elementary calculus:** 42, 48, 55

(ii) **Further Mathematics for Technicians,** the optional unit for **National Certificate/Diploma** courses in Engineering, to include all or part of the following chapters:
1. **Advanced graphical techniques:** 29–31
2. **Algebraic techniques:** 15, 35, 38
2. **Trigonometry:** 23, 26, 27, 34
3. **Calculus:** 42–44, 48, 55–56

(iii) **The mathematical contents of Electrical and Electronic Principles units of the City & Guilds Level 3 Certificate in Engineering (2800).**

(iv) Any **introductory/access/foundation course** involving Engineering Mathematics **at University, Colleges of Further and Higher education and in schools**.

Each topic considered in the text is presented in a way that assumes in the reader little previous knowledge of that topic.

Engineering Mathematics 5th Edition provides a follow-up to **Basic Engineering Mathematics** and a lead into **Higher Engineering Mathematics 5th Edition**.

This textbook contains over **1000 worked problems**, followed by some **1750 further problems** (all **with** answers). The further problems are contained within some **220 Exercises**; each Exercise follows on directly from the relevant section of work, every two or three pages. In addition, the text contains **238 multiple-choice questions**. Where at all possible, the problems mirror practical situations found in engineering and science. **500 line diagrams** enhance the understanding of the theory.

At regular intervals throughout the text are some **18 Revision tests** to check understanding. For example, Revision test 1 covers material contained in Chapters 1 to 4, Revision test 2 covers the material in Chapters 5 to 8, and so on. These Revision Tests do not have answers given since it is envisaged that lecturers could set the tests for students to attempt as part of their course structure. Lecturers' may obtain a complimentary set of solutions of the Revision Tests in an **Instructor's Manual** available from the publishers via the internet – see below.

A list of **Essential Formulae** is included in the Instructor's Manual for convenience of reference. **Learning by Example** is at the heart of **Engineering Mathematics 5th Edition**.

JOHN BIRD
Royal Naval School of Marine Engineering,
HMS Sultan,
formerly University of Portsmouth and
Highbury College,
Portsmouth

Free web downloads

Additional material on statistics

Chapters on Linear correlation, Linear regression and Sampling and estimation theories are available for free to students and lecturers at http://books.elsevier.com/companions/9780750685559

In addition, a suite of support material is available to lecturers only from Elsevier's textbook website.

Solutions manual

Within the text are some 1750 further problems arranged within 220 Exercises. A sample of over 1250 worked solutions has been prepared for lecturers.

Instructor's manual

This manual provides full worked solutions and mark scheme for all 18 Revision Tests in this book.

Illustrations

Lecturers can also download electronic files for all illustrations in this fifth edition.

To access the lecturer support material, please go to http://textbooks.elsevier.com and search for the book. On the book web page, you will see a link to the Instructor Manual on the right. If you do not have an account for the textbook website already, you will need to register and request access to the book's subject area. If you already have an account but do not have access to the right subject area, please follow the 'Request Access to this Subject Area' link at the top of the subject area homepage.

Section 1

Number and Algebra

Chapter 1

Revision of fractions, decimals and percentages

1.1 Fractions

When 2 is divided by 3, it may be written as $\frac{2}{3}$ or 2/3 or 2/3. $\frac{2}{3}$ is called a **fraction**. The number above the line, i.e. 2, is called the **numerator** and the number below the line, i.e. 3, is called the **denominator**.

When the value of the numerator is less than the value of the denominator, the fraction is called a **proper fraction**; thus $\frac{2}{3}$ is a proper fraction. When the value of the numerator is greater than the denominator, the fraction is called an **improper fraction**. Thus $\frac{7}{3}$ is an improper fraction and can also be expressed as a **mixed number**, that is, an integer and a proper fraction. Thus the improper fraction $\frac{7}{3}$ is equal to the mixed number $2\frac{1}{3}$.

When a fraction is simplified by dividing the numerator and denominator by the same number, the process is called **cancelling**. Cancelling by 0 is not permissible.

Problem 1. Simplify $\frac{1}{3} + \frac{2}{7}$

The lowest common multiple (i.e. LCM) of the two denominators is 3×7, i.e. 21.

Expressing each fraction so that their denominators are 21, gives:

$$\frac{1}{3} + \frac{2}{7} = \frac{1}{3} \times \frac{7}{7} + \frac{2}{7} \times \frac{3}{3} = \frac{7}{21} + \frac{6}{21}$$

$$= \frac{7+6}{21} = \mathbf{\frac{13}{21}}$$

Alternatively:

$$\underset{\uparrow}{\underset{\text{Step (1)}}{\frac{1}{3} + \frac{2}{7}}} = \frac{\overset{\text{Step (2) Step (3)}}{\overset{\downarrow \qquad \downarrow}{(7 \times 1) + (3 \times 2)}}}{21}$$

Step 1: the LCM of the two denominators;

Step 2: for the fraction $\frac{1}{3}$, 3 into 21 goes 7 times, $7 \times$ the numerator is 7×1;

Step 3: for the fraction $\frac{2}{7}$, 7 into 21 goes 3 times, $3 \times$ the numerator is 3×2.

Thus $\frac{1}{3} + \frac{2}{7} = \frac{7+6}{21} = \mathbf{\frac{13}{21}}$ as obtained previously.

Problem 2. Find the value of $3\frac{2}{3} - 2\frac{1}{6}$

One method is to split the mixed numbers into integers and their fractional parts. Then

$$3\frac{2}{3} - 2\frac{1}{6} = \left(3 + \frac{2}{3}\right) - \left(2 + \frac{1}{6}\right)$$

$$= 3 + \frac{2}{3} - 2 - \frac{1}{6}$$

$$= 1 + \frac{4}{6} - \frac{1}{6} = 1\frac{3}{6} = \mathbf{1\frac{1}{2}}$$

Another method is to express the mixed numbers as improper fractions.

Since $3 = \frac{9}{3}$, then $3\frac{2}{3} = \frac{9}{3} + \frac{2}{3} = \frac{11}{3}$

Similarly, $2\frac{1}{6} = \frac{12}{6} + \frac{1}{6} = \frac{13}{6}$

Thus $3\frac{2}{3} - 2\frac{1}{6} = \frac{11}{3} - \frac{13}{6} = \frac{22}{6} - \frac{13}{6} = \frac{9}{6} = 1\frac{1}{2}$

as obtained previously.

Problem 3. Determine the value of

$$4\frac{5}{8} - 3\frac{1}{4} + 1\frac{2}{5}$$

$$4\frac{5}{8} - 3\frac{1}{4} + 1\frac{2}{5} = (4 - 3 + 1) + \left(\frac{5}{8} - \frac{1}{4} + \frac{2}{5}\right)$$

$$= 2 + \frac{5 \times 5 - 10 \times 1 + 8 \times 2}{40}$$

$$= 2 + \frac{25 - 10 + 16}{40}$$

$$= 2 + \frac{31}{40} = 2\frac{31}{40}$$

Problem 4. Find the value of $\frac{3}{7} \times \frac{14}{15}$

Dividing numerator and denominator by 3 gives:

$$\frac{{}^1\cancel{3}}{7} \times \frac{14}{\cancel{15}_5} = \frac{1}{7} \times \frac{14}{5} = \frac{1 \times 14}{7 \times 5}$$

Dividing numerator and denominator by 7 gives:

$$\frac{1 \times \cancel{14}^2}{{}_1\cancel{7} \times 5} = \frac{1 \times 2}{1 \times 5} = \frac{2}{5}$$

This process of dividing both the numerator and denominator of a fraction by the same factor(s) is called **cancelling**.

Problem 5. Evaluate $1\frac{3}{5} \times 2\frac{1}{3} \times 3\frac{3}{7}$

Mixed numbers **must** be expressed as improper fractions before multiplication can be performed. Thus,

$$1\frac{3}{5} \times 2\frac{1}{3} \times 3\frac{3}{7}$$

$$= \left(\frac{5}{5} + \frac{3}{5}\right) \times \left(\frac{6}{3} + \frac{1}{3}\right) \times \left(\frac{21}{7} + \frac{3}{7}\right)$$

$$= \frac{8}{5} \times \frac{{}^1\cancel{7}}{{}_1\cancel{3}} \times \frac{\cancel{24}^8}{\cancel{7}_1} = \frac{8 \times 1 \times 8}{5 \times 1 \times 1}$$

$$= \frac{64}{5} = 12\frac{4}{5}$$

Problem 6. Simplify $\frac{3}{7} \div \frac{12}{21}$

$$\frac{3}{7} \div \frac{12}{21} = \frac{\dfrac{3}{7}}{\dfrac{12}{21}}$$

Multiplying both numerator and denominator by the reciprocal of the denominator gives:

$$\frac{\dfrac{3}{7}}{\dfrac{12}{21}} = \frac{\dfrac{{}^1\cancel{3}}{{}_1\cancel{7}} \times \dfrac{\cancel{21}^3}{\cancel{12}_4}}{\dfrac{{}^1\cancel{12}}{{}_1\cancel{21}} \times \dfrac{\cancel{21}^1}{\cancel{12}_1}} = \frac{\dfrac{3}{4}}{1} = \frac{3}{4}$$

This method can be remembered by the rule: invert the second fraction and change the operation from division to multiplication. Thus:

$$\frac{3}{7} \div \frac{12}{21} = \frac{{}^1\cancel{3}}{{}_1\cancel{7}} \times \frac{\cancel{21}^3}{\cancel{12}_4} = \frac{3}{4} \text{ as obtained previously.}$$

Problem 7. Find the value of $5\frac{3}{5} \div 7\frac{1}{3}$

The mixed numbers must be expressed as improper fractions. Thus,

$$5\frac{3}{5} \div 7\frac{1}{3} = \frac{28}{5} \div \frac{22}{3} = \frac{{}^{14}\cancel{28}}{5} \times \frac{3}{\cancel{22}_{11}} = \frac{42}{55}$$

Problem 8. Simplify

$$\frac{1}{3} - \left(\frac{2}{5} + \frac{1}{4}\right) \div \left(\frac{3}{8} \times \frac{1}{3}\right)$$

The order of precedence of operations for problems containing fractions is the same as that for integers, i.e. remembered by **BODMAS** (**B**rackets, **O**f, **D**ivision, **M**ultiplication, **A**ddition and **S**ubtraction). Thus,

$$\frac{1}{3} - \left(\frac{2}{5} + \frac{1}{4}\right) \div \left(\frac{3}{8} \times \frac{1}{3}\right)$$

$$= \frac{1}{3} - \frac{4 \times 2 + 5 \times 1}{20} \div \frac{\cancel{3}^{\,1}}{\cancel{24}_{\,8}} \qquad (B)$$

$$= \frac{1}{3} - \frac{13}{\,_5\cancel{20}} \times \frac{\cancel{8}^{\,2}}{1} \qquad (D)$$

$$= \frac{1}{3} - \frac{26}{5} \qquad (M)$$

$$= \frac{(5 \times 1) - (3 \times 26)}{15} \qquad (S)$$

$$= \frac{-73}{15} = -4\frac{13}{15}$$

Problem 9. Determine the value of

$$\frac{7}{6} \text{ of } \left(3\frac{1}{2} - 2\frac{1}{4}\right) + 5\frac{1}{8} \div \frac{3}{16} - \frac{1}{2}$$

$$\frac{7}{6} \text{ of } \left(3\frac{1}{2} - 2\frac{1}{4}\right) + 5\frac{1}{8} \div \frac{3}{16} - \frac{1}{2}$$

$$= \frac{7}{6} \text{ of } 1\frac{1}{4} + \frac{41}{8} \div \frac{3}{16} - \frac{1}{2} \qquad (B)$$

$$= \frac{7}{6} \times \frac{5}{4} + \frac{41}{8} \div \frac{3}{16} - \frac{1}{2} \qquad (O)$$

$$= \frac{7}{6} \times \frac{5}{4} + \frac{41}{\,_1\cancel{8}} \times \frac{\cancel{16}^{\,2}}{3} - \frac{1}{2} \qquad (D)$$

$$= \frac{35}{24} + \frac{82}{3} - \frac{1}{2} \qquad (M)$$

$$= \frac{35 + 656}{24} - \frac{1}{2} \qquad (A)$$

$$= \frac{691}{24} - \frac{1}{2} \qquad (A)$$

$$= \frac{691 - 12}{24} \qquad (S)$$

$$= \frac{679}{24} = 28\frac{7}{24}$$

Now try the following exercise

Exercise 1   Further problems on fractions

Evaluate the following:

1. (a) $\dfrac{1}{2} + \dfrac{2}{5}$ (b) $\dfrac{7}{16} - \dfrac{1}{4}$

$$\left[\text{(a) } \frac{9}{10} \quad \text{(b) } \frac{3}{16} \right]$$

2. (a) $\dfrac{2}{7} + \dfrac{3}{11}$ (b) $\dfrac{2}{9} - \dfrac{1}{7} + \dfrac{2}{3}$

$$\left[\text{(a) } \frac{43}{77} \quad \text{(b) } \frac{47}{63} \right]$$

3. (a) $10\dfrac{3}{7} - 8\dfrac{2}{3}$ (b) $3\dfrac{1}{4} - 4\dfrac{4}{5} + 1\dfrac{5}{6}$

$$\left[\text{(a) } 1\frac{16}{21} \quad \text{(b) } \frac{17}{60} \right]$$

4. (a) $\dfrac{3}{4} \times \dfrac{5}{9}$ (b) $\dfrac{17}{35} \times \dfrac{15}{119}$

$$\left[\text{(a) } \frac{5}{12} \quad \text{(b) } \frac{3}{49} \right]$$

5. (a) $\dfrac{3}{5} \times \dfrac{7}{9} \times 1\dfrac{2}{7}$ (b) $\dfrac{13}{17} \times 4\dfrac{7}{11} \times 3\dfrac{4}{39}$

$$\left[\text{(a) } \frac{3}{5} \quad \text{(b) } 11 \right]$$

6. (a) $\dfrac{3}{8} \div \dfrac{45}{64}$ (b) $1\dfrac{1}{3} \div 2\dfrac{5}{9}$

$$\left[\text{(a) } \frac{8}{15} \quad \text{(b) } \frac{12}{23} \right]$$

7. $\dfrac{1}{2} + \dfrac{3}{5} \div \dfrac{8}{15} - \dfrac{1}{3}$ $\qquad \left[1\dfrac{7}{24} \right]$

8. $\dfrac{7}{15} \text{ of } \left(15 \times \dfrac{5}{7}\right) + \left(\dfrac{3}{4} \div \dfrac{15}{16}\right)$ $\qquad \left[5\dfrac{4}{5} \right]$

9. $\dfrac{1}{4} \times \dfrac{2}{3} - \dfrac{1}{3} \div \dfrac{3}{5} + \dfrac{2}{7}$ $\qquad \left[-\dfrac{13}{126} \right]$

10. $\left(\dfrac{2}{3} \times 1\dfrac{1}{4}\right) \div \left(\dfrac{2}{3} + \dfrac{1}{4}\right) + 1\dfrac{3}{5}$ $\qquad \left[2\dfrac{28}{55} \right]$

11. If a storage tank is holding 450 litres when it is three-quarters full, how much will it contain when it is two-thirds full?

$$[\text{400 litres}]$$

12. Three people, P, Q and R contribute to a fund. P provides 3/5 of the total, Q provides 2/3 of the remainder, and R provides £8. Determine (a) the total of the fund, (b) the contributions of P and Q.   [(a) £60 (b) £36, £16]

1.2   Ratio and proportion

The ratio of one quantity to another is a fraction, and is the number of times one quantity is contained in another quantity **of the same kind**. If one quantity is

directly proportional to another, then as one quantity doubles, the other quantity also doubles. When a quantity is **inversely proportional** to another, then as one quantity doubles, the other quantity is halved.

Problem 10. A piece of timber 273 cm long is cut into three pieces in the ratio of 3 to 7 to 11. Determine the lengths of the three pieces

---

The total number of parts is $3 + 7 + 11$, that is, 21. Hence 21 parts correspond to 273 cm

$$1 \text{ part corresponds to } \frac{273}{21} = 13 \text{ cm}$$

$$3 \text{ parts correspond to } 3 \times 13 = 39 \text{ cm}$$

$$7 \text{ parts correspond to } 7 \times 13 = 91 \text{ cm}$$

$$11 \text{ parts correspond to } 11 \times 13 = 143 \text{ cm}$$

i.e. **the lengths of the three pieces are 39 cm, 91 cm and 143 cm**.

(Check: $39 + 91 + 143 = 273$)

Problem 11. A gear wheel having 80 teeth is in mesh with a 25 tooth gear. What is the gear ratio?

$$\text{Gear ratio} = 80 : 25 = \frac{80}{25} = \frac{16}{5} = 3.2$$

i.e. gear ratio $= \mathbf{16 : 5}$ or $\mathbf{3.2 : 1}$

Problem 12. An alloy is made up of metals A and B in the ratio 2.5 : 1 by mass. How much of A has to be added to 6 kg of B to make the alloy?

Ratio A : B : :2.5 : 1 (i.e. A is to B as 2.5 is to 1) or
$$\frac{A}{B} = \frac{2.5}{1} = 2.5$$

When B $= 6$ kg, $\dfrac{A}{6} = 2.5$ from which,

$$A = 6 \times 2.5 = \mathbf{15 \, kg}$$

Problem 13. If 3 people can complete a task in 4 hours, how long will it take 5 people to complete the same task, assuming the rate of work remains constant

---

The more the number of people, the more quickly the task is done, hence inverse proportion exists.

3 people complete the task in 4 hours.

1 person takes three times as long, i.e.

$4 \times 3 = 12$ hours,

5 people can do it in one fifth of the time that one person takes, that is $\dfrac{12}{5}$ hours or **2 hours 24 minutes**.

Now try the following exercise

Exercise 2   Further problems on ratio and proportion

1. Divide 621 cm in the ratio of 3 to 7 to 13.
   [81 cm to 189 cm to 351 cm]

2. When mixing a quantity of paints, dyes of four different colours are used in the ratio of $7 : 3 : 19 : 5$. If the mass of the first dye used is $3\frac{1}{2}$ g, determine the total mass of the dyes used. [17 g]

3. Determine how much copper and how much zinc is needed to make a 99 kg brass ingot if they have to be in the proportions copper : zinc: :8 : 3 by mass. [72 kg : 27 kg]

4. It takes 21 hours for 12 men to resurface a stretch of road. Find how many men it takes to resurface a similar stretch of road in 50 hours 24 minutes, assuming the work rate remains constant. [5]

5. It takes 3 hours 15 minutes to fly from city A to city B at a constant speed. Find how long the journey takes if

   (a) the speed is $1\frac{1}{2}$ times that of the original speed and

   (b) if the speed is three-quarters of the original speed.
   [(a) 2 h 10 min   (b) 4 h 20 min]

1.3   Decimals

The decimal system of numbers is based on the **digits** 0 to 9. A number such as 53.17 is called a **decimal fraction**, a decimal point separating the integer part, i.e. 53, from the fractional part, i.e. 0.17.

A number which can be expressed exactly as a decimal fraction is called a **terminating decimal** and those which cannot be expressed exactly as a decimal fraction are called **non-terminating decimals**. Thus, $\frac{3}{2}=1.5$ is a terminating decimal, but $\frac{4}{3}=1.33333\ldots$ is a non-terminating decimal. $1.33333\ldots$ can be written as $1.\dot{3}$, called 'one point-three recurring'.

The answer to a non-terminating decimal may be expressed in two ways, depending on the accuracy required:

(i) correct to a number of **significant figures**, that is, figures which signify something, and
(ii) correct to a number of **decimal places**, that is, the number of figures after the decimal point.

The last digit in the answer is unaltered if the next digit on the right is in the group of numbers 0, 1, 2, 3 or 4, but is increased by 1 if the next digit on the right is in the group of numbers 5, 6, 7, 8 or 9. Thus the non-terminating decimal 7.6183... becomes 7.62, correct to 3 significant figures, since the next digit on the right is 8, which is in the group of numbers 5, 6, 7, 8 or 9. Also 7.6183... becomes 7.618, correct to 3 decimal places, since the next digit on the right is 3, which is in the group of numbers 0, 1, 2, 3 or 4.

Problem 14. Evaluate $42.7+3.04+8.7+0.06$

The numbers are written so that the decimal points are under each other. Each column is added, starting from the right.

$$\begin{array}{r} 42.7 \\ 3.04 \\ 8.7 \\ 0.06 \\ \hline 54.50 \end{array}$$

Thus $\mathbf{42.7+3.04+8.7+0.06=54.50}$

Problem 15. Take 81.70 from 87.23

The numbers are written with the decimal points under each other.

$$\begin{array}{r} 87.23 \\ -81.70 \\ \hline 5.53 \end{array}$$

Thus $\mathbf{87.23-81.70=5.53}$

Problem 16. Find the value of
$$23.4-17.83-57.6+32.68$$

The sum of the positive decimal fractions is
$$23.4+32.68=56.08$$

The sum of the negative decimal fractions is
$$17.83+57.6=75.43$$

Taking the sum of the negative decimal fractions from the sum of the positive decimal fractions gives:
$$56.08-75.43$$
i.e. $-(75.43-56.08)=\mathbf{-19.35}$

Problem 17. Determine the value of 74.3×3.8

When multiplying decimal fractions: (i) the numbers are multiplied as if they are integers, and (ii) the position of the decimal point in the answer is such that there are as many digits to the right of it as the sum of the digits to the right of the decimal points of the two numbers being multiplied together. Thus

(i)
$$\begin{array}{r} 743 \\ 38 \\ \hline 5\,944 \\ 22\,290 \\ \hline 28\,234 \end{array}$$

(ii) As there are $(1+1)=2$ digits to the right of the decimal points of the two numbers being multiplied together, $(74.\underline{3}\times3.\underline{8})$, then

$$\mathbf{74.3\times3.8=282.34}$$

Problem 18. Evaluate $37.81\div1.7$, correct to (i) 4 significant figures and (ii) 4 decimal places

$$37.81\div1.7=\frac{37.81}{1.7}$$

The denominator is changed into an integer by multiplying by 10. The numerator is also multiplied by 10 to keep the fraction the same. Thus

$$37.81\div1.7=\frac{37.81\times10}{1.7\times10}$$
$$=\frac{378.1}{17}$$

The long division is similar to the long division of integers and the first four steps are as shown:

$$
\begin{array}{r}
22.24117.. \\
17 \overline{)\,378.100000} \\
34 \\
\overline{38} \\
34 \\
\overline{41} \\
34 \\
\overline{70} \\
68 \\
\overline{20}
\end{array}
$$

(i) **37.81 ÷ 1.7 = 22.24, correct to 4 significant figures,** and

(ii) **37.81 ÷ 1.7 = 22.2412, correct to 4 decimal places.**

Problem 19. Convert (a) 0.4375 to a proper fraction and (b) 4.285 to a mixed number

(a) 0.4375 can be written as $\dfrac{0.4375 \times 10\,000}{10\,000}$ without changing its value,

i.e. $0.4375 = \dfrac{4375}{10\,000}$

By cancelling

$$\dfrac{4375}{10\,000} = \dfrac{875}{2000} = \dfrac{175}{400} = \dfrac{35}{80} = \dfrac{7}{16}$$

i.e. $\mathbf{0.4375 = \dfrac{7}{16}}$

(b) Similarly, $\mathbf{4.285 = 4\dfrac{285}{1000} = 4\dfrac{57}{200}}$

Problem 20. Express as decimal fractions:

$$\text{(a) } \dfrac{9}{16} \text{ and (b) } 5\dfrac{7}{8}$$

(a) To convert a proper fraction to a decimal fraction, the numerator is divided by the denominator. Division by 16 can be done by the long division method, or, more simply, by dividing by 2 and then 8:

$$
\begin{array}{r}
4.50 \\
2\overline{)\,9.00}
\end{array}
\qquad
\begin{array}{r}
0.5625 \\
8\overline{)\,4.5000}
\end{array}
$$

Thus $\dfrac{9}{16} = \mathbf{0.5625}$

(b) For mixed numbers, it is only necessary to convert the proper fraction part of the mixed number to a decimal fraction. Thus, dealing with the $\frac{7}{8}$ gives:

$$
\begin{array}{r}
0.875 \\
8\overline{)\,7.000}
\end{array}
\qquad \text{i.e.} \quad \dfrac{7}{8} = 0.875
$$

Thus $5\dfrac{7}{8} = \mathbf{5.875}$

Now try the following exercise

Exercise 3   Further problems on decimals

In Problems 1 to 6, determine the values of the expressions given:

1. $23.6 + 14.71 - 18.9 - 7.421$ [11.989]

2. $73.84 - 113.247 + 8.21 - 0.068$
   [−31.265]

3. $3.8 \times 4.1 \times 0.7$ [10.906]

4. 374.1×0.006 [2.2446]

5. $421.8 \div 17$, (a) correct to 4 significant figures and (b) correct to 3 decimal places.
   [(a) 24.81   (b) 24.812]

6. $\dfrac{0.0147}{2.3}$, (a) correct to 5 decimal places and
   (b) correct to 2 significant figures.
   [(a) 0.00639   (b) 0.0064]

7. Convert to proper fractions:
   (a) 0.65 (b) 0.84 (c) 0.0125 (d) 0.282 and (e) 0.024
   $$\left[\text{(a) } \dfrac{13}{20} \text{ (b) } \dfrac{21}{25} \text{ (c) } \dfrac{1}{80} \text{ (d) } \dfrac{141}{500} \text{ (e) } \dfrac{3}{125} \right]$$

8. Convert to mixed numbers:
   (a) 1.82 (b) 4.275 (c) 14.125 (d) 15.35 and (e) 16.2125
   $$\left[\begin{array}{l} \text{(a) } 1\dfrac{41}{50} \text{ (b) } 4\dfrac{11}{40} \text{ (c) } 14\dfrac{1}{8} \\ \text{(d) } 15\dfrac{7}{20} \text{ (e) } 16\dfrac{17}{80} \end{array} \right]$$

In Problems 9 to 12, express as decimal fractions to the accuracy stated:

9. $\dfrac{4}{9}$, correct to 5 significant figures.
   [0.44444]

10. $\dfrac{17}{27}$, correct to 5 decimal places.

[0.62963]

11. $1\dfrac{9}{16}$, correct to 4 significant figures.

[1.563]

12. $13\dfrac{31}{37}$, correct to 2 decimal places.

[13.84]

13. Determine the dimension marked x in the length of shaft shown in Figure 1.1. The dimensions are in millimetres.

[12.52 mm]

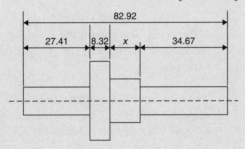

Figure 1.1

14. A tank contains 1800 litres of oil. How many tins containing 0.75 litres can be filled from this tank? [2400]

1.4 Percentages

Percentages are used to give a common standard and are fractions having the number 100 as their denominators. For example, 25 per cent means $\dfrac{25}{100}$ i.e. $\dfrac{1}{4}$ and is written 25%.

Problem 21. Express as percentages:
(a) 1.875   and   (b) 0.0125

A decimal fraction is converted to a percentage by multiplying by 100. Thus,

(a) 1.875 corresponds to $1.875 \times 100\%$, i.e. **187.5%**

(b) 0.0125 corresponds to $0.0125 \times 100\%$, i.e. **1.25%**

Problem 22. Express as percentages:
$$\text{(a) } \dfrac{5}{16} \quad \text{and} \quad \text{(b) } 1\dfrac{2}{5}$$

To convert fractions to percentages, they are (i) converted to decimal fractions and (ii) multiplied by 100

(a) By division, $\dfrac{5}{16} = 0.3125$, hence $\dfrac{5}{16}$ corresponds to $0.3125 \times 100\%$, i.e. **31.25%**

(b) Similarly, $1\dfrac{2}{5} = 1.4$ when expressed as a decimal fraction.

Hence $1\dfrac{2}{5} = 1.4 \times 100\% = \textbf{140\%}$

Problem 23. It takes 50 minutes to machine a certain part, Using a new type of tool, the time can be reduced by 15%. Calculate the new time taken

$$15\% \text{ of 50 minutes} = \dfrac{15}{100} \times 50 = \dfrac{750}{100}$$
$$= 7.5 \text{ minutes.}$$

hence the **new time taken** is

$$50 - 7.5 = \textbf{42.5 minutes.}$$

Alternatively, if the time is reduced by 15%, then it now takes 85% of the original time, i.e. 85% of $50 = \dfrac{85}{100} \times 50 = \dfrac{4250}{100} = \textbf{42.5 minutes}$, as above.

Problem 24. Find 12.5% of £378

12.5% of £378 means $\dfrac{12.5}{100} \times 378$, since per cent means 'per hundred'.

Hence 12.5% of £378 $= \dfrac{12.5^{1}}{100_{8}} \times 378 = \dfrac{1}{8} \times 378 = \dfrac{378}{8} = \textbf{£47.25}$

Problem 25. Express 25 minutes as a percentage of 2 hours, correct to the nearest 1%

Working in minute units, 2 hours $= 120$ minutes.
Hence 25 minutes is $\dfrac{25}{120}$ ths of 2 hours. By cancelling,

$$\dfrac{25}{120} = \dfrac{5}{24}$$

Expressing $\dfrac{5}{24}$ as a decimal fraction gives $0.208\dot{3}$

Multiplying by 100 to convert the decimal fraction to a percentage gives:

$$0.208\dot{3} \times 100 = 20.83\%$$

Thus **25 minutes is 21% of 2 hours,** correct to the nearest 1%.

> **Problem 26.** A German silver alloy consists of 60% copper, 25% zinc and 15% nickel. Determine the masses of the copper, zinc and nickel in a 3.74 kilogram block of the alloy

By direct proportion:

100% corresponds to 3.74 kg

1% corresponds to $\dfrac{3.74}{100} = 0.0374$ kg

60% corresponds to $60 \times 0.0374 = 2.244$ kg

25% corresponds to $25 \times 0.0374 = 0.935$ kg

15% corresponds to $15 \times 0.0374 = 0.561$ kg

Thus, the masses of the copper, zinc and nickel are **2.244 kg, 0.935 kg and 0.561 kg**, respectively.

(Check: $2.244 + 0.935 + 0.561 = 3.74$)

Now try the following exercise

Exercise 4 Further problems percentages

1. Convert to percentages:
   (a) 0.057 (b) 0.374 (c) 1.285
   $\qquad$ [(a) 5.7% (b) 37.4% (c) 128.5%]

2. Express as percentages, correct to 3 significant figures:

   (a) $\dfrac{7}{33}$ (b) $\dfrac{19}{24}$ (c) $1\dfrac{11}{16}$
   $\qquad$ [(a) 21.2% (b) 79.2% (c) 169%]

3. Calculate correct to 4 significant figures:
   (a) 18% of 2758 tonnes (b) 47% of 18.42 grams (c) 147% of 14.1 seconds
   $\qquad$ [(a) 496.4 t (b) 8.657 g (c) 20.73 s]

4. When 1600 bolts are manufactured, 36 are unsatisfactory. Determine the percentage unsatisfactory. [2.25%]

5. Express: (a) 140 kg as a percentage of 1 t (b) 47 s as a percentage of 5 min (c) 13.4 cm as a percentage of 2.5 m
   $\qquad$ [(a) 14% (b) 15.67% (c) 5.36%]

6. A block of monel alloy consists of 70% nickel and 30% copper. If it contains 88.2 g of nickel, determine the mass of copper in the block. [37.8 g]

7. A drilling machine should be set to 250 rev/min. The nearest speed available on the machine is 268 rev/min. Calculate the percentage over speed. [7.2%]

8. Two kilograms of a compound contains 30% of element A, 45% of element B and 25% of element C. Determine the masses of the three elements present.
   $\qquad$ [A 0.6 kg, B 0.9 kg, C 0.5 kg]

9. A concrete mixture contains seven parts by volume of ballast, four parts by volume of sand and two parts by volume of cement. Determine the percentage of each of these three constituents correct to the nearest 1% and the mass of cement in a two tonne dry mix, correct to 1 significant figure.
   $\qquad$ [54%, 31%, 15%, 0.3 t]

10. In a sample of iron ore, 18% is iron. How much ore is needed to produce 3600 kg of iron? [20 000 kg]

11. A screws' dimension is $12.5 \pm 8\%$ mm. Calculate the possible maximum and minimum length of the screw.
    $\qquad$ [13.5 mm, 11.5 mm]

12. The output power of an engine is 450 kW. If the efficiency of the engine is 75%, determine the power input. [600 kW]

Chapter 2

Indices, standard form and engineering notation

2.1 Indices

The lowest factors of 2000 are $2 \times 2 \times 2 \times 2 \times 5 \times 5 \times 5$. These factors are written as $2^4 \times 5^3$, where 2 and 5 are called **bases** and the numbers 4 and 5 are called **indices**.

When an index is an integer it is called a **power**. Thus, 2^4 is called 'two to the power of four', and has a base of 2 and an index of 4. Similarly, 5^3 is called 'five to the power of 3' and has a base of 5 and an index of 3.

Special names may be used when the indices are 2 and 3, these being called 'squared' and 'cubed', respectively. Thus 7^2 is called 'seven squared' and 9^3 is called 'nine cubed'. When no index is shown, the power is 1, i.e. 2 means 2^1.

Reciprocal

The **reciprocal** of a number is when the index is -1 and its value is given by 1, divided by the base. Thus the reciprocal of 2 is 2^{-1} and its value is $\frac{1}{2}$ or 0.5. Similarly, the reciprocal of 5 is 5^{-1} which means $\frac{1}{5}$ or 0.2.

Square root

The **square root** of a number is when the index is $\frac{1}{2}$, and the square root of 2 is written as $2^{1/2}$ or $\sqrt{2}$. The value of a square root is the value of the base which when multiplied by itself gives the number. Since $3 \times 3 = 9$, then $\sqrt{9} = 3$. However, $(-3) \times (-3) = 9$, so $\sqrt{9} = -3$. There are always two answers when finding the square root of a number and this is shown by putting both a $+$ and a $-$ sign in front of the answer to a square root problem. Thus $\sqrt{9} = \pm 3$ and $4^{1/2} = \sqrt{4} = \pm 2$, and so on.

Laws of indices

When simplifying calculations involving indices, certain basic rules or laws can be applied, called the **laws of indices**. These are given below.

(i) When multiplying two or more numbers having the same base, the indices are added. Thus

$$3^2 \times 3^4 = 3^{2+4} = 3^6$$

(ii) When a number is divided by a number having the same base, the indices are subtracted. Thus

$$\frac{3^5}{3^2} = 3^{5-2} = 3^3$$

(iii) When a number which is raised to a power is raised to a further power, the indices are multiplied. Thus

$$(3^5)^2 = 3^{5 \times 2} = 3^{10}$$

(iv) When a number has an index of 0, its value is 1. Thus $3^0 = 1$

(v) A number raised to a negative power is the reciprocal of that number raised to a positive power. Thus $3^{-4} = \frac{1}{3^4}$ Similarly, $\frac{1}{2^{-3}} = 2^3$

(vi) When a number is raised to a fractional power the denominator of the fraction is the root of the number and the numerator is the power.

Thus $8^{2/3} = \sqrt[3]{8^2} = (2)^2 = 4$

and $25^{1/2} = \sqrt[2]{25^1} = \sqrt{25^1} = \pm 5$

(Note that $\sqrt{} \equiv \sqrt[2]{}$)

2.2 Worked problems on indices

Problem 1. Evaluate (a) $5^2 \times 5^3$, (b) $3^2 \times 3^4 \times 3$ and (c) $2 \times 2^2 \times 2^5$

From law (i):

(a) $5^2 \times 5^3 = 5^{(2+3)} = 5^5 = 5 \times 5 \times 5 \times 5 \times 5 = \mathbf{3125}$

(b) $3^2 \times 3^4 \times 3 = 3^{(2+4+1)} = 3^7$
$$= 3 \times 3 \times \cdots \text{ to 7 terms}$$
$$= \mathbf{2187}$$

(c) $2 \times 2^2 \times 2^5 = 2^{(1+2+5)} = 2^8 = \mathbf{256}$

Problem 2. Find the value of:
$$\text{(a) } \frac{7^5}{7^3} \quad \text{and} \quad \text{(b) } \frac{5^7}{5^4}$$

From law (ii):

(a) $\dfrac{7^5}{7^3} = 7^{(5-3)} = 7^2 = \mathbf{49}$

(b) $\dfrac{5^7}{5^4} = 5^{(7-4)} = 5^3 = \mathbf{125}$

Problem 3. Evaluate: (a) $5^2 \times 5^3 \div 5^4$ and (b) $(3 \times 3^5) \div (3^2 \times 3^3)$

From laws (i) and (ii):

(a) $5^2 \times 5^3 \div 5^4 = \dfrac{5^2 \times 5^3}{5^4} = \dfrac{5^{(2+3)}}{5^4}$
$$= \frac{5^5}{5^4} = 5^{(5-4)} = 5^1 = \mathbf{5}$$

(b) $(3 \times 3^5) \div (3^2 \times 3^3) = \dfrac{3 \times 3^5}{3^2 \times 3^3} = \dfrac{3^{(1+5)}}{3^{(2+3)}}$
$$= \frac{3^6}{3^5} = 3^{(6-5)} = 3^1 = \mathbf{3}$$

Problem 4. Simplify: (a) $(2^3)^4$ (b) $(3^2)^5$, expressing the answers in index form.

From law (iii):

(a) $(2^3)^4 = 2^{3\times4} = \mathbf{2^{12}}$ (b) $(3^2)^5 = 3^{2\times5} = \mathbf{3^{10}}$

Problem 5. Evaluate: $\dfrac{(10^2)^3}{10^4 \times 10^2}$

From the laws of indices:
$$\frac{(10^2)^3}{10^4 \times 10^2} = \frac{10^{(2\times3)}}{10^{(4+2)}} = \frac{10^6}{10^6}$$
$$= 10^{6-6} = 10^0 = \mathbf{1}$$

Problem 6. Find the value of
$$\text{(a) } \frac{2^3 \times 2^4}{2^7 \times 2^5} \quad \text{and} \quad \text{(b) } \frac{(3^2)^3}{3 \times 3^9}$$

From the laws of indices:

(a) $\dfrac{2^3 \times 2^4}{2^7 \times 2^5} = \dfrac{2^{(3+4)}}{2^{(7+5)}} = \dfrac{2^7}{2^{12}} = 2^{7-12} = 2^{-5}$
$$= \frac{1}{2^5} = \mathbf{\frac{1}{32}}$$

(b) $\dfrac{(3^2)^3}{3 \times 3^9} = \dfrac{3^{2\times3}}{3^{1+9}} = \dfrac{3^6}{3^{10}} = 3^{6-10} = 3^{-4}$
$$= \frac{1}{3^4} = \mathbf{\frac{1}{81}}$$

Now try the following exercise

Exercise 5 Further problems on indices

In Problems 1 to 10, simplify the expressions given, expressing the answers in index form and with positive indices:

1. (a) $3^3 \times 3^4$ (b) $4^2 \times 4^3 \times 4^4$
   $\qquad$ [(a) 3^7 (b) 4^9]

2. (a) $2^3 \times 2 \times 2^2$ (b) $7^2 \times 7^4 \times 7 \times 7^3$
   $\qquad$ [(a) 2^6 (b) 7^{10}]

3. (a) $\dfrac{2^4}{2^3}$ (b) $\dfrac{3^7}{3^2}$
   $\qquad$ [(a) 2 (b) 3^5]

4. (a) $5^6 \div 5^3$ (b) $7^{13}/7^{10}$
   $\qquad$ [(a) 5^3 (b) 7^3]

5. (a) $(7^2)^3$ (b) $(3^3)^2$ $\qquad$ [(a) 7^6 (b) 3^6]

6.　(a) $\dfrac{2^2 \times 2^3}{2^4}$ (b) $\dfrac{3^7 \times 3^4}{3^5}$

$$[(a)\ 2\quad (b)\ 3^6]$$

7.　(a) $\dfrac{5^7}{5^2 \times 5^3}$ (b) $\dfrac{13^5}{13 \times 13^2}$

$$[(a)\ 5^2\quad (b)\ 13^2]$$

8.　(a) $\dfrac{(9 \times 3^2)^3}{(3 \times 27)^2}$ (b) $\dfrac{(16 \times 4)^2}{(2 \times 8)^3}$

$$[(a)\ 3^4\quad (b)\ 1]$$

9.　(a) $\dfrac{5^{-2}}{5^{-4}}$ (b) $\dfrac{3^2 \times 3^{-4}}{3^3}$

$$\left[(a)\ 5^2\quad (b)\ \dfrac{1}{3^5}\right]$$

10.　(a) $\dfrac{7^2 \times 7^{-3}}{7 \times 7^{-4}}$ (b) $\dfrac{2^3 \times 2^{-4} \times 2^5}{2 \times 2^{-2} \times 2^6}$

$$\left[(a)\ 7^2\quad (b)\ \dfrac{1}{2}\right]$$

2.3 Further worked problems on indices

Problem 7. Evaluate: $\dfrac{3^3 \times 5^7}{5^3 \times 3^4}$

The laws of indices only apply to terms **having the same base**. Grouping terms having the same base, and then applying the laws of indices to each of the groups independently gives:

$$\frac{3^3 \times 5^7}{5^3 \times 3^4} = \frac{3^3}{3^4} \times \frac{5^7}{5^3} = 3^{(3-4)} \times 5^{(7-3)}$$

$$= 3^{-1} \times 5^4 = \frac{5^4}{3^1} = \frac{625}{3} = \mathbf{208\frac{1}{3}}$$

Problem 8. Find the value of

$$\frac{2^3 \times 3^5 \times (7^2)^2}{7^4 \times 2^4 \times 3^3}$$

$$\frac{2^3 \times 3^5 \times (7^2)^2}{7^4 \times 2^4 \times 3^3} = 2^{3-4} \times 3^{5-3} \times 7^{2\times2-4}$$

$$= 2^{-1} \times 3^2 \times 7^0$$

$$= \frac{1}{2} \times 3^2 \times 1 = \frac{9}{2} = \mathbf{4\frac{1}{2}}$$

Problem 9. Evaluate:

(a) $4^{1/2}$ (b) $16^{3/4}$ (c) $27^{2/3}$ (d) $9^{-1/2}$

(a)　$4^{1/2} = \sqrt{4} = \mathbf{\pm 2}$

(b)　$16^{3/4} = \sqrt[4]{16^3} = (\pm 2)^3 = \mathbf{\pm 8}$

(Note that it does not matter whether the 4th root of 16 is found first or whether 16 cubed is found first — the same answer will result).

(c)　$27^{2/3} = \sqrt[3]{27^2} = (3)^2 = \mathbf{9}$

(d)　$9^{-1/2} = \dfrac{1}{9^{1/2}} = \dfrac{1}{\sqrt{9}} = \dfrac{1}{\pm 3} = \mathbf{\pm\dfrac{1}{3}}$

Problem 10. Evaluate: $\dfrac{4^{1.5} \times 8^{1/3}}{2^2 \times 32^{-2/5}}$

$$4^{1.5} = 4^{3/2} = \sqrt{4^3} = 2^3 = 8$$

$$8^{1/3} = \sqrt[3]{8} = 2,\ 2^2 = 4$$

and $\quad 32^{-2/5} = \dfrac{1}{32^{2/5}} = \dfrac{1}{\sqrt[5]{32^2}} = \dfrac{1}{2^2} = \dfrac{1}{4}$

Hence $\quad \dfrac{4^{1.5} \times 8^{1/3}}{2^2 \times 32^{-2/5}} = \dfrac{8 \times 2}{4 \times \frac{1}{4}} = \dfrac{16}{1} = \mathbf{16}$

Alternatively,

$$\frac{4^{1.5} \times 8^{1/3}}{2^2 \times 32^{-2/5}} = \frac{[(2)^2]^{3/2} \times (2^3)^{1/3}}{2^2 \times (2^5)^{-2/5}} = \frac{2^3 \times 2^1}{2^2 \times 2^{-2}}$$

$$= 2^{3+1-2-(-2)} = 2^4 = \mathbf{16}$$

Problem 11. Evaluate: $\dfrac{3^2 \times 5^5 + 3^3 \times 5^3}{3^4 \times 5^4}$

Dividing each term by the HCF (i.e. highest common factor) of the three terms, i.e. $3^2 \times 5^3$, gives:

$$\frac{3^2 \times 5^5 + 3^3 \times 5^3}{3^4 \times 5^4} = \frac{\dfrac{3^2 \times 5^5}{3^2 \times 5^3} + \dfrac{3^3 \times 5^3}{3^2 \times 5^3}}{\dfrac{3^4 \times 5^4}{3^2 \times 5^3}}$$

$$= \frac{3^{(2-2)} \times 5^{(5-3)} + 3^{(3-2)} \times 5^0}{3^{(4-2)} \times 5^{(4-3)}}$$

$$= \frac{3^0 \times 5^2 + 3^1 \times 5^0}{3^2 \times 5^1}$$

$$= \frac{1 \times 25 + 3 \times 1}{9 \times 5} = \mathbf{\frac{28}{45}}$$

Problem 12. Find the value of

$$\frac{3^2 \times 5^5}{3^4 \times 5^4 + 3^3 \times 5^3}$$

To simplify the arithmetic, each term is divided by the HCF of all the terms, i.e. $3^2 \times 5^3$. Thus

$$\frac{3^2 \times 5^5}{3^4 \times 5^4 + 3^3 \times 5^3}$$

$$= \frac{\dfrac{3^2 \times 5^5}{3^2 \times 5^3}}{\dfrac{3^4 \times 5^4}{3^2 \times 5^3} + \dfrac{3^3 \times 5^3}{3^2 \times 5^3}}$$

$$= \frac{3^{(2-2)} \times 5^{(5-3)}}{3^{(4-2)} \times 5^{(4-3)} + 3^{(3-2)} \times 5^{(3-3)}}$$

$$= \frac{3^0 \times 5^2}{3^2 \times 5^1 + 3^1 \times 5^0} = \frac{25}{45 + 3} = \mathbf{\frac{25}{48}}$$

Problem 13. Simplify: $\dfrac{\left(\dfrac{4}{3}\right)^3 \times \left(\dfrac{3}{5}\right)^{-2}}{\left(\dfrac{2}{5}\right)^{-3}}$

giving the answer with positive indices

A fraction raised to a power means that both the numerator and the denominator of the fraction are raised to that power, i.e. $\left(\dfrac{4}{3}\right)^3 = \dfrac{4^3}{3^3}$

A fraction raised to a negative power has the same value as the inverse of the fraction raised to a positive power.

Thus, $\left(\dfrac{3}{5}\right)^{-2} = \dfrac{1}{\left(\dfrac{3}{5}\right)^2} = \dfrac{1}{\dfrac{3^2}{5^2}} = 1 \times \dfrac{5^2}{3^2} = \dfrac{5^2}{3^2}$

Similarly, $\left(\dfrac{2}{5}\right)^{-3} = \left(\dfrac{5}{2}\right)^3 = \dfrac{5^3}{2^3}$

Thus, $\dfrac{\left(\dfrac{4}{3}\right)^3 \times \left(\dfrac{3}{5}\right)^{-2}}{\left(\dfrac{2}{5}\right)^{-3}} = \dfrac{\dfrac{4^3}{3^3} \times \dfrac{5^2}{3^2}}{\dfrac{5^3}{2^3}}$

$$= \frac{4^3}{3^3} \times \frac{5^2}{3^2} \times \frac{2^3}{5^3}$$

$$= \frac{(2^2)^3 \times 2^3}{3^{(3+2)} \times 5^{(3-2)}}$$

$$= \mathbf{\frac{2^9}{3^5 \times 5}}$$

Now try the following exercise

Exercise 6 Further problems on indices

In Problems 1 and 2, simplify the expressions given, expressing the answers in index form and with positive indices:

1. (a) $\dfrac{3^3 \times 5^2}{5^4 \times 3^4}$ (b) $\dfrac{7^{-2} \times 3^{-2}}{3^5 \times 7^4 \times 7^{-3}}$

$$\left[\text{(a) } \frac{1}{3 \times 5^2} \quad \text{(b) } \frac{1}{7^3 \times 3^7}\right]$$

2. (a) $\dfrac{4^2 \times 9^3}{8^3 \times 3^4}$ (b) $\dfrac{8^{-2} \times 5^2 \times 3^{-4}}{25^2 \times 2^4 \times 9^{-2}}$

$$\left[\text{(a) } \frac{3^2}{2^5} \quad \text{(b) } \frac{1}{2^{10} \times 5^2}\right]$$

3. Evaluate (a) $\left(\dfrac{1}{3^2}\right)^{-1}$ (b) $81^{0.25}$

  (c) $16^{(-1/4)}$ (d) $\left(\dfrac{4}{9}\right)^{1/2}$

$$\left[\text{(a) } 9 \quad \text{(b) } \pm 3 \quad \text{(c) } \pm\frac{1}{2} \quad \text{(d) } \pm\frac{2}{3}\right]$$

In Problems 4 to 8, evaluate the expressions given.

4. $\dfrac{9^2 \times 7^4}{3^4 \times 7^4 + 3^3 \times 7^2}$ $\left[\dfrac{147}{148}\right]$

5. $\dfrac{(2^4)^2 - 3^{-2} \times 4^4}{2^3 \times 16^2}$ $\left[\dfrac{1}{9}\right]$

6. $\dfrac{\left(\dfrac{1}{2}\right)^3 - \left(\dfrac{2}{3}\right)^{-2}}{\left(\dfrac{3}{5}\right)^2}$ $\left[-5\dfrac{65}{72}\right]$

7. $\dfrac{\left(\frac{4}{3}\right)^4}{\left(\frac{2}{9}\right)^2}$ [64]

8. $\dfrac{(3^2)^{3/2} \times (8^{1/3})^2}{(3)^2 \times (4^3)^{1/2} \times (9)^{-1/2}}$ $\left[4\frac{1}{2}\right]$

2.4 Standard form

A number written with one digit to the left of the decimal point and multiplied by 10 raised to some power is said to be written in **standard form**. Thus: 5837 is written as 5.837×10^3 in standard form, and 0.0415 is written as 4.15×10^{-2} in standard form.

When a number is written in standard form, the first factor is called the **mantissa** and the second factor is called the **exponent**. Thus the number 5.8×10^3 has a mantissa of 5.8 and an exponent of 10^3.

(i) Numbers having the same exponent can be added or subtracted in standard form by adding or subtracting the mantissae and keeping the exponent the same. Thus:

$$2.3 \times 10^4 + 3.7 \times 10^4$$
$$= (2.3 + 3.7) \times 10^4 = 6.0 \times 10^4$$

and $5.9 \times 10^{-2} - 4.6 \times 10^{-2}$

$$= (5.9 - 4.6) \times 10^{-2} = 1.3 \times 10^{-2}$$

When the numbers have different exponents, one way of adding or subtracting the numbers is to express one of the numbers in non-standard form, so that both numbers have the same exponent. Thus:

$$2.3 \times 10^4 + 3.7 \times 10^3$$
$$= 2.3 \times 10^4 + 0.37 \times 10^4$$
$$= (2.3 + 0.37) \times 10^4 = 2.67 \times 10^4$$

Alternatively,

$$2.3 \times 10^4 + 3.7 \times 10^3$$
$$= 23\,000 + 3700 = 26\,700$$
$$= 2.67 \times 10^4$$

(ii) The laws of indices are used when multiplying or dividing numbers given in standard form. For example,

$$(2.5 \times 10^3) \times (5 \times 10^2)$$
$$= (2.5 \times 5) \times (10^{3+2})$$
$$= 12.5 \times 10^5 \text{ or } 1.25 \times 10^6$$

Similarly,

$$\frac{6 \times 10^4}{1.5 \times 10^2} = \frac{6}{1.5} \times (10^{4-2}) = 4 \times 10^2$$

2.5 Worked problems on standard form

Problem 14. Express in standard form:
(a) 38.71 (b) 3746 (c) 0.0124

For a number to be in standard form, it is expressed with only one digit to the left of the decimal point. Thus:

(a) 38.71 must be divided by 10 to achieve one digit to the left of the decimal point and it must also be multiplied by 10 to maintain the equality, i.e.

$$38.71 = \frac{38.71}{10} \times 10 = \textbf{3.871} \times \textbf{10 in standard form}$$

(b) $3746 = \dfrac{3746}{1000} \times 1000 = \textbf{3.746} \times \textbf{10}^3$ in standard form

(c) $0.0124 = 0.0124 \times \dfrac{100}{100} = \dfrac{1.24}{100}$

$$= \textbf{1.24} \times \textbf{10}^{-2} \text{ in standard form}$$

Problem 15. Express the following numbers, which are in standard form, as decimal numbers:
(a) 1.725×10^{-2} (b) 5.491×10^4 (c) 9.84×10^0

(a) $1.725 \times 10^{-2} = \dfrac{1.725}{100} = \textbf{0.01725}$

(b) $5.491 \times 10^4 = 5.491 \times 10\,000 = \textbf{54\,910}$

(c) $9.84 \times 10^0 = 9.84 \times 1 = \textbf{9.84}$ (since $10^0 = 1$)

Problem 16. Express in standard form, correct to 3 significant figures:

(a) $\dfrac{3}{8}$ (b) $19\dfrac{2}{3}$ (c) $741\dfrac{9}{16}$

(a) $\dfrac{3}{8} = 0.375$, and expressing it in standard form gives: $0.375 = \mathbf{3.75 \times 10^{-1}}$

(b) $19\dfrac{2}{3} = 19.\dot{6} = \mathbf{1.97 \times 10}$ in standard form, correct to 3 significant figures

(c) $741\dfrac{9}{16} = 741.5625 = \mathbf{7.42 \times 10^2}$ in standard form, correct to 3 significant figures

Problem 17. Express the following numbers, given in standard form, as fractions or mixed numbers: (a) 2.5×10^{-1} (b) 6.25×10^{-2} (c) 1.354×10^2

(a) $2.5 \times 10^{-1} = \dfrac{2.5}{10} = \dfrac{25}{100} = \dfrac{\mathbf{1}}{\mathbf{4}}$

(b) $6.25 \times 10^{-2} = \dfrac{6.25}{100} = \dfrac{625}{10\,000} = \dfrac{\mathbf{1}}{\mathbf{16}}$

(c) $1.354 \times 10^2 = 135.4 = 135\dfrac{4}{10} = \mathbf{135}\dfrac{\mathbf{2}}{\mathbf{5}}$

Now try the following exercise

Exercise 7   Further problems on standard form

In Problems 1 to 4, express in standard form:

1. (a) 73.9   (b) 28.4   (c) 197.72

   $\begin{bmatrix}(a)\ 7.39 \times 10 & (b)\ 2.84 \times 10 \\ (c)\ 1.9772 \times 10^2 \end{bmatrix}$

2. (a) 2748   (b) 33 170   (c) 274 218

   $\begin{bmatrix}(a)\ 2.748 \times 10^3 & (b)\ 3.317 \times 10^4 \\ (c)\ 2.74218 \times 10^5 \end{bmatrix}$

3. (a) 0.2401   (b) 0.0174   (c) 0.00923

   $\begin{bmatrix}(a)\ 2.401 \times 10^{-1} & (b)\ 1.74 \times 10^{-2} \\ (c)\ 9.23 \times 10^{-3} \end{bmatrix}$

4. (a) $\dfrac{1}{2}$ (b) $11\dfrac{7}{8}$ (c) $130\dfrac{3}{5}$ (d) $\dfrac{1}{32}$

   $\begin{bmatrix}(a)\ 5 \times 10^{-1} & (b)\ 1.1875 \times 10 \\ (c)\ 1.306 \times 10^2 & (d)\ 3.125 \times 10^{-2} \end{bmatrix}$

In Problems 5 and 6, express the numbers given as integers or decimal fractions:

5. (a) 1.01×10^3 (b) 9.327×10^2
    (c) 5.41×10^4 (d) 7×10^0

   [(a) 1010 (b) 932.7 (c) 54 100 (d) 7]

6. (a) 3.89×10^{-2} (b) 6.741×10^{-1}
    (c) 8×10^{-3}

   [(a) 0.0389 (b) 0.6741 (c) 0.008]

2.6 Further worked problems on standard form

Problem 18. Find the value of:

(a) $7.9 \times 10^{-2} - 5.4 \times 10^{-2}$
(b) $8.3 \times 10^3 + 5.415 \times 10^3$ and
(c) $9.293 \times 10^2 + 1.3 \times 10^3$
   expressing the answers in standard form.

Numbers having the same exponent can be added or subtracted by adding or subtracting the mantissae and keeping the exponent the same. Thus:

(a) $7.9 \times 10^{-2} - 5.4 \times 10^{-2}$

$= (7.9 - 5.4) \times 10^{-2} = \mathbf{2.5 \times 10^{-2}}$

(b) $8.3 \times 10^3 + 5.415 \times 10^3$

$= (8.3 + 5.415) \times 10^3 = 13.715 \times 10^3$

$= \mathbf{1.3715 \times 10^4}$ in standard form

(c) Since only numbers having the same exponents can be added by straight addition of the mantissae, the numbers are converted to this form before adding. Thus:

$$9.293 \times 10^2 + 1.3 \times 10^3$$

$$= 9.293 \times 10^2 + 13 \times 10^2$$

$$= (9.293 + 13) \times 10^2$$

$$= 22.293 \times 10^2 = \mathbf{2.2293 \times 10^3}$$

in standard form.

Alternatively, the numbers can be expressed as decimal fractions, giving:

$$9.293 \times 10^2 + 1.3 \times 10^3$$

$$= 929.3 + 1300 = 2229.3$$

$$= \mathbf{2.2293 \times 10^3}$$

in standard form as obtained previously. This method is often the 'safest' way of doing this type of problem.

Problem 19. Evaluate (a) $(3.75 \times 10^3)(6 \times 10^4)$ and (b) $\dfrac{3.5 \times 10^5}{7 \times 10^2}$ expressing answers in standard form

---

(a) $(3.75 \times 10^3)(6 \times 10^4) = (3.75 \times 6)(10^{3+4})$

$$= 22.50 \times 10^7$$

$$= \mathbf{2.25 \times 10^8}$$

(b) $\dfrac{3.5 \times 10^5}{7 \times 10^2} = \dfrac{3.5}{7} \times 10^{5-2}$

$$= 0.5 \times 10^3 = \mathbf{5 \times 10^2}$$

Now try the following exercise

Exercise 8 Further problems on standard form

In Problems 1 to 4, find values of the expressions given, stating the answers in standard form:

1. (a) $3.7 \times 10^2 + 9.81 \times 10^2$
   (b) $1.431 \times 10^{-1} + 7.3 \times 10^{-1}$
   [(a) 1.351×10^3 (b) 8.731×10^{-1}]

2. (a) $4.831 \times 10^2 + 1.24 \times 10^3$
   (b) $3.24 \times 10^{-3} - 1.11 \times 10^{-4}$
   [(a) 1.7231×10^3 (b) 3.129×10^{-3}]

3. (a) $(4.5 \times 10^{-2})(3 \times 10^3)$
   (b) $2 \times (5.5 \times 10^4)$
   [(a) 1.35×10^2 (b) 1.1×10^5]

4. (a) $\dfrac{6 \times 10^{-3}}{3 \times 10^{-5}}$ (b) $\dfrac{(2.4 \times 10^3)(3 \times 10^{-2})}{(4.8 \times 10^4)}$
   [(a) 2×10^2 (b) 1.5×10^{-3}]

5. Write the following statements in standard form:
   (a) The density of aluminium is $2710\,kg\,m^{-3}$
   [$2.71 \times 10^3\,kg\,m^{-3}$]
   (b) Poisson's ratio for gold is 0.44
   [4.4×10^{-1}]
   (c) The impedance of free space is $376.73\,\Omega$
   [$3.7673 \times 10^2\,\Omega$]
   (d) The electron rest energy is 0.511 MeV
   [5.11×10^{-1} MeV]
   (e) Proton charge-mass ratio is $9\,5789\,700\,C\,kg^{-1}$
   [$9.57897 \times 10^7\,C\,kg^{-1}$]
   (f) The normal volume of a perfect gas is $0.02241\,m^3\,mol^{-1}$
   [$2.241 \times 10^{-2}\,m^3\,mol^{-1}$]

2.7 Engineering notation and common prefixes

Engineering notation is similar to scientific notation except that the power of ten is always a multiple of 3.

For example, $0.00035 = 3.5 \times 10^{-4}$ in scientific notation,

but $0.00035 = 0.35 \times 10^{-3}$ or 350×10^{-6} in engineering notation.

Units used in engineering and science may be made larger or smaller by using **prefixes** that denote multiplication or division by a particular amount. The eight most common multiples, with their meaning, are listed in Table 2.1, where it is noticed that the prefixes involve powers of ten which are all multiples of 3:
For example,

5 MV means $5 \times 1\,000\,000 = 5 \times 10^6$
$$= \mathbf{5\,000\,000\ volts}$$

3.6 kΩ means $3.6 \times 1000 = 3.6 \times 10^3$
$$= \mathbf{3600\ ohms}$$

Table 2.1

Prefix	Name	Meaning	
T	tera	multiply by 1 000 000 000 000	(i.e. $\times 10^{12}$)
G	giga	multiply by 1 000 000 000	(i.e. $\times 10^{9}$)
M	mega	multiply by 1 000 000	(i.e. $\times 10^{6}$)
k	kilo	multiply by 1000	(i.e. $\times 10^{3}$)
m	milli	divide by 1000	(i.e. $\times 10^{-3}$)
μ	micro	divide by 1 000 000	(i.e. $\times 10^{-6}$)
n	nano	divide by 1 000 000 000	(i.e. $\times 10^{-9}$)
p	pico	divide by 1 000 000 000 000	(i.e. $\times 10^{-12}$)

7.5 μC means $7.5 \div 1\,000\,000 = \dfrac{7.5}{10^6}$ or

$$7.5 \times 10^{-6} = \textbf{0.0000075 coulombs}$$

and **4 mA** means 4×10^{-3} or $= \dfrac{4}{10^3}$

$$= \dfrac{4}{1000} = \textbf{0.004 amperes}$$

Similarly,

$$\textbf{0.00006 J} = \textbf{0.06 mJ or 60 μJ}$$
$$\textbf{5 620 000 N} = \textbf{5620 kN or 5.62 MN}$$
$$\textbf{47} \times \textbf{10}^4\,\Omega = 470\,000\,\Omega = \textbf{470 kΩ or 0.47 MΩ}$$

and $\textbf{12} \times \textbf{10}^{-5}\textbf{A} = 0.00012\,A = \textbf{0.12 mA or 120 μA}$

A calculator is needed for many engineering calculations, and having a calculator which has an 'EXP' and 'ENG' function is most helpful.
For example, to calculate: $3 \times 10^4 \times 0.5 \times 10^{-6}$ volts, input your calculator in the following order:
(a) Enter '3' (b) Press 'EXP' (or $\times 10^x$) (c) Enter '4' (d) Press '×' (e) Enter '0.5' (f) Press 'EXP' (or $\times 10^x$) (g) Enter '−6' (h) Press '='
The answer is **0.015 V** $\left(\text{or } \dfrac{7}{200}\right)$ Now press the 'ENG' button, and the answer changes to **15 × 10⁻³ V**.
The 'ENG' or 'Engineering' button ensures that the value is stated to a power of 10 that is a multiple of 3, enabling you, in this example, to express the answer as **15 mV**.

Now try the following exercise

Exercise 9 Further problems on engineering notation and common prefixes

1. Express the following in engineering notation and in prefix form:
   (a) 100 000 W (b) 0.00054 A
   (c) $15 \times 10^5\,\Omega$ (d) 225×10^{-4} V
   (e) 35 000 000 000 Hz (f) 1.5×10^{-11} F
   (g) 0.000017 A (h) 46200 Ω
   [(a) 100 kW (b) 0.54 mA or 540 μA
   (c) 1.5 MΩ (d) 22.5 mV (e) 35 GHz (f) 15 pF
   (g) 17 μA (h) 46.2 kΩ]

2. Rewrite the following as indicated:
   (a) 0.025 mA = ...μA
   (b) 1000 pF = ...nF
   (c) 62×10^4 V = ...kV
   (d) 1 250 000 Ω = ...MΩ
   [(a) 25 μA (b) 1 nF (c) 620 kV (d) 1.25 MΩ]

3. Use a calculator to evaluate the following in engineering notation:
   (a) $4.5 \times 10^{-7} \times 3 \times 10^4$
   (b) $\dfrac{(1.6 \times 10^{-5})(25 \times 10^3)}{(100 \times 10^6)}$
   [(a) 13.5×10^{-3} (b) 4×10^3]

Chapter 3

Computer numbering systems

The system of numbers in everyday use is the **denary** or **decimal** system of numbers, using the digits 0 to 9. It has ten different digits (0, 1, 2, 3, 4, 5, 6, 7, 8 and 9) and is said to have a **radix** or **base** of 10.

The **binary** system of numbers has a radix of 2 and uses only the digits 0 and 1.

3.2 Conversion of binary to decimal

The decimal number 234.5 is equivalent to

$$2 \times 10^2 + 3 \times 10^1 + 4 \times 10^0 + 5 \times 10^{-1}$$

i.e. is the sum of term comprising: (a digit) multiplied by (the base raised to some power).

In the binary system of numbers, the base is 2, so 1101.1 is equivalent to:

$$1 \times 2^3 + 1 \times 2^2 + 0 \times 2^1 + 1 \times 2^0 + 1 \times 2^{-1}$$

Thus the decimal number equivalent to the binary number 1101.1 is

$$8 + 4 + 0 + 1 + \frac{1}{2}, \text{ that is } 13.5$$

i.e. $\mathbf{1101.1_2 = 13.5_{10}}$, the suffixes 2 and 10 denoting binary and decimal systems of number respectively.

Problem 1. Convert 11011_2 to a decimal number

From above: $11011_2 = 1 \times 2^4 + 1 \times 2^3 + 0 \times 2^2$
$$+ 1 \times 2^1 + 1 \times 2^0$$
$$= 16 + 8 + 0 + 2 + 1$$
$$= \mathbf{27_{10}}$$

Problem 2. Convert 0.1011_2 to a decimal fraction

$$0.1011_2 = 1 \times 2^{-1} + 0 \times 2^{-2} + 1 \times 2^{-3}$$
$$+ 1 \times 2^{-4}$$
$$= 1 \times \frac{1}{2} + 0 \times \frac{1}{2^2} + 1 \times \frac{1}{2^3}$$
$$+ 1 \times \frac{1}{2^4}$$
$$= \frac{1}{2} + \frac{1}{8} + \frac{1}{16}$$
$$= 0.5 + 0.125 + 0.0625$$
$$= \mathbf{0.6875_{10}}$$

Problem 3. Convert 101.0101_2 to a decimal number

$$101.0101_2 = 1 \times 2^2 + 0 \times 2^1 + 1 \times 2^0$$
$$+ 0 \times 2^{-1} + 1 \times 2^{-2}$$
$$+ 0 \times 2^{-3} + 1 \times 2^{-4}$$
$$= 4 + 0 + 1 + 0 + 0.25$$
$$+ 0 + 0.0625$$
$$= \mathbf{5.3125_{10}}$$

Now try the following exercise

Exercise 10 Further problems on conversion of binary to decimal numbers

In Problems 1 to 4, convert the binary number given to decimal numbers.

1. (a) 110 (b) 1011 (c) 1110 (d) 1001

   $[$(a) 6_{10} (b) 11_{10} (c) 14_{10} (d) $9_{10}]$

2. (a) 10101 (b) 11001 (c) 101101 (d) 110011

   $[$(a) 21_{10} (b) 25_{10} (c) 45_{10} (d) $51_{10}]$

3. (a) 0.1101 (b) 0.11001 (c) 0.00111 (d) 0.01011

   $$\begin{bmatrix} \text{(a) } 0.8125_{10} & \text{(b) } 0.78125_{10} \\ \text{(c) } 0.21875_{10} & \text{(d) } 0.34375_{10} \end{bmatrix}$$

4. (a) 11010.11 (b) 10111.011 (c) 110101.0111 (d) 11010101.10111

   $$\begin{bmatrix} \text{(a) } 26.75_{10} & \text{(b) } 23.375_{10} \\ \text{(c) } 53.4375_{10} & \text{(d) } 213.71875_{10} \end{bmatrix}$$

3.3 Conversion of decimal to binary

An integer decimal number can be converted to a corresponding binary number by repeatedly dividing by 2 and noting the remainder at each stage, as shown below for 39_{10}

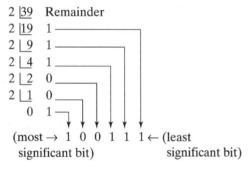

(most → 1 0 0 1 1 1 ← (least
significant bit)   significant bit)

The result is obtained by writing the top digit of the remainder as the least significant bit, (a bit is a **b**inary dig**it** and the least significant bit is the one on the right). The bottom bit of the remainder is the most significant bit, i.e. the bit on the left.

Thus $39_{10} = 100111_2$

The fractional part of a decimal number can be converted to a binary number by repeatedly multiplying by 2, as shown below for the fraction 0.625

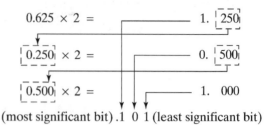

(most significant bit) .1 0 1 (least significant bit)

For fractions, the most significant bit of the result is the top bit obtained from the integer part of multiplication by 2. The least significant bit of the result is the bottom bit obtained from the integer part of multiplication by 2.

Thus $0.625_{10} = 0.101_2$

Problem 4. Convert 47_{10} to a binary number

From above, repeatedly dividing by 2 and noting the remainder gives:

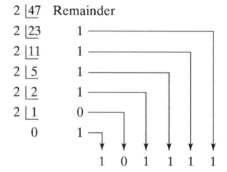

Thus $47_{10} = 101111_2$

Problem 5. Convert 0.40625_{10} to a binary number

From above, repeatedly multiplying by 2 gives:

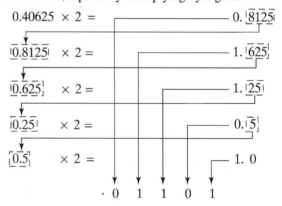

i.e. $\mathbf{040625_{10} = 0.01101_2}$

Problem 6. Convert 58.3125_{10} to a binary number

The integer part is repeatedly divided by 2, giving:

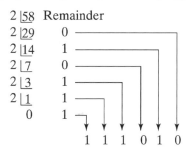

The fractional part is repeatedly multiplied by 2 giving:

$$0.3125 \times 2 = \qquad 0.625$$
$$0.625 \times 2 = \qquad 1.25$$
$$0.25 \times 2 = \qquad 0.5$$
$$0.5 \times 2 = \qquad 1.0$$
$$\qquad .0\ 1\ 0\ 1$$

Thus $58.3125_{10} = 111010.0101_2$

Now try the following exercise

Exercise 11   Further problems on conversion of decimal to binary numbers

In Problem 1 to 4, convert the decimal numbers given to binary numbers.

1. (a) 5 (b) 15 (c) 19 (d) 29

$$\begin{bmatrix} \text{(a) } 101_2 & \text{(b) } 1111_2 \\ \text{(c) } 10011_2 & \text{(d) } 11101_2 \end{bmatrix}$$

2. (a) 31 (b) 42 (c) 57 (d) 63

$$\begin{bmatrix} \text{(a) } 11111_2 & \text{(b) } 101010_2 \\ \text{(c) } 111001_2 & \text{(d) } 111111_2 \end{bmatrix}$$

3. (a) 0.25 (b) 0.21875 (c) 0.28125
   (d) 0.59375

$$\begin{bmatrix} \text{(a) } 0.01_2 & \text{(b) } 0.00111_2 \\ \text{(c) } 0.01001_2 & \text{(d) } 0.10011_2 \end{bmatrix}$$

4. (a) 47.40625 (b) 30.8125
   (c) 53.90625 (d) 61.65625

$$\begin{bmatrix} \text{(a) } 101111.01101_2 & \text{(b) } 11110.1101_2 \\ \text{(c) } 110101.11101_2 & \text{(d) } 111101.10101_2 \end{bmatrix}$$

3.4   Conversion of decimal to binary via octal

For decimal integers containing several digits, repeatedly dividing by 2 can be a lengthy process. In this case, it is usually easier to convert a decimal number to a binary number via the octal system of numbers. This system has a radix of 8, using the digits 0, 1, 2, 3, 4, 5, 6 and 7. The denary number equivalent to the octal number 4317_8 is

$$4 \times 8^3 + 3 \times 8^2 + 1 \times 8^1 + 7 \times 8^0$$

i.e. $4 \times 512 + 3 \times 64 + 1 \times 8 + 7 \times 1$ or 2255

An integer decimal number can be converted ﬨ responding octal number by repeatedly divi and noting the remainder at each stage, as shown for 493_{10}

$$\begin{array}{r|l}
8 \underline{|493} & \text{Remainder} \\
8 \underline{|\ 61} & 5 \\
8 \underline{|\ \ 7} & 5 \\
0 & 7
\end{array}$$

$$7\ 5\ 5$$

Thus $493_{10} = 755_8$

The fractional part of a decimal number can be converted to an octal number by repeatedly multiplying by 8, as shown below for the fraction 0.4375_{10}

$$0.4375 \times 8 = \quad 3 . \lceil 5 \rceil$$
$$\lceil 0.5 \rceil \times 8 = \quad 4 . 0$$
$$.3\ 4$$

For fractions, the most significant bit is the top integer obtained by multiplication of the decimal fraction by 8, thus

$$0.4375_{10} = 0.34_8$$

The natural binary code for digits 0 to 7 is shown in Table 3.1, and an octal number can be converted to a binary number by writing down the three bits corresponding to the octal digit.

Thus $437_8 = 100\ 011\ 111_2$

and $26.35_8 = 010\ 110.011\ 101_2$

Table 3.1

Octal digit	Natural binary number
0	000
1	001
2	010
3	011
4	100
5	101
6	110
7	111

The '0' on the extreme left does not signify anything, thus $26.35_8 = 10\ 110.011\ 101_2$

Conversion of decimal to binary via octal is demonstrated in the following worked problems.

Problem 7. Convert 3714_{10} to a binary number, via octal

Dividing repeatedly by 8, and noting the remainder gives:

```
8 │3714    Remainder
8 │ 464      2 ────────────┐
8 │  58      0 ─────────┐   │
8 │   7      2 ──────┐   │   │
    0        7 ───┐   │   │   │
                  ↓   ↓   ↓   ↓
                  7   2   0   2
```

From Table 3.1, $7202_8 = 111\ 010\ 000\ 010_2$

i.e. $\mathbf{3714_{10} = 111\ 010\ 000\ 010_2}$

Problem 8. Convert 0.59375_{10} to a binary number, via octal

Multiplying repeatedly by 8, and noting the integer values, gives:

$$0.59375 \times 8 = \quad\text{4.75}$$
$$0.75 \quad \times 8 = \quad\text{6.00}$$
$$.4\ 6$$

Thus $0.59375_{10} = 0.46_8$

From Table 3.1. $\qquad 0.46_8 = 0.100\ 110_2$

i.e. $\qquad \mathbf{0.59375_{10} = 0.100\ 11_2}$

Problem 9. Convert 5613.90625_{10} to a binary number, via octal

The integer part is repeatedly divided by 8, noting the remainder, giving:

```
8 │5613    Remainder
8 │ 701      5 ──────────────┐
8 │  87      5 ───────────┐   │
8 │  10      7 ────────┐   │   │
8 │   1      2 ─────┐   │   │   │
    0        1 ──┐   │   │   │   │
                 ↓   ↓   ↓   ↓   ↓
                 1   2   7   5   5
```

This octal number is converted to a binary number, (see Table 3.1)

$$12755_8 = 001\ 010\ 111\ 101\ 101_2$$

i.e. $5613_{10} = 1\ 010\ 111\ 101\ 101_2$

The fractional part is repeatedly multiplied by 8, and noting the integer part, giving:

$$0.90625 \times 8 = \quad\text{7.25}$$
$$0.25 \quad \times 8 = \quad\text{2.00}$$
$$.7\ 2$$

This octal fraction is converted to a binary number, (see Table 3.1)

$$0.72_8 = 0.111\ 010_2$$

i.e. $0.90625_{10} = 0.111\ 01_2$

Thus, $5613.90625_{10} = 1\ 010\ 111\ 101\ 101.111\ 01_2$

Problem 10. Convert $11\ 110\ 011.100\ 01_2$ to a decimal number via octal

Grouping the binary number in three's from the binary point gives: $011\ 110\ 011.100\ 010_2$

Using Table 3.1 to convert this binary number to an octal number gives: 363.42_8 and

$$363.42_8 = 3 \times 8^2 + 6 \times 8^1 + 3 \times 8^0$$
$$+ 4 \times 8^{-1} + 2 \times 8^{-2}$$
$$= 192 + 48 + 3 + 0.5 + 0.03125$$
$$= 243.53125_{10}$$

Now try the following exercise

Exercise 12   Further problems on conversion between decimal and binary numbers via octal

In Problems 1 to 3, convert the decimal numbers given to binary numbers, via octal.

1. (a) 343   (b) 572   (c) 1265

> (a) 101010111_2 (b) 1000111100_2
> (c) 10011110001_2

2. (a) 0.46875   (b) 0.6875   (c) 0.71875

> (a) 0.01111_2 (b) 0.1011_2
> (c) 0.10111_2

3. (a) 247.09375   (b) 514.4375

   (c) 1716.78125

> (a) 11110111.00011_2
> (b) 1000000010.0111_2
> (c) $11010110100.11.001_2$

4. Convert the following binary numbers to decimal numbers via octal:

   (a) 111.011 1   (b) 101 001.01

   (c) 1 110 011 011 010.001 1

> (a) 7.4375_{10} (b) 41.25_{10}
> (c) 7386.1875_{10}

3.5   Hexadecimal numbers

The complexity of computers requires higher order numbering systems such as octal (base 8) and hexadecimal (base 16), which are merely extensions of the binary system. A **hexadecimal numbering system** has a radix of 16 and uses the following 16 distinct digits:

0, 1, 2, 3, 4, 5, 6, 7, 8, 9, A, B, C, D, E and F

'A' corresponds to 10 in the denary system, B to 11, C to 12, and so on.

To convert from hexadecimal to decimal:

For example

$$1A_{16} = 1 \times 16^1 + A \times 16^0$$
$$= 1 \times 16^1 + 10 \times 1 = 16 + 10 = 26$$

i.e. $\mathbf{1A_{16} = 26_{10}}$

Similarly,

$$\mathbf{2E_{16}} = 2 \times 16^1 + E \times 16^0$$
$$= 2 \times 16^1 + 14 \times 16^0 = 32 + 14 = \mathbf{46_{10}}$$

and $\mathbf{1BF_{16}} = 1 \times 16^2 + B \times 16^1 + F \times 16^0$
$$= 1 \times 16^2 + 11 \times 16^1 + 15 \times 16^0$$
$$= 256 + 176 + 15 = \mathbf{447_{10}}$$

Table 3.2 compares decimal, binary, octal and hexadecimal numbers and shows, for example, that

$$23_{10} = 10111_2 = 27_8 = 17_{16}$$

Problem 11.   Convert the following hexadecimal numbers into their decimal equivalents: (a) $7A_{16}$ (b) $3F_{16}$

(a) $7A_{16} = 7 \times 16^1 + A \times 16^0 = 7 \times 16 + 10 \times 1$
$$= 112 + 10 = 122$$
   Thus $\mathbf{7A_{16} = 122_{10}}$

(b) $3F_{16} = 3 \times 16^1 + F \times 16^0 = 3 \times 16 + 15 \times 1$
$$= 48 + 15 = 63$$
   Thus, $\mathbf{3F_{16} = 63_{10}}$

Problem 12.   Convert the following hexadecimal numbers into their decimal equivalents: (a) $C9_{16}$ (b) BD_{16}

Table 3.2

Decimal	Binary	Octal	Hexadecimal
0	0000	0	0
1	0001	1	1
2	0010	2	2
3	0011	3	3
4	0100	4	4
5	0101	5	5
6	0110	6	6
7	0111	7	7
8	1000	10	8
9	1001	11	9
10	1010	12	A
11	1011	13	B
12	1100	14	C
13	1101	15	D
14	1110	16	E
15	1111	17	F
16	10000	20	10
17	10001	21	11
18	10010	22	12
19	10011	23	13
20	10100	24	14
21	10101	25	15
22	10110	26	16
23	10111	27	17
24	11000	30	18
25	11001	31	19
26	11010	32	1A
27	11011	33	1B
28	11100	34	1C
29	11101	35	1D
30	11110	36	1E
31	11111	37	1F
32	100000	40	20

(a) $C9_{16} = C \times 16^1 + 9 \times 16^0 = 12 \times 16 + 9 \times 1$
$$= 192 + 9 = 201$$
Thus $C9_{16} = 201_{10}$

(b) $BD_{16} = B \times 16^1 + D \times 16^0 = 11 \times 16 + 13 \times 1$
$$= 176 + 13 = 189$$
Thus, $BD_{16} = 189_{10}$

Problem 13. Convert $1A4E_{16}$ into a denary number

$1A4E_{16}$

$= 1 \times 16^3 + A \times 16^2 + 4 \times 16^1 + E \times 16^0$

$= 1 \times 16^3 + 10 \times 16^2 + 4 \times 16^1 + 14 \times 16^0$

$= 1 \times 4096 + 10 \times 256 + 4 \times 16 + 14 \times 1$

$= 4096 + 2560 + 64 + 14 = 6734$

Thus, $1A4E_{16} = 6734_{10}$

To convert from decimal to hexadecimal:

This is achieved by repeatedly dividing by 16 and noting the remainder at each stage, as shown below for 26_{10}.

```
16 ⌊26    Remainder
16 ⌊ 1    10 ≡ A₁₆
    0      1 ≡ 1₁₆
```

most significant bit → 1 A ← least significant bit

Hence $26_{10} = 1A_{16}$

Similarly, for 447_{10}

```
16 ⌊447   Remainder
16 ⌊ 27   15 ≡ F₁₆
16 ⌊  1   11 ≡ B₁₆
     0     1 ≡ 1₁₆
                  1 B F
```

Thus $447_{10} = 1BF_{16}$

Problem 14. Convert the following decimal numbers into their hexadecimal equivalents:
(a) 37_{10} (b) 108_{10}

(a) 16 ⌊37   Remainder
     16 ⌊2 $5 = 5_{16}$ ⌐
       0      $2 = 2_{16}$ ⌐

most significant bit $\rightarrow$ 2   5 $\leftarrow$ least significant bit

Hence $37_{10} = \mathbf{25_{16}}$

(b) 16 ⌊108   Remainder
     16 ⌊6 $12 = C_{16}$ ⌐
       0      $6 = 6_{16}$ ⌐
                 6 C

Hence $\mathbf{108_{10} = 6C_{16}}$

> **Problem 15.** Convert the following decimal numbers into their hexadecimal equivalents:
> (a) 162_{10} (b) 239_{10}

(a) 16 ⌊162   Remainder
     16 ⌊10 $2 = 2_{16}$ ⌐
       0     $10 = A_{16}$ ⌐
                 A 2

Hence $162_{10} = \mathbf{A2_{16}}$

(b) 16 ⌊239   Remainder
     16 ⌊14 $15 = F_{16}$ ⌐
       0     $14 = E_{16}$ ⌐
                 E F

Hence $239_{10} = \mathbf{EF_{16}}$

To convert from binary to hexadecimal:

The binary bits are arranged in groups of four, starting from right to left, and a hexadecimal symbol is assigned to each group. For example, the binary number 1110011110101001 is initially grouped in

fours as: 1110   0111   1010   1001

and a hexadecimal symbol

assigned to each group as   E   7   A   9

from Table 3.2

Hence $\mathbf{1110011110101001_2 = E7A9_{16}}$

To convert from hexadecimal to binary:

The above procedure is reversed, thus, for example,

$6CF3_{16} = 0110 \;\; 1100 \;\; 1111 \;\; 0011$

from Table 3.2

i.e. $\mathbf{6CF3_{16} = 110110011110011_2}$

> **Problem 16.** Convert the following binary numbers into their hexadecimal equivalents:
> (a) 11010110_2 (b) 1100111_2

(a) Grouping bits in fours from the
    right gives: 1101   0110
    and assigning hexadecimal symbols
    to each group gives: D   6
                      from Table 3.2

Thus, $\mathbf{11010110_2 = D6_{16}}$

(b) Grouping bits in fours from the
    right gives: 0110   0111
    and assigning hexadecimal symbols
    to each group gives: 6   7
                      from Table 3.2

Thus, $\mathbf{1100111_2 = 67_{16}}$

> **Problem 17.** Convert the following binary numbers into their hexadecimal equivalents:
> (a) 11001111_2 (b) 110011110_2

(a) Grouping bits in fours from the
    right gives: 1100   1111
    and assigning hexadecimal
    symbols to each group gives: C   F
                      from Table 3.2

Thus, $\mathbf{11001111_2 = CF_{16}}$

(b) Grouping bits in fours from
    the right gives: 0001   1001   1110
    and assigning hexadecimal
    symbols to each group gives: 1   9   E
                      from Table 3.2

Thus, $\mathbf{110011110_2 = 19E_{16}}$

> **Problem 18.** Convert the following hexadecimal numbers into their binary equivalents: (a) $3F_{16}$
> (b) $A6_{16}$

(a) Spacing out hexadecimal digits gives: 3   F

and converting each into

binary gives: 0011   1111
                       from Table 3.2

Thus, $\mathbf{3F_{16}} = \mathbf{111111_2}$

(b) Spacing out hexadecimal digits

gives: A   6

and converting each into binary

gives: 1010   0110
                       from Table 3.2

Thus, $\mathbf{A6_{16}} = \mathbf{10100110_2}$

Problem 19. Convert the following hexadecimal numbers into their binary equivalents: (a) $7B_{16}$ (b) $17D_{16}$

(a) Spacing out hexadecimal

digits gives: 7   B

and converting each into

binary gives: 0111   1011
                       from Table 3.2

Thus, $\mathbf{7B_{16}} = \mathbf{1111011_2}$

(b) Spacing out hexadecimal

digits gives: 1   7   D

and converting each into

binary gives: 0001   0111   1101
                       from Table 3.2

Thus, $\mathbf{17D_{16}} = \mathbf{101111101_2}$

Now try the following exercise

Exercise 13   Further problems on hexadecimal numbers

In Problems 1 to 4, convert the given hexadecimal numbers into their decimal equivalents.

1. $E7_{16}$ $[231_{10}]$ 2. $2C_{16}$ $[44_{10}]$

3. 98_{16} $[152_{10}]$ 4. $2F1_{16}$ $[753_{10}]$

In Problems 5 to 8, convert the given decimal numbers into their hexadecimal equivalents.

5. 54_{10} $[36_{16}]$ 6. 200_{10} $[C8_{16}]$

7. 91_{10} $[5B_{16}]$ 8. 238_{10} $[EE_{16}]$

In Problems 9 to 12, convert the given binary numbers into their hexadecimal equivalents.

9. 11010111_2 $[D7_{16}]$

10. 11101010_2 $[EA_{16}]$

11. 10001011_2 $[8B_{16}]$

12. 10100101_2 $[A5_{16}]$

In Problems 13 to 16, convert the given hexadecimal numbers into their binary equivalents.

13. 37_{16} $[110111_2]$

14. ED_{16} $[11101101_2]$

15. $9F_{16}$ $[10011111_2]$

16. $A21_{16}$ $[101000100001_2]$

Chapter 4

Calculations and evaluation of formulae

4.1 Errors and approximations

(i)   In all problems in which the measurement of distance, time, mass or other quantities occurs, an exact answer cannot be given; only an answer which is correct to a stated degree of accuracy can be given. To take account of this an **error due to measurement** is said to exist.

(ii) To take account of measurement errors it is usual to limit answers so that the result given is **not more than one significant figure greater than the least accurate number given in the data**.

(iii) **Rounding-off errors** can exist with decimal fractions. For example, to state that $\pi = 3.142$ is not strictly correct, but '$\pi = 3.142$ correct to 4 significant figures' is a true statement. (Actually, $\pi = 3.14159265\ldots$)

(iv) It is possible, through an incorrect procedure, to obtain the wrong answer to a calculation. This type of error is known as **a blunder**.

(v)   An **order of magnitude error** is said to exist if incorrect positioning of the decimal point occurs after a calculation has been completed.

(vi) Blunders and order of magnitude errors can be reduced by determining **approximate values of calculations**. Answers which do not seem feasible must be checked and the calculation must be repeated as necessary.

An engineer will often need to make a quick mental approximation for a calculation. For example, $\dfrac{49.1 \times 18.4 \times 122.1}{61.2 \times 38.1}$ may be approximated

to $\dfrac{50 \times 20 \times 120}{60 \times 40}$ and then, by cancelling,

$\dfrac{50 \times {}^1\cancel{20} \times \cancel{120}^{41}}{{}_1\cancel{60} \times \cancel{40}_{71}} = 50$. An accurate answer

somewhere between 45 and 55 could therefore be expected. Certainly an answer around 500 or 5 would not be expected. Actually, by calculator $\dfrac{49.1 \times 18.4 \times 122.1}{61.2 \times 38.1} = 47.31$, correct to 4 significant figures.

Problem 1.   The area A of a triangle is given by $A = \dfrac{1}{2}bh$. The base b when measured is found to be 3.26 cm, and the perpendicular height h is 7.5 cm. Determine the area of the triangle.

Area of triangle $= \dfrac{1}{2}bh = \dfrac{1}{2} \times 3.26 \times 7.5$
$= 12.225\,\text{cm}^2$ (by calculator).

The approximate values is $\dfrac{1}{2} \times 3 \times 8 = 12\,\text{cm}^2$, so there are no obvious blunder or magnitude errors. However, it is not usual in a measurement type problem to state the answer to an accuracy greater than 1 significant figure more than the least accurate number in the data: this is 7.5 cm, so the result should not have more than 3 significant figures
Thus, **area of triangle $= 12.2\,\text{cm}^2$**

Problem 2.   State which type of error has been made in the following statements:

(a) $72 \times 31.429 = 2262.9$
(b) $16 \times 0.08 \times 7 = 89.6$

(c) $11.714 \times 0.0088 = 0.3247$ correct to 4 decimal places.

(d) $\dfrac{29.74 \times 0.0512}{11.89} = 0.12$, correct to 2 significant figures.

(a) $72 \times 31.429 = 2262.888$ (by calculator), hence a **rounding-off error** has occurred. The answer should have stated:

$72 \times 31.429 = 2262.9$, correct to 5 significant figures or 2262.9, correct to 1 decimal place.

(b) $16 \times 0.08 \times 7 = 16 \times \dfrac{8}{100} \times 7 = \dfrac{32 \times 7}{25}$

$\qquad = \dfrac{224}{25} = 8\dfrac{24}{25} = 8.96$

Hence an **order of magnitude** error has occurred.

(c) 11.714×0.0088 is approximately equal to $12 \times 9 \times 10^{-3}$, i.e. about 108×10^{-3} or 0.108. Thus a **blunder** has been made.

(d) $\dfrac{29.74 \times 0.0512}{11.89} \approx \dfrac{30 \times 5 \times 10^{-2}}{12}$

$\qquad = \dfrac{150}{12 \times 10^2} = \dfrac{15}{120} = \dfrac{1}{8}$ or 0.125

hence no order of magnitude error has occurred. However, $\dfrac{29.74 \times 0.0512}{11.89} = 0.128$ correct to 3 significant figures, which equals 0.13 correct to 2 significant figures.

Hence a **rounding-off error** has occurred.

Problem 3. Without using a calculator, determine an approximate value of:

(a) $\dfrac{11.7 \times 19.1}{9.3 \times 5.7}$ (b) $\dfrac{2.19 \times 203.6 \times 17.91}{12.1 \times 8.76}$

(a) $\dfrac{11.7 \times 19.1}{9.3 \times 5.7}$ is approximately equal to $\dfrac{10 \times 20}{10 \times 5}$

i.e. about **4**

(By calculator, $\dfrac{11.7 \times 19.1}{9.3 \times 5.7} = \mathbf{4.22}$, correct to 3 significant figures.)

(b) $\dfrac{2.19 \times 203.6 \times 17.91}{12.1 \times 8.76} \approx \dfrac{2 \times \overset{20}{200} \times \overset{2}{20^2}}{\underset{1}{10} \times \underset{1}{10}}$

$= 2 \times 20 \times 2$ after cancelling,

i.e. $\dfrac{2.19 \times 203.6 \times 17.91}{12.1 \times 8.76} \approx \mathbf{80}$

(By calculator, $\dfrac{2.19 \times 203.6 \times 17.91}{12.1 \times 8.76} \approx \mathbf{75.3}$, correct to 3 significant figures.)

Now try the following exercise

Exercise 14 Further problems on errors

In Problems 1 to 5 state which type of error, or errors, have been made:

1. $25 \times 0.06 \times 1.4 = 0.21$
   [order of magnitude error]

2. $137 \times 6.842 = 937.4$
   $\begin{bmatrix} \text{Rounding-off error–should add 'correct} \\ \text{to 4 significant figures' or 'correct to} \\ \text{1 decimal place'} \end{bmatrix}$

3. $\dfrac{24 \times 0.008}{12.6} = 10.42$ [Blunder]

4. For a gas $pV = c$. When pressure $p = 1\,03\,400$ Pa and $V = 0.54\,\text{m}^3$ then $c = 55\,836\,\text{Pa}\,\text{m}^3$.
   $\begin{bmatrix} \text{Measured values, hence} \\ c = 55\,800\,\text{Pa}\,\text{m}^3 \end{bmatrix}$

5. $\dfrac{4.6 \times 0.07}{52.3 \times 0.274} = 0.225$
   $\begin{bmatrix} \text{Order of magnitude error and rounding-} \\ \text{off error–should be 0.0225, correct to 3} \\ \text{significant figures or 0.0225,} \\ \text{correct to 4 decimal places} \end{bmatrix}$

In Problems 6 to 8, evaluate the expressions approximately, without using a calculator.

6. 4.7×6.3 [≈ 30 (29.61, by calculator)]

7. $\dfrac{2.87 \times 4.07}{6.12 \times 0.96}$
   $\begin{bmatrix} \approx 2 \ (1.988, \text{correct to 4 s.f., by} \\ \text{calculator)} \end{bmatrix}$

8. $\dfrac{72.1 \times 1.96 \times 48.6}{139.3 \times 5.2}$
   $\begin{bmatrix} \approx 10 \ (9.481, \text{correct to 4 s.f., by} \\ \text{calculator)} \end{bmatrix}$

4.2   Use of calculator

The most modern aid to calculations is the pocket-sized electronic calculator. With one of these, calculations can be quickly and accurately performed, correct to about 9 significant figures. The scientific type of calculator has made the use of tables and logarithms largely redundant.

To help you to become competent at using your calculator check that you agree with the answers to the following problems:

Problem 4.   Evaluate the following, correct to 4 significant figures:

(a) $4.7826 + 0.02713$ (b) $17.6941 - 11.8762$
(c) 21.93×0.012981

(a) $4.7826 + 0.02713 = 4.80973 = \mathbf{4.810}$, correct to 4 significant figures

(b) $17.6941 - 11.8762 = 5.8179 = \mathbf{5.818}$, correct to 4 significant figures

(c) $21.93 \times 0.012981 = 0.2846733\ldots = \mathbf{0.2847}$, correct to 4 significant figures

---

Problem 5.   Evaluate the following, correct to 4 decimal places:

(a) $46.32 \times 97.17 \times 0.01258$ (b) $\dfrac{4.621}{23.76}$

(c) $\dfrac{1}{2}(62.49 \times 0.0172)$

(a) $46.32 \times 97.17 \times 0.01258 = 56.6215031\ldots$
$= \mathbf{56.6215}$, correct to 4 decimal places

(b) $\dfrac{4.621}{23.76} = 0.19448653\ldots = \mathbf{0.1945}$, correct to 4 decimal places

(c) $\dfrac{1}{2}(62.49 \times 0.0172) = 0.537414 = \mathbf{0.5374}$, correct to 4 decimal places

---

Problem 6.   Evaluate the following, correct to 3 decimal places:

(a) $\dfrac{1}{52.73}$ (b) $\dfrac{1}{0.0275}$ (c) $\dfrac{1}{4.92} + \dfrac{1}{1.97}$

(a) $\dfrac{1}{52.73} = 0.01896453\ldots = \mathbf{0.019}$, correct to 3 decimal places

(b) $\dfrac{1}{0.0275} = 36.3636363\ldots = \mathbf{36.364}$, correct to 3 decimal places

(c) $\dfrac{1}{4.92} + \dfrac{1}{1.97} = 0.71086624\ldots = \mathbf{0.711}$, correct to 3 decimal places

---

Problem 7.   Evaluate the following, expressing the answers in standard form, correct to 4 significant figures.

(a) $(0.00451)^2$ (b) $631.7 - (6.21 + 2.95)^2$
(c) $46.27^2 - 31.79^2$

(a) $(0.00451)^2 = 2.03401 \times 10^{-5} = \mathbf{2.034 \times 10^{-5}}$, correct to 4 significant figures

(b) $631.7 - (6.21 + 2.95)^2 = 547.7944$
$= 5.477944 \times 10^2 = \mathbf{5.478 \times 10^2}$, correct to 4 significant figures

(c) $46.27^2 - 31.79^2 = 1130.3088 = \mathbf{1.130 \times 10^3}$, correct to 4 significant figures

---

Problem 8.   Evaluate the following, correct to 3 decimal places:

(a) $\dfrac{(2.37)^2}{0.0526}$ (b) $\left(\dfrac{3.60}{1.92}\right)^2 + \left(\dfrac{5.40}{2.45}\right)^2$

(c) $\dfrac{15}{7.6^2 - 4.8^2}$

(a) $\dfrac{(2.37)^2}{0.0526} = 106.785171\ldots = \mathbf{106.785}$, correct to 3 decimal places

(b) $\left(\dfrac{3.60}{1.92}\right)^2 + \left(\dfrac{5.40}{2.45}\right)^2 = 8.37360084\ldots = \mathbf{8.374}$, correct to 3 decimal places

(c) $\dfrac{15}{7.6^2 - 4.8^2} = 0.43202764\ldots = \mathbf{0.432}$, correct to 3 decimal places

---

Problem 9.   Evaluate the following, correct to 4 significant figures:

(a) $\sqrt{5.462}$ (b) $\sqrt{54.62}$ (c) $\sqrt{546.2}$

Section 1

(a) $\sqrt{5.462} = 2.3370922\ldots = \mathbf{2.337}$, correct to 4 significant figures

(b) $\sqrt{54.62} = 7.39053448\ldots = \mathbf{7.391}$, correct to 4 significant figures

(c) $\sqrt{546.2} = 23.370922\ldots = \mathbf{23.37}$, correct to 4 significant figures

Problem 10. Evaluate the following, correct to 3 decimal places:

(a) $\sqrt{0.007328}$ (b) $\sqrt{52.91} - \sqrt{31.76}$

(c) $\sqrt{1.6291 \times 10^4}$

(a) $\sqrt{0.007328} = 0.08560373 = \mathbf{0.086}$, correct to 3 decimal places

(b) $\sqrt{52.91} - \sqrt{31.76} = 1.63832491\ldots = \mathbf{1.638}$, correct to 3 decimal places

(c) $\sqrt{1.6291 \times 10^4} = \sqrt{16291} = 127.636201\ldots$
$\qquad = \mathbf{127.636}$, correct to 3 decimal places

Problem 11. Evaluate the following, correct to 4 significant figures:

(a) 4.72^3 (b) $(0.8316)^4$ (c) $\sqrt{76.21^2 - 29.10^2}$

(a) $4.72^3 = 105.15404\ldots = \mathbf{105.2}$, correct to 4 significant figures

(b) $(0.8316)^4 = 0.47825324\ldots = \mathbf{0.4783}$, correct to 4 significant figures

(c) $\sqrt{76.21^2 - 29.10^2} = 70.4354605\ldots = \mathbf{70.44}$, correct to 4 significant figures

Problem 12. Evaluate the following, correct to 3 significant figures:

(a) $\sqrt{\dfrac{6.09^2}{25.2 \times \sqrt{7}}}$ (b) $\sqrt[3]{47.291}$

(c) $\sqrt{7.213^2 + 6.418^3 + 3.291^4}$

(a) $\sqrt{\dfrac{6.09^2}{25.2 \times \sqrt{7}}} = 0.74583457\ldots = \mathbf{0.746}$, correct

to 3 significant figures

(b) $\sqrt[3]{47.291} = 3.61625876\ldots = \mathbf{3.62}$, correct to 3 significant figures

(c) $\sqrt{7.213^2 + 6.418^3 + 3.291^4} = 20.8252991\ldots,$
$\qquad = \mathbf{20.8}$ correct to 3 significant figures

Problem 13. Evaluate the following, expressing the answers in standard form, correct to 4 decimal places:

(a) $(5.176 \times 10^{-3})^2$

(b) $\left(\dfrac{1.974 \times 10^1 \times 8.61 \times 10^{-2}}{3.462}\right)^4$

(c) $\sqrt{1.792 \times 10^{-4}}$

(a) $(5.176 \times 10^{-3})^2 = 2.679097\ldots \times 10^{-5}$
$\qquad = \mathbf{2.6791 \times 10^{-5}}$, correct to 4 decimal places

(b) $\left(\dfrac{1.974 \times 10^1 \times 8.61 \times 10^{-2}}{3.462}\right)^4 = 0.05808887\ldots$
$\qquad = \mathbf{5.8089 \times 10^{-2}}$, correct to 4 decimal places

(c) $\sqrt{1.792 \times 10^{-4}} = 0.0133865\ldots = \mathbf{1.3387 \times 10^{-2}}$, correct to 4 decimal places

Now try the following exercise

Exercise 15   Further problems on the use of a calculator

In Problems 1 to 9, use a calculator to evaluate the quantities shown correct to 4 significant figures:

1. (a) 3.249^2 (b) 73.78^2 (c) 311.4^2 (d) 0.0639^2
$$\begin{bmatrix} \text{(a) } 10.56 & \text{(b) } 5443 & \text{(c) } 96970 \\ \text{(d) } 0.004083 & & \end{bmatrix}$$

2. (a) $\sqrt{4.735}$ (b) $\sqrt{35.46}$ (c) $\sqrt{73\,280}$
(d) $\sqrt{0.0256}$
$$\begin{bmatrix} \text{(a) } 2.176 & \text{(b) } 5.955 & \text{(c) } 270.7 \\ \text{(d) } 0.1600 & & \end{bmatrix}$$

3. (a) $\dfrac{1}{7.768}$ (b) $\dfrac{1}{48.46}$ (c) $\dfrac{1}{0.0816}$
(d) $\dfrac{1}{1.118}$
$$\begin{bmatrix} \text{(a) } 0.1287 & \text{(b) } 0.02064 \\ \text{(c) } 12.25 & \text{(d) } 0.8945 \end{bmatrix}$$

4. (a) $127.8 \times 0.0431 \times 19.8$
(b) $15.76 \div 4.329$
$$[\text{(a) } 109.1 \quad \text{(b) } 3.641]$$

5. (a) $\dfrac{137.6}{552.9}$ (b) $\dfrac{11.82 \times 1.736}{0.041}$
$$[\text{(a) } 0.2489 \text{ (b) } 500.5]$$

6. (a) 13.6^3 (b) 3.476^4 (c) 0.124^5

[(a) 2515   (b) 146.0   (c) 0.00002932]

7. (a) $\left(\dfrac{24.68 \times 0.0532}{7.412}\right)^3$

(b) $\left(\dfrac{0.2681 \times 41.2^2}{32.6 \times 11.89}\right)^4$

[(a) 0.005559   (b) 1.900]

8. (a) $\dfrac{14.32^3}{21.68^2}$ (b) $\dfrac{4.821^3}{17.33^2 - 15.86 \times 11.6}$

[(a) 6.248   (b) 0.9630]

9. (a) $\sqrt{\dfrac{(15.62)^2}{29.21 \times \sqrt{10.52}}}$

(b) $\sqrt{6.921^2 + 4.816^3 - 2.161^4}$

[(a) 1.605   (b) 11.74]

10. Evaluate the following, expressing the answers in standard form, correct to 3 decimal places: (a) $(8.291 \times 10^{-2})^2$

(b) $\sqrt{7.623 \times 10^{-3}}$

[(a) 6.874×10^{-3} (b) 8.731×10^{-2}]

4.3   Conversion tables and charts

It is often necessary to make calculations from various conversion tables and charts. Examples include currency exchange rates, imperial to metric unit conversions, train or bus timetables, production schedules and so on.

Problem 14.   Currency exchange rates for five countries are shown in Table 4.1

Table 4.1

France	£1 = 1.46 euros
Japan	£1 = 220 yen
Norway	£1 = 12.10 kronor
Switzerland	£1 = 2.30 francs
U.S.A.	£1 = 1.95 dollars ($)

Calculate:

(a)   how many French euros £27.80 will buy

(b)   the number of Japanese yen which can be bought for £23

(c)   the pounds sterling which can be exchanged for 7114.80 Norwegian kronor

(d)   the number of American dollars which can be purchased for £90, and

(e)   the pounds sterling which can be exchanged for 2990 Swiss francs

(a)   £1 = 1.46 euros, hence
£27.80 = 27.80 × 1.46 euros = **40.59 euros**

(b)   £1 = 220 yen, hence
£23 = 23 × 220 yen = **5060 yen**

(c)   £1 = 12.10 kronor, hence
7114.80 kronor = £$\dfrac{7114.80}{12.10}$ = **£588**

(d)   £1 = 1.95 dollars, hence
£90 = 90 × 1.95 dollars = **$175.50**

(e)   £1 = 2.30 Swiss francs, hence
2990 franc = £$\dfrac{2990}{2.30}$ = **£1300**

Problem 15.   Some approximate imperial to metric conversions are shown in Table 4.2

Table 4.2

length	1 inch = 2.54 cm 1 mile = 1.61 km
weight	2.2 lb = 1 kg (1 lb = 16 oz)
capacity	1.76 pints = 1 litre (8 pints = 1 gallon)

Use the table to determine:

(a)   the number of millimetres in 9.5 inches,
(b)   a speed of 50 miles per hour in kilometres per hour,
(c)   the number of miles in 300 km,
(d)   the number of kilograms in 30 pounds weight,
(e)   the number of pounds and ounces in 42 kilograms (correct to the nearest ounce),
(f)   the number of litres in 15 gallons, and
(g)   the number of gallons in 40 litres.

(a) 9.5 inches $= 9.5 \times 2.54$ cm $= 24.13$ cm

     24.13 cm $= 24.13 \times 10$ mm $= \textbf{241.3 mm}$

(b) 50 m.p.h. $= 50 \times 1.61$ km/h $= \textbf{80.5 km/h}$

(c) 300 km $= \dfrac{300}{1.61}$ miles $= \textbf{186.3 miles}$

(d) 30 lb $= \dfrac{30}{2.2}$ kg $= \textbf{13.64 kg}$

(e) 42 kg $= 42 \times 2.2$ lb $= 92.4$ lb

    0.4 lb $= 0.4 \times 16$ oz $= 6.4$ oz $= 6$ oz, correct to the nearest ounce

    Thus 42 kg $= \textbf{92 lb 6 oz}$, correct to the nearest ounce.

(f) 15 gallons $= 15 \times 8$ pints $= 120$ pints

    120 pints $= \dfrac{120}{1.76}$ litres $= \textbf{68.18 litres}$

(g) 40 litres $= 40 \times 1.76$ pints $= 70.4$ pints

    70.4 pints $= \dfrac{70.4}{8}$ gallons $= \textbf{8.8 gallons}$

Now try the following exercise

Exercise 16   Further problems conversion tables and charts

1. Currency exchange rates listed in a newspaper included the following:

    Italy   £1 = 1.48 euro
    Japan   £1 = 225 yen
    Australia   £1 = 2.50 dollars
    Canada   £1 = \$2.20
    Sweden   £1 = 13.25 kronor

    Calculate (a) how many Italian euros £32.50 will buy, (b) the number of Canadian dollars that can be purchased for £74.80, (c) the pounds sterling which can be exchanged for 14 040 yen, (d) the pounds sterling which can be exchanged for 1754.30 Swedish kronor, and (e) the Australian dollars which can be bought for £55

             [(a) 48.10 euros   (b) \$164.56
               (c) £62.40   (d) £132.40
               (e) 137.50 dollars]

2. Below is a list of some metric to imperial conversions.

    Length   2.54 cm = 1 inch
                  1.61 km = 1 mile
    Weight   1 kg = 2.2 lb (1 lb = 16 ounces)
    Capacity   1 litre = 1.76 pints
                  (8 pints = 1 gallon)

    Use the list to determine (a) the number of millimetres in 15 inches, (b) a speed of 35 mph in km/h, (c) the number of kilometres in 235 miles, (d) the number of pounds and ounces in 24 kg (correct to the nearest ounce), (e) the number of kilograms in 15 lb, (f) the number of litres in 12 gallons and (g) the number of gallons in 25 litres.

        [(a) 381 mm   (b) 56.35 km/h
         (c) 378.35 km   (d) 52 lb 13 oz
         (e) 6.82 kg   (f) 54.55 litre
         (g) 5.5 gallons]

3. Deduce the following information from the train timetable shown in Table 4.3:

  (a) At what time should a man catch a train at Mossley Hill to enable him to be in Manchester Piccadilly by 8.15 a.m.?

  (b) A girl leaves Hunts Cross at 8.17 a.m. and travels to Manchester Oxford Road. How long does the journey take? What is the average speed of the journey?

  (c) A man living at Edge Hill has to be at work at Trafford Park by 8.45 a.m. It takes him 10 minutes to walk to his work from Trafford Park station. What time train should he catch from Edge Hill?

        [(a) 7.09 a.m.
         (b) 52 minutes, 31.15 m.p.h.
         (c) 7.04 a.m.]

4.4   Evaluation of formulae

The statement $v = u + at$ is said to be a **formula** for v in terms of u, a and t.

v, u, a and t are called **symbols**.

The single term on the left-hand side of the equation, v, is called the **subject of the formulae**.

Table 4.3 Liverpool, Hunt's Cross and Warrington → Manchester

Miles	Station			MX	MO (A SS)	SX ◇C	SO. ◇C	SX BHX	SX BHX	BHX ◇D	BHX	SX BHX	BHX	BHX ◇E	SX BHX	BHX	BHX ◇C	SX BHX ◇	BHX
0	Liverpool Lime Street	82, 99	d	05 25		05 37		06 03	06 23		07 00		07 17	07 30		07 52	08 00	08 23		08 30
1½	Edge Hill	82, 99	d							06 34		07 04		07 34		08 04			08 34
3½	Mossley Hill	82	d							06 39		07 09		07 39		08 09			08 39
4½	West Allerton	82	d							06 41		07 11		07 41		08 11			08 41
5½	Allerton	82	d							06 43		07 13		07 43		08 13			08 43
—	Liverpool Central	101	d						*06 45*		*07 15*		*07 45*			*08 15*			
—	Garston (Merseyside)	101	d						*06 26*	*06 56*		*07 26*	*07 56*				*08 26*		
7½	Hunt's Cross		d	05u38			05u50	06 17		06 47		07 17		07 47		08u05	08 17		08 47
8½	Halewood		d					06 20		06 50		07 20		07 50		08 20			08 50
10½	Hough Green		d					06 24		06 54		07 24		07 54		08 24			08 54
12½	Widnes		d					06 27		06 57	07 35	07 27		07 57	08 12	08 27			08 57
16	Sankey for Penketh		d	00 02				06 32		07 02		07 32		08 02		08 32			09 02
18½	Warrington Central		a	00 07		06 02		06 37		07 07	07 43	07 37		08 07	08 20	08 37			09 07
—			d			06 03		06 46	07 20	07 00	07 43	07 19		08 00	08 20	08 46			09 07
20½	Padgate		d							07 03	08 03	07 33			08 33		08 46		09 00 / 09 03
21½	Birchwood		d	05 56		06 08		06 36	07 25	07 06	08 06	07 36		08 06	08 25	08 34		08 51	09 06
24½	Glazebrook		d					06 41		07 11	08 11	07 41		08 11		08 41			09 11
25½	Irlam		d	06 02				06 44		07 14	07 54	07 44		08 14	08 34	08 44			09 14
28	Flixton		d	06 06				06 48		07 18		07 48		08 15	08 38	08 48			09 18
28½	Chassen Road		d	06 08				06 50		07 20		07 50		08 20	08 40	08 50			09 20
29	Urmston		d	06 10	00 03			06 52		07 22		07 52	08 09	08 22	08 42	08 52		09 09	09 22
30½	Humphrey Park		d	06 13	00 13			06 55		07 25	08 11	07 55		08 25	08 45	08 55		09 11	09 25
31	Trafford Park		d	06 15				06 57		07 27		07 57		08 27	08 47	08 57			09 27
34	Deansgate	81	d		00 23			07 03		07 33	08 32	08 03		08 33	08 52	09 03			09 33
34½	Manchester Oxford Road	81	a	06 22	00 27			07 05	07 40	07 35	08 08	08 05		08 35	08 54	09 05	09 08		09 35
—			d	06 23	00 27			07 08	07 41		08 09	08 41		08 37		09 05	09 09		
35	Manchester Piccadilly	81	a	06 25	00 34			07 11	07 43	07 43	08 11	08 43		08 39	08 57		09 11		
—	Stockport	81, 90	a	06 34				*07 32*	*07 54*		*08 32*		*08 43*	*08 54*	*09 19*		*09 32*		
—	Sheffield	90	a	07 30					*08 42*				*08 42*						

(*Continued*)

Table 4.3   Continued

The following transcribes Table 4.3, a railway timetable. Reading along each station row gives the successive train times (24-hour clock). Trains marked *BHX* and *◇* in the column headings. Times shown in italics denote the Liverpool Central branch services.

Station			Train times (read left to right)
Liverpool Lime Street	82, 99	d	08 54, 09 00, 09 23, 09 30, 09 56, 10 00, 10 23, 10 30, 10 56, 11 00, 11 23, 11 30, 11 56, 12 00, 12 23, 12 30, 12 56, 13 00
Edge Hill	82, 99	d	09 04, 09 34, 10 04, 10 34, 11 04, 11 34, 12 04, 12 34, 13 04
Mossley Hill	82	d	09 09, 09 39, 10 09, 10 39, 11 09, 11 39, 12 09, 12 39, 13 09
West Allerton	82	d	09 11, 09 41, 10 11, 10 41, 11 11, 11 41, 12 11, 12 41, 13 11
Allerton	82	d	09 13, 09 43, 10 13, 10 43, 11 13, 11 43, 12 13, 12 43, 13 13
Liverpool Central	101	d	*09 15, 09 45, 10 15, 10 45, 11 15, 11 45, 12 15, 12 45*
Garston (Merseyside)	101	d	*09 26, 09 56, 10 26, 10 56, 11 26, 11 56, 12 26, 12 56*
Hunt's Cross		d	09u09, 09 17, 09 47, 10u09, 10 17, 10 47, 11u09, 11 17, 11 47, 12u09, 12 17, 12 47, 13u09, 13 17
Halewood		d	09 20, 09 50, 10 20, 10 50, 11 20, 11 50, 12 20, 12 50, 13 20
Hough Green		d	09 24, 09 54, 10 24, 10 54, 11 24, 11 54, 12 24, 12 54, 13 24
Widnes		d	09 27, 09 57, 10 27, 10 57, 11 27, 11 57, 12 27, 12 57, 13 27
Sankey for Penketh		d	09 32, 10 02, 10 32, 11 02, 11 32, 12 02, 12 32, 13 02, 13 32
Warrington Central		a	09 21, 09 37, 09 46, 10 07, 10 21, 10 37, 10 46, 11 07, 11 21, 11 37, 11 46, 12 07, 12 21, 12 37, 12 46, 13 07, 13 21, 13 37
		d	09 22, 09 46, 10 00, 10 22, 10 46, 11 00, 11 22, 11 46, 12 00, 12 22, 12 46, 13 00, 13 22
Padgate		d	09 33, 10 03, 11 03, 12 03, 13 03
Birchwood		d	09 36, 09 51, 10 06, 10 51, 11 06, 11 51, 12 06, 12 51, 13 04
Glazebrook		d	09 41, 10 11, 11 11, 12 11, 13 11
Irlam		d	09 44, 10 14, 11 14, 12 14, 13 14
Flixton		d	09 48, 10 18, 11 18, 12 18, 13 18
Chassen Road		d	09 50, 10 20, 11 20, 12 20, 13 20
Urmston		d	09 52, 10 22, 11 22, 12 22, 13 22
Humphrey Park		d	09 55, 10 25, 11 25, 12 25, 13 25
Trafford Park		d	09 57, 10 27, 11 27, 12 27, 13 27
Deansgate	81	d	10 03, 10 33, 11 33, 12 33, 13 33
Manchester Oxford Road	81	a	09 40, 10 05, 10 35, 11 08, 11 35, 11 40, 12 08, 12 35, 12 40, 13 08, 13 35, 13 40
		d	09 41, 10 06, 10 09, 11 09, 11 41, 12 09, 12 41, 13 09, 13 41
Manchester Piccadilly	81	a	09 43, 10 08, 10 11, 11 11, 11 43, 12 11, 12 43, 13 11, 13 43
Stockport	81, 90	a	*09 54, 10 25, 10 54, 11 32, 11 54, 12 32, 12 54, 13 32, 13 54*
Sheffield	90	a	10 42, 11 42, 12 41, 13 42, 14 39

Reproduced with permission of British Rail

Provided values are given for all the symbols in a formula except one, the remaining symbol can be made the subject of the formula and may be evaluated by using a calculator.

Problem 16. In an electrical circuit the voltage V is given by Ohm's law, i.e. $V = IR$. Find, correct to 4 significant figures, the voltage when $I = 5.36$ A and $R = 14.76\,\Omega$.

$$V = IR = (5.36)(14.76)$$

Hence, **voltage $V = 79.11$ V, correct to 4 significant figures**.

Problem 17. The surface area A of a hollow cone is given by $A = \pi r l$. Determine, correct to 1 decimal place, the surface area when $r = 3.0$ cm and $l = 8.5$ cm.

$$A = \pi r l = \pi\ (3.0)(8.5)\ \text{cm}^2$$

Hence, **surface area $A = 80.1$ cm^2**, correct to 1 decimal place.

Problem 18. Velocity v is given by $v = u + at$. If $u = 9.86$ m/s, $a = 4.25$ m/s^2 and $t = 6.84$ s, find v, correct to 3 significant figures.

$$v = u + at = 9.86 + (4.25)(6.84)$$
$$= 9.86 + 29.07 = 38.93$$

Hence, **velocity $v = 38.9$ m/s, correct to 3 significant figures.**

Problem 19. The power, P watts, dissipated in an electrical circuit may be expressed by the formula $P = \dfrac{V^2}{R}$. Evaluate the power, correct to 3 significant figures, given that $V = 17.48$ V and $R = 36.12\,\Omega$.

$$P = \frac{V^2}{R} = \frac{(17.48)^2}{36.12} = \frac{305.5504}{36.12}$$

Hence **power, $P = 8.46$ W, correct to 3 significant figures.**

Problem 20. The volume V cm^3 of a right circular cone is given by $V = \dfrac{1}{3}\pi r^2 h$. Given that $r = 4.321$ cm and $h = 18.35$ cm, find the volume, correct to 4 significant figures.

$$V = \frac{1}{3}\pi r^2 h = \frac{1}{3}\pi(4.321)^2(18.35)$$
$$= \frac{1}{3}\pi(18.671041)(18.35)$$

Hence volume, $V = 358.8$ cm^3, correct to 4 significant figures.

Problem 21. Force F newtons is given by the formula $F = \dfrac{Gm_1 m_2}{d^2}$, where m_1 and m_2 are masses, d their distance apart and G is a constant. Find the value of the force given that $G = 6.67 \times 10^{-11}$, $m_1 = 7.36$, $m_2 = 15.5$ and $d = 22.6$. Express the answer in standard form, correct to 3 significant figures.

$$F = \frac{Gm_1 m_2}{d^2} = \frac{(6.67 \times 10^{-11})(7.36)(15.5)}{(22.6)^2}$$
$$= \frac{(6.67)(7.36)(15.5)}{(10^{11})(510.76)} = \frac{1.490}{10^{11}}$$

Hence force $F = 1.49 \times 10^{-11}$ newtons, correct to 3 significant figures.

Problem 22. The time of swing t seconds, of a simple pendulum is given by $t = 2\pi\sqrt{\dfrac{l}{g}}$ Determine the time, correct to 3 decimal places, given that $l = 12.0$ and $g = 9.81$

$$t = 2\pi\sqrt{\frac{l}{g}} = (2)\pi\sqrt{\frac{12.0}{9.81}}$$
$$= (2)\pi\sqrt{1.22324159}$$
$$= (2)\pi(1.106002527)$$

Hence time $t = 6.950$ seconds, correct to 3 decimal places.

Problem 23. Resistance, $R\Omega$, varies with temperature according to the formula $R = R_0(1 + \alpha t)$. Evaluate R, correct to 3 significant figures, given $R_0 = 14.59$, $\alpha = 0.0043$ and $t = 80$.

$$R = R_0(1 + \alpha t) = 14.59[1 + (0.0043)(80)]$$
$$= 14.59(1 + 0.344)$$
$$= 14.59(1.344)$$

Hence, **resistance, $R = 19.6\ \Omega$, correct to 3 significant figures.**

Now try the following exercise

Exercise 17 Further problems on evaluation of formulae

1. A formula used in connection with gases is $R = (PV)/T$. Evaluate R when $P = 1500$, $V = 5$ and $T = 200$. [$R = 37.5$]

2. The velocity of a body is given by $v = u + at$. The initial velocity u is measured when time t is 15 seconds and found to be 12 m/s. If the acceleration a is 9.81 m/s² calculate the final velocity v. [159 m/s]

3. Find the distance s, given that $s = \frac{1}{2}gt^2$, time $t = 0.032$ seconds and acceleration due to gravity $g = 9.81$ m/s².
   [0.00502 m or 5.02 mm]

4. The energy stored in a capacitor is given by $E = \frac{1}{2}CV^2$ joules. Determine the energy when capacitance $C = 5 \times 10^{-6}$ farads and voltage $V = 240V$. [0.144 J]

5. Resistance R_2 is given by $R_2 = R_1(1 + \alpha t)$. Find R_2, correct to 4 significant figures, when $R_1 = 220$, $\alpha = 0.00027$ and $t = 75.6$.
   [224.5]

6. Density $= \dfrac{\text{mass}}{\text{volume}}$. Find the density when the mass is 2.462 kg and the volume is 173 cm³. Give the answer in units of kg/m³.
   [14 230 kg/m³]

7. Velocity $=$ frequency $\times$ wavelength. Find the velocity when the frequency is 1825 Hz and the wavelength is 0.154 m. [281.1 m/s]

8. Evaluate resistance R_T, given
   $$\frac{1}{R_T} = \frac{1}{R_1} + \frac{1}{R_2} + \frac{1}{R_3} \quad \text{when} \quad R_1 = 5.5\ \Omega,$$
   $R_2 = 7.42\ \Omega$ and $R_3 = 12.6\ \Omega$. [$2.526\ \Omega$]

9. Power $= \dfrac{\text{force} \times \text{distance}}{\text{time}}$. Find the power when a force of 3760 N raises an object a distance of 4.73 m in 35 s. [508.1 W]

10. The potential difference, V volts, available at battery terminals is given by $V = E - Ir$. Evaluate V when $E = 5.62$, $I = 0.70$ and $R = 4.30$ [$V = 2.61\ V$]

11. Given force $F = \frac{1}{2}m(v^2 - u^2)$, find F when $m = 18.3$, $v = 12.7$ and $u = 8.24$
    [$F = 854.5$]

12. The current I amperes flowing in a number of cells is given by $I = \dfrac{nE}{R + nr}$. Evaluate the current when $n = 36$. $E = 2.20$, $R = 2.80$ and $r = 0.50$ [$I = 3.81$ A]

13. The time, t seconds, of oscillation for a simple pendulum is given by $t = 2\pi\sqrt{\dfrac{l}{g}}$. Determine the time when $\pi = 3.142$, $l = 54.32$ and $g = 9.81$ [$t = 14.79$ s]

14. Energy, E joules, is given by the formula $E = \frac{1}{2}LI^2$. Evaluate the energy when $L = 5.5$ and $I = 1.2$ [$E = 3.96$ J]

15. The current I amperes in an a.c. circuit is given by $I = \dfrac{V}{\sqrt{R^2 + X^2}}$. Evaluate the current when $V = 250$, $R = 11.0$ and $X = 16.2$ [$I = 12.77$ A]

16. Distance s metres is given by the formula $s = ut + \frac{1}{2}at^2$. If $u = 9.50$, $t = 4.60$ and $a = -2.50$, evaluate the distance. [$s = 17.25$ m]

17. The area, A, of any triangle is given by $A = \sqrt{s(s-a)(s-b)(s-c)}$ where $s = \dfrac{a+b+c}{2}$. Evaluate the area given $a = 3.60$ cm, $b = 4.00$ cm and $c = 5.20$ cm.
    [$A = 7.184$ cm²]

18. Given that $a = 0.290$, $b = 14.86$, $c = 0.042$, $d = 31.8$ and $e = 0.650$, evaluate v, given that $v = \sqrt{\dfrac{ab}{c} - \dfrac{d}{e}}$ [$v = 7.327$]

This Revision test covers the material contained in Chapters 1 to 4. *The marks for each question are shown in brackets at the end of each question.*

1. Simplify (a) $2\frac{2}{3} \div 3\frac{1}{3}$

   (b) $\dfrac{1}{\left(\frac{4}{7} \times 2\frac{1}{4}\right)} \div \left(\frac{1}{3} + \frac{1}{5}\right) + 2\frac{7}{24}$ (9)

2. A piece of steel, 1.69 m long, is cut into three pieces in the ratio 2 to 5 to 6. Determine, in centimetres, the lengths of the three pieces. (4)

3. Evaluate $\dfrac{576.29}{19.3}$

   (a) correct to 4 significant figures

   (b) correct to 1 decimal place   (2)

4. Determine, correct to 1 decimal places, 57% of 17.64 g   (2)

5. Express 54.7 mm as a percentage of 1.15 m, correct to 3 significant figures.   (3)

6. Evaluate the following:

   (a) $\dfrac{2^3 \times 2 \times 2^2}{2^4}$ (b) $\dfrac{(2^3 \times 16)^2}{(8 \times 2)^3}$

   (c) $\left(\dfrac{1}{4^2}\right)^{-1}$ (d) $(27)^{-\frac{1}{3}}$

   (e) $\dfrac{\left(\frac{3}{2}\right)^{-2} - \frac{2}{9}}{\left(\frac{2}{3}\right)^2}$ (14)

7. Express the following in both standard form and engineering notation

   (a) 1623   (b) 0.076   (c) $145\frac{2}{5}$ (3)

8. Determine the value of the following, giving the answer in both standard form and engineering notation

   (a) $5.9 \times 10^2 + 7.31 \times 10^2$

   (b) $2.75 \times 10^{-2} - 2.65 \times 10^{-3}$ (4)

9. Convert the following binary numbers to decimal form:

   (a) 1101   (b) 101101.0101   (5)

10. Convert the following decimal number to binary form:

    (a) 27   (b) 44.1875   (6)

11. Convert the following decimal numbers to binary, via octal:

    (a) 479   (b) 185.2890625   (6)

12. Convert (a) $5F_{16}$ into its decimal equivalent (b) 132_{10} into its hexadecimal equivalent (c) 110101011_2 into its hexadecimal equivalent   (6)

13. Evaluate the following, each correct to 4 significant figures:

    (a) 61.22^2 (b) $\dfrac{1}{0.0419}$ (c) $\sqrt{0.0527}$ (3)

14. Evaluate the following, each correct to 2 decimal places:

    (a) $\left(\dfrac{36.2^2 \times 0.561}{27.8 \times 12.83}\right)^3$

    (b) $\sqrt{\dfrac{14.69^2}{\sqrt{17.42} \times 37.98}}$ (7)

15. If 1.6 km = 1 mile, determine the speed of 45 miles/hour in kilometres per hour.   (3)

16. Evaluate B, correct to 3 significant figures, when $W = 7.20$, $v = 10.0$ and $g = 9.81$, given that $B = \dfrac{Wv^2}{2g}$.   (3)

Chapter 5

Algebra

5.1 Basic operations

Algebra is that part of mathematics in which the relations and properties of numbers are investigated by means of general symbols. For example, the area of a rectangle is found by multiplying the length by the breadth; this is expressed algebraically as $A = l \times b$, where A represents the area, l the length and b the breadth.

The basic laws introduced in arithmetic are generalised in algebra.

Let a, b, c and d represent any four numbers. Then:

(i) $a + (b + c) = (a + b) + c$

(ii) $a(bc) = (ab)c$

(iii) $a + b = b + a$

(iv) $ab = ba$

(v) $a(b + c) = ab + ac$

(vi) $\dfrac{a + b}{c} = \dfrac{a}{c} + \dfrac{b}{c}$

(vii) $(a + b)(c + d) = ac + ad + bc + bd$

Problem 1. Evaluate: $3ab - 2bc + abc$ when $a = 1$, $b = 3$ and $c = 5$

Replacing a, b and c with their numerical values gives:

$$3ab - 2bc + abc = 3 \times 1 \times 3 - 2 \times 3 \times 5$$
$$+ 1 \times 3 \times 5$$
$$= 9 - 30 + 15 = \mathbf{-6}$$

Problem 2. Find the value of $4p^2qr^3$, given the $p = 2$, $q = \dfrac{1}{2}$ and $r = 1\dfrac{1}{2}$

Replacing p, q and r with their numerical values gives:

$$4p^2qr^3 = 4(2)^2 \left(\frac{1}{2}\right) \left(\frac{3}{2}\right)^3$$

$$= 4 \times 2 \times 2 \times \frac{1}{2} \times \frac{3}{2} \times \frac{3}{2} \times \frac{3}{2} = \mathbf{27}$$

Problem 3. Find the sum of: $3x$, $2x$, $-x$ and $-7x$

The sum of the positive term is: $3x + 2x = 5x$

The sum of the negative terms is: $x + 7x = 8x$

Taking the sum of the negative terms from the sum of the positive terms gives:

$$5x - 8x = \mathbf{-3x}$$

Alternatively

$$3x + 2x + (-x) + (-7x) = 3x + 2x - x - 7x$$
$$= \mathbf{-3x}$$

Problem 4. Find the sum of $4a$, $3b$, c, $-2a$, $-5b$ and $6c$

Each symbol must be dealt with individually.

For the 'a' terms: $+4a - 2a = 2a$

For the 'b' terms: $+3b - 5b = -2b$

For the 'c' terms: $+c + 6c = 7c$

Thus
$$4a + 3b + c + (-2a) + (-5b) + 6c$$
$$= 4a + 3b + c - 2a - 5b + 6c$$
$$= \mathbf{2a - 2b + 7c}$$

Problem 5. Find the sum of: $5a - 2b$, $2a + c$, $4b - 5d$ and $b - a + 3d - 4c$

The algebraic expressions may be tabulated as shown below, forming columns for the a's, b's, c's and d's. Thus:

$$
\begin{array}{l}
+5a - 2b \\
+2a \qquad + c \\
\quad\; + 4b \qquad - 5d \\
-a + \; b - 4c + 3d \\
\hline
\textbf{6a + 3b - 3c - 2d}
\end{array}
$$

Adding gives:

Problem 6. Subtract $2x + 3y - 4z$ from $x - 2y + 5z$

$$
\begin{array}{l}
x \; - 2y + 5z \\
2x \; + 3y - 4z \\
\hline
\textbf{-x - 5y + 9z}
\end{array}
$$

Subtracting gives:

(Note that $+5z - -4z = +5z + 4z = 9z$)

An alternative method of subtracting algebraic expressions is to 'change the signs of the bottom line and add'. Hence:

$$
\begin{array}{l}
x - 2y + 5z \\
-2x - 3y + 4z \\
\hline
\textbf{-x - 5y + 9z}
\end{array}
$$

Adding gives:

Problem 7. Multiply $2a + 3b$ by $a + b$

Each term in the first expression is multiplied by a, then each term in the first expression is multiplied by b, and the two results are added. The usual layout is shown below.

$$
\begin{array}{l}
2a \; + \; 3b \\
a \; + \; b \\
\hline
\end{array}
$$

Multiplying by $a \rightarrow$ $\quad 2a^2 + 3ab$

Multiplying by $b \rightarrow$ $\qquad\; + 2ab + 3b^2$

Adding gives: $\quad\; \textbf{2a}^2 + \textbf{5ab} + \textbf{3b}^2$

Problem 8. Multiply $3x - 2y^2 + 4xy$ by $2x - 5y$

$$
\begin{array}{l}
3x \; - \; 2y^2 \; + 4xy \\
2x \; - \; 5y \\
\hline
\end{array}
$$

Multiplying by $2x \rightarrow$ $\quad 6x^2 - 4xy^2 + 8x^2y$

Multiplying by $-5y \rightarrow$ $\quad\quad\; - 20xy^2 \qquad - 15xy + 10y^3$

Adding gives: $\quad \textbf{6x}^2 - \textbf{24xy}^2 + \textbf{8x}^2\textbf{y} - \textbf{15xy} + \textbf{10y}^3$

Problem 9. Simplify: $2p \div 8pq$

$2p \div 8pq$ means $\dfrac{2p}{8pq}$. This can be reduced by cancelling as in arithmetic.

Thus: $\qquad \dfrac{2p}{8pq} = \dfrac{1\!\!\!\not2 \times \not p^1}{\not8_4 \times \not p_1 \times q} = \dfrac{1}{4q}$

Now try the following exercise

Exercise 18   Further problems on basic operations

1. Find the value of $2xy + 3yz - xyz$, when $x = 2, y = -2$ and $z = 4$ $[-16]$

2. Evaluate $3pq^3r^3$ when $p = \dfrac{2}{3}$, $q = -2$ and $r = -1$ $[-8]$

3. Find the sum of $3a, -2a, -6a, 5a$ and $4a$ $[4a]$

4. Add together $2a + 3b + 4c$, $-5a - 2b + c$, $4a - 5b - 6c$ $[a - 4b - c]$

5. Add together $3d + 4e$, $-2e + f$, $2d - 3f$, $4d - e + 2f - 3e$ $[9d - 2e]$

6. From $4x - 3y + 2z$ subtract $x + 2y - 3z$ $[3x - 5y + 5z]$

7. Subtract $\dfrac{3}{2}a - \dfrac{b}{3} + c$ from $\dfrac{b}{2} - 4a - 3c$ $\left[-5\dfrac{1}{2}a + \dfrac{5}{6}b - 4c \right]$

8. Multiply $3x + 2y$ by $x - y$ $[3x^2 - xy - 2y^2]$

9. Multiply $2a - 5b + c$ by $3a + b$ $[6a^2 - 13ab + 3ac - 5b^2 + bc]$

10. Simplify (i) $3a \div 9ab$ (ii) $4a^2b \div 2a$ $\left[\text{(i) } \dfrac{1}{3b} \quad \text{(ii) } 2ab \right]$

5.2 Laws of Indices

The laws of indices are:

(i) $a^m \times a^n = a^{m+n}$ (ii) $\dfrac{a^m}{a^n} = a^{m-n}$

(iii) $(a^m)^n = a^{mn}$ (iv) $a^{m/n} = \sqrt[n]{a^m}$

(v) $a^{-n} = \dfrac{1}{a^n}$ (vi) $a^0 = 1$

Problem 10. Simplify: $a^3 b^2 c \times a b^3 c^5$

Grouping like terms gives:

$$a^3 \times a \times b^2 \times b^3 \times c \times c^5$$

Using the first law of indices gives:

$$a^{3+1} \times b^{2+3} \times c^{1+5}$$

i.e. $a^4 \times b^5 \times c^6 = \boldsymbol{a^4 b^5 c^6}$

Problem 11. Simplify:

$$a^{1/2} b^2 c^{-2} \times a^{1/6} b^{1/2} c$$

Using the first law of indices,

$$a^{1/2} b^2 c^{-2} \times a^{(1/6)} b^{(1/2)} c$$

$$= a^{(1/2)+(1/6)} \times b^{2+(1/2)} \times c^{-2+1}$$

$$= \boldsymbol{a^{2/3} b^{5/2} c^{-1}}$$

Problem 12. Simplify: $\dfrac{a^3 b^2 c^4}{abc^{-2}}$ and evaluate when $a = 3$, $b = \dfrac{1}{8}$ and $c = 2$

Using the second law of indices,

$$\frac{a^3}{a} = a^{3-1} = a^2, \quad \frac{b^2}{b} = b^{2-1} = b$$

and $\dfrac{c^4}{c^{-2}} = c^{4-(-2)} = c^6$

Thus $\dfrac{a^3 b^2 c^4}{abc^{-2}} = \boldsymbol{a^2 b c^6}$

When $a = 3$, $b = \dfrac{1}{8}$ and $c = 2$,

$$a^2 b c^6 = (3)^2 \left(\frac{1}{8}\right)(2)^6 = (9)\left(\frac{1}{8}\right)(64) = \boldsymbol{72}$$

Problem 13. Simplify: $\dfrac{p^{1/2} q^2 r^{2/3}}{p^{1/4} q^{1/2} r^{1/6}}$ and evaluate when $p = 16$, $q = 9$ and $r = 4$, taking positive roots only

Using the second law of indices gives:

$$p^{(1/2)-(1/4)} q^{2-(1/2)} r^{(2/3)-(1/6)} = \boldsymbol{p^{1/4} q^{3/2} r^{1/2}}$$

When $p = 16$, $q = 9$ and $r = 4$,

$$p^{1/4} q^{3/2} r^{1/2} = (16)^{1/4} (9)^{3/2} (4)^{1/2}$$

$$= (\sqrt[4]{16})(\sqrt{9^3})(\sqrt{4})$$

$$= (2)(3^3)(2) = \boldsymbol{108}$$

Problem 14. Simplify: $\dfrac{x^2 y^3 + xy^2}{xy}$

Algebraic expressions of the form $\dfrac{a+b}{c}$ can be split into $\dfrac{a}{c} + \dfrac{b}{c}$. Thus

$$\frac{x^2 y^3 + xy^2}{xy} = \frac{x^2 y^3}{xy} + \frac{xy^2}{xy}$$

$$= x^{2-1} y^{3-1} + x^{1-1} y^{2-1}$$

$$= \boldsymbol{xy^2 + y}$$

(since $x^0 = 1$, from the sixth law of indices)

Problem 15. Simplify: $\dfrac{x^2 y}{xy^2 - xy}$

The highest common factor (HCF) of each of the three terms comprising the numerator and denominator is xy. Dividing each term by xy gives:

$$\frac{x^2 y}{xy^2 - xy} = \frac{\dfrac{x^2 y}{xy}}{\dfrac{xy^2}{xy} - \dfrac{xy}{xy}} = \boldsymbol{\frac{x}{y-1}}$$

Problem 16. Simplify: $(p^3)^{1/2}(q^2)^4$

Using the third law of indices gives:

$$p^{3 \times (1/2)} q^{2 \times 4} = \boldsymbol{p^{(3/2)} q^8}$$

Problem 17. Simplify: $\dfrac{(mn^2)^3}{(m^{1/2}n^{1/4})^4}$

The brackets indicate that each letter in the bracket must be raised to the power outside. Using the third law of indices gives:

$$\frac{(mn^2)^3}{(m^{1/2}n^{1/4})^4} = \frac{m^{1\times3}n^{2\times3}}{m^{(1/2)\times4}n^{(1/4)\times4}} = \frac{m^3n^6}{m^2n^1}$$

Using the second law of indices gives:

$$\frac{m^3n^6}{m^2n^1} = m^{3-2}n^{6-1} = \boldsymbol{mn^5}$$

Problem 18. Simplify:
$(a^3\sqrt{b}\sqrt{c^5})(\sqrt{a}\sqrt[3]{b^2}\,c^3)$ and evaluate when $a = \dfrac{1}{4}$, $b = 6$ and $c = 1$

Using the fourth law of indices, the expression can be written as:

$$(a^3b^{1/2}c^{5/2})(a^{1/2}b^{2/3}c^3)$$

Using the first law of indices gives:

$$a^{3+(1/2)}b^{(1/2)+(2/3)}c^{(5/2)+3} = a^{7/2}b^{7/6}c^{11/2}$$

It is usual to express the answer in the same form as the question. Hence

$$a^{7/2}b^{7/6}c^{11/2} = \sqrt{a^7}\sqrt[6]{b^7}\sqrt{c^{11}}$$

When $a = \dfrac{1}{4}$, $b = 64$ and $c = 1$,

$$\sqrt{a^7}\sqrt[6]{b^7}\sqrt{c^{11}} = \sqrt{\left(\frac{1}{4}\right)^7}\left(\sqrt[6]{64^7}\right)\left(\sqrt{1^{11}}\right)$$

$$= \left(\frac{1}{2}\right)^7(2)^7(1) = \boldsymbol{1}$$

Problem 19. Simplify: $\dfrac{d^2e^2f^{1/2}}{(d^{3/2}ef^{5/2})^2}$ expressing the answer with positive indices only

Using the third law of indices gives:

$$\frac{d^2e^2f^{1/2}}{(d^{3/2}ef^{5/2})^2} = \frac{d^2e^2f^{1/2}}{d^3e^2f^5}$$

Using the second law of indices gives:

$$d^{2-3}e^{2-2}f^{(1/2)-5} = d^{-1}e^0f^{-9/2}$$

$$= d^{-1}f^{(-9/2)}\text{ since }e^0 = 1$$

from the sixth law of indices

$$= \frac{1}{df^{9/2}}$$

from the fifth law of indices

Problem 20. Simplify: $\dfrac{(x^2y^{1/2})(\sqrt{x}\sqrt[3]{y^2})}{(x^5y^3)^{1/2}}$

Using the third and fourth laws of indices gives:

$$\frac{(x^2y^{1/2})(\sqrt{x}\sqrt[3]{y^2})}{(x^5y^3)^{1/2}} = \frac{(x^2y^{1/2})(x^{1/2}y^{2/3})}{x^{5/2}y^{3/2}}$$

Using the first and second laws of indices gives:

$$x^{2+(1/2)-(5/2)}y^{(1/2)+(2/3)-(3/2)} = x^0y^{-1/3}$$

$$= \boldsymbol{y^{-1/3}}$$

$$\text{or }\frac{1}{y^{1/3}}\text{ or }\frac{1}{\sqrt[3]{y}}$$

from the fifth and sixth law of indices.

Now try the following exercise

Exercise 19   Further problems on laws of indices

1. Simplify $(x^2y^3z)(x^3yz^2)$ and evaluate when $x = \dfrac{1}{2}$, $y = 2$ and $z = 3$

$$\left[x^5y^4z^3, \quad 13\frac{1}{2}\right]$$

2. Simplify $(a^{3/2}bc^{-3})(a^{1/2}bc^{-1/2}c)$ and evaluate when $a = 3$, $b = 4$ and $c = 2$

$$\left[a^2b^{1/2}c^{-2}, \quad \pm4\frac{1}{2}\right]$$

3. Simplify: $\dfrac{a^5bc^3}{a^2b^3c^2}$ and evaluate when $a=\dfrac{3}{2}$, $b=\dfrac{1}{2}$ and $c=\dfrac{2}{3}$ $\qquad [a^3b^{-2}c,\ 9]$

In Problems 4 to 10, simplify the given expressions:

4. $\dfrac{x^{1/5}y^{1/2}z^{1/3}}{x^{-1/2}y^{1/3}z^{-1/6}}$ $\qquad [x^{7/10}y^{1/6}z^{1/2}]$

5. $\dfrac{a^2b+a^3b}{a^2b^2}$ $\qquad \left[\dfrac{1+a}{b}\right]$

6. $\dfrac{p^3q^2}{pq^2-p^2q}$ $\qquad \left[\dfrac{p^2q}{q-p}\right]$

7. $(a^2)^{1/2}(b^2)^3(c^{1/2})^3$ $\qquad [ab^6c^{3/2}]$

8. $\dfrac{(abc)^2}{(a^2b^{-1}c^{-3})^3}$ $\qquad [a^{-4}b^5c^{11}]$

9. $(\sqrt{x}\sqrt{y^3}\sqrt[3]{z^2})(\sqrt{x}\sqrt{y^3}\sqrt{z^3})$ $\qquad [xy^3\sqrt[6]{z^{13}}]$

10. $\dfrac{(a^3b^{1/2}c^{-1/2})(ab)^{1/3}}{(\sqrt{a^3}\sqrt{bc})}$

$\qquad \left[a^{11/6}b^{1/3}c^{-3/2} \text{ or } \dfrac{\sqrt[6]{a^{11}}\sqrt[3]{b}}{\sqrt{c^3}}\right]$

5.3 Brackets and factorisation

When two or more terms in an algebraic expression contain a common factor, then this factor can be shown outside of a **bracket**. For example

$$ab+ac=a(b+c)$$

which is simply the reverse of law (v) of algebra on page 34, and

$$6px+2py-4pz=2p(3x+y-2z)$$

This process is called **factorisation**.

Problem 21. Remove the brackets and simplify the expression:

$$(3a+b)+2(b+c)-4(c+d)$$

Both b and c in the second bracket have to be multiplied by 2, and c and d in the third bracket by -4 when the brackets are removed. Thus:

$$(3a+b)+2(b+c)-4(c+d)$$
$$=3a+b+2b+2c-4c-4d$$

Collecting similar terms together gives:

$$\mathbf{3a+3b-2c-4d}$$

Problem 22. Simplify:

$$a^2-(2a-ab)-a(3b+a)$$

When the brackets are removed, both $2a$ and $-ab$ in the first bracket must be multiplied by -1 and both $3b$ and a in the second bracket by $-a$. Thus

$$a^2-(2a-ab)-a(3b+a)$$
$$=a^2-2a+ab-3ab-a^2$$

Collecting similar terms together gives: $-2a-2ab$ Since $-2a$ is a common factor, the answer can be expressed as: $\mathbf{-2a(1+b)}$

Problem 23. Simplify: $(a+b)(a-b)$

Each term in the second bracket has to be multiplied by each term in the first bracket. Thus:

$$(a+b)(a-b)=a(a-b)+b(a-b)$$
$$=a^2-ab+ab-b^2$$
$$=\mathbf{a^2-b^2}$$

Alternatively

$$
\begin{array}{r}
a + b \\
a - b \\
\hline
\end{array}
$$

Multiplying by $a\rightarrow \quad a^2+ab$

Multiplying by $-b\rightarrow \quad -ab-b^2$

Adding gives: $\quad \dfrac{a^2 \qquad -b^2}{}$

Problem 24. Simplify: $(3x-3y)^2$

$$(2x-3y)^2=(2x-3y)(2x-3y)$$
$$=2x(2x-3y)-3y(2x-3y)$$
$$=4x^2-6xy-6xy+9y^2$$
$$=\mathbf{4x^2-12xy+9y^2}$$

Alternatively,

$$2x - 3y$$
$$2x - 3y$$

Multiplying by $2x \rightarrow$ $\dfrac{4x^2 - 6xy}{}$

Multiplying by $-3y \rightarrow$ $\dfrac{\qquad - 6xy + 9y^2}{}$

Adding gives: $4x^2 - 12xy + 9y^2$

Problem 25. Remove the brackets from the expression: $2[p^2 - 3(q+r) + q^2]$

In this problem there are two brackets and the 'inner' one is removed first.

Hence, $2[p^2 - 3(q+r) + q^2]$

$$= 2[p^2 - 3q - 3r + q^2]$$
$$= 2p^2 - 6q - 6r + 2q^2$$

Problem 26. Remove the brackets and simplify the expression:

$$2a - [3\{2(4a - b) - 5(a + 2b)\} + 4a]$$

Removing the innermost brackets gives:

$$2a - [3\{8a - 2b - 5a - 10b\} + 4a]$$

Collecting together similar terms gives:

$$2a - [3\{3a - 12b\} + 4a]$$

Removing the 'curly' brackets gives:

$$2a - [9a - 36b + 4a]$$

Collecting together similar terms gives:

$$2a - [13a - 36b]$$

Removing the outer brackets gives:

$$2a - 13a - 36b$$

i.e. $-11a + 36b$ or $36b - 11a$

(see law (iii), page 38)

Problem 27. Simplify:

$$x(2x - 4y) - 2x(4x + y)$$

Removing brackets gives:

$$2x^2 - 4xy - 8x^2 - 2xy$$

Collecting together similar terms gives:

$$-6x^2 - 6xy$$

Factorising gives:

$$-6x(x + y)$$

since $-6x$ is common to both terms

Problem 28. Factorise: (a) $xy - 3xz$
(b) $4a^2 + 16ab^3$ (c) $3a^2b - 6ab^2 + 15ab$

For each part of this problem, the HCF of the terms will become one of the factors. Thus:

(a) $xy - 3xz = x(y - 3z)$

(b) $4a^2 + 16ab^3 = 4a(a + 4b^3)$

(c) $3a^2b - 6ab^2 + 15ab = 3ab(a - 2b + 5)$

Problem 29. Factorise: $ax - ay + bx - by$

The first two terms have a common factor of a and the last two terms a common factor of b. Thus:

$$ax - ay + bx - by = a(x - y) + b(x - y)$$

The two newly formed terms have a common factor of $(x - y)$. Thus:

$$a(x - y) + b(x - y) = (x - y)(a + b)$$

Problem 30. Factorise:

$$2ax - 3ay + 2bx - 3by$$

---

a is a common factor of the first two terms and b a common factor of the last two terms. Thus:

$$2ax - 3ay + 2bx - 3by$$
$$= a(2x - 3y) + b(2x - 3y)$$

$(2x - 3y)$ is now a common factor thus:

$$a(2x - 3y) + b(2x - 3y)$$
$$= (2x - 3y)(a + b)$$

Alternatively, $2x$ is a common factor of the original first and third terms and $-3y$ is a common factor of the second and fourth terms. Thus:

$$2ax - 3ay + 2bx - 3by$$
$$= 2x(a + b) - 3y(a + b)$$

$(a+b)$ is now a common factor thus:

$$2x(a+b) - 3y(a+b) = \mathbf{(a+b)(2x-3y)}$$

as before

Problem 31. Factorise $x^3 + 3x^2 - x - 3$

x^2 is a common factor of the first two terms, thus:

$$x^3 + 3x^2 - x - 3 = x^2(x+3) - x - 3$$

-1 is a common factor of the last two terms, thus:

$$x^2(x+3) - x - 3 = x^2(x+3) - 1(x+3)$$

$(x+3)$ is now a common factor, thus:

$$x^2(x+3) - 1(x+3) = \mathbf{(x+3)(x^2-1)}$$

Now try the following exercise

Exercise 20 Further problems on brackets and factorisation

In Problems 1 to 9, remove the brackets and simplify where possible:

1. $(x+2y)+(2x-y)$ $\quad [3x+y]$
2. $2(x-y)-3(y-x)$ $\quad [5(x-y)]$
3. $2(p+3q-r)-4(r-q+2p)+p$ $\quad [-5p+10q-6r]$
4. $(a+b)(a+2b)$ $\quad [a^2+3ab+2b^2]$
5. $(p+q)(3p-2q)$ $\quad [3p^2+pq-2q^2]$
6. (i) $(x-2y)^2$ (ii) $(3a-b)^2$ $\quad$ [(i) $x^2-4xy+4y^2$ (ii) $9a^2-6ab+b^2$]
7. $3a+2[a-(3a-2)]$ $\quad [4-a]$
8. $2-5[a(a-2b)-(a-b)^2]$ $\quad [2+5b^2]$
9. $24p-[2\{3(5p-q)-2(p+2q)\}+3q]$ $\quad [11q-2p]$

In Problem 10 to 12, factorise:

10. (i) $pb+2pc$ (ii) $2q^2+8qn$ $\quad$ [(i) $p(b+2c)$ (ii) $2q(q+4n)$]
11. (i) $21a^2b^2-28ab$ (ii) $2xy^2+6x^2y+8x^3y$ $\quad$ [(i) $7ab(3ab-4)$ (ii) $2xy(y+3x+4x^2)$]
12. (i) $ay+by+a+b$ (ii) $px+qx+py+qy$ (iii) $2ax+3ay-4bx-6by$ $\quad$ [(i) $(a+b)(y+1)$ (ii) $(p+q)(x+y)$ (iii) $(a-2b)(2x+3y)$]

5.4 Fundamental laws and precedence

The **laws of precedence** which apply to arithmetic also apply to algebraic expressions. The order is Brackets, Of, Division, Multiplication, Addition and Subtraction (i.e. **BODMAS**).

Problem 32. Simplify: $2a+5a\times 3a-a$

Multiplication is performed before addition and subtraction thus:

$$2a + 5a \times 3a - a = 2a + 15a^2 - a$$
$$= \mathbf{a + 15a^2} \text{ or } \mathbf{a(1+15a)}$$

Problem 33. Simplify: $(a+5a)\times 2a-3a$

The order of precedence is brackets, multiplication, then subtraction. Hence

$$(a+5a)\times 2a - 3a = 6a \times 2a - 3a$$
$$= \mathbf{12a^2 - 3a}$$
$$\text{or } \mathbf{3a(4a-1)}$$

Problem 34. Simplify: $a+5a\times(2a-3a)$

The order of precedence is brackets, multiplication, then subtraction. Hence

$$a + 5a \times (2a - 3a) = a + 5a \times -a$$
$$= a + -5a^2$$
$$= \mathbf{a - 5a^2} \text{ or } \mathbf{a(1-5a)}$$

Problem 35. Simplify: $a\div 5a+2a-3a$

The order of precedence is division, then addition and subtraction. Hence

$$a \div 5a + 2a - 3a = \frac{a}{5a} + 2a - 3a$$

$$= \frac{1}{5} + 2a - 3a = \mathbf{\frac{1}{5} - a}$$

Problem 36. Simplify:

$$a \div (5a + 2a) - 3a$$

The order of precedence is brackets, division and subtraction. Hence

$$a \div (5a + 2a) - 3a = a \div 7a - 3a$$

$$= \frac{a}{7a} - 3a = \mathbf{\frac{1}{7} - 3a}$$

Problem 37. Simplify:

$$3c + 2c \times 4c + c \div 5c - 8c$$

The order of precedence is division, multiplication, addition and subtraction. Hence:

$$3c + 2c \times 4c + c \div 5c - 8c$$

$$= 3c + 2c \times 4c + \frac{c}{5c} - 8c$$

$$= 3c + 8c^2 + \frac{1}{5} - 8c$$

$$= \mathbf{8c^2 - 5c + \frac{1}{5}} \quad \text{or} \quad \mathbf{c(8c - 5) + \frac{1}{5}}$$

Problem 38. Simplify:

$$3c + 2c \times 4c + c \div (5c - 8c)$$

The order of precedence is brackets, division, multiplication and addition. Hence,

$$3c + 2c \times 4c + c \div (5c - 8c)$$

$$= 3c + 2c \times 4c + c \div -3c$$

$$= 3c + 2c \times 4c + \frac{c}{-3c}$$

Now $\dfrac{c}{-3c} = \dfrac{1}{-3}$

Multiplying numerator and denominator by -1 gives:

$$\frac{1 \times -1}{-3 \times -1} \quad \text{i.e.} \quad -\frac{1}{3}$$

Hence:

$$3c + 2c \times 4c + \frac{c}{-3c}$$

$$= 3c + 2c \times 4c - \frac{1}{3}$$

$$= \mathbf{3c + 8c^2 - \frac{1}{3}} \quad \text{or} \quad \mathbf{c(3 + 8c) - \frac{1}{3}}$$

Problem 39. Simplify:

$$(3c + 2c)(4c + c) \div (5c - 8c)$$

The order of precedence is brackets, division and multiplication. Hence

$$(3c + 2c)(4c + c) \div (5c - 8c)$$

$$= 5c \times 5c \div -3c = 5c \times \frac{5c}{-3c}$$

$$= 5c \times -\frac{5}{3} = \mathbf{-\frac{25}{3}c}$$

Problem 40. Simplify:

$$(2a - 3) \div 4a + 5 \times 6 - 3a$$

The bracket around the $(2a - 3)$ shows that both $2a$ and -3 have to be divided by $4a$, and to remove the bracket the expression is written in fraction form.

Hence, $\qquad (2a - 3) \div 4a + 5 \times 6 - 3a$

$$= \frac{2a - 3}{4a} + 5 \times 6 - 3a$$

$$= \frac{2a - 3}{4a} + 30 - 3a$$

$$= \frac{2a}{4a} - \frac{3}{4a} + 30 - 3a$$

$$= \frac{1}{2} - \frac{3}{4a} + 30 - 3a$$

$$= \mathbf{30\frac{1}{2} - \frac{3}{4a} - 3a}$$

Problem 41. Simplify:

$$\frac{1}{3} \text{ of } 3p + 4p(3p - p)$$

Applying BODMAS, the expression becomes

$$\frac{1}{3} \text{ of } 3p + 4p \times 2p$$

and changing 'of' to '×' gives:

$$\frac{1}{3} \times 3p + 4p \times 2p$$

i.e. $p + 8p^2$ or $p(1 + 8p)$

Now try the following exercise

Exercise 21   Further problems on fundamental laws and precedence

Simplify the following:

1. $2x \div 4x + 6x$ $\left[\dfrac{1}{2} + 6x\right]$

2. $2x \div (4x + 6x)$ $\left[\dfrac{1}{5}\right]$

3. $3a - 2a \times 4a + a$ $[4a(1 - 2a)]$

4. $3a - 2a(4a + a)$ $[a(3 - 10a)]$

5. $2y + 4 \div 6y + 3 \times 4 - 5y$ $\left[\dfrac{2}{3y} - 3y + 12\right]$

6. $2y + 4 \div 6y + 3(4 - 5y)$ $\left[\dfrac{2}{3y} + 12 - 13y\right]$

7. $3 \div y + 2 \div y + 1$ $\left[\dfrac{5}{y} + 1\right]$

8. $p^2 - 3pq \times 2p \div 6q + pq$ $[pq]$

9. $(x + 1)(x - 4) \div (2x + 2)$ $\left[\dfrac{1}{2}(x - 4)\right]$

10. $\dfrac{1}{4}$ of $2y + 3y(2y - y)$ $\left[y\left(\dfrac{1}{2} + 3y\right)\right]$

5.5   Direct and inverse proportionality

An expression such as $y = 3x$ contains two variables. For every value of x there is a corresponding value of y. The variable x is called the **independent variable** and y is called the **dependent variable**.

When an increase or decrease in an independent variable leads to an increase or decrease of the same proportion in the dependent variable this is termed **direct proportion**. If $y = 3x$ then y is directly proportional to x, which may be written as $y \alpha x$ or $y = kx$, where k is called the **coefficient of proportionality** (in this case, k being equal to 3).

When an increase in an independent variable leads to a decrease of the same proportion in the dependent variable (or vice versa) this is termed **inverse proportion**. If y is inversely proportional to x then $y \alpha \dfrac{1}{x}$ or $y = k/x$. Alternatively, $k = xy$, that is, for inverse proportionality the product of the variable is constant.

Examples of laws involving direct and inverse proportional in science include:

(i)   **Hooke's law**, which states that within the elastic limit of a material, the strain ε produced is directly proportional to the stress, σ, producing it, i.e. $\varepsilon \alpha \sigma$ or $\varepsilon = k\sigma$.

(ii) **Charles's law**, which states that for a given mass of gas at constant pressure the volume V is directly proportional to its thermodynamic temperature T, i.e. $V \alpha T$ or $V = kT$.

(iii) **Ohm's law**, which states that the current I flowing through a fixed resistor is directly proportional to the applied voltage V, i.e. $I \alpha V$ or $I = kV$.

(iv) **Boyle's law**, which states that for a gas at constant temperature, the volume V of a fixed mass of a gas is inversely proportional to its absolute pressure p, i.e. $p \alpha (1/V)$ or $p = k/V$, i.e. $pV = k$

Problem 42.   If y is directly proportional to x and $y = 2.48$ when $x = 0.4$, determine (a) the coefficient of proportionality and (b) the value of y when $x = 0.65$

(a) $y \alpha x$, i.e. $y = kx$. If $y = 2.48$ when $x = 0.4$, $2.48 = k(0.4)$
     Hence the coefficient of proportionality,

$$k = \frac{2.48}{0.4} = \mathbf{6.2}$$

(b) $y = kx$, hence, when $x = 0.65$,
     $y = (6.2)(0.65) = \mathbf{4.03}$

Problem 43.   Hooke's law states that stress σ is directly proportional to strain ε within the elastic limit of a material. When, for mild steel, the stress is

25×10^6 Pascals, the strain is 0.000125. Determine (a) the coefficient of proportionality and (b) the value of strain when the stress is 18×10^6 Pascals

(a) $\sigma \alpha \varepsilon$, i.e. $\sigma = k\varepsilon$, from which $k = \sigma/\varepsilon$. Hence the coefficient of proportionality,

$$k = \frac{25 \times 10^6}{0.000125} = \mathbf{200 \times 10^9\, pascals}$$

(The coefficient of proportionality k in this case is called Young's Modulus of Elasticity)

(b) Since $\sigma = k\varepsilon$, $\varepsilon = \sigma/k$
Hence when $\sigma = 18 \times 10^6$,

$$\text{strain } \varepsilon = \frac{18 \times 10^6}{200 \times 10^9} = \mathbf{0.00009}$$

Problem 44. The electrical resistance R of a piece of wire is inversely proportional to the cross-sectional area A. When $A = 5\,\text{mm}^2$, $R = 7.02$ ohms. Determine (a) the coefficient of proportionality and (b) the cross-sectional area when the resistance is 4 ohms

(a) $R \alpha \dfrac{1}{A}$, i.e. $R = k/A$ or $k = RA$. Hence, when $R = 7.2$ and $A = 5$, the coefficient of proportionality, $k = (7.2)(5) = \mathbf{36}$

(b) Since $k = RA$ then $A = k/R$
When $R = 4$, the cross-sectional area,

$$A = \frac{36}{4} = \mathbf{9\,mm^2}$$

Problem 45. Boyle's law states that at constant temperature, the volume V of a fixed mass of gas is inversely proportional to its absolute pressure p. If a gas occupies a volume of $0.08\,\text{m}^3$ at a pressure of 1.5×10^6 Pascals determine (a) the coefficient of proportionality and (b) the volume if the pressure is changed to 4×10^6 Pascals

(a) $V \alpha \dfrac{1}{p}$ i.e. $V = k/p$ or $k = pV$
Hence the coefficient of proportionality,

$$k = (1.5 \times 10^6)(0.08) = \mathbf{0.12 \times 10^6}$$

(b) Volume $V = \dfrac{k}{p} = \dfrac{0.12 \times 10^6}{4 \times 10^6} = \mathbf{0.03\,m^3}$

Now try the following exercise

Exercise 22   Further problems on direct and inverse proportionality

1. If p is directly proportional to q and $p = 37.5$ when $q = 2.5$, determine (a) the constant of proportionality and (b) the value of p when q is 5.2   [(a) 15   (b) 78]

2. Charles's law states that for a given mass of gas at constant pressure the volume is directly proportional to its thermodynamic tempera-ture. A gas occupies a volume of 2.25 litres at 300 K. Determine (a) the constant of propor-tionality, (b) the volume at 420 K, and (c) the temperature when the volume is 2.625 litres.
   [(a) 0.0075   (b) 3.15 litres   (c) 350 K]

3. Ohm's law states that the current flowing in a fixed resistor is directly proportional to the applied voltage. When 30 volts is applied across a resistor the current flowing through the resistor is 2.4×10^{-3} amperes. Deter-mine (a) the constant of proportionality, (b) the current when the voltage is 52 volts and (c) the voltage when the current is 3.6×10^{-3} amperes.
   $$\begin{bmatrix} \text{(a) } 0.00008 \quad \text{(b) } 4.16 \times 10^{-3}\,\text{A} \\ \text{(c) } 45\,\text{V} \end{bmatrix}$$

4. If y is inversely proportional to x and $y = 15.3$ when $x = 0.6$, determine (a) the coefficient of proportionality, (b) the value of y when x is 1.5, and (c) the value of x when y is 27.2
   [(a) 9.18   (b) 6.12   (c) 0.3375]

5. Boyle's law states that for a gas at constant temperature, the volume of a fixed mass of gas is inversely proportional to its absolute pres-sure. If a gas occupies a volume of $1.5\,\text{m}^3$ at a pressure of 200×10^3 Pascals, determine (a) the constant of proportionality, (b) the volume when the pressure is 800×10^3 Pascals and (c) the pressure when the volume is $1.25\,\text{m}^3$.
   $$\begin{bmatrix} \text{(a) } 300 \times 10^3 \quad \text{(b) } 0.375\,\text{m}^2 \\ \text{(c) } 240 \times 10^3\,\text{Pa} \end{bmatrix}$$

Chapter 6

Further algebra

6.1 Polynominal division

Before looking at long division in algebra let us revise long division with numbers (we may have forgotten, since calculators do the job for us!)

For example, $\dfrac{208}{16}$ is achieved as follows:

$$16\overline{)208}$$
with quotient 13, 16, 48, 48, $\therefore$

(1) 16 divided into 2 won't go
(2) 16 divided into 20 goes 1
(3) Put 1 above the zero
(4) Multiply 16 by 1 giving 16
(5) Subtract 16 from 20 giving 4
(6) Bring down the 8
(7) 16 divided into 48 goes 3 times
(8) Put the 3 above the 8
(9) $3 \times 16 = 48$
(10) $48 - 48 = 0$

Hence $\dfrac{208}{16} = \mathbf{13}$ exactly

Similarly, $\dfrac{172}{15}$ is laid out as follows:

$$15\overline{)172}$$
with quotient 11, 15, 22, 15, 7

Hence $\dfrac{175}{15} = 11$ remainder 7 or $11 + \dfrac{7}{15} = \mathbf{11\dfrac{7}{15}}$

Below are some examples of division in algebra, which in some respects, is similar to long division with numbers.

(Note that a **polynomial** is an expression of the form

$$f(x) = a + bx + cx^2 + dx^3 + \cdots$$

and **polynomial division** is sometimes required when resolving into partial fractions — see Chapter 7).

Problem 1. Divide $2x^2 + x - 3$ by $x - 1$

$2x^2 + x - 3$ is called the **dividend** and $x - 1$ the **divisor**. The usual layout is shown below with the dividend and divisor both arranged in descending powers of the symbols.

$$x - 1\overline{)2x^2 + x - 3}$$
with quotient $2x + 3$, $2x^2 - 2x$, $3x - 3$, $3x - 3$, $\therefore$

Dividing the first term of the dividend by the first term of the divisor, i.e. $\dfrac{2x^2}{x}$ gives $2x$, which is put above the first term of the dividend as shown. The divisor is then multiplied by $2x$, i.e. $2x(x - 1) = 2x^2 - 2x$, which is placed under the dividend as shown. Subtracting gives $3x - 3$. The process is then repeated, i.e. the first term of the divisor, x, is divided into $3x$, giving $+3$, which is placed above the dividend as shown. Then $3(x - 1) = 3x - 3$ which is placed under the $3x - 3$. The remainder, on subtraction, is zero, which completes the process.

Thus $(2x^2 + x - 3) \div (x - 1) = (2x + 3)$

[A check can be made on this answer by multiplying $(2x+3)$ by $(x-1)$ which equals $2x^2+x-3$]

Problem 2.   Divide $3x^3+x^2+3x+5$ by $x+1$

$$
\begin{array}{r}
(1)\quad(4)\quad(7) \\
3x^2-2x+5 \\
x+1\,\overline{)\,3x^3+x^2+3x+5} \\
\underline{3x^3+3x^2} \\
-2x^2+3x+5 \\
\underline{-2x^2-2x} \\
5x+5 \\
\underline{5x+5} \\
\cdot\quad\cdot
\end{array}
$$

(1) x into $3x^3$ goes $3x^2$. Put $3x^2$ above $3x^3$
(2) $3x^2(x+1)=3x^3+3x^2$
(3)   Subtract
(4) x into $-2x^2$ goes $-2x$. Put $-2x$ above the dividend
(5) $-2x(x+1)=-2x^2-2x$
(6)   Subtract
(7) x into $5x$ goes 5. Put 5 above the dividend
(8) $5(x+1)=5x+5$
(9)   Subtract

Thus $\dfrac{3x^3+x^2+3x+5}{x+1}=3x^2-2x+5$

Problem 3.   Simplify $\dfrac{x^3+y^3}{x+y}$

$$
\begin{array}{r}
(1)\quad(4)\quad\quad(7) \\
x^2-xy\;+\;y^2 \\
x+y\,\overline{)\,x^3+\;0\;+\;0\;+\;y^3} \\
\underline{x^3+x^2y} \\
-x^2y+y^3 \\
\underline{-x^2y-xy^2} \\
xy^2+y^3 \\
\underline{xy^2+y^3} \\
\cdot\quad\cdot
\end{array}
$$

(1) x into x^3 goes x^2. Put x^2 above x^3 of dividend
(2) $x^2(x+y)=x^3+x^2y$
(3)   Subtract
(4) x into $-x^2y$ goes $-xy$. Put $-xy$ above dividend
(5) $-xy(x+y)=-x^2y-xy^2$
(6)   Subtract

(7) x into xy^2 goes y^2. Put y^2 above divided
(8) $y^2(x+y)=xy^2+y^3$
(9)   Subtract

Thus $\dfrac{x^3+y^3}{x+y}=x^2-xy+y^2$

The zero's shown in the dividend are not normally shown, but are included to clarify the subtraction process and to keep similar terms in their respective columns.

Problem 4.   Divide (x^2+3x-2) by $(x-2)$

$$
\begin{array}{r}
x+5 \\
x-2\,\overline{)\,x^2+3x-2} \\
\underline{x^2-2x} \\
5x-2 \\
\underline{5x-10} \\
8
\end{array}
$$

Hence $\dfrac{x^2+3x-2}{x-2}=x+5+\dfrac{8}{x-2}$

Problem 5.   Divided $4a^3-6a^2b+5b^3$ by $2a-b$

$$
\begin{array}{r}
2a^2-2ab-b^2 \\
2a-b\,\overline{)\,4a^3-6a^2b+5b^3} \\
\underline{4a^3-2a^2b} \\
-4a^2b+5b^3 \\
\underline{-4a^2b+2ab^2} \\
-2ab^2+5b^3 \\
\underline{-2ab^2+b^3} \\
4b^3
\end{array}
$$

Thus

$$\frac{4a^3-6a^2b+5b^3}{2a-b}=2a^2-2ab-b^2+\frac{4b^3}{2a-b}$$

Now try the following exercise

Exercise 23   Further problem on polynomial division

1.   Divide $(2x^2+xy-y^2)$ by $(x+y)$ $[2x-y]$

2.   Divide $(3x^2+5x-2)$ by $(x+2)$ $[3x-1]$

3. Determine $(10x^2 + 11x - 6) \div (2x + 3)$

$$[5x - 2]$$

4. Find: $\dfrac{14x^2 - 19x - 3}{2x - 3}$ $\quad [7x + 1]$

5. Divide $(x^3 + 3x^2 y + 3xy^2 + y^3)$ by $(x + y)$

$$[x^2 + 2xy + y^2]$$

6. Find $(5x^2 - x + 4) \div (x - 1)$

$$\left[5x + 4 + \dfrac{8}{x - 1} \right]$$

7. Divide $(3x^3 + 2x^2 - 5x + 4)$ by $(x + 2)$

$$\left[3x^3 - 4x + 3 - \dfrac{2}{x + 2} \right]$$

8. Determine: $\dfrac{5x^4 + 3x^3 - 2x + 1}{x - 3}$

$$\left[5x^3 + 18x^2 + 54x + 160 + \dfrac{481}{x - 3} \right]$$

6.2 The factor theorem

There is a simple relationship between the factors of a quadratic expression and the roots of the equation obtained by equating the expression to zero.

For example, consider the quadratic equation $x^2 + 2x - 8 = 0$

To solve this we may factorise the quadratic expression $x^2 + 2x - 8$ giving $(x - 2)(x + 4)$

Hence $(x - 2)(x + 4) = 0$

Then, if the product of two number is zero, one or both of those numbers must equal zero. Therefore,

either $(x - 2) = 0$, from which, $x = 2$

or $\quad (x + 4) = 0$, from which, $x = -4$

It is clear then that a factor of $(x - 2)$ indicates a root of $+2$, while a factor of $(x + 4)$ indicates a root of -4. In general, we can therefore say that:

a factor of $(x - a)$ corresponds to a root of $x = a$

In practice, we always deduce the roots of a simple quadratic equation from the factors of the quadratic expression, as in the above example. However, we could reverse this process. If, by trial and error, we could determine that $x = 2$ is a root of the equation $x^2 + 2x - 8 = 0$

we could deduce at once that $(x - 2)$ is a factor of the expression $x^2 + 2x - 8$. We wouldn't normally solve quadratic equations this way — but suppose we have to factorise a cubic expression (i.e. one in which the highest power of the variable is 3). A cubic equation might have three simple linear factors and the difficulty of discovering all these factors by trial and error would be considerable. It is to deal with this kind of case that we use the **factor theorem**. This is just a generalised version of what we established above for the quadratic expression. The factor theorem provides a method of factorising any polynomial, $f(x)$, which has simple factors.

A statement of the **factor theorem** says:

> **'if $x = a$ is a root of the equation $f(x) = 0$, then $(x - a)$ is a factor of $f(x)$'**

The following worked problems show the use of the factor theorem.

Problem 6. Factorise $x^3 - 7x - 6$ and use it to solve the cubic equation: $x^3 - 7x - 6 = 0$

Let $f(x) = x^3 - 7x - 6$

If $x = 1$, then $f(1) = 1^3 - 7(1) - 6 = -12$

If $x = 2$, then $f(2) = 2^3 - 7(2) - 6 = -12$

If $x = 3$, then $f(3) = 3^3 - 7(3) - 6 = 0$

If $f(3) = 0$, then $(x - 3)$ is a factor — from the factor theorem.

We have a choice now. We can divide $x^3 - 7x - 6$ by $(x - 3)$ or we could continue our 'trial and error' by substituting further values for x in the given expression — and hope to arrive at $f(x) = 0$.

Let us do both ways. Firstly, dividing out gives:

$$
\begin{array}{r}
x^2 + 3x + 2 \\
x - 3 \overline{\smash{)}\ x^3 + 0\ \ - 7x - 6} \\
\underline{x^3 - 3x^2} \\
3x^2 - 7x - 6 \\
\underline{3x^2 - 9x} \\
2x - 6 \\
\underline{2x - 6} \\
\cdot \quad \cdot
\end{array}
$$

Hence $\dfrac{x^3 - 7x - 6}{x - 3} = x^2 + 3x + 2$

i.e. $x^3 - 7x - 6 = (x - 3)(x^2 + 3x + 2)$

$x^3 + 3x + 2$ factorises 'on sight' as $(x + 1)(x + 2)$

Therefore

$$x^3 - 7x - 6 = (x - 3)(x + 1)(x + 2)$$

A second method is to continue to substitute values of x into $f(x)$.

Our expression for $f(3)$ was $3^3 - 7(3) - 6$. We can see that if we continue with positive values of x the first term will predominate such that $f(x)$ will not be zero. Therefore let us try some negative values for x:

$f(-1) = (-1)^3 - 7(-1) - 6 = 0$; hence $(x + 1)$ is a factor (as shown above).

Also, $f(-2) = (-2)^3 - 7(-2) - 6 = 0$; hence $(x + 2)$ is a factor (also as shown above).

To solve $x^3 - 7x - 6 = 0$, we substitute the factors, i.e.

$$(x - 3)(x + 1)(x + 2) = 0$$

from which, $x = 3, x = -1$ and $x = -2$

Note that the values of x, i.e. 3, -1 and -2, are all factors of the constant term, i.e. the 6. This can give us a clue as to what values of x we should consider.

Problem 7. Solve the cubic equation $x^3 - 2x^2 - 5x + 6 = 0$ by using the factor theorem

---

Let $f(x) = x^3 - 2x^2 - 5x + 6$ and let us substitute simple values of x like 1, 2, 3, -1, -2, and so on.

$$f(1) = 1^3 - 2(1)^2 - 5(1) + 6 = 0,$$

hence $(x - 1)$ is a factor

$$f(2) = 2^3 - 2(2)^2 - 5(2) + 6 \neq 0$$

$$f(3) = 3^3 - 2(3)^2 - 5(3) + 6 = 0,$$

hence $(x - 3)$ is a factor

$$f(-1) = (-1)^3 - 2(-1)^2 - 5(-1) + 6 \neq 0$$

$$f(-2) = (-2)^3 - 2(-2)^2 - 5(-2) + 6 = 0,$$

hence $(x + 2)$ is a factor

Hence, $x^3 - 2x^2 - 5x + 6 = (x - 1)(x - 3)(x + 2)$

Therefore if $x^3 - 2x^2 - 5x + 6 = 0$

then $(x - 1)(x - 3)(x + 2) = 0$

from which, $x = 1, x = 3$ and $x = -2$

Alternatively, having obtained one factor, i.e. $(x - 1)$ we could divide this into $(x^3 - 2x^2 - 5x + 6)$ as follows:

$$
\begin{array}{r}
x^2 - x - 6 \\
x - 1 \overline{\smash{)}\ x^3 - 2x^2 - 5x + 6} \\
\underline{x^3 - x^2} \\
-x^2 - 5x + 6 \\
\underline{-x^2 + x} \\
-6x + 6 \\
\underline{-6x + 6} \\
\cdot \quad \cdot
\end{array}
$$

Hence $x^3 - 2x^2 - 5x + 6$

$$= (x - 1)(x^2 - x - 6)$$

$$= (x - 1)(x - 3)(x + 2)$$

Summarising, the factor theorem provides us with a method of factorising simple expressions, and an alternative, in certain circumstances, to polynomial division.

Now try the following exercise

Exercise 24   Further problems on the factor theorem

Use the factor theorem to factorise the expressions given in problems 1 to 4.

1. $x^2 + 2x - 3$ $[(x - 1)(x + 3)]$

2. $x^3 + x^2 - 4x - 4$ $[(x + 1)(x + 2)(x - 2)]$

3. $2x^3 + 5x^2 - 4x - 7$

   $[(x + 1)(2x^2 + 3x - 7)]$

4. $2x^3 - x^2 - 16x + 15$

   $[(x - 1)(x + 3)(2x - 5)]$

5. Use the factor theorem to factorise $x^3 + 4x^2 + x - 6$ and hence solve the cubic equation $x^3 + 4x^2 + x - 6 = 0$
$$\left[\begin{array}{l} x^3 + 4x^2 + x - 6 \\ = (x - 1)(x + 3)(x + 2); \\ x = 1, x = -3 \text{ and } x = -2 \end{array}\right]$$

6. Solve the equation $x^3 - 2x^2 - x + 2 = 0$
   $[x = 1, x = 2 \text{ and } x = -1]$

Section 1

6.3 The remainder theorem

Dividing a general quadratic expression $(ax^2 + bx + c)$ by $(x - p)$, where p is any whole number, by long division (see Section 6.1) gives:

$$x - p \overline{\smash{\big)}\, \begin{array}{l} ax + (b + ap) \\ \hline ax^2 + bx \qquad\quad + c \\ \underline{ax^2 - apx} \\ \qquad (b + ap)x + c \\ \qquad \underline{(b + ap)x - (b + ap)p} \\ \qquad\qquad\quad c + (b + ap)p \end{array}}$$

The remainder, $c + (b + ap)p = c + bp + ap^2$ or $ap^2 + bp + c$. This is, in fact, what the **remainder theorem** states, i.e.

'if $(ax^2 + bx + c)$ is divided by $(x - p)$,

the remainder will be $ap^2 + bp + c$'

If, in the dividend $(ax^2 + bx + c)$, we substitute p for x we get the remainder $ap^2 + bp + c$

For example, when $(3x^2 - 4x + 5)$ is divided by $(x - 2)$ the remainder is $ap^2 + bp + c$, (where $a = 3$, $b = -4$, $c = 5$ and $p = 2$),
i.e. the remainder is:

$$3(2)^2 + (-4)(2) + 5 = 12 - 8 + 5 = \mathbf{9}$$

We can check this by dividing $(3x^2 - 4x + 5)$ by $(x - 2)$ by long division:

$$x - 2 \overline{\smash{\big)}\, \begin{array}{l} 3x + 2 \\ \hline 3x^2 - 4x + 5 \\ \underline{3x^2 - 6x} \\ \qquad 2x + 5 \\ \qquad \underline{2x - 4} \\ \qquad\qquad 9 \end{array}}$$

Similarly, when $(4x^2 - 7x + 9)$ is divided by $(x + 3)$, the remainder is $ap^2 + bp + c$, (where $a = 4$, $b = -7$, $c = 9$ and $p = -3$) i.e. the remainder is: $4(-3)^2 + (-7)(-3) + 9 = 36 + 21 + 9 = \mathbf{66}$

Also, when $(x^2 + 3x - 2)$ is divided by $(x - 1)$, the remainder is $1(1)^2 + 3(1) - 2 = \mathbf{2}$

It is not particularly useful, on its own, to know the remainder of an algebraic division. However, if the remainder should be zero then $(x - p)$ is a factor. This is very useful therefore when factorising expressions.

For example, when $(2x^2 + x - 3)$ is divided by $(x - 1)$, the remainder is $2(1)^2 + 1(1) - 3 = 0$, which means that $(x - 1)$ is a factor of $(2x^2 + x - 3)$.

In this case the other factor is $(2x + 3)$, i.e.

$$(2x^2 + x - 3) = (x - 1)(2x - 3).$$

The **remainder theorem** may also be stated for a **cubic equation** as:

'if $(ax^3 + bx^2 + cx + d)$ is divided by $(x - p)$, the remainder will be $ap^3 + bp^2 + cp + d$'

As before, the remainder may be obtained by substituting p for x in the dividend.

For example, when $(3x^3 + 2x^2 - x + 4)$ is divided by $(x - 1)$, the remainder is: $ap^3 + bp^2 + cp + d$ (where $a = 3$, $b = 2$, $c = -1$, $d = 4$ and $p = 1$), i.e. the remainder is: $3(1)^3 + 2(1)^2 + (-1)(1) + 4 = 3 + 2 - 1 + 4 = \mathbf{8}$.

Similarly, when $(x^3 - 7x - 6)$ is divided by $(x - 3)$, the remainder is: $1(3)^3 + 0(3)^2 - 7(3) - 6 = 0$, which mean that $(x - 3)$ is a factor of $(x^3 - 7x - 6)$.

Here are some more examples on the remainder theorem.

Problem 8. Without dividing out, find the remainder when $2x^2 - 3x + 4$ is divided by $(x - 2)$

By the remainder theorem, the remainder is given by: $ap^2 + bp + c$, where $a = 2$, $b = -3$, $c = 4$ and $p = 2$.

Hence **the remainder is:**

$$2(2)^2 + (-3)(2) + 4 = 8 - 6 + 4 = \mathbf{6}$$

Problem 9. Use the remainder theorem to determine the remainder when $(3x^3 - 2x^2 + x - 5)$ is divided by $(x + 2)$

By the remainder theorem, the remainder is given by: $ap^3 + bp^2 + cp + d$, where $a = 3$, $b = -2$, $c = 1$, $d = -5$ and $p = -2$

Hence **the remainder is:**

$$3(-2)^3 + (-2)(-2)^2 + (1)(-2) + (-5)$$
$$= -24 - 8 - 2 - 5 = \mathbf{-39}$$

Problem 10. Determine the remainder when $(x^3 - 2x^2 - 5x + 6)$ is divided by (a) $(x - 1)$ and (b) $(x + 2)$. Hence factorise the cubic expression

(a) When $(x^3 - 2x^2 - 5x + 6)$ is divided by $(x - 1)$, the remainder is given by $ap^3 + bp^2 + cp + d$, where $a = 1, b = -2, c = -5, d = 6$ and $p = 1$,

   i.e.   **the remainder** $= (1)(1)^3 + (-2)(1)^2$
   $$+ (-5)(1) + 6$$
   $$= 1 - 2 - 5 + 6 = \mathbf{0}$$

   Hence $(x - 1)$ is a factor of $(x^3 - 2x^2 - 5x + 6)$

(b) When $(x^3 - 2x^2 - 5x + 6)$ is divided by $(x + 2)$, **the remainder is** given by

   $$(1)(-2)^3 + (-2)(-2)^2 + (-5)(-2) + 6$$
   $$= -8 - 8 + 10 + 6 = \mathbf{0}$$

   Hence $(x + 2)$ is also a factor of $(x^3 - 2x^2 - 5x + 6)$

   Therefore $(x - 1)(x + 2)(\quad) = x^3 - 2x^2 - 5x + 6$

To determine the third factor (shown blank) we could

   (i) divide $(x^3 - 2x^2 - 5x + 6)$ by $(x - 1)$ $(x + 2)$

or (ii) use the factor theorem where $f(x) = x^3 - 2x^2 - 5x + 6$ and hoping to choose a value of x which makes $f(x) = 0$

or (iii) use the remainder theorem, again hoping to choose a factor $(x - p)$ which makes the remainder zero

(i) Dividing $(x^3 - 2x^2 - 5x + 6)$ by $(x^2 + x - 2)$ gives:

$$\begin{array}{r} x - 3 \\ x^2 + x - 2 \overline{\smash{\big)}\ x^3 - 2x^2 - 5x + 6} \\ \underline{x^3 + x^2 - 2x} \\ -3x^2 - 3x + 6 \\ \underline{-3x^2 - 3x + 6} \\ \cdot \qquad \cdot \qquad \cdot \end{array}$$

   Thus $(x^3 - 2x^2 - 5x + 6)$
   $$= (x - 1)(x + 2)(x - 3)$$

(ii) Using the factor theorem, we let

   $$f(x) = x^3 - 2x^2 - 5x + 6$$
   Then $f(3) = 3^3 - 2(3)^2 - 5(3) + 6$
   $$= 27 - 18 - 15 + 6 = 0$$

Hence $(x - 3)$ is a factor.

(iii) Using the remainder theorem, when $(x^3 - 2x^2 - 5x + 6)$ is divided by $(x - 3)$, the remainder is given by $ap^3 + bp^2 + cp + d$, where $a = 1, b = -2, c = -5, d = 6$ and $p = 3$.

Hence the remainder is:

   $$1(3)^3 + (-2)(3)^2 + (-5)(3) + 6$$
   $$= 27 - 18 - 15 + 6 = 0$$

Hence $(x - 3)$ is a factor.

Thus $(x^3 - 2x^2 - 5x + 6)$
   $$= (x - 1)(x + 2)(x - 3)$$

Now try the following exercise

> **Exercise 25   Further problems on the remainder theorem**
>
> 1. Find the remainder when $3x^2 - 4x + 2$ is divided by:
>   (a) $(x - 2)$ (b) $(x + 1)$ [(a) 6 (b) 9]
>
> 2. Determine the remainder when $x^3 - 6x^2 + x - 5$ is divided by:
>   (a) $(x + 2)$ (b) $(x - 3)$
>   [(a) -39 (b) -29]
>
> 3. Use the remainder theorem to find the factors of $x^3 - 6x^2 + 11x - 6$
>   [$(x - 1)(x - 2)(x - 3)$]
>
> 4. Determine the factors of $x^3 + 7x^2 + 14x + 8$ and hence solve the cubic equation:
> $x^3 + 7x^2 + 14x + 8 = 0$
>   [$x = -1, x = -2$ and $x = -4$]
>
> 5. Determine the value of 'a' if $(x + 2)$ is a factor of $(x^3 - ax^2 + 7x + 10)$ [$a = -3$]
>
> 6. Using the remainder theorem, solve the equation: $2x^3 - x^2 - 7x + 6 = 0$
>   [$x = 1, x = -2$ and $x = 1.5$]

Chapter 7

Partial fractions

7.1 Introduction to partial fractions

By algebraic addition,

$$\frac{1}{x-2} + \frac{3}{x+1} = \frac{(x+1) + 3(x-2)}{(x-2)(x+1)}$$

$$= \frac{4x-5}{x^2-x-2}$$

The reverse process of moving from $\dfrac{4x-5}{x^2-x-2}$ to $\dfrac{1}{x-2} + \dfrac{3}{x+1}$ is called resolving into **partial fractions**.

In order to resolve an algebraic expression into partial fractions:

(i) the denominator must factorise (in the above example, $x^2 - x - 2$ factorises as $(x-2)(x+1)$, and

(ii) the numerator must be at least one degree less than the denominator (in the above example $(4x-5)$ is of degree 1 since the highest powered x term is x^1 and $(x^2 - x - 2)$ is of degree 2)

When the degree of the numerator is equal to or higher than the degree of the denominator, the numerator must be divided by the denominator (see Problems 3 and 4).

There are basically three types of partial fraction and the form of partial fraction used is summarised in Table 7.1 where $f(x)$ is assumed to be of less degree than the relevant denominator and A, B and C are constants to be determined.

(In the latter type in Table 7.1, $ax^2 + bx + c$ is a quadratic expression which does not factorise without containing surds or imaginary terms.)

Resolving an algebraic expression into partial fractions is used as a preliminary to integrating certain functions (see Chapter 51).

7.2 Worked problems on partial fractions with linear factors

Problem 1. Resolve $\dfrac{11-3x}{x^2+2x-3}$ into partial fractions

The denominator factorises as $(x-1)(x+3)$ and the numerator is of less degree than the denominator.

Table 7.1

Type	Denominator containing	Expression	Form of partial fraction
1	Linear factors (see Problems 1 to 4)	$\dfrac{f(x)}{(x+a)(x-b)(x+c)}$	$\dfrac{A}{(x+a)} + \dfrac{B}{(x-b)} + \dfrac{C}{(x+c)}$
2	Repeated linear factors (see Problems 5 to 7)	$\dfrac{f(x)}{(x+a)^3}$	$\dfrac{A}{(x+a)} + \dfrac{B}{(x+a)^2} + \dfrac{C}{(x+a)^3}$
3	Quadratic factors (see Problems 8 and 9)	$\dfrac{f(x)}{(ax^2+bx+c)(x+d)}$	$\dfrac{Ax+B}{(ax^2+bx+c)} + \dfrac{C}{(x+d)}$

Thus $\dfrac{11 - 3x}{x^2 + 2x - 3}$ may be resolved into partial fractions.
Let

$$\frac{11 - 3x}{x^2 + 2x - 3} \equiv \frac{11 - 3x}{(x - 1)(x + 3)} \equiv \frac{A}{(x - 1)} + \frac{B}{(x + 3)},$$

where A and B are constants to be determined,

i.e. $\dfrac{11 - 3x}{(x - 1)(x + 3)} \equiv \dfrac{A(x + 3) + B(x - 1)}{(x - 1)(x + 3)}$

by algebraic addition.

Since the denominators are the same on each side of the identity then the numerators are equal to each other.

Thus, $11 - 3x \equiv A(x + 3) + B(x - 1)$

To determine constants A and B, values of x are chosen to make the term in A or B equal to zero.

When $x = 1$, then $11 - 3(1) \equiv A(1 + 3) + B(0)$

i.e. $8 = 4A$

i.e. $A = 2$

When $x = -3$, then $11 - 3(-3) \equiv A(0) + B(-3 - 1)$

i.e. $20 = -4B$

i.e. $B = -5$

Thus $\dfrac{11 - 3x}{x^2 + 2x - 3} \equiv \dfrac{2}{(x - 1)} + \dfrac{-5}{(x + 3)}$

$$\equiv \frac{2}{(x - 1)} - \frac{5}{(x + 3)}$$

$$\left[\text{Check:} \quad \frac{2}{(x - 1)} - \frac{5}{(x + 3)} \right.$$

$$= \frac{2(x + 3) - 5(x - 1)}{(x - 1)(x + 3)}$$

$$\left. = \frac{11 - 3x}{x^2 + 2x - 3} \right]$$

Problem 2. Convert $\dfrac{2x^2 - 9x - 35}{(x + 1)(x - 2)(x + 3)}$ into the sum of three partial fractions

Let $\dfrac{2x^2 - 9x - 35}{(x + 1)(x - 2)(x + 3)}$

$$\equiv \frac{A}{(x + 1)} + \frac{B}{(x - 2)} + \frac{C}{(x + 3)}$$

$$\equiv \frac{\begin{array}{c} A(x - 2)(x + 3) + B(x + 1)(x + 3) \\ + C(x + 1)(x - 2) \end{array}}{(x + 1)(x - 2)(x + 3)}$$

by algebraic addition

Equating the numerators gives:

$2x^2 - 9x - 35 \equiv A(x - 2)(x + 3) + B(x + 1)(x + 3)$
$\qquad\qquad\qquad\qquad + C(x + 1)(x - 2)$

Let $x = -1$. Then

$$2(-1)^2 - 9(-1) - 35 \equiv A(-3)(2) + B(0)(2)$$
$$+ C(0)(-3)$$

i.e. $-24 = -6A$

i.e. $A = \dfrac{-24}{-6} = 4$

Let $x = 2$. Then

$$2(2)^2 - 9(2) - 35 \equiv A(0)(5) + B(3)(5)$$
$$+ C(3)(0)$$

i.e. $-45 = 15B$

i.e. $B = \dfrac{-45}{15} = -3$

Let $x = -3$. Then

$$2(-3)^2 - 9(-3) - 35 \equiv A(-5)(0) + B(-2)(0)$$
$$+ C(-2)(-5)$$

i.e. $10 = 10C$

i.e. $C = 1$

Thus $\dfrac{2x^2 - 9x - 35}{(x + 1)(x - 2)(x + 3)}$

$$\equiv \frac{4}{(x + 1)} - \frac{3}{(x - 2)} + \frac{1}{(x + 3)}$$

Problem 3. Resolve $\dfrac{x^2 + 1}{x^2 - 3x + 2}$ into partial fractions

The denominator is of the same degree as the numerator. Thus dividing out gives:

$$x^2 - 3x + 2 \overline{\smash{\big)}\ \begin{aligned} &\ \ \ \ \ \ \ 1 \\ &x^2 \ \ \ \ \ \ \ \ +1 \\ &\underline{x^2 - 3x + 2} \\ &\ \ \ \ \ \ \ 3x - 1 \end{aligned}}$$

For more on polynomial division, see Section 6.1, page 48.

Hence $\dfrac{x^2 + 1}{x^2 - 3x + 2} \equiv 1 + \dfrac{3x - 1}{x^2 - 3x + 2}$

$$\equiv 1 + \dfrac{3x - 1}{(x - 1)(x - 2)}$$

Let $\dfrac{3x - 1}{(x - 1)(x - 2)} \equiv \dfrac{A}{(x - 1)} + \dfrac{B}{(x - 2)}$

$$\equiv \dfrac{A(x - 2) + B(x - 1)}{(x - 1)(x - 2)}$$

Equating numerators gives:

$$3x - 1 \equiv A(x - 2) + B(x - 1)$$

Let $x = 1$.   Then $2 = -A$

i.e. $A = -2$

Let $x = 2$.   Then $5 = B$

Hence $\dfrac{3x - 1}{(x - 1)(x - 2)} \equiv \dfrac{-2}{(x - 1)} + \dfrac{5}{(x - 2)}$

Thus $\dfrac{x^2 + 1}{x^2 - 3x + 2} \equiv 1 - \dfrac{2}{(x - 1)} + \dfrac{5}{(x - 2)}$

Problem 4. Express $\dfrac{x^3 - 2x^2 - 4x - 4}{x^2 + x - 2}$ in partial fractions

The numerator is of higher degree than the denominator. Thus dividing out gives:

$$x^2 + x - 2 \overline{\smash{\big)}\ \begin{aligned} &\ \ \ \ \ \ \ \ \ \ \ x - 3 \\ &x^3 - 2x^2 - 4x - 4 \\ &\underline{x^3 + x^2 - 2x} \\ &\ \ \ \ -3x^2 - 2x - 4 \\ &\ \ \ \ \underline{-3x^2 - 3x + 6} \\ &\ \ \ \ \ \ \ \ \ \ \ \ \ \ \ x - 10 \end{aligned}}$$

Thus $\dfrac{x^3 - 2x^2 - 4x - 4}{x^2 + x - 2} \equiv x - 3 + \dfrac{x - 10}{x^2 + x - 2}$

$$\equiv x - 3 + \dfrac{x - 10}{(x + 2)(x - 1)}$$

Let $\dfrac{x - 10}{(x + 2)(x - 1)} \equiv \dfrac{A}{(x + 2)} + \dfrac{B}{(x - 1)}$

$$\equiv \dfrac{A(x - 1) + B(x + 2)}{(x + 2)(x - 1)}$$

Equating the numerators gives:

$$x - 10 \equiv A(x - 1) + B(x + 2)$$

Let $x = -2$.   Then $-12 = -3A$

i.e.   **A = 4**

Let $x = 1$.   Then $-9 = 3B$

i.e.   **B = −3**

Hence $\dfrac{x - 10}{(x + 2)(x - 1)} \equiv \dfrac{4}{(x + 2)} - \dfrac{3}{(x - 1)}$

Thus $\dfrac{x^3 - 2x^2 - 4x - 4}{x^2 + x - 2}$

$$\equiv x - 3 + \dfrac{4}{(x + 2)} - \dfrac{3}{(x - 1)}$$

Now try the following exercise

Exercise 26   Further problems on partial fractions with linear factors

Resolve the following into partial fractions:

1. $\dfrac{12}{x^2 - 9}$ $\qquad \left[\dfrac{2}{(x - 3)} - \dfrac{2}{(x + 3)} \right]$

2. $\dfrac{4(x - 4)}{x^2 - 2x - 3}$ $\qquad \left[\dfrac{5}{(x + 1)} - \dfrac{1}{(x - 3)} \right]$

3. $\dfrac{x^2 - 3x + 6}{x(x - 2)(x - 1)}$ $\left[\dfrac{3}{x} + \dfrac{2}{(x - 2)} - \dfrac{4}{(x - 1)} \right]$

4. $\dfrac{3(2x^2 - 8x - 1)}{(x + 4)(x + 1)(2x - 1)}$

$$\left[\dfrac{7}{(x + 4)} - \dfrac{3}{(x + 1)} - \dfrac{2}{(2x - 1)} \right]$$

5. $\dfrac{x^2 + 9x + 8}{x^2 + x - 6}$ $\qquad \left[1 + \dfrac{2}{(x + 3)} + \dfrac{6}{(x - 2)} \right]$

6. $\dfrac{x^2 - x - 14}{x^2 - 2x - 3}$ $\left[1 - \dfrac{2}{(x-3)} + \dfrac{3}{(x+1)}\right]$

7. $\dfrac{3x^3 - 2x^2 - 16x + 20}{(x-2)(x+2)}$

$$\left[3x - 2 + \dfrac{1}{(x-2)} - \dfrac{5}{(x+2)}\right]$$

7.3   Worked problems on partial fractions with repeated linear factors

Problem 5.   Resolve $\dfrac{2x+3}{(x-2)^2}$ into partial fractions

The denominator contains a repeated linear factor, $(x-2)^2$

Let $\dfrac{2x+3}{(x-2)^2} \equiv \dfrac{A}{(x-2)} + \dfrac{B}{(x-2)^2}$

$\equiv \dfrac{A(x-2) + B}{(x-2)^2}$

Equating the numerators gives:

$$2x + 3 \equiv A(x-2) + B$$

Let $x = 2$.   Then $7 = A(0) + B$

i.e. $\mathbf{B = 7}$

$2x + 3 \equiv A(x-2) + B$

$\equiv Ax - 2A + B$

Since an identity is true for all values of the unknown, the coefficients of similar terms may be equated.

Hence, equating the coefficients of x gives: $\mathbf{2 = A}$
[Also, as a check, equating the constant terms gives:
$3 = -2A + B$. When $A = 2$ and $B = 7$,
RHS $= -2(2) + 7 = 3 =$ LHS]

Hence $\dfrac{2x+3}{(x-2)^2} \equiv \dfrac{2}{(x-2)} + \dfrac{7}{(x-2)^2}$

Problem 6.   Express $\dfrac{5x^2 - 2x - 19}{(x+3)(x-1)^2}$ as the sum of three partial fractions

The denominator is a combination of a linear factor and a repeated linear factor.

Let $\dfrac{5x^2 - 2x - 19}{(x+3)(x-1)^2}$

$\equiv \dfrac{A}{(x+3)} + \dfrac{B}{(x-1)} + \dfrac{C}{x-1)^2}$

$\equiv \dfrac{A(x-1)^2 + B(x+3)(x-1) + C(x+3)}{(x+3)(x-1)^2}$

by algebraic addition

Equating the numerators gives:

$$5x^2 - 2x - 19 \equiv A(x-1)^2 + B(x+3)(x-1)$$
$$+ C(x+3) \quad (1)$$

Let $x = -3$. Then

$5(-3)^2 - 2(-3) - 19 \equiv A(-4)^2 + B(0)(-4) + C(0)$

i.e. $32 = 16A$

i.e. $\mathbf{A = 2}$

Let $x = 1$. Then

$5(1)^2 - 2(1) - 19 \equiv A(0)^2 + B(4)(0) + C(4)$

i.e. $-16 = 4C$

i.e. $\mathbf{C = -4}$

Without expanding the RHS of equation (1) it can be seen that equating the coefficients of x^2 gives:
$5 = A + B$, and since $A = 2$, $\mathbf{B = 3}$
[Check: Identity (1) may be expressed as:

$$5x^2 - 2x - 19 \equiv A(x^2 - 2x + 1)$$
$$+ B(x^2 + 2x - 3) + C(x+3)$$

i.e. $5x^2 - 2x - 19 \equiv Ax^2 - 2Ax + A + Bx^2$
$$+ 2Bx - 3B + Cx + 3C$$

Equating the x term coefficients gives:

$$-2 \equiv -2A + 2B + C$$

When $A = 2$, $B = 3$ and $C = -4$ then $-2A + 2B + C = -2(2) + 2(3) - 4 = -2 =$ LHS
Equating the constant term gives:

$$-19 \equiv A - 3B + 3C$$
$$\text{RHS} = 2 - 3(3) + 3(-4) = 2 - 9 - 12$$
$$= -19 = \text{LHS}]$$

Hence $\dfrac{5x^2 - 2x - 19}{(x+3)(x-1)^2}$

$$\equiv \frac{2}{(x+3)} + \frac{3}{(x-1)} - \frac{4}{(x-1)^2}$$

Problem 7. Resolve $\dfrac{3x^2 + 16x + 15}{(x+3)^3}$ into partial fractions

Let

$$\frac{3x^2 + 16x + 15}{(x+3)^3} \equiv \frac{A}{(x+3)} + \frac{B}{(x+3)^2} + \frac{C}{(x+3)^3}$$

$$\equiv \frac{A(x+3)^2 + B(x+3) + C}{(x+3)^3}$$

Equating the numerators gives:

$$3x^2 + 16x + 15 \equiv A(x+3)^2 + B(x+3) + C \quad (1)$$

Let $x = -3$. Then

$$3(-3)^2 + 16(-3) + 15 \equiv A(0)^2 + B(0) + C$$

i.e. $\qquad\qquad\qquad -6 = C$

Identity (1) may be expanded as:

$$3x^2 + 16x + 15 \equiv A(x^2 + 6x + 9) + B(x+3) + C$$

i.e. $\quad 3x^2 + 16x + 15 \equiv Ax^2 + 6Ax + 9A$
$$+ Bx + 3B + C$$

Equating the coefficients of x^2 terms gives:

$$3 = A$$

Equating the coefficients of x terms gives:

$$16 = 6A + B$$

Since $\quad A = 3, \quad \mathbf{B = -2}$

[Check: equating the constant terms gives:

$$15 = 9A + 3B + C$$

When $A = 3$, $B = -2$ and $C = -6$,

$$9A + 3B + C = 9(3) + 3(-2) + (-6)$$
$$= 27 - 6 - 6 = 15 = \text{LHS}]$$

Thus $\dfrac{3x^2 + 16x + 15}{(x+3)^3}$

$$\equiv \frac{3}{(x+3)} - \frac{2}{(x+3)^2} - \frac{6}{(x+3)^3}$$

Now try the following exercise

Exercise 27 Further problems on partial fractions with repeated linear factors

1. $\dfrac{4x - 3}{(x+1)^2}$ $\qquad \left[\dfrac{4}{(x+1)} - \dfrac{7}{(x+1)^2}\right]$

2. $\dfrac{x^2 + 7x + 3}{x^2(x+3)}$ $\qquad \left[\dfrac{1}{x^2} + \dfrac{2}{x} - \dfrac{1}{(x+3)}\right]$

3. $\dfrac{5x^2 - 30x + 44}{(x-2)^3}$

$$\left[\frac{5}{(x-2)} - \frac{10}{(x-2)^2} + \frac{4}{(x-2)^3}\right]$$

4. $\dfrac{18 + 21x - x^2}{(x-5)(x+2)^2}$

$$\left[\frac{2}{(x-5)} - \frac{3}{(x+2)} + \frac{4}{(x+2)^2}\right]$$

7.4 Worked problems on partial fractions with quadratic factors

Problem 8. Express $\dfrac{7x^2 + 5x + 13}{(x^2+2)(x+1)}$ in partial fractions

The denominator is a combination of a quadratic factor, $(x^2 + 2)$, which does not factorise without introducing imaginary surd terms, and a linear factor, $(x+1)$. Let

$$\frac{7x^2 + 5x + 13}{(x^2+2)(x+1)} \equiv \frac{Ax + B}{(x^2+2)} + \frac{C}{(x+1)}$$

$$\equiv \frac{(Ax+B)(x+1) + C(x^2+2)}{(x^2+2)(x+1)}$$

Equating numerators gives:

$$7x^2 + 5x + 13 \equiv (Ax+B)(x+1) + C(x^2+2) \quad (1)$$

Let $x = -1$. Then

$$7(-1)^2 + 5(-1) + 13 \equiv (Ax+B)(0) + C(1+2)$$

i.e. $\qquad 15 = 3C$

i.e. $\qquad \mathbf{C = 5}$

Identity (1) may be expanded as:

$$7x^2 + 5x + 13 \equiv Ax^2 + Ax + Bx + B + Cx^2 + 2C$$

Equating the coefficients of x^2 terms gives:

$$7 = A + C, \text{and since C } = 5, \mathbf{A = 2}$$

Equating the coefficients of x terms gives:

$$5 = A + B, \text{and since A } = 2, \mathbf{B = 3}$$

[Check: equating the constant terms gives:

$$13 = B + 2C$$

When B $= 3$ and C $= 5$, B $+ 2C = 3 + 10 = 13 = $LHS]

Hence $\qquad \dfrac{7x^2 + 5x + 13}{(x^2 + 2)(x + 1)} \equiv \dfrac{2x + 3}{(x^2 + 2)} + \dfrac{5}{(x + 1)}$

Problem 9. Resolve $\dfrac{3 + 6x + 4x^2 - 2x^3}{x^2(x^2 + 3)}$ into partial fractions

Terms such as x^2 may be treated as $(x + 0)^2$, i.e. they are repeated linear factors

Let $\qquad \dfrac{3 + 6x + 4x^2 - 2x^3}{x^2(x^2 + 3)}$

$$\equiv \dfrac{A}{x} + \dfrac{B}{x^2} + \dfrac{Cx + D}{(x^2 + 3)}$$

$$\equiv \dfrac{Ax(x^2 + 3) + B(x^2 + 3) + (Cx + D)x^2}{x^2(x^2 + 3)}$$

Equating the numerators gives:

$$3 + 6x + 4x^2 - 2x^3 \equiv Ax(x^2 + 3)$$
$$+ B(x^2 + 3) + (Cx + D)x^2$$
$$\equiv Ax^3 + 3Ax + Bx^2 + 3B$$
$$+ Cx^3 + Dx^2$$

Let $x = 0$. Then $3 = 3B$

i.e. $\qquad \mathbf{B = 1}$

Equating the coefficients of x^3 terms gives:

$$-2 = A + C \qquad (1)$$

Equating the coefficients of x^2 terms gives:

$$4 = B + D$$

Since B $= 1$, $\mathbf{D = 3}$
Equating the coefficients of x terms gives:

$$6 = 3A$$

i.e. $\mathbf{A = 2}$

From equation (1), since A $= 2$, $\mathbf{C = -4}$

Hence $\qquad \dfrac{3 + 6x + 4x^2 - 2x^3}{x^2(x^2 + 3)}$

$$\equiv \dfrac{2}{x} + \dfrac{1}{x^2} + \dfrac{-4x + 3}{x^2 + 3}$$

$$\equiv \dfrac{2}{x} + \dfrac{1}{x^2} + \dfrac{3 - 4x}{x^2 + 3}$$

Now try the following exercise

Exercise 28 Further problems on partial fractions with quadratic factors

1. $\dfrac{x^2 - x - 13}{(x^2 + 7)(x - 2)} \qquad \left[\dfrac{2x + 3}{(x^2 + 7)} - \dfrac{1}{(x - 2)}\right]$

2. $\dfrac{6x - 5}{(x - 4)(x^2 + 3)} \qquad \left[\dfrac{1}{(x - 4)} + \dfrac{2 - x}{(x^2 + 3)}\right]$

3. $\dfrac{15 + 5x + 5x^2 - 4x^3}{x^2(x^2 + 5)} \qquad \left[\dfrac{1}{x} + \dfrac{3}{x^2} + \dfrac{2 - 5x}{(x^2 + 5)}\right]$

4. $\dfrac{x^3 + 4x^2 + 20x - 7}{(x - 1)^2(x^2 + 8)}$

$$\left[\dfrac{3}{(x - 1)} + \dfrac{2}{(x - 1)^2} + \dfrac{1 - 2x}{(x^2 + 8)}\right]$$

5. When solving the differential equation $\dfrac{d^2\theta}{dt^2} - 6\dfrac{d\theta}{dt} - 10\theta = 20 - e^{2t}$ by Laplace transforms, for given boundary conditions, the following expression for $\mathcal{L}\{\theta\}$ results:

$$\mathcal{L}\{\theta\} = \dfrac{4s^3 - \dfrac{39}{2}s^2 + 42s - 40}{s(s - 2)(s^2 - 6s + 10)}$$

Show that the expression can be resolved into partial fractions to give:

$$\mathcal{L}\{\theta\} = \dfrac{2}{s} - \dfrac{1}{2(s - 2)} + \dfrac{5s - 3}{2(s^2 - 6s + 10)}$$

Chapter 8

Simple equations

8.1 Expressions, equations and identities

$(3x - 5)$ is an example of an **algebraic expression**, whereas $3x - 5 = 1$ is an example of an **equation** (i.e. it contains an 'equals' sign).

An equation is simply a statement that two quantities are equal. For example, $1\,\text{m} = 1000\,\text{mm}$ or $F = \dfrac{9}{5}C + 32$ or $y = mx + c$.

An **identity** is a relationship that is true for all values of the unknown, whereas an equation is only true for particular values of the unknown. For example, $3x - 5 = 1$ is an equation, since it is only true when $x = 2$, whereas $3x \equiv 8x - 5x$ is an identity since it is true for all values of x. (Note '$\equiv$' means 'is identical to').

Simple linear equations (or equations of the first degree) are those in which an unknown quantity is raised only to the power 1.

To **'solve an equation'** means 'to find the value of the unknown'.

Any arithmetic operation may be applied to an equation **as long as the equality of the equation is maintained**.

8.2 Worked problems on simple equations

Problem 1. Solve the equation: $4x = 20$

Dividing each side of the equation by 4 gives: $\dfrac{4x}{4} = \dfrac{20}{4}$

(Note that the same operation has been applied to both the left-hand side (LHS) and the right-hand side (RHS) of the equation so the equality has been maintained).

Cancelling gives: $x = 5$, which is the solution to the equation.

Solutions to simple equations should always be checked and this is accomplished by substituting the solution into the original equation. In this case, $\text{LHS} = 4(5) = 20 = \text{RHS}$.

Problem 2. Solve: $\dfrac{2x}{5} = 6$

The LHS is a fraction and this can be removed by multiplying both sides of the equation by 5.

Hence, $5\left(\dfrac{2x}{5}\right) = 5(6)$

Cancelling gives: $2x = 30$

Dividing both sides of the equation by 2 gives:

$$\dfrac{2x}{2} = \dfrac{30}{2} \quad \text{i.e.} \quad x = 15$$

Problem 3. Solve: $a - 5 = 8$

Adding 5 to both sides of the equation gives:

$$a - 5 + 5 = 8 + 5$$

i.e. $\qquad\qquad\qquad a = 13$

The result of the above procedure is to move the '-5' from the LHS of the original equation, across the equals sign, to the RHS, but the sign is changed to $+$.

Problem 4. Solve: $x + 3 = 7$

Subtracting 3 from both sides of the equation gives:

$$x + 3 - 3 = 7 - 3$$

i.e. $\qquad\qquad\qquad x = 4$

The result of the above procedure is to move the '+3' from the LHS of the original equation, across the equals sign, to the RHS, but the sign is changed to −. Thus a term can be moved from one side of an equation to the other as long as a change in sign is made.

Problem 5. Solve: $6x + 1 = 2x + 9$

In such equations the terms containing x are grouped on one side of the equation and the remaining terms grouped on the other side of the equation. As in Problems 3 and 4, changing from one side of an equation to the other must be accompanied by a change of sign.

Thus since $\quad 6x + 1 = 2x + 9$

then $\quad\quad 6x - 2x = 9 - 1$

$$4x = 8$$

$$\frac{4x}{4} = \frac{8}{4}$$

i.e. $\quad\quad\quad\quad x = 2$

Check: LHS of original equation $= 6(2) + 1 = 13$

$\quad\quad$ RHS of original equation $= 2(2) + 9 = 13$

Hence the solution $x = 2$ is correct.

Problem 6. Solve: $4 - 3p = 2p - 11$

In order to keep the p term positive the terms in p are moved to the RHS and the constant terms to the LHS.

Hence $\quad 4 + 11 = 2p + 3p$

$$15 = 5p$$

$$\frac{15}{5} = \frac{5p}{5}$$

Hence $\quad\quad 3 = p \quad$ or $\quad p = 3$

Check: LHS $= 4 - 3(3) = 4 - 9 = -5$

$\quad\quad$ RHS $= 2(3) - 11 = 6 - 11 = -5$

Hence the solution $p = 3$ is correct.

If, in this example, the unknown quantities had been grouped initially on the LHS instead of the RHS then:

$$-3p - 2p = -11 - 4$$

i.e. $\quad\quad -5p = -15$

$$\frac{-5p}{-5} = \frac{-15}{-5}$$

and $\quad\quad\quad p = 3$, as before

It is often easier, however, to work with positive values where possible.

Problem 7. Solve: $3(x - 2) = 9$

Removing the bracket gives: $\quad 3x - 6 = 9$

Rearranging gives: $\quad\quad\quad\quad 3x = 9 + 6$

$$3x = 15$$

$$\frac{3x}{3} = \frac{15}{3}$$

i.e. $\quad\quad\quad\quad\quad x = 5$

Check: LHS $= 3(5 - 2) = 3(3) = 9 = $ RHS
Hence the solution $x = 5$ is correct.

Problem 8. Solve:

$$4(2r - 3) - 2(r - 4) = 3(r - 3) - 1$$

Removing brackets gives:

$$8r - 12 - 2r + 8 = 3r - 9 - 1$$

Rearranging gives:

$$8r - 2r - 3r = -9 - 1 + 12 - 8$$

i.e. $\quad\quad\quad 3r = -6$

$$r = \frac{-6}{3} = -2$$

Check:

LHS $= 4(-4 - 3) - 2(-2 - 4) = -28 + 12 = -16$

RHS $= 3(-2 - 3) - 1 = -15 - 1 = -16$

Hence the solution $r = -2$ is correct.

Now try the following exercise

Exercise 29 Further problems on simple equations

Solve the following equations:

1. $2x + 5 = 7$ $\quad\quad$ [1]

2. $8 - 3t = 2$ $\quad\quad$ [2]

3. $2x - 1 = 5x + 11$ $\quad\quad$ [−4]

4. $7 - 4p = 2p - 3$ $\quad\quad \left[1\frac{2}{3}\right]$

5. $2a + 6 - 5a = 0$ [2]

6. $3x - 2 - 5x = 2x - 4$ $\left[\dfrac{1}{2}\right]$

7. $20d - 3 + 3d = 11d + 5 - 8$ [0]

8. $5(f - 2) - 3(2f + 5) + 15 = 0$ [−10]

9. $2x = 4(x - 3)$ [6]

10. $6(2 - 3y) - 42 = -2(y - 1)$ [−2]

11. $2(3g - 5) - 5 = 0$ $\left[2\dfrac{1}{2}\right]$

12. $4(3x + 1) = 7(x + 4) - 2(x + 5)$ [2]

13. $10 + 3(r - 7) = 16 - (r + 2)$ $\left[6\dfrac{1}{4}\right]$

14. $8 + 4(x - 1) - 5(x - 3) = 2(5 - 2x)$ [−3]

8.3 Further worked problems on simple equations

Problem 9. Solve: $\dfrac{3}{x} = \dfrac{4}{5}$

The lowest common multiple (LCM) of the denominators, i.e. the lowest algebraic expression that both x and 5 will divide into, is $5x$.

Multiplying both sides by $5x$ gives:

$$5x\left(\frac{3}{x}\right) = 5x\left(\frac{4}{5}\right)$$

Cancelling gives:

$$15 = 4x \tag{1}$$

$$\frac{15}{4} = \frac{4x}{4}$$

i.e. $\qquad x = \dfrac{15}{4} \quad \text{or} \quad 3\dfrac{3}{4}$

Check:

$$\text{LHS} = \frac{3}{3\dfrac{3}{4}} = \frac{3}{\dfrac{15}{4}} = 3\left(\frac{4}{15}\right) = \frac{12}{15} = \frac{4}{5} = \text{RHS}$$

(Note that when there is only one fraction on each side of an equation 'cross-multiplication' can be applied. In this example, if $\dfrac{3}{x} = \dfrac{4}{5}$ then $(3)(5) = 4x$, which is a quicker way of arriving at equation (1) above.)

Problem 10. Solve: $\dfrac{2y}{5} + \dfrac{3}{4} + 5 = \dfrac{1}{20} - \dfrac{3y}{2}$

The LCM of the denominators is 20. Multiplying each term by 20 gives:

$$20\left(\frac{2y}{5}\right) + 20\left(\frac{3}{4}\right) + 20(5)$$

$$= 20\left(\frac{1}{20}\right) - 20\left(\frac{3y}{2}\right)$$

Cancelling gives:

$$4(2y) + 5(3) + 100 = 1 - 10(3y)$$

i.e. $\qquad 8y + 15 + 100 = 1 - 30y$

Rearranging gives:

$$8y + 30y = 1 - 15 - 100$$

$$38y = -114$$

$$y = \frac{-114}{38} = \mathbf{-3}$$

Check: $\text{LHS} = \dfrac{2(-3)}{5} + \dfrac{3}{4} + 5 = \dfrac{-6}{5} + \dfrac{3}{4} + 5$

$$= \frac{-9}{20} + 5 = 4\frac{11}{20}$$

$$\text{RHS} = \frac{1}{20} - \frac{3(-3)}{2} = \frac{1}{20} + \frac{9}{2} = 4\frac{11}{20}$$

Hence the solution $y = -3$ is correct.

Problem 11. Solve: $\dfrac{3}{t - 2} = \dfrac{4}{3t + 4}$

By 'cross-multiplication': $3(3t + 4) = 4(t - 2)$

Removing brackets gives: $9t + 12 = 4t - 8$

Rearranging gives: $9t - 4t = -8 - 12$

i.e. $5t = -20$

$$t = \frac{-20}{5} = \mathbf{-4}$$

Check: LHS $= \dfrac{3}{-4-2} = \dfrac{3}{-6} = -\dfrac{1}{2}$

RHS $= \dfrac{4}{3(-4)+4} = \dfrac{4}{-12+4}$

$= \dfrac{4}{-8} = -\dfrac{1}{2}$

Hence the solution $t=-4$ is correct.

Problem 12. Solve: $\sqrt{x}=2$

[$\sqrt{x}=2$ is not a 'simple equation' since the power of x is $\frac{1}{2}$ i.e $\sqrt{x}=x^{(1/2)}$; however, it is included here since it occurs often in practise].

Wherever square root signs are involved with the unknown quantity, both sides of the equation must be squared. Hence

$$(\sqrt{x})^2 = (2)^2$$

i.e. $x = 4$

Problem 13. Solve: $2\sqrt{2}=8$

To avoid possible errors it is usually best to arrange the term containing the square root on its own. Thus

$$\dfrac{2\sqrt{d}}{2} = \dfrac{8}{2}$$

i.e. $\sqrt{d} = 4$

Squaring both sides gives: $d=16$, which may be checked in the original equation

Problem 14. Solve: $x^2 = 25$

This problem involves a square term and thus is not a simple equation (it is, in fact, a quadratic equation). However the solution of such an equation is often required and is therefore included here for completeness. Whenever a square of the unknown is involved, the square root of both sides of the equation is taken. Hence

$$\sqrt{x^2} = \sqrt{25}$$

i.e. $x = 5$

However, $x=-5$ is also a solution of the equation because $(-5)\times(-5)=+25$. Therefore, whenever the square root of a number is required there are always two answers, one positive, the other negative.

The solution of $x^2=25$ is thus written as $x=\pm 5$

Problem 15. Solve: $\dfrac{15}{4t^2} = \dfrac{2}{3}$

'Cross-multiplying' gives: $15(3) = 2(4t^2)$

i.e. $45 = 8t^2$

$\dfrac{45}{8} = t^2$

i.e. $t^2 = 5.625$

Hence $t = \sqrt{5.625} = \pm 2.372$, correct to 4 significant figures.

Now try the following exercise

Exercise 30 Further problems on simple equations

Solve the following equations:

1. $2+\dfrac{3}{4}y = 1+\dfrac{2}{3}y+\dfrac{5}{6}$ [-2]

2. $\dfrac{1}{4}(2x-1)+3 = \dfrac{1}{2}$ $\left[-4\frac{1}{2}\right]$

3. $\dfrac{1}{5}(2f-3)+\dfrac{1}{6}(f-4)+\dfrac{2}{15} = 0$ [2]

4. $\dfrac{1}{3}(3m-6)-\dfrac{1}{4}(5m+4)+\dfrac{1}{5}(2m-9) = -3$ [12]

5. $\dfrac{x}{3}-\dfrac{x}{5} = 2$ [15]

6. $1-\dfrac{y}{3} = 3+\dfrac{y}{3}-\dfrac{y}{6}$ [-4]

7. $\dfrac{1}{3n}+\dfrac{1}{4n} = \dfrac{7}{24}$ [2]

8. $\dfrac{x+3}{4} = \dfrac{x-3}{5}+2$ [13]

9. $\dfrac{y}{5}+\dfrac{7}{20} = \dfrac{5-y}{4}$ [2]

10. $\dfrac{v-2}{2v-3} = \dfrac{1}{3}$ [3]

11. $\dfrac{2}{a-3} = \dfrac{3}{2a+1}$ [−11]

12. $\dfrac{x}{4} - \dfrac{x+6}{5} = \dfrac{x+3}{2}$ [−6]

13. $3\sqrt{t} = 9$ [9]

14. $\dfrac{3\sqrt{x}}{1-\sqrt{x}} = -6$ [4]

15. $10 = 5\sqrt{\dfrac{x}{2} - 1}$ [10]

16. $16 = \dfrac{t^2}{9}$ [±12]

17. $\sqrt{\dfrac{y+2}{y-2}} = \dfrac{1}{2}$ $\left[-3\dfrac{1}{3}\right]$

18. $\dfrac{11}{2} = 5 + \dfrac{8}{x^2}$ [±4]

8.4 Practical problems involving simple equations

Problem 16. A copper wire has a length l of 1.5 km, a resistance R of 5 Ω and a resistivity of 17.2×10^{-6} Ω mm. Find the cross-sectional area, a, of the wire, given that $R = \rho l/a$

Since $R = \rho l/a$
then

$$5\,\Omega = \frac{(17.2 \times 10^{-6}\ \Omega\ \text{mm})(1500 \times 10^3\ \text{mm})}{a}$$

From the units given, a is measured in mm^2.

Thus $\qquad 5a = 17.2 \times 10^{-6} \times 1500 \times 10^3$

and $\qquad a = \dfrac{17.2 \times 10^{-6} \times 1500 \times 10^3}{5}$

$\qquad\qquad = \dfrac{17.2 \times 1500 \times 10^3}{10^6 \times 5}$

$\qquad\qquad = \dfrac{17.2 \times 15}{10 \times 5} = 5.16$

Hence the cross-sectional area of the wire is 5.16 mm^2

Problem 17. The temperature coefficient of resistance α may be calculated from the formula $R_t = R_0(1 + \alpha t)$. Find α given $R_t = 0.928$, $R_0 = 0.8$ and $t = 40$

Since $R_t = R_0(1 + \alpha t)$ then

$$0.928 = 0.8[1 + \alpha(40)]$$

$$0.928 = 0.8 + (0.8)(\alpha)(40)$$

$$0.928 - 0.8 = 32\alpha$$

$$0.128 = 32\alpha$$

Hence $\qquad \alpha = \dfrac{0.128}{32} = \mathbf{0.004}$

Problem 18. The distance s metres travelled in time t seconds is given by the formula: $s = ut + \frac{1}{2}at^2$, where u is the initial velocity in m/s and a is the acceleration in m/s^2. Find the acceleration of the body if it travels 168 m in 6 s, with an initial velocity of 10 m/s

$s = ut + \dfrac{1}{2}at^2$, and $s = 168$, $u = 10$ and $t = 6$

Hence $\qquad 168 = (10)(6) + \dfrac{1}{2}a(6)^2$

$$168 = 60 + 18a$$

$$168 - 60 = 18a$$

$$108 = 18a$$

$$a = \dfrac{108}{18} = 6$$

Hence the acceleration of the body is 6 m/s^2

Problem 19. When three resistors in an electrical circuit are connected in parallel the total resistance R_T is given by:

$$\frac{1}{R_T} = \frac{1}{R_1} + \frac{1}{R_2} + \frac{1}{R_3}.$$

Find the total resistance when $R_1 = 5\,\Omega$, $R_2 = 10\,\Omega$ and $R_3 = 30\,\Omega$

$$\frac{1}{R_T} = \frac{1}{5} + \frac{1}{10} + \frac{1}{30}$$

$$= \frac{6+3+1}{30} = \frac{10}{30} = \frac{1}{3}$$

Taking the reciprocal of both sides gives: $R_T = 3\,\Omega$

Alternatively, if $\frac{1}{R_T} = \frac{1}{5} + \frac{1}{10} + \frac{1}{30}$ the LCM of the denominators is $30R_T$

Hence

$$30R_T\left(\frac{1}{R_T}\right) = 30R_T\left(\frac{1}{5}\right) + 30R_T\left(\frac{1}{10}\right)$$
$$+ 30R_T\left(\frac{1}{30}\right)$$

Cancelling gives:

$$30 = 6R_T + 3R_T + R_T$$
$$30 = 10R_T$$
$$R_T = \frac{30}{10} = 3\,\Omega, \text{ as above}$$

Now try the following exercise

Exercise 31   Practical problems involving simple equations

1. A formula used for calculating resistance of a cable is $R = (\rho l)/a$. Given $R = 1.25$, $l = 2500$ and $a = 2 \times 10^{-4}$ find the value of ρ. [10^{-7}]

2. Force F newtons is given by $F = ma$, where m is the mass in kilograms and a is the acceleration in metres per second squared. Find the acceleration when a force of 4 kN is applied to a mass of 500 kg. [$8\,\text{m/s}^2$]

3. $PV = mRT$ is the characteristic gas equation. Find the value of m when $P = 100 \times 10^3$, $V = 3.00$, $R = 288$ and $T = 300$. [3.472]

4. When three resistors R_1, R_2 and R_3 are connected in parallel the total resistance R_T is determined from $\frac{1}{R_T} = \frac{1}{R_1} + \frac{1}{R_2} + \frac{1}{R_3}$
   (a) Find the total resistance when $R_1 = 3\,\Omega$, $R_2 = 6\,\Omega$ and $R_3 = 18\,\Omega$.

(b) Find the value of R_3 given that $R_T = 3\,\Omega$, $R_1 = 5\,\Omega$ and $R_2 = 10\,\Omega$. [(a) $1.8\,\Omega$ (b) $30\,\Omega$]

5. Ohm's law may be represented by $I = V/R$, where I is the current in amperes, V is the voltage in volts and R is the resistance in ohms. A soldering iron takes a current of 0.30 A from a 240 V supply. Find the resistance of the element. [$800\,\Omega$]

8.5   Further practical problems involving simple equations

Problem 20. The extension x m of an aluminium tie bar of length l m and cross-sectional area A m^2 when carrying a load of F newtons is given by the modulus of elasticity $E = Fl/Ax$. Find the extension of the tie bar (in mm) if $E = 70 \times 10^9\,\text{N/m}^2$, $F = 20 \times 10^6\,\text{N}$, $A = 0.1\,\text{m}^2$ and $l = 1.4\,\text{m}$

$E = Fl/Ax$, hence

$$70 \times 10^9\,\frac{\text{N}}{\text{m}^2} = \frac{(20 \times 10^6\,\text{N})(1.4\,\text{m})}{(0.1\,\text{m}^2)(x)}$$

(the unit of x is thus metres)

$$70 \times 10^9 \times 0.1 \times x = 20 \times 10^6 \times 1.4$$

$$x = \frac{20 \times 10^6 \times 1.4}{70 \times 10^9 \times 0.1}$$

Cancelling gives: $\quad x = \frac{2 \times 1.4}{7 \times 100}\,\text{m}$

$$= \frac{2 \times 1.4}{7 \times 100} \times 1000\,\text{mm}$$

Hence the extension of the tie bar, $x = 4\,\text{mm}$

Problem 21. Power in a d.c. circuit is given by $P = \frac{V^2}{R}$ where V is the supply voltage and R is the circuit resistance. Find the supply voltage if the circuit resistance is $1.25\,\Omega$ and the power measured is 320 W

Since $P = \dfrac{V^2}{R}$ then $320 = \dfrac{V^2}{1.25}$

$(320)(1.25) = V^2$

i.e. $V^2 = 400$

Supply voltage, $V = \sqrt{400} = \pm 20\,\text{V}$

Problem 22. A formula relating initial and final states of pressures, P_1 and P_2, volumes V_1 and V_2, and absolute temperatures, T_1 and T_2, of an ideal gas is $\dfrac{P_1 V_1}{T_1} = \dfrac{P_2 V_2}{T_2}$. Find the value of P_2 given $P_1 = 100 \times 10^3$, $V_1 = 1.0$, $V_2 = 0.266$, $T_1 = 423$ and $T_2 = 293$

Since $\dfrac{P_1 V_1}{T_1} = \dfrac{P_2 V_2}{T_2}$

then $\dfrac{(100 \times 10^3)(1.0)}{423} = \dfrac{P_2(0.266)}{293}$

'Cross-multiplying' gives:

$100 \times 10^3)(1.0)(293) = P_2(0.266)(423)$

$P_2 = \dfrac{(100 \times 10^3)(1.0)(293)}{(0.266)(423)}$

Hence $P_2 = 260 \times 10^3$ or 2.6×10^5

Problem 23. The stress f in a material of a thick cylinder can be obtained from $\dfrac{D}{d} = \sqrt{\dfrac{f+p}{f-p}}$
Calculate the stress, given that $D = 21.5$, $d = 10.75$ and $p = 1800$

Since $\dfrac{D}{d} = \sqrt{\dfrac{f+p}{f-p}}$

then $\dfrac{21.5}{10.75} = \sqrt{\dfrac{f+1800}{f-1800}}$

i.e. $2 = \sqrt{\dfrac{f+1800}{f-1800}}$

Squaring both sides gives:

$4 = \dfrac{f+1800}{f-1800}$

$4(f - 1800) = f + 1800$

$4f - 7200 = f + 1800$

$4f - f = 1800 + 7200$

$3f = 9000$

$f = \dfrac{9000}{3} = 3000$

Hence **stress,** $f = 3000$

Now try the following exercise

Exercise 32   Practical problems involving simple equations

1. Given $R_2 = R_1(1 + \alpha t)$, find α given $R_1 = 5.0$, $R_2 = 6.03$ and $t = 51.5$ [0.004]

2. If $v^2 = u^2 + 2as$, find u given $v = 24$, $a = -40$ and $s = 4.05$ [30]

3. The relationship between the temperature on a Fahrenheit scale and that on a Celsius scale is given by $F = \dfrac{9}{5}C + 32$. Express $113°F$ in degrees Celsius.   [45°C]

4. If $t = 2\pi\sqrt{w/Sg}$, find the value of S given $w = 1.219$, $g = 9.81$ and $t = 0.3132$ [50]

5. Applying the principle of moments to a beam results in the following equation:

    $F \times 3 = (5 - F) \times 7$

    where F is the force in newtons. Determine the value of F.   [3.5 N]

6. A rectangular laboratory has a length equal to one and a half times its width and a perimeter of 40 m. Find its length and width.
                                  [12 m, 8 m]

This Revision test covers the material contained in Chapters 5 to 8. *The marks for each question are shown in brackets at the end of each question.*

1. Evaluate: $3xy^2z^3 - 2yz$ when $x = \dfrac{4}{3}$, $y = 2$ and $z = \dfrac{1}{2}$ (3)

2. Simplify the following:

   (a) $\dfrac{8a^2b\sqrt{c^3}}{(2a)^2\sqrt{b}\sqrt{c}}$

   (b) $3x + 4 \div 2x + 5 \times 2 - 4x$ (6)

3. Remove the brackets in the following expressions and simplify:

   (a) $(2x - y)^2$

   (b) $4ab - [3\{2(4a - b) + b(2 - a)\}]$ (5)

4. Factorise: $3x^2y + 9xy^2 + 6xy^3$ (3)

5. If x is inversely proportional to y and $x = 12$ when $y = 0.4$, determine

   (a) the value of x when y is 3, and

   (b) the value of y when $x = 2$. (4)

6. Factorise $x^3 + 4x^2 + x - 6$ using the factor theorem. Hence solve the equation $x^3 + 4x^2 + x - 6 = 0$ (6)

7. Use the remainder theorem to find the remainder when $2x^3 + x^2 - 7x - 6$ is divided by

   (a) $(x - 2)$ (b) $(x + 1)$

   Hence factorise the cubic expression. (7)

8. Simplify $\dfrac{6x^2 + 7x - 5}{2x - 1}$ by dividing out. (5)

9. Resolve the following into partial fractions:

   (a) $\dfrac{x - 11}{x^2 - x - 2}$ (b) $\dfrac{3 - x}{(x^2 + 3)(x + 3)}$

   (c) $\dfrac{x^3 - 6x + 9}{x^2 + x - 2}$ (24)

10. Solve the following equations:

    (a) $3t - 2 = 5t + 4$

    (b) $4(k - 1) - 2(3k + 2) + 14 = 0$

    (c) $\dfrac{a}{2} - \dfrac{2a}{5} = 1$ (d) $\sqrt{\dfrac{s + 1}{s - 1}} = 2$ (13)

11. A rectangular football pitch has its length equal to twice its width and a perimeter of 360 m. Find its length and width. (4)

Chapter 9

Simultaneous equations

9.1 Introduction to simultaneous equations

Only one equation is necessary when finding the value of a **single unknown quantity** (as with simple equations in Chapter 8). However, when an equation contains **two unknown quantities** it has an infinite number of solutions. When two equations are available connecting the same two unknown values then a unique solution is possible. Similarly, for three unknown quantities it is necessary to have three equations in order to solve for a particular value of each of the unknown quantities, and so on.

Equations that have to be solved together to find the unique values of the unknown quantities, which are true for each of the equations, are called **simultaneous equations**.

Two methods of solving simultaneous equations analytically are:

(a) by **substitution**, and (b) by **elimination**.

(A graphical solution of simultaneous equations is shown in Chapter 31 and determinants and matrices are used to solve simultaneous equations in Chapter 62).

9.2 Worked problems on simultaneous equations in two unknowns

Problem 1. Solve the following equations for x and y, (a) by substitution, and (b) by elimination:

$$x + 2y = -1 \qquad (1)$$
$$4x - 3y = 18 \qquad (2)$$

(a) **By substitution**

From equation (1): $x = -1 - 2y$

Substituting this expression for x into equation (2) gives:

$$4(-1 - 2y) - 3y = 18$$

This is now a simple equation in y.
Removing the bracket gives:

$$-4 - 8y - 3y = 18$$
$$-11y = 18 + 4 = 22$$
$$y = \frac{22}{-11} = -2$$

Substituting $y = -2$ into equation (1) gives:

$$x + 2(-2) = -1$$
$$x - 4 = -1$$
$$x = -1 + 4 = 3$$

Thus $x = 3$ and $y = -2$ is the solution to the simultaneous equations.

(Check: In equation (2), since $x = 3$ and $y = -2$, LHS $= 4(3) - 3(-2) = 12 + 6 = 18 =$ RHS)

(b) **By elimination**

$$x + 2y = -1 \qquad (1)$$
$$4x - 3y = 18 \qquad (2)$$

If equation (1) is multiplied throughout by 4 the coefficient of x will be the same as in equation (2), giving:

$$4x + 8y = -4 \qquad (3)$$

Subtracting equation (3) from equation (2) gives:

$$4x - 3y = 18 \qquad (2)$$
$$4x + 8y = -4 \qquad (3)$$
$$\overline{0 - 11y = 22}$$

Hence $y = \dfrac{22}{-11} = -2$

(Note, in the above subtraction,
$$18 - (-4) = 18 + 4 = 22).$$

Substituting $y = -2$ into either equation (1) or equation (2) will give $x = 3$ as in method (a). The solution $x = 3, y = -2$ is the only pair of values that satisfies both of the original equations.

> **Problem 2.** Solve, by a substitution method, the simultaneous equations:
> $$3x - 2y = 12 \qquad (1)$$
> $$x + 3y = -7 \qquad (2)$$

From equation (2), $x = -7 - 3y$

Substituting for x in equation (1) gives:

$$3(-7 - 3y) - 2y = 12$$
i.e. $-21 - 9y - 2y = 12$
$$-11y = 12 + 21 = 33$$
Hence $y = \dfrac{33}{-11} = -3$

Substituting $y = -3$ in equation (2) gives:

$$x + 3(-3) = -7$$
i.e. $x - 9 = -7$
Hence $x = -7 + 9 = 2$

Thus $x = 2, y = -3$ is the solution of the simultaneous equations.

(Such solutions should always be checked by substituting values into each of the original two equations.)

> **Problem 3.** Use an elimination method to solve the simultaneous equations:
> $$3x + 4y = 5 \qquad (1)$$
> $$2x - 5y = -12 \qquad (2)$$

If equation (1) is multiplied throughout by 2 and equation (2) by 3, then the coefficient of x will be the same in the newly formed equations. Thus

$2 \times$ equation (1) gives: $6x + 8y = 10 \qquad (3)$
$3 \times$ equation (2) gives: $6x - 15y = -36 \qquad (4)$

Equation (3) − equation (4) gives:

$$0 + 23y = 46$$
i.e. $y = \dfrac{46}{23} = 2$

(Note $+8y - -15y = 8y + 15y = 23y$ and $10 - (-36) = 10 + 36 = 46$. Alternatively, 'change the signs of the bottom line and add'.)

Substituting $y = 2$ in equation (1) gives:

$$3x + 4(2) = 5$$
from which $3x = 5 - 8 = -3$
and $x = -1$

Checking in equation (2), left-hand side $= 2(-1) - 5(2) = -2 - 10 = -12 =$ right-hand side.

Hence $x = -1$ and $y = 2$ is the solution of the simultaneous equations.

The elimination method is the most common method of solving simultaneous equations.

> **Problem 4.** Solve :
> $$7x - 2y = 26 \qquad (1)$$
> $$6x + 5y = 29 \qquad (2)$$

When equation (1) is multiplied by 5 and equation (2) by 2 the coefficients of y in each equation are numerically the same, i.e. 10, but are of opposite sign.

$5 \times$ equation (1) gives: $35x - 10y = 130 \qquad (3)$
$2 \times$ equation (2) gives: $12x + 10y = 58 \qquad (4)$
Adding equation (3) and (4) gives: $47x + 0 = 188 \qquad (5)$

Hence $x = \dfrac{188}{47} = 4$

[Note that when the signs of common coefficients are **different** the two equations are **added**, and when the signs of common coefficients are the **same** the two equations are **subtracted** (as in Problems 1 and 3).]

Substituting $x = 4$ in equation (1) gives:

$$7(4) - 2y = 26$$

$$28 - 2y = 26$$

$$28 - 26 = 2y$$

$$2 = 2y$$

Hence $\qquad y = 1$

Checking, by substituting $x = 4$ and $y = 1$ in equation (2), gives:

$$\text{LHS} = 6(4) + 5(1) = 24 + 5 = 29 = \text{RHS}$$

Thus the solution is $x = 4$, $y = 1$, since these values maintain the equality when substituted in both equations.

___

Now try the following exercise

Exercise 33 Further problems on simultaneous equations

Solve the following simultaneous equations and verify the results.

1. $a + b = 7$
    $a - b = 3$ $\qquad\qquad$ $[a = 5, \quad b = 2]$

2. $2x + 5y = 7$
    $x + 3y = 4$ $\qquad\qquad$ $[x = 1, \quad y = 1]$

3. $3s + 2t = 12$
    $4s - t = 5$ $\qquad\qquad$ $[s = 2, \quad t = 3]$

4. $3x - 2y = 13$
    $2x + 5y = -4$ $\qquad\qquad$ $[x = 3, \quad y = -2]$

5. $5x = 2y$
    $3x + 7y = 41$ $\qquad\qquad$ $[x = 2, \quad y = 5]$

6. $5c = 1 - 3d$
    $2d + c + 4 = 0$ $\qquad\qquad$ $[c = 2, \quad d = -3]$

9.3 Further worked problems on simultaneous equations

Problem 5. Solve

$$3p = 2q \qquad\qquad\qquad (1)$$

$$4p + q + 11 = 0 \qquad\qquad (2)$$

Rearranging gives:

$$3p - 2q = 0 \qquad\qquad\qquad (3)$$

$$4p + q = -11 \qquad\qquad\qquad (4)$$

Multiplying equation (4) by 2 gives:

$$8p + 2q = -22 \qquad\qquad\qquad (5)$$

Adding equations (3) and (5) gives:

$$11p + 0 = -22$$

$$p = \frac{-22}{11} = -2$$

Substituting $p = -2$ into equation (1) gives:

$$3(-2) = 2q$$

$$-6 = 2q$$

$$q = \frac{-6}{2} = -3$$

Checking, by substituting $p = -2$ and $q = -3$ into equation (2) gives:

$$\text{LHS} = 4(-2) + (-3) + 11 = -8 - 3 + 11$$

$$= 0 = \text{RHS}$$

Hence the solution is $p = -2$, $q = -3$

Problem 6. Solve

$$\frac{x}{8} + \frac{5}{2} = y \qquad\qquad\qquad (1)$$

$$13 - \frac{y}{3} = 3x \qquad\qquad\qquad (2)$$

Whenever fractions are involved in simultaneous equation it is usual to firstly remove them. Thus, multiplying equation (1) by 8 gives:

$$8\left(\frac{x}{8}\right) + 8\left(\frac{5}{2}\right) = 8y$$

i.e. $\qquad\qquad x + 20 = 8y \qquad\qquad (3)$

Multiplying equation (2) by 3 gives:

$$39 - y = 9x \qquad (4)$$

Rearranging equation (3) and (4) gives:

$$x - 8y = -20 \qquad (5)$$
$$9x + y = 39 \qquad (6)$$

Multiplying equation (6) by 8 gives:

$$72x + 8y = 312 \qquad (7)$$

Adding equations (5) and (7) gives:

$$73x + 0 = 292$$
$$x = \frac{292}{73} = 4$$

Substituting $x = 4$ into equation (5) gives:

$$4 - 8y = -20$$
$$4 + 20 = 8y$$
$$24 = 8y$$
$$y = \frac{24}{8} = 3$$

Checking: substituting $x = 4$, $y = 3$ in the original equations, gives:

Equation (1):   LHS $= \dfrac{4}{8} + \dfrac{5}{2} = \dfrac{1}{2} + 2\dfrac{1}{2} = 3$

$\qquad\qquad = y =$ RHS

Equation (2):   LHS $= 13 - \dfrac{3}{3} = 13 - 1 = 12$

$\qquad\qquad$ RHS $= 3x = 3(4) = 12$

Hence the solution is $x = 4, y = 3$

Problem 7.   Solve

$$2.5x + 0.75 - 3y = 0$$
$$1.6x = 1.08 - 1.2y$$

It is often easier to remove decimal fractions. Thus multiplying equations (1) and (2) by 100 gives:

$$250x + 75 - 300y = 0 \qquad (1)$$
$$160x = 108 - 120y \qquad (2)$$

Rearranging gives:

$$250x - 300y = -75 \qquad (3)$$
$$160x + 120y = 108 \qquad (4)$$

Multiplying equation (3) by 2 gives:

$$500x - 600y = -150 \qquad (5)$$

Multiplying equation (4) by 5 gives:

$$800x + 600y = 540 \qquad (6)$$

Adding equations (5) and (6) gives:

$$1300x + 0 = 390$$
$$x = \frac{390}{1300} = \frac{39}{130} = \frac{3}{10} = \mathbf{0.3}$$

Substituting $x = 0.3$ into equation (1) gives:

$$250(0.3) + 75 - 300y = 0$$
$$75 + 75 = 300y$$
$$150 = 300y$$
$$y = \frac{150}{300} = \mathbf{0.5}$$

Checking $x = 0.3$, $y = 0.5$ in equation (2) gives:

$$\text{LHS} = 160(0.3) = 48$$
$$\text{RHS} = 108 - 120(0.5)$$
$$= 108 - 60 = 48$$

Hence the solution is $x = 0.3, y = 0.5$

Now try the following exercise

Exercise 34   Further problems on simultanuous equations

Solve the following simultaneous equations and verify the results.

1. $7p + 11 + 2q = 0$

    $-1 = 3q - 5p \qquad\qquad [p = -1, \quad q = -2]$

2. $\dfrac{x}{2} + \dfrac{y}{3} = 4$

    $\dfrac{x}{6} - \dfrac{y}{9} = 0 \qquad\qquad [x = 4, \quad y = 6]$

3. $\dfrac{a}{2} - 7 = -2b$

    $12 = 5a + \dfrac{2}{3}b$ $\qquad\qquad$ $[a = 2, \quad b = 3]$

4. $\dfrac{x}{5} + \dfrac{2y}{3} = \dfrac{49}{15}$

    $\dfrac{3x}{7} - \dfrac{y}{2} + \dfrac{5}{7} = 0$ $\qquad\qquad$ $[x = 3, \quad y = 4]$

5. $1.5x - 2.2y = -18$

    $2.4x + 0.6y = 33$ $\qquad\qquad$ $[x = 10, \quad y = 15]$

6. $3b - 2.5a = 0.45$

    $1.6a + 0.8b = 0.8$ $\qquad$ $[a = 0.30, \quad b = 0.40]$

9.4 More difficult worked problems on simultaneous equations

Problem 8. Solve

$$\dfrac{2}{x} + \dfrac{3}{y} = 7 \qquad\qquad (1)$$

$$\dfrac{1}{x} - \dfrac{4}{y} = -2 \qquad\qquad (2)$$

In this type of equation the solutions is easier if a substitution is initially made. Let $\dfrac{1}{x} = a$ and $\dfrac{1}{y} = b$

Thus equation (1) becomes: $2a + 3b = 7 \qquad (3)$

and equation (2) becomes: $\quad a - 4b = -2 \qquad (4)$

Multiplying equation (4) by 2 gives:

$$2a - 8b = -4 \qquad\qquad (5)$$

Subtracting equation (5) from equation (3) gives:

$$0 + 11b = 11$$

i.e. $\qquad\qquad\qquad b = 1$

Substituting $b = 1$ in equation (3) gives:

$$2a + 3 = 7$$

$$2a = 7 - 3 = 4$$

i.e. $\qquad\qquad\qquad a = 2$

Checking, substituting $a = 2$ and $b = 1$ in equation (4) gives:

$$\text{LHS} = 2 - 4(1) = 2 - 4 = -2 = \text{RHS}$$

Hence $a = 2$ and $b = 1$

However, since $\qquad \dfrac{1}{x} = a \quad$ then $\quad x = \dfrac{1}{a} = \dfrac{1}{2}$

and since $\qquad \dfrac{1}{y} = b \quad$ then $\quad y = \dfrac{1}{b} = \dfrac{1}{1} = 1$

Hence the solutions is $x = \dfrac{1}{2}, y = 1,$

which may be checked in the original equations.

Problem 9. Solve

$$\dfrac{1}{2a} + \dfrac{3}{5b} = 4 \qquad\qquad (1)$$

$$\dfrac{4}{a} + \dfrac{1}{2b} = 10.5 \qquad\qquad (2)$$

Let $\dfrac{1}{a} = x \quad$ and $\quad \dfrac{1}{b} = y$

then $\qquad\qquad \dfrac{x}{2} + \dfrac{3}{5}y = 4 \qquad\qquad (3)$

$$4x + \dfrac{1}{2}y = 10.5 \qquad\qquad (4)$$

To remove fractions, equation (3) is multiplied by 10 giving:

$$10\left(\dfrac{x}{2}\right) + 10\left(\dfrac{3}{5}y\right) = 10(4)$$

i.e. $\qquad\qquad 5x + 6y = 40 \qquad\qquad (5)$

Multiplying equation (4) by 2 gives:

$$8x + y = 21 \qquad\qquad (6)$$

Multiplying equation (6) by 6 gives:

$$48x + 6y = 126 \qquad\qquad (7)$$

Subtracting equation (5) from equation (7) gives:

$$43x + 0 = 86$$

$$x = \dfrac{86}{43} = 2$$

Substituting $x = 2$ into equation (3) gives:

$$\frac{2}{2} + \frac{3}{5}y = 4$$

$$\frac{3}{5}y = 4 - 1 = 3$$

$$y = \frac{5}{3}(3) = 5$$

Since $\frac{1}{a} = x$ then $a = \frac{1}{x} = \frac{1}{2}$

and since $\frac{1}{b} = y$ then $b = \frac{1}{y} = \frac{1}{5}$

Hence the solutions is $a = \dfrac{1}{2}$, $b = \dfrac{1}{5}$

which may be checked in the original equations.

Problem 10. Solve

$$\frac{1}{x+y} = \frac{4}{27} \qquad (1)$$

$$\frac{1}{2x-y} = \frac{4}{33} \qquad (2)$$

To eliminate fractions, both sides of equation (1) are multiplied by $27(x + y)$ giving:

$$27(x+y)\left(\frac{1}{x+y}\right) = 27(x+y)\left(\frac{4}{27}\right)$$

i.e. $\qquad\qquad 27(1) = 4(x + y)$

$$27 = 4x + 4y \qquad (3)$$

Similarly, in equation (2): $33 = 4(2x - y)$

i.e. $\qquad\qquad 33 = 8x - 4y \qquad (4)$

Equation (3) + equation (4) gives:

$$60 = 12x, \quad \text{i.e. } x = \frac{60}{12} = 5$$

Substituting $x = 5$ in equation (3) gives:

$$27 = 4(5) + 4y$$

from which $4y = 27 - 20 = 7$

and $y = \dfrac{7}{4} = 1\dfrac{3}{4}$

Hence $x = 5$, $y = 1\dfrac{3}{4}$ is the required solution, which may be checked in the original equations.

Now try the following exercise

Exercise 35   Further more difficult problems on simultaneous equations

In Problems 1 to 5, solve the simultaneous equations and verify the results

1. $\dfrac{3}{x} + \dfrac{2}{y} = 14$

   $\dfrac{5}{x} - \dfrac{3}{y} = -2$ $\qquad \left[x = \dfrac{1}{2}, \quad y = \dfrac{1}{4}\right]$

2. $\dfrac{4}{a} - \dfrac{3}{b} = 18$

   $\dfrac{2}{a} + \dfrac{5}{b} = -4$ $\qquad \left[a = \dfrac{1}{3}, \quad b = -\dfrac{1}{2}\right]$

3. $\dfrac{1}{2p} + \dfrac{3}{5q} = 5$

   $\dfrac{5}{p} - \dfrac{1}{2q} = \dfrac{35}{2}$ $\qquad \left[p = \dfrac{1}{4}, \quad q = \dfrac{1}{5}\right]$

4. $\dfrac{c+1}{4} - \dfrac{d+2}{3} + 1 = 0$

   $\dfrac{1-c}{5} + \dfrac{3-d}{4} + \dfrac{13}{20} = 0$ $\qquad [c = 3, \quad d = 4]$

5. $\dfrac{3r+2}{5} - \dfrac{2s-1}{4} = \dfrac{11}{5}$

   $\dfrac{3+2r}{4} + \dfrac{5-s}{3} = \dfrac{15}{4}$ $\qquad \left[r = 3, \quad s = \dfrac{1}{2}\right]$

6. If $5x - \dfrac{3}{y} = 1$ and $x + \dfrac{4}{y} = \dfrac{5}{2}$ find the value of

   $\dfrac{xy+1}{y}$ $\qquad\qquad\qquad\qquad\qquad [1]$

9.5 Practical problems involving simultaneous equations

There are a number of situations in engineering and science where the solution of simultaneous equations is required. Some are demonstrated in the following worked problems.

Problem 11. The law connecting friction F and load L for an experiment is of the form $F = aL + b$, where a and b are constants. When $F = 5.6, L = 8.0$ and when $F = 4.4, L = 2.0$. Find the values of a and b and the value of F when $L = 6.5$

Substituting $F = 5.6, L = 8.0$ into $F = aL + b$ gives:

$$5.6 = 8.0a + b \qquad (1)$$

Substituting $F = 4.4, L = 2.0$ into $F = aL + b$ gives:

$$4.4 = 2.0a + b \qquad (2)$$

Subtracting equation (2) from equation (1) gives:

$$1.2 = 6.0a$$

$$a = \frac{1.2}{6.0} = \frac{1}{5}$$

Substituting $a = \frac{1}{5}$ into equation (1) gives:

$$5.6 = 8.0 \left(\frac{1}{5} \right) + b$$

$$5.6 = 1.6 + b$$

$$5.6 - 1.6 = b$$

i.e. $b = 4$

Checking, substituting $a = \frac{1}{5}$ and $b = 4$ in equation (2), gives:

$$\text{RHS} = 2.0 \left(\frac{1}{5} \right) + 4 = 0.4 + 4 = 4.4 = \text{LHS}$$

Hence $a = \dfrac{1}{5}$ and $b = 4$

When $L = 6.5$, $F = al + b = \frac{1}{5}(6.5) + 4$

$$= 1.3 + 4, \quad \text{i.e. } F = 5.3$$

Problem 12.   The equation of a straight line, of gradient m and intercept on the y-axis c, is $y = mx + c$. If a straight line passes through the point where $x = 1$ and $y = -2$, and also through the point where $x = 3\frac{1}{2}$ and $y = 10\frac{1}{2}$, find the values of the gradient and the y-axis intercept

Substituting $x = 1$ and $y = -2$ into $y = mx + c$ gives:

$$-2 = m + c \qquad (1)$$

Substituting $x = 3\frac{1}{2}$ and $y = 10\frac{1}{2}$ into $y = mx + c$ gives:

$$10\frac{1}{2} = 3\frac{1}{2}m + c \qquad (2)$$

Subtracting equation (1) from equation (2) gives:

$$12\frac{1}{2} = 2\frac{1}{2}m \quad \text{from which,} \quad m = \frac{12\frac{1}{2}}{2\frac{1}{2}} = 5$$

Substituting $m = 5$ into equation (1) gives:

$$-2 = 5 + c$$

$$c = -2 - 5 = -7$$

Checking, substituting $m = 5$ and $c = -7$ in equation (2), gives:

$$\text{RHS} = \left(3\frac{1}{2} \right)(5) + (-7) = 17\frac{1}{2} - 7$$

$$= 10\frac{1}{2} = \text{LHS}$$

Hence the gradient, $m = 5$ and the y-axis intercept, $c = -7$

Problem 13.   When Kirchhoff's laws are applied to the electrical circuit shown in Fig. 9.1 the currents I_1 and I_2 are connected by the equations:

$$27 = 1.5I_1 + 8(I_1 - I_2) \qquad (1)$$

$$-26 = 2I_2 - 8(I_1 - I_2) \qquad (2)$$

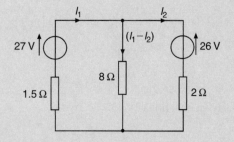

Figure 9.1

Solve the equations to find the values of currents I_1 and I_2

Removing the brackets from equation (1) gives:

$$27 = 1.5I_1 + 8I_1 - 8I_2$$

Rearranging gives:

$$9.5I_1 - 8I_2 = 27 \qquad (3)$$

Removing the brackets from equation (2) gives:

$$-26 = 2I_2 - 8I_1 + 8I_2$$

Rearranging gives:

$$-8I_1 + 10I_2 = -26 \qquad (4)$$

Multiplying equation (3) by 5 gives:

$$47.5I_1 - 40I_2 = 135 \qquad (5)$$

Multiplying equation (4) by 4 gives:

$$-32I_1 + 40I_2 = -104 \qquad (6)$$

Adding equations (5) and (6) gives:

$$15.5I_1 + 0 = 31$$

$$I_1 = \frac{31}{15.5} = 2$$

Substituting $I_1 = 2$ into equation (3) gives:

$$9.5(2) - 8I_2 = 27$$

$$19 - 8I_2 = 27$$

$$19 - 27 = 8I_2$$

$$-8 = 8I_2$$

$$I_2 = -1$$

Hence the solution is $I_1 = 2$ and $I_2 = -1$

(which may be checked in the original equations).

Problem 14. The distance s metres from a fixed point of a vehicle travelling in a straight line with constant acceleration, a m/s^2, is given by $s = ut + \frac{1}{2}at^2$, where u is the initial velocity in m/s and t the time in seconds. Determine the initial velocity and the acceleration given that $s = 42$ m when $t = 2$ s and $s = 144$ m when $t = 4$ s. Find also the distance travelled after 3 s

---

Substituting $s = 42, t = 2$ into $s = ut + \frac{1}{2}at^2$ gives:

$$42 = 2u + \frac{1}{2}a(2)^2$$

i.e. $$42 = 2u + 2a \qquad (1)$$

Substituting $s = 144, t = 4$ into $s = ut + \frac{1}{2}at^2$ gives:

$$144 = 4u + \frac{1}{2}a(4)^2$$

i.e. $$144 = 4u + 8a \qquad (2)$$

Multiplying equation (1) by 2 gives:

$$84 = 4u + 4a \qquad (3)$$

Subtracting equation (3) from equation (2) gives:

$$60 = 0 + 4a$$

$$a = \frac{60}{15} = 15$$

Substituting $a = 15$ into equation (1) gives:

$$42 = 2u + 2(15)$$

$$42 - 30 = 2u$$

$$u = \frac{12}{2} = 6$$

Substituting $a = 15, u = 6$ in equation (2) gives:

$$\text{RHS} = 4(6) + 8(15) = 24 + 120 = 144 = \text{LHS}$$

Hence the initial velocity, $u = 6$ m/s and the acceleration, $a = 15$ m/s^2.

Distance travelled after 3 s is given by $s = ut + \frac{1}{2}at^2$ where $t = 3, u = 6$ and $a = 15$

Hence $s = (6)(3) + \frac{1}{2}(15)(3)^2 = 18 + 67.5$

i.e. distance travelled after 3 s $= 85.5$ m

Problem 15. The resistance $R\,\Omega$ of a length of wire at $t°$C is given by $R = R_0(1 + \alpha t)$, where R_0 is the resistance at $0°$C and α is the temperature coefficient of resistance in /$°$C. Find the values of α and R_0 if $R = 30\,\Omega$ at $50°$C and $R = 35\,\Omega$ at $100°$C

Substituting $R = 30, t = 50$ into $R = R_0(1 + \alpha t)$ gives:

$$30 = R_0(1 + 50\alpha) \qquad (1)$$

Substituting $R = 35, t = 100$ into $R = R_0(1 + \alpha t)$ gives:

$$35 = R_0(1 + 100\alpha) \qquad (2)$$

Although these equations may be solved by the conventional substitution method, an easier way is to eliminate R_0 by division. Thus, dividing equation (1) by equation (2) gives:

$$\frac{30}{35} = \frac{R_0(1 + 50\alpha)}{R_0(1 + 100\alpha)} = \frac{1 + 50\alpha}{1 + 100\alpha}$$

'Cross-multiplying' gives:

$$30(1 + 100\alpha) = 35(1 + 50\alpha)$$

$$30 + 3000\alpha = 35 + 1750\alpha$$

$$3000\alpha - 1750\alpha = 35 - 30$$

$$1250\alpha = 5$$

i.e. $\quad\quad \alpha = \dfrac{5}{1250} = \dfrac{1}{250} \quad$ or $\quad \mathbf{0.004}$

Substituting $\alpha = \dfrac{1}{250}$ into equation (1) gives:

$$30 = R_0 \left\{ 1 + (50) \left(\frac{1}{250} \right) \right\}$$

$$30 = R_0 (1.2)$$

$$R_0 = \frac{30}{1.2} = \mathbf{25}$$

Checking, substituting $\alpha = \dfrac{1}{250}$ and $R_0 = 25$ in equation (2) gives:

$$\text{RHS} = 25 \left\{ 1 + (100) \left(\frac{1}{250} \right) \right\}$$

$$= 25(1.4) = 35 = \text{LHS}$$

Thus the solution is $\alpha = 0.004/°\text{C}$ and $R_0 = 25\ \Omega$.

Problem 16. The molar heat capacity of a solid compound is given by the equation $c = a + bT$, where a and b are constants. When $c = 52, T = 100$ and when $c = 172, T = 400$. Determine the values of a and b

When $c = 52, \quad T = 100$, hence

$$52 = a + 100b \tag{1}$$

When $c = 172, \quad T = 400$, hence

$$172 = a + 400b \tag{2}$$

Equation (2) – equation (1) gives:

$$120 = 300b$$

from which, $\quad b = \dfrac{120}{300} = \mathbf{0.4}$

Substituting $b = 0.4$ in equation (1) gives:

$$52 = a + 100(0.4)$$

$$a = 52 - 40 = \mathbf{12}$$

Hence $a = 12$ and $b = 0.4$

Now try the following exercise

Exercise 36 Further practical problems involving simultaneous equations

1. In a system of pulleys, the effort P required to raise a load W is given by $P = aW + b$, where a and b are constants.

   If $W = 40$ when $P = 12$ and $W = 90$ when $P = 22$, find the values of a and b.

   $$\left[a = \frac{1}{5}, \quad b = 4 \right]$$

2. Applying Kirchhoff's laws to an electrical circuit produces the following equations:

   $$5 = 0.2I_1 + 2(I_1 - I_2)$$

   $$12 = 3I_2 + 0.4I_2 - 2(I_1 - I_2)$$

   Determine the values of currents I_1 and I_2
   $$[I_1 = 6.47, \quad I_2 = 4.62]$$

3. Velocity v is given by the formula $v = u + at$. If $v = 20$ when $t = 2$ and $v = 40$ when $t = 7$, find the values of u and a. Hence find the velocity when $t = 3.5$.

   $$[u = 12, \quad a = 4, \quad v = 26]$$

4. $y = mx + c$ is the equation of a straight line of slope m and y-axis intercept c. If the line passes through the point where $x = 2$ and $y = 2$, and also through the point where $x = 5$ and $y = \frac{1}{2}$, find the slope and y-axis intercept of the straight line. $\quad \left[m = -\dfrac{1}{2}, \quad c = 3 \right]$

5. The resistance R ohms of copper wire at $t°\text{C}$ is given by $R = R_0(1 + \alpha t)$, where R_0 is the resistance at $0°\text{C}$ and α is the temperature coefficient of resistance. If $R = 25.44\ \Omega$ at $30°\text{C}$ and $R = 32.17\ \Omega$ at $100°\text{C}$, find α and R_0. $\quad [\alpha = 0.00426, \quad R_0 = 22.56\ \Omega]$

6. The molar heat capacity of a solid compound is given by the equation $c = a + bT$. When $c = 52, T = 100$ and when $c = 172, T = 400$. Find the values of a and b.
   $$[a = 12, \quad b = 0.40]$$

Chapter 10

Transposition of formulae

10.1 Introduction to transposition of formulae

When a symbol other than the subject is required to be calculated it is usual to rearrange the formula to make a new subject. This rearranging process is called **transposing the formula** or **transposition**.

The rules used for transposition of formulae are the same as those used for the solution of simple equations (see Chapter 8)—basically, **that the equality of an equation must be maintained**.

10.2 Worked problems on transposition of formulae

Problem 1. Transpose $p = q + r + s$ to make r the subject

The aim is to obtain r on its own on the left-hand side (LHS) of the equation. Changing the equation around so that r is on the LHS gives:

$$q + r + s = p \quad \quad (1)$$

Substracting $(q + s)$ from both sides of the equation gives:

$$q + r + s - (q + s) = p - (q + s)$$

Thus $\quad q + r + s - q - s = p - q - s$

i.e. $\quad\quad\quad\quad\quad\quad \boldsymbol{r = p - q - s} \quad \quad (2)$

It is shown with simple equations, that a quantity can be moved from one side of an equation to the other with an appropriate change of sign. Thus equation (2) follows immediately from equation (1) above.

Problem 2. If $a + b = w - x + y$, express x as the subject

Rearranging gives:

$$w - x + y = a + b \quad \text{and} \quad -x = a + b - w - y$$

Multiplying both sides by -1 gives:

$$(-1)(-x) = (-1)(a + b - w - y)$$

i.e. $\quad\quad\quad\quad\quad x = -a - b + w + y$

The result of multiplying each side of the equation by -1 is to change all the signs in the equation.

It is conventional to express answers with positive quantities first. Hence rather than $x = -a - b + w + y$, $\boldsymbol{x = w + y - a - b}$, since the order of terms connected by $+$ and $-$ signs is immaterial.

Problem 3. Transpose $v = f\lambda$ to make λ the subject

Rearranging gives: $\quad\quad\quad\quad f\lambda = v$

Dividing both sides by f gives: $\quad \dfrac{f\lambda}{f} = \dfrac{v}{f}$

i.e. $\quad\quad\quad\quad\quad\quad\quad \boldsymbol{\lambda = \dfrac{v}{f}}$

Problem 4. When a body falls freely through a height h, the velocity v is given by $v^2 = 2gh$. Express this formula with h as the subject

Rearranging gives: $\quad\quad\quad\quad 2gh = v^2$

Dividing both sides by $2g$ gives: $\quad \dfrac{2gh}{2g} = \dfrac{v^2}{2g}$

i.e. $\quad\quad\quad\quad\quad\quad\quad \boldsymbol{h = \dfrac{v^2}{2g}}$

Problem 5. If $I = \dfrac{V}{R}$, rearrange to make V the subject

Rearranging gives: $\dfrac{V}{R} = I$

Multiplying both sides by R gives:

$$R\left(\dfrac{V}{R}\right) = R(I)$$

Hence $\qquad V = IR$

Problem 6. Transpose: $a = \dfrac{F}{m}$ for m

Rearranging gives: $\dfrac{F}{m} = a$

Multiplying both sides by m gives:

$$m\left(\dfrac{F}{m}\right) = m(a) \quad \text{i.e.} \quad F = ma$$

Rearranging gives: $\qquad ma = F$

Dividing both sides by a gives: $\dfrac{ma}{a} = \dfrac{F}{a}$

i.e. $\qquad m = \dfrac{F}{a}$

Problem 7. Rearrange the formula: $R = \dfrac{\rho l}{a}$ to make (i) a the subject, and (ii) l the subject

(i) Rearranging gives: $\dfrac{\rho l}{a} = R$

Multiplying both sides by a gives:

$$a\left(\dfrac{\rho l}{a}\right) = a(R) \quad \text{i.e.} \quad \rho l = aR$$

Rearranging gives: $aR = \rho l$
Dividing both sides by R gives:

$$\dfrac{aR}{R} = \dfrac{\rho l}{R}$$

i.e. $\qquad a = \dfrac{\rho l}{R}$

(ii) Multiplying both sides of $\dfrac{\rho l}{a} = R$ by a gives:

$$\rho l = aR$$

Dividing both sides by ρ gives: $\dfrac{\rho l}{\rho} = \dfrac{aR}{\rho}$

i.e. $\qquad l = \dfrac{aR}{\rho}$

Now try the following exercise

Exercise 37 Further problems on transposition of formulae

Make the symbol indicated the subject of each of the formulae shown and express each in its simplest form.

1. $a + b = c - d - e$ (d) $[d = c - a - b]$

2. $x + 3y = t$ (y) $\left[y = \dfrac{1}{3}(t - x)\right]$

3. $c = 2\pi r$ (r) $\left[r = \dfrac{c}{2\pi}\right]$

4. $y = mx + c$ (x) $\left[x = \dfrac{y - c}{m}\right]$

5. $I = PRT$ (T) $\left[T = \dfrac{I}{PR}\right]$

6. $I = \dfrac{E}{R}$ (R) $\left[R = \dfrac{E}{I}\right]$

7. $S = \dfrac{a}{1 - r}$ (r) $\left[R = \dfrac{S - a}{S} \text{ or } 1 - \dfrac{a}{S}\right]$

8. $F = \dfrac{9}{5}C + 32$ (C) $\left[C = \dfrac{5}{9}(F - 32)\right]$

10.3 Further worked problems on transposition of formulae

Problem 8. Transpose the formula: $v = u + \dfrac{ft}{m}$ to make f the subject

Rearranging gives: $u + \dfrac{ft}{m} = v$ and $\dfrac{ft}{m} = v - u$

Multiplying each side by m gives:

$$m\left(\dfrac{ft}{m}\right) = m(v - u) \quad \text{i.e.} \quad ft = m(v - u)$$

Dividing both sides by t gives:

$$\dfrac{ft}{t} = \dfrac{m}{t}(v - u) \quad \text{i.e.} \quad f = \dfrac{m}{t}(v - u)$$

Problem 9. The final length, l_2 of a piece of wire heated through $\theta°C$ is given by the formula $l_2 = l_1(1 + \alpha\theta)$. Make the coefficient of expansion, α, the subject

Rearranging gives: $\qquad\qquad l_1(1 + \alpha\theta) = l_2$

Removing the bracket gives: $\qquad l_1 + l_1\alpha\theta = l_2$

Rearranging gives: $\qquad\qquad\quad l_1\alpha\theta = l_2 - l_1$

Dividing both sides by $l_1\theta$ gives:

$$\frac{l_1\alpha\theta}{l_1\theta} = \frac{l_2 - l_1}{l_1\theta} \quad \text{i.e.} \quad \boldsymbol{\alpha = \frac{l_2 - l_1}{l_1\theta}}$$

Problem 10. A formula for the distance moved by a body is given by: $s = \dfrac{1}{2}(v + u)t$. Rearrange the formula to make u the subject

Rearranging gives: $\qquad\qquad \dfrac{1}{2}(v + u)t = s$

Multiplying both sides by 2 gives: $\quad (v + u)t = 2s$

Dividing both sides by t gives:

$$\frac{(v + u)t}{t} = \frac{2s}{t}$$

i.e. $\qquad\qquad\qquad v + u = \dfrac{2s}{t}$

Hence $\qquad \boldsymbol{u = \dfrac{2s}{t} - v} \quad$ or $\quad \boldsymbol{u = \dfrac{2s - vt}{t}}$

Problem 11. A formula for kinetic energy is $k = \dfrac{1}{2}mv^2$. Transpose the formula to make v the subject

Rearranging gives: $\dfrac{1}{2}mv^2 = k$

Whenever the prospective new subject is a squared term, that term is isolated on the LHS, and then the square root of both sides of the equation is taken.

Multiplying both sides by 2 gives: $\quad mv^2 = 2k$

Dividing both sides by m gives: $\quad \dfrac{mv^2}{m} = \dfrac{2k}{m}$

i.e. $\qquad\qquad\qquad\qquad\qquad v^2 = \dfrac{2k}{m}$

Taking the square root of both sides gives:

$$\sqrt{v^2} = \sqrt{\frac{2k}{m}}$$

i.e. $\qquad\qquad \boldsymbol{v = \sqrt{\dfrac{2k}{m}}}$

Problem 12. In a right-angled triangle having sides x, y and hypotenuse z, Pythagoras' theorem states $z^2 = x^2 + y^2$. Transpose the formula to find x

Rearranging gives: $\quad x^2 + y^2 = z^2$

and $\qquad\qquad\qquad x^2 = z^2 - y^2$

Taking the square root of both sides gives:

$$\boldsymbol{x = \sqrt{z^2 - y^2}}$$

Problem 13. Given $t = 2\pi\sqrt{\dfrac{l}{g}}$ find g in terms of t, l and π

Whenever the prospective new subject is within a square root sign, it is best to isolate that term on the LHS and then to square both sides of the equation.

Rearranging gives: $2\pi\sqrt{\dfrac{l}{g}} = t$

Dividing both sides by 2π gives: $\sqrt{\dfrac{l}{g}} = \dfrac{t}{2\pi}$

Squaring both sides gives: $\dfrac{l}{g} = \left(\dfrac{t}{2\pi}\right)^2 = \dfrac{t^2}{4\pi^2}$

Cross-multiplying, i.e. multiplying each term by $4\pi^2 g$, gives:

$$4\pi^2 l = gt^2$$

or $\qquad\qquad\qquad\qquad\qquad gt^2 = 4\pi^2 l$

Dividing both sides by t^2 gives: $\quad \dfrac{gt^2}{t^2} = \dfrac{4\pi^2 l}{t^2}$

i.e. $\qquad\qquad\qquad\qquad \boldsymbol{g = \dfrac{4\pi^2 l}{t^2}}$

Problem 14. The impedance of an a.c. circuit is given by $Z = \sqrt{R^2 + X^2}$. Make the reactance, X, the subject

Rearranging gives: $\sqrt{R^2 + X^2} = Z$

Squaring both sides gives: $R^2 + X^2 = Z^2$

Rearranging gives: $X^2 = Z^2 - R^2$

Taking the square root of both sides gives:

$$X = \sqrt{Z^2 - R^2}$$

Problem 15. The volume V of a hemisphere is given by $V = \dfrac{2}{3}\pi r^3$. Find r in terms of V

Rearranging gives: $\dfrac{2}{3}\pi r^3 = V$

Multiplying both sides by 3 gives: $2\pi r^3 = 3V$

Dividing both sides by 2π gives:

$$\frac{2\pi r^3}{2\pi} = \frac{3V}{2\pi} \quad \text{i.e.} \quad r^3 = \frac{3V}{2\pi}$$

Taking the cube root of both sides gives:

$$\sqrt[3]{r^3} = \sqrt[3]{\frac{3V}{2\pi}} \quad \text{i.e.} \quad r = \sqrt[3]{\frac{3V}{2\pi}}$$

Now try the following exercise

Exercise 38   Further problems on transposition of formulae

Make the symbol indicated the subject of each of the formulae shown and express each in its simplest form.

1. $y = \dfrac{\lambda(x-d)}{d}$ (x)

$$\left[x = \frac{d}{\lambda}(y+\lambda) \quad \text{or} \quad x = d + \frac{yd}{\lambda} \right]$$

2. $A = \dfrac{3(F-f)}{L}$ (f)

$$\left[f = \frac{3F - AL}{3} \quad \text{or} \quad f = F - \frac{AL}{3} \right]$$

3. $y = \dfrac{Ml^2}{8EI}$ (E) $\quad \left[E = \dfrac{Ml^2}{8yI} \right]$

4. $R = R_0(1 + \alpha t)$ (t) $\quad \left[t = \dfrac{R - R_0}{R_0 \alpha} \right]$

5. $\dfrac{1}{R} = \dfrac{1}{R_1} + \dfrac{1}{R_2}$ (R_2) $\quad \left[R_2 = \dfrac{RR_1}{R_1 - R} \right]$

6. $I = \dfrac{E - e}{R + r}$ (R)

$$\left[R = \frac{E - e - Ir}{I} \quad \text{or} \quad R = \frac{E - e}{I} - r \right]$$

7. $y = 4ab^2 c^2$ (b) $\quad \left[b = \sqrt{\dfrac{y}{4ac^2}} \right]$

8. $\dfrac{a^2}{x^2} + \dfrac{b^2}{y^2} = 1$ (x) $\quad \left[x = \dfrac{ay}{\sqrt{y^2 - b^2}} \right]$

9. $t = 2\pi \sqrt{\dfrac{l}{g}}$ (l) $\quad \left[l = \dfrac{t^2 g}{4\pi^2} \right]$

10. $v^2 = u^2 + 2as$ (u) $\quad \left[u = \sqrt{v^2 - 2as} \right]$

11. $A = \dfrac{\pi R^2 \theta}{360}$ (R) $\quad \left[R = \sqrt{\dfrac{360A}{\pi\theta}} \right]$

12. $N = \sqrt{\dfrac{a + x}{y}}$ (a) $\quad [a = N^2 y - x]$

13. $Z = \sqrt{R^2 + (2\pi f L)^2}$ (L) $\quad \left[L = \dfrac{\sqrt{Z^2 - R^2}}{2\pi f} \right]$

10.4   Harder worked problems on transposition of formulae

Problem 16. Transpose the formula $p = \dfrac{a^2 x + a^2 y}{r}$ to make a the subject

Rearranging gives: $\dfrac{a^2 x + a^2 y}{r} = p$

Multiplying both sides by r gives: $a^2 x + a^2 y = rp$

Factorising the LHS gives: $a^2(x + y) = rp$

Dividing both sides by $(x + y)$ gives:

$$\frac{a^2(x + y)}{(x + y)} = \frac{rp}{(x + y)} \quad \text{i.e.} \quad a^2 = \frac{rp}{(x + y)}$$

Taking the square root of both sides gives:

$$a = \sqrt{\frac{rp}{x+y}}$$

Problem 17. Make b the subject of the formula

$$a = \frac{x-y}{\sqrt{bd+be}}$$

Rearranging gives: $\dfrac{x-y}{\sqrt{bd+be}} = a$

Multiplying both sides by $\sqrt{bd+be}$ gives:

$$x - y = a\sqrt{bd+be}$$

or $\qquad a\sqrt{bd+be} = x - y$

Dividing both sides by a gives:

$$\sqrt{bd+be} = \frac{x-y}{a}$$

Squaring both sides gives:

$$bd + be = \left(\frac{x-y}{a}\right)^2$$

Factorising the LHS gives:

$$b(d+e) = \left(\frac{x-y}{a}\right)^2$$

Dividing both sides by $(d+e)$ gives:

$$b = \frac{\left(\dfrac{x-y}{a}\right)^2}{(d+e)}$$

i.e. $\qquad b = \dfrac{(x-y)^2}{a^2(d+e)}$

Problem 18. If $a = \dfrac{b}{1+b}$ make b the subject of the formula

Rearranging gives: $\dfrac{b}{1+b} = a$

Multiplying both sides by $(1+b)$ gives:

$$b = a(1+b)$$

Removing the bracket gives: $b = a + ab$

Rearranging to obtain terms in b on the LHS gives:

$$b - ab = a$$

Factorising the LHS gives: $b(1-a) = a$

Dividing both sides by $(1-a)$ gives:

$$b = \frac{a}{1-a}$$

Problem 19. Transpose the formula $V = \dfrac{Er}{R+r}$ to make r the subject

Rearranging gives: $\dfrac{Er}{R+r} = V$

Multiplying both sides by $(R+r)$ gives:

$$Er = V(R+r)$$

Removing the bracket gives: $Er = VR + Vr$

Rearranging to obtain terms in r on the LHS gives:

$$Er - Vr = VR$$

Factorising gives: $r(E-V) = VR$

Dividing both sides by $(E-V)$ gives:

$$r = \frac{VR}{E-V}$$

Problem 20. Given that: $\dfrac{D}{d} = \sqrt{\dfrac{f+p}{f-p}}$ express p in terms of D, d and f

Rearranging gives: $\sqrt{\dfrac{f+p}{f-p}} = \dfrac{D}{d}$

Squaring both sides gives: $\left(\dfrac{f+p}{f-p}\right) = \dfrac{D^2}{d^2}$

Cross-multiplying, i.e. multiplying each term by $d^2(f-p)$, gives:

$$d^2(f+p) = D^2(f-p)$$

Removing brackets gives: $d^2 f + d^2 p = D^2 f - D^2 p$

Rearranging, to obtain terms in p on the LHS gives:

$$d^2 p + D^2 p = D^2 f - d^2 f$$

Factorising gives: $p(d^2 + D^2) = f(D^2 - d^2)$

Dividing both sides by $(d^2 + D^2)$ gives:

$$p = \frac{f(D^2 - d^2)}{(d^2 + D^2)}$$

Now try the following exercise

Exercise 39 Further problems on transposition of formulae

Make the symbol indicated the subject of each of the formulae shown in Problems 1 to 7, and express each in its simplest form.

1. $y = \dfrac{a^2 m - a^2 n}{x}$ (a) $\left[a = \sqrt{\dfrac{xy}{m-n}} \right]$

2. $M = \pi(R^4 - r^4)$ (R) $\left[R = \sqrt[4]{\dfrac{M}{\pi} + r^4} \right]$

3. $x + y = \dfrac{r}{3+r}$ (r) $\left[r = \dfrac{3(x+y)}{(1-x-y)} \right]$

4. $m = \dfrac{\mu L}{L + rCR}$ (L) $\left[L = \dfrac{mrCR}{\mu - m} \right]$

5. $a^2 = \dfrac{b^2 - c^2}{b^2}$ (b) $\left[b = \dfrac{c}{\sqrt{1 - a^2}} \right]$

6. $\dfrac{x}{y} = \dfrac{1 + r^2}{1 - r^2}$ (r) $\left[r = \sqrt{\dfrac{x-y}{x+y}} \right]$

7. $\dfrac{p}{q} = \sqrt{\dfrac{a + 2b}{a - 2b}}$ (b) $\left[b = \dfrac{a(p^2 - q^2)}{2(p^2 + q^2)} \right]$

8. A formula for the focal length, f, of a convex lens is $\dfrac{1}{f} = \dfrac{1}{u} + \dfrac{1}{v}$. Transpose the formula to make v the subject and evaluate v when $f = 5$ and $u = 6$.

$$\left[v = \dfrac{uf}{u - f}, \quad 30 \right]$$

9. The quantity of heat, Q, is given by the formula $Q = mc(t_2 - t_1)$. Make t_2 the subject of the formula and evaluate t_2 when $m = 10$, $t_1 = 15$, $c = 4$ and $Q = 1600$.

$$\left[t_2 = t_1 + \dfrac{Q}{mc}, \quad 55 \right]$$

10. The velocity, v, of water in a pipe appears in the formula $h = \dfrac{0.03 L v^2}{2dg}$. Express v as the subject of the formula and evaluate v when $h = 0.712$, $L = 150$, $d = 0.30$ and $g = 9.81$

$$\left[v = \sqrt{\dfrac{2dgh}{0.03L}}, \quad 0.965 \right]$$

11. The sag S at the centre of a wire is given by the formula: $S = \sqrt{\dfrac{3d(l - d)}{8}}$. Make l the subject of the formula and evaluate l when $d = 1.75$ and $S = 0.80$

$$\left[l = \dfrac{8S^2}{3d} + d, \quad 2.725 \right]$$

12. In an electrical alternating current circuit the impedance Z is given by:

$$Z = \sqrt{R^2 + \left(\omega L - \dfrac{1}{\omega C} \right)^2}$$

Transpose the formula to make C the subject and hence evaluate C when $Z = 130$, $R = 120$, $\omega = 314$ and $L = 0.32$

$$\left[C = \dfrac{1}{\omega \left\{ \omega L - \sqrt{Z^2 - R^2} \right\}}, \quad 63.1 \times 10^{-6} \right]$$

13. An approximate relationship between the number of teeth, T, on a milling cutter, the diameter of cutter, D, and the depth of cut, d, is given by: $T = \dfrac{12.5 D}{D + 4d}$. Determine the value of D when $T = 10$ and $d = 4$ mm.   [64 mm]

14. Make λ, the wavelength of X-rays, the subject of the following formula:

$$\dfrac{\mu}{\rho} = \dfrac{CZ^4 \sqrt{\lambda^5}\, n}{a}$$

$$\left[\lambda = \sqrt[5]{\left(\dfrac{a\mu}{\rho C Z^4 n} \right)^2} \right]$$

Chapter 11

Quadratic equations

11.1 Introduction to quadratic equations

As stated in Chapter 8, an **equation** is a statement that two quantities are equal and to **'solve an equation'** means 'to find the value of the unknown'. The value of the unknown is called the **root** of the equation.

A **quadratic equation** is one in which the highest power of the unknown quantity is 2. For example, $x^2 - 3x + 1 = 0$ is a quadratic equation.

There are four methods of **solving quadratic equations**.

These are: (i)   **by factorisation** (where possible)
           (ii)   **by 'completing the square'**
           (iii)   **by using the 'quadratic formula'**
   or   (iv)   **graphically** (see Chapter 31).

11.2 Solution of quadratic equations by factorisation

Multiplying out $(2x + 1)(x - 3)$ gives $2x^2 - 6x + x - 3$, i.e. $2x^2 - 5x - 3$. The reverse process of moving from $2x^2 - 5x - 3$ to $(2x + 1)(x - 3)$ is called **factorising**.

If the quadratic expression can be factorised this provides the simplest method of solving a quadratic equation.

For example, if $2x^2 - 5x - 3 = 0$, then,

by factorising: $(2x + 1)(x - 3) = 0$

Hence either $(2x + 1) = 0$ i.e. $x = -\dfrac{1}{2}$

or $(x - 3) = 0$ i.e. $x = 3$

The technique of factorising is often one of 'trial and error'.

Problem 1. Solve the equations:
(a) $x^2 + 2x - 8 = 0$ (b) $3x^2 - 11x - 4 = 0$ by factorisation

(a) $x^2 + 2x - 8 = 0$. The factors of x^2 are x and x. These are placed in brackets thus: $(x \quad)(x \quad)$

The factors of -8 are $+8$ and -1, or -8 and $+1$, or $+4$ and -2, or -4 and $+2$. The only combination to given a middle term of $+2x$ is $+4$ and -2, i.e.

$$x^2 + 2x - 8 = (x + 4)(x - 2)$$

(Note that the product of the two inner terms added to the product of the two outer terms must equal to middle term, $+2x$ in this case.)

The quadratic equation $x^2 + 2x - 8 = 0$ thus becomes $(x + 4)(x - 2) = 0$.

Since the only way that this can be true is for either the first or the second, or both factors to be zero, then

either $(x + 4) = 0$ i.e. $x = -4$

or $(x - 2) = 0$ i.e. $x = 2$

Hence the roots of $x^2 + 2x - 8 = 0$ are $x = -4$ and 2

(b) $3x^2 - 11x - 4 = 0$
The factors of $3x^2$ are $3x$ and x. These are placed in brackets thus: $(3x \quad)(x \quad)$

The factors of -4 are -4 and $+1$, or $+4$ and -1, or -2 and 2.

Remembering that the product of the two inner terms added to the product of the two outer terms must equal $-11x$, the only combination to give this is $+1$ and -4, i.e.

$$3x^2 - 11x - 4 = (3x + 1)(x - 4)$$

The quadratic equation $3x^2 - 11x - 4 = 0$ thus becomes $(3x + 1)(x - 4) = 0$.

Hence, either $(3x + 1) = 0$ i.e. $x = -\dfrac{1}{3}$

or $(x - 4) = 0$ i.e. $x = 4$

and both solutions may be checked in the original equation.

Problem 2.   Determine the roots of:
(a) $x^2 - 6x + 9 = 0$, and (b) $4x^2 - 25 = 0$, by factorisation

(a) $x^2 - 6x + 9 = 0$. Hence $(x - 3)(x - 3) = 0$, i.e. $(x - 3)^2 = 0$ (the left-hand side is known as **a perfect square**). Hence $x = 3$ is the only root of the equation $x^2 - 6x + 9 = 0$.

(b) $4x^2 - 25 = 0$ (the left-hand side is **the difference of two squares**, $(2x)^2$ and $(5)^2$). Thus $(2x + 5)(2x - 5) = 0$.

Hence either $(2x + 5) = 0$ i.e. $x = -\dfrac{5}{2}$

or $(2x - 5) = 0$ i.e. $x = \dfrac{5}{2}$

Problem 3.   Solve the following quadratic equations by factorising: (a) $4x^2 + 8x + 3 = 0$
(b) $15x^2 + 2x - 8 = 0$.

(a) $4x^2 + 8x + 3 = 0$. The factors of $4x^2$ are $4x$ and x or $2x$ and $2x$. The factors of 3 are 3 and 1, or -3 and -1. Remembering that the product of the inner terms added to the product of the two outer terms must equal $+8x$, the only combination that is true (by trial and error) is:

$$(4x^2 + 8x + 3) = (\underbrace{2x + 3)(2x} + 1)$$

Hence $(2x + 3)(2x + 1) = 0$ from which, either

$(2x + 3) = 0$ or $(2x + 1) = 0$

Thus, $2x = -3$, from which, $x = -\dfrac{3}{2}$

or $2x = -1$, from which, $x = -\dfrac{1}{2}$

which may be checked in the original equation.

(b) $15x^2 + 2x - 8 = 0$. The factors of $15x^2$ are $15x$ and x or $5x$ and $3x$. The factors of -8 are -4 and $+2$, or 4 and -2, or -8 and $+1$, or 8 and -1. By trial and error the only combination that works is:

$$15x^2 + 2x - 8 = (5x + 4)(3x - 2)$$

Hence $(5x + 4)(3x - 2) = 0$ from which

either $5x + 4 = 0$

or $3x - 2 = 0$

Hence $x = -\dfrac{4}{5}$ or $x = \dfrac{2}{3}$

which may be checked in the original equation.

Problem 4.   The roots of quadratic equation are $\dfrac{1}{3}$ and -2. Determine the equation

If the roots of a quadratic equation are α and β then $(x - \alpha)(x - \beta) = 0$.

Hence if $\alpha = \dfrac{1}{3}$ and $\beta = -2$, then

$$\left(x - \frac{1}{3}\right)(x - (-2)) = 0$$

$$\left(x - \frac{1}{3}\right)(x + 2) = 0$$

$$x^2 - \frac{1}{3}x + 2x - \frac{2}{3} = 0$$

$$x^2 + \frac{5}{3}x - \frac{2}{3} = 0$$

Hence $\mathbf{3x^2 + 5x - 2 = 0}$

Problem 5.   Find the equations is x whose roots are: (a)   5 and -5 (b)   1.2 and -0.4

(a)   If 5 and -5 are the roots of a quadratic equation then:

$$(x - 5)(x + 5) = 0$$

i.e. $x^2 - 5x + 5x - 25 = 0$

i.e. $\mathbf{x^2 - 25 = 0}$

(b) If 1.2 and -0.4 are the roots of a quadratic equation then:

$$(x - 1.2)(x + 0.4) = 0$$

i.e. $x^2 - 1.2x + 0.4x - 0.48 = 0$

i.e. $\mathbf{x^2 - 0.8x - 0.48 = 0}$

Now try the following exercise

Exercise 40   Further problems on solving quadratic equations by factorisation

In Problems 1 to 10, solve the given equations by factorisation.

1. $x^2 + 4x - 32 = 0$ [4, -8]

2. $x^2 - 16 = 0$ [4, -4]

3. $(x + 2)^2 = 16$ [2, -6]

4. $2x^2 - x - 3 = 0$ $\left[-1, 1\frac{1}{2}\right]$

5. $6x^2 - 5x + 1 = 0$ $\left[\frac{1}{2}, \frac{1}{3}\right]$

6. $10x^2 + 3x - 4 = 0$ $\left[\frac{1}{2}, -\frac{4}{5}\right]$

7. $x^2 - 4x + 4 = 0$ [2]

8. $21x^2 - 25x = 4$ $\left[1\frac{1}{3}, -\frac{1}{7}\right]$

9. $6x^2 - 5x - 4 = 0$ $\left[\frac{4}{3}, -\frac{1}{2}\right]$

10. $8x^2 + 2x - 15 = 0$ $\left[\frac{5}{4}, -\frac{3}{2}\right]$

In Problems 11 to 16, determine the quadratic equations in x whose roots are:

11. 3 and 1 $[x^2 - 4x + 3 = 0]$

12. 2 and -5 $[x^2 + 3x - 10 = 0]$

13. -1 and -4 $[x^2 + 5x + 4 = 0]$

14. $2\frac{1}{2}$ and $-\frac{1}{2}$ $[4x^2 - 8x - 5 = 0]$

15. 6 and -6 $[x^2 - 36 = 0]$

16. 2.4 and -0.7 $[x^2 - 1.7x - 1.68 = 0]$

11.3 Solution of quadratic equations by 'completing the square'

An expression such as x^2 or $(x + 2)^2$ or $(x - 3)^2$ is called a perfect square.

If $x^2 = 3$ then $x = \pm\sqrt{3}$

If $(x + 2)^2 = 5$ then $x + 2 = \pm\sqrt{5}$ and $x = -2 \pm\sqrt{5}$

If $(x - 3)^2 = 8$ then $x - 3 = \pm\sqrt{8}$ and $x = 3 \pm\sqrt{8}$

Hence if a quadratic equation can be rearranged so that one side of the equation is a perfect square and the other side of the equation is a number, then the solution of the equation is readily obtained by taking the square roots of each side as in the above examples. The process of rearranging one side of a quadratic equation into a perfect square before solving is called 'completing the square'.

$$(x + a)^2 = x^2 + 2ax + a^2$$

Thus in order to make the quadratic expression $x^2 + 2ax$ into a perfect square it is necessary to add (half the coefficient of x)^2 i.e. $\left(\dfrac{2a}{2}\right)^2$ or a^2

For example, $x^2 + 3x$ becomes a perfect square by adding $\left(\dfrac{3}{2}\right)^2$, i.e.

$$x^2 + 3x + \left(\frac{3}{2}\right)^2 = \left(x + \frac{3}{2}\right)^2$$

The method is demonstrated in the following worked problems.

Problem 6.   Solve $2x^2 + 5x = 3$ by 'completing the square'

The procedure is as follows:

1. Rearrange the equations so that all terms are on the same side of the equals sign (and the coefficient of the x^2 term is positive).

Hence $2x^2 + 5x - 3 = 0$

2. Make the coefficient of the x^2 term unity. In this case this is achieved by dividing throughout by 2. Hence

$$\frac{2x^2}{2} + \frac{5x}{2} - \frac{3}{2} = 0$$

i.e. $x^2 + \dfrac{5}{2}x - \dfrac{3}{2} = 0$

3. Rearrange the equations so that the x^2 and x terms are on one side of the equals sign and the constant is on the other side, Hence

$$x^2 + \frac{5}{2}x = \frac{3}{2}$$

4. Add to both sides of the equation (half the coefficient of x)^2. In this case the coefficient of x is $\frac{5}{2}$. Half the coefficient squared is therefore $\left(\frac{5}{4}\right)^2$.

Thus, $x^2 + \frac{5}{2}x + \left(\frac{5}{4}\right)^2 = \frac{3}{2} + \left(\frac{5}{4}\right)^2$

The LHS is now a perfect square, i.e.

$$\left(x + \frac{5}{4}\right)^2 = \frac{3}{2} + \left(\frac{5}{4}\right)^2$$

5. Evaluate the RHS. Thus

$$\left(x + \frac{5}{4}\right)^2 = \frac{3}{2} + \frac{25}{16} = \frac{24 + 25}{16} = \frac{49}{16}$$

6. Taking the square root of both sides of the equation (remembering that the square root of a number gives a $\pm$ answer). Thus

$$\sqrt{\left(x + \frac{5}{4}\right)^2} = \sqrt{\frac{49}{16}}$$

i.e. $x + \frac{5}{4} = \pm\frac{7}{4}$

7. Solve the simple equation. Thus

$$x = -\frac{5}{4} \pm \frac{7}{4}$$

i.e. $x = -\frac{5}{4} + \frac{7}{4} = \frac{2}{4} = \frac{1}{2}$

and $x = -\frac{5}{4} - \frac{7}{4} = -\frac{12}{4} = -3$

Hence $x = \frac{1}{2}$ or -3 are the roots of the equation $2x^2 + 5x = 3$

Problem 7. Solve $2x^2 + 9x + 8 = 0$, correct to 3 significant figures, by 'completing the square'

Making the coefficient of x^2 unity gives:

$$x^2 + \frac{9}{2}x + 4 = 0$$

and rearranging gives: $x^2 + \frac{9}{2}x = -4$

Adding to both sides (half the coefficient of $x)^2$ gives:

$$x^2 + \frac{9}{2}x + \left(\frac{9}{4}\right)^2 = \left(\frac{9}{4}\right)^2 - 4$$

The LHS is now a perfect square, thus:

$$\left(x + \frac{9}{4}\right)^2 = \frac{81}{16} - 4 = \frac{17}{16}$$

Taking the square root of both sides gives:

$$x + \frac{9}{4} = \sqrt{\frac{17}{16}} = \pm 1.031$$

Hence $x = -\frac{9}{4} \pm 1.031$

i.e. $x = -1.22$ or -3.28, correct to 3 significant figures.

Problem 8. By 'completing the square', solve the quadratic equation $4.6y^2 + 3.5y - 1.75 = 0$, correct to 3 decimal places

Making the coefficient of y^2 unity gives:

$$y^2 + \frac{3.5}{4.6}y - \frac{1.75}{4.6} = 0$$

and rearranging gives: $y^2 + \frac{3.5}{4.6}y = \frac{1.75}{4.6}$

Adding to both sides (half the coefficient of $y)^2$ gives:

$$y^2 + \frac{3.5}{4.6}y + \left(\frac{3.5}{9.2}\right)^2 = \frac{1.75}{4.6} + \left(\frac{3.5}{9.2}\right)^2$$

The LHS is now a perfect square, thus:

$$\left(y + \frac{3.5}{9.2}\right)^2 = 0.5251654$$

Taking the square root of both sides gives:

$$y + \frac{3.5}{9.2} = \sqrt{0.5251654} = \pm 0.7246830$$

Hence, $y = -\frac{3.5}{9.2} \pm 0.7246830$

i.e $y = 0.344$ or -1.105

Now try the following exercise

Solve the following equations by completing the
square, each correct to 3 decimal places.

1. $x^2 + 4x + 1 = 0$ [$-3.732, -0.268$]

2. $2x^2 + 5x - 4 = 0$ [$-3.137, 0.637$]

3. $3x^2 - x - 5 = 0$ [$1.468, -1.135$]

4. $5x^2 - 8x + 2 = 0$ [$1.290, 0.310$]

5. $4x^2 - 11x + 3 = 0$ [$2.443, 0.307$]

11.4 Solution of quadratic equations by formula

Let the general form of a quadratic equation be given by:

$$ax^2 + bx + c = 0$$

where a, b and c are constants.

Dividing $ax^2 + bx + c = 0$ by a gives:

$$x^2 + \frac{b}{a}x + \frac{c}{a} = 0$$

Rearranging gives:

$$x^2 + \frac{b}{a}x = -\frac{c}{a}$$

Adding to each side of the equation the square of half the
coefficient of the terms in x to make the LHS a perfect
square gives:

$$x^2 + \frac{b}{a}x + \left(\frac{b}{2a}\right)^2 = \left(\frac{b}{2a}\right)^2 - \frac{c}{a}$$

Rearranging gives:

$$\left(x + \frac{b}{a}\right)^2 = \frac{b^2}{4a^2} - \frac{c}{a} = \frac{b^2 - 4ac}{4a^2}$$

Taking the square root of both sides gives:

$$x + \frac{b}{2a} = \sqrt{\frac{b^2 - 4ac}{4a^2}} = \frac{\pm\sqrt{b^2 - 4ac}}{2a}$$

Hence $\quad x = -\frac{b}{2a} \pm \frac{\sqrt{b^2 - 4ac}}{2a}$

i.e. the quadratic formula is: $x = \dfrac{-b \pm \sqrt{b^2 - 4ac}}{2a}$

(This method of solution is 'completing the square' – as
shown in Section 10.3.). Summarising:

if $ax^2 + bx + c = 0$

then $x = \dfrac{-b \pm \sqrt{b^2 - 4ac}}{2a}$

This is known as the **quadratic formula.**

Problem 9. Solve (a) $x^2 + 2x - 8 = 0$ and
(b) $3x^2 - 11x - 4 = 0$ by using the quadratic
formula

(a)   Comparing $x^2 + 2x - 8 = 0$ with $ax^2 + bx + c = 0$
gives $a = 1$, $b = 2$ and $c = -8$.

Substituting these values into the quadratic formula

$$x = \frac{-b \pm \sqrt{b^2 - 4ac}}{2a} \text{ gives}$$

$$x = \frac{-2 \pm \sqrt{2^2 - 4(1)(-8)}}{2(1)}$$

$$= \frac{-2 \pm \sqrt{4 + 32}}{2} = \frac{-2 \pm \sqrt{36}}{2}$$

$$= \frac{-2 \pm 6}{2} = \frac{-2 + 6}{2} \text{ or } \frac{-2 - 6}{2}$$

Hence $x = \dfrac{4}{2} = \mathbf{2}$ or $\dfrac{-8}{2} = \mathbf{-4}$ (as   in
Problem 1(a)).

(b)   Comparing $3x^2 - 11x - 4 = 0$ with $ax^2 + bx + c = 0$
gives $a = 3$, $b = -11$ and $c = -4$. Hence,

$$x = \frac{-(-11) \pm \sqrt{(-11)^2 - 4(3)(-4)}}{2(3)}$$

$$= \frac{-11 \pm \sqrt{121 + 48}}{6} = \frac{11 \pm \sqrt{169}}{6}$$

$$= \frac{11 \pm 13}{6} = \frac{11 + 13}{6} \text{ or } \frac{11 - 13}{6}$$

Hence $x = \dfrac{24}{6} = \mathbf{4}$ or $\dfrac{-2}{6} = -\dfrac{1}{3}$ (as in Problem
1(b)).

Problem 10. Solve $4x^2 + 7x + 2 = 0$ giving the
roots correct to 2 decimal places

Comparing $4x^2 + 7x + 2 = 0$ with $ax^2 + bx + c = 0$ gives $a = 4$, $b = 7$ and $c = 2$. Hence,

$$x = \frac{-7 \pm \sqrt{7^2 - 4(4)(2)}}{2(4)}$$

$$= \frac{-7 \pm \sqrt{17}}{8} = \frac{-7 \pm 4.123}{8}$$

$$= \frac{-7 \pm 4.123}{8} \quad \text{or} \quad \frac{-7 - 4.123}{8}$$

Hence, $x = -0.36$ or -1.39, **correct to 2 decimal places.**

Now try the following exercise

Exercise 42 Further problems on solving quadratic equations by formula

Solve the following equations by using the quadratic formula, correct to 3 decimal places.

1. $2x^2 + 5x - 4 = 0$ [0.637, −3.137]

2. $5.76x^2 + 2.86x - 1.35 = 0$ [0.296, −0.792]

3. $2x^2 - 7x + 4 = 0$ [2.781, 0.719]

4. $4x + 5 = \dfrac{3}{x}$ [0.443, −1.693]

5. $(2x + 1) = \dfrac{5}{x - 3}$ [3.608, −1.108]

11.5 Practical problems involving quadratic equations

There are many **practical problems** where a quadratic equation has first to be obtained, from given information, before it is solved.

Problem 11. Calculate the diameter of a solid cylinder which has a height of 82.0 cm and a total surface area of 2.0 m²

Total surface area of a cylinder

= curved surface area

           + 2 circular ends (from Chapter 20)

$= 2\pi rh = 2\pi r^2$

(where r = radius and h = height)

Since the total surface area $= 2.0\,\text{m}^2$ and the height $h = 82$ cm or 0.82 m, then

$$2.0 = 2\pi r(0.82) + 2\pi r^2$$

i.e. $2\pi r^2 + 2\pi r(0.82) - 2.0 = 0$

Dividing throughout by 2π gives:

$$r^2 + 0.82r - \frac{1}{\pi} = 0$$

Using the quadratic formula:

$$r = \frac{-0.82 \pm \sqrt{(0.82)^2 - 4(1)\left(-\dfrac{1}{\pi}\right)}}{2(1)}$$

$$= \frac{-0.82 \pm \sqrt{1.9456}}{2} = \frac{-0.82 \pm 1.3948}{2}$$

$$= 0.2874 \quad \text{or} \quad -1.1074$$

Thus the radius r of the cylinder is 0.2874 m (the negative solution being neglected).
Hence the diameter of the cylinder

$$= 2 \times 0.2874$$

$$= \mathbf{0.5748\,m} \quad \textbf{or} \quad \mathbf{57.5\,cm}$$

       correct to 3 significant figures

Problem 12. The height s metres of a mass projected vertically upward at time t seconds is $s = ut - \dfrac{1}{2}gt^2$. Determine how long the mass will take after being projected to reach a height of 16 m (a) on the ascent and (b) on the descent, when $u = 30$ m/s and $g = 9.81$ m/s²

When height $s = 16$ m, $16 = 30\,t - \dfrac{1}{2}(9.81)t^2$

i.e. $\qquad\qquad 4.905t^2 - 30t + 16 = 0$

Using the quadratic formula:

$$t = \frac{-(-30) \pm \sqrt{(-30)^2 - 4(4.905)(16)}}{2(4.905)}$$

$$= \frac{30 \pm \sqrt{586.1}}{9.81} = \frac{30 \pm 24.21}{9.81}$$

$$= 5.53 \quad \text{or} \quad 0.59$$

Hence the mass will reach a height of 16 m after 0.59 s on the ascent and after 5.53 s on the descent.

Problem 13. A shed is 4.0 m long and 2.0 m wide. A concrete path of constant width is laid all the way around the shed. If the area of the path is 9.50 m² calculate its width to the nearest centimetre

Figure 11.1 shows a plan view of the shed with its surrounding path of width t metres.

Area of path $= 2(2.0 \times t) + 2t(4.0 + 2t)$

i.e. $9.50 = 4.0t + 8.0t + 4t^2$

or $4t^2 + 12.0t - 9.50 = 0$

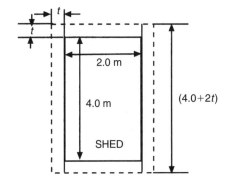

Figure 11.1

Hence $t = \dfrac{-(12.0) \pm \sqrt{(12.0)^2 - 4(4)(-9.50)}}{2(4)}$

$= \dfrac{-12.0 \pm \sqrt{296.0}}{8}$

$= \dfrac{-12.0 \pm 17.20465}{8}$

Hence $t = 0.6506$ m or -3.65058 m

Neglecting the negative result which is meaningless, the width of the path, $t = \mathbf{0.651\,m}$ or **65 cm**, correct to the nearest centimetre.

Problem 14. If the total surface area of a solid cone is 486.2 cm² and its slant height is 15.3 cm, determine its base diameter

From Chapter 20, page 157, the total surface area A of a solid cone is given by: $A = \pi r l + \pi r^2$ where l is the slant height and r the base radius.

If $A = 482.2$ and $l = 15.3$, then

$$482.2 = \pi r(15.3) + \pi r^2$$

i.e. $\pi r^2 + 15.3\pi r - 482.2 = 0$

or $r^2 + 15.3r - \dfrac{482.2}{\pi} = 0$

Using the quadratic formula,

$$r = \dfrac{-15.3 \pm \sqrt{(15.3)^2 - 4\left(\dfrac{-482.2}{\pi}\right)}}{2}$$

$$= \dfrac{-15.3 \pm \sqrt{848.0461}}{2}$$

$$= \dfrac{-15.3 \pm 29.12123}{2}$$

Hence radius $r = 6.9106$ cm (or -22.21 cm, which is meaningless, and is thus ignored).

Thus **the diameter of the base**

$$= 2r = 2(6.9106) = \mathbf{13.82\,cm}$$

Now try the following exercise

Exercise 43 Further practical problems involving quadratic equations

1. The angle a rotating shaft turns through in t seconds is given by: $\theta = \omega t + \dfrac{1}{2}\alpha t^2$. Determine the time taken to complete 4 radians if ω is 3.0 rad/s and α is 0.60 rad/s². [1.191 s]

2. The power P developed in an electrical circuit is given by $P = 10I - 8I^2$, where I is the current in amperes. Determine the current necessary to produce a power of 2.5 watts in the circuit. [0.345 A or 0.905 A]

3. The sag l metres in a cable stretched between two supports, distance x m apart is given by: $l = \dfrac{12}{x} + x$. Determine the distance between supports when the sag is 20 m. [0.619 m or 19.38 m]

4. The acid dissociation constant K_a of ethanoic acid is 1.8×10^{-5} mol dm⁻³ for a particular solution. Using the Ostwald dilution law $K_a = \dfrac{x^2}{v(1-x)}$ determine x, the degree of ionization, given that $v = 10$ dm³. [0.0133]

5. A rectangular building is 15 m long by 11 m wide. A concrete path of constant width is laid all the way around the building. If the area of the path is 60.0 m², calculate its width correct to the neareast millimetre. [1.066 m]

Section 1

6. The total surface area of a closed cylindrical container is 20.0 m². Calculate the radius of the cylinder if its height is 2.80 m.

[86.78 cm]

7. The bending moment M at a point in a beam is given by $M = \dfrac{3x(20-x)}{2}$ where x metres is the distance from the point of support. Determine the value of x when the bending moment is 50 Nm. [1.835 m or 18.165 m]

8. A tennis court measures 24 m by 11 m. In the layout of a number of courts an area of ground must be allowed for at the ends and at the sides of each court. If a border of constant width is allowed around each court and the total area of the court and its border is 950 m², find the width of the borders.

[7 m]

9. Two resistors, when connected in series, have a total resistance of 40 ohms. When connected in parallel their total resistance is 8.4 ohms. If one of the resistors has a resistance R_x ohms:

  (a) show that $R_x^2 - 40R_x + 336 = 0$ and
  (b) calculated the resistance of each.

[(b) 12 ohms, 28 ohms]

11.6 The solution of linear and quadratic equations simultaneously

Sometimes a linear equation and a quadratic equation need to be solved simultaneously. An algebraic method of solution is shown in Problem 15; a graphical solution is shown in Chapter 31, page 281.

Problem 15. Determine the values of x and y which simultaneously satisfy the equations: $y = 5x - 4 - 2x^2$ and $y = 6x - 7$

For a simultaneous solution the values of y must be equal, hence the RHS of each equation is equated.

Thus $5x - 4 - 2x^2 = 6x - 7$

Rearranging gives:

$$5x^2 - 4 - 2x^2 - 6x + 7 = 0$$

i.e. $-x + 3 - 2x^2 = 0$

or $2x^2 + x - 3 = 0$

Factorising gives: $(2x + 3)(x - 1) = 0$

i.e. $x = -\dfrac{3}{2}$ or $x = 1$

In the equation $y = 6x - 7$

when $x = -\dfrac{3}{2}$, $y = 6\left(\dfrac{-3}{2}\right) - 7 = -16$

and when $x = 1$, $y = 6 - 7 = -1$

[Checking the result in $y = 5x - 4 - 2x^2$:

when $x = -\dfrac{3}{2}$, $y = 5\left(-\dfrac{3}{2}\right) - 4 - 2\left(-\dfrac{3}{2}\right)^2$

$= -\dfrac{15}{2} - 4 - \dfrac{9}{2} = -16$

as above; and when $x = 1$, $y = 5 - 4 - 2 = -1$ as above.]

Hence the simultaneous solutions occur when

$$x = -\dfrac{3}{2}, \quad y = -16$$

and when $x = 1$, $y = -1$

Now try the following exercise

Exercise 44   Further problems on solving linear and quadratic equations simultaneously

In Problems 1 to 3 determine the solutions of the simulations equations.

1. $y = x^2 + x + 1$
   $y = 4 - x$

   $[x = 1, y = 3 \text{ and } x = -3, y = 7]$

2. $y = 15x^2 + 21x - 11$
   $y = 2x - 1$

   $\left[x = \dfrac{2}{5}, y = -\dfrac{1}{5} \text{ and } x = -1\dfrac{2}{3}, y = -4\dfrac{1}{3}\right]$

3. $2x^2 + y = 4 + 5x$
   $x + y = 4$

   $[x = 0, y = 4 \text{ and } x = 3, y = 1]$

Chapter 12

Inequalities

12.1 Introduction in inequalities

An **inequality** is any expression involving one of the symbols $<$, $>$ $\leq$ or $\geq$

$p < q$ means p is less than q
$p > q$ means p is greater than q
$p \leq q$ means p is less than or equal to q
$p \geq q$ means p is greater than or equal to q

Some simple rules

(i) When a quantity is **added or subtracted** to both sides of an inequality, the inequality still remains.

For example, if $p < 3$

then $\quad p + 2 < 3 + 2$ (adding 2 to both sides)

and $\quad p - 2 < 3 - 2$ (subtracting 2 from both sides)

(ii) When **multiplying or dividing** both sides of an inequality by a **positive** quantity, say 5, the inequality **remains the same**. For example,

if $p > 4$ then $5p > 20$ and $\dfrac{p}{5} > \dfrac{4}{5}$

(iii) When **multiplying or dividing** both sides of an inequality by a **negative** quantity, say -3, **the inequality is reversed**. For example,

if $p > 1$ then $-3p < -3$ and $\dfrac{p}{-3} < \dfrac{1}{-3}$

(Note $>$ has changed to $<$ in each example.)

To **solve an inequality** means finding all the values of the variable for which the inequality is true. Knowledge of simple equations and quadratic equations are needed in this chapter.

12.2 Simple inequalities

The solution of some simple inequalities, using only the rules given in section 12.1, is demonstrated in the following worked problems.

Problem 1. Solve the following inequalities:
(a) $3 + x > 7$ (b) $3t < 6$
(c) $z - 2 \geq 5$ (d) $\dfrac{p}{3} \leq 2$

(a) Subtracting 3 from both sides of the inequality: $3 + x > 7$ gives:

$$3 + x - 3 > 7 - 3, \text{ i.e. } \boldsymbol{x > 4}$$

Hence, all values of x greater than 4 satisfy the inequality.

(b) Dividing both sides of the inequality: $3t < 6$ by 3 gives:

$$\frac{3t}{3} < \frac{6}{3}, \text{ i.e. } \boldsymbol{t < 2}$$

Hence, all values of t less than 2 satisfy the inequality.

(c) Adding 2 to both sides of the inequality $z - 2 \geq 5$ gives:

$$z - 2 + 2 \geq 5 + 2, \text{ i.e. } \boldsymbol{z \geq 7}$$

Hence, all values of z greater than or equal to 7 satisfy the inequality.

(d) Multiplying both sides of the inequality $\dfrac{p}{3} \leq 2$ by 3 gives:

$$(3)\frac{p}{3} \leq (3)2, \text{ i.e. } \boldsymbol{p \leq 6}$$

Hence, all values of p less than or equal to 6 satisfy the inequality.

Problem 2. Solve the inequality: $4x + 1 > x + 5$

Subtracting 1 from both sides of the inequality: $4x + 1 > x + 5$ gives:

$$4x > x + 4$$

Subtracting x from both sides of the inequality: $4x > x + 4$ gives:

$$3x > 4$$

Dividing both sides of the inequality: $3x > 4$ by 3 gives:

$$x > \frac{4}{3}$$

Hence all values of x greater than $\frac{4}{3}$ satisfy the inequality:

$$4x + 1 > x + 5$$

Problem 3. Solve the inequality: $3 - 4t \le 8 + t$

Subtracting 3 from both sides of the inequality: $3 - 4t \le 8 + t$ gives:

$$-4t \le 5 + t$$

Subtracting t from both sides of the inequality: $-4t \le 5 + t$ gives:

$$-5t \le 5$$

Dividing both sides of the inequality $-5t \le 5$ by -5 gives:

$$t \ge -1 \text{ (remembering to reverse the inequality)}$$

Hence, all values of t greater than or equal to -1 satisfy the inequality.

Now try the following exercise

Exercise 45   Further problems on simple inequalitites

Solve the following inequalities:

1. (a) $3t > 6$ (b) $2x < 10$
$$[(a)\ t > 2 \quad (b)\ x < 5]$$

2. (a) $\dfrac{x}{2} > 1.5$ (b) $x + 2 \ge 5$
$$[(a)\ x > 3 \quad (b)\ x \ge 3]$$

3. (a) $4t - 1 \le 3$ (b) $5 - x \ge -1$
$$[(a)\ t \le 1 \quad (b)\ x \le 6]$$

4. (a) $\dfrac{7 - 2k}{4} \le 1$ (b) $3z + 2 > z + 3$
$$\left[(a)\ k \ge \frac{3}{2} \quad (b)\ z > \frac{1}{2}\right]$$

5. (a) $5 - 2y \le 9 + y$ (b) $1 - 6x \le 5 + 2x$
$$\left[(a)\ y \ge -\frac{4}{3} \quad (b)\ x \ge -\frac{1}{2}\right]$$

12.3   Inequalities involving a modulus

The **modulus** of a number is the size of the number, regardless of sign. Vertical lines enclosing the number denote a modulus.

For example, $|4| = 4$ and $|-4| = 4$ (the modulus of a number is never negative),

The inequality: $|t| < 1$ means that all numbers whose actual size, regardless of sign, is less than 1, i.e. any value between -1 and $+1$.

Thus $|t| < 1$ **means** $-1 < t < 1$.

Similarly, $|x| > 3$ means all numbers whose actual size, regardless of sign, is greater than 3, i.e. any value greater than 3 and any value less than -3.

Thus $|x| > 3$ **means** $x > 3$ **and** $x < -3$.

Inequalities involving a modulus are demonstrated in the following worked problems.

Problem 4. Solve the following inequality:
$$|3x + 1| < 4$$

Since $|3x + 1| < 4$ then $-4 < 3x + 1 < 4$

Now $-4 < 3x + 1$ becomes $-5 < 3x$, i.e. $-\dfrac{5}{3} < x$

and $3x + 1 < 4$ becomes $3x < 3$, i.e. $x < 1$

Hence, these two results together become $-\dfrac{5}{3} < x < 1$ and mean that the inequality $|3x + 1| < 4$ is satisfied for any value of x greater than $-\dfrac{5}{3}$ but less than 1.

Problem 5. Solve the inequality: $|1 + 2t| \le 5$

Since $|1 + 2t| \le 5$ then $-5 \le 1 + 2t \le 5$

Now $-5 \le 1 + 2t$ becomes $-6 \le 2t$, i.e. $-3 \le t$

and $1 + 2t \le 5$ becomes $2t \le 4$ i.e. $t \le 2$

Hence, these two results together become: $-3 \le t \le 2$

Problem 6. Solve the inequality: $|3z - 4| > 2$

$|3z - 4| > 2$ means $3z - 4 > 2$ and $3z - 4 < -2$,

i.e. $3z > 6$ and $3z < 2$,

i.e. the inequality: $|3z - 4| > 2$ is satisfied when

$$z > 2 \text{ and } z < \frac{2}{3}$$

Now try the following exercise

Exercise 46 Further problems on inequalities involving a modulus

Solve the following inequalities:

1. $|t + 1| < 4$ $[-5 < t < 3]$

2. $|y + 3| \leq 2$ $[-5 \leq y \leq -1]$

3. $|2x - 1| < 4$ $\left[-\dfrac{3}{2} < x < \dfrac{5}{2}\right]$

4. $|3t - 5| > 4$ $\left[t > 3 \text{ and } t < \dfrac{1}{3}\right]$

5. $|1 - k| \geq 3$ $[k \geq 4 \text{ and } k \leq -2]$

12.4 Inequalities involving quotients

If $\dfrac{p}{q} > 0$ then $\dfrac{p}{q}$ must be a **positive** value.

For $\dfrac{p}{q}$ to be positive, **either** p is positive **and** q is positive **or** p is negative **and** q is negative.

i.e. $\dfrac{+}{+} = +$ and $\dfrac{-}{-} = +$

If $\dfrac{p}{q} < 0$ then $\dfrac{p}{q}$ must be a **negative** value.

For $\dfrac{p}{q}$ to be negative, **either** p is positive **and** q is negative **or** p is negative **and** q is positive.

i.e. $\dfrac{+}{-} = -$ and $\dfrac{-}{+} = -$

This reasoning is used when solving inequalities involving quotients as demonstrated in the following worked problems.

Problem 7. Solve the inequality: $\dfrac{t + 1}{3t - 6} > 0$

Since $\dfrac{t + 1}{3t - 6} > 0$ then $\dfrac{t + 1}{3t - 6}$ must be **positive.**

For $\dfrac{t + 1}{3t - 6}$ to be positive,

     either   (i) $t + 1 > 0$ **and** $3t - 6 > 0$

     or   (ii) $t + 1 < 0$ **and** $3t - 6 < 0$

(i) If $t + 1 > 0$ then $t > -1$ and if $3t - 6 > 0$ then $3t > 6$ and $t > 2$

     Both of the inequalities $t > -1$ **and** $t > 2$ are only true when $t > 2$,

     i.e. the fraction $\dfrac{t + 1}{3t - 6}$ is positive when $t > 2$

(ii) If $t + 1 < 0$ then $t < -1$ and if $3t - 6 < 0$ then $3t < 6$ and $t < 2$

     Both of the inequalities $t < -1$ **and** $t < 2$ are only true when $t < -1$,

     i.e. the fraction $\dfrac{t + 1}{3t - 6}$ is positive when $t < -1$

Summarising, $\dfrac{t + 1}{3t - 6} > 0$ when $t > 2$ **or** $t < -1$

Problem 8. Solve the inequality: $\dfrac{2x + 3}{x + 2} \leq 1$

Since $\dfrac{2x + 3}{x + 2} \leq 1$ then $\dfrac{2x + 3}{x + 2} - 1 \leq 0$

i.e. $\dfrac{2x + 3}{x + 2} - \dfrac{x + 2}{x + 2} \leq 0$,

i.e. $\dfrac{2x + 3 - (x + 2)}{x + 2} \leq 0$ or $\dfrac{x + 1}{x + 2} \leq 0$

For $\dfrac{x + 1}{x + 2}$ to be negative or zero,

     either   (i) $x + 1 \leq 0$ **and** $x + 2 > 0$

     or   (ii) $x + 1 \geq 0$ **and** $x + 2 < 0$

(i) If $x + 1 \leq 0$ then $x \leq -1$ and if $x + 2 > 0$ then $x > -2$ (Note that $>$ is used for the denominator, not $\geq$; a zero denominator gives a value for the fraction which is impossible to evaluate.)

     Hence, the inequality $\dfrac{x + 1}{x + 2} \leq 0$ is true when x is greater than -2 and less than or equal to -1, which may be written as $-2 < x \leq -1$

(ii) If $x + 1 \geq 0$ then $x \geq -1$ and if $x + 2 < 0$ then $x < -2$

     It is not possible to satisfy both $x \geq -1$ and $x < -2$ thus no values of x satisfies (ii).

Summarising, $\dfrac{2x+3}{x+2} \le 1$ when $-2 < x \le -1$

Now try the following exercise

> **Exercise 47   Further problems on inequalities involving quotients**
>
> Solve the following inequalitites:
>
> 1. $\dfrac{x+4}{6-2x} \ge 0$ $[-4 \le x < 3]$
>
> 2. $\dfrac{2t+4}{t-5} > 1$ $[t > 5 \text{ or } t < -9]$
>
> 3. $\dfrac{3z-4}{z+5} \le 2$ $[-5 < z \le 14]$
>
> 4. $\dfrac{2-x}{x+3} \ge 4$ $[-3 < x \le -2]$

12.5   Inequalities involving square functions

The following two general rules apply when inequalities involve square functions:

(i) **if $x^2 > k$ then $x > \sqrt{k}$ or $x < -\sqrt{k}$**   (1)
(ii) **if $x^2 > k$ then $-\sqrt{k} < x < \sqrt{k}$**   (2)

These rules are demonstrated in the following worked problems.

Problem 9. Solve the inequality: $t^2 > 9$

Since $t^2 > 9$ then $t^2 - 9 > 0$, i.e. $(t+3)(t-3) > 0$ by factorising
For $(t+3)(t-3)$ to be positive,

     either   (i) $(t+3) > 0$ **and** $(t-3) > 0$
        or   (ii) $(t+3) < 0$ **and** $(t-3) < 0$

(i) If $(t+3) > 0$ then $t > -3$ and if $(t-3) > 0$ then $t > 3$
     Both of these are true only when $t > 3$

(ii) If $(t+3) < 0$ then $t < -3$ and if $(t-3) < 0$ then $t < 3$
     Both of these are true only when $t < -3$

Summarising, $t^2 > 9$ when $t > 3$ **or** $t < -3$

This demonstrates the general rule:

    if $x^2 > k$ then $x > \sqrt{k}$ or $x < -\sqrt{k}$ (1)

Problem 10. Solve the inequality: $x^2 > 4$

From the general rule stated above in equation (1): if $x^2 > 4$ then $x > \sqrt{4}$ or $x < -\sqrt{4}$

i.e. the inequality: $x^2 > 4$ is satisfied when **$x > 2$ or $x < -2$**

Problem 11. Solve the inequality: $(2z+1)^2 > 9$

From equation (1), if $(2z+1)^2 > 9$ then

         $2z+1 > \sqrt{9}$ or $2z+1 < -\sqrt{9}$
   i.e. $2z+1 > 3$ or $2z+1 < -3$
   i.e. $2z > 2$ or $2z < -4$,
   i.e.   **$z > 1$**   or   **$z < -2$**

Problem 12. Solve the inequality: $t^2 < 9$

Since $t^2 < 9$ then $t^2 - 9 < 0$, i.e. $(t+3)(t-3) < 0$ by factorising. For $(t+3)(t-3)$ to be negative,

     either   (i) $(t+3) > 0$ **and** $(t-3) < 0$
        or   (ii) $(t+3) < 0$ **and** $(t-3) > 0$

(i) If $(t+3) > 0$ then $t > -3$ and if $(t-3) < 0$ then $t < 3$
     Hence (i) is satisfied when $t > -3$ and $t < 3$ which may be written as: **$-3 < t < 3$**

(ii) If $(t+3) < 0$ then $t < -3$ and if $(t-3) > 0$ then $t > 3$
     It is not possible to satisfy both $t < -3$ and $t > 3$, thus no values of t satisfies (ii).

Summarising, $t^2 < 9$ when $-3 < t < 3$ which means that all values of t between -3 and $+3$ will satisfy the inequality.

This demonstrates the general rule:

   if $x^2 < k$ then $-\sqrt{k} < x < \sqrt{k}$ (2)

Problem 13. Solve the inequality: $x^2 < 4$

From the general rule stated above in equation (2): if $x^2 < 4$ then $-\sqrt{4} < x < \sqrt{4}$

i.e. the inequality: $x^2 < 4$ is satisfied when:
$$-2 < x < 2$$

Problem 14. Solve the inequality: $(y-3)^2 \le 16$

From equation (2), $\quad -\sqrt{16} \le (y-3) \le \sqrt{16}$

i.e. $\qquad\qquad\qquad -4 \le (y-3) \le 4$

from which, $\qquad\qquad 3-4 \le y \le 4+3,$

i.e. $\qquad\qquad\qquad\qquad \mathbf{-1 \le y \le 7}$

Now try the following exercise

Exercise 48 Further problems on inequalities involving square functions

Solve the following inequalities:

1. $z^2 > 16$ $\qquad\qquad\qquad$ $[z > 4 \text{ or } z < -4]$

2. $z^2 < 16$ $\qquad\qquad\qquad$ $[-4 < z < 4]$

3. $2x^2 \ge 6$ $\qquad\qquad$ $[x \ge \sqrt{3} \text{ or } x \le -\sqrt{3}]$

4. $3k^2 - 2 \le 10$ $\qquad\qquad$ $[-2 \le k \le 2]$

5. $(t-1)^2 \le 36$ $\qquad\qquad$ $[-5 \le t \le 7]$

6. $(t-1)^2 \ge 36$ $\qquad\qquad$ $[t \ge 7 \text{ or } t \le -5]$

7. $7 - 3y^2 \le -5$ $\qquad\qquad$ $[y \ge 2 \text{ or } y \le -2]$

8. $(4k+5)^2 > 9$ $\qquad$ $\left[k > -\dfrac{1}{2} \text{ or } k < -2\right]$

12.6 Quadratic inequalities

Inequalities involving quadratic expressions are solved using either **factorisation** or **'completing the square'**. For example,

$$x^2 - 2x - 3 \text{ is factorised as } (x+1)(x-3)$$
$$\text{and } 6x^2 + 7x - 5 \text{ is factorised as } (2x-1)(3x+5)$$

If a quadratic expression does not factorise, then the technique of 'completing the square' is used. In general, the procedure for $x^2 + bx + c$ is:

$$x^2 + bx + c \equiv \left(x + \frac{b}{2}\right)^2 + c - \left(\frac{b}{2}\right)^2$$

For example, $x^2 + 4x - 7$ does not factorise; completing the square gives:

$$x^2 + 4x - 7 \equiv (x+2)^2 - 7 - 2^2 \equiv (x+2)^2 - 11$$

Similarly,

$$x^2 + 6x - 5 \equiv (x+3)^2 - 5 - 3^2 \equiv (x-3)^2 - 14$$

Solving quadratic inequalities is demonstrated in the following worked problems.

Problem 15. Solve the inequality:
$$x^2 + 2x - 3 > 0$$

Since $x^2 + 2x - 3 > 0$ then $(x-1)(x+3) > 0$ by factorising. For the product $(x-1)(x+3)$ to be positive,

$\quad$ **either** $\quad$ (i) $(x-1) > 0$ $\quad$ **and** $\quad$ $(x+3) > 0$

$\quad$ **or** $\quad\quad$ (ii) $(x-1) < 0$ $\quad$ **and** $\quad$ $(x+3) < 0$

(i) $\quad$ Since $(x-1) > 0$ then $x > 1$ and since $(x+3) > 0$ then $x > -3$
$\quad\quad$ Both of these inequalities are satisfied only when
$\quad\quad$ $\boldsymbol{x > 1}$

(ii) $\quad$ Since $(x-1) < 0$ then $x < 1$ and since $(x+3) < 0$ then $x < -3$
$\quad\quad$ Both of these inequalities are satisfied only when
$\quad\quad$ $\boldsymbol{x < -3}$

Summarising, $x^2 + 2x - 3 > 0$ is satisfied when either $\boldsymbol{x > 1}$ **or** $\boldsymbol{x < -3}$

Problem 16. Solve the inequality: $t^2 - 2t - 8 < 0$

Since $\quad t^2 - 2t - 8 < 0$ $\quad$ then $\quad$ $(t-4)(t+2) < 0$ $\quad$ by factorising.
For the product $(t-4)(t+2)$ to be negative,

$\quad$ **either** $\quad$ (i) $(t-4) > 0$ $\quad$ **and** $\quad$ $(t+2) < 0$

$\quad$ **or** $\quad$ (ii) $(t-4) < 0$ $\quad$ **and** $\quad$ $(t+2) > 0$

(i) $\quad$ Since $(t-4) > 0$ then $t > 4$ and since $(t+2) < 0$ then $t < -2$
$\quad\quad$ It is not possible to satisfy both $t > 4$ and $t < -2$, thus no values of t satisfies the inequality (i)

(ii) $\quad$ Since $(t-4) < 0$ then $t < 4$ and since $(t+2) > 0$ then $t > -2$
$\quad\quad$ Hence, (ii) is satisfied when $-2 < t < 4$

Summarising, $\quad t^2 - 2t - 8 < 0$ $\quad$ is $\quad$ satisfied $\quad$ when $\boldsymbol{-2 < t < 4}$

Problem 17. Solve the inequality:
$$x^2 + 6x + 3 < 0$$

$x^2 + 6x + 3$ does not factorise; completing the square gives:

$$x^2 + 6x + 3 \equiv (x + 3)^2 + 3 - 3^2$$
$$\equiv (x + 3)^2 - 6$$

The inequality thus becomes: $(x + 3)^2 - 6 < 0$ or $(x + 3)^2 < 6$

From equation (2), $-\sqrt{6} < (x + 3) < \sqrt{6}$

from which, $(-\sqrt{6} - 3) < x < (\sqrt{6} - 3)$

Hence, $x^2 + 6x + 3 < 0$ is satisfied when **−5.45 < x < −0.55** correct to 2 decimal places.

Problem 18. Solve the inequality:
$$y^2 - 8y - 10 \geq 0$$

$y^2 - 8y - 10$ does not factorise; completing the square gives:

$$y^2 - 8y - 10 \equiv (y - 4)^2 - 10 - 4^2$$
$$\equiv (y - 4)^2 - 26$$

The inequality thus becomes: $(y - 4)^2 - 26 \geq 0$ or $(y - 4)^2 \geq 26$

From equation (1), $(y - 4) \geq \sqrt{26}$ or $(y - 4) \leq -\sqrt{26}$

from which, $\quad y \geq 4 + \sqrt{26}$ **or** $y \leq 4 - \sqrt{26}$

Hence, $y^2 - 8y - 10 \geq 0$ is satisfied when **y ≥ 9.10 or y ≤ −1.10** correct to 2 decimal places.

Now try the following exercise

Exercise 49 Further problems on quadratic inequalities

Solve the following inequalities:

1. $x^2 - x - 6 > 0$ $\qquad\qquad$ $[x > 3 \text{ or } x < -2]$

2. $t^2 + 2t - 8 \leq 0$ $\qquad\qquad$ $[-4 \leq t \leq 2]$

3. $2x^2 + 3x - 2 < 0$ $\qquad\qquad$ $\left[-2 < x < \dfrac{1}{2}\right]$

4. $y^2 - y - 20 \geq 0$ $\qquad\qquad$ $[y \geq 5 \text{ or } y \leq -4]$

5. $z^2 + 4z + 4 \leq 4$ $\qquad\qquad$ $[-4 \leq z \leq 0]$

6. $x^2 + 6x - 6 \leq 0$
$$\left[(-\sqrt{3} - 3) \leq x \leq (\sqrt{3} - 3)\right]$$

7. $t^2 - 4t - 7 \geq 0$
$$[t \geq (\sqrt{11} + 2) \text{ or } t \leq (2 - \sqrt{11})]$$

8. $k^2 + k - 3 \geq 0$
$$\left[k \geq \left(\sqrt{\dfrac{13}{4}} - \dfrac{1}{2}\right) \text{ or } k \leq \left(-\sqrt{\dfrac{13}{4}} - \dfrac{1}{2}\right)\right]$$

Chapter 13

Logarithms

13.1 Introduction to logarithms

With the use of calculators firmly established, logarithmic tables are now rarely used for calculation. However, the theory of logarithms is important, for there are several scientific and engineering laws that involve the rules of logarithms.

If a number y can be written in the form a^x, then the index x is called the 'logarithm of y to the base of a',

i.e.　　　**if $y = a^x$　then　$x = \log_a y$**

Thus, since $1000 = 10^3$, then $3 = \log_{10} 1000$

z Check this using the 'log' button on your calculator.

(a) Logarithms having a base of 10 are called **common logarithms** and $\log_{10}$ is usually abbreviated to lg. The following values may be checked by using a calculator:

$$\lg 17.9 = 1.2528\ldots,$$

$$\lg 462.7 = 2.6652\ldots$$

and　$\lg 0.0173 = -1.7619\ldots$

(b) Logarithms having a base of e (where 'e' is a mathematical constant approximately equal to 2.7183) are called **hyperbolic**, **Napierian** or **natural logarithms**, and $\log_e$ is usually abbreviated to ln.

The following values may be checked by using a calculator:

$$\ln 3.15 = 1.1474\ldots,$$

$$\ln 362.7 = 5.8935\ldots$$

and　$\ln 0.156 = -1.8578\ldots$

For more on Napierian logarithms see Chapter 14.

13.2 Laws of logarithms

There are three laws of logarithms, which apply to any base:

(i) To multiply two numbers:

$$\log (A \times B) = \log A + \log B$$

The following may be checked by using a calculator:

$$\lg 10 = 1,$$

also　　　$\lg 5 + \lg 2 = 0.69897\ldots$

$$+ 0.301029\ldots = 1$$

Hence　　$\lg (5 \times 2) = \lg 10 = \lg 5 + \lg 2$

(ii) To divide two numbers:

$$\log \left(\frac{A}{B}\right) = \log A - \log B$$

The following may be checked using a calculator:

$$\ln \left(\frac{5}{2}\right) = \ln 2.5 = 0.91629\ldots$$

Also　　$\ln 5 - \ln 2 = 1.60943\ldots - 0.69314\ldots$

$$= 0.91629\ldots$$

Hence　　$\ln \left(\frac{5}{2}\right) = \ln 5 - \ln 2$

(iii) To raise a number to a power:

$$\lg A^n = n \log A$$

The following may be checked using a calculator:

$$\lg 5^2 = \lg 25 = 1.39794\ldots$$

Also　　$2 \lg 5 = 2 \times 0.69897\ldots = 1.39794\ldots$

Hence　$\lg 5^2 = 2 \lg 5$

Problem 1. Evaluate
(a) $\log_3 9$ (b) $\log_{10} 10$ (c) $\log_{16} 8$

(a) Let $x = \log_3 9$ then $3^x = 9$ from the definition of a logarithm, i.e. $3^x = 3^2$, from which $x = 2$.
Hence **$\log_3 9 = 2$**

(b) Let $x = \log_{10} 10$ then $10^x = 10$ from the definition of a logarithm, i.e. $10^x = 10^1$, from which $x = 1$.
Hence **$\log_{10} 10 = 1$** (which may be checked by a calculator)

(c) Let $x = \log_{16} 8$ then $16^x = 8$, from the definition of a logarithm, i.e. $(2^4)^x = 2^3$, i.e. $2^{4x} = 2^3$ from the laws of indices, from which, $4x = 3$ and $x = \dfrac{3}{4}$
Hence   **$\log_{16} 8 = \dfrac{3}{4}$**

Problem 2. Evaluate
(a) $\lg 0.001$ (b) $\ln e$ (c) $\log_3 \dfrac{1}{81}$

(a) Let $x = \lg 0.001 = \log_{10} 0.001$ then $10^x = 0.001$, i.e. $10^x = 10^{-3}$, from which $x = -3$
Hence **$\lg 0.001 = -3$** (which may be checked by a calculator).

(b) Let $x = \ln e = \log_e e$ then $e^x = e$, i.e. $e^x = e^1$ from which $x = 1$
Hence **$\ln e = 1$** (which may be checked by a calculator).

(c) Let $x = \log_3 \dfrac{1}{81}$ then $3^x = \dfrac{1}{81} = \dfrac{1}{3^4} = 3^{-4}$, from which $x = -4$
Hence **$\log_3 \dfrac{1}{81} = -4$**

Problem 3. Solve the following equations:
(a) $\lg x = 3$ (b) $\log_2 x = 3$ (c) $\log_5 x = -2$

(a) If $\lg x = 3$ then $\log_{10} x = 3$ and $x = 10^3$, i.e. **$x = 1000$**

(b) If $\log_2 x = 3$ then $x = 2^3 = \mathbf{8}$

(c) If $\log_5 x = -2$ then $x = 5^{-2} = \dfrac{1}{5^2} = \mathbf{\dfrac{1}{25}}$

Problem 4. Write (a) $\log 30$ (b) $\log 450$ in terms of $\log 2$, $\log 3$ and $\log 5$ to any base

(a) $\log 30 = \log(2 \times 15) = \log(2 \times 3 \times 5)$
$$= \mathbf{\log 2 + \log 3 + \log 5}$$
by the first law of logarithms

(b) $\log 450 = \log(2 \times 225) = \log(2 \times 3 \times 75)$
$$= \log(2 \times 3 \times 3 \times 25)$$
$$= \log(2 \times 3^2 \times 5^2)$$
$$= \log 2 + \log 3^2 + \log 5^2$$
by the first law of logarithms

i.e $\log 450 = \mathbf{\log 2 + 2\log 3 + 2\log 5}$
by the third law of logarithms

Problem 5. Write $\log\left(\dfrac{8 \times \sqrt[4]{5}}{81}\right)$ in terms of $\log 2$, $\log 3$ and $\log 5$ to any base

$$\log\left(\frac{8 \times \sqrt[4]{5}}{81}\right) = \log 8 + \log \sqrt[4]{5} - \log 81,$$
by the first and second laws of logarithms
$$= \log 2^3 + \log 5^{(1/4)} - \log 3^4$$
by the law of indices

i.e $\log\left(\dfrac{8 \times \sqrt[4]{5}}{81}\right) = \mathbf{3\log 2 + \dfrac{1}{4}\log 5 - 4\log 3}$
by the third law of logarithms.

Problem 6. Simplify $\log 64 - \log 128 + \log 32$

$64 = 2^6$, $128 = 2^7$ and $32 = 2^5$

Hence $\log 64 - \log 128 + \log 32$
$$= \log 2^6 - \log 2^7 + \log 2^5$$
$$= 6\log 2 - 7\log 2 + 5\log 2$$
by the third law of logarithms
$$= \mathbf{4\log 2}$$

Problem 7. Evaluate
$$\frac{\log 25 - \log 125 + \dfrac{1}{2}\log 625}{3\log 5}$$

$$\frac{\log 25 - \log 125 + \frac{1}{2} \log 625}{3 \log 5}$$

$$= \frac{\log 5^2 - \log 5^3 + \frac{1}{2} \log 5^4}{3 \log 5}$$

$$= \frac{2 \log 5 - 3 \log 5 + \frac{4}{2} \log 5}{3 \log 5}$$

$$= \frac{1 \log 5}{3 \log 5} = \frac{1}{3}$$

Problem 8. Solve the equation:

$$\log (x - 1) + \log (x + 1) = 2 \log (x + 2)$$

$$\log (x - 1) + \log (x + 1) = \log (x - 1)(x + 1)$$

from the first

law of logarithms

$$= \log (x^2 - 1)$$

$$2 \log (x + 2) = \log (x + 2)^2$$

$$= \log (x^2 + 4x + 4)$$

Hence if $\qquad \log (x^2 - 1) = \log (x^2 + 4x + 4)$

then $\qquad x^2 - 1 = x^2 + 4x + 4$

i.e. $\qquad -1 = 4x + 4$

i.e. $\qquad -5 = 4x$

i.e. $\qquad x = -\frac{5}{4} \quad \text{or} \quad -1\frac{1}{4}$

Now try the following exercise

Exercise 50   Further problems on the laws of logarithms

In Problems 1 to 11, evaluate the given expression:

1. $\log_{10} 10000$ [4]   2. $\log_2 16$ [4]

3. $\log_5 125$ [3]   4. $\log_2 \frac{1}{8}$ [−3]

5. $\log_8 2$ $\left[\frac{1}{3}\right]$ 6. $\log_7 343$ [3]

7. $\lg 100$ [2]   8. $\lg 0.01$ [−2]

9. $\log_4 8$ $\left[1\frac{1}{2}\right]$ 10. $\log_{27} 3$ $\left[\frac{1}{3}\right]$

11. $\ln e^2$ [2]

In Problems 12 to 18 solve the equations:

12. $\log_{10} x = 4$ [10 000]

13. $\lg x = 5$ [100 000]

14. $\log_3 x = 2$ [9]

15. $\log_4 x = -2\frac{1}{2}$ $\left[\pm \frac{1}{32}\right]$

16. $\lg x = -2$ [0.01]

17. $\log_8 x = -\frac{4}{3}$ $\left[\frac{1}{16}\right]$

18. $\ln x = 3$ $[e^3]$

In Problems 19 to 22 write the given expressions in terms of log 2, log 3 and log 5 to any base:

19. $\log 60$ $\qquad$ [2 log 2 + log 3 + log 5]

20. $\log 300$ $\qquad$ [2 log 2 + log 3 + 2 log 5]

21. $\log \left(\frac{16 \times \sqrt[4]{5}}{27} \right)$

$$\left[4 \log 2 + \frac{1}{4} \log 5 - 3 \log 3 \right]$$

22. $\log \left(\frac{125 \times \sqrt[4]{16}}{\sqrt[4]{81}^3} \right)$

$$[\log 2 - 3 \log 3 + 3 \log 5]$$

Simplify the expressions given in Problems 23 to 25:

23. $\log 27 - \log 9 + \log 81$ $\qquad$ [5 log 3]

24. $\log 64 + \log 32 - \log 128$ $\qquad$ [4 log 2]

25. $\log 8 - \log 4 + \log 32$ $\qquad$ [6 log 2]

Evaluate the expression given in Problems 26 and 27:

26. $\dfrac{\frac{1}{2} \log 16 - \frac{1}{3} \log 8}{\log 4}$ $\qquad$ $\left[\frac{1}{2}\right]$

27. $\dfrac{\log 9 - \log 3 + \frac{1}{2} \log 81}{2 \log 3}$ $\qquad$ $\left[\frac{3}{2}\right]$

Solve the equations given in Problems 28 to 30:

28. $\log x^4 - \log x^3 = \log 5x - \log 2x$
$$[x = 2.5]$$

29. $\log 2t^3 - \log t = \log 16 + \log t$ $\quad [t = 8]$

30. $2 \log b^2 - 3 \log b = \log 8b - \log 4b$
$$[b = 2]$$

13.3 Indicial equations

The laws of logarithms may be used to solve certain equations involving powers — called **indicial equations**. For example, to solve, say, $3^x = 27$, logarithms to base of 10 are taken of both sides,

i.e. $\qquad \log_{10} 3^x = \log_{10} 27$

and $\qquad x \log_{10} 3 = \log_{10} 27$ by the third law of logarithms.

Rearranging gives $\qquad x = \dfrac{\log_{10} 27}{\log_{10} 3}$

$$= \dfrac{1.43136\ldots}{0.4771\ldots} = 3$$

which may be readily checked.

(Note, $(\log 8/\log 2)$ is **not** equal to $\lg(8/2)$)

Problem 9. Solve the equation $2^x = 3$, correct to 4 significant figures

Taking logarithms to base 10 of both sides of $2^x = 3$ gives:

$$\log_{10} 2^x = \log_{10} 3$$

i.e. $\qquad x \log_{10} 2 = \log_{10} 3$

Rearranging gives:

$$x = \frac{\log_{10} 3}{\log_{10} 2} = \frac{0.47712125\ldots}{0.30102999\ldots} = \textbf{1.585}$$

correct to 4 significant figures.

Problem 10. Solve the equation $2^{x+1} = 3^{2x-5}$ correct to 2 decimal places

Taking logarithms to base 10 of both sides gives:

$$\log_{10} 2^{x+1} = \log_{10} 3^{2x-5}$$

i.e. $\qquad (x+1)\log_{10} 2 = (2x-5)\log_{10} 3$

$$x \log_{10} 2 + \log_{10} 2 = 2x \log_{10} 3 - 5 \log_{10} 3$$

$$x(0.3010) + (0.3010) = 2x(0.4771) - 5(0.4771)$$

i.e. $\qquad 0.3010x + 0.3010 = 0.9542x - 2.3855$

Hence $\qquad 2.3855 + 0.3010 = 0.9542x - 0.3010x$

$$2.6865 = 0.6532x$$

from which $\qquad x = \dfrac{2.6865}{0.6532} = \textbf{4.11}$

correct to 2 decimal places.

Problem 11. Solve the equation $x^{3.2} = 41.15$, correct to 4 significant figures

Taking logarithms to base 10 of both sides gives:

$$\log_{10} x^{3.2} = \log_{10} 41.15$$

$$3.2 \log_{10} x = \log_{10} 41.15$$

Hence $\qquad \log_{10} x = \dfrac{\log_{10} 41.15}{3.2} = 0.50449$

Thus $x =$ antilog $0.50449 = 10^{0.50449} = \textbf{3.195}$ correct to 4 significant figures.

Now try the following exercise

Exercise 51 Indicial equations

Solve the following indicial equations for x, each correct to 4 significant figures:

1. $3^x = 6.4$ $\qquad\qquad\qquad\qquad$ [1.690]

2. $2^x = 9$ $\qquad\qquad\qquad\qquad\quad$ [3.170]

3. $2^{x-1} = 3^{2x-1}$ $\qquad\qquad\qquad$ [0.2696]

4. $x^{1.5} = 14.91$ $\qquad\qquad\qquad\quad$ [6.058]

5. $25.28 = 4.2^x$ $\qquad\qquad\qquad\quad$ [2.251]

6. $4^{2x-1} = 5^{x+2}$ $\qquad\qquad\qquad$ [3.959]

7. $x^{-0.25} = 0.792$ $\qquad\qquad\qquad$ [2.542]

8. $0.027^x = 3.26$ [−0.3272]

9. The decibel gain n of an amplifier is given by: $n = 10 \log_{10} \left(\dfrac{P_2}{P_1} \right)$ where P_1 is the power input and P_2 is the power output. Find the power gain $\dfrac{P_2}{P_1}$ when $n = 25$ decibels.

                                   [316.2]

13.4   Graphs of logarithmic functions

A graph of $y = \log_{10} x$ is shown in Fig. 13.1 and a graph of $y = \log_e x$ is shown in Fig. 13.2. Both are seen to be of similar shape; in fact, the same general shape occurs for a logarithm to any base.

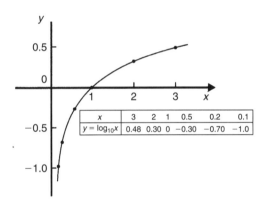

x	3	2	1	0.5	0.2	0.1
$y = \log_{10} x$	0.48	0.30	0	−0.30	−0.70	−1.0

Figure 13.1

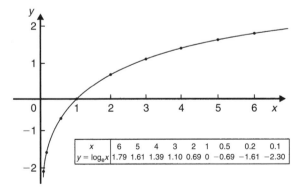

x	6	5	4	3	2	1	0.5	0.2	0.1
$y = \log_e x$	1.79	1.61	1.39	1.10	0.69	0	−0.69	−1.61	−2.30

Figure 13.2

In general, with a logarithm to any base a, it is noted that:

(i)   **$\log_a 1 = 0$**

Let $\log_a = x$, then $a^x = 1$ from the definition of the logarithm.

If $a^x = 1$ then $x = 0$ from the laws of logarithms.

Hence $\log_a 1 = 0$. In the above graphs it is seen that $\log_{10} 1 = 0$ and $\log_e 1 = 0$

(ii)   **$\log_a a = 1$**

Let $\log_a a = x$, then $a^x = a$, from the definition of a logarithm.

If $a^x = a$ then $x = 1$

Hence $\log_a a = 1$. (Check with a calculator that $\log_{10} 10 = 1$ and $\log_e e = 1$)

(iii)   **$\log_a 0 \to -\infty$**

Let $\log_a 0 = x$ then $a^x = 0$ from the definition of a logarithm.

If $a^x = 0$, and a is a positive real number, then x must approach minus infinity. (For example, check with a calculator, $2^{-2} = 0.25$, $2^{-20} = 9.54 \times 10^{-7}$, $2^{-200} = 6.22 \times 10^{-61}$, and so on.)

Hence $\log_a 0 \to -\infty$

Revision Test 3

This Revision test covers the material contained in Chapters 9 and 13. *The marks for each question are shown in brackets at the end of each question.*

1. Solve the following pairs of simultaneous equations:

   (a) $7x - 3y = 23$

   $2x + 4y = -8$

   (b) $3a - 8 + \dfrac{b}{8} = 0$

   $b + \dfrac{a}{2} = \dfrac{21}{4}$ (12)

2. In an engineering process two variables x and y are related by the equation $y = ax + \dfrac{b}{x}$ where a and b are constants. Evaluate a and b if $y = 15$ when $x = 1$ and $y = 13$ when $x = 3$ (4)

3. Transpose the following equations:

   (a) $y = mx + c$ for m

   (b) $x = \dfrac{2(y - z)}{t}$ for z

   (c) $\dfrac{1}{R_T} = \dfrac{1}{R_A} + \dfrac{1}{R_B}$ for R_A

   (d) $x^2 - y^2 = 3ab$ for y

   (e) $K = \dfrac{p - q}{1 + pq}$ for q (17)

4. The passage of sound waves through walls is governed by the equation:

   $$v = \sqrt{\dfrac{K + \dfrac{4}{3}G}{\rho}}$$

   Make the shear modulus G the subject of the formula. (4)

5. Solve the following equations by factorisation:

   (a) $x^2 - 9 = 0$ (b) $2x^2 - 5x - 3 = 0$ (6)

6. Determine the quadratic equation in x whose roots are 1 and -3 (4)

7. Solve the equation $4x^2 - 9x + 3 = 0$ correct to 3 decimal places. (5)

8. The current i flowing through an electronic device is given by:

   $$i = 0.005\,v^2 + 0.014\,v$$

   where v is the voltage. Calculate the values of v when $i = 3 \times 10^{-3}$ (5)

9. Solve the following inequalities:

   (a) $2 - 5x \leq 9 + 2x$ (b) $|3 + 2t| \leq 6$

   (c) $\dfrac{x - 1}{3x + 5} > 0$ (d) $(3t + 2)^2 > 16$

   (e) $2x^2 - x - 3 < 0$ (14)

10. Evaluate $\log_{16} 8$ (3)

11. Solve

    (a) $\log_3 x = -2$

    (b) $\log 2x^2 + \log x = \log 32 - \log x$ (6)

12. Solve the following equations, each correct to 3 significant figures:

    (a) $2^x = 5.5$

    (b) $3^{2t-1} = 7^{t+2}$

    (c) $3e^{2x} = 4.2$ (10)

Chapter 14

Exponential functions

14.1 The exponential function

An exponential function is one which contains e^x, e being a constant called the exponent and having an approximate value of 2.7183. The exponent arises from the natural laws of growth and decay and is used as a base for natural or Napierian logarithms.

14.2 Evaluating exponential functions

The value of e^x may be determined by using:

(a) a calculator, or
(b) the power series for e^x (see Section 14.3), or
(c) tables of exponential functions.

The most common method of evaluating an exponential function is by using a scientific notation **calculator**, this now having replaced the use of tables. Most scientific notation calculators contain an e^x function which enables all practical values of e^x and e^{-x} to be determined, correct to 8 or 9 significant figures. For example,

$$e^1 = 2.7182818$$

$$e^{2.4} = 11.023176$$

$$e^{-1.618} = 0.19829489$$

each correct to 8 significant figures.

In practical situations the degree of accuracy given by a calculator is often far greater than is appropriate. The accepted convention is that the final result is stated to one significant figure greater than the least significant measured value. Use your calculator to check the following values:

$$e^{0.12} = 1.1275, \text{ correct to 5 significant figures}$$

$$e^{-1.47} = 0.22993, \text{ correct to 5 decimal places}$$

$$e^{-0.431} = 0.6499, \text{ correct to 4 decimal places}$$

$$e^{9.32} = 11159, \text{ correct to 5 significant figures}$$

$$e^{-2.785} = 0.0617291, \text{ correct to 7 decimal places}$$

Problem 1. Using a calculator, evaluate, correct to 5 significant figures:

(a) $e^{2.731}$ (b) $e^{-3.162}$ (c) $\dfrac{5}{3}e^{5.253}$

(a) $e^{2.731} = 15.348227\ldots = \mathbf{15.348}$, correct to 5 significant figures.

(b) $e^{-3.162} = 0.04234097\ldots = \mathbf{0.042341}$, correct to 5 significant figures.

(c) $\dfrac{5}{3}e^{5.253} = \dfrac{5}{3}(191.138825\ldots) = \mathbf{318.56}$, correct to 5 significant figures.

Problem 2. Use a calculator to determine the following, each correct to 4 significant figures:

(a) $3.72e^{0.18}$ (b) $53.2e^{-1.4}$ (c) $\dfrac{5}{122}e^7$

(a) $3.72e^{0.18} = (3.72)(1.197217\ldots) = \mathbf{4.454}$, correct to 4 significant figures.

(b) $53.2e^{-1.4} = (53.2)(0.246596\ldots) = \mathbf{13.12}$, correct to 4 significant figures.

(c) $\dfrac{5}{122}e^7 = \dfrac{5}{122}(1096.6331\ldots) = \mathbf{44.94}$, correct to 4 significant figures.

Problem 3. Evaluate the following correct to 4 decimal places, using a calculator:

(a) $0.0256(e^{5.21} - e^{2.49})$

(b) $5\left(\dfrac{e^{0.25} - e^{-0.25}}{e^{0.25} + e^{-0.25}}\right)$

(a) $0.0256(e^{5.21} - e^{2.49})$

$\quad = 0.0256(183.094058\ldots - 12.0612761\ldots)$

$\quad = \mathbf{4.3784},$ correct to 4 decimal places

(b) $5\left(\dfrac{e^{0.25} - e^{-0.25}}{e^{0.25} + e^{-0.25}}\right)$

$\quad = 5\left(\dfrac{1.28402541\ldots - 0.77880078\ldots}{1.28402541\ldots + 0.77880078\ldots}\right)$

$\quad = 5\left(\dfrac{0.5052246\ldots}{2.0628261\ldots}\right)$

$\quad = \mathbf{1.2246},$ correct to 4 decimal places

Problem 4. The instantaneous voltage v in a capacitive circuit is related to time t by the equation $v = Ve^{-t/CR}$ where V, C and R are constants. Determine v, correct to 4 significant figures, when $t = 30 \times 10^{-3}$ seconds, $C = 10 \times 10^{-6}$ farads, $R = 47 \times 10^3$ ohms and $V = 200$ volts

$v = Ve^{-t/CR} = 200e^{(-30 \times 10^{-3})/(10 \times 10^{-6} \times 47 \times 10^3)}$

Using a calculator, $v = 200e^{-0.0638297\ldots}$

$\quad\quad\quad\quad\quad\quad = 200(0.9381646\ldots)$

$\quad\quad\quad\quad\quad\quad = \mathbf{187.6\ volts}$

Now try the following exercise

Exercise 52   Further problems on evaluating exponential functions

In Problems 1 and 2 use a calculator to evaluate the given functions correct to 4 significant figures:

1.   (a) $e^{4.4}$ (b) $e^{-0.25}$ (c) $e^{0.92}$

$\quad\quad\quad$ [(a) 81.45   (b) 0.7788   (c) 2.509]

2.   (a) $e^{-1.8}$ (b) $e^{-0.78}$ (c) e^{10}

$\quad\quad\quad$ [(a) 0.1653   (b) 0.4584   (c) 22030]

3.   Evaluate, correct to 5 significant figures:

$\quad$ (a) $3.5e^{2.8}$ (b) $-\dfrac{6}{5}e^{-1.5}$ (c) $2.16e^{5.7}$

$\quad\quad\quad$ [(a) 57.556   (b) -0.26776 (c) 645.55]

4.   Use a calculator to evaluate the following, correct to 5 significant figures:

$\quad$ (a) $e^{1.629}$ (b) $e^{-2.7483}$ (c) $0.62e^{4.178}$

$\quad\quad\quad$ [(a) 5.0988   (b) 0.064037   (c) 40.446]

In Problems 5 and 6, evaluate correct to 5 decimal places:

5.   (a) $\dfrac{1}{7}e^{3.4629}$ (b) $8.52e^{-1.2651}$ (c) $\dfrac{5e^{2.6921}}{3e^{1.1171}}$

$\quad\quad\quad$ $\begin{bmatrix}\text{(a) } 4.55848 \quad \text{(b) } 2.40444 \\ \text{(c) } 8.05124\end{bmatrix}$

6.   (a) $\dfrac{5.6823}{e^{-2.1347}}$ (b) $\dfrac{e^{2.1127} - e^{-2.1127}}{2}$

$\quad$ (c) $\dfrac{4(e^{-1.7295} - 1)}{e^{3.6817}}$

$\quad\quad\quad$ $\begin{bmatrix}\text{(a) } 48.04106 \quad \text{(b) } 4.07482 \\ \text{(c) } -0.08286\end{bmatrix}$

7.   The length of bar, l, at a temperature θ is given by $l = l_0 e^{\alpha\theta}$, where l_0 and α are constants. Evaluate l, correct to 4 significant figures, when $l_0 = 2.587$, $\theta = 321.7$ and $\alpha = 1.771 \times 10^{-4}$.   [2.739]

14.3   The power series for e^x

The value of e^x can be calculated to any required degree of accuracy since it is defined in terms of the following **power series**:

$$e^x = 1 + x + \frac{x^2}{2!} + \frac{x^3}{3!} + \frac{x^4}{4!} + \cdots \quad (1)$$

(where $3! = 3 \times 2 \times 1$ and is called 'factorial 3')
The series is valid for all values of x.
The series is said to **converge**, i.e. if all the terms are added, an actual value for e^x (where x is a real number) is obtained. The more terms that are taken, the closer will be the value of e^x to its actual value. The value of

the exponent e, correct to say 4 decimal places, may be determined by substituting $x = 1$ in the power series of equation (1). Thus

$$e^1 = 1 + 1 + \frac{(1)^2}{2!} + \frac{(1)^3}{3!} + \frac{(1)^4}{4!} + \frac{(1)^5}{5!}$$

$$+ \frac{(1)^6}{6!} + \frac{(1)^7}{7!} + \frac{(1)^8}{8!} + \cdots$$

$$= 1 + 1 + 0.5 + 0.16667 + 0.04167$$

$$+ 0.00833 + 0.00139 + 0.00020$$

$$+ 0.00002 + \cdots$$

$$= 2.71828$$

i.e. $e = 2.71828$ correct to 4 decimal places.

The value of $e^{0.05}$, correct to say 8 significant figures, is found by substituting $x = 0.05$ in the power series for e^x. Thus

$$e^{0.05} = 1 + 0.05 + \frac{(0.05)^2}{2!} + \frac{(0.05)^3}{3!}$$

$$+ \frac{(0.05)^4}{4!} + \frac{(0.05)^5}{5!} + \cdots$$

$$= 1 + 0.05 + 0.00125 + 0.000020833$$

$$+ 0.000000260 + 0.000000003$$

and by adding,

$$e^{0.05} = 1.0512711,$$

correct to 8 significant figures

In this example, successive terms in the series grow smaller very rapidly and it is relatively easy to determine the value of $e^{0.05}$ to a high degree of accuracy. However, when x is nearer to unity or larger than unity, a very large number of terms are required for an accurate result.

If in the series of equation (1), x is replaced by $-x$, then

$$e^{-x} = 1 + (-x) + \frac{(-x)^2}{2!} + \frac{(-x)^3}{3!} + \cdots$$

$$e^{-x} = 1 - x + \frac{x^2}{2!} - \frac{x^3}{3!} + \cdots$$

In a similar manner the power series for e^x may be used to evaluate any exponential function of the form ae^{kx}, where a and k are constants. In the series of equation (1),

let x be replaced by kx. Then

$$ae^{kx} = a\left\{1 + (kx) + \frac{(kx)^2}{2!} + \frac{(kx)^3}{3!} + \cdots\right\}$$

Thus $5e^{2x} = 5\left\{1 + (2x) + \frac{(2x)^2}{2!} + \frac{(2x)^3}{3!} + \cdots\right\}$

$$= 5\left\{1 + 2x + \frac{4x^2}{2} + \frac{8x^3}{6} + \cdots\right\}$$

i.e. $5e^{2x} = 5\left\{1 + 2x + 2x^2 + \frac{4}{3}x^3 + \cdots\right\}$

Problem 5. Determine the value of $5e^{0.5}$, correct to 5 significant figures by using the power series for e^x

$$e^x = 1 + x + \frac{x^2}{2!} + \frac{x^3}{3!} + \frac{x^4}{4!} + \cdots$$

Hence $e^{0.5} = 1 + 0.5 + \frac{(0.5)^2}{(2)(1)} + \frac{(0.5)^3}{(3)(2)(1)}$

$$+ \frac{(0.5)^4}{(4)(3)(2)(1)} + \frac{(0.5)^5}{(5)(4)(3)(2)(1)}$$

$$+ \frac{(0.5)^6}{(6)(5)(4)(3)(2)(1)}$$

$$= 1 + 0.5 + 0.125 + 0.020833$$

$$+ 0.0026042 + 0.0002604$$

$$+ 0.0000217$$

i.e. $e^{0.5} = 1.64872$ correct to 6 significant figures

Hence $5e^{0.5} = 5(1.64872) = \mathbf{8.2436}$, correct to 5 significant figures.

Problem 6. Determine the value of $3e^{-1}$, correct to 4 decimal places, using the power series for e^x

Substituting $x = -1$ in the power series

$$e^x = 1 + x + \frac{x^2}{2!} + \frac{x^3}{3!} + \frac{x^4}{4!} + \cdots$$

gives $e^{-1} = 1 + (-1) + \frac{(-1)^2}{2!} + \frac{(-1)^3}{3!}$

$$+ \frac{(-1)^4}{4!} + \cdots$$

$$= 1 - 1 + 0.5 - 0.166667 + 0.041667$$
$$- 0.008333 + 0.001389$$
$$- 0.000198 + \cdots$$
$$= 0.367858 \text{ correct to 6 decimal places}$$

Hence $3e^{-1} = (3)(0.367858) = \mathbf{1.1036}$ correct to 4 decimal places.

Problem 7. Expand $e^x(x^2 - 1)$ as far as the term in x^5

The power series for e^x is:

$$e^x = 1 + x + \frac{x^2}{2!} + \frac{x^3}{3!} + \frac{x^4}{4!} + \frac{x^5}{5!} + \cdots$$

Hence:

$$e^x(x^2 - 1)$$

$$= \left(1 + x + \frac{x^2}{2!} + \frac{x^3}{3!} + \frac{x^4}{4!} + \frac{x^5}{5!} + \cdots\right)(x^2 - 1)$$

$$= \left(x^2 + x^3 + \frac{x^4}{2!} + \frac{x^5}{3!} + \cdots\right)$$

$$- \left(1 + x + \frac{x^2}{2!} + \frac{x^3}{3!} + \frac{x^4}{4!} + \frac{x^5}{5!} + \cdots\right)$$

Grouping like terms gives:

$$e^x(x^2 - 1)$$

$$= -1 - x + \left(x^2 - \frac{x^2}{2!}\right) + \left(x^3 - \frac{x^3}{3!}\right)$$

$$+ \left(\frac{x^4}{2!} - \frac{x^4}{4!}\right) + \left(\frac{x^5}{3!} - \frac{x^5}{5!}\right) + \cdots$$

$$= \mathbf{-1 - x + \frac{1}{2}x^2 + \frac{5}{6}x^3 + \frac{11}{24}x^4 + \frac{19}{120}x^5}$$

when expanded as far as the term in x^5

Now try the following exercise

Exercise 53 Further problems on the power series for e^x

1. Evaluate $5.6e^{-1}$, correct to 4 decimal places, using the power series for e^x. [2.0601]

2. Use the power series for e^x to determine, correct to 4 significant figures, (a) e^2 (b) $e^{-0.3}$ and check your result by using a calculator. [(a) 7.389 (b) 0.7408]

3. Expand $(1 - 2x)e^{2x}$ as far as the term in x^4.
$$\left[1 - 2x^2 - \frac{8}{3}x^3 - 2x^4\right]$$

4. Expand $(2e^{x^2})(x^{1/2})$ to six terms.
$$\left[2x^{1/2} + 2x^{5/2} + x^{9/2} + \frac{1}{3}x^{13/2} + \frac{1}{12}x^{17/2} + \frac{1}{60}x^{21/2}\right]$$

14.4 Graphs of exponential functions

Values of e^x and e^{-x} obtained from a calculator, correct to 2 decimal places, over a range $x = -3$ to $x = 3$, are shown in the following table.

x	−3.0	−2.5	−2.0	−1.5	−1.0	−0.5	0
e^x	0.05	0.08	0.14	0.22	0.37	0.61	1.00
e^{-x}	20.09	12.18	7.9	4.48	2.72	1.65	1.00

x	0.5	1.0	1.5	2.0	2.5	3.0
e^x	1.65	2.72	4.48	7.39	12.18	20.09
e^{-x}	0.61	0.37	0.22	0.14	0.08	0.05

Figure 14.1 shows graphs of $y = e^x$ and $y = e^{-x}$

Problem 8. Plot a graph of $y = 2e^{0.3x}$ over a range of $x = -2$ to $x = 3$. Hence determine the value of y when $x = 2.2$ and the value of x when $y = 1.6$

A table of values is drawn up as shown below.

x	−3	−2	−1	0	1	2	3
$0.3x$	−0.9	−0.6	−0.3	0	0.3	0.6	0.9
$e^{0.3x}$	0.407	0.549	0.741	1.000	1.350	1.822	2.460
$2e^{0.3x}$	0.81	1.10	1.48	2.00	2.70	3.64	4.92

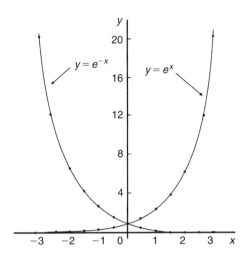

Figure 14.1

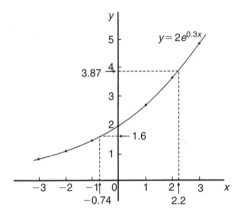

Figure 14.2

A graph of $y = 2e^{0.3x}$ is shown plotted in Fig. 14.2.

From the graph, **when $x = 2.2$, $y = 3.87$** and **when $y = 1.6$, $x = -0.74$**

Problem 9. Plot a graph of $y = \dfrac{1}{3}e^{-2x}$ over the range $x = -1.5$ to $x = 1.5$. Determine from the graph the value of y when $x = -1.2$ and the value of x when $y = 1.4$

A table of values is drawn up as shown below.

x	−1.5	−1.0	−0.5	0	0.5	1.0	1.5
$-2x$	3	2	1	0	−1	−2	−3
e^{-2x}	20.086	7.389	2.718	1.00	0.368	0.135	0.050
$\dfrac{1}{3}e^{-2x}$	6.70	2.46	0.91	0.33	0.12	0.05	0.02

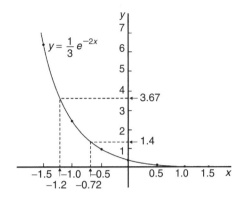

Figure 14.3

A graph of $\dfrac{1}{3}e^{-2x}$ is shown in Fig. 14.3.
From the graph, **when $x = -1.2$, $y = 3.67$** and **when $y = 1.4$, $x = -0.72$**

Problem 10. The decay of voltage, v volts, across a capacitor at time t seconds is given by $v = 250e^{-t/3}$. Draw a graph showing the natural decay curve over the first 6 seconds. From the graph, find (a) the voltage after 3.4 s, and (b) the time when the voltage is 150 V

A table of values is drawn up as shown below.

t	0	1	2	3
$e^{-t/3}$	1.00	0.7165	0.5134	0.3679
$v = 250e^{-t/3}$	250.0	179.1	128.4	91.97

t	4	5	6
$e^{-t/3}$	0.2636	0.1889	0.1353
$v = 250e^{-t/3}$	65.90	47.22	33.83

The natural decay curve of $v = 250e^{-t/3}$ is shown in Fig. 14.4.
From the graph:

(a) **when time $t = 3.4$ s, voltage $v = 80$ volts** and

(b) **when voltage $v = 150$ volts, time $t = 1.5$ seconds.**

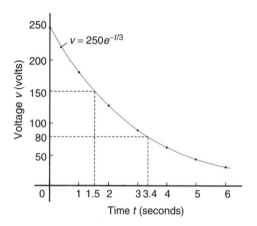

Figure 14.4

Now try the following exercise

Exercise 54 Further problems on exponential graphs

1. Plot a graph of $y = 3e^{0.2x}$ over the range $x = -3$ to $x = 3$. Hence determine the value of y when $x = 1.4$ and the value of x when $y = 4.5$
   [3.95, 2.05]

2. Plot a graph of $y = \frac{1}{2}e^{-1.5x}$ over a range $x = -1.5$ to $x = 1.5$ and hence determine the value of y when $x = -0.8$ and the value of x when $y = 3.5$
   [1.65, −1.30]

3. In a chemical reaction the amount of starting material C cm^3 left after t minutes is given by $C = 40e^{-0.006t}$. Plot a graph of C against t and determine (a) the concentration C after 1 hour, and (b) the time taken for the concentration to decrease by half.
   [(a) 28 cm^3 (b) 116 min]

4. The rate at which a body cools is given by $\theta = 250e^{-0.05t}$ where the excess of temperature of a body above its surroundings at time t minutes is $\theta°$ C. Plot a graph showing the natural decay curve for the first hour of cooling. Hence determine (a) the temperature after 25 minutes, and (b) the time when the temperature is 195°C
   [(a) 70°C (b) 5 minutes]

14.5 Napierian logarithms

Logarithms having a base of e are called **hyperbolic, Napierian** or **natural logarithms** and the Napierian logarithm of x is written as $\log_e x$, or more commonly, $\ln x$.

14.6 Evaluating Napierian logarithms

The value of a Napierian logarithm may be determined by using:

(a) a calculator, or

(b) a relationship between common and Napierian logarithms, or

(c) Napierian logarithm tables.

The most common method of evaluating a Napierian logarithm is by a scientific notation **calculator**, this now having replaced the use of four-figure tables, and also the relationship between common and Napierian logarithms,

$$\log_e y = 2.3026 \log_{10} y$$

Most scientific notation calculators contain a '$\ln x$' function which displays the value of the Napierian logarithm of a number when the appropriate key is pressed.
   Using a calculator,

$$\ln 4.692 = 1.5458589\ldots = 1.5459,$$

correct to 4 decimal places

and $\ln 35.78 = 3.57738907\ldots = 3.5774,$

correct to 4 decimal places

Use your calculator to check the following values:

   $\ln 1.732 = 054928$, correct to 5 significant figures

   $\ln 1 = 0$

   $\ln 593 = 6.3852$, correct to 5 significant figures

   $\ln 1750 = 7.4674$, correct to 4 decimal places

   $\ln 0.17 = -1.772$, correct to 4 significant figures

   $\ln 0.00032 = -8.04719$, correct to 6 significant figures

   $\ln e^3 = 3$

   $\ln e^1 = 1$

From the last two examples we can conclude that

$$\log_e e^x = x$$

This is useful when solving equations involving exponential functions.

For example, to solve $e^{3x} = 8$, take Napierian logarithms of both sides, which gives

$$\ln e^{3x} = \ln 8$$

i.e. $\qquad\qquad 3x = \ln 8$

from which $\qquad x = \dfrac{1}{3} \ln 8 = \mathbf{0.6931}$,

correct to 4 decimal places

Problem 11. Using a calculator evaluate correct to 5 significant figures:

(a) ln 47.291 (b) ln 0.06213
(c) 3.2 ln 762.923

(a) ln 47.291 = 3.8563200... = **3.8563**, correct to 5 significant figures.

(b) ln 0.06213 = −2.7785263... = **−2.7785**, correct to 5 significant figures.

(c) 3.2 ln 762.923 = 3.2(6.6371571...) = **21.239**, correct to 5 significant figures.

Problem 12. Use a calculator to evaluate the following, each correct to 5 significant figures:

(a) $\dfrac{1}{4} \ln 4.7291$ (b) $\dfrac{\ln 7.8693}{7.8693}$

(c) $\dfrac{5.29 \ln 24.07}{e^{-0.1762}}$

(a) $\dfrac{1}{4} \ln 4.7291 = \dfrac{1}{4}(1.5537349\ldots) = \mathbf{0.38843}$,

correct to 5 significant figures.

(b) $\dfrac{\ln 7.8693}{7.8693} = \dfrac{2.06296911\ldots}{7.8693} = \mathbf{0.26215}$,

correct to 5 significant figures.

(c) $\dfrac{5.29 \ln 24.07}{e^{-0.1762}} = \dfrac{5.29(3.18096625\ldots)}{0.83845027\ldots}$

$= \mathbf{20.070}$, correct to 5 significant figures.

Problem 13. Evaluate the following:

(a) $\dfrac{\ln e^{2.5}}{\lg 10^{0.5}}$ (b) $\dfrac{4e^{2.23} \lg 2.23}{\ln 2.23}$ (correct to 3 decimal places)

(a) $\dfrac{\ln e^{2.5}}{\lg 10^{0.5}} = \dfrac{2.5}{0.5} = \mathbf{5}$

(b) $\dfrac{4e^{2.23} \lg 2.23}{\ln 2.23}$

$= \dfrac{4(9.29986607\ldots)(0.34830486\ldots)}{0.80200158\ldots}$

$= \mathbf{16.156}$, correct to 3 decimal places

Problem 14. Solve the equation $7 = 4e^{-3x}$ to find x, correct to 4 significant figures

Rearranging $7 = 4e^{-3x}$ gives: $\dfrac{7}{4} = e^{-3x}$

Taking the reciprocal of both sides gives:

$$\frac{4}{7} = \frac{1}{e^{-3x}} = e^{3x}$$

Taking Napierian logarithms of both sides gives:

$$\ln \left(\frac{4}{7}\right) = \ln (e^{3x})$$

Since $\log_e e^\alpha = \alpha$, then $\ln \left(\dfrac{4}{7}\right) = 3x$

Hence $x = \dfrac{1}{3} \ln \left(\dfrac{4}{7}\right) = \dfrac{1}{3}(-0.55962) = \mathbf{-0.1865}$, correct to 4 significant figures.

Problem 15. Given $20 = 60(1 - e^{-t/2})$ determine the value of t, correct to 3 significant figures

Rearranging $20 = 60(1 - e^{-t/2})$ gives:

$$\frac{20}{60} = 1 - e^{-1/2}$$

and

$$e^{-t/2} = 1 - \frac{20}{60} = \frac{2}{3}$$

Taking the reciprocal of both sides gives:

$$e^{t/2} = \frac{3}{2}$$

Taking Napierian logarithms of both sides gives:

$$\ln e^{t/2} = \ln \frac{3}{2}$$

i.e. $$\frac{t}{2} = \ln \frac{3}{2}$$

from which, $t = 2 \ln \dfrac{3}{2} = \mathbf{0.881}$, correct to 3 significant figures.

Problem 16.   Solve the equation $3.72 = \ln\left(\dfrac{5.14}{x}\right)$ to find x

---

From the definition of a logarithm, since
$$3.72 = \left(\frac{5.14}{x}\right) \text{ then } e^{3.72} = \frac{5.14}{x}$$

Rearranging gives: $$x = \frac{5.14}{e^{3.72}} = 5.14e^{-3.72}$$

i.e. $$x = \mathbf{0.1246},$$

correct to 4 significant figures

Now try the following exercise

Exercise 55   Further problems on evaluating Napierian logarithms

In Problems 1 to 3 use a calculator to evaluate the given functions, correct to 4 decimal places

1.   (a) ln 1.73   (b) ln 5.413   (c) ln 9.412
     [(a) 0.5481   (b) 1.6888   (c) 2.2420]

2.   (a) ln 17.3   (b) ln 541.3   (c) ln 9412
     [(a) 2.8507   (b) 6.2940   (c) 9.1497]

3.   (a) ln 0.173   (b) ln 0.005413   (c) ln 0.09412
     [(a) −1.7545   (b) −5.2190   (c) −2.3632]

In Problems 4 and 5, evaluate correct to 5 significant figures:

4.   (a) $\dfrac{1}{6} \ln 5.2932$ (b) $\dfrac{\ln 82.473}{4.829}$

     (c) $\dfrac{5.62 \ln 321.62}{e^{1.2942}}$

     [(a) 0.27774 (b) 0.91374 (c) 8.8941]

5.   (a) $\dfrac{2.946 \ln e^{1.76}}{\lg 10^{1.41}}$ (b) $\dfrac{5e^{-0.1629}}{2 \ln 0.00165}$

     (c) $\dfrac{\ln 4.8629 - \ln 2.4711}{5.173}$

     [(a) 3.6773   (b) −0.33154   (c) 0.13087]

In Problems 6 to 10 solve the given equations, each correct to 4 significant figures.

6. $1.5 = 4e^{2t}$ [−0.4904]

7. $7.83 = 2.91e^{-1.7x}$ [−0.5822]

8. $16 = 24(1 - e^{-t/2})$ [2.197]

9. $5.17 = \ln\left(\dfrac{x}{4.64}\right)$ [816.2]

10. $3.72 \ln\left(\dfrac{1.59}{x}\right) = 2.43$ [0.8274]

11. The work done in an isothermal expansion of a gas from pressure p_1 to p_2 is given by:

$$w = w_0 \ln\left(\frac{p_1}{p_2}\right)$$

If the initial pressure $p_1 = 7.0\,\text{kPa}$, calculate the final pressure p_2 if $w = 3\,w_0$
     [$p_2 = 348.5\,\text{Pa}$]

14.7   Laws of growth and decay

The laws of exponential growth and decay are of the form $y = Ae^{-kx}$ and $y = A(1 - e^{-kx})$, where A and k are constants. When plotted, the form of each of these equations is as shown in Fig. 14.5. The laws occur frequently in engineering and science and examples of quantities related by a natural law include:

(i)   Linear expansion $l = l_0e^{\alpha\theta}$

(ii)   Change in electrical resistance with
        temperature $R_\theta = R_0e^{\alpha\theta}$

(iii)   Tension in belts $T_1 = T_0e^{\mu\theta}$

(iv)   Newton's law of cooling $\theta = \theta_0e^{-kt}$

(v)   Biological growth $y = y_0e^{kt}$

(vi)   Discharge of a capacitor $q = Qe^{-t/CR}$

(vii)   Atmospheric pressure $p = p_0e^{-h/c}$

(viii) Radioactive decay $N = N_0e^{-\lambda t}$

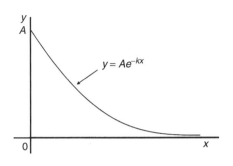

$y = Ae^{-kx}$

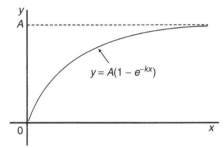

$y = A(1 - e^{-kx})$

Figure 14.5

(ix) Decay of current in an inductive
circuit $i = Ie^{-Rt/L}$

(x)   Growth of current in a
capacitive circuit $i = I(1 - e^{-t/CR})$

Problem 17. The resistance R of an electrical
conductor at temperature $\theta°C$ is given by $R = R_0 e^{\alpha\theta}$,
where α is a constant and $R_0 = 5 \times 10^3$ ohms.
Determine the value of α, correct to 4 significant
figures, when $R = 6 \times 10^3$ ohms and $\theta = 1500°C$.
Also, find the temperature, correct to the nearest
degree, when the resistance R is 5.4×10^3 ohms

Transposing $R = R_0 e^{\alpha\theta}$ gives $\dfrac{R}{R_0} = e^{\alpha\theta}$

Taking Napierian logarithms of both sides gives:

$$\ln \frac{R}{R_0} = \ln e^{\alpha\theta} = \alpha\theta$$

Hence

$$\alpha = \frac{1}{\theta} \ln \frac{R}{R_0} = \frac{1}{1500} \ln \left(\frac{6 \times 10^3}{5 \times 10^3} \right)$$

$$= \frac{1}{1500}(0.1823215\ldots)$$

$$= 1.215477\ldots \times 10^{-4}$$

Hence $\alpha = \mathbf{1.215 \times 10^{-4}}$, correct to 4 significant
figures.

From above, $\ln \dfrac{R}{R_0} = \alpha\theta$ hence $\theta = \dfrac{1}{\alpha} \ln \dfrac{R}{R_0}$

When $R = 5.4 \times 10^3$, $\alpha = 1.215477\ldots \times 10^{-4}$ and
$R_0 = 5 \times 10^3$

$$\theta = \frac{1}{1.215477\ldots \times 10^{-4}} \ln \left(\frac{5.4 \times 10^3}{5 \times 10^3} \right)$$

$$= \frac{104}{1.215477\ldots}(7.696104\ldots \times 10^{-2})$$

$$= \mathbf{633°C} \quad \text{correct to the nearest degree.}$$

Problem 18. In an experiment involving
Newton's law of cooling, the temperature $\theta(°C)$ is
given by $\theta = \theta_0 e^{-kt}$. Find the value of constant k
when $\theta_0 = 56.6°C$, $\theta = 16.5°C$ and $t = 83.0$ seconds

Transposing $\theta = \theta_0 e^{-kt}$ gives $\dfrac{\theta}{\theta_0} = e^{-kt}$ from which

$$\frac{\theta_0}{\theta} = \frac{1}{e^{-kt}} = e^{kt}$$

Taking Napierian logarithms of both sides gives:

$$\ln \frac{\theta_0}{\theta} = kt$$

from which,

$$k = \frac{1}{t} \ln \frac{\theta_0}{\theta} = \frac{1}{83.0} \ln \left(\frac{56.6}{16.5} \right)$$

$$= \frac{1}{83.0}(1.2326486\ldots)$$

Hence $k = \mathbf{1.485 \times 10^{-2}}$

Problem 19. The current i amperes flowing in a
capacitor at time t seconds is given by
$i = 8.0(1 - e^{-t/CR})$, where the circuit resistance R is
25×10^3 ohms and capacitance C is 16×10^{-6}
farads. Determine (a) the current i after 0.5 seconds
and (b) the time, to the nearest millisecond, for the
current to reach 6.0 A. Sketch the graph of current
against time

(a)   Current $i = 8.0(1 - e^{-t/CR})$

$$= 8.0[1 - e^{0.5/(16 \times 10^{-6})(25 \times 10^3)}]$$

$$= 8.0(1 - e^{-1.25})$$

$$= 8.0(1 - 0.2865047\ldots)$$

$$= 8.0(0.7134952\ldots)$$

$$= \mathbf{5.71\ amperes}$$

(b)   Transposing $i = 8.0(1 - e^{-t/CR})$ gives:

$$\frac{i}{8.0} = 1 - e^{-t/CR}$$

from which, $e^{-t/CR} = 1 - \dfrac{i}{8.0} = \dfrac{8.0 - i}{8.0}$

Taking the reciprocal of both sides gives:

$$e^{t/CR} = \frac{8.0}{8.0 - i}$$

Taking Napierian logarithms of both sides gives:

$$\frac{t}{CR} = \ln\left(\frac{8.0}{8.0 - i}\right)$$

Hence

$$t = CR\ln\left(\frac{8.0}{8.0 - i}\right)$$

$$= (16 \times 10^{-6})(25 \times 10^3)\ln\left(\frac{8.0}{8.0 - 6.0}\right)$$

when $i = 6.0$ amperes,

i.e. $t = \dfrac{400}{10^3}\ln\left(\dfrac{8.0}{2.0}\right) = 0.4\ln 4.0$

$$= 0.4(1.3862943\ldots) = 0.5545\,\text{s}$$

$$= \mathbf{555\,ms},$$

to the nearest millisecond

A graph of current against time is shown in Fig. 14.6.

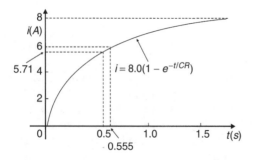

Figure 14.6

Problem 20.   The temperature θ_2 of a winding which is being heated electrically at time t is given by: $\theta_2 = \theta_1(1 - e^{-t/\tau})$ where θ_1 is the temperature (in degrees Celsius) at time $t = 0$ and τ is a constant. Calculate

(a) θ_1, correct to the nearest degree, when θ_2 is 50°C, t is 30 s and τ is 60 s

(b)   the time t, correct to 1 decimal place, for θ_2 to be half the value of θ_1

(a)   Transposing the formula to make θ_1 the subject gives:

$$\theta_1 = \frac{\theta_2}{(1 - e^{-t/\tau})} = \frac{50}{1 - e^{-30/60}}$$

$$= \frac{50}{1 - e^{-0.5}} = \frac{50}{0.393469\ldots}$$

i.e. $\boldsymbol{\theta_1 = 127°C}$, correct to the nearest degree

(b)   Transposing to make t the subject of the formula gives:

$$\frac{\theta_2}{\theta_1} = 1 - e^{-t/\tau}$$

from which, $e^{-t/\tau} = 1 - \dfrac{\theta_2}{\theta_1}$

Hence $-\dfrac{t}{\tau} = \ln\left(1 - \dfrac{\theta_2}{\theta_1}\right)$

i.e. $t = -\tau\ln\left(1 - \dfrac{\theta_2}{\theta_1}\right)$

Since $\theta_2 = \dfrac{1}{2}\theta_1$

$$t = -60\ln\left(1 - \frac{1}{2}\right) = -60\ln 0.5$$

$$= 41.59\,\text{s}$$

Hence the time for the temperature θ_2 to be one half of the value of θ_1 is 41.6 s, correct to 1 decimal place.

Now try the following exercise

Exercise 56   Further problems on the laws of growth and decay

1.   The pressure p pascals at height h metres above ground level is given by $p = p_0 e^{-h/C}$, where p_0 is the pressure at ground level and C is a constant. Find pressure p when

$p_0 = 1.012 \times 10^5$ Pa, height $h = 1420$ m and $C = 71500$. [9.921×10^4 Pa]

2. The voltage drop, v volts, across an inductor L henrys at time t seconds is given by $v = 200e^{-Rt/L}$, where $R = 150\,\Omega$ and $L = 12.5 \times 10^{-3}$ H. Determine (a) the voltage when $t = 160 \times 10^{-6}$ s, and (b) the time for the voltage to reach 85 V.

[(a) 29.32 volts (b) 71.31×10^{-6} s]

3. The length l metres of a metal bar at temperature $t°C$ is given by $l = l_0 e^{\alpha t}$, when l_0 and α are constants. Determine (a) the value of l when $l_0 = 1.894$, $\alpha = 2.038 \times 10^{-4}$ and $t = 250°C$, and (b) the value of l_0 when $l = 2.416$, $t = 310°C$ and $\alpha = 1.682 \times 10^{-4}$

[(a) 1.993 m (b) 2.293 m]

4. The temperature $\theta_2°C$ of an electrical conductor at time t seconds is given by $\theta_2 = \theta_1(1 - e^{-t/T})$, when θ_1 is the initial temperature and T seconds is a constant. Determine (a) θ_2 when $\theta_1 = 159.9°C$, $t = 30$ s and $T = 80$ s, and (b) the time t for θ_2 to fall to half the value of θ_1 if T remains at 80 s.

[(a) 50°C (b) 55.45 s]

5. A belt is in contact with a pulley for a sector of $\theta = 1.12$ radians and the coefficient of friction between these two surfaces is $\mu = 0.26$. Determine the tension on the taut side of the belt, T newtons, when tension on the slack side is given by $T_0 = 22.7$ newtons, given that these quantities are related by the law $T = T_0 e^{\mu\theta}$. Determine also the value of θ when $T = 28.0$ newtons

[30.4 N, 0.807 rad]

6. The instantaneous current i at time t is given by:

$$i = 10e^{-t/CR}$$

when a capacitor is being charged. The capacitance C is 7×10^{-6} farads and the resistance R is 0.3×10^6 ohms. Determine:

(a) the instantaneous current when t is 2.5 seconds, and

(b) the time for the instantaneous current to fall to 5 amperes.

Sketch a curve of current against time from $t = 0$ to $t = 6$ seconds.

[(a) 3.04 A (b) 1.46 s]

7. The amount of product x (in mol/cm^3) found in a chemical reaction starting with 2.5 mol/cm^3 of reactant is given by $x = 2.5(1 - e^{-4t})$ where t is the time, in minutes, to form product x. Plot a graph at 30 second intervals up to 2.5 minutes and determine x after 1 minute. [2.45 mol/cm^3]

8. The current i flowing in a capacitor at time t is given by:

$$i = 12.5(1 - e^{-t/CR})$$

where resistance R is 30 kilohms and the capacitance C is 20 microfarads. Determine

(a) the current flowing after 0.5 seconds, and

(b) the time for the current to reach 10 amperes.

[(a) 7.07 A (b) 0.966 s]

9. The amount A after n years of a sum invested P is given by the compound interest law: $A = Pe^{rn/100}$ when the per unit interest rate r is added continuously. Determine, correct to the nearest pound, the amount after 8 years for a sum of £1500 invested if the interest rate is 6% per annum. [£2424]

Chapter 15

Number sequences

15.1 Arithmetic progressions

When a sequence has a constant difference between successive terms it is called an **arithmetic progression** (often abbreviated to *AP*).
Examples include:

(i)   1, 4, 7, 10, 13, ... where the **common difference** is 3

and   (ii) $a, a+d, a+2d, a+3d, \ldots$ where the common difference is d.

If the 1st term of an *AP* is 'a' and the common difference is 'd' then

the n'th term is: $a + (n-1)d$

In example (i) above, the 7th term is given by $1 + (7-1)3 = \mathbf{19}$, which may be readily checked.

The sum S of an *AP* can be obtained by multiplying the average of all the terms by the number of terms.
The average of all the terms $= \dfrac{a+1}{2}$, where 'a' is the 1st term and l is the last term, i.e. $l = a + (n-1)d$, for n terms.
Hence the sum of n terms,

$$S_n = n\left(\frac{a+1}{2}\right) = \frac{n}{2}\{a + [a + (n-1)d]\}$$

i.e. $S_n = \dfrac{n}{2}[2a + (n-1)d]$

For example, the sum of the first 7 terms of the series 1, 4, 7, 10, 13, ... is given by

$$S_7 = \frac{7}{7}[2(1) + (7-1)3] \quad \text{since } a = 1 \quad \text{and} \quad d = 3$$

$$= \frac{7}{2}[2+18] = \frac{7}{2}[20] = \mathbf{70}$$

15.2 Worked problems on arithmetic progressions

Problem 1. Determine (a) the 9th, and (b) the 16th term of the series 2, 7, 12, 17,...

2, 7, 12, 17, ... is an arithmetic progression with a common difference, d, of 5

(a)   The n'th term of an *AP* is given by $a + (n-1)d$
Since the first term $a = 2$, $d = 5$ and $n = 9$ then the 9th term is:

$$2 + (9-1)5 = 2 + (8)(5) = 2 + 40 = \mathbf{42}$$

(b)   The 16th term is:

$$2 + (16-1)5 = 2 + (15)(5) = 2 + 75 = \mathbf{77}$$

Problem 2. The 6th term of an *AP* is 17 and the 13th term is 38. Determine the 19th term.

The n'th term of an *AP* is $a + (n-1)d$
The 6th term is: $a + 5d = 17$ (1)
The 13th term is: $a + 12d = 38$ (2)

Equation (2) − equation (1) gives: $7d = 21$, from which, $d = \dfrac{21}{7} = 3$

Substituting in equation (1) gives: $a + 15 = 17$, from which, $a = 2$

Hence the 19th term is:

$$a + (n-1)d = 2 + (19-1)3 = 2 + (18)(3)$$

$$= 2 + 54 = \mathbf{56}$$

Problem 3. Determine the number of the term whose value is 22 is the series

$$2\frac{1}{2}, 4, 5\frac{1}{2}, 7, \ldots$$

$2\frac{1}{2}, 4, 5\frac{1}{2}, 7, \ldots$ is an AP

where $a = 2\frac{1}{2}$ and $d = 1\frac{1}{2}$

Hence if the n'th term is 22 then: $a + (n-1)d = 22$

i.e. $$2\frac{1}{2} + (n-1)\left(1\frac{1}{2}\right) = 22$$

$$(n-1)\left(1\frac{1}{2}\right) = 22 - 2\frac{1}{2} = 19\frac{1}{2}$$

$$n - 1 = \frac{19\frac{1}{2}}{1\frac{1}{2}} = 13$$

and $$n = 13 + 1 = 14$$

i.e. **the 14th term of the AP is 22**

Problem 4. Find the sum of the first 12 terms of the series 5, 9, 13, 17, ...

5, 9, 13, 17, ... is an AP where $a = 5$ and $d = 4$

The sum of n terms of an AP,

$$S_n = \frac{n}{2}[2a + (n-1)d]$$

Hence the sum of the first 12 terms,

$$S_{12} = \frac{12}{2}[2(5) + (12-1)4]$$

$$= 6[10 + 44] = 6(54) = \mathbf{324}$$

Now try the following exercise

Exercise 57   Further problems on arithmetic progressions

1. Find the 11th term of the series 8, 14, 20, 26, ...   [68]

2. Find the 17th term of the series 11, 10.7, 10.4, 10.1, ...   [6.2]

3. The 7th term of a series is 29 and the 11th term is 54. Determine the sixteenth term.   [85.25]

4. Find the 15th term of an arithmetic progression of which the first term is 2.5 and the 10th term is 16.   [23.5]

5. Determine the number of the term which is 29 in the series 7, 9.2, 11.4, 13.6, ...   [11]

6. Find the sum of the first 11 terms of the series 4, 7, 10, 13, ...   [209]

7. Determine the sum of the series 6.5, 8.0, 9.5, 11.0, ..., 32   [346.5]

15.3 Further worked problems on arithmetic progressions

Problem 5. The sum of 7 terms of an AP is 35 and the common difference is 1.2. Determine the 1st term of the series

$n = 7$, $d = 1.2$ and $S_7 = 35$

Since the sum of n terms of an AP is given by

$$S_n = \frac{n}{2}[2a + (n-1)]d$$

then $$35 = \frac{7}{2}[2a + (7-1)1.2] = \frac{7}{2}[2a + 7.2]$$

Hence $$\frac{35 \times 2}{7} = 2a + 7.2$$

$$10 = 2a + 7.2$$

Thus $$2a = 10 - 7.2 = 2.8$$

from which $$a = \frac{2.8}{2} = 14$$

i.e. **the first term, $a = 1.4$**

Problem 6. Three numbers are in arithmetic progression. Their sum is 15 and their product is 80. Determine the three numbers

Let the three numbers be $(a - d)$, a and $(a + d)$

Then $(a - d) + a + (a + d) = 15$, i.e. $3a = 15$, from which, $a = 5$

Also, $a(a-d)(a+d) = 80$, i.e. $a(a^2 - d^2) = 80$

Since a = 5, $5(5^2 - d^2) = 80$
$$125 - 5d^2 = 80$$
$$125 - 80 = 5d^2$$
$$45 = 5d^2$$

from which, $d^2 = \dfrac{45}{5} = 9$. Hence, $d = \sqrt{9} = \pm 3$

The three numbers are thus $(5-3)$, 5 and $(5+3)$, i.e. **2, 5 and 8**

Problem 7. Find the sum of all the numbers between 0 and 207 which are exactly divisible by 3

The series 3, 6, 9, 12, ... 207 is an *AP* whose first term $a = 3$ and common difference $d = 3$

The last term is $a + (n-1)d = 207$

i.e. $\quad\quad\quad 3 + (n-1)3 = 207$

from which $\quad (n-1) = \dfrac{207 - 3}{3} = 68$

Hence $\quad\quad\quad\quad n = 68 + 1 = 69$

The sum of all 69 terms is given by

$$S_{69} = \frac{n}{2}[2a + (n-1)d]$$

$$= \frac{69}{2}[2(3) + (69-1)3]$$

$$= \frac{69}{2}[6 + 204] = \frac{69}{2}(210) = \mathbf{7245}$$

Problem 8. The 1st, 12th and last term of an arithmetic progression are 4, 31.5, and 376.5 respectively. Determine (a) the number of terms in the series, (b) the sum of all the terms and (c) the 80th term

---

(a) Let the *AP* be a, $a+d$, $a+2d$, $\dots$, $a+(n-1)d$, where $a = 4$

The 12th term is: $a + (12-1)d = 31.5$

i.e. $\quad 4 + 11d = 31.5$, from which,
$$11d = 31.5 - 4 = 27.5$$

Hence $\quad\quad\quad d = \dfrac{27.5}{11} = 2.5$

The last term is $a + (n-1)d$

i.e. $\quad 4 + (n-1)(2.5) = 376.5$

$$(n-1) = \frac{376.5 - 4}{2.5}$$

$$= \frac{372.5}{2.5} = 149$$

Hence the number of terms in the series, $n = 149 + 1 = 150$

(b) Sum of all the terms,

$$S_{150} = \frac{n}{2}[2a + (n-1)d]$$

$$= \frac{150}{2}[2(4) + (150-1)(2.5)]$$

$$= 75[8 + (149)(2.5)]$$

$$= 85[8 + 372.5]$$

$$= 75(380.5) = \mathbf{28537.5}$$

(c) The 80th term is:

$$a + (n-1)d = 4 + (80-1)(2.5)$$

$$= 4 + (79)(2.5)$$

$$= 4 + 197.5 = \mathbf{201.5}$$

Now try the following exercise

Exercise 58 Further problems on arithmetic progressions

1. The sum of 15 terms of an arithmetic progression is 202.5 and the common difference is 2. Find the first term of the series. $[-0.5]$

2. Three numbers are in arithmetic progression. Their sum is 9 and their product is 20.25. Determine the three numbers. [1.5, 3, 4.5]

3. Find the sum of all the numbers between 5 and 250 which are exactly divisible by 4. [7808]

4. Find the number of terms of the series 5, 8, 11, ... of which the sum is 1025. [25]

5. Insert four terms between 5 and 22.5 to form an arithmetic progression.
    [8.5, 12, 15.5, 19]

6. The 1st, 10th and last terms of an arithmetic progression are 9, 40.5, and 425.5 respectively.

Find (a) the number of terms, (b) the sum of all terms and (c) the 70th term.

[(a) 120   (b) 26070   (c) 250.5]

7. On commencing employment a man is paid a salary of £7200 per annum and receives annual increments of £350. Determine his salary in the 9th year and calculate the total he will have received in the first 12 years.

[£10 000, £109 500]

8. An oil company bores a hole 80 m deep. Estimate the cost of boring if the cost is £30 for drilling the first metre with an increase in cost of £2 per metre for each succeeding metre.

[£8720]

15.4 Geometric progressions

When a sequence has a constant ratio between successive terms it is called a **geometric progression** (often abbreviated to *GP*). The constant is called the **common ratio, *r***

Examples include

(i) 1, 2, 4, 8, ... where the common ratio is 2

and (ii) $a, ar, ar^2, ar^3, \dots$ where the common ratio is r

If the first term of a *GP* is '*a*' and the common ratio is *r*, then

the *n*'th term is: ar^{n-1}

which can be readily checked from the above examples. For example, the 8th term of the *GP* 1, 2, 4, 8, ... is $(1)(2)^7 = \mathbf{128}$, since $a = 1$ and $r = 2$

Let a *GP* be $a, ar, ar^2, ar^3, \dots ar^{n-1}$

then the sum of *n* terms,

$$S_n = a + ar + ar^2 + ar^3 + \cdots + ar^{n-1} \dots \quad (1)$$

Multiplying throughout by *r* gives:

$$rS_n = ar + ar^2 + ar^3 + ar^4 + \cdots + ar^{n-1} + ar^n \dots \quad (2)$$

Subtracting equation (2) from equation (1) gives:

$$S_n - rS_n = a - ar^n$$

i.e. $S_n(1 - r) = a(1 - r^n)$

Thus the sum of *n* terms,

$$S_n = \frac{a(1 - r^n)}{(1 - r)} \quad \text{which is valid when } r < 1$$

Subtracting equation (1) from equation (2) gives

$$S_n = \frac{a(r^n - 1)}{(r - 1)} \quad \text{which is valid when } r > 1$$

For example, the sum of the first 8 terms of the *GP* 1, 2, 4, 8, 16, ... is given by:

$$S_8 = \frac{1(2^8 - 1)}{(2 - 1)} \quad \text{since } a = 1 \text{ and } r = 2$$

i.e. $S_8 = \frac{1(256 - 1)}{1} = \mathbf{255}$

When the common ratio *r* of a *GP* is less than unity, the sum of *n* terms,

$$S_n = \frac{a(1 - r^n)}{(1 - r)}, \text{which may be written as}$$

$$S_n = \frac{a}{(1 - r)} - \frac{ar^n}{(1 - r)}$$

Since, $r < 1$, r^n becomes less as *n* increases,

i.e. $r^n \to 0$ as $n \to \infty$

Hence $\dfrac{ar^n}{(1 - r)} \to 0$ as $n \to \infty$.

Thus $S_n \to \dfrac{a}{(1 - r)}$ as $n \to \infty$

The quantity $\dfrac{a}{(1 - r)}$ is called the **sum to infinity**, S_∞, and is the limiting value of the sum of an infinite number of terms,

i.e. $S_\infty = \dfrac{a}{(1 - r)}$ which is valid when $-1 < r < 1$

For example, the sum to infinity of the GP $1 + \dfrac{1}{2} + \dfrac{1}{4} + \cdots$ is

$$S_\infty = \frac{1}{1 - \dfrac{1}{2}}, \text{since } a = 1 \text{ and } r = \frac{1}{2} \text{ i.e. } S_\infty = 2$$

15.5 Worked problems on geometric progressions

Problem 9. Determine the 10th term of the series 3, 6, 12, 24, ...

3, 6, 12, 24, ... is a geometric progression with a common ratio r of 2.

The n'th term of a GP is ar^{n-1}, where a is the first term. Hence the 10th term is:

$$(3)(2)^{10-1} = (3)(2)^9 = 3(512) = \mathbf{1536}$$

Problem 10. Find the sum of the first 7 terms of the series, $\dfrac{1}{2}, 1\dfrac{1}{2}, 4\dfrac{1}{2}, 13\dfrac{1}{2}, \ldots$

$$\frac{1}{2}, 1\frac{1}{2}, 4\frac{1}{2}, 13\frac{1}{2}, \ldots$$

is a GP with a common ratio $r = 3$

The sum of n terms, $S_n = \dfrac{a(r^n - 1)}{(r - 1)}$

Hence $S_7 = \dfrac{\frac{1}{2}(3^7 - 1)}{(3 - 1)} = \dfrac{\frac{1}{2}(2187 - 1)}{2} = \mathbf{546\dfrac{1}{2}}$

Problem 11. The first term of a geometric progression is 12 and the 5th term is 55. Determine the 8th term and the 11th term

The 5th term is given by $ar^4 = 55$, where the first term $a = 12$

Hence $r^4 = \dfrac{55}{a} = \dfrac{55}{12}$ and

$r = \sqrt[4]{\dfrac{55}{12}} = 1.4631719\ldots$

The 8th term is

$$ar^7 = (12)(1.4631719\ldots)^7 = \mathbf{172.3}$$

The 11th term is

$$ar^{10} = (12)(1.4631719\ldots)^{10} = \mathbf{539.7}$$

Problem 12. Which term of the series: 2187, 729, 243, ... is $\dfrac{1}{9}$?

2187, 729, 243, ... is a GP with a common ratio $r = \dfrac{1}{3}$ and first term $a = 2187$

The n'th term of a GP is given by: ar^{n-1}

Hence $\dfrac{1}{9} = (2187)\left(\dfrac{1}{3}\right)^{n-1}$ from which

$$\left(\frac{1}{3}\right)^{n-1} = \frac{1}{(9)(2187)} = \frac{1}{3^2 3^7} = \frac{1}{3^9} = \left(\frac{1}{3}\right)^9$$

Thus $(n - 1) = 9$, from which, $n = 9 + 1 = 10$

i.e. $\dfrac{1}{9}$ **is the 10th term of the** GP

Problem 13. Find the sum of the first 9 terms of the series: 72.0, 57.6, 46.08, ...

The common ratio,

$$r = \frac{ar}{a} = \frac{57.6}{72.0} = 0.8$$

$$\left(\text{also}\frac{ar^2}{ar} = \frac{46.08}{57.6} = 0.8\right)$$

The sum of 9 terms,

$$S_9 = \frac{a(1 - r^n)}{(1 - r)} = \frac{72.0(1 - 0.8^9)}{(1 - 0.8)}$$

$$= \frac{72.0(1 - 0.1342)}{0.2} = \mathbf{311.7}$$

Problem 14. Find the sum to infinity of the series 3, 1, $\dfrac{1}{3}$, ...

3, 1, $\dfrac{1}{3}$, ... is a GP of common ratio, $r = \dfrac{1}{3}$

The sum to infinity,

$$S_\infty = \frac{a}{1 - r} = \frac{3}{1 - \frac{1}{3}} = \frac{3}{\frac{2}{3}} = \frac{9}{2} = \mathbf{4\frac{1}{2}}$$

Now try the following exercise

---

1. Find the 10th term of the series 5, 10, 20, 40, ...   [2560]

2. Determine the sum to the first 7 terms of the series 0.25, 0.75, 2.25, 6.75, ...   [273.25]

3. The first term of a geometric progression is 4 and the 6th term is 128. Determine the 8th and 11th terms.   [512, 4096]

4. Find the sum of the first 7 terms of the series 2, 5, $12\frac{1}{2}$, ... (correct to 4 significant figures).
                                                [812.5]

5. Determine the sum to infinity of the series 4, 2, 1, ...   [8]

6. Find the sum to infinity of the series $2\frac{1}{2}$, $-1\frac{1}{4}$, $\frac{5}{8}$, ...
                                        $\left[1\frac{2}{3}\right]$

15.6 Further worked problems on geometric progressions

Problem 15. In a geometric progression the 6th term is 8 times the 3rd term and the sum of the 7th and 8th terms is 192. Determine (a) the common ratio, (b) the 1st term, and (c) the sum of the 5th to 11th term, inclusive

(a) Let the *GP* be a, ar, ar^2, ar^3, ... , ar^{n-1}

The 3rd term $= ar^2$ and the 6th term $= ar^5$

The 6th term is 8 times the 3rd

Hence $ar^5 = 8\,ar^2$ from which, $r^3 = 8$ and $r = \sqrt[3]{8}$

i.e. **the common ratio $r = 2$**

(b) The sum of the 7th and 8th terms is 192. Hence $ar^6 + ar^7 = 192$. Since $r = 2$, then

$$64a + 128a = 192$$

$$192a = 192,$$

from which, **a, the first term $= 1$**

(c) The sum of the 5th to 11th terms (inclusive) is given by:

$$S_{11} - S_4 = \frac{a(r^{11} - 1)}{(r - 1)} - \frac{a(r^4 - 1)}{(r - 1)}$$

$$= \frac{1(2^{11} - 1)}{(2 - 1)} - \frac{1(2^4 - 1)}{(2 - 1)}$$

$$= (2^{11} - 1) - (2^4 - 1)$$

$$= 2^{11} - 2^4 = 2408 - 16 = \mathbf{2032}$$

Problem 16. A hire tool firm finds that their net return from hiring tools is decreasing by 10% per annum. If their net gain on a certain tool this year is £400, find the possible total of all future profits from this tool (assuming the tool lasts for ever)

The net gain forms a series:

$$£400 + £400 \times 0.9 + £400 \times 0.9^2 + \cdots,$$

which is a *GP* with $a = 400$ and $r = 0.9$

The sum to infinity,

$$S_\infty = \frac{a}{(1 - r)} = \frac{400}{(1 - 0.9)}$$

$$= \mathbf{£4000 = total\ future\ profits}$$

Problem 17. If £100 is invested at compound interest of 8% per annum, determine (a) the value after 10 years, (b) the time, correct to the nearest year, it takes to reach more than £300

(a) Let the *GP* be a, ar, ar^2, ... ar^n

The first term $a = £100$ and

the common ratio $r = 1.08$

Hence the second term is $ar = (100)(1.08) = £108$, which is the value after 1 year, the 3rd term is $ar^2 = (100)(1.08)^2 = £116.64$, which is the value after 2 years, and so on.

Thus the value after 10 years
$= ar^{10} = (100)(1.08)^{10} = \mathbf{£215.89}$

(b) When £300 has been reached, $300 = ar^n$

i.e. $300 = 100(1.08)^n$

and $3 = (1.08)^n$

Taking logarithms to base 10 of both sides gives:

$$\lg 3 = \lg (1.08)^n = n\,\lg(1.08),$$

by the laws of logarithms from which,

$$n = \frac{\lg 3}{\lg 1.08} = 14.3$$

Hence it will take 15 years to reach more than £300

Problem 18. A drilling machine is to have 6 speeds ranging from 50 rev/min to 750 rev/min. If the speeds form a geometric progression determine their values, each correct to the nearest whole number

---

Let the *GP* of n terms by given by a, ar, ar^2, ... ar^{n-1}

The 1st term $a = 50$ rev/min.

The 6th term is given by ar^{6-1}, which is 750 rev/min, i.e., $ar^5 = 750$

from which $r^5 = \dfrac{750}{a} = \dfrac{750}{50} = 15$

Thus the common ratio, $r = \sqrt[5]{15} = 1.7188$

The 1st term is $a = 50$ rev/min.

the 2nd term is $ar = (50)(1.7188) = 85.94$,

the 3rd term is $ar^2 = (50)(1.7188)^2 = 147.71$,

the 4th term is $ar^3 = (50)(1.7188)^3 = 253.89$,

the 5th term is $ar^4 = (50)(1.7188)^4 = 436.39$,

the 6th term is $ar^5 = (50)(1.7188)^5 = 750.06$.

Hence, correct to the nearest whole number, the 6 speeds of the drilling machine are: **50, 86, 148, 254, 436 and 750 rev/min.**

Now try the following exercise

Exercise 60   Further problems on geometric progressions

1. In a geometric progression the 5th term is 9 times the 3rd term and the sum of the 6th and 7th terms is 1944. Determine (a) the common ratio, (b) the 1st term and (c) the sum of the 4th to 10th terms inclusive.
    [(a) 3   (b) 2   (c) 59 022]

2. Which term of the series 3, 9, 27, ... is 59 049?
    [10th]

3. The value of a lathe originally valued at £3000 depreciates 15% per annum. Calculate its value after 4 years. The machine is sold when its value is less than £550. After how many years is the lathe sold?   [£1566, 11 years]

4. If the population of Great Britain is 55 million and is decreasing at 2.4% per annum, what will be the population in 5 years time?
    [48.71 M]

5. 100 g of a radioactive substance disintegrates at a rate of 3% per annum. How much of the substance is left after 11 years?   [71.53 g]

6. If £250 is invested at compound interest of 6% per annum determine (a) the value after 15 years, (b) the time, correct to the nearest year, it takes to reach £750.
    [(a) £599.14   (b) 19 years]

7. A drilling machine is to have 8 speeds ranging from 100 rev/min to 1000 rev/min. If the speeds form a geometric progression determine their values, each correct to the nearest whole number.

$$\begin{bmatrix} 100,\ 139,\ 193,\ 268,\ 373,\ 518, \\ 720,\ 1000\ \text{rev/min} \end{bmatrix}$$

15.7 Combinations and permutations

A **combination** is the number of selections of r different items from n distinguishable items when order of selection is ignored. A combination is denoted by ^nC_r or $\dbinom{n}{r}$

where

$$^nC_r = \frac{n!}{r!(n-r)!}$$

where, for example, 4! denotes $4 \times 3 \times 2 \times 1$ and is termed 'factorial 4'.

Thus,

$$^5C_3 = \frac{5!}{3!(5-3)!} = \frac{5 \times 4 \times 3 \times 2 \times 1}{(3 \times 2 \times 1)(2 \times 1)}$$

$$= \frac{120}{6 \times 2} = 10$$

For example, the five letters A, B, C, D, E can be arranged in groups of three as follows: ABC, ABD, ABE, ACD, ACE, ADE, BCD, BCE, BDE, CDE, i.e. there are ten groups. The above calculation $^5C_3$

produces the answer of 10 combinations without having to list all of them.

A **permutation** is the number of ways of selecting $r \leq n$ objects from n distinguishable objects when order of selection is important. A permutation is denoted by ^nP_r or $_nP_r$

where $^nP_r = n(n-1)(n-2)\ldots(n-r+1)$

or $^nP_r = \dfrac{n!}{(n-r)!}$

Thus, $^4P_2 = \dfrac{4!}{(4-2)!} = \dfrac{4!}{2!}$

$$= \frac{4 \times 3 \times 2}{2} = 12$$

Problem 19.   Evaluate: (a) $^7C_4$ (b) $^{10}C_6$

(a) $^7C_4 = \dfrac{7!}{4!(7-4)!} = \dfrac{7!}{4!3!}$

$$= \frac{7 \times 6 \times 5 \times 4 \times 3 \times 2}{(4 \times 3 \times 2)(3 \times 2)} = 35$$

(b) $^{10}C_6 = \dfrac{10!}{6!(10-6)!} = \dfrac{10!}{6!4!} = 210$

Problem 20.   Evaluate: (a) $^6P_2$ (b) $^9P_5$

(a) $^6P_2 = \dfrac{6!}{(6-2)!} = \dfrac{6!}{4!}$

$$= \frac{6 \times 5 \times 4 \times 3 \times 2}{4 \times 3 \times 2} = 30$$

(b) $^9P_5 = \dfrac{9!}{(9-5)!} = \dfrac{9!}{4!}$

$$= \frac{9 \times 8 \times 7 \times 6 \times 5 \times 4!}{4!} = 15\,120$$

Now try the following exercise

Exercise 61   Further problems on permutations and combinations

Evaluate the following:

1. (a) $^9C_6$ (b) $^3C_1$ [(a) 84   (b) 3]

2. (a) $^6C_2$ (b) $^8C_5$ [(a) 15   (b) 56]

3. (a) $^4P_2$ (b) $^7P_4$ [(a) 12   (b) 840]

4. (a) $^{10}P_3$ (b) $^8P_5$ [(a) 720   (b) 6720]

Chapter 16

The binomial series

16.1 Pascal's triangle

A **binomial expression** is one that contains two terms connected by a plus or minus sign. Thus $(p+q)$, $(a+x)^2$, $(2x+y)^3$ are examples of binomial expression. Expanding $(a+x)^n$ for integer values of n from 0 to 6 gives the results shown at the bottom of the page.

From the results the following patterns emerge:

(i) 'a' decreases in power moving from left to right.

(ii) 'x' increases in power moving from left to right.

(iii) The coefficients of each term of the expansions are symmetrical about the middle coefficient when n is even and symmetrical about the two middle coefficients when n is odd.

(iv) The coefficients are shown separately in Table 16.1 and this arrangement is known as **Pascal's triangle**. A coefficient of a term may be obtained by adding the two adjacent coefficients immediately above in the previous row. This is shown by the triangles in Table 16.1, where, for example, $1+3=4$, $10+5=15$, and so on.

(v) Pasal's triangle method is used for expansions of the form $(a+x)^n$ for integer values of n less than about 8

Table 16.1

From Table 16.1 the row the Pascal's triangle corresponding to $(a+x)^6$ is as shown in (1) below. Adding adjacent coefficients gives the coefficients of $(a+x)^7$ as shown in (2) below.

The first and last terms of the expansion of $(a+x)^7$ and a^7 and x^7 respectively. The powers of 'a' decrease and the powers of 'x' increase moving from left to right. Hence,

$$(a+x)^7 = a^7 + 7a^6x + 21a^5x^2 + 35a^4x^3 + 35a^3x^4$$
$$+ 21a^2x^5 + 7ax^6 + x^7$$

Problem 1. Use the Pascal's triangle method to determine the expansion of $(a+x)^7$

$$
\begin{array}{ll}
(a+x)^0 = & 1 \\
(a+x)^1 = & a+x \\
(a+x)^2 = (a+x)(a+x) = & a^2 + 2ax + x^2 \\
(a+x)^3 = (a+x)^2(a+x) = & a^3 + 3a^2x + 3ax^2 + x^3 \\
(a+x)^4 = (a+x)^3(a+x) = & a^4 + 4a^3x + 6a^2x^2 + 4ax^3 + x^4 \\
(a+x)^5 = (a+x)^4(a+x) = & a^5 + 5a^4x + 10a^3x^2 + 10a^2x^3 + 5ax^4 + x^5 \\
(a+x)^6 = (a+x)^5(a+x) = & a^6 + 6a^5x + 15a^4x^2 + 20a^3x^3 + 15a^2x^4 + 6ax^5 + x^6 \\
\end{array}
$$

Problem 2. Determine, using Pascal's triangle method, the expansion of $(2p - 3q)^5$

Comparing $(2p - 3q)^5$ with $(a + x)^5$ shows that $a = 2p$ and $x = -3q$

Using Pascal's triangle method:

$(a + x)^5 = a^5 + 5a^4x + 10a^3x^2 + 10a^2x^3 + \cdots$

Hence

$$(2p - 3q)^5 = (2p)^5 + 5(2p)^4(-3q)$$
$$+ 10(2p)^3(-3q)^2$$
$$+ 10(2p)^2(-3q)^3$$
$$+ 5(2p)(-3q)^4 + (-3q)^5$$

i.e. $(2p - 3q)^5 = 32p^5 - 240p^4q + 720p^3q^2$
$$- 1080p^2q^3 + 810pq^4 - 243q^5$$

Now try the following exercise

Exercise 62 Further problems on Pascal's triangle

1. Use Pascal's triangle to expand $(x - y)^7$
$$\left[\begin{array}{l} x^7 - 7x^6y + 21x^5y^2 - 35x^4y^3 \\ + 35x^3y^4 - 21x^2y^5 + 7xy^6 - y^7 \end{array}\right]$$

2. Expand $(2a + 3b)^5$ using Pascal's triangle.
$$\left[\begin{array}{l} 32a^5 + 240a^4b + 720a^3b^2 \\ + 1080a^2b^3 + 810ab^4 + 243b^5 \end{array}\right]$$

16.2 The binomial series

The **binomial series** or **binomial theorem** is a formula for raising a binomial expression to any power without lengthy multiplication. The general binomial expansion of $(a + x)^n$ is given by:

$$(a + x)^n = a^n + na^{n-1}x + \frac{n(n-1)}{2!}a^{n-2}x^2$$
$$+ \frac{n(n-1)(n-2)}{3!}a^{n-3}x^3$$
$$+ \cdots + x^n$$

where, for example, 3! denote $3 \times 2 \times 1$ and is termed 'factorial 3'.

With the binomial theorem n may be a fraction, a decimal fraction or a positive or negative integer.

In the general expansion of $(a + x)^n$ it is noted that the 4th term is:
$$\frac{n(n-1)(n-2)}{3!}a^{n-3}x^3$$

The number 3 is very evident in this expression.

For any term in a binomial expansion, say the r'th term, $(r - 1)$ is very evident. It may therefore be reasoned that **the r'th term of the expansion $(a + x)^n$ is:**
$$\frac{n(n-1)(n-2)\dots \text{to}(r-1)\text{terms}}{(r-1)!}a^{n-(r-1)}x^{r-1}$$

If $a = 1$ in the binomial expansion of $(a + x)^n$ then:
$$(1 + x)^n = 1 + nx + \frac{n(n-1)}{2!}x^2$$
$$+ \frac{n(n-1)(n-2)}{3!}x^3 + \cdots$$

which is valid for $-1 < x < 1$

When x is small compared with 1 then:
$$(1 + x)^n \approx 1 + nx$$

16.3 Worked problems on the binomial series

Problem 3. Use the binomial series to determine the expansion of $(2 + x)^7$

The binomial expansion is given by:
$$(a + x)^n = a^n + na^{n-1}x + \frac{n(n-1)}{2!}a^{n-2}x^2$$
$$+ \frac{n(n-1)(n-2)}{3!}a^{n-3}x^3 + \cdots$$

When $a = 2$ and $n = 7$:
$$(2 + x)^7 = 2^7 + 7(2)^6 + \frac{(7)(6)}{(2)(1)}(2)^5x^2$$
$$+ \frac{(7)(6)(5)}{(3)(2)(1)}(2)^4x^3 + \frac{(7)(6)(5)(4)}{(4)(3)(2)(1)}(2)^3x^4$$
$$+ \frac{(7)(6)(5)(4)(3)}{(5)(4)(3)(2)(1)}(2)^2x^5$$
$$+ \frac{(7)(6)(5)(4)(3)(2)}{(6)(5)(4)(3)(2)(1)}(2)x^6$$
$$+ \frac{(7)(6)(5)(4)(3)(2)(1)}{(7)(6)(5)(4)(3)(2)(1)}x^7$$

i.e. $(2 + x)^7 = 128 + 448x + 672x^2 + 560x^3$
$$+ 280x^4 + 84x^5 + 14x^6 + x^7$$

Problem 4. Expand $\left(c - \dfrac{1}{c}\right)^5$ using the binomial series

$$\left(c - \frac{1}{c}\right)^5 = c^5 + 5c^4\left(-\frac{1}{c}\right) + \frac{(5)(4)}{(2)(1)}c^3\left(-\frac{1}{c}\right)^2$$
$$+ \frac{(5)(4)(3)}{(3)(2)(1)}c^2\left(-\frac{1}{c}\right)^3$$
$$+ \frac{(5)(4)(3)(2)}{(4)(3)(2)(1)}c\left(-\frac{1}{c}\right)^4$$
$$+ \frac{(5)(4)(3)(2)(1)}{(5)(4)(3)(2)(1)}\left(-\frac{1}{c}\right)^5$$

i.e. $\left(c - \dfrac{1}{c}\right)^5 = c^5 - 5c^4 + 10c - \dfrac{10}{c} + \dfrac{5}{c^3} - \dfrac{1}{c^5}$

Problem 5. Without fully expanding $(3 + x)^7$, determine the fifth term

The r'th term of the expansion $(a + x)^n$ is given by:
$$\frac{n(n-1)(n-2)\dots\text{to }(r-1)\text{ terms}}{(r-1)!}a^{n-(r-1)}x^{r-1}$$
Substituting $n = 7$, $a = 3$ and $r - 1 = 5 - 1 = 4$ gives:
$$\frac{(7)(6)(5)(4)}{(4)(3)(2)(1)}(3)^{7-4}x^4$$
i.e. the fifth term of $(3 + x)^7 = 35(3)^3x^4 = \mathbf{945x^4}$

Problem 6. Find the middle term of $\left(2p - \dfrac{1}{2q}\right)^{10}$

In the expansion of $(a + x)^{10}$ there are $10 + 1$, i.e. 11 terms. Hence the middle term is the sixth. Using the general expression for the r'th term where $a = 2p$, $x = -\dfrac{1}{2q}$, $n = 10$ and $r - 1 = 5$ gives:
$$\frac{(10)(9)(8)(7)(6)}{(5)(4)(3)(2)(1)}(2p)^{10-5}\left(-\frac{1}{2q}\right)^5$$
$$= 252(32p^5)\left(-\frac{1}{32q^5}\right)$$

Hence the middle term of $\left(2q - \dfrac{1}{2q}\right)^{10}$ is $-252\dfrac{p^5}{q^5}$

Problem 7. Evaluate $(1.002)^9$ using the binomial theorem correct to (a) 3 decimal places and (b) 7 significant figures

$$(1 + x)^n = 1 + nx + \frac{n(n-1)}{2!}x^2$$
$$+ \frac{n(n-1)(n-2)}{3!}x^3 + \cdots$$
$$(1.002)^9 = (1 + 0.002)^9$$

Substituting $x = 0.002$ and $n = 9$ in the general expansion for $(1 + x)^n$ gives:
$$(1 + 0.002)^9 = 1 + 9(0.002) + \frac{(9)(8)}{(2)(1)}(0.002)^2$$
$$+ \frac{(9)(8)(7)}{(3)(2)(1)}(0.002)^3 + \cdots$$
$$= 1 + 0.018 + 0.000144$$
$$+ 0.000000672 + \cdots$$
$$= 1.018144672\dots$$

Hence, $(1.002)^9 = \mathbf{1.018}$, **correct to 3 decimal places**
$$= \mathbf{1.018145}\text{, \textbf{correct to}}$$
7 significant figures

Problem 8. Determine the value of $(3.039)^4$, correct to 6 significant figures using the binomial theorem

$(3.039)^4$ may be written in the form $(1 + x)^n$ as:
$$(3.039)^4 = (3 + 0.039)^4$$
$$= \left[3\left(1 + \frac{0.039}{3}\right)\right]^4$$
$$= 3^4(1 + 0.013)^4$$
$$(1 + 0.013)^4 = 1 + 4(0.013) + \frac{(4)(3)}{(2)(1)}(0.013)^2$$
$$+ \frac{(4)(3)(2)}{(3)(2)(1)}(0.013)^3 + \cdots$$
$$= 1 + 0.052 + 0.001014$$
$$+ 0.000008788 + \cdots$$
$$= 1.0530228$$
correct to 8 significant figures

Hence $(3.039)^4 = 3^4(1.0530228)$
$$= \mathbf{85.2948}\text{, \textbf{correct to}}$$
6 significant figures

Now try the following exercise

Exercise 63   Further problems on the binomial series

1. Use the binomial theorem to expand $(a + 2x)^4$
$$\left[\begin{array}{c} a^4 + 8a^3x + 24a^2x^2 \\ + 32ax^3 + 16x^4 \end{array}\right]$$

2. Use the binomial theorem to expand $(2 - x)^6$
$$\left[\begin{array}{c} 64 - 192x + 240x^2 - 160x^3 \\ + 60x^4 - 12x^5 + x^6 \end{array}\right]$$

3. Expand $(2x - 3y)^4$
$$\left[\begin{array}{c} 16x^4 - 96x^3y + 216x^2y^2 \\ -216xy^3 + 81y^4 \end{array}\right]$$

4. Determine the expansion of $\left(2x + \dfrac{2}{x}\right)^5$
$$\left[\begin{array}{c} 32x^5 + 160x^3 + 320x + \dfrac{320}{x} \\ + \dfrac{160}{x^3} + \dfrac{32}{x^5} \end{array}\right]$$

5. Expand $(p + 2q)^{11}$ as far as the fifth term
$$\left[\begin{array}{c} p^{11} + 22p^{10}q + 220p^9q^2 \\ + 1320p^8q^3 + 5280p^7q^4 \end{array}\right]$$

6. Determine the sixth term of $\left(3p + \dfrac{q}{3}\right)^{13}$
$$[34\,749\,p^8q^5]$$

7. Determine the middle term of $(2a - 5b)^8$
$$[700\,000\,a^4b^4]$$

8. Use the binomial theorem to determine, correct to 4 decimal places:
    (a) $(1.003)^8$ (b) $(0.98)^7$
    $$[\text{(a) } 1.0243 \quad \text{(b) } 0.8681]$$

9. Evaluate $(4.044)^6$ correct to 3 decimal places.
    $$[4373.880]$$

16.4   Further worked problems on the binomial series

Problem 9.

(a) Expand $\dfrac{1}{(1 + 2x)^3}$ in ascending powers of x as far as the term in x^3, using the binomial series.

(b) State the limits of x for which the expansion is valid

(a) Using the binomial expansion of $(1 + x)^n$, where $n = -3$ and x is replaced by $2x$ gives:

$$\frac{1}{(1 + 2x)^3} = (1 + 2x)^{-3}$$

$$= 1 + (-3)(2x) + \frac{(-3)(-4)}{2!}(2x)^2$$

$$+ \frac{(-3)(-4)(-5)}{3!}(2x)^3 + \cdots$$

$$= 1 - 6x + 24x^2 - 80x^3 +$$

(b) The expansion is valid provided $|2x| < 1$,
    i.e. $|x| < \dfrac{1}{2}$ or $-\dfrac{1}{2} < x < \dfrac{1}{2}$

Problem 10.

(a) Expand $\dfrac{1}{(4 - x)^2}$ in ascending powers of x as far as the term in x^3, using the binomial theorem.

(b) What are the limits of x for which the expansion in (a) is true?

(a) $\dfrac{1}{(4 - x)^2} = \dfrac{1}{\left[4\left(1 - \dfrac{x}{4}\right)\right]^2} = \dfrac{1}{4^2\left(1 - \dfrac{x}{4}\right)^2}$

$$= \frac{1}{16}\left(1 - \frac{x}{4}\right)^{-2}$$

Using the expansion of $(1 + x)^n$

$$\frac{1}{(4 - x)^2} = \frac{1}{16}\left(1 - \frac{x}{4}\right)^{-2}$$

$$= \frac{1}{16}\left[1 + (-2)\left(-\frac{x}{4}\right)\right.$$

$$+ \frac{(-2)(-3)}{2!}\left(-\frac{x}{4}\right)^2$$

$$\left. + \frac{(-2)(-3)(-4)}{3!}\left(-\frac{x}{4}\right)^3 + \ldots\right]$$

$$= \frac{1}{16}\left(1 + \frac{x}{2} + \frac{3x^2}{16} + \frac{x^3}{16} + \cdots\right)$$

(b) The expansion in (a) is true provided $\left|\dfrac{x}{4}\right| < 1$,
    i.e. $|x| < 4$ or $-4 < x < 4$

Problem 11. Use the binomial theorem to expand $\sqrt{4+x}$ in ascending powers of x to four terms. Give the limits of x for which the expansion is valid

$$\sqrt{4+x} = \sqrt{4\left(1+\frac{x}{4}\right)}$$

$$= \sqrt{4}\sqrt{1+\frac{x}{4}}$$

$$= 2\left(1+\frac{x}{4}\right)^{\frac{1}{2}}$$

Using the expansion of $(1+x)^n$,

$$2\left(1+\frac{x}{4}\right)^{\frac{1}{2}}$$

$$= 2\left[1+\left(\frac{1}{2}\right)\left(\frac{x}{4}\right)+\frac{(1/2)(-1/2)}{2!}\left(\frac{x}{4}\right)^2\right.$$

$$\left.+\frac{(1/2)(-1/2)(-3/2)}{3!}\left(\frac{x}{4}\right)^3+\cdots\right]$$

$$= 2\left(1+\frac{x}{8}-\frac{x^2}{128}+\frac{x^3}{1024}-\cdots\right)$$

$$= 2+\frac{x}{4}-\frac{x^2}{64}+\frac{x^3}{512}-\cdots$$

This is valid when $\left|\frac{x}{4}\right| < 1$,

i.e. $\left|\frac{x}{4}\right| < 4$ or $-4 < x < 4$

Problem 12. Expand $\dfrac{1}{\sqrt{1-2t}}$ in ascending powers of t as far as the term in t^3. State the limits of t for which the expression is valid

$$\frac{1}{\sqrt{1-2t}}$$

$$= (1-2t)^{-\frac{1}{2}}$$

$$= 1+\left(-\frac{1}{2}\right)(-2t)+\frac{(-1/2)(-3/2)}{2!}(-2t)^2$$

$$+\frac{(-1/2)(-3/2)(-5/2)}{3!}(-2t)^3+\cdots$$

$$\text{using the expansion for}(1+x)^n$$

$$= 1+t+\frac{3}{2}t^2+\frac{5}{2}t^3+\cdots$$

The expression is valid when $|2t| < 1$,

i.e. $|t| < \dfrac{1}{2}$ or $-\dfrac{1}{2} < t < \dfrac{1}{2}$

Problem 13. Simplify $\dfrac{\sqrt[3]{1-3x}\sqrt{1+x}}{\left(1+\frac{x}{2}\right)^3}$ given that powers of x above the first may be neglected

$$\frac{\sqrt[3]{1-3x}\sqrt{1+x}}{\left(1+\frac{x}{2}\right)^3}$$

$$= (1-3x)^{\frac{1}{3}}(1+x)^{\frac{1}{2}}\left(1+\frac{x}{2}\right)^{-3}$$

$$\approx \left[1+\left(\frac{1}{3}\right)(-3x)\right]\left[1+\left(\frac{1}{2}\right)(x)\right]\left[1+(-3)\left(\frac{x}{2}\right)\right]$$

when expanded by the binomial theorem as far as the x term only,

$$= (1-x)\left(1+\frac{x}{2}\right)\left(1-\frac{3x}{2}\right)$$

$$= \left(1-x+\frac{x}{2}-\frac{3x}{2}\right)\quad \begin{array}{l}\text{when powers of } x \text{ higher}\\ \text{than unity are neglected}\end{array}$$

$$= (1-2x)$$

Problem 14. Express $\dfrac{\sqrt{1+2x}}{\sqrt[3]{1-3x}}$ as a power series as far as the term in x^2. State the range of values of x for which the series is convergent

$$\frac{\sqrt{1+2x}}{\sqrt[3]{1-3x}} = (1+2x)^{\frac{1}{2}}(1-3x)^{-\frac{1}{3}}$$

$$(1+2x)^{\frac{1}{2}} = 1+\left(\frac{1}{2}\right)(2x)$$

$$+\frac{(1/2)(-1/2)}{2!}(2x)^2+\cdots$$

$$= 1+x-\frac{x^2}{2}+\cdots \text{ which is valid for}$$

$$|2x| < 1, \text{ i.e. } |x| < \frac{1}{2}$$

$$(1-3x)^{-\frac{1}{3}} = 1+(-1/3)(-3x)$$

$$+\frac{(-1/3)(-4/3)}{2!}(-3x)^2+\cdots$$

$$= 1+x+2x^2+\cdots \text{ which is valid for}$$

$$|3x| < 1, \text{ i.e. } |x| < \frac{1}{3}$$

Hence $\dfrac{\sqrt{1+2x}}{\sqrt[3]{1-3x}}$

$= (1+2x)^{\frac{1}{2}}(1-3x)^{\frac{1}{3}}$

$= \left(1 + x - \dfrac{x^2}{2} + \cdots\right)(1 + x + 2x^2 + \cdots)$

$= 1 + x + 2x^2 + x + x^2 - \dfrac{x^2}{2}$

neglecting terms of higher power than 2

$= \mathbf{1 + 2x + \dfrac{5}{2}x^2}$

The series is convergent if $-\dfrac{1}{3} < x < \dfrac{1}{3}$

Now try the following exercise

Exercise 64   Further problems on the binomial series

In Problems 1 to 5 expand in ascending powers of x as far as the term in x^3, using the binomial theorem. State in each case the limits of x for which the series is valid.

1. $\dfrac{1}{(1-x)}$ $[1 + x + x^2 + x^3 + \cdots, \; |x| < 1]$

2. $\dfrac{1}{(1+x)^2}$ $\left[\begin{array}{l} 1 - 2x + 3x^2 - 4x^3 + \cdots, \\ |x| < 1 \end{array}\right]$

3. $\dfrac{1}{(2+x)^3}$

$\left[\begin{array}{l} \dfrac{1}{8}\left(1 - \dfrac{3x}{2} + \dfrac{3x^2}{2} - \dfrac{5x^3}{4} + \cdots\right) \\ |x| < 2 \end{array}\right]$

4. $\sqrt{2+x}$

$\left[\begin{array}{l} \sqrt{2}\left(1 + \dfrac{x}{4} - \dfrac{x^2}{32} + \dfrac{x^3}{128} - \cdots\right) \\ |x| < 2 \end{array}\right]$

5. $\dfrac{1}{\sqrt{1+3x}}$

$\left[\begin{array}{l} \left(1 - \dfrac{3}{2}x + \dfrac{27}{8}x^2 - \dfrac{135}{16}x^3 + \cdots\right) \\ |x| < \dfrac{1}{3} \end{array}\right]$

6. Expand $(2+3x)^{-6}$ to three terms. For what values of x is the expansion valid?

$\left[\begin{array}{l} \dfrac{1}{64}\left(1 - 9x + \dfrac{189}{4}x^2\right) \\ |x| < \dfrac{2}{3} \end{array}\right]$

7. When x is very small show that:

(a) $\dfrac{1}{(1-x)^2\sqrt{1-x}} \approx 1 + \dfrac{5}{2}x$

(b) $\dfrac{(1-2x)}{(1-3x)^4} \approx 1 + 10x$

(c) $\dfrac{\sqrt{1+5x}}{\sqrt[3]{1-2x}} \approx 1 + \dfrac{19}{6}x$

8. If x very small such that x^2 and higher powers may be neglected, determine the power series for $\dfrac{\sqrt{x+4}\;\sqrt[3]{8-x}}{\sqrt[5]{(1+x)^3}}$ $\left[4 - \dfrac{31}{15}x\right]$

9. Express the following as power series in ascending powers of x as far as the term in x^2. State in each case the range of x for which the series is valid.

(a) $\sqrt{\dfrac{1-x}{1+x}}$ (b) $\dfrac{(1+x)\;\sqrt[3]{(1-3x)^2}}{\sqrt{1+x^2}}$

$\left[\begin{array}{l} \text{(a)}\;\; 1 - x + \dfrac{1}{2}x^2, \;\; |x| < 1 \\[2mm] \text{(b)}\;\; 1 - x - \dfrac{7}{2}x^2, \;\; |x| < \dfrac{1}{3} \end{array}\right]$

16.5 Practical problems involving the binomial theorem

Binomial expansions may be used for numerical approximations, for calculations with small variations and in probability theory.

Problem 15.   The radius of a cylinder is reduced by 4% and its height is increased by 2%. Determine the approximate percentage change in (a) its volume and (b) its curved surface area, (neglecting the products of small quantities)

Volume of cylinder $= \pi r^2 h$

Let r and h be the original values of radius and height.

The new values are $0.96r$ or $(1 - 0.04)r$ and $1.02\,h$ or $(1 + 0.02)h$

(a) New volume $= \pi[(1 - 0.04)r]^2[(1 + 0.02)h]$

$\qquad = \pi r^2 h(1 - 0.04)^2(1 + 0.02)$

Now $(1 - 0.04)^2 = 1 - 2(0.04) + (0.04)^2$

$\qquad\qquad = (1 - 0.08)$, neglecting powers of small terms

Hence new volume

$\qquad \approx \pi r^2 h(1 - 0.08)(1 + 0.02)$

$\qquad \approx \pi r^2 h(1 - 0.08 + 0.02)$, neglecting products of small terms

$\qquad \approx \pi r^2 h(1 - 0.06)$ or $0.94\pi r^2 h$, i.e. 94% of the original volume

Hence the volume is reduced by approximately 6%

(b) Curved surface area of cylinder $= 2\pi rh$.

New surface area

$\qquad = 2\pi[(1 - 0.04)r][(1 + 0.02)h]$

$\qquad = 2\pi rh(1 - 0.04)(1 + 0.02)$

$\qquad \approx 2\pi rh(1 - 0.04 + 0.02)$, neglecting products of small terms

$\qquad \approx 2\pi rh(1 - 0.02)$ or $0.98(2\pi rh)$, i.e. 98% of the original surface area

Hence the curved surface area is reduced by approximately 2%

Problem 16. The second moment of area of a rectangle through its centroid is given by $\dfrac{bl^3}{12}$. Determine the approximate change in the second moment of area if b is increased by 3.5% and l is reduced by 2.5%

New values of b and l are $(1 + 0.035)b$ and $(1 - 0.025)l$ respectively.
New second moment of area

$= \dfrac{1}{2}[(1 + 0.035)b][(1 - 0.025)l]^3$

$= \dfrac{bl^3}{12}(1 + 0.035)(1 - 0.025)^3$

$\approx \dfrac{bl^3}{12}(1 + 0.035)(1 - 0.075)$, neglecting powers of small terms

$\approx \dfrac{bl^3}{12}(1 + 0.035 - 0.075)$, neglecting products of small terms

$\approx \dfrac{bl^3}{12}(1 - 0.040)$ or $(0.96)\dfrac{bl^3}{12}$, i.e. 96% of the original second moment of area

Hence the second moment of area is reduced by approximately 4%

Problem 17. The resonant frequency of a vibrating shaft is given by: $f = \dfrac{1}{2\pi}\sqrt{\dfrac{k}{I}}$, where k is the stiffness and I is the inertia of the shaft. Use the binomial theorem to determine the approximate percentage error in determining the frequency using the measured values of k and I when the measured value of k is 4% too large and the measured value of I is 2% too small

Let f, k and I be the true values of frequency, stiffness and inertia respectively. Since the measured value of stiffness, k_1, is 4% too large, then

$$k_1 = \dfrac{104}{100}k = (1 + 0.04)k$$

The measured value of inertia, I_1, is 2% too small, hence

$$I_1 = \dfrac{98}{100}I = (1 - 0.02)I$$

The measured value of frequency,

$$f_1 = \dfrac{1}{2\pi}\sqrt{\dfrac{k_1}{I_1}} = \dfrac{1}{2\pi}k_1^{\frac{1}{2}}I_1^{-\frac{1}{2}}$$

$$= \dfrac{1}{2\pi}[(1 + 0.04)k]^{\frac{1}{2}}[(1 - 0.02)I]^{-\frac{1}{2}}$$

$$= \dfrac{1}{2\pi}(1 + 0.04)^{\frac{1}{2}}k^{\frac{1}{2}}(1 - 0.02)^{-\frac{1}{2}}I^{-\frac{1}{2}}$$

$$= \dfrac{1}{2\pi}k^{\frac{1}{2}}I^{-\frac{1}{2}}(1 + 0.04)^{\frac{1}{2}}(1 - 0.02)^{-\frac{1}{2}}$$

i.e. $f_1 = f(1 + 0.04)^{\frac{1}{2}}(1 - 0.02)^{-\frac{1}{2}}$

$$\approx f\left[1 + \left(\frac{1}{2}\right)(0.04)\right]\left[1 + \left(-\frac{1}{2}\right)(-0.02)\right]$$

$$\approx f(1 + 0.02)(1 + 0.01)$$

Neglecting the products of small terms,

$$f_1 \approx (1 + 0.02 + 0.01)f \approx 1.03f$$

Thus the percentage error in f based on the measured values of k and I is approximately $[(1.03)(100) - 100]$, i.e. **3% too large**

Now try the following exercise

Exercise 65 Further practical problems involving the binomial theorem

1. Pressure p and volume v are related by $pv^3 = c$, where c is a constant. Determine the approximate percentage change in c when p is increased by 3% and v decreased by 1.2%.
   [0.6% decrease]

2. Kinetic energy is given by $\frac{1}{2}mv^2$. Determine the approximate change in the kinetic energy when mass m is increased by 2.5% and the velocity v is reduced by 3%.
   [3.5% decrease]

3. An error of $+1.5\%$ was made when measuring the radius of a sphere. Ignoring the products of small quantities determine the approximate error in calculating (a) the volume, and (b) the surface area.
   [(a) 4.5% increase (b) 3.0% increase]

4. The power developed by an engine is given by $I = k$ PLAN, where k is a constant. Determine the approximate percentage change in the power when P and A are each increased by 2.5% and L and N are each decreased by 1.4%.
   [2.2% increase]

5. The radius of a cone is increased by 2.7% and its height reduced by 0.9%. Determine the approximate percentage change in its volume, neglecting the products of small terms.
   [4.5% increase]

6. The electric field strength H due to a magnet of length $2l$ and moment M at a point on its axis distance x from the centre is given by:

   $$H = \frac{M}{2l}\left\{\frac{1}{(x - l)^2} - \frac{1}{(x + l)^2}\right\}$$

   Show that is l is very small compared with x, then $H \approx \dfrac{2M}{x^3}$

7. The shear stress τ in a shaft of diameter D under a torque T is given by: $\tau = \dfrac{kT}{\pi D^3}$. Determine the approximate percentage error in calculating τ if T is measured 3% too small and D 1.5% too large.
   [7.5% decrease]

8. The energy W stored in a flywheel is given by: $W = kr^5N^2$, where k is a constant, r is the radius and N the number of revolutions. Determine the approximate percentage change in W when r is increased by 1.3% and N is decreased by 2%.
   [2.5% increase]

9. In a series electrical circuit containing inductance L and capacitance C the resonant frequency is given by: $f_r = \dfrac{1}{2\pi\sqrt{LC}}$. If the values of L and C used in the calculation are 2.6% too large and 0.8% too small respectively, determine the approximate percentage error in the frequency.
   [0.9% too small]

10. The viscosity η of a liquid is given by: $\eta = \dfrac{kr^4}{vl}$, where k is a constant. If there is an error in r of $+2\%$, in v of $+4\%$ and l of -3%, what is the resultant error in η?
    [+7%]

11. A magnetic pole, distance x from the plane of a coil of radius r, and on the axis of the coil, is subject to a force F when a current flows in the coil. The force is given by: $F = \dfrac{kx}{\sqrt{(r^2 + x^2)^5}}$, where k is a constant. Use the binomial theorem to show that when x is small compared to r, then $F \approx \dfrac{kx}{r^5} - \dfrac{5kx^3}{2r^7}$

12. The flow of water through a pipe is given by: $G = \sqrt{\dfrac{(3d)^5H}{L}}$. If d decreases by 2% and H by 1%, use the binomial theorem to estimate the decrease in G.
    [5.5%]

Chapter 17

Solving equations by iterative methods

17.1 Introduction to iterative methods

Many equations can only be solved graphically or by methods of successive approximation to the roots, called **iterative methods**. Three methods of successive approximations are (i) by using the Newton-Raphson formula, given in Section 16.2, (ii) the bisection methods, and (iii) an algebraic methods. The latter two methods are discussed in *Higher Engineering Mathematics, fifth edition*.

Each successive approximation method relies on a reasonably good first estimate of the value of a root begin made. One way of determining this is to sketch a graph of the function, say $y = f(x)$, and determine the approximate values of roots from the points where the graph cuts the x-axis. Another way is by using a functional notation method. This method uses the property that the value of the graph of $f(x) = 0$ changes sign for values of x just before and just after the value of a root. For example, one root of the equation $x^2 - x - 6 = 0$ is $x = 3$.

Using functional notation:

$$f(x) = x^2 - x - 6$$
$$f(2) = 2^2 - 2 - 6 = -4$$
$$f(4) = 4^2 - 4- = +6$$

It can be seen from these results that the value of $f(x)$ changes from -4 at $f(2)$ to $+6$ at $f(4)$, indicating that a root lies between 2 and 4. This is shown more clearly in Fig. 17.1.

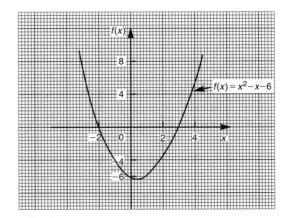

Figure 17.1

17.2 The Newton–Raphson method

The Newton–Raphson formula, often just referred to as **Newton's method,** may be stated as follows:

if r_1 is the approximate value of a real root of the equation $f(x) = 0$, then a closer approximation to the root r_2 is given by:

$$r_2 = r_1 - \frac{f(r_1)}{f'(r_1)}$$

The advantages of Newton's method over other methods of successive approximations is that it can be used for any type of mathematical equation (i.e. ones containing trigonometric, exponential, logarithmic, hyperbolic and algebraic functions), and it is usually easier to apply than other methods. The method is demonstrated in the following worked problems.

17.3 Worked problems on the Newton–Raphson method

Problem 1. Use Newton's method to determine the positive root of the quadratic equation $5x^2 + 11x - 17 = 0$, correct to 3 significant figures. Check the value of the root by using the quadratic formula.

The functional notation method is used to determine the first approximation to the root:

$$f(x) = 5x^2 + 11x - 17$$
$$f(0) = 5(0)^2 + 11(0) - 17 = -17$$
$$f(1) = 5(1)^2 + 11(1) - 17 = -1$$
$$f(2) = 5(2)^2 + 11(2) - 17 = 25$$

This shows that the value of the root is close to $x = 1$. Let the first approximation to the root, r_1, be 1. Newton's formula states that a closer approximation,

$$r_2 = r_1 - \frac{f(r_1)}{f'(r_1)}$$

$$f(x) = 5x^2 + 11x - 17, \quad \text{thus,}$$
$$f(r) = 5(r_1)^2 + 11(r_1) - 17$$
$$= 5(1)^2 + 11(1) - 17 = -1$$

$f'(x)$ is the differential coefficient of $f(x)$, i.e. $f'(x) = 10x + 11$ (see Chapter 42). Thus $f'(r_1) = 10(r_1) + 11 = 10(1) + 11 = 21$
By Newton's formula, a better approximation to the root is:

$$r_2 = 1 - \frac{-1}{21} = 1 - (-0.048) = 1.05,$$

correct to 3 significant figures

A still better approximation to the root, r_3, is given by:

$$r_3 = r_2 - \frac{f(r_2)}{f'(r_2)}$$
$$= 1.05 - \frac{[5(1.05)^2 + 11(1.05) - 17]}{[10(1.05) + 11]}$$
$$= 1.05 - \frac{0.0625}{21.5}$$
$$= 1.05 - 0.003 = 1.047,$$

i.e. 1.05, correct to 3 significant figures.
  Since the value of r_2 and r_3 are the same when expressed to the required degree of accuracy, the required root is **1.05**, correct to 3 significant figures.

Checking, using the quadratic equation formula,

$$x = \frac{-11 \pm \sqrt{121 - 4(5)(-17)}}{(2)(5)}$$
$$= \frac{-11 \pm 21.47}{10}$$

The positive root is 1.047, i.e. **1.05**, correct to 3 significant figures

Problem 2. Taking the first approximation as 2, determine the root of the equation $x^2 - 3\sin x + 2\ln(x+1) = 3.5$, correct to 3 significant figures, by using Netwon's method

Newton's formula state that $r_2 = r_1 - \frac{f(r_1)}{f'(r_1)}$, where r_1 is a first approximation to the root and r_2 is a better approximation to the root.

Since
$$fx = x^2 - 3\sin x + 2\ln(x+1) - 3.5$$
$$f(r_1) = f(2) = 2^2 - 3\sin 2 + 2\ln 3 - 3.5,$$
where $\sin 2$ means the sin of 2 radians
$$= 4 - 2.7279 + 2.1972 - 3.5 = -0.0307$$
$$f'x = 2x - 3\cos x + \frac{2}{x+1}$$
$$f'(r_1) = f'(2) = 2(2) - 3\cos 2 + \frac{2}{3}$$
$$= 4 + 1.2484 + 0.6667 = 5.9151$$

Hence,
$$r_2 = r_1 - \frac{f(r_1)}{f'(r_1)}$$
$$= 2 - \frac{-0.0307}{5.9151} = 2.005 \text{ or } 2.01,$$

correct to 3 significant figures.

A still better approximation to the root, r_3, is given by:

$$r_3 = r_2 - \frac{f(r_2)}{f'(r_2)}$$
$$= 2.005 - \frac{[(2.005)^2 - 3\sin 2.005 + 2\ln 3.005 - 3.5]}{\left[2(2.005) - 3\cos 2.005 + \frac{2}{2.005+1}\right]}$$
$$= 2.005 - \frac{(-0.00104)}{5.9376} = 2.005 + 0.000175$$

i.e. $r_3 = 2.01$, correct to 3 significant figures.

Since the values of r_2 and r_3 are the same when expressed to the required degree of accuracy, then the required root is **2.01**, correct to 3 significant figures.

Problem 3. Use Newton's method to find the positive root of:

$$(x+4)^3 - e^{1.92x} + 5\cos\frac{x}{3} = 9,$$

correct to 3 significant figures

The function notational method is used to determine the approximate value of the root:

$$f(x) = (x+4)^3 - e^{1.92x} + 5\cos\frac{x}{3} - 9$$

$$f(0) = (0+4)^3 - e^0 + 5\cos 0 - 9 = 59$$

$$f(1) = 5^3 - e^{1.92} + 5\cos\frac{1}{3} - 9 \approx 114$$

$$f(2) = 6^3 - e^{3.84} + 5\cos\frac{2}{3} - 9 \approx 164$$

$$f(3) = 7^3 - e^{5.76} + 5\cos 1 - 9 \approx 19$$

$$f(4) = 8^3 - e^{7.68} + 5\cos\frac{4}{3} - 9 \approx -1660$$

From these results, let a first approximation to the root be $r_1 = 3$.

Newton's formula states that a better approximation to the root,

$$r_2 = r_1 - \frac{f(r_1)}{f'(r_1)}$$

$$f(r_1) = f(3) = 7^3 - e^{5.76} + 5\cos 1 - 9$$

$$= 19.35$$

$$f'(x) = 3(x+4)^2 - 1.92e^{1.92x} - \frac{5}{3}\sin\frac{x}{3}$$

$$f'(r_1) = f'(3) = 3(7)^2 - 1.92e^{5.76} - \frac{5}{3}\sin 1$$

$$= -463.7$$

Thus, $r_3 = 3 - \dfrac{19.35}{-463.7} = 3 + 0.042 = 3.042 = 3.04$, correct to 3 significant figures

Similarly, $r_3 = 3.042 - \dfrac{f(3.042)}{f'(3.042)}$

$$= 3.042 - \frac{-1.146}{(-513.1)}$$

$$= 3.042 - 0.0022 = 3.0398 = 3.04,$$

correct to 3 significant figures.

Since r_2 and r_3 are the same when expressed to the required degree of accuracy, then the required root is **3.04,** correct to 3 significant figures.

Now try the following exercise

Exercise 66 Further problems on Newton's method

In Problems 1 to 7, use **Newton's method** to solve the equations given to the accuracy stated.

1. $x^2 - 2x - 13 = 0$, correct to 3 decimal places. [−2.742, 4.742]

2. $3x^3 - 10x = 14$, correct to 4 significant figures. [2.313]

3. $x^4 - 3x^3 + 7x - 12 = 0$, correct to 3 decimal places. [−1.721, 2.648]

4. $3x^4 - 4x^3 + 7x = 12$, correct to 3 decimal places. [−1.386, 1.491]

5. $3\ln x + 4x = 5$, correct to 3 decimal places. [1.147]

6. $x^3 = 5\cos 2x$, correct to 3 significant figures. [−1.693, −0.846, 0.744]

7. $300e^{-2\theta} + \dfrac{\theta}{2} = 6$, correct to 3 significant figures. [2.05]

8. A Fourier analysis of the instantaneous value of a waveform can be represented by:
$$y = \left(t + \frac{\pi}{4}\right) + \sin t + \frac{1}{8}\sin 3t$$
Use Newton's method to determine the value of t near to 0.04, correct to 4 decimal places, when the amplitude, y, is 0.880 [0.0399]

9. A damped oscillation of a system is given by the equation: $y = -7.4e^{0.5t}\sin 3t$. Determine the value of t near to 4.2, correct to 3 significant figure, when the magnitude y of the oscillation is zero. [4.19]

Revision Test 4

This Revision test covers the material contained in Chapters 14 to 17. *The marks for each question are shown in brackets at the end of each question.*

1. Evaluate the following, each correct to 4 significant figures:

   (a) $e^{-0.683}$ (b) $\dfrac{5(e^{-2.73} - 1)}{e^{1.68}}$ (3)

2. Expand xe^{3x} to six terms (5)

3. Plot a graph of $y = \frac{1}{2}e^{-1.2x}$ over the range $x = -2$ to $x = +1$ and hence determine, correct to 1 decimal place, (a) the value of y when $x = -0.75$, and (b) the value of x when $y = 4.0$. (6)

4. Evaluate the following, each correct to 3 decimal places:

   (a) $\ln 0.0753$ (b) $\dfrac{\ln 3.68 - \ln 2.91}{4.63}$ (2)

5. Two quantities x and y are related by the equation $y = ae^{-kx}$, where a and k are constants. Determine, correct to 1 decimal place, the value of y when $a = 2.114$, $k = -3.20$ and $x = 1.429$ (3)

6. Determine the 20th term of the series 15.6, 15, 14.4, 13.8, ... (3)

7. The sum of 13 terms of an arithmetic progression is 286 and the common difference is 3. Determine the first term of the series. (4)

8. Determine the 11th term of the series 1.5, 3, 6, 12, ... (2)

9. A machine is to have seven speeds ranging from 25 rev/min to 500 rev/min. If the speeds form a geometric progression, determine their value, each correct to the nearest whole number. (8)

10. Use the binomial series to expand $(2a - 3b)^6$ (7)

11. Expand the following in ascending powers of t as far as the term in t^3

    (a) $\dfrac{1}{1+t}$ (b) $\dfrac{1}{\sqrt{1 - 3t}}$

    For each case, state the limits for which the expansion is valid. (10)

12. The modulus of rigidity G is given by $G = \dfrac{R^4\theta}{L}$ where R is the radius, θ the angle of twist and L the length. Find the approximate percentage error in G when R is measured 1.5% too large, θ is measure 3% too small and L is measured 1% too small. (6)

13. The solution to a differential equation associated with the path taken by a projectile for which the resistance to motion is proportional to the velocity is given by: $y = 2.5(e^x - e^{-x}) + x - 25$

    Use Newton's method to determine the value of x, correct to 2 decimal places, for which the value of y is zero. (11)

All questions have only one correct answer (answers on page 570).

1. The relationship between the temperature in degrees Fahrenheit (F) and the temperature in degrees Celsius (C) is given by: $F = \frac{9}{5}C + 32$. 135°F is equivalent to:

   (a) 43°C (b) 57.2°C
   (c) 185.4°C (d) 184°C

2. Transposing $I = \frac{V}{R}$ for resistance R gives:

   (a) $I - V$ (b) $\frac{V}{I}$ (c) $\frac{I}{V}$ (d) VI

3. 11 mm expressed as a percentage of 41 mm is:

   (a) 2.68, correct to 3 significant figures

   (b) 2.6, correct to 2 significant figures

   (c) 26.83, correct to 2 decimal places

   (d) 0.2682, correct to 4 decimal places

4. When two resistors R_1 and R_2 are connected in parallel the formula $\frac{1}{R_T} = \frac{1}{R_1} + \frac{1}{R_2}$ is used to determine the total resistance R_T. If $R_1 = 470\,\Omega$ and $R_2 = 2.7\,\text{k}\Omega$, R_T (correct to 3 significant figures) is equal to:

   (a) 2.68 Ω (b) 400 Ω
   (c) 473 Ω (d) 3170 Ω

5. $1\frac{1}{3} + 1\frac{2}{3} \div 2\frac{2}{3} - \frac{1}{3}$ is equal to:

   (a) $1\frac{5}{8}$ (b) $\frac{19}{24}$ (c) $2\frac{1}{21}$ (d) $1\frac{2}{7}$

6. Transposing $v = f\lambda$ to make wavelength λ the subject gives:

   (a) $\frac{v}{f}$ (b) $v + f$ (c) $f - v$ (d) $\frac{f}{v}$

7. The value of $\frac{2^{-3}}{2^{-4}} - 1$ is equal to:

   (a) 1 (b) 2 (c) $-\frac{1}{2}$ (d) $\frac{1}{2}$

8. Four engineers can complete a task in 5 hours. Assuming the rate of work remains constant, six engineers will complete the task in:

   (a) 126 h (b) 4 h 48 min
   (c) 3 h 20 min (d) 7 h 30 min

9. In an engineering equation $\frac{3^4}{3^r} = \frac{1}{9}$. The value of r is:

   (a) −6 (b) 2 (c) 6 (d) −2

10. Transposing the formula $R = R_0(1 + \alpha t)$ for t gives:

    (a) $\frac{R - R_0}{(1 + \alpha)}$ (b) $\frac{R - R_0 - 1}{\alpha}$

    (c) $\frac{R - R_0}{\alpha R_0}$ (d) $\frac{R}{R_0 \alpha}$

11. $2x^2 - (x - xy) - x(2y - x)$ simplifies to:

    (a) $x(3x - 1 - y)$ (b) $x^2 - 3xy - xy$
    (c) $x(xy - y - 1)$ (d) $3x^2 - x + xy$

12. The current I in an a.c. circuit is given by:

    $$I = \frac{V}{\sqrt{R^2 + X^2}}$$

    When $R = 4.8$, $X = 10.5$ and $I = 15$, the value of voltage V is:

    (a) 173.18 (b) 1.30
    (c) 0.98 (d) 229.50

13. The height s of a mass projected vertically upwards at time t is given by: $s = ut - \frac{1}{2}gt^2$. When $g = 10$, $t = 1.5$ and $s = 3.75$, the value of u is:

    (a) 10 (b) −5 (c) +5 (d) −10

14. The quantity of heat Q is given by the formula $Q = mc(t_2 - t_1)$. When $m = 5$, $t_1 = 20$, $c = 8$ and $Q = 1200$, the value of t_2 is:

    (a) 10 (b) 1.5 (c) 21.5 (d) 50

15. When $p = 3$, $q = -\frac{1}{2}$ and $r = -2$, the engineering expression $2p^2q^3r^4$ is equal to:

    (a) −36 (b) 1296 (c) 36 (d) 18

16. Electrical resistance $R = \frac{\rho l}{a}$; transposing this equation for l gives:

    (a) $\frac{Ra}{\rho}$ (b) $\frac{R}{a\rho}$ (c) $\frac{a}{R\rho}$ (d) $\frac{\rho a}{R}$

17. $\frac{3}{4} \div 1\frac{3}{4}$ is equal to:

    (a) $\frac{3}{7}$ (b) $1\frac{9}{16}$ (c) $1\frac{5}{16}$ (d) $2\frac{1}{2}$

18. $(2e - 3f)(e + f)$ is equal to:

    (a) $2e^2 - 3f^2$ (b) $2e^2 - 5ef - 3f^2$
    (c) $2e^2 + 3f^2$ (d) $2e^2 - ef - 3f^2$

19. The solution of the simultaneous equations $3x - 2y = 13$ and $2x + 5y = -4$ is:

    (a) $x = -2, y = 3$ (b) $x = 1, y = -5$
    (c) $x = 3, y = -2$ (d) $x = -7, y = 2$

20. $16^{-3/4}$ is equal to:

    (a) 8 (b) $-\dfrac{1}{2^3}$ (c) 4 (d) $\dfrac{1}{8}$

21. A formula for the focal length f of a convex lens is $\dfrac{1}{f} = \dfrac{1}{u} + \dfrac{1}{v}$. When $f = 4$ and $u = 6$, v is:

    (a) -2 (b) $\frac{1}{12}$ (c) 12 (d) $-\frac{1}{2}$

22. If $x = \dfrac{57.06 \times 0.0711}{\sqrt{0.0635}}$ cm, which of the following statements is correct?

    (a) $x = 16$ cm, correct to 2 significant figures
    (b) $x = 16.09$ cm, correct to 4 significant figures
    (c) $x = 1.61 \times 10^1$ cm, correct to 3 decimal places
    (d) $x = 16.099$ cm, correct to 3 decimal places

23. Volume $= \dfrac{\text{mass}}{\text{density}}$. The density (in kg/m^3) when the mass is 2.532 kg and the volume is 162 cm^3 is:

    (a) 0.01563 kg/m^3 (b) 410.2 kg/m^3
    (c) $15\,630$ kg/m^3 (d) 64.0 kg/m^3

24. $(5.5 \times 10^2)(2 \times 10^3)$ cm in standard form is equal to:

    (a) 11×10^6 cm   (b) 1.1×10^6 cm
    (c) 11×10^5 cm   (d) 1.1×10^5 cm

25. $PV = mRT$ is the characteristic gas equation. When $P = 100 \times 10^3$, $V = 4.0$, $R = 288$ and $T = 300$, the value of m is:

    (a) 4.630 (b) $313\,600$
    (c) 0.216 (d) $100\,592$

26. $\log_{16} 8$ is equal to:

    (a) $\frac{1}{2}$ (b) 144 (c) $\frac{3}{4}$ (d) 2

27. The quadratic equation in x whose roots are -2 and $+5$ is:

    (a) $x^2 - 3x - 10 = 0$ (b) $x^2 + 7x + 10 = 0$
    (c) $x^2 + 3x - 10 = 0$ (d) $x^2 - 7x - 10 = 0$

28. The area A of a triangular piece of land of sides a, b and c may be calculated using $A = \sqrt{s(s-a)(s-b)(s-c)}$ where $s = \dfrac{a+b+c}{2}$.

    When $a = 15$ m, $b = 11$ m and $c = 8$ m, the area, correct to the nearest square metre, is:

    (a) 1836 m^2 (b) 648 m^2
    (c) 445 m^2 (d) 43 m^2

29. The engineering expression $\dfrac{(16 \times 4)^2}{(8 \times 2)^4}$ is equal to:

    (a) 4 (b) 2^{-4} (c) $\dfrac{1}{2^2}$ (d) 1

30. In a system of pulleys, the effort P required to raise a load W is given by $P = aW + b$, where a and b are constants. If $W = 40$ when $P = 12$ and $W = 90$ when $P = 22$, the values of a and b are:

    (a) $a = 5, b = \frac{1}{4}$ (b) $a = 1, b = -28$
    (c) $a = \frac{1}{3}, b = -8$ (d) $a = \frac{1}{5}, b = 4$

31. $(16^{-\frac{1}{4}} - 27^{-\frac{2}{3}})$ is equal to:

    (a) $\frac{7}{18}$ (b) -7 (c) $1\frac{8}{9}$ (d) $-8\frac{1}{2}$

32. Resistance R ohms varies with temperature t according to the formula $R = R_0(1 + \alpha t)$. Given $R = 21\,\Omega$, $\alpha = 0.004$ and $t = 100$, R_0 has a value of:

    (a) $21.4\,\Omega$ (b) $29.4\,\Omega$
    (c) $15\,\Omega$ (d) $0.067\,\Omega$

33. $(p + x)^4 = p^4 + 4p^3x + 6p^2x^2 + 4px^3 + x^4$. Using Pascal's triangle, the third term of $(p + x)^5$ is:

    (a) $10p^2x^3$ (b) $5p^4x$
    (c) $5p^3x^2$ (d) $10p^3x^2$

34. The value of $\frac{2}{5}$ of $(4\frac{1}{2} - 3\frac{1}{4}) + 5 \div \frac{5}{16} - \frac{1}{4}$ is:

    (a) $17\frac{7}{20}$ (b) $80\frac{1}{2}$ (c) $16\frac{1}{4}$ (d) 88

35. $\log_2 \frac{1}{8}$ is equal to:

    (a) -3 (b) $\frac{1}{4}$ (c) 3 (d) 16

36. The value of $\dfrac{\ln 2}{e^2 \lg 2}$, correct to 3 significant figures, is:

    (a) 0.0588

    (b) 0.312

    (c) 17.0

    (d) 3.209

37. $8x^2 + 13x - 6 = (x + p)(qx - 3)$. The values of p and q are:

    (a) $p = -2, q = 4$

    (b) $p = 3, q = 2$

    (c) $p = 2, q = 8$

    (d) $p = 1, q = 8$

38. If $\log_2 x = 3$ then:

    (a) $x = 8$

    (b) $x = \frac{3}{2}$

    (c) $x = 9$

    (d) $x = \frac{2}{3}$

39. The pressure p Pascals at height h metres above ground level is given by $p = p_0 e^{-h/k}$, where p_0 is the pressure at ground level and k is a constant. When p_0 is 1.01×10^5 Pa and the pressure at a height of 1500 m is 9.90×10^4 Pa, the value of k, correct to 3 significant figures is:

    (a) 1.33×10^{-5}

    (b) 75 000

    (c) 173 000

    (d) 197

40. The fifth term of an arithmetic progression is 18 and the twelfth term is 46.

    The eighteenth term is:

    (a) 72   (b) 74   (c) 68   (d) 70

41. The height S metres of a mass thrown vertically upwards at time t seconds is given by $S = 80t - 16t^2$. To reach a height of 50 metres on the descent will take the mass:

    (a) 0.73 s

    (b) 5.56 s

    (c) 4.27 s

    (d) 81.77 s

42. $(2x - y)^2$ is equal to:

    (a) $4x^2 + y^2$

    (b) $2x^2 - 2xy + y^2$

    (c) $4x^2 - y^2$

    (d) $4x^2 - 4xy + y^2$

43. The final length l_2 of a piece of wire heated through $\theta°$C is given by the formula $l_2 = l_1(1 + \alpha\theta)$. Transposing, the coefficient of expansion α is given by:

    (a) $\dfrac{l_2}{l_1} - \dfrac{1}{\theta}$

    (b) $\dfrac{l_2 - l_1}{l_1 \theta}$

    (c) $l_2 - l_1 - l_1 \theta$

    (d) $\dfrac{l_1 - l_2}{l_1 \theta}$

44. The roots of the quadratic equation $8x^2 + 10x - 3 = 0$ are:

    (a) $-\frac{1}{4}$ and $\frac{3}{2}$

    (b) 4 and $\frac{2}{3}$

    (c) $-\frac{3}{2}$ and $\frac{1}{4}$

    (d) $\frac{2}{3}$ and -4

45. The current i amperes flowing in a capacitor at time t seconds is given by $i = 10(1 - e^{-t/CR})$, where resistance R is 25×10^3 ohms and capacitance C is 16×10^{-6} farads. When current i reaches 7 amperes, the time t is:

    (a) -0.48 s

    (b) 0.14 s

    (c) 0.21 s

    (d) 0.48 s

46. The value of $\dfrac{3.67 \ln 21.28}{e^{-0.189}}$, correct to 4 significant figures, is:

    (a) 9.289

    (b) 13.56

    (c) 13.5566

    (d) -3.844×10^9

47. The volume V_2 of a material when the temperature is increased is given by $V_2 = V_1[1 + \gamma(t_2 - t_1)]$. The value of t_2 when $V_2 = 61.5\,\text{cm}^3$, $V_1 = 60\,\text{cm}^3$, $\gamma = 54 \times 10^{-6}$ and $t_1 = 250$ is:

    (a) 213

    (b) 463

    (c) 713

    (d) 28 028

48. A formula used for calculating the resistance of a cable is $R = \dfrac{\rho l}{a}$. A cable's resistance $R = 0.50\,\Omega$, its length l is 5000 m and its cross-sectional area a is $4 \times 10^{-4}\,\text{m}^2$. The resistivity ρ of the material is:

    (a) $6.25 \times 10^7\,\Omega\text{m}$

    (b) $4 \times 10^{-8}\,\Omega\text{m}$

    (c) $2.5 \times 10^7\,\Omega\text{m}$

    (d) $3.2 \times 10^{-7}\,\Omega\text{m}$

49. In the equation $5.0 = 3.0 \ln\left(\dfrac{2.9}{x}\right)$, x has a value correct to 3 significant figures of:

    (a) 1.59

    (b) 0.392

    (c) 0.548

    (d) 0.0625

50. Current I in an electrical circuit is given by $I = \dfrac{E - e}{R + r}$. Transposing for R gives:

    (a) $\dfrac{E - e - Ir}{I}$

    (b) $\dfrac{E - e}{I + r}$

    (c) $(E - e)(I + r)$

    (d) $\dfrac{E - e}{Ir}$

51. $(\sqrt{x})(y^{3/2})(x^2 y)$ is equal to:

    (a) $\sqrt{(xy)^5}$

    (b) $x^{\sqrt{2}} y^{5/2}$

    (c) $xy^{5/2}$

    (d) $x\sqrt{y^3}$

52. The roots of the quadratic equation $2x^2 - 5x + 1 = 0$, correct to 2 decimal places, are:

    (a) -0.22 and -2.28

    (b) 2.69 and -0.19

    (c) 0.19 and -2.69

    (d) 2.28 and 0.22

53. Transposing $t = 2\pi \sqrt{\dfrac{l}{g}}$ for g gives:

    (a) $\dfrac{(t - 2\pi)^2}{l}$

    (b) $\left(\dfrac{2\pi}{t}\right) l^2$

    (c) $\dfrac{\sqrt{\dfrac{t}{2\pi}}}{l}$

    (d) $\dfrac{4\pi^2 l}{t^2}$

54. $\log_3 9$ is equal to:

    (a) 3   (b) 27   (c) $\frac{1}{3}$ (d) 2

55. The second moment of area of a rectangle through its centroid is given by $\dfrac{bl^3}{12}$.

    Using the binomial theorem, the approximate percentage change in the second moment of area if b is increased by 3% and l is reduced by 2% is:

    (a) -6% (b) $+1\%$ (c) $+3\%$ (d) -3%

56. The equation $x^4 - 3x^2 - 3x + 1 = 0$ has:

    (a) 1 real root   (b) 2 real roots
    (c) 3 real roots   (d) 4 real roots

57. The motion of a particle in an electrostatic field is described by the equation $y = x^3 + 3x^2 + 5x - 28$. When $x = 2$, y is approximately zero. Using one iteration of the Newton–Raphson method, a better approximation (correct to 2 decimal places) is:

    (a) 1.89   (b) 2.07   (c) 2.11   (d) 1.93

58. In hexadecimal, the decimal number 123 is:

    (a) 1111011   (b) 123
    (c) 173   (d) 7B

59. $6x^2 - 5x - 6$ divided by $2x - 3$ gives:

    (a) $2x - 1$ (b) $3x + 2$
    (c) $3x - 2$ (d) $6x + 1$

60. The first term of a geometric progression is 9 and the fourth term is 45. The eighth term is:

    (a) 225   (b) 150.5
    (c) 384.7   (d) 657.9

61. The solution of the inequality $\dfrac{3t + 2}{t + 1} \le 1$ is:

    (a) $t \ge -2\frac{1}{2}$ (b) $-1 < t \le \frac{1}{2}$
    (c) $t < -1$ (d) $-\frac{1}{2} < t \le 1$

62. The solution of the inequality $x^2 - x - 2 < 0$ is:

    (a) $1 < x < -2$ (b) $x > 2$
    (c) $-1 < x < 2$ (d) $x < -1$

Section 2

Mensuration

Chapter 18

Areas of plane figures

18.1 Mensuration

Mensuration is a branch of mathematics concerned with the determination of lengths, areas and volumes.

18.2 Properties of quadrilaterals

Polygon

A **polygon** is a closed plane figure bounded by straight lines. A polygon, which has:

(i)   3 sides is called a **triangle**
(ii) 4 sides is called a **quadrilateral**
(iii) 5 sides is called a **pentagon**
(iv) 6 sides is called a **hexagon**
(v)   7 sides is called a **heptagon**
(vi) 8 sides is called an **octagon**

There are five types of **quadrilateral**, these being:

(i)   rectangle
(ii) square
(iii) parallelogram
(iv) rhombus
(v)   trapezium

(The properties of these are given below).

If the opposite corners of any quadrilateral are joined by a straight line, two triangles are produced. Since the sum of the angles of a triangle is 180°, the sum of the angles of a quadrilateral is 360°.

In a **rectangle**, shown in Fig. 18.1:

(i)   all four angles are right angles,
(ii) opposite sides are parallel and equal in length, and
(iii) diagonals AC and BD are equal in length and bisect one another.

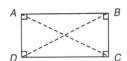

Figure 18.1

In a **square**, shown in Fig. 18.2:

(i)   all four angles are right angles,
(ii) opposite sides are parallel,
(iii) all four sides are equal in length, and
(iv) diagonals PR and QS are equal in length and bisect one another at right angles.

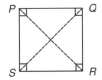

Figure 18.2

In a **parallelogram**, shown in Fig. 18.3:

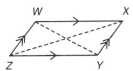

Figure 18.3

(i)   opposite angles are equal,
(ii) opposite sides are parallel and equal in length, and
(iii) diagonals WY and XZ bisect one another.

In a **rhombus**, shown in Fig. 18.4:

(i)   opposite angles are equal,
(ii) opposite angles are bisected by a diagonal,
(iii) opposite sides are parallel,

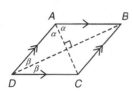

Figure 18.4

(iv) all four sides are equal in length, and
(v) diagonals AC and BD bisect one another at right angles.

In a **trapezium**, shown in Fig. 18.5:

(i) only one pair of sides is parallel

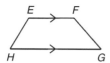

Figure 18.5

18.3 Worked problems on areas of plane figures

Table 18.1 summarises the areas of common plane figures.

Table 18.1

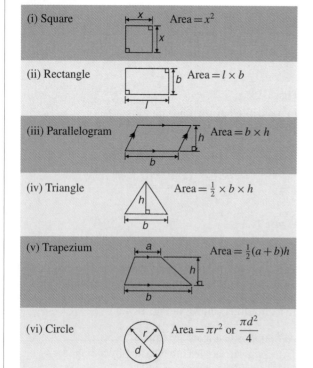

(i) Square	Area $= x^2$
(ii) Rectangle	Area $= l \times b$
(iii) Parallelogram	Area $= b \times h$
(iv) Triangle	Area $= \frac{1}{2} \times b \times h$
(v) Trapezium	Area $= \frac{1}{2}(a+b)h$
(vi) Circle	Area $= \pi r^2$ or $\dfrac{\pi d^2}{4}$

Table 18.1 *(Continued)*

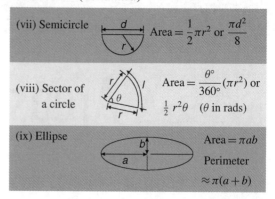

(vii) Semicircle	Area $= \frac{1}{2}\pi r^2$ or $\dfrac{\pi d^2}{8}$
(viii) Sector of a circle	Area $= \dfrac{\theta°}{360°}(\pi r^2)$ or $\frac{1}{2} r^2\theta$ (θ in rads)
(ix) Ellipse	Area $= \pi ab$ Perimeter $\approx \pi(a+b)$

Problem 1. State the types of quadrilateral shown in Fig. 18.6 and determine the angles marked a to l

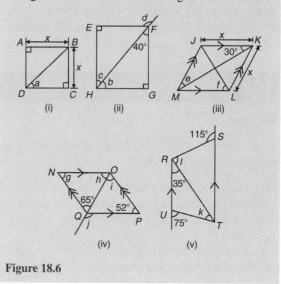

Figure 18.6

(iii) **ABCD is a square**
The diagonals of a square bisect each of the right angles, hence

$$a = \frac{90°}{2} = 45°$$

(ii) **EFGH is a rectangle**
In triangle FGH, $40° + 90° + b = 180°$ (angles in a triangle add up to $180°$) from which, $b = 50°$. Also $c = 40°$ (alternate angles between parallel lines EF and HG).
(Alternatively, b and c are complementary, i.e. add up to $90°$)
$d = 90° + c$ (external angle of a triangle equals the sum of the interior opposite angles), hence

$$d = 90° + 40° = 130°$$

(iii) *JKLM* is a rhombus

The diagonals of a rhombus bisect the interior angles and opposite internal angles are equal. Thus $\angle JKM = \angle MKL = \angle JMK = \angle LMK = 30°$, hence, $e = 30°$

In triangle *KLM*, $30° + \angle KLM + 30° = 180°$ (angles in a triangle add up to 180°), hence $\angle KLM = 120°$.

The diagonal *JL* bisects $\angle KLM$, hence

$$f = \frac{120°}{2} = 60°$$

(iv) *NOPQ* is a parallelogram

$g = 52°$ (since opposite interior angles of a parallelogram are equal).

In triangle *NOQ*, $g + h + 65° = 180°$ (angles in a triangle add up to 180°), from which,

$$h = 180° - 65° - 52° = 63°$$

$i = 65°$ (alternate angles between parallel lines *NQ* and *OP*).

$j = 52° + i = 52° + 65° = 117°$ (external angle of a triangle equals the sum of the interior opposite angles).

(v) *RSTU* is a trapezium

$35° + k = 75°$ (external angle of a triangle equals the sum of the interior opposite angles), hence $k = 40°$

$\angle STR = 35°$ (alternate angles between parallel lines *RU* and *ST*).

$l + 35° = 115°$ (external angle of a triangle equals the sum of the interior opposite angles), hence

$$l = 115° - 35° = 80°$$

Problem 2. A rectangular tray is 820 mm long and 400 mm wide. Find its area in (a) mm², (b) cm², (c) m²

(a) Area = length × width = $820 × 400$
$$= 328\,000\,\text{mm}^2$$

(b) $1\,\text{cm}^2 = 100\,\text{mm}^2$. Hence

$$328\,000\,\text{mm}^2 = \frac{328\,000}{100}\,\text{cm}^2 = 3280\,\text{cm}^2$$

(c) $1\,\text{m}^2 = 10\,000\,\text{cm}^2$. Hence

$$3280\,\text{cm}^2 = \frac{3280}{10\,000}\,\text{m}^2 = 0.3280\,\text{m}^2$$

Problem 3. Find (a) the cross-sectional area of the girder shown in Fig. 18.7(a) and (b) the area of the path shown in Fig. 18.7(b)

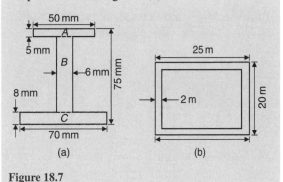

Figure 18.7

(a) The girder may be divided into three separate rectangles as shown.

Area of rectangle $A = 50 × 5 = 250\,\text{mm}^2$

Area of rectangle $B = (75 - 8 - 5) × 6$
$$= 62 × 6 = 372\,\text{mm}^2$$

Area of rectangle $C = 70 × 8 = 560\,\text{mm}^2$

Total area of girder $= 250 + 372 + 560 = 1182\,\text{mm}^2$ or $11.82\,\text{cm}^2$

(b) Area of path = area of large rectangle − area of small rectangle
$$= (25 × 20) - (21 × 16) = 500 - 336$$
$$= 164\,\text{m}^2$$

Problem 4. Find the area of the parallelogram shown in Fig. 18.8 (dimensions are in mm)

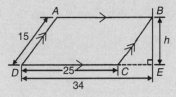

Figure 18.8

Area of parallelogram = base × perpendicular height. The perpendicular height h is found using Pythagoras' theorem.

$$BC^2 = CE^2 + h^2$$

i.e. $15^2 = (34 - 25)^2 + h^2$

$$h^2 = 15^2 - 9^2 = 225 - 81 = 144$$

Hence, $h = \sqrt{144} = 12$ mm (-12 can be neglected).

Hence, area of $ABCD = 25 \times 12 = \textbf{300 mm}^2$

Problem 5. Figure 18.9 shows the gable end of a building. Determine the area of brickwork in the gable end

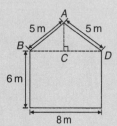

Figure 18.9

The shape is that of a rectangle and a triangle.

Area of rectangle $= 6 \times 8 = 48$ m^2

Area of triangle $= \frac{1}{2} \times$ base $\times$ height.

$CD = 4$ m, $AD = 5$ m, hence $AC = 3$ m (since it is a 3, 4, 5 triangle).

Hence, area of triangle $ABD = \frac{1}{2} \times 8 \times 3 = 12$ m^2

Total area of brickwork $= 48 + 12 = \textbf{60 m}^2$

Problem 6. Determine the area of the shape shown in Fig. 18.10

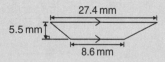

Figure 18.10

The shape shown is a trapezium.

Area of trapezium

$$= \frac{1}{2} \text{ (sum of parallel sides)(perpendicular distance between them)}$$

$$= \frac{1}{2}(27.4 + 8.6)(5.5)$$

$$= \frac{1}{2} \times 36 \times 5.5 = \textbf{99 mm}^2$$

Now try the following exercise

Exercise 67   Further problems on areas of plane figures

1. A rectangular plate is 85 mm long and 42 mm wide. Find its area in square centimetres.
    [35.7 cm^2]

2. A rectangular field has an area of 1.2 hectares and a length of 150 m. Find (a) its width and (b) the length of a diagonal (1 hectare $= 10\,000$ m^2).
    [(a) 80 m   (b) 170 m]

3. Determine the area of each of the angle iron sections shown in Fig. 18.11.
    [(a) 29 cm^2 (b) 650 mm^2]

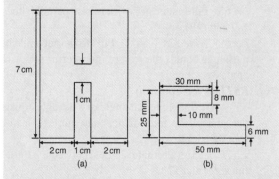

Figure 18.11

4. A rectangular garden measures 40 m by 15 m. A 1 m flower border is made round the two shorter sides and one long side. A circular swimming pool of diameter 8 m is constructed in the middle of the garden. Find, correct to the nearest square metre, the area remaining.
    [482 m^2]

5. The area of a trapezium is 13.5 cm^2 and the perpendicular distance between its parallel sides is 3 cm. If the length of one of the parallel sides is 5.6 cm, find the length of the other parallel side.
    [3.4 cm]

6. Find the angles p, q, r, s and t in Fig. 18.12(a) to (c).

$$\begin{bmatrix} p = 105°, \ q = 35°, \ r = 142°, \\ s = 95°, \ t = 146° \end{bmatrix}$$

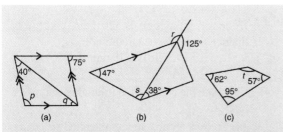

Figure 18.12

7. Name the types of quadrilateral shown in Fig. 18.13(i) to (iv), and determine (a) the area, and (b) the perimeter of each.

$$\begin{bmatrix} \text{(i) rhombus (a) } 14\,\text{cm}^2 \text{ (b) } 16\,\text{cm} \\ \text{(ii) parallelogram (a) } 180\,\text{cm}^2 \text{ (b) } 80\,\text{mm} \\ \text{(iii) rectangle (a) } 3600\,\text{mm}^2 \text{ (b) } 300\,\text{mm} \\ \text{(iv) trapezium (a) } 190\,\text{cm}^2 \text{ (b) } 62.91\,\text{cm} \end{bmatrix}$$

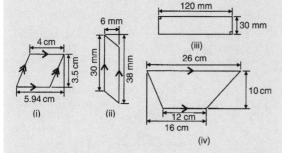

Figure 18.13

8. Calculate the area of the steel plate shown in Fig. 18.14   [6750 mm²]

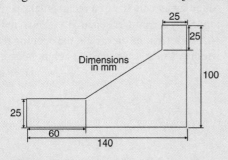

Figure 18.14

18.4 Further worked problems on areas of plane figures

Problem 7. Find the areas of the circles having (a) a radius of 5 cm, (b) a diameter of 15 mm, (c) a circumference of 70 mm

Area of a circle $= \pi r^2$ or $\dfrac{\pi d^2}{4}$

(a) Area $= \pi r^2 = \pi(5)^2 = 25\pi = \mathbf{78.54\,cm^2}$

(b) Area $= \dfrac{\pi d^2}{4} = \dfrac{\pi(15)^2}{4} = \dfrac{225\pi}{4} = \mathbf{176.7\,mm^2}$

(c) Circumference, $c = 2\pi r$, hence

$$r = \frac{c}{2\pi} = \frac{70}{2\pi} = \frac{35}{\pi}\,\text{mm}$$

$$\text{Area of circle} = \pi r^2 = \pi\left(\frac{35}{\pi}\right)^2 = \frac{35^2}{\pi}$$

$$= \mathbf{389.9\,mm^2} \text{ or } \mathbf{3.899\,cm^2}$$

Problem 8. Calculate the areas of the following sectors of circles having:

(a) radius 6 cm with angle subtended at centre 50°
(b) diameter 80 mm with angle subtended at centre 107°42′
(c) radius 8 cm with angle subtended at centre 1.15 radians

Area of sector of a circle $= \dfrac{\theta^2}{360}(\pi r^2)$

$$\text{or } \frac{1}{2}r^2\theta \quad (\theta \text{ in radians}).$$

(a) Area of sector

$$= \frac{50}{360}(\pi 6^2) = \frac{50 \times \pi \times 36}{360} = 5\pi$$

$$= \mathbf{15.71\,cm^2}$$

(b) If diameter $= 80$ mm, then radius, $r = 40$ mm, and area of sector

$$= \frac{107°42'}{360}(\pi 40^2) = \frac{107\frac{42}{60}}{360}(\pi 40^2)$$

$$= \frac{107.7}{360}(\pi 40^2) = \mathbf{1504\,mm^2} \quad \text{or} \quad \mathbf{15.04\,cm^2}$$

(c) Area of sector $= \frac{1}{2}r^2\theta = \frac{1}{2} \times 8^2 \times 1.15$
$$= \mathbf{36.8\,cm^2}$$

Problem 9. A hollow shaft has an outside diameter of 5.45 cm and an inside diameter of 2.25 cm. Calculate the cross-sectional area of the shaft

The cross-sectional area of the shaft is shown by the shaded part in Fig. 18.15 (often called an **annulus**).

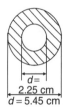

Figure 18.15

Area of shaded part = area of large circle − area of small circle

$$= \frac{\pi D^2}{4} - \frac{\pi d^2}{4} = \frac{\pi}{4}(D^2 - d^2)$$
$$= \frac{\pi}{4}(5.45^2 - 2.25^2) = \mathbf{19.35 \ cm^2}$$

Problem 10. The major axis of an ellipse is 15.0 cm and the minor axis is 9.0 cm. Find its area and approximate perimeter

If the major axis = 15.0 cm, then the semi-major axis = 7.5 cm.
If the minor axis = 9.0 cm, then the semi-minor axis = 4.5 cm.
Hence, from Table 18.1(ix),

$$\textbf{area} = \pi ab = \pi(7.5)(4.5) = \mathbf{106.0 \ cm^2}$$

and $\textbf{perimeter} \approx \pi(a+b) = \pi(7.5+4.5)$

$$= 12.0\pi = \mathbf{37.7 \ cm}$$

Now try the following exercise

Exercise 68   Further problems on areas of plane figures

1. Determine the area of circles having a (a) radius of 4 cm (b) diameter of 30 mm (c) circumference of 200 mm.

$$\left[\begin{array}{ll} \text{(a) } 50.27 \ cm^2 & \text{(b) } 706.9 \ mm^2 \\ \text{(c) } 3183 \ mm^2 \end{array} \right]$$

2. An annulus has an outside diameter of 60 mm and an inside diameter of 20 mm. Determine its area. [2513 mm²]

3. If the area of a circle is 320 mm², find (a) its diameter, and (b) its circumference.
   [(a) 20.19 mm (b) 63.41 mm]

4. Calculate the areas of the following sectors of circles:
   (a) radius 9 cm, angle subtended at centre 75°
   (b) diameter 35 mm, angle subtended at centre 48°37′
   (c) diameter 5 cm, angle subtended at centre 2.19 radians

$$\left[\begin{array}{ll} \text{(a) } 53.01 \ cm^2 & \text{(b) } 129.9 \ mm^2 \\ \text{(c) } 6.84 \ cm^2 \end{array} \right]$$

5. Determine the area of the shaded template shown in Fig. 18.16   [5773 mm²]

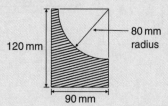

Figure 18.16

6. An archway consists of a rectangular opening topped by a semi-circular arch as shown in Fig. 18.17. Determine the area of the opening if the width is 1 m and the greatest height is 2 m.
   [1.89 m²]

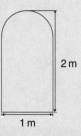

Figure 18.17

7. The major axis of an ellipse is 200 mm and the minor axis 100 mm. Determine the area and approximate perimeter of the ellipse.
   [15 710 mm², 471 mm]

8. If fencing costs £15 per metre, find the cost (to the nearest pound) of enclosing an elliptical plot of land which has major and minor diameter lengths of 120 m and 80 m.

[£4712]

9. A cycling track is in the form of an ellipse, the axes being 250 m and 150 m respectively for the inner boundary, and 270 m and 170 m for the outer boundary. Calculate the area of the track.

[6597 m²]

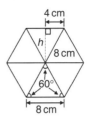

Figure 18.19

are equal to each other. Hence each of the triangles is equilateral with each angle 60° and each side 8 cm.

$$\text{Area of one triangle} = \frac{1}{2} \times \text{base} \times \text{height}$$

$$= \frac{1}{2} \times 8 \times h$$

h is calculated using Pythagoras' theorem:

$$8^2 = h^2 + 4^2$$

from which, $h = \sqrt{8^2 - 4^2} = 6.928$ cm

Hence area of one triangle

$$= \frac{1}{2} \times 8 \times 6.928 = 27.71 \text{ cm}^2$$

Area of hexagon $= 6 \times 27.71 = \textbf{166.3 cm}^2$

18.5 Worked problems on areas of composite figures

Problem 11. Calculate the area of a regular octagon, if each side is 5 cm and the width across the flats is 12 cm

An octagon is an 8-sided polygon. If radii are drawn from the centre of the polygon to the vertices then 8 equal triangles are produced (see Fig. 18.18).

$$\text{Area of one triangle} = \frac{1}{2} \times \text{base} \times \text{height}$$

$$= \frac{1}{2} \times 5 \times \frac{12}{2} = 15 \text{ cm}^2$$

$$\text{Area of octagon} = 8 \times 15 = \textbf{120 cm}^2$$

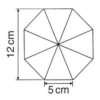

Figure 18.18

Problem 12. Determine the area of a regular hexagon that has sides 8 cm long

A hexagon is a 6-sided polygon which may be divided into 6 equal triangles as shown in Fig. 18.19. The angle subtended at the centre of each triangle is $360°/6 = 60°$. The other two angles in the triangle add up to 120° and

Problem 13. Figure 18.20 shows a plan of a floor of a building that is to be carpeted. Calculate the area of the floor in square metres. Calculate the cost, correct to the nearest pound, of carpeting the floor with carpet costing £16.80 per m², assuming 30% extra carpet is required due to wastage in fitting

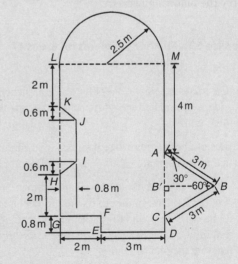

Figure 18.20

Area of floor plan

= area of triangle ABC + area of semicircle

+ area of rectangle $CGLM$

+ area of rectangle $CDEF$

− area of trapezium $HIJK$

Triangle ABC is equilateral since $AB = BC = 3$ m and hence angle $B'CB = 60°$
$\sin B'CB = BB'/3$, i.e. $BB' = 3\sin 60° = 2.598$ m

Area of triangle $ABC = \frac{1}{2}(AC)(BB')$

$\qquad = \frac{1}{2}(3)(2.598) = 3.897\,\text{m}^2$

Area of semicircle $= \frac{1}{2}\pi r^2 = \frac{1}{2}\pi(2.5)^2 = 9.817\,\text{m}^2$

Area of CGLM $= 5 \times 7 = 35\,\text{m}^2$

Area of CDEF $= 0.8 \times 3 = 2.4\,\text{m}^2$

Area of HIJK $= \frac{1}{2}(KH + IJ)(0.8)$

Since MC $= 7$ m then LG $= 7$ m, hence
JI $= 7 - 5.2 = 1.8$ m

Hence area of HIJK $= \frac{1}{2}(3 + 1.8)(0.8) = 1.92\,\text{m}^2$

Total floor area $= 3.897 + 9.817 + 35 + 2.4 - 1.92$
$\qquad\qquad = 49.194\,\text{m}^2$

To allow for 30% wastage, amount of carpet required $= 1.3 \times 49.194 = 63.95\,\text{m}^2$

Cost of carpet at £16.80 per m^2 $= 63.95 \times 16.80 =$ £1074, correct to the nearest pound.

Now try the following exercise

Exercise 69   Further problems on areas of plane figures

1. Calculate the area of a regular octagon if each side is 20 mm and the width across the flats is 48.3 mm.   [1932 mm²]

2. Determine the area of a regular hexagon which has sides 25 mm.   [1624 mm²]

3. A plot of land is in the shape shown in Fig. 18.21. Determine (a) its area in hectares (1 ha = 10^4 m²), and (b) the length of fencing required, to the nearest metre, to completely enclose the plot of land.
    [(a) 0.918 ha (b) 456 m]

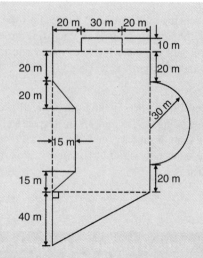

Figure 18.21

4. If paving slabs are produced in 250 mm × 250 mm squares, determine the number of slabs required to cover an area of 2 m²   [32]

18.6   Areas of similar shapes

The areas of similar shapes are proportional to the squares of corresponding linear dimensions.

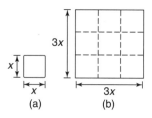

Figure 18.22

For example, Fig. 18.22 shows two squares, one of which has sides three times as long as the other.

$$\text{Area of Fig. 18.22(a)} = (x)(x) = x^2$$

$$\text{Area of Fig. 18.22(b)} = (3x)(3x) = 9x^2$$

Hence Fig. 18.22(b) has an area $(3)^2$, i.e. 9 times the area of Fig. 18.22(a).

Problem 14. A rectangular garage is shown on a building plan having dimensions 10 mm by 20 mm.

If the plan is drawn to a scale of 1 to 250, determine the true area of the garage in square metres

Area of garage on the plan $= 10\,\text{mm} \times 20\,\text{mm}$
$$= 200\,\text{mm}^2$$
Since the areas of similar shapes are proportional to the squares of corresponding dimensions then:

$$\text{true area of garage} = 200 \times (250)^2$$
$$= 12.5 \times 10^6\,\text{mm}^2$$
$$= \frac{12.5 \times 10^6}{10^6}\,\text{m}^2 = \mathbf{12.5\,m^2}$$

Now try the following exercise

Exercise 70   Further problems on areas of similar shapes

1. The area of a park on a map is $500\,\text{mm}^2$. If the scale of the map is 1 to 40 000 determine the true area of the park in hectares (1 hectare $= 10^4\,\text{m}^2$).   [80 ha]

2. A model of a boiler is made having an overall height of 75 mm corresponding to an overall height of the actual boiler of 6 m. If the area of metal required for the model is $12\,500\,\text{mm}^2$ determine, in square metres, the area of metal required for the actual boiler.   [$80\,\text{m}^2$]

3. The scale of an Ordnance Survey map is 1:2500. A circular sports field has a diameter of 8 cm on the map. Calculate its area in hectares, giving your answer correct to 3 significant figures. (1 hectare $= 10^4\,\text{m}^2$)   [3.14 ha]

Chapter 19

The circle and its properties

19.1 Introduction

A **circle** is a plain figure enclosed by a curved line, every point on which is equidistant from a point within, called the **centre**.

19.2 Properties of circles

(i) The distance from the centre to the curve is called the **radius**, r, of the circle (see OP in Fig. 19.1).

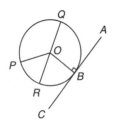

Figure 19.1

(ii) The boundary of a circle is called the **circumference**, c.

(iii) Any straight line passing through the centre and touching the circumference at each end is called the **diameter**, d (see QR in Fig. 19.1). Thus $d = 2r$

(iv) The ratio $\dfrac{\text{circumference}}{\text{diameter}} = $ a constant for any circle.

This constant is denoted by the Greek letter π (pronounced 'pie'), where $\pi = 3.14159$, correct to 5 decimal places.

Hence $c/d = \pi$ or $c = \pi d$ or $c = 2\pi r$

(v) A **semicircle** is one half of the whole circle.

(vi) A **quadrant** is one quarter of a whole circle.

(vii) A **tangent** to a circle is a straight line that meets the circle in one point only and does not cut the circle when produced. AC in Fig. 19.1 is a tangent to the circle since it touches the curve at point B only. If radius OB is drawn, then angle ABO is a right angle.

(viii) A **sector** of a circle is the part of a circle between radii (for example, the portion OXY of Fig. 19.2 is a sector). If a sector is less than a semicircle it is called a **minor sector**, if greater than a semicircle it is called a **major sector**.

Figure 19.2

(ix) A **chord** of a circle is any straight line that divides the circle into two parts and is terminated at each end by the circumference. ST, in Fig. 19.2 is a chord.

(x) A **segment** is the name given to the parts into which a circle is divided by a chord. If the segment is less than a semicircle it is called a **minor segment** (see shaded area in Fig. 19.2). If the segment is greater than a semicircle it is called a **major segment** (see the unshaded area in Fig. 19.2).

(xi) An **arc** is a portion of the circumference of a circle. The distance SRT in Fig. 19.2 is called a **minor arc** and the distance $SXYT$ is called a **major arc**.

(xii) The angle at the centre of a circle, subtended by an arc, is double the angle at the circumference subtended by the same arc. With reference to Fig. 19.3,

Angle AOC = 2 × angle ABC

Figure 19.3

(xiii) The angle in a semicircle is a right angle (see angle *BQP* in Fig. 19.3).

Problem 1. Find the circumference of a circle of radius 12.0 cm

Circumference,

$$c = 2 \times \pi \times \text{radius} = 2\pi r = 2\pi(12.0)$$
$$= \mathbf{75.40\, cm}$$

Problem 2. If the diameter of a circle is 75 mm, find its circumference

Circumference,

$$c = \pi \times \text{diameter} = \pi d = \pi(75) = \mathbf{235.6\, mm}$$

Problem 3. Determine the radius of a circle if its perimeter is 112 cm

Perimeter = circumference, $c = 2\pi r$

Hence **radius** $r = \dfrac{c}{2\pi} = \dfrac{112}{2\pi} = \mathbf{17.83\, cm}$

Problem 4. In Fig. 19.4, *AB* is a tangent to the circle at *B*. If the circle radius is 40 mm and *AB* = 150 mm, calculate the length *AO*

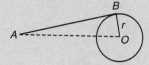

Figure 19.4

A tangent to a circle is at right angles to a radius drawn from the point of contact, i.e. $ABO = 90°$. Hence, using Pythagoras' theorem (see page 187):

$$AO^2 = AB^2 + OB^2$$

from which, $$AO = \sqrt{AB^2 + OB^2}$$
$$= \sqrt{150^2 + 40^2} = \mathbf{155.2\, mm}$$

Now try the following exercise

Exercise 71 Further problems on properties of a circle

1. Calculate the length of the circumference of a circle of radius 7.2 cm. [45.24 cm]

2. If the diameter of a circle is 82.6 mm, calculate the circumference of the circle. [259.5 mm]

3. Determine the radius of a circle whose circumference is 16.52 cm. [2.629 cm]

4. Find the diameter of a circle whose perimeter is 149.8 cm. [47.68 cm]

5. A crank mechanism is shown in Fig. 19.5, where XY is a tangent to the circle at point X. If the circle radius OX is 10 cm and length OY is 40 cm, determine the length of the connecting rod XY. [38.73 cm]

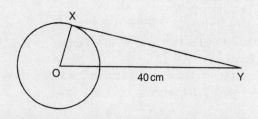

Figure 19.5

6. If the circumference of the earth is 40 000 km at the equator, calculate its diameter. [12730 km]

7. Calculate the length of wire in the paper clip shown in Figure 19.6. The dimensions are in millimetres. [97.13 mm]

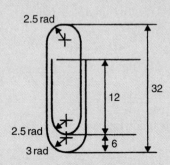

Figure 19.6

19.3 Arc length and area of a sector

One **radian** is defined as the angle subtended at the centre of a circle by an arc equal in length to the radius. With reference to Fig. 19.7, for arc length, s,

θ radians $= s/r$ or **arc length,** $s = r\theta$ (1)

where θ is in radians.

Figure 19.7

When $s =$ whole circumference $(=2\pi r)$ then

$$\theta = s/r = 2\pi r/r = 2\pi$$

i.e. 2π radians $= 360°$ or

$$\pi \text{ radians} = 180°$$

Thus 1 rad $= 180°/\pi = 57.30°$, correct to 2 decimal places.

Since π rad $= 180°$, then $\pi/2 = 90°$, $\pi/3 = 60°$, $\pi/4 = 45°$, and so on.

$$\textbf{Area of a sector} = \frac{\theta}{360}(\pi r^2)$$

when θ is in degrees

$$= \frac{\theta}{2\pi}(\pi r^2) = \frac{1}{2}r^2\theta \quad\quad (2)$$

when θ is in radians

Problem 5. Convert to radians: (a) 125°
(b) 69° 47′

(a) Since $180° = \pi$ rad then $1° = \pi/180$ rad, therefore

$$125° = 125\left(\frac{\pi}{180}\right)^c = \textbf{2.182 radians}$$

(Note that ^{c} means 'circular measure' and indicates radian measure.)

(b) $69°47' = 69\dfrac{47°}{60} = 69.783°$

$$69.783° = 69.783\left(\frac{\pi}{180}\right)^c = \textbf{1.218 radians}$$

Problem 6. Convert to degrees and minutes:
(a) 0.749 radians (b) $3\pi/4$ radians

(a) Since π rad $= 180°$ then 1 rad $= 180°/\pi$, therefore

$$0.749 \text{ rad} = 0.749\left(\frac{180}{\pi}\right)^° = 42.915°$$

$0.915° = (0.915 \times 60)' = 55'$, correct to the nearest minute, hence

$$\textbf{0.749 radians} = \textbf{42°55}'$$

(b) Since 1 rad $= \left(\dfrac{180}{\pi}\right)^°$ then

$$\frac{3\pi}{4} \text{ rad} = \frac{3\pi}{4}\left(\frac{180}{\pi}\right)^°$$

$$= \frac{3}{4}(180)° = \textbf{135}°$$

Problem 7. Express in radians, in terms of π:
(a) 150° (b) 270° (c) 37.5°

Since $180° = \pi$ rad then $1° = 180/\pi$, hence

(a) $150° = 150\left(\dfrac{\pi}{180}\right) \text{rad} = \dfrac{5\pi}{6}\textbf{rad}$

(b) $270° = 270\left(\dfrac{\pi}{180}\right) \text{rad} = \dfrac{3\pi}{2}\textbf{rad}$

(c) $37.5° = 37.5\left(\dfrac{\pi}{180}\right) \text{rad} = \dfrac{75\pi}{360}\text{rad} = \dfrac{5\pi}{24}\textbf{rad}$

Section 2

Now try the following exercise

Exercise 72 Further problems on radians and degrees

1. Convert to radians in terms of π: (a) 30° (b) 75° (c) 225°

$$\left[\text{(a)} \ \frac{\pi}{6} \quad \text{(b)} \ \frac{5\pi}{12} \quad \text{(c)} \ \frac{5\pi}{4} \right]$$

2. Convert to radians: (a) 48° (b) 84° 51′ (c) 232° 15′

[(a) 0.838 (b) 1.481 (c) 4.054]

3. Convert to degrees: (a) $\frac{5\pi}{6}$ rad (b) $\frac{4\pi}{9}$ rad (c) $\frac{7\pi}{12}$ rad

[(a) 150° (b) 80° (c) 105°]

4. Convert to degrees and minutes:
   (a) 0.0125 rad (b) 2.69 rad (c) 7.241 rad

[(a) 0° 43′ (b) 154° 8′ (c) 414° 53′]

19.4 Worked problems on arc length and sector of a circle

Problem 8. Find the length of arc of a circle of radius 5.5 cm when the angle subtended at the centre is 1.20 radians

From equation (1), length of arc, $s = r\theta$, where θ is in radians, hence

$$s = (5.5)(1.20) = \textbf{6.60 cm}$$

Problem 9. Determine the diameter and circumference of a circle if an arc of length 4.75 cm subtends an angle of 0.91 radians

Since $s = r\theta$ then $r = \dfrac{s}{\theta} = \dfrac{4.75}{0.91} = 5.22$ cm.

Diameter $= 2 \times$ radius $= 2 \times 5.22 = \textbf{10.44 cm}$.

Circumference, $c = \pi d = \pi(10.44) = \textbf{32.80 cm}$.

Problem 10. If an angle of 125° is subtended by an arc of a circle of radius 8.4 cm, find the length of (a) the minor arc, and (b) the major arc, correct to 3 significant figures

Since $180° = \pi$ rad then $1° = \left(\dfrac{\pi}{180}\right)$ rad

and $125° = 125\left(\dfrac{\pi}{180}\right)$ rad

Length of minor arc,

$$s = r\theta = (8.4)(125)\left(\frac{\pi}{180}\right) = \textbf{18.3 cm}$$

correct to 3 significant figures.

Length of major arc = (circumference − minor arc) $= 2\pi(8.4) - 18.3 = \textbf{34.5 cm}$, correct to 3 significant figures.
(Alternatively, major arc $= r\theta = 8.4(360 - 125)(\pi/180) = \textbf{34.5 cm}$.)

Problem 11. Determine the angle, in degrees and minutes, subtended at the centre of a circle of diameter 42 mm by an arc of length 36 mm. Calculate also the area of the minor sector formed

Since length of arc, $s = r\theta$ then $\theta = s/r$

Radius, $r = \dfrac{\text{diameter}}{2} = \dfrac{42}{2} = 21$ mm

hence $\theta = \dfrac{s}{r} = \dfrac{36}{21} = 1.7143$ radians

1.7143 rad $= 1.7143 \times (180/\pi)° = 98.22° = \textbf{98° 13}′ =$ angle subtended at centre of circle.
From equation (2),

$$\textbf{area of sector} = \frac{1}{2}r^2\theta = \frac{1}{2}(21)^2(1.7143)$$

$$= \textbf{378 mm}^2$$

Problem 12. A football stadiums floodlights can spread its illumination over an angle of 45° to a distance of 55 m. Determine the maximum area that is floodlit

$$\textbf{Floodlit area} = \text{area of sector} = \frac{1}{2}r^2\theta$$

$$= \frac{1}{2}(55)^2\left(45 \times \frac{\pi}{180}\right)$$

from equation (2)

$$= \textbf{1188 m}^2$$

Problem 13. An automatic garden spray produces a spray to a distance of 1.8 m and revolves through an angle α which may be varied. If the desired spray catchment area is to be 2.5 m², to what should angle α be set, correct to the nearest degree

Area of sector $= \frac{1}{2}r^2\theta$, hence $2.5 = \frac{1}{2}(1.8)^2\alpha$ from

which, $\alpha = \dfrac{2.5 \times 2}{1.8^2} = 1.5432$ radians

$1.5432 \text{ rad} = \left(1.5432 \times \dfrac{180}{\pi}\right)^{\circ} = 88.42°$

Hence **angle $\alpha = 88°$**, correct to the nearest degree.

Problem 14. The angle of a tapered groove is checked using a 20 mm diameter roller as shown in Fig. 19.8. If the roller lies 2.12 mm below the top of the groove, determine the value of angle θ

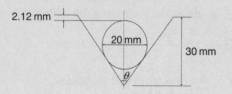

2.12 mm

20 mm

30 mm

θ

Figure 19.8

In Fig. 19.9, triangle ABC is right-angled at C (see Section 19.2(vii), page 150).

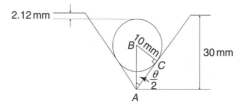

2.12 mm

10 mm

B

C

$\dfrac{\theta}{2}$

A

30 mm

Figure 19.9

Length BC $= 10$ mm (i.e. the radius of the circle), and AB $= 30 - 10 - 2.12 = 17.88$ mm from Fig. 19.8

Hence $\sin\dfrac{\theta}{2} = \dfrac{10}{17.88}$ and

$\dfrac{\theta}{2} = \sin^{-1}\left(\dfrac{10}{17.88}\right) = 34°$

and **angle $\theta = 68°$**

Now try the following exercise

Exercise 73   Further problems on arc length and area of a sector

1. Find the length of an arc of a circle of radius 8.32 cm when the angle subtended at the centre is 2.14 radians. Calculate also the area of the minor sector formed.
   [17.80 cm, 74.07 cm²]

2. If the angle subtended at the centre of a circle of diameter 82 mm is 1.46 rad, find the lengths of the (a) minor arc   (b) major arc.
   [(a) 59.86 mm   (b) 197.8 mm]

3. A pendulum of length 1.5 m swings through an angle of 10° in a single swing. Find, in centimetres, the length of the arc traced by the pendulum bob.   [26.2 cm]

4. Determine the length of the radius and circumference of a circle if an arc length of 32.6 cm subtends an angle of 3.76 radians.
   [8.67 cm, 54.48 cm]

5. Determine the angle of lap, in degrees and minutes, if 180 mm of a belt drive are in contact with a pulley of diameter 250 mm.
   [82.5°]

6. Determine the number of complete revolutions a motorcycle wheel will make in travelling 2 km, if the wheel's diameter is 85.1 cm.
   [748]

7. The floodlights at a sports ground spread its illumination over an angle of 40° to a distance of 48 m. Determine (a) the angle in radians, and (b) the maximum area that is floodlit.
   [(a) 0.698 rad   (b) 804.2 m²]

8. Find the area swept out in 50 minutes by the minute hand of a large floral clock, if the hand is 2 m long.   [10.47 m²]

9. Determine (a) the shaded area in Fig. 19.10 (b) the percentage of the whole sector that the area of the shaded portion represents.
   [(a) 396 mm² (b) 42.24%]

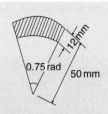

Figure 19.10

10. Determine the length of steel strip required to make the clip shown in Fig. 19.11
    [701.8 mm]

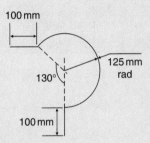

Figure 19.11

11. A 50° tapered hole is checked with a 40 mm diameter ball as shown in Fig. 19.12. Determine the length shown as x. [7.74 mm]

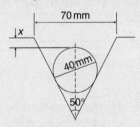

Figure 19.12

19.5 The equation of a circle

The simplest equation of a circle, centre at the origin, radius r, is given by:

$$x^2 + y^2 = r^2$$

For example, Fig. 19.13 shows a circle $x^2 + y^2 = 9$.

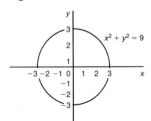

Figure 19.13

More generally, the equation of a circle, centre (a, b), radius r, is given by:

$$(x - a)^2 + (y - b)^2 = r^2 \qquad (1)$$

Figure 19.14 shows a circle $(x - 2)^2 + (y - 3)^2 = 4$

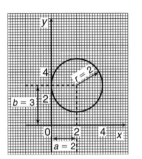

Figure 19.14

The general equation of a circle is:

$$x^2 + y^2 + 2ex + 2fy + c = 0 \qquad (2)$$

Multiplying out the bracketed terms in equation (1) gives:

$$x^2 - 2ax + a^2 + y^2 - 2by + b^2 = r^2$$

Comparing this with equation (2) gives:

$$2e = -2a, \quad \text{i.e.} \quad a = -\frac{2e}{2}$$

$$\text{and} \quad 2f = -2b, \quad \text{i.e.} \quad b = -\frac{2f}{2}$$

$$\text{and} \quad c = a^2 + b^2 - r^2, \quad \text{i.e.} \quad r = \sqrt{a^2 + b^2 - c}$$

Thus, for example, the equation

$$x^2 + y^2 - 4x - 6y + 9 = 0$$

represents a circle with centre

$$a = -\left(\tfrac{-4}{2}\right), b = -\left(\tfrac{-6}{2}\right)$$

i.e. at (2, 3) and radius $r = \sqrt{2^2 + 3^2 - 9} = 2$
Hence $x^2 + y^2 - 4x - 6y + 9 = 0$ is the circle shown in Fig. 19.14, which may be checked by multiplying out the brackets in the equation

$$(x - 2)^2 + (y - 3)^2 = 4$$

Problem 15. Determine (a) the radius, and (b) the co-ordinates of the centre of the circle given by the equation: $x^2 + y^2 + 8x - 2y + 8 = 0$

156 Engineering Mathematics

$x^2 + y^2 + 8x - 2y + 8 = 0$ is of the form shown in equation (2),

where $a = -\left(\frac{8}{2}\right) = -4$, $b = -\left(\frac{-2}{2}\right) = 1$

and $r = \sqrt{(-4)^2 + 1^2 - 8} = \sqrt{9} = 3$

Hence $x^2 + y^2 + 8x - 2y + 8 = 0$ represents a circle **centre (−4, 1)** and **radius 3**, as shown in Fig. 19.15.

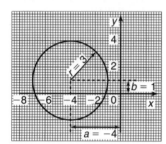

Figure 19.15

Alternatively, $x^2 + y^2 + 8x - 2y + 8 = 0$ may be rearranged as:

$$(x + 4)^2 + (y - 1)^2 - 9 = 0$$

i.e. $(x + 4)^2 + (y - 1)^2 = 3^2$

which represents a circle, **centre (−4, 1)** and **radius 3**, as stated above.

> **Problem 16.** Sketch the circle given by the equation: $x^2 + y^2 - 4x + 6y - 3 = 0$

The equation of a circle, centre (a, b), radius r is given by:

$$(x - a)^2 + (y - b)^2 = r^2$$

The general equation of a circle is

$$x^2 + y^2 + 2ex + 2fy + c = 0$$

From above $a = -\dfrac{2e}{2}, b = -\dfrac{2f}{2}$

and $r = \sqrt{a^2 + b^2 - c}$

Hence if $x^2 + y^2 - 4x + 6y - 3 = 0$

then $a = -\left(\frac{-4}{2}\right) = 2$, $b = -\left(\frac{6}{2}\right) = -3$

and $r = \sqrt{2^2 + (-3)^2 - (-3)} = \sqrt{16} = 4$

Thus **the circle has centre (2, −3)** and **radius 4**, as shown in Fig. 19.16

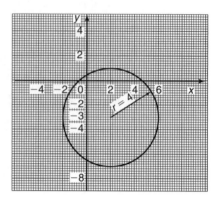

Figure 19.16

Alternatively, $x^2 + y^2 - 4x + 6y - 3 = 0$ may be rearranged as:

$$(x - 2)^2 + (y + 3)^2 - 3 - 13 = 0$$

i.e. $(x - 2)^2 + (y + 3)^2 = 4^2$

which represents a circle, **centre (2, −3)** and **radius 4**, as stated above.

Now try the following exercise

> **Exercise 74 Further problems on the equation of a circle**
>
> 1. Determine (a) the radius, and (b) the co-ordinates of the centre of the circle given by the equation $x^2 + y^2 - 6x + 8y + 21 = 0$
>
>   [(a) 2   (b) (3 −4)]
>
> 2. Sketch the circle given by the equation $x^2 + y^2 - 6x + 4y - 3 = 0$
>
>   [Centre at (3, −2), radius 4]
>
> 3. Sketch the curve $x^2 + (y - 1)^2 - 25 = 0$
>
>   [Circle, centre (0,1), radius 5]
>
> 4. Sketch the curve $x = 6\sqrt{1 - \left(\dfrac{y}{6}\right)^2}$
>
>   [Circle, centre (0, 0), radius 6]

Chapter 20

Volumes and surface areas of common solids

20.1 Volumes and surface areas of regular solids

A summary of volumes and surface areas of regular solids is shown in Table 20.1.

Table 20.1

(i) Rectangular prism (or cuboid)

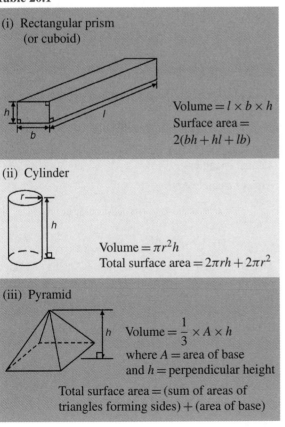

$$\text{Volume} = l \times b \times h$$
$$\text{Surface area} = 2(bh + hl + lb)$$

(ii) Cylinder

$$\text{Volume} = \pi r^2 h$$
$$\text{Total surface area} = 2\pi rh + 2\pi r^2$$

(iii) Pyramid

$$\text{Volume} = \frac{1}{3} \times A \times h$$

where A = area of base
and h = perpendicular height

Total surface area = (sum of areas of triangles forming sides) + (area of base)

(iv) Cone

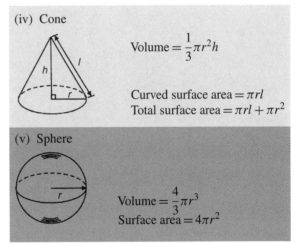

$$\text{Volume} = \frac{1}{3}\pi r^2 h$$

$$\text{Curved surface area} = \pi rl$$
$$\text{Total surface area} = \pi rl + \pi r^2$$

(v) Sphere

$$\text{Volume} = \frac{4}{3}\pi r^3$$
$$\text{Surface area} = 4\pi r^2$$

20.2 Worked problems on volumes and surface areas of regular solids

Problem 1. A water tank is the shape of a rectangular prism having length 2 m, breadth 75 cm and height 50 cm. Determine the capacity of the tank in (a) m^3 (b) cm^3 (c) litres

Volume of rectangular prism = $l \times b \times h$ (see Table 20.1)

(a) Volume of tank = $2 \times 0.75 \times 0.5 = \mathbf{0.75\,m^3}$
(b) $1\,m^3 = 10^6\,cm^3$, hence
$0.75\,m^3 = 0.75 \times 10^6\,cm^3 = \mathbf{750\,000\,cm^3}$
(c) 1 litre = $1000\,cm^3$, hence
$750\,000\,cm^3 = \dfrac{750\,000}{1000}$ litres = **750 litres**

Problem 2. Find the volume and total surface area of a cylinder of length 15 cm and diameter 8 cm

Volume of cylinder $= \pi r^2 h$ (see Table 20.1)

Since diameter $= 8$ cm, then radius $r = 4$ cm

Hence volume $= \pi \times 4^2 \times 15 = \mathbf{754\,cm^3}$

Total surface area (i.e. including the two ends)

$$= 2\pi rh + 2\pi r^2 = (2 \times \pi \times 4 \times 15)$$
$$+ (2 \times \pi \times 4^2) = \mathbf{477.5\,cm^2}$$

Problem 3. Determine the volume (in cm^3) of the shape shown in Fig. 20.1.

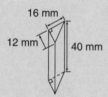

16 mm

12 mm

40 mm

Figure 20.1

The solid shown in Fig. 20.1 is a triangular prism. The volume V of any prism is given by: $V = Ah$, where A is the cross-sectional area and h is the perpendicular height.

Hence volume $= \frac{1}{2} \times 16 \times 12 \times 40$

$$= 3840\,mm^3 = \mathbf{3.840\,cm^3}$$

(since $1\,cm^3 = 1000\,mm^3$)

Problem 4. Calculate the volume and total surface area of the solid prism shown in Fig. 20.2

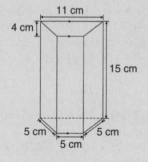

11 cm

4 cm

15 cm

5 cm   5 cm

5 cm

Figure 20.2

The solid shown in Fig. 20.2 is a trapezoidal prism.

Volume $=$ cross-sectional area $\times$ height

$$= \frac{1}{2}(11 + 5)4 \times 15 = 32 \times 15 = \mathbf{480\,cm^3}$$

Surface area $=$ sum of two trapeziums $+$ 4 rectangles

$$= (2 \times 32) + (5 \times 15) + (11 \times 15)$$
$$+ 2(5 \times 15)$$
$$= 64 + 75 + 165 + 150 = \mathbf{454\,cm^2}$$

Problem 5. Determine the volume and the total surface area of the square pyramid shown in Fig. 20.3 if its perpendicular height is 12 cm.

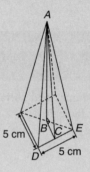

A

B C E

5 cm

D 5 cm

Figure 20.3

Volume of pyramid

$$= \frac{1}{3}(\text{area of base}) \times (\text{perpendicular height})$$

$$= \frac{1}{3}(5 \times 5) \times 12 = \mathbf{100\,cm^3}$$

The total surface area consists of a square base and 4 equal triangles.

Area of triangle ADE

$$= \frac{1}{2} \times \text{base} \times \text{perpendicular height}$$

$$= \frac{1}{2} \times 5 \times AC$$

The length AC may be calculated using Pythagoras' theorem on triangle ABC, where $AB = 12$ cm, $BC = \frac{1}{2} \times 5 = 2.5$ cm

Hence, $\quad AC = \sqrt{AB^2 + BC^2} = \sqrt{12^2 + 2.5^2}$

$$= 12.26\,cm$$

Hence area of triangle $ADE \quad = \frac{1}{2} \times 5 \times 12.26$

$$= 30.65\,cm^2$$

Total surface area of pyramid $= (5 \times 5) + 4(30.65)$

$$= 147.6 \, \text{cm}^2$$

Problem 6. Determine the volume and total surface area of a cone of radius 5 cm and perpendicular height 12 cm

The cone is shown in Fig. 20.4.

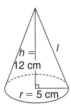

Figure 20.4

Volume of cone $= \frac{1}{3}\pi r^2 h = \frac{1}{3} \times \pi \times 5^2 \times 12$

$$= 314.2 \, \text{cm}^3$$

Total surface area $=$ curved surface area

$\qquad\qquad\qquad\qquad + \text{area of base}$

$$= \pi r l + \pi r^2$$

From Fig. 20.4, slant height l may be calculated using Pythagoras' theorem

$$l = \sqrt{12^2 + 5^2} = 13 \, \text{cm}$$

Hence total surface area $= (\pi \times 5 \times 13) + (\pi \times 5^2)$

$$= 282.7 \, \text{cm}^2$$

Problem 7. Find the volume and surface area of a sphere of diameter 8 cm

Since diameter $= 8$ cm, then radius, $r = 4$ cm.

Volume of sphere $= \dfrac{4}{3}\pi r^3 = \dfrac{4}{3} \times \pi \times 4^3$

$$= 268.1 \, \text{cm}^3$$

Surface area of sphere $= 4\pi r^2 = 4 \times \pi \times 4^2$

$$= 201.1 \, \text{cm}^2$$

Now try the following exercise

Exercise 75   Further problems on volumes and surface areas of regular solids

1. A rectangular block of metal has dimensions of 40 mm by 25 mm by 15 mm. Determine its volume. Find also its mass if the metal has a density of 9 g/cm³.   [15 cm³, 135 g]

2. Determine the maximum capacity, in litres, of a fish tank measuring 50 cm by 40 cm by 2.5 m (1 litre $= 1000$ cm³).   [500 litre]

2. Determine how many cubic metres of concrete are required for a 120 m long path, 150 mm wide and 80 mm deep.   [1.44 m³]

4. Calculate the volume of a metal tube whose outside diameter is 8 cm and whose inside diameter is 6 cm, if the length of the tube is 4 m.   [8796 cm³]

5. The volume of a cylinder is 400 cm³. If its radius is 5.20 cm, find its height. Determine also its curved surface area.
    [4.709 cm, 153.9 cm²]

6. If a cone has a diameter of 80 mm and a perpendicular height of 120 mm calculate its volume in cm³ and its curved surface area.
    [201.1 cm³, 159.0 cm²]

7. A cylinder is cast from a rectangular piece of alloy 5 cm by 7 cm by 12 cm. If the length of the cylinder is to be 60 cm, find its diameter.
    [2.99 cm]

8. Find the volume and the total surface area of a regular hexagonal bar of metal of length 3 m if each side of the hexagon is 6 cm.
    [28 060 cm³, 1.099 m²]

9. A square pyramid has a perpendicular height of 4 cm. If a side of the base is 2.4 cm long find the volume and total surface area of the pyramid.   [7.68 cm³, 25.81 cm²]

10. A sphere has a diameter of 6 cm. Determine its volume and surface area.
    [113.1 cm³, 113.1 cm²]

11.   Find the total surface area of a hemisphere of diameter 50 mm.   [5890 mm² or 58.90 cm²]

12.   How long will it take a tap dripping at a rate of 800 mm³/s to fill a 3-litre can?

[62.5 minutes]

20.3   Further worked problems on volumes and surface areas of regular solids

Problem 8.   A wooden section is shown in Fig. 20.5. Find (a) its volume (in m³), and (b) its total surface area.

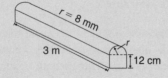

Figure 20.5

The section of wood is a prism whose end comprises a rectangle and a semicircle. Since the radius of the semicircle is 8 cm, the diameter is 16 cm.

Hence the rectangle has dimensions 12 cm by 16 cm.

Area of end $= (12 \times 16) + \frac{1}{2}\pi 8^2 = 292.5 \, \text{cm}^2$

Volume of wooden section

$= \text{area of end} \times \text{perpendicular height}$

$= 292.5 \times 300 = 87\,750 \, \text{cm}^3 = \dfrac{87\,750 \, \text{m}^3}{10^6}$

$$= \mathbf{0.08775 \, m^3}$$

The total surface area comprises the two ends (each of area 292.5 cm²), three rectangles and a curved surface (which is half a cylinder), hence

total surface area $= (2 \times 292.5) + 2(12 \times 300)$

$+ (16 \times 300) + \frac{1}{2}(2\pi \times 8 \times 300)$

$= 585 + 7200 + 4800 + 2400\pi$

$= \mathbf{20\,125 \, cm^2} \quad \text{or} \quad \mathbf{2.0125 \, m^2}$

Problem 9.   A pyramid has a rectangular base 3.60 cm by 5.40 cm. Determine the volume and total surface area of the pyramid if each of its sloping edges is 15.0 cm

The pyramid is shown in Fig. 20.6. To calculate the volume of the pyramid the perpendicular height EF is required. Diagonal BD is calculated using Pythagoras' theorem,

i.e. $BD = \sqrt{3.60^2 + 5.40^2} = 6.490 \, \text{cm}$

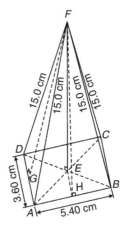

Figure 20.6

Hence $EB = \dfrac{1}{2}BD = \dfrac{6.490}{2} = 3.245 \, \text{cm}$

Using Pythagoras' theorem on triangle BEF gives

$$BF^2 = EB^2 + EF^2$$

from which, $EF = \sqrt{BF^2 - EB^2}$

$$= \sqrt{15.0^2 - 3.245^2} = 14.64 \, \text{cm}$$

Volume of pyramid

$= \frac{1}{3}(\text{area of base})(\text{perpendicular height})$

$= \frac{1}{3}(3.60 \times 5.40)(14.64) = \mathbf{94.87 \, cm^3}$

Area of triangle ADF (which equals triangle BCF) $= \frac{1}{2}(AD)(FG)$, where G is the midpoint of AD. Using Pythagoras' theorem on triangle FGA gives:

$$FG = \sqrt{15.0^2 - 1.80^2} = 14.89 \, \text{cm}$$

Hence area of triangle $ADF = \frac{1}{2}(3.60)(14.89)$

$$= 26.80 \, \text{cm}^2$$

Similarly, if H is the mid-point of AB, then

$$FH = \sqrt{15.0^2 - 2.70^2} = 14.75 \, \text{cm},$$

hence area of triangle ABF (which equals triangle CDF)

$$= \frac{1}{2}(5.40)(14.75) = 39.83 \, \text{cm}^2$$

Total surface area of pyramid

$$= 2(26.80) + 2(39.83) + (3.60)(5.40)$$

$$= 53.60 + 79.66 + 19.44$$

$$= \mathbf{152.7\,cm^2}$$

Problem 10. Calculate the volume and total surface area of a hemisphere of diameter 5.0 cm

Volume of hemisphere $= \frac{1}{2}$(volume of sphere)

$$= \frac{2}{3}\pi r^3 = \frac{2}{3}\pi \left(\frac{5.0}{2}\right)^3$$

$$= \mathbf{32.7\,cm^3}$$

Total surface area

$$= \text{curved surface area} + \text{area of circle}$$

$$= \frac{1}{2}(\text{surface area of sphere}) + \pi r^2$$

$$= \frac{1}{2}(4\pi r^2) + \pi r^2$$

$$= 2\pi r^2 + \pi r^2 = 3\pi r^2 = 3\pi \left(\frac{5.0}{2}\right)^2$$

$$= \mathbf{58.9\,cm^2}$$

Problem 11. A rectangular piece of metal having dimensions 4 cm by 3 cm by 12 cm is melted down and recast into a pyramid having a rectangular base measuring 2.5 cm by 5 cm. Calculate the perpendicular height of the pyramid

Volume of rectangular prism of metal $= 4 \times 3 \times 12$

$$= 144\,cm^3$$

Volume of pyramid

$$= \frac{1}{3}(\text{area of base})(\text{perpendicular height})$$

Assuming no waste of metal,

$$144 = \frac{1}{3}(2.5 \times 5)(\text{height})$$

i.e.   perpendicular height $= \dfrac{144 \times 3}{2.5 \times 5} = \mathbf{34.56\,cm}$

Problem 12. A rivet consists of a cylindrical head, of diameter 1 cm and depth 2 mm, and a shaft of diameter 2 mm and length 1.5 cm. Determine the volume of metal in 2000 such rivets

Radius of cylindrical head $= \frac{1}{2}$ cm $= 0.5$ cm and height of cylindrical head $= 2$ mm $= 0.2$ cm

Hence, volume of cylindrical head

$$= \pi r^2 h = \pi (0.5)^2 (0.2) = 0.1571\,cm^3$$

Volume of cylindrical shaft

$$= \pi r^2 h = \pi \left(\frac{0.2}{2}\right)^2 (1.5) = 0.0471\,cm^3$$

Total volume of 1 rivet $= 0.1571 + 0.0471$

$$= 0.2042\,cm^3$$

Volume of metal in 2000 such rivets

$$= 2000 \times 0.2042 = \mathbf{408.4\,cm^3}$$

Problem 13. A solid metal cylinder of radius 6 cm and height 15 cm is melted down and recast into a shape comprising a hemisphere surmounted by a cone. Assuming that 8% of the metal is wasted in the process, determine the height of the conical portion, if its diameter is to be 12 cm

Volume of cylinder $= \pi r^2 h = \pi \times 6^2 \times 15$

$$= 540\pi\,cm^3$$

If 8% of metal is lost then 92% of 540π gives the volume of the new shape (shown in Fig. 20.7).

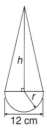

Figure 20.7

Hence the volume of (hemisphere + cone)

$$= 0.92 \times 540\pi\,cm^3,$$

i.e. $\frac{1}{2}\left(\frac{4}{3}\pi r^3\right) + \frac{1}{3}\pi r^2 h = 0.92 \times 540\pi$

Dividing throughout by π gives:

$$\frac{2}{3}r^3 + \frac{1}{3}r^2 h = 0.92 \times 540$$

Since the diameter of the new shape is to be 12 cm, then radius $r = 6$ cm,

hence $\frac{2}{3}(6)^3 + \frac{1}{3}(6)^2 h = 0.92 \times 540$

$$144 + 12h = 496.8$$

i.e. height of conical portion,

$$h = \frac{496.8 - 144}{12} = \textbf{29.4 cm}$$

Problem 14. A block of copper having a mass of 50 kg is drawn out to make 500 m of wire of uniform cross-section. Given that the density of copper is 8.91 g/cm³, calculate (a) the volume of copper, (b) the cross-sectional area of the wire, and (c) the diameter of the cross-section of the wire

(a) A density of 8.91 g/cm³ means that 8.91 g of copper has a volume of 1 cm³, or 1 g of copper has a volume of (1/8.91) cm³

Hence 50 kg, i.e. 50 000 g, has a volume

$$\frac{50\,000}{8.91} \text{cm}^3 = \textbf{5612 cm}^3$$

(b) Volume of wire

$$= \text{area of circular cross-section}$$

$$\times \text{length of wire.}$$

Hence 5612 cm³ = area × (500 × 100 cm),

from which, area $= \dfrac{5612}{500 \times 100}$ cm²

$$= \textbf{0.1122 cm}^2$$

(c) Area of circle $= \pi r^2$ or $\dfrac{\pi d^2}{4}$, hence

$$0.1122 = \frac{\pi d^2}{4}\text{from which}$$

$$d = \sqrt{\frac{4 \times 0.1122}{\pi}} = 0.3780 \text{ cm}$$

i.e. diameter of cross-section is 3.780 mm

Problem 15. A boiler consists of a cylindrical section of length 8 m and diameter 6 m, on one end of which is surmounted a hemispherical section of diameter 6 m, and on the other end a conical section of height 4 m and base diameter 6 m. Calculate the volume of the boiler and the total surface area

The boiler is shown in Fig. 20.8

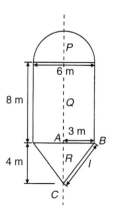

Figure 20.8

Volume of hemisphere, P

$$= \tfrac{2}{3}\pi r^3 = \tfrac{2}{3} \times \pi \times 3^3 = 18\pi \text{ m}^3$$

Volume of cylinder, Q

$$= \pi r^2 h = \pi \times 3^2 \times 8 = 72\pi \text{ m}^3$$

Volume of cone, R

$$= \tfrac{1}{3}\pi r^2 h = \tfrac{1}{3} \times \pi \times 3^2 \times 4 = 12\pi \text{ m}^3$$

Total volume of boiler $= 18\pi + 72\pi + 12\pi$

$$= 102\pi = \textbf{320.4 m}^3$$

Surface area of hemisphere, P

$$= \tfrac{1}{2}(4\pi r^2) = 2 \times \pi \times 3^2 = 18\pi \text{ m}^2$$

Curved surface area of cylinder, Q

$$= 2\pi rh = 2 \times \pi \times 3 \times 8 = 48\pi \text{ m}^2$$

The slant height of the cone, l, is obtained by Pythagoras' theorem on triangle ABC, i.e.

$$l = \sqrt{4^2 + 3^2} = 5$$

Curved surface area of cone, R

$$= \pi rl = \pi \times 3 \times 5 = 15\pi \text{ m}^2$$

Total surface area of boiler $= 18\pi + 48\pi + 15\pi$

$$= 81\pi = \textbf{254.5 m}^2$$

Now try the following exercise

Exercise 76   Further problems on volumes and surface areas of regular solids

1. Determine the mass of a hemispherical copper container whose external and internal radii are 12 cm and 10 cm, assuming that 1 cm^3 of copper weighs 8.9 g.   [13.57 kg]

2. If the volume of a sphere is 566 cm^3, find its radius.   [5.131 cm]

3. A metal plumb bob comprises a hemisphere surmounted by a cone. If the diameter of the hemisphere and cone are each 4 cm and the total length is 5 cm, find its total volume.   [29.32 cm^3]

4. A marquee is in the form of a cylinder surmounted by a cone. The total height is 6 m and the cylindrical portion has a height of 3.5 m, with a diameter of 15 m. Calculate the surface area of material needed to make the marquee assuming 12% of the material is wasted in the process.   [393.4 m^2]

5. Determine (a) the volume and (b) the total surface area of the following solids:
    (i)   a cone of radius 8.0 cm and perpendicular height 10 cm
    (ii) a sphere of diameter 7.0 cm
    (iii) a hemisphere of radius 3.0 cm
    (iv) a 2.5 cm by 2.5 cm square pyramid of perpendicular height 5.0 cm
    (v)   a 4.0 cm by 6.0 cm rectangular pyramid of perpendicular height 12.0 cm
    (vi) a 4.2 cm by 4.2 cm square pyramid whose sloping edges are each 15.0 cm
    (vii) a pyramid having an octagonal base of side 5.0 cm and perpendicular height 20 cm.

$$\begin{bmatrix} \text{(i)} & \text{(a) } 670\,\text{cm}^3 & \text{(b) } 523\,\text{cm}^2 \\ \text{(ii)} & \text{(a) } 180\,\text{cm}^3 & \text{(b) } 154\,\text{cm}^2 \\ \text{(iii)} & \text{(a) } 56.5\,\text{cm}^3 & \text{(b) } 84.8\,\text{cm}^2 \end{bmatrix}$$

$$\begin{bmatrix} \text{(iv)} & \text{(a) } 10.4\,\text{cm}^3 & \text{(b) } 32.0\,\text{cm}^2 \\ \text{(v)} & \text{(a) } 96.0\,\text{cm}^3 & \text{(b) } 146\,\text{cm}^2 \\ \text{(vi)} & \text{(a) } 86.5\,\text{cm}^3 & \text{(b) } 142\,\text{cm}^2 \\ \text{(vii)} & \text{(a) } 805\,\text{cm}^3 & \text{(b) } 539\,\text{cm}^2 \end{bmatrix}$$

6. The volume of a sphere is 325 cm^3. Determine its diameter.   [8.53 cm]

7. A metal sphere weighing 24 kg is melted down and recast into a solid cone of base radius 8.0 cm. If the density of the metal is 8000 kg/m^3 determine (a) the diameter of the metal sphere and (b) the perpendicular height of the cone, assuming that 15% of the metal is lost in the process.
    [(a) 17.9 cm (b) 38.0 cm]

8. Find the volume of a regular hexagonal pyramid if the perpendicular height is 16.0 cm and the side of base is 3.0 cm.   [125 cm^3]

9. A buoy consists of a hemisphere surmounted by a cone. The diameter of the cone and hemisphere is 2.5 m and the slant height of the cone is 4.0 m. Determine the volume and surface area of the buoy.   [10.3 m^3, 25.5 m^2]

10. A petrol container is in the form of a central cylindrical portion 5.0 m long with a hemispherical section surmounted on each end. If the diameters of the hemisphere and cylinder are both 1.2 m determine the capacity of the tank in litres (1 litre $=$ 1000 cm^3).   [6560 litre]

11. Figure 20.9 shows a metal rod section. Determine its volume and total surface area.   [657.1 cm^3, 1027 cm^2]

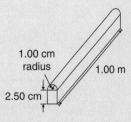

1.00 cm radius

1.00 m

2.50 cm

Figure 20.9

12. Find the volume (in cm³) of the die-casting shown in Figure 20.10. The dimensions are in millimetres. [220.7 cm³]

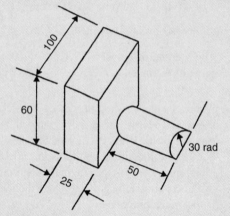

Figure 20.10

13. The cross-section of part of a circular ventilation shaft is shown in Figure 20.11, ends AB and CD being open. Calculate (a) the volume of the air, correct to the nearest litre, contained in the part of the system shown, neglecting the sheet metal thickness, (given 1 litre = 1000 cm³), (b) the cross-sectional area of the sheet metal used to make the system, in square metres, and (c) the cost of the sheet metal if the material costs £11.50 per square metre, assuming that 25% extra metal is required due to wastage.

[(a) 1458 litre (b) 9.77 m² (c) £140.45]

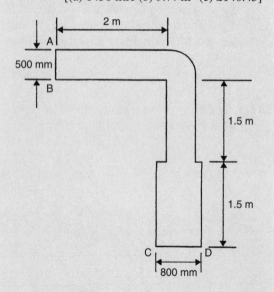

Figure 20.11

20.4 Volumes and surface areas of frusta of pyramids and cones

The **frustum** of a pyramid or cone is the portion remaining when a part containing the vertex is cut off by a plane parallel to the base.

The **volume of a frustum of a pyramid or cone** is given by the volume of the whole pyramid or cone minus the volume of the small pyramid or cone cut off.

The **surface area of the sides of a frustum of a pyramid or cone** is given by the surface area of the whole pyramid or cone minus the surface area of the small pyramid or cone cut off. This gives the lateral surface area of the frustum. If the total surface area of the frustum is required then the surface area of the two parallel ends are added to the lateral surface area.

There is an alternative method for finding the volume and surface area of a **frustum of a cone**. With reference to Fig. 20.12:

Figure 20.12

$$\text{Volume} = \tfrac{1}{3}\pi h(R^2 + Rr + r^2)$$
$$\text{Curved surface area} = \pi l(R + r)$$
$$\text{Total surface area} = \pi l(R + r) + \pi r^2 + \pi R^2$$

Problem 16. Determine the volume of a frustum of a cone if the diameter of the ends are 6.0 cm and 4.0 cm and its perpendicular height is 3.6 cm

Method 1

A section through the vertex of a complete cone is shown in Fig. 20.13

Using similar triangles

$$\frac{AP}{DP} = \frac{DR}{BR}$$

Hence

$$\frac{AP}{2.0} = \frac{3.6}{1.0}$$

from which

$$AP = \frac{(2.0)(3.6)}{1.0} = 7.2 \text{ cm}$$

The height of the large cone = 3.6 + 7.2 = 10.8 cm.

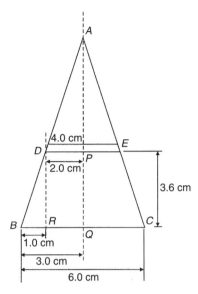

Figure 20.13

Volume of frustum of cone

  = volume of large cone

    − volume of small cone cut off

$$= \tfrac{1}{3}\pi(3.0)^2(10.8) - \tfrac{1}{3}\pi(2.0)^2(7.2)$$

$$= 101.79 - 30.16 = \mathbf{71.6\,cm^3}$$

Method 2

From above, volume of the frustum of a cone

$$= \tfrac{1}{3}\pi h(R^2 + Rr + r^2),$$

where $R = 3.0\,cm,$

    $r = 2.0\,cm$ and $h = 3.6\,cm$

Hence volume of frustum

$$= \tfrac{1}{3}\pi(3.6)[(3.0)^2 + (3.0)(2.0) + (2.0)^2]$$

$$= \tfrac{1}{3}\pi(3.6)(19.0) = \mathbf{71.6\,cm^3}$$

> **Problem 17.** Find the total surface area of the frustum of the cone in Problem 16

Method 1

Curved surface area of frustum = curved surface area of large cone—curved surface area of small cone cut off.

From Fig. 20.13, using Pythagoras' theorem:

$$AB^2 = AQ^2 + BQ^2 \quad \text{from which}$$

$$AB = \sqrt{10.8^2 + 3.0^2} = 11.21\,cm$$

and $AD^2 = AP^2 + DP^2$ from which

$$AD = \sqrt{7.2^2 + 2.0^2} = 7.47\,cm$$

Curved surface area of large cone

$$= \pi r l = \pi(BQ)(AB) = \pi(3.0)(11.21)$$

$$= 105.65\,cm^2$$

and curved surface area of small cone

$$= \pi(DP)(AD) = \pi(2.0)(7.47) = 46.94\,cm^2$$

Hence, curved surface area of frustum

$$= 105.65 - 46.94$$

$$= 58.71\,cm^2$$

Total surface area of frustum

  = curved surface area

    + area of two circular ends

$$= 58.71 + \pi(2.0)^2 + \pi(3.0)^2$$

$$= 58.71 + 12.57 + 28.27 = \mathbf{99.6\,cm^2}$$

Method 2

From page 164, total surface area of frustum

$$= \pi l(R + r) + \pi r^2 + \pi R^2$$

where $l = BD = 11.21 - 7.47 = 3.74\,cm,$ $R = 3.0\,cm$ and $r = 2.0\,cm.$

Hence total surface area of frustum

$$= \pi(3.74)(3.0 + 2.0) + \pi(2.0)^2 + \pi(3.0)^2$$

$$= \mathbf{99.6\,cm^2}$$

> **Problem 18.** A storage hopper is in the shape of a frustum of a pyramid. Determine its volume if the ends of the frustum are squares of sides 8.0 m and 4.6 m, respectively, and the perpendicular height between its ends is 3.6 m

The frustum is shown shaded in Fig. 20.14(a) as part of a complete pyramid. A section perpendicular to the base through the vertex is shown in Fig. 20.14(b)

By similar triangles: $\dfrac{CG}{BG} = \dfrac{BH}{AH}$

Height $CG = BG\left(\dfrac{BH}{AH}\right) = \dfrac{(2.3)(3.6)}{1.7} = 4.87\,\text{m}$

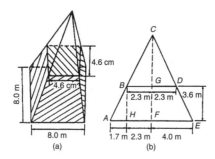

Figure 20.14

Height of complete pyramid $= 3.6 + 4.87 = 8.47\,\text{m}$

Volume of large pyramid $= \frac{1}{3}(8.0)^2(8.47)$

$= 180.69\,\text{m}^3$

Volume of small pyramid cut off

$= \frac{1}{3}(4.6)^2(4.87) = 34.35\,\text{m}^3$

Hence volume of storage hopper

$= 180.69 - 34.35 = \mathbf{146.3\,m^3}$

Problem 19. Determine the lateral surface area of the storage hopper in Problem 18

The lateral surface area of the storage hopper consists of four equal trapeziums.
From Fig. 20.15, area of trapezium $PRSU$

$= \frac{1}{2}(PR + SU)(QT)$

$OT = 1.7\,\text{m}$ (same as AH in Fig. 20.14(b)) and $OQ = 3.6\,\text{m}$.
By Pythagoras' theorem,

$QT = \sqrt{OQ^2 + OT^2} = \sqrt{3.6^2 + 1.7^2} = 3.98\,\text{m}$

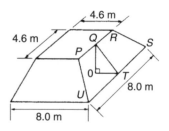

Figure 20.15

Area of trapezium $PRSU = \frac{1}{2}(4.6 + 8.0)(3.98)$

$= 25.07\,\text{m}^2$

Lateral surface area of hopper $= 4(25.07)$

$= \mathbf{100.3\,m^2}$

Problem 20. A lampshade is in the shape of a frustum of a cone. The vertical height of the shade is 25.0 cm and the diameters of the ends are 20.0 cm and 10.0 cm, respectively. Determine the area of the material needed to form the lampshade, correct to 3 significant figures

The curved surface area of a frustum of a cone $= \pi l(R + r)$ from page 164.
Since the diameters of the ends of the frustum are 20.0 cm and 10.0 cm, then from Fig. 20.16,

$r = 5.0\,\text{cm},\ R = 10.0\,\text{cm}$

and $l = \sqrt{25.0^2 + 5.0^2} = 25.50\,\text{cm}$,

from Pythagoras' theorem.

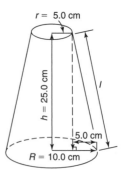

Figure 20.16

Hence curved surface area

$= \pi(25.50)(10.0 + 5.0) = 1201.7\,\text{cm}^2$

i.e. the area of material needed to form the lampshade is **1200 cm²**, correct to 3 significant figures.

Problem 21. A cooling tower is in the form of a cylinder surmounted by a frustum of a cone as shown in Fig. 20.17. Determine the volume of air space in the tower if 40% of the space is used for pipes and other structures

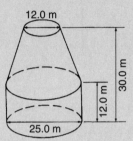

12.0 m

30.0 m

12.0 m

25.0 m

Figure 20.17

Volume of cylindrical portion

$$= \pi r^2 h = \pi \left(\frac{25.0}{2} \right)^2 (12.0) = 5890 \, \text{m}^3$$

Volume of frustum of cone

$$= \frac{1}{3} \pi h (R^2 + Rr + r^2)$$

where $h = 30.0 - 12.0 = 18.0 \, \text{m}$,

$R = 25.0/2 = 12.5 \, \text{m}$ and $r = 12.0/2 = 6.0 \, \text{m}$

Hence volume of frustum of cone

$$= \frac{1}{3} \pi (18.0)[(12.5)^2 + (12.5)(6.0) + (6.0)^2]$$

$$= 5038 \, \text{m}^3$$

Total volume of cooling tower $= 5890 + 5038$

$$= 10\,928 \, \text{m}^3$$

If 40% of space is occupied then volume of air space $= 0.6 \times 10\,928 = \mathbf{6557 \, m^3}$

Now try the following exercise

Exercise 77 Further problems on volumes and surface areas of frustra of pyramids and cones

1. The radii of the faces of a frustum of a cone are 2.0 cm and 4.0 cm and the thickness of the frustum is 5.0 cm. Determine its volume and total surface area. [147 cm³, 164 cm²]

2. A frustum of a pyramid has square ends, the squares having sides 9.0 cm and 5.0 cm, respectively. Calculate the volume and total surface area of the frustum if the perpendicular distance between its ends is 8.0 cm. [403 cm³, 337 cm²]

3. A cooling tower is in the form of a frustum of a cone. The base has a diameter of 32.0 m, the top has a diameter of 14.0 m and the vertical height is 24.0 m. Calculate the volume of the tower and the curved surface area. [10 480 m³, 1852 m²]

4. A loudspeaker diaphragm is in the form of a frustum of a cone. If the end diameters are 28.0 cm and 6.00 cm and the vertical distance between the ends is 30.0 cm, find the area of material needed to cover the curved surface of the speaker. [1707 cm²]

5. A rectangular prism of metal having dimensions 4.3 cm by 7.2 cm by 12.4 cm is melted down and recast into a frustum of a square pyramid, 10% of the metal being lost in the process. If the ends of the frustum are squares of side 3 cm and 8 cm respectively, find the thickness of the frustum. [10.69 cm]

6. Determine the volume and total surface area of a bucket consisting of an inverted frustum of a cone, of slant height 36.0 cm and end diameters 55.0 cm and 35.0 cm. [55 910 cm³, 6051 cm²]

7. A cylindrical tank of diameter 2.0 m and perpendicular height 3.0 m is to be replaced by a tank of the same capacity but in the form of a frustum of a cone. If the diameters of the ends of the frustum are 1.0 m and 2.0 m, respectively, determine the vertical height required. [5.14 m]

20.5 The frustum and zone of a sphere

Volume of sphere $= \frac{4}{3} \pi r^3$ and the surface area of sphere $= 4\pi r^2$

A **frustum of a sphere** is the portion contained between two parallel planes. In Fig. 20.18, PQRS is a frustum of the sphere. A **zone of a sphere** is the curved surface of a frustum. With reference to Fig. 20.18:

Surface area of a zone of a sphere $= 2\pi rh$

Volume of frustum of sphere

$$= \frac{\pi h}{6}(h^2 + 3r_1^2 + 3r_2^2)$$

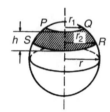

Figure 20.18

Problem 22. Determine the volume of a frustum of a sphere of diameter 49.74 cm if the diameter of the ends of the frustum are 24.0 cm and 40.0 cm, and the height of the frustum is 7.00 cm

From above, volume of frustum of a sphere

$$= \frac{\pi h}{6}(h^2 + 3r_1^2 + 3r_2^2)$$

where $h = 7.00$ cm, $r_1 = 24.0/2 = 12.0$ cm and $r_2 = 40.0/2 = 20.0$ cm.

Hence volume of frustum

$$= \frac{\pi(7.00)}{6}[(7.00)^2 + 3(12.0)^2 + 3(20.0)^2]$$

$$= \mathbf{6161\ cm^3}$$

Problem 23. Determine for the frustum of Problem 22 the curved surface area of the frustum

The curved surface area of the frustum = surface area of zone $= 2\pi rh$ (from above), where $r =$ radius of sphere $= 49.74/2 = 24.87$ cm and $h = 7.00$ cm. Hence, surface area of zone $= 2\pi(24.87)(7.00) = \mathbf{1094\ cm^2}$

Problem 24. The diameters of the ends of the frustum of a sphere are 14.0 cm and 26.0 cm respectively, and the thickness of the frustum is 5.0 cm. Determine, correct to 3 significant figures (a) the volume of the frustum of the sphere, (b) the radius of the sphere and (c) the area of the zone formed

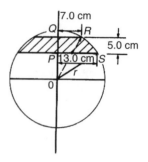

Figure 20.19

The frustum is shown shaded in the cross-section of Fig. 20.19

(a) Volume of frustum of sphere

$$= \frac{\pi h}{6}(h^2 + 3r_1^2 + 3r_2^2)$$

from above, where $h = 5.0$ cm, $r_1 = 14.0/2 = 7.0$ cm and $r_2 = 26.0/2 = 13.0$ cm.

Hence volume of frustum of sphere

$$= \frac{\pi(5.0)}{6}[(5.0)^2 + 3(7.0)^2 + 3(13.0)^2]$$

$$= \frac{\pi(5.0)}{6}[25.0 + 147.0 + 507.0]$$

$$= \mathbf{1780\ cm^3}$$ correct to 3 significant figures.

(b) The radius, r, of the sphere may be calculated using Fig. 20.19. Using Pythagoras' theorem:

$$OS^2 = PS^2 + OP^2$$

i.e. $$r^2 = (13.0)^2 + OP^2 \qquad (1)$$

$$OR^2 = QR^2 + OQ^2$$

i.e. $$r^2 = (7.0)^2 + OQ^2$$

However $OQ = QP + OP = 5.0 + OP$, therefore

$$r^2 = (7.0)^2 + (5.0 + OP)^2 \qquad (2)$$

Equating equations (1) and (2) gives:

$$(13.0)^2 + OP^2 = (7.0)^2 + (5.0 + OP)^2$$

$$169.0 + OP^2 = 49.0 + 25.0$$

$$+ 10.0(OP) + OP^2$$

$$169.0 = 74.0 + 10.0(OP)$$

Hence

$$OP = \frac{169.0 - 74.0}{10.0} = 9.50\,\text{cm}$$

Substituting $OP = 9.50\,\text{cm}$ into equation (1) gives:

$$r^2 = (13.0)^2 + (9.50)^2$$

from which $r = \sqrt{13.0^2 + 9.50^2}$

i.e. **radius of sphere, $r = 16.1$ cm**

(c) Area of zone of sphere

$$= 2\pi rh = 2\pi(16.1)(5.0)$$

$$= \mathbf{506\,cm^2}, \text{ correct to 3 significant figures.}$$

Problem 25. A frustum of a sphere of diameter 12.0 cm is formed by two parallel planes, one through the diameter and the other distance h from the diameter. The curved surface area of the frustum is required to be $\frac{1}{4}$ of the total surface area of the sphere. Determine (a) the volume and surface area of the sphere, (b) the thickness h of the frustum, (c) the volume of the frustum and (d) the volume of the frustum expressed as a percentage of the sphere

---

(a) Volume of sphere,

$$V = \frac{4}{3}\pi r^3 = \frac{4}{3}\pi\left(\frac{12.0}{2}\right)^3$$

$$= \mathbf{904.8\,cm^3}$$

Surface area of sphere

$$= 4\pi r^2 = 4\pi\left(\frac{12.0}{2}\right)^2$$

$$= \mathbf{452.4\,cm^2}$$

(b) Curved surface area of frustum

$$= \tfrac{1}{4} \times \text{surface area of sphere}$$

$$= \tfrac{1}{4} \times 452.4 = 113.1\,\text{cm}^2$$

From above,

$$113.1 = 2\pi rh = 2\pi\left(\frac{12.0}{2}\right)h$$

Hence thickness of frustum

$$h = \frac{113.1}{2\pi(6.0)} = \mathbf{3.0\,cm}$$

(c) Volume of frustum,

$$V = \frac{\pi h}{6}(h^2 + 3r_1^2 + 3r_2^2)$$

where $h = 3.0\,\text{cm}$, $r_2 = 6.0\,\text{cm}$ and

$$r_1 = \sqrt{OQ^2 - OP^2}$$

from Fig. 20.20,

i.e. $r_1 = \sqrt{6.0^2 - 3.0^2} = 5.196\,\text{cm}$

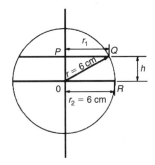

Figure 20.20

Hence volume of frustum

$$= \frac{\pi(3.0)}{6}[(3.0)^2 + 3(5.196)^2 + 3(6.0)^2]$$

$$= \frac{\pi}{2}[9.0 + 81 + 108.0] = \mathbf{311.0\,cm^3}$$

(d)

$$\frac{\text{Volume of frustum}}{\text{Volume of sphere}} = \frac{311.0}{904.8} \times 100\%$$

$$= \mathbf{34.37\%}$$

Problem 26. A spherical storage tank is filled with liquid to a depth of 20 cm. If the internal diameter of the vessel is 30 cm, determine the number of litres of liquid in the container (1 litre = 1000 cm^3)

The liquid is represented by the shaded area in the section shown in Fig. 20.21. The volume of liquid comprises a hemisphere and a frustum of thickness 5 cm. Hence volume of liquid

$$= \frac{2}{3}\pi r^3 + \frac{\pi h}{6}[h^2 + 3r_1^2 + 3r_2^2]$$

where $r_2 = 30/2 = 15\,\text{cm}$ and

$$r_1 = \sqrt{15^2 - 5^2} = 14.14\,\text{cm}$$

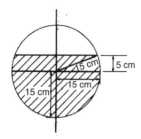

Figure 20.21

Volume of liquid

$$= \frac{2}{3}\pi(15)^3 + \frac{\pi(5)}{6}[5^2 + 3(14.14)^2 + 3(15)^2]$$

$$= 7069 + 3403 = 10\,470 \, \text{cm}^3$$

Since 1 litre $= 1000 \, \text{cm}^3$, the number of litres of liquid

$$= \frac{10\,470}{1000} = \textbf{10.47 litres}$$

Now try the following exercise

Exercise 78   Further problems on frustums and zones of spheres

1. Determine the volume and surface area of a frustum of a sphere of diameter 47.85 cm, if the radii of the ends of the frustum are 14.0 cm and 22.0 cm and the height of the frustum is 10.0 cm.   [11 210 cm³, 1503 cm²]

2. Determine the volume (in cm³) and the surface area (in cm²) of a frustum of a sphere if the diameter of the ends are 80.0 mm and 120.0 mm and the thickness is 30.0 mm.
[259.2 cm³, 118.3 cm²]

3. A sphere has a radius of 6.50 cm. Determine its volume and surface area. A frustum of the sphere is formed by two parallel planes, one through the diameter and the other at a distance h from the diameter. If the curved surface area of the frustum is to be $\frac{1}{5}$ of the surface area of the sphere, find the height h and the volume of the frustum.
$$\left[\begin{array}{l} 1150\,\text{cm}^3, 531\,\text{cm}^2, \\ 2.60\,\text{cm}, 326.7\,\text{cm}^3 \end{array}\right]$$

4. A sphere has a diameter of 32.0 mm. Calculate the volume (in cm³) of the frustum of the sphere contained between two parallel planes distances 12.0 mm and 10.00 mm from the centre and on opposite sides of it.
[14.84 cm³]

5. A spherical storage tank is filled with liquid to a depth of 30.0 cm. If the inner diameter of the vessel is 45.0 cm determine the number of litres of liquid in the container (1 litre $= 1000 \, \text{cm}^3$).   [35.34 litres]

20.6   Prismoidal rule

The prismoidal rule applies to a solid of length x divided by only three equidistant plane areas, A_1, A_2 and A_3 as shown in Fig. 20.22 and is merely an extension of Simpson's rule (see Chapter 21)—but for volumes.

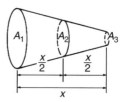

Figure 20.22

With reference to Fig. 20.22

$$\textbf{Volume, } V = \frac{x}{6}[A_1 + 4A_2 + A_3]$$

The prismoidal rule gives precise values of volume for regular solids such as pyramids, cones, spheres and prismoids.

Problem 27.   A container is in the shape of a frustum of a cone. Its diameter at the bottom is 18 cm and at the top 30 cm. If the depth is 24 cm determine the capacity of the container, correct to the nearest litre, by the prismoidal rule. (1 litre $= 1000 \, \text{cm}^3$)

The container is shown in Fig. 20.23. At the mid-point, i.e. at a distance of 12 cm from one end, the radius r_2 is $(9 + 15)/2 = 12$ cm, since the sloping side changes uniformly.

Volume of container by the prismoidal rule

$$= \frac{x}{6}[A_1 + 4A_2 + A_3]$$

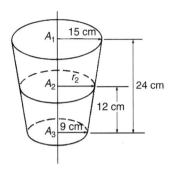

Figure 20.23

from above, where $x = 24\,\text{cm}$, $A_1 = \pi(15)^2\,\text{cm}^2$, $A_2 = \pi(12)^2\,\text{cm}^2$ and $A_3 = \pi(9)^2\,\text{cm}^2$
Hence volume of container

$$= \frac{24}{6}[\pi(15)^2 + 4\pi(12)^2 + \pi(9)^2]$$

$$= 4[706.86 + 1809.56 + 254.47]$$

$$= 11\,080\,\text{cm}^3 = \frac{11\,080}{1000}\,\text{litres}$$

$$= \textbf{11 litres, correct to the nearest litre}$$

(Check: Volume of frustum of cone

$$= \tfrac{1}{3}\pi h[R^2 + Rr + r^2] \quad \text{from Section 20.4}$$

$$= \tfrac{1}{3}\pi(24)[(15)^2 + (15)(9) + (9)^2]$$

$$= 11\,080\,\text{cm}^3\,(\text{as shown above})$$

> **Problem 28.** A frustum of a sphere of radius 13 cm is formed by two parallel planes on opposite sides of the centre, each at distance of 5 cm from the centre. Determine the volume of the frustum (a) by using the prismoidal rule, and (b) by using the formula for the volume of a frustum of a sphere

The frustum of the sphere is shown by the section in Fig. 20.24.

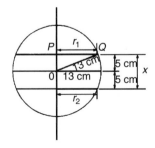

Figure 20.24

Radius $r_1 = r_2 = PQ = \sqrt{13^2 - 5^2} = 12\,\text{cm}$, by Pythagoras' theorem.

(a) Using the prismoidal rule, volume of frustum,

$$V = \frac{x}{6}[A_1 + 4A_2 + A_3]$$

$$= \frac{10}{6}[\pi(12)^2 + 4\pi(13)^2 + \pi(12)^2]$$

$$= \frac{10\pi}{6}[144 + 676 + 144] = \textbf{5047 cm}^3$$

(b) Using the formula for the volume of a frustum of a sphere:

$$\text{Volume} \quad V = \frac{\pi h}{6}(h^2 + 3r_1^2 + 3r_2^2)$$

$$= \frac{\pi\,(10)}{6}[10^2 + 3(12)^2 + 3(12)^2]$$

$$= \frac{10\pi}{6}(100 + 432 + 432)$$

$$= \textbf{5047 cm}^3$$

> **Problem 29.** A hole is to be excavated in the form of a prismoid. The bottom is to be a rectangle 16 m long by 12 m wide; the top is also a rectangle, 26 m long by 20 m wide. Find the volume of earth to be removed, correct to 3 significant figures, if the depth of the hole is 6.0 m

The prismoid is shown in Fig. 20.25. Let A_1 represent the area of the top of the hole, i.e. $A_1 = 20 \times 26 = 520\,\text{m}^2$. Let A_3 represent the area of the bottom of the hole, i.e. $A_3 = 16 \times 12 = 192\,\text{m}^2$. Let A_2 represent the rectangular area through the middle of the hole parallel to areas A_1 and A_2. The length of this rectangle is $(26 + 16)/2 = 21\,\text{m}$ and the width is $(20 + 12)/2 = 16\,\text{m}$, assuming the sloping edges are uniform. Thus area $A_2 = 21 \times 16 = 336\,\text{m}^2$.

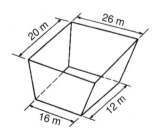

Figure 20.25

Using the prismoidal rule,

$$\text{volume of hole} = \frac{x}{6}[A_1 + 4A_2 + A_3]$$

$$= \frac{6}{6}[520 + 4(336) + 192]$$

$$= 2056\,\text{m}^3 = \mathbf{2060\,m^3},$$

correct to 3 significant figures.

Problem 30. The roof of a building is in the form of a frustum of a pyramid with a square base of side 5.0 m. The flat top is a square of side 1.0 m and all the sloping sides are pitched at the same angle. The vertical height of the flat top above the level of the eaves is 4.0 m. Calculate, using the prismoidal rule, the volume enclosed by the roof

---

Let area of top of frustum be $A_1 = (1.0)^2 = 1.0\,\text{m}^2$
Let area of bottom of frustum be $A_3 = (5.0)^2 = 25.0\,\text{m}^2$
Let area of section through the middle of the frustum parallel to A_1 and A_3 be A_2. The length of the side of the square forming A_2 is the average of the sides forming A_1 and A_3, i.e. $(1.0+5.0)/2 = 3.0\,\text{m}$. Hence $A_2 = (3.0)^2 = 9.0\,\text{m}^2$
Using the prismoidal rule,

$$\text{volume of frustum} = \frac{x}{6}[A_1 + 4A_2 + A_3]$$

$$= \frac{4.0}{6}[1.0 + 4(9.0) + 25.0]$$

Hence, **volume enclosed by roof $= 41.3\,\text{m}^3$**

Now try the following exercise

Exercise 79 Further problems on the prismoidal rule

1. Use the prismoidal rule to find the volume of a frustum of a sphere contained between two parallel planes on opposite sides of the centre each of radius 7.0 cm and each 4.0 cm from the centre. [1500 cm³]

2. Determine the volume of a cone of perpendicular height 16.0 cm and base diameter 10.0 cm by using the prismoidal rule. [418.9 cm³]

3. A bucket is in the form of a frustum of a cone. The diameter of the base is 28.0 cm and the diameter of the top is 42.0 cm. If the height is 32.0 cm, determine the capacity of the bucket (in litres) using the prismoidal rule (1 litre = 1000 cm³). [31.20 litres]

4. Determine the capacity of a water reservoir, in litres, the top being a 30.0 m by 12.0 m rectangle, the bottom being a 20.0 m by 8.0 m rectangle and the depth being 5.0 m (1 litre = 1000 cm³). [1.267×10^6 litre]

20.7 Volumes of similar shapes

The volumes of similar bodies are proportional to the cubes of corresponding linear dimensions.

For example, Fig. 20.26 shows two cubes, one of which has sides three times as long as those of the other.

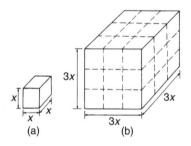

Figure 20.26

Volume of Fig. 20.26(a) $= (x)(x)(x) = x^3$

Volume of Fig. 20.26(b) $= (3x)(3x)(3x) = 27x^3$

Hence Fig. 20.26(b) has a volume $(3)^3$, i.e. 27 times the volume of Fig. 20.26(a).

Problem 31. A car has a mass of 1000 kg. A model of the car is made to a scale of 1 to 50. Determine the mass of the model if the car and its model are made of the same material

---

$$\frac{\text{Volume of model}}{\text{Volume of car}} = \left(\frac{1}{50}\right)^3$$

since the volume of similar bodies are proportional to the cube of corresponding dimensions.

Mass = density × volume, and since both car and model are made of the same material then:

$$\frac{\text{Mass of model}}{\text{Mass of car}} = \left(\frac{1}{50}\right)^3$$

Hence mass of model = (mass of car) $\left(\dfrac{1}{50}\right)^3$

$$= \frac{1000}{50^3}$$

$$= \mathbf{0.008\,kg} \quad \text{or} \quad \mathbf{8\,g}$$

Now try the following exercise

Exercise 80   Further problems on volumes of similar shapes

1. The diameter of two spherical bearings are in the ratio 2:5. What is the ratio of their volumes?   [8:125]

2. An engineering component has a mass of 400 g. If each of its dimensions are reduced by 30% determine its new mass.   [137.2 g]

Chapter 21

Irregular areas and volumes and mean values of waveforms

21.1   Area of irregular figures

Area of irregular plane surfaces may be approximately determined by using (a) a planimeter, (b) the trapezoidal rule, (c) the mid-ordinate rule, and (d) Simpson's rule. Such methods may be used, for example, by engineers estimating areas of indicator diagrams of steam engines, surveyors estimating areas of plots of land or naval architects estimating areas of water planes or transverse sections of ships.

(a)   A **planimeter** is an instrument for directly measuring small areas bounded by an irregular curve.

(b)   **Trapezoidal rule**
To determine the areas *PQRS* in Fig. 21.1:

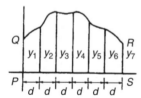

Figure 21.1

(i)   Divide base *PS* into any number of equal intervals, each of width *d* (the greater the number of intervals, the greater the accuracy).
(ii)   Accurately measure ordinates y_1, y_2, y_3, etc.

(iii)   Area *PQRS*

$$= d\left[\frac{y_1 + y_7}{2} + y_2 + y_3 + y_4 + y_5 + y_6\right]$$

In general, the trapezoidal rule states:

$$\text{Area} = \left(\begin{array}{c}\text{width of}\\\text{interval}\end{array}\right)\left[\frac{1}{2}\left(\begin{array}{c}\text{first + last}\\\text{ordinate}\end{array}\right) + \left(\begin{array}{c}\text{sum of}\\\text{remaining}\\\text{ordinates}\end{array}\right)\right]$$

(c)   **Mid-ordinate rule**
To determine the area *ABCD* of Fig. 21.2:

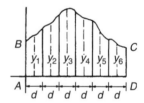

Figure 21.2

(i)   Divide base *AD* into any number of equal intervals, each of width *d* (the greater the number of intervals, the greater the accuracy).
(ii)   Erect ordinates in the middle of each interval (shown by broken lines in Fig. 21.2).
(iii)   Accurately measure ordinates y_1, y_2, y_3, etc.
(iv)   Area *ABCD*

$$= d(y_1 + y_2 + y_3 + y_4 + y_5 + y_6).$$

In general, the mid-ordinate rule states:

$$\text{Area} = \begin{pmatrix} \text{width of} \\ \text{interval} \end{pmatrix} \begin{pmatrix} \text{sum of} \\ \text{mid-ordinates} \end{pmatrix}$$

(d) **Simpson's rule**

To determine the area $PQRS$ of Fig. 21.1:

(i)   Divide base PS into an **even** number of intervals, each of width d (the greater the number of intervals, the greater the accuracy).

(ii) Accurately measure ordinates y_1, y_2, y_3, etc.

(iii) Area $PQRS$

$$= \frac{d}{3}[(y_1 + y_7) + 4(y_2 + y_4 + y_6) + 2(y_3 + y_5)]$$

In general, Simpson's rule states:

$$\text{Area} = \frac{1}{3}\begin{pmatrix} \text{width of} \\ \text{interval} \end{pmatrix} \times \left[\begin{pmatrix} \text{first + last} \\ \text{ordinate} \end{pmatrix} \right.$$
$$+ 4\begin{pmatrix} \text{sum of even} \\ \text{ordinates} \end{pmatrix}$$
$$\left. + 2\begin{pmatrix} \text{sum of remaining} \\ \text{odd ordinates} \end{pmatrix} \right]$$

Problem 1. A car starts from rest and its speed is measured every second for 6 s:

Time t(s)	0	1	2	3	4	5	6
Speed v (m/s)	0	2.5	5.5	8.75	12.5	17.5	24.0

Determine the distance travelled in 6 seconds (i.e. the are under the v/t graph), by (a) the trapezoidal rule, (b) the mid-ordinate rule, and (c) Simpson's rule

---

A graph of speed/time is shown in Fig. 21.3.

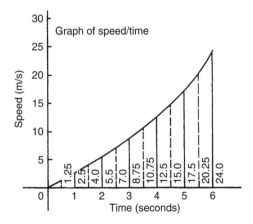

Figure 21.3

(a) **Trapezodial rule** (see para. (b) above).

The time base is divided into 6 strips each of width 1 s, and the length of the ordinates measured. Thus

$$\text{area} = (1)\left[\left(\frac{0 + 24.0}{2}\right) + 2.5 + 5.5 + 8.75 \right.$$
$$\left. + 12.5 + 17.5\right] = \mathbf{58.75 \ m}$$

(b) **Mid-ordinate rule** (see para. (c) above).

The time base is divided into 6 strips each of width 1 second. Mid-ordinates are erected as shown in Fig. 21.3 by the broken lines. The length of each mid-ordinate is measured. Thus

$$\text{area} = (1)[1.25 + 4.0 + 7.0 + 10.75$$
$$+ 15.0 + 20.25] = \mathbf{58.25 \ m}$$

(c) **Simpson's rule** (see para. (d) above).

The time base is divided into 6 strips each of width 1 s, and the length of the ordinates measured. Thus

$$\text{area} = \frac{1}{3}(1)[(0 + 24.0) + 4(2.5 + 8.75$$
$$+ 17.5) + 2(5.5 + 12.5)] = \mathbf{58.33 \ m}$$

Problem 2. A river is 15 m wide. Soundings of the depth are made at equal intervals of 3 m across the river and are as shown below.

Depth (m)	0	2.2	3.3	4.5	4.2	2.4	0

Calculate the cross-sectional area of the flow of water at this point using Simpson's rule

---

From para. (d) above,

$$\text{Area} = \frac{1}{3}(3)[(0 + 0) + 4(2.2 + 4.5 + 2.4)$$
$$+ 2(3.3 + 4.2)]$$
$$= (1)[0 + 36.4 + 15] = \mathbf{51.4 \ m^2}$$

Now try the following exercise

Exercise 81   Further problems on areas of irregular figures

1. Plot a graph of $y = 3x - x^2$ by completing a table of values of y from $x = 0$ to $x = 3$. Determine the area enclosed by the curve, the

x-axis and ordinate $x = 0$ and $x = 3$ by (a) the trapezoidal rule, (b) the mid-ordinate rule and (c) by Simpson's rule. [4.5 square units]

2. Plot the graph of $y = 2x^2 + 3$ between $x = 0$ and $x = 4$. Estimate the area enclosed by the curve, the ordinates $x = 0$ and $x = 4$, and the x-axis by an approximate method.
[54.7 square units]

3. The velocity of a car at one second intervals is given in the following table:

time t(s)	0	1	2	3	4	5	6
velocity v (m/s)	0	2.0	4.5	8.0	14.0	21.0	29.0

Determine the distance travelled in 6 seconds (i.e. the area under the v/t graph) using Simpson's rule. [63.33 m]

4. The shape of a piece of land is shown in Fig. 21.4. To estimate the area of the land, a surveyor takes measurements at intervals of 50 m, perpendicular to the straight portion with the results shown (the dimensions being in metres). Estimate the area of the land in hectares (1 ha $= 10^4$ m^2). [4.70 ha]

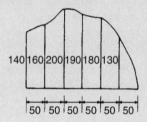

140 160 200 190 180 130

50 50 50 50 50 50

Figure 21.4

5. The deck of a ship is 35 m long. At equal intervals of 5 m the width is given by the following table:

Width (m)	0	2.8	5.2	6.5	5.8	4.1	3.0	2.3

Estimate the area of the deck. [143 m^2]

21.2 Volumes of irregular solids

If the cross-sectional areas $A_1, A_2, A_3, \ldots$ of an irregular solid bounded by two parallel planes are known at equal

intervals of width d (as shown in Fig. 21.5), then by Simpson's rule:

$$\text{Volume, } V = \frac{d}{3}\left[\begin{array}{l}(A_1 + A_7) + 4(A_2 + A_4 + A_6) \\ \qquad\qquad + 2(A_3 + A_5)\end{array}\right]$$

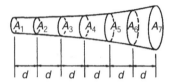

Figure 21.5

Problem 3. A tree trunk is 12 m in length and has a varying cross-section. The cross-sectional areas at intervals of 2 m measured from one end are:

0.52, 0.55, 0.59, 0.63, 0.72, 0.84, 0.97 m^2

Estimate the volume of the tree trunk

A sketch of the tree trunk is similar to that shown in Fig. 21.5, where $d = 2$ m, $A_1 = 0.52$ m^2, $A_2 = 0.55$ m^2, and so on.
Using Simpson's rule for volumes gives:

$$\text{Volume} = \frac{2}{3}[(0.52 + 0.97) + 4(0.55$$

$$+ 0.63 + 0.84) + 2(0.59 + 0.72)]$$

$$= \frac{2}{3}[1.49 + 8.08 + 2.62] = \mathbf{8.13\,m^3}$$

Problem 4. The areas of seven horizontal cross-sections of a water reservoir at intervals of 10 m are:

210, 250, 320, 350, 290, 230, 170 m^2

Calculate the capacity of the reservoir in litres

Using Simpson's rule for volumes gives:

$$\text{Volume} = \frac{10}{3}[(210 + 170) + 4(250$$

$$+ 350 + 230) + 2(320 + 290)]$$

$$= \frac{10}{3}[380 + 3320 + 1220]$$

$$= \mathbf{16\,400\,m^3}$$

$$16\,400\,\text{m} = 16\,400 \times 10^6\,\text{cm}^3$$

Since 1 litre $= 1000 \, \text{cm}^3$, capacity of reservoir

$$= \frac{16\,400 \times 10^6}{1000} \, \text{litres}$$

$$= 16\,400\,000 = \mathbf{1.64 \times 10^7 \, litres}$$

Now try the following exercise

Exercise 82 Further problems on volumes of irregular solids

1. The areas of equidistantly spaced sections of the underwater form of a small boat are as follows:

    1.76, 2.78, 3.10, 3.12, 2.61, 1.24, 0.85 m^2

    Determine the underwater volume if the sections are 3 m apart.   [42.59 m^3].

2. To estimate the amount of earth to be removed when constructing a cutting the cross-sectional area at intervals of 8 m were estimated as follows:

    0, 2.8, 3.7, 4.5, 4.1, 2.6, 0 m^3

    Estimate the volume of earth to be excavated.   [147 m^3]

3. The circumference of a 12 m long log of timber of varying circular cross-section is measured at intervals of 2 m along its length and the results are:

Distance from one end (m)	Circumference (m)
0	2.80
2	3.25
4	3.94
6	4.32
8	5.16
10	5.82
12	6.36

    Estimate the volume of the timber in cubic metres.   [20.42 m^3]

21.3 The mean or average value of a waveform

The mean or average value, y, of the waveform shown in Fig. 21.6 is given by:

$$y = \frac{\text{area under curve}}{\text{length of base, } b}$$

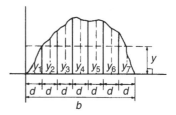

Figure 21.6

If the mid-ordinate rule is used to find the area under the curve, then:

$$y = \frac{\text{sum of mid-ordinates}}{\text{number of mid-ordinates}}$$

$$\left(= \frac{y_1 + y_2 + y_3 + y_4 + y_5 + y_6 + y_7}{7} \right.$$

$$\left. \text{for Fig. 21.6} \right)$$

For a **sine wave**, the mean or average value:

(i) over one complete cycle is zero (see Fig. 21.7(a)),

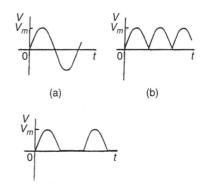

Figure 21.7

(ii) over half a cycle is **0.637 × maximum value**, or **2/π × maximum value**,

(iii) of a full-wave rectified waveform (see Fig. 21.7(b)) is **0.637 × maximum value**,

(iv)   of a half-wave rectified waveform (see Fig. 21.7(c)) is **0.318 × maximum value, or $\dfrac{1}{\pi}$ × maximum value,**

Problem 5.   Determine the average values over half a cycle of the periodic waveforms shown in Fig. 21.8

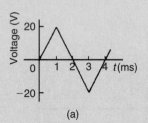

(a)

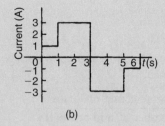

(b)

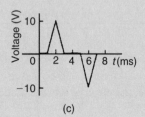

(c)

Figure 21.8

(a)   Area under triangular waveform (a) for a half cycle is given by:

$$\text{Area} = \frac{1}{2}(\text{base})(\text{perpendicular height})$$

$$= \frac{1}{2}(2 \times 10^{-3})(20)$$

$$= 20 \times 10^{-3}\,\text{Vs}$$

Average value of waveform

$$= \frac{\text{area under curve}}{\text{length of base}}$$

$$= \frac{20 \times 10^{-3}\,\text{Vs}}{2 \times 10^{-3}\,\text{s}} = \mathbf{10\,V}$$

(b)   Area under waveform (b) for a half cycle

$$= (1 \times 1) + (3 \times 2) = 7\,\text{As}$$

Average value of waveform

$$= \frac{\text{area under curve}}{\text{length of base}}$$

$$= \frac{7\,\text{As}}{3\,\text{s}} = \mathbf{2.33\,A}$$

(c)   A half cycle of the voltage waveform (c) is completed in 4 ms.

Area under curve

$$= \frac{1}{2}\{(3 - 1)10^{-3}\}(10)$$

$$= 10 \times 10^{-3}\,\text{Vs}$$

Average value of waveform

$$= \frac{\text{area under curve}}{\text{length of base}}$$

$$= \frac{10 \times 10^{-3}\,\text{Vs}}{4 \times 10^{-3}\,\text{s}} = \mathbf{2.5\,V}$$

Problem 6.   Determine the mean value of current over one complete cycle of the periodic waveforms shown in Fig. 21.9

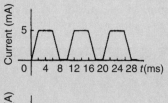

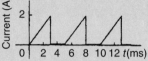

Figure 21.9

(a)   One cycle of the trapezoidal waveform (a) is completed in 10 ms (i.e. the periodic time is 10 ms).

Area under curve = area of trapezium

$$= \frac{1}{2}(\text{sum of parallel sides})(\text{perpendicular}$$

distance between parallel sides)

$$= \frac{1}{2}\{(4+8) \times 10^{-3}\}(5 \times 10^{-3})$$

$$= 30 \times 10^{-6}\,\text{As}$$

Mean value over one cycle

$$= \frac{\text{area under curve}}{\text{length of base}}$$

$$= \frac{30 \times 10^{-6}\,\text{As}}{10 \times 10^{-3}\,\text{s}} = \mathbf{3\,mA}$$

(b) One cycle of the sawtooth waveform (b) is completed in 5 ms.
Area under curve

$$= \frac{1}{2}(3 \times 10^{-3})(2) = 3 \times 10^{-3}\,\text{As}$$

Mean value over one cycle

$$= \frac{\text{area under curve}}{\text{length of base}}$$

$$= \frac{3 \times 10^{-3}\,\text{As}}{5 \times 10^{-3}\,\text{s}} = \mathbf{0.6\,A}$$

> **Problem 7.** The power used in a manufacturing process during a 6 hour period is recorded at intervals of 1 hour as shown below
>
> | Time (h) | 0 | 1 | 2 | 3 | 4 | 5 | 6 |
> |------------|---|----|----|----|----|----|---|
> | Power (kW) | 0 | 14 | 29 | 51 | 45 | 23 | 0 |
>
> Plot a graph of power against time and, by using the mid-ordinate rule, determine (a) the area under the curve and (b) the average value of the power

The graph of power/time is shown in Fig. 21.10.

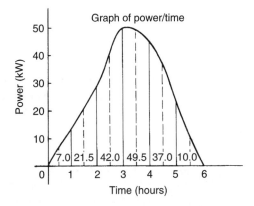

Figure 21.10

(a) The time base is divided into 6 equal intervals, each of width 1 hour. Mid-ordinates are erected (shown by broken lines in Fig. 21.10) and measured. The values are shown in Fig. 21.10.

Area under curve

$$= (\text{width of interval})(\text{sum of mid-ordinates})$$

$$= (1)[7.0 + 21.5 + 42.0 + 49.5 + 37.0 + 10.0]$$

$$= \mathbf{167\,kWh}$$

(i.e. a measure of electrical energy)

(b) Average value of waveform

$$= \frac{\text{area under curve}}{\text{length of base}}$$

$$= \frac{167\,\text{kWh}}{6\,\text{h}} = \mathbf{27.83\,kW}$$

Alternatively, average value

$$= \frac{\text{Sum of mid-ordinates}}{\text{number of mid-ordinate}}$$

> **Problem 8.** Figure 21.11 shows a sinusoidal output voltage of a full-wave rectifier. Determine, using the mid-ordinate rule with 6 intervals, the mean output voltage

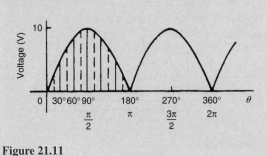

Figure 21.11

One cycle of the output voltage is completed in π radians or $180°$. The base is divided into 6 intervals, each of width $30°$. The mid-ordinate of each interval will lie at $15°$, $45°$, $75°$, etc.

At $15°$ the height of the mid-ordinate is $10 \sin 15° = 2.588\,\text{V}$

At $45°$ the height of the mid-ordinate is $10 \sin 45° = 7.071\,\text{V}$, and so on.

The results are tabulated below:

Mid-ordinate	Height of mid-ordinate
15°	$10 \sin 15° = 2.588$ V
45°	$10 \sin 45° = 7.071$ V
75°	$10 \sin 75° = 9.659$ V
105°	$10 \sin 105° = 9.659$ V
135°	$10 \sin 135° = 7.071$ V
165°	$10 \sin 165° = 2.588$ V
Sum of mid-ordinates $= 38.636$ V	

Mean or average value of output voltage

$$= \frac{\text{sum of mid-ordinates}}{\text{number of mid-ordinate}}$$

$$= \frac{38.636}{6} = \mathbf{6.439 \text{ V}}$$

(With a larger number of intervals a more accurate answer may be obtained.)

For a sine wave the actual mean value is $0.637 \times$ maximum value, which in this problem gives 6.37 V

Problem 9. An indicator diagram for a steam engine is shown in Fig. 21.12. The base line has been divided into 6 equally spaced intervals and the lengths of the 7 ordinates measured with the results shown in centimetres. Determine (a) the area of the indicator diagram using Simpson's rule, and (b) the mean pressure in the cylinder given that 1 cm represents 100 kPa

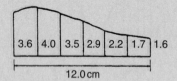

| 3.6 | 4.0 | 3.5 | 2.9 | 2.2 | 1.7 | 1.6 |

12.0 cm

Figure 21.12

(a) The width of each interval is $\dfrac{12.0}{6}$ cm. Using Simpson's rule,

$$\text{area} = \frac{1}{3}(2.0)[(3.6 + 1.6) + 4(4.0$$
$$+ 2.9 + 1.7) + 2(3.5 + 2.2)]$$

$$= \frac{2}{3}[5.2 + 34.4 + 11.4]$$

$$= \mathbf{34 \text{ cm}^2}$$

(b) Mean height of ordinates

$$= \frac{\text{area of diagram}}{\text{length of base}} = \frac{34}{12} = 2.83 \text{ cm}$$

Since 1 cm represents 100 kPa, the mean pressure in the cylinder

$$= 2.83 \text{ cm} \times 100 \text{ kPa/cm} = \mathbf{283 \text{ kPa}}$$

Now try the following exercise

Exercise 83   Further problems on mean or average values of waveforms

1. Determine the mean value of the periodic waveforms shown in Fig. 21.13 over a half cycle.   [(a) 2 A   (b) 50 V   (c) 2.5 A]

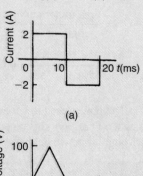

(a)

(b)

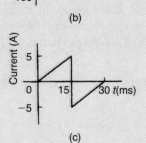

(c)

Figure 21.13

2. Find the average value of the periodic waveform shown in Fig. 21.14 over one complete cycle.   [(a) 2.5 V   (b) 3 A]

3. An alternating current has the following values at equal intervals of 5 ms.

Time (ms)	0	5	10	15	20	25	30
Current (A)	0	0.9	2.6	4.9	5.8	3.5	0

Plot a graph of current against time and estimate the area under the curve over the 30 ms period using the mid-ordinate rule and determine its mean value.   [0.093 As, 3.1 A]

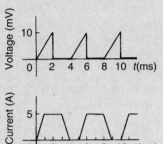

Figure 21.14

4. Determine, using an approximate method, the average value of a sine wave of maximum value 50 V for (a) a half cycle and (b) a complete cycle.

    [(a) 31.83 V   (b) 0]

5. An indicator diagram of a steam engine is 12 cm long. Seven evenly spaced ordinates, including the end ordinates, are measured as follows:

    5.90, 5.52, 4.22, 3.63, 3.32, 3.24, 3.16 cm

    Determine the area of the diagram and the mean pressure in the cylinder if 1 cm represents 90 kPa.   [49.13 cm^2, 368.5 kPa]

This Revision test covers the material contained in Chapters 18 to 21. *The marks for each question are shown in brackets at the end of each question.*

1. A swimming pool is 55 m long and 10 m wide. The perpendicular depth at the deep end is 5 m and at the shallow end is 1.5 m, the slope from one end to the other being uniform. The inside of the pool needs two coats of a protective paint before it is filled with water. Determine how many litres of paint will be needed if 1 litre covers 10 m². (7)

2. A steel template is of the shape shown in Fig. R5.1, the circular area being removed. Determine the area of the template, in square centimetres, correct to 1 decimal place. (7)

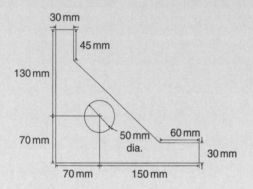

Figure R5.1

3. The area of a plot of land on a map is 400 mm². If the scale of the map is 1 to 50 000, determine the true area of the land in hectares (1 hectare = 10^4 m²). (3)

4. Determine the shaded area in Fig. R5.2, correct to the nearest square centimetre. (3)

5. Determine the diameter of a circle whose circumference is 178.4 cm. (2)

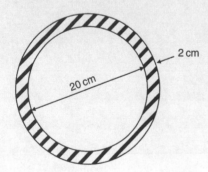

Figure R5.2

6. Convert
   (a) 125°47′ to radians
   (b) 1.724 radians to degrees and minutes (2)

7. Calculate the length of metal strip needed to make the clip shown in Fig. R5.3. (6)

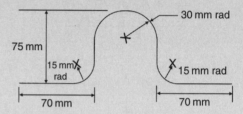

Figure R5.3

8. A lorry has wheels of radius 50 cm. Calculate the number of complete revolutions a wheel makes (correct to the nearest revolution) when travelling 3 miles (assume 1 mile = 1.6 km). (5)

9. The equation of a circle is:
$$x^2 + y^2 + 12x - 4y + 4 = 0.$$
Determine (a) the diameter of the circle, and (b) the coordinates of the centre of the circle. (5)

10. Determine the volume (in cubic metres) and the total surface area (in square metres) of a solid metal cone of base radius 0.5 m and perpendicular height 1.20 m. Give answers correct to 2 decimal places. (5)

11. Calculate the total surface area of a 10 cm by 15 cm rectangular pyramid of height 20 cm. (5)

12. A water container is of the form of a central cylindrical part 3.0 m long and diameter 1.0 m, with a hemispherical section surmounted at each end as shown in Fig. R5.4. Determine the maximum capacity of the container, correct to the nearest litre. (1 litre = 1000 cm³.) (5)

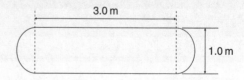

Figure R5.4

13. Find the total surface area of a bucket consisting of an inverted frustum of a cone, of slant height 35.0 cm and end diameters 60.0 cm and 40.0 cm. (4)

14. A boat has a mass of 20 000 kg. A model of the boat is made to a scale of 1 to 8. If the model is made of the same material as the boat, determine the mass of the model (in grams). (3)

15. Plot a graph of $y = 3x^2 + 5$ from $x = 1$ to $x = 4$. Estimate, correct to 2 decimal places, using 6 intervals, the area enclosed by the curve, the ordinates $x = 1$ and $x = 4$, and the x-axis by (a) the trapezoidal rule, (b) the mid-ordinate rule, and (c) Simpson's rule. (12)

16. A vehicle starts from rest and its velocity is measured every second for 6 seconds, with the following results:

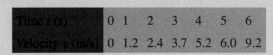

Time t (s)	0	1	2	3	4	5	6
Velocity v (m/s)	0	1.2	2.4	3.7	5.2	6.0	9.2

Using Simpson's rule, calculate (a) the distance travelled in 6 s (i.e. the area under the v/t graph) and (b) the average speed over this period. (6)

Section 2

Section 3

Trigonometry

Chapter 22

Introduction to trigonometry

22.1 Trigonometry

Trigonometry is the branch of mathematics that deals with the measurement of sides and angles of triangles, and their relationship with each other. There are many applications in engineering where knowledge of trigonometry is needed.

22.2 The theorem of Pythagoras

With reference to Fig. 22.1, the side opposite the right angle (i.e. side b) is called the **hypotenuse**. The **theorem of Pythagoras** states:

'In any right-angle triangle, the square on the hypotenuse is equal to the sum of the squares on the other two sides.'

Hence $$b^2 = a^2 + c^2$$

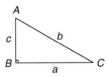

Figure 22.1

Problem 1. In Fig. 22.2, find the length of *EF*.

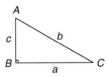

Figure 22.2

By Pythagoras' theorem: $e^2 = d^2 + f^2$

Hence $13^2 = d^2 + 5^2$

$$169 = d^2 + 25$$

$$d^2 = 169 - 25 = 144$$

Thus $d = \sqrt{144} = 12\,\text{cm}$

i.e. $EF = \mathbf{12\ cm}$

Problem 2. Two aircraft leave an airfield at the same time. One travels due north at an average speed of 300 km/h and the other due west at an average speed of 220 km/h. Calculate their distance apart after 4 hours

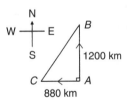

Figure 22.3

After 4 hours, the first aircraft has travelled $4 \times 300 = 1{,}200$ km, due north, and the second aircraft has travelled $4 \times 220 = 880$ km due west, as shown in Fig. 22.3. Distance apart after 4 hour $= BC$.
From Pythagoras' theorem:

$$BC^2 = 1200^2 + 880^2$$

$$= 1440000 + 774400 \text{ and}$$

$$BC = \sqrt{2214400}$$

Hence distance apart after 4 hours $= \mathbf{1488\ km}$

Now try the following exercise

Exercise 84   Further problems on the theorem of Pythagoras

1. In a triangle CDE, $D = 90°$, $CD = 14.83$ mm and $CE = 28.31$ mm. Determine the length of DE. [24.11 mm]

2. Triangle PQR is isosceles, Q being a right angle. If the hypotenuse is 38.47 cm find (a) the lengths of sides PQ and QR, and (b) the value of $\angle QPR$. [(a) 27.20 cm each (b) 45°]

3. A man cycles 24 km due south and then 20 km due east. Another man, starting at the same time as the first man, cycles 32 km due east and then 7 km due south. Find the distance between the two men. [20.81 km]

4. A ladder 3.5 m long is placed against a perpendicular wall with its foot 1.0 m from the wall. How far up the wall (to the nearest centimetre) does the ladder reach? If the foot of the ladder is now moved 30 cm further away from the wall, how far does the top of the ladder fall? [3.35 m, 10 cm]

5. Two ships leave a port at the same time. One travels due west at 18.4 km/h and the other due south at 27.6 km/h. Calculate how far apart the two ships are after 4 hours. [132.7 km]

6. Figure 22.4 shows a bolt rounded off at one end. Determine the dimension h. [2.94 mm]

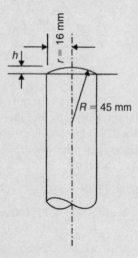

Figure 22.4

7. Figure 22.5 shows a cross-section of a component that is to be made from a round bar. If the diameter of the bar is 74 mm, calculate the dimension x. [24 mm]

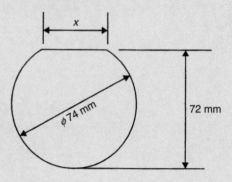

Figure 22.5

22.3   Trigonometric ratios of acute angles

(a) With reference to the right-angled triangle shown in Fig. 22.6:

(i) $\quad \sin\theta = \dfrac{\text{opposite side}}{\text{hypotenuse}}$

$\quad$ i.e. $\sin\theta = \dfrac{b}{c}$

(ii) $\quad \cos\text{ine }\theta = \dfrac{\text{adjacent side}}{\text{hypotenuse}}$

$\quad$ i.e. $\cos\theta = \dfrac{a}{c}$

(iii) $\quad \tan\text{gent }\theta = \dfrac{\text{opposite side}}{\text{adjacent side}}$

$\quad$ i.e. $\tan\theta = \dfrac{b}{a}$

(iv) $\quad \sec\text{ant }\theta = \dfrac{\text{hypotenuse}}{\text{adjacent side}}$

$\quad$ i.e. $\sec\theta = \dfrac{c}{a}$

(v) $\quad \csc\text{ant }\theta = \dfrac{\text{hypotenuse}}{\text{opposite side}}$

$\quad$ i.e. $\csc\theta = \dfrac{c}{b}$

(vi) cotangent $\theta = \dfrac{\text{adjacent side}}{\text{opposite side}}$

i.e. $\cot\theta = \dfrac{a}{b}$

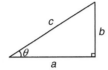

Figure 22.6

(b) From above,

(i) $\dfrac{\sin\theta}{\cos\theta} = \dfrac{\frac{b}{c}}{\frac{a}{c}} = \dfrac{b}{a} = \tan\theta,$

i.e. $\tan\theta = \dfrac{\sin\theta}{\cos\theta}$

(ii) $\dfrac{\cos\theta}{\sin\theta} = \dfrac{\frac{a}{c}}{\frac{b}{c}} = \dfrac{a}{b} = \cot\theta,$

i.e. $\cot\theta = \dfrac{\cos\theta}{\sin\theta}$

(iii) $\sec\theta = \dfrac{1}{\cos\theta}$

(iv) $\operatorname{cosec}\theta = \dfrac{1}{\sin\theta}$

(Note 's' and 'c' go together)

(v) $\cot\theta = \dfrac{1}{\tan\theta}$

Secants, cosecants and cotangents are called the **reciprocal ratios**.

Problem 3. If $\cos X = \dfrac{9}{41}$ determine the value of the other five trigonometric ratios

Figure 22.7 shows a right-angled triangle XYZ.

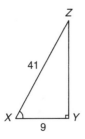

Figure 22.7

Since $\cos X = \dfrac{9}{41}$, then $XY = 9$ units and $XZ = 41$ units.

Using Pythagoras' theorem: $41^2 = 9^2 + YZ^2$ from which $YZ = \sqrt{41^2 - 9^2} = 40$ unts.

Thus, $\sin X = \dfrac{40}{41},\ \tan X = \dfrac{40}{9} = 4\dfrac{4}{9},$

$\operatorname{cosec} X = \dfrac{41}{40} = 1\dfrac{1}{40},\ \sec X = \dfrac{41}{9} = 4\dfrac{5}{9}$

and $\cot X = \dfrac{9}{40}$

Problem 4. If $\sin\theta = 0.625$ and $\cos\theta = 0.500$ determine the values of $\operatorname{cosec}\theta$, $\sec\theta$, $\tan\theta$ and $\cot\theta$

$\operatorname{cosec}\theta = \dfrac{1}{\sin\theta} = \dfrac{1}{0.625} = \mathbf{1.60}$

$\sec\theta = \dfrac{1}{\cos\theta} = \dfrac{1}{0.500} = \mathbf{2.00}$

$\tan\theta = \dfrac{\sin\theta}{\cos\theta} = \dfrac{0.625}{0.500} = \mathbf{1.25}$

$\cot\theta = \dfrac{\cos\theta}{\sin\theta} = \dfrac{0.500}{0.625} = \mathbf{0.80}$

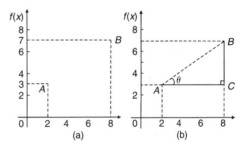

Figure 22.8

Problem 5. Point A lies at co-ordinate (2, 3) and point B at (8, 7). Determine (a) the distance AB, (b) the gradient of the straight line AB, and (c) the angle AB makes with the horizontal

(a) Points A and B are shown in Fig. 22.8(a). In Fig. 22.8(b), the horizontal and vertical lines AC and BC are constructed. Since ABC is a right-angled triangle, and $AC = (8 - 2) = 6$ and $BC = (7 - 3) = 4$, then by Pythagoras' theorem:

$$AB^2 = AC^2 + BC^2 = 6^2 + 4^2$$

and $AB = \sqrt{6^2 + 4^2} = \sqrt{52} = \mathbf{7.211},$

correct to 3 decimal places

(b) The gradient of AB is given by $\tan \theta$,

i.e. **gradient** $= \tan \theta = \dfrac{BC}{AC} = \dfrac{4}{6} = \dfrac{2}{3}$

(c) **The angle AB makes with the horizontal** is given by: $\tan^{-1} \dfrac{2}{3} = \mathbf{33.69°}$

Now try the following exercise

Exercise 85   Further problems on trigonometric ratios of acute angles

1. In triangle ABC shown in Fig. 22.9, find $\sin A$, $\cos A$, $\tan A$, $\sin B$, $\cos B$ and $\tan B$.

Figure 22.9

$$\left[\begin{array}{l} \sin A = \dfrac{3}{5}, \cos A = \dfrac{4}{5}, \tan A = \dfrac{3}{4} \\[2mm] \sin B = \dfrac{4}{5}, \cos B = \dfrac{3}{5}, \tan B = \dfrac{4}{3} \end{array}\right]$$

2. For the right-angled triangle shown in Fig. 22.10, find:

(a) $\sin \alpha$ (b) $\cos \theta$ (c) $\tan \theta$

Figure 22.10

$$\left[\text{(a)} \ \dfrac{15}{17} \quad \text{(b)} \ \dfrac{15}{17} \quad \text{(c)} \ \dfrac{8}{15}\right]$$

3. If $\cos A = \dfrac{12}{13}$ find $\sin A$ and $\tan A$, in fraction form.

$$\left[\sin A = \dfrac{5}{13}, \quad \tan A = \dfrac{5}{12}\right]$$

4. Point P lies at co-ordinate $(-3, 1)$ and point Q at $(5, -4)$. Determine (a) the distance PQ, (b) the gradient of the straight line PQ, and (c) the angle PQ makes with the horizontal.

$$[\text{(a)} \ 9.434 \quad \text{(b)} \ -0.625 \quad \text{(c)} \ -32°]$$

22.4 Fractional and surd forms of trigonometric ratios

In Fig. 22.11, ABC is an equilateral triangle of side 2 units. AD bisects angle A and bisects the side BC. Using Pythagoras' theorem on triangle ABD gives:

$$AD = \sqrt{2^2 - 1^2} = \sqrt{3}.$$

Hence, $\sin 30° = \dfrac{BD}{AB} = \dfrac{1}{2}$, $\cos 30° = \dfrac{AD}{AB} = \dfrac{\sqrt{3}}{2}$

and $\tan 30° = \dfrac{BD}{AD} = \dfrac{1}{\sqrt{3}}$

$\sin 60° = \dfrac{AD}{AB} = \dfrac{\sqrt{3}}{2}$, $\cos 60° = \dfrac{BD}{AB} = \dfrac{1}{2}$

and $\tan 60° = \dfrac{AD}{BD} = \sqrt{3}$

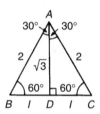

Figure 22.11

Figure 22.12

In Fig. 22.12, PQR is an isosceles triangle with $PQ = QR = 1$ unit. By Pythagoras' theorem, $PR = \sqrt{1^2 + 1^2} = \sqrt{2}$

Hence,

$$\sin 45° = \dfrac{1}{\sqrt{2}}, \quad \cos 45° = \dfrac{1}{\sqrt{2}} \text{ and } \tan 45° = 1$$

A quantity that is not exactly expressible as a rational number is called a **surd**. For example, $\sqrt{2}$ and $\sqrt{3}$ are called surds because they cannot be expressed as a fraction and the decimal part may be continued indefinitely. For example,

$$\sqrt{2} = 1.4142135\ldots, \text{ and } \sqrt{3} = 1.7320508\ldots$$

From above,

$$\sin 30° = \cos 60°, \sin 45° = \cos 45° \text{ and}$$

$$\sin 60° = \cos 30°.$$

In general,

$$\sin \theta = \cos(90° - \theta) \quad \text{and} \quad \cos \theta = \sin(90° - \theta)$$

For example, it may be checked by calculator that $\sin 25° = \cos 65°$, $\sin 42° = \cos 48°$ and $\cos 84° \, 10' = \sin 5° \, 50'$, and so on.

Problem 6. Using surd forms, evaluate:

$$\frac{3 \tan 60° - 2 \cos 30°}{\tan 30°}$$

From above, $\tan 60° = \sqrt{3}$, $\cos 30° = \dfrac{\sqrt{3}}{2}$ and $\tan 30° = \dfrac{1}{\sqrt{3}}$, hence

$$\frac{3 \tan 60° - 2 \cos 30°}{\tan 30°} = \frac{3\left(\sqrt{3}\right) - 2\left(\dfrac{\sqrt{3}}{2}\right)}{\dfrac{1}{\sqrt{3}}}$$

$$= \frac{3\sqrt{3} - \sqrt{3}}{\dfrac{1}{\sqrt{3}}} = \frac{2\sqrt{3}}{\dfrac{1}{\sqrt{3}}}$$

$$= 2\sqrt{3}\left(\dfrac{\sqrt{3}}{1}\right) = 2(3) = \mathbf{6}$$

Now try the following exercise

Exercise 86   Further problems on fractional and surd form of trigonometrical ratios

Evaluate the following, without using calculators, leaving where necessary in surd form:

1. $3 \sin 30° - 2 \cos 60°$ $\left[\dfrac{1}{2}\right]$

2. $5 \tan 60° - 3 \sin 60°$ $\left[\dfrac{7}{2}\sqrt{3}\right]$

3. $\dfrac{\tan 60°}{3 \tan 30°}$ $[1]$

4. $(\tan 45°)(4 \cos 60° - 2 \sin 60°)$ $[2 - \sqrt{3}]$

5. $\dfrac{\tan 60° - \tan 30°}{1 + \tan 30° \tan 60°}$ $\left[\dfrac{1}{\sqrt{3}}\right]$

22.5   Solution of right-angled triangles

To 'solve a right-angled triangle' means 'to find the unknown sides and angles'. This is achieved by using (i) the theorem of Pythagoras, and/or (ii) trigonometric ratios. This is demonstrated in the following problems.

Problem 7. In triangle PQR shown in Fig. 22.13, find the lengths of PQ and PR.

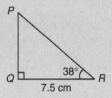

Figure 22.13

$$\tan 38° = \frac{PQ}{QR} = \frac{PQ}{7.5} \text{ hence}$$

$$PQ = 7.5 \tan 38° = 7.5(0.7813)$$

$$= \mathbf{5.860 \, cm}$$

$$\cos 38° = \frac{QR}{PR} = \frac{7.5}{PR} \text{ hence}$$

$$PR = \frac{7.5}{\cos 38°} = \frac{7.5}{0.7880} = \mathbf{9.518 \, cm}$$

[Check: Using Pythagoras' theorem $(7.5)^2 + (5.860)^2 = 90.59 = (9.518)^2$]

Problem 8. Solve the triangle ABC shown in Fig. 22.14

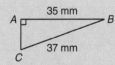

Figure 22.14

To 'solve triangle ABC' means 'to find the length AC and angles B and C'.

$$\sin C = \frac{35}{37} = 0.94595 \text{ hence}$$

$$\angle C = \sin^{-1} 0.94595 = \mathbf{71.08°} \text{ or } \mathbf{71°5'}$$

$\angle B = 180° - 90° - 71.08° = \mathbf{18.92°} \text{ or } \mathbf{18°55'}$ (since angles in a triangle add up to $180°$)

$$\sin B = \frac{AC}{37} \text{ hence}$$

$$\mathbf{AC} = 37 \sin 18.92° = 37(0.3242)$$

$$= \mathbf{12.0\,mm},$$

or, using Pythagoras' theorem, $37^2 = 35^2 + AC^2$, from which, $\mathbf{AC} = \sqrt{37^2 - 35^2} = \mathbf{12.0\,mm}$

Problem 9. Solve triangle XYZ given $\angle X = 90°$, $\angle Y = 23°17'$ and $YZ = 20.0$ mm. Determine also its area

It is always advisable to make a reasonably accurate sketch so as to visualize the expected magnitudes of unknown sides and angles. Such as sketch is shown in Fig. 22.15

$$\angle Z = 180° - 90° - 23°17' = \mathbf{66°43'}$$

$$\sin 23°17' = \frac{XZ}{20.0} \text{ hence } \mathbf{XZ} = 20.0 \sin 23°17'$$

$$= 20.0(0.3953)$$

$$= \mathbf{7.906\,mm}$$

$$\cos 23°17' = \frac{XY}{20.0} \text{ hence } \mathbf{XY} = 20.0 \cos 23°17'$$

$$= 20.0(0.9186)$$

$$= \mathbf{18.37\,mm}$$

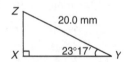

Figure 22.15

[Check: Using Pythagoras' theorem
$(18.37)^2 + (7.906)^2 = 400.0 = (20.0)^2$]

Area of triangle XYZ

$$= \frac{1}{2}(\text{base})(\text{perpendicular height})$$

$$= \frac{1}{2}(XY)(XZ) = \frac{1}{2}(18.37)(7.906)$$

$$= \mathbf{72.62\,mm^2}$$

Now try the following exercise

Exercise 87   Further problems on the solution of right-angled triangles

1. Solve triangle ABC in Fig. 22.16 (i).

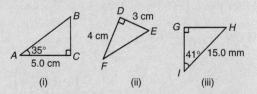

Figure 22.16

$[BC = 3.50\,cm, AB = 6.10\,cm, \angle B = 55°]$

2. Solve triangle DEF in Fig. 22.16 (ii).
$[FE = 5\,cm, \angle E = 53.13°, \angle F = 36.87°]$

3. Solve triangle GHI in Fig. 22.16(iii).
$[GH = 9.841\,mm, GI = 11.32\,mm, \angle H = 49°]$

4. Solve the triangle JKL in Fig. 22.17 (i) and find its area.

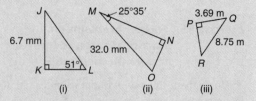

Figure 22.17

$$\begin{bmatrix} KL = 5.43\,cm,\ JL = 8.62\,cm, \\ \angle J = 39°,\ \text{area} = 18.19\,cm^2 \end{bmatrix}$$

5. Solve the triangle MNO in Fig. 22.17(ii) and find its area.

$$\begin{bmatrix} MN = 28.86\,mm,\ NO = 13.82\,mm. \\ \angle O = 64.42°,\ \text{area} = 199.4\,mm^2 \end{bmatrix}$$

6. Solve the triangle PQR in Fig. 22.17(iii) and find its area.

$$\begin{bmatrix} PR = 7.934\,\text{m}, \angle Q = 65.06°, \\ \angle R = 24.94°, \ \text{area} = 14.64\,\text{m}^2 \end{bmatrix}$$

7. A ladder rests against the top of the perpendicular wall of a building and makes an angle of 73° with the ground. If the foot of the ladder is 2 m from the wall, calculate the height of the building.   [6.54 m]

8. Determine the length x in Fig. 22.18.   [9.40 mm]

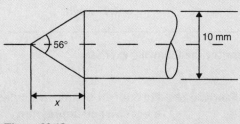

Figure 22.18

22.6   Angle of elevation and depression

(a) If, in Fig. 22.19, BC represents horizontal ground and AB a vertical flagpole, then the **angle of elevation** of the top of the flagpole, A, from the point C is the angle that the imaginary straight line AC must be raised (or elevated) from the horizontal CB, i.e. angle θ.

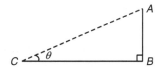

Figure 22.19

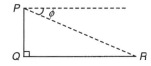

Figure 22.20

(b) If, in Fig. 22.20, PQ represents a vertical cliff and R a ship at sea, then the **angle of depression** of the ship from point P is the angle through which the imaginary straight line PR must be lowered (or depressed) from the horizontal to the ship, i.e. angle ϕ.

(Note, $\angle PRQ$ is also ϕ – alternate angles between parallel lines.)

> **Problem 10.** An electricity pylon stands on horizontal ground. At a point 80 m from the base of the pylon, the angle of elevation of the top of the pylon is 23°. Calculate the height of the pylon to the nearest metre

Figure 22.21 shows the pylon AB and the angle of elevation of A from point C is 23° and

$$\tan 23° = \frac{AB}{BC} = \frac{AB}{80}$$

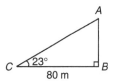

Figure 22.21

Hence, height of pylon $AB = 80 \tan 23°$

$$= 80(0.4245) = 33.96\,\text{m}$$

$$= \textbf{34 m to the nearest metre}.$$

> **Problem 11.** A surveyor measures the angle of elevation of the top of a perpendicular building as 19°. He move 120 m nearer the building and finds the angle of elevation is now 47°. Determine the height of the building

The building PQ and the angles of elevation are shown in Fig. 22.22

In triangle PQS, $\tan 19° = \dfrac{h}{x + 120}$

hence $h = \tan 19°(x + 120)$

i.e. $h = 0.3443(x + 120)$ (1)

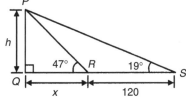

Figure 22.22

In triangle PQR, $\tan 47° = \dfrac{h}{x}$

hence $\quad h = \tan 47°(x)$, i.e. $h = 1.0724x$ $\qquad$ (2)

Equating equations (1) and (2) gives:

$$0.3443(x + 120) = 1.0724x$$

$$0.3443x + (0.3443)(120) = 1.0724x$$

$$(0.3443)(120) = (1.0724 - 0.3443)x$$

$$41.316 = 0.7281x$$

$$x = \frac{41.316}{0.7281} = 56.74\,\text{m}$$

From equation (2), **height of building**,

$$h = 1.0724x = 1.0724(56.74) = \mathbf{60.85\,m}$$

Problem 12. The angle of depression of a ship viewed at a particular instant from the top of a 75 m vertical cliff is 30°. Find the distance of the ship from the base of the cliff at this instant. The ship is sailing away from the cliff at constant speed and 1 minute later its angle of depression from the top of the cliff is 20°. Determine the speed of the ship in km/h

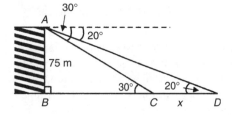

Figure 22.23

Figure 22.23 shows the cliff AB, the initial position of the ship at C and the final position at D. Since the angle of depression is initially 30° then $\angle ACB = 30°$ (alternate angles between parallel lines).

$$\tan 30° = \frac{AB}{BC} = \frac{75}{BC}$$

hence $\quad BC = \dfrac{75}{\tan 30°} = \dfrac{75}{0.5774}$

$$= \mathbf{129.9\,m}$$

$$= \textbf{initial position of ship from base of cliff}$$

In triangle ABD,

$$\tan 20° = \frac{AB}{BD} = \frac{75}{BC + CD} = \frac{75}{129.9 + x}$$

Hence

$$129.9 + x = \frac{75}{\tan 20°} = \frac{75}{0.3640} = 206.0\,\text{m}$$

from which,

$$x = 206.0 - 129.9 = 76.1\,\text{m}$$

Thus the ship sails 76.1 m in 1 minute, i.e. 60 s, hence,

$$\textbf{speed of ship} = \frac{\text{distance}}{\text{time}} = \frac{76.1}{60}\text{m/s}$$

$$= \frac{76.1 \times 60 \times 60}{60 \times 1,000}\text{km/h}$$

$$= \mathbf{4.57\,km/h}$$

Now try the following exercise

Exercise 88 Further problems on angles of elevation and depression

1. If the angle of elevation of the top of a vertical 30 m high aerial is 32°, how far is it to the aerial? [48 m]

2. From the top of a vertical cliff 80.0 m high the angles of depression of two buoys lying due west of the cliff are 23° and 15°, respectively. How far are the buoys apart? [110.1 m]

3. From a point on horizontal ground a surveyor measures the angle of elevation of the top of a flagpole as 18°40′. He moves 50 m nearer to the flagpole and measures the angle of elevation as 26°22′. Determine the height of the flagpole. [53.0 m]

4. A flagpole stands on the edge of the top of a building. At a point 200 m from the building the angles of elevation of the top and bottom of the pole are 32° and 30° respectively. Calculate the height of the flagpole. [9.50 m]

5. From a ship at sea, the angle of elevation of the top and bottom of a vertical lighthouse standing on the edge of a vertical cliff are 31° and 26°, respectively. If the lighthouse is 25.0 m high, calculate the height of the cliff. [107.8 m]

6. From a window 4.2 m above horizontal ground the angle of depression of the foot of a building

across the road is 24° and the angle of elevation of the top of the building is 34°. Determine, correct to the nearest centimetre, the width of the road and the height of the building.

[9.43 m, 10.56 m]

7. The elevation of a tower from two points, one due east of the tower and the other due west of it are 20° and 24°, respectively, and the two points of observation are 300 m apart. Find the height of the tower to the nearest metre.

[60 m]

22.7 Evaluating trigonometric ratios of any angles

The easiest method of evaluating trigonometric functions of any angle is by using a **calculator**. The following values, correct to 4 decimal places, may be checked:

$$\begin{aligned}
\text{sine } 18° &= 0.3090, \\
\text{sine } 172° &= 0.1392 \\
\text{sine } 241.63° &= -0.8799, \\
\text{cosine } 56° &= 0.5592 \\
\text{cosine } 115° &= -0.4226, \\
\text{cosine } 331.78° &= 0.8811 \\
\text{tangent } 29° &= 0.5543, \\
\text{tangent } 178° &= -0.0349 \\
\text{tangent } 296.42° &= -2.0127
\end{aligned}$$

To evaluate, say, sine 42°23′ using a calculating means finding sine $42\dfrac{23°}{60}$ since there are 60 minutes in 1 degree.

$\dfrac{23}{60} = 0.383\dot{3}$, thus $42°23′ = 42.383\dot{3}°$

Thus sine 42°23′ = sine 42.383\dot{3}° = 0.6741, correct to 4 decimal places.

Similarly, cosine 72°38′ = cosine $72\dfrac{38°}{60} = 0.2985$, correct to 4 decimal places.

Most calculators contain only sine, cosine and tangent functions. Thus to evaluate secants, cosecants and cotangents, reciprocals need to be used.

The following values, correct to 4 decimal places, may be checked:

$$\text{secant } 32° = \frac{1}{\cos 32°} = 1.1792$$

$$\text{cosecant } 75° = \frac{1}{\sin 75°} = 1.0353$$

$$\text{cotangent } 41° = \frac{1}{\tan 41°} = 1.1504$$

$$\text{secant } 215.12° = \frac{1}{\cos 215.12°} = -1.2226$$

$$\text{cosecant } 321.62° = \frac{1}{\sin 321.62°} = -1.6106$$

$$\text{cotangent } 263.59° = \frac{1}{\tan 263.59°} = 0.1123$$

Problem 13. Evaluate correct to 4 decimal places:

(a) sine 168°14′   (b) cosine 271.41°
(c) tangent 98°4′

(a) sine 168°14′ = sine $168\dfrac{14°}{60}$ = **0.2039**

(b) cosine 271.41° = **0.0246**

(c) tangent 98°4′ = $\tan 98\dfrac{4°}{60}$ = **−7.0558**

Problem 14. Evaluate, correct to 4 decimal places:

(a) secant 161°   (b) secant 302°29′

(a) sec 161° = $\dfrac{1}{\cos 161°}$ = **−1.0576**

(b) sec 302°29′ = $\dfrac{1}{\cos 302°29′}$ = $\dfrac{1}{\cos 302\dfrac{29°}{60}}$

= **1.8620**

Problem 15. Evaluate, correct to 4 significant figures:

(a) cosecant 279.16°   (b) cosecant 49°7′

(a) cosec 279.16° = $\dfrac{1}{\sin 279.16°}$ = **−1.013**

(b) cosec 49°7′ = $\dfrac{1}{\sin 49°7′}$ = $\dfrac{1}{\sin 49\dfrac{7°}{60}}$ = **1.323**

Problem 16. Evaluate, correct to 4 decimal places:
(a) cotangent 17.49°   (b) cotangent 163°52′

(a) $\cot 17.49° = \dfrac{1}{\tan 17.49°} = \mathbf{3.1735}$

(b) $\cot 163°52′ = \dfrac{1}{\tan 163°52′} = \dfrac{1}{\tan 163\frac{52°}{60}}$

$= \mathbf{-3.4570}$

Problem 17. Evaluate, correct to 4 significant figures:
(a) sin 1.481   (b) cos (3π/5)   (c) tan 2.93

(a)   sin 1.481 means the sine of 1.481 radians. Hence a calculator needs to be on the radian function.

Hence sin 1.481 = **0.9960**

(b) $\cos(3\pi/5) = \cos 1.884955\ldots = \mathbf{-0.3090}$

(c)   tan 2.93 = **−0.2148**

Problem 18. Evaluate, correct to 4 decimal places:
(a) secant 5.37   (b) cosecant π/4
(c) cotangent π/24

(a)   Again, with no degrees sign, it is assumed that 5.37 means 5.37 radians.

Hence $\sec 5.37 = \dfrac{1}{\cos 5.37} = \mathbf{1.6361}$

(b) $\csc(\pi/4) = \dfrac{1}{\sin(\pi/4)} = \dfrac{1}{\sin 0.785398\ldots}$

$= \mathbf{1.4142}$

(c) $\cot(5\pi/24) = \dfrac{1}{\tan(5\pi/24)} = \dfrac{1}{\tan 0.654498\ldots}$

$= \mathbf{1.3032}$

Problem 19. Determine the acute angles:
(a) $\sec^{-1} 2.3164$ (b) $\csc^{-1} 1.1784$
(c) $\cot^{-1} 2.1273$

(a) $\sec^{-1} 2.3164 = \cos^{-1}\left(\dfrac{1}{2.3164}\right)$

$= \cos^{-1} 0.4317\ldots$

$= \mathbf{64.42°}$ or **64°25′** or **1.124 radians**

(b) $\csc^{-1} 1.1784 = \sin^{-1}\left(\dfrac{1}{1.1784}\right)$

$= \sin^{-1} 0.8486\ldots$

$= \mathbf{58.06°}$ or **58°4′** or **1.013 radians**

(c) $\cot^{-1} 2.1273 = \tan^{-1}\left(\dfrac{1}{2.1273}\right)$

$= \tan^{-1} 0.4700\ldots$

$= \mathbf{25.18°}$ or **25°11′** or **0.439 radians**

Problem 20. Evaluate the following expression, correct to 4 significant figures:

$$\frac{4 \sec 32°10′ - 2 \cot 15°19′}{3 \csc 63°8′ \tan 14°57′}$$

By calculator:

$\sec 32°10′ = 1.1813, \quad \cot 15°19′ = 3.6512$

$\csc 63°8′ = 1.1210, \quad \tan 14°57′ = 0.2670$

Hence

$$\frac{4 \sec 32°10′ - 2 \cot 15°19′}{3 \csc 63°8′ \tan 14°57′}$$

$$= \frac{4(1.1813) - 2(3.6512)}{3(1.1210)(0.2670)}$$

$$= \frac{4.7252 - 7.3024}{0.8979} = \frac{-2.5772}{0.8979}$$

$= \mathbf{-2.870}$, correct to 4 significant figures.

Problem 21. Evaluate correct to 4 decimal places:
(a) sec (−115°)   (b) cosec (−95°17′)

(a)   Positive angles are considered by convention to be anticlockwise and negative angles as clockwise.

Hence −115° is actually the same as 245° (i.e. 360° − 115°).

Hence $\sec(-115°) = \sec 245° = \dfrac{1}{\cos 245°}$

$= \mathbf{-2.3662}$

(b) $\csc(-95°47′) = \dfrac{1}{\sin\left(-95\frac{47°}{60}\right)} = \mathbf{-1.0051}$

Section 3

Now try the following exercise

Exercise 89   Further problems on evaluation trigonometric ratios

In Problems 1 to 8, evaluate correct to 4 decimal places:

1. (a) sine 27° (b) sine 172.41° (c) sine 302°52′
    [(a) 0.4540 (b) 0.1321 (c) −0.8399]

2. (a) cosine 124° (b) cosine 21.46°
    (c) cosine 284°10′
    [(a) −0.5592 (b) 0.9307 (c) 0.2447]

3. (a) tangent 145° (b) tangent 310.59°
    (c) tangent 49°16′
    [(a) −0.7002 (b) −1.1671 (c) 1.1612]

4. (a) secant 73° (b) secant 286.45°
    (c) secant 155°41′
    [(a) 3.4203 (b) 3.5313 (c) −1.0974]

5. (a) cosecant 213° (b) cosecant 15.62°
    (c) cosecant 311°50′
    [(a) −1.8361 (b) 3.7139 (c) −1.3421]

6. (a) cotangent 71° (b) cotangent 151.62°
    (c) cotangent 321°23′
    [(a) 0.3443 (b) −1.8510 (c) −1.2519]

7. (a) sine $\dfrac{2\pi}{3}$ (b) cos 1.681 (c) tan 3.672
    [(a) 0.8660 (b) −0.1010 (c) 0.5865]

8. (a) sine $\dfrac{\pi}{8}$ (b) cosec 2.961 (c) cot 2.612
    [(a) 1.0824 (b) 5.5675 (c) −1.7083]

In Problems 9 to 14, determine the acute angle in degrees (correct to 2 decimal places), degrees and minutes, and in radians (correct to 3 decimal places).

9. $\sin^{-1} 0.2341$ [13.54°, 13°32′, 0.236 rad]

10. $\cos^{-1} 0.8271$ [34.20°, 34°12′, 0.597 rad]

11. $\tan^{-1} 0.8106$ [39.03°, 39°2′, 0.681 rad]

12. $\sec^{-1} 1.6214$ [51.92°, 51°55′, 0.906 rad]

13. $\csc^{-1} 2.4891$
    [23.69° , 23°41′, 0.413 rad]

14. $\cot^{-1} 1.9614$ [27.01°, 27°1′, 0.471 rad]

In Problems 15 to 18, evaluate correct to 4 significant figures.

15. $4 \cos 56°19′ − 3 \sin 21°57′$ [1.097]

16. $\dfrac{11.5 \tan 49°11′ − \sin 90°}{3 \cos 45°}$ [5.805]

17. $\dfrac{5 \sin 86°3′}{3 \tan 14°29′ − 2 \cos 31°9′}$ [−5.325]

18. $\dfrac{6.4 \csc 29°5′ − \sec 81°}{2 \cot 12°}$ [0.7199]

19. Determine the acute angle, in degrees and minutes, correct to the nearest minute, given by: $\sin^{-1}\left(\dfrac{4.32 \sin 42°16′}{7.86}\right)$ [21°42′]

20. If $\tan x = 1.5276$, determine $\sec x$, cosec x, and $\cot x$. (Assume x is an acute angle)   [1.8258, 1.1952, 0.6546]

In Problems 21 to 23 evaluate correct to 4 significant figures.

21. $\dfrac{(\sin 34°27′)(\cos 69°2′)}{(2 \tan 53°39′)}$ [0.07448]

22. $3 \cot 14°15′ \sec 23°9′$ [12.85]

23. $\dfrac{\csc 27°19′ + \sec 45°29′}{1 − \csc 27°19′ \sec 45°29′}$ [−1.710]

24. Evaluate correct to 4 decimal places:
    (a) $\sin(−125°)$ (b) $\tan(−241°)$
    (c) $\cos(−49°15′)$
    [(a) −0.8192 (b) −1.8040 (c) 0.6528]

25. Evaluate correct to 5 significant figures:
    (a) cosec $(−143°)$ (b) $\cot(−252°)$
    (c) $\sec(−67°22′)$
    [(a) −1.6616 (b) −0.32492 (c) 2.5985]

22.8   Trigonometric approximations for small angles

If angle x is a small angle (i.e. less than about 5°) and is expressed in radians, then the following trigonometric approximations may be shown to be true:

(i) $\sin x \approx x$
(ii) $\tan x \approx x$
(iii) $\cos x \approx 1 − \dfrac{x^2}{2}$

For example, let $x = 1°$, i.e. $1 \times \dfrac{\pi}{180} = 0.01745$ radians, correct to 5 decimal places. By calculator, $\sin 1° = 0.01745$ and $\tan 1° = 0.01746$, showing that: $\sin x = x \approx \tan x$ when $x = 0.01745$ radians. Also, $\cos 1° = 0.99985$; when $x = 1°$, i.e. 0.001745 radians,

$$1 - \frac{x^2}{2} = 1 - \frac{0.01745^2}{2} = 0.99985,$$

correct to 5 decimal places, showing that

$$\cos x = 1 - \frac{x^2}{2} \text{ when } x = 0.01745 \text{ radians.}$$

Similarly, let $x = 5°$, i.e. $5 \times \dfrac{\pi}{180} = 0.08727$ radians, correct to 5 decimal places.

By calculator, $\sin 5° = 0.08716$, thus $\sin x \approx x$,

$\tan 5° = 0.08749$, thus $\tan x \approx x$,

and $\cos 5° = 0.99619$;

since $x = 0.08727$ radians,

$$1 - \frac{x^2}{2} = 1 - \frac{0.08727^2}{2} = 0.99619 \text{ showing that:}$$

$$\cos x = 1 - \frac{x^2}{2} \text{ when } x = 0.0827 \text{ radians.}$$

If $\sin x \approx x$ for small angles, then $\dfrac{\sin x}{x} \approx 1$, and this relationship can occur in engineering considerations.

Chapter 23

Trigonometric waveforms

23.1 Graphs of trigonometric functions

By drawing up tables of values from 0° to 360°, graphs of $y = \sin A$, $y = \cos A$ and $y = \tan A$ may be plotted. Values obtained with a calculator (correct to 3 decimal places — which is more than sufficient for plotting graphs), using 30° intervals, are shown below, with the respective graphs shown in Fig. 23.1.

(a) $y = \sin A$

A	0	30°	60°	90°	120°	150°	180°
sin A	0	0.500	0.866	1.000	0.866	0.500	0

A	210°	240°	270°	300°	330°	360°
sin A	−0.500	−0.866	−1.000	−0.866	−0.500	0

(b) $y = \cos A$

A	0	30°	60°	90°	120°	150°	180°
cos A	1.000	0.866	0.500	0	−0.500	−0.866	−1.000

A	210°	240°	270°	300°	330°	360°
cos A	−0.866	−0.500	0	0.500	0.866	1.000

(c) $y = \tan A$

A	0	30°	60°	90°	120°	150°	180°
tan A	0	0.577	1.732	∞	−1.732	−0.577	0

A	210°	240°	270°	300°	330°	360°
tan A	0.577	1.732	∞	−1.732	−0.577	0

From Fig. 23.1 it is seen that:

(i) Sine and cosine graphs oscillate between peak values of ±1.

(ii) The cosine curve is the same shape as the sine curve but displaced by 90°.

(iii) The sine and cosine curves are continuous and they repeat at intervals of 360°; the tangent curve appears to be discontinuous and repeats at intervals of 180°.

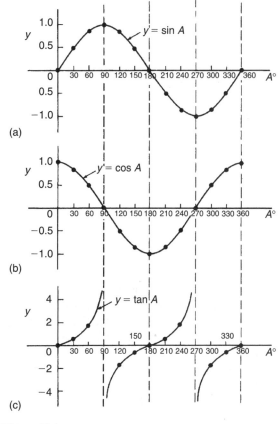

(a)

(b)

(c)

Figure 23.1

23.2 Angles of any magnitude

Figure 23.2 shows rectangular axes XX' and YY' intersecting at origin 0. As with graphical work, measurements made to the right and above 0 are positive, while those to the left and downwards are negative. Let 0A be free to rotate about 0. By convention, when 0A moves anticlockwise angular measurement is considered positive, and vice versa. Let 0A be rotated anticlockwise

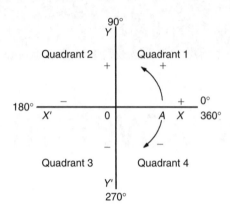

Figure 23.2

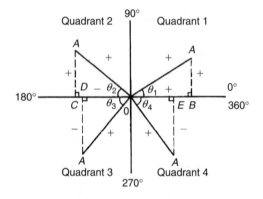

Figure 23.3

so that θ_1 is any angle in the first quadrant and left perpendicular AB be constructed to form the right-angled triangle $0AB$ in Fig. 23.3. Since all three sides of the triangle are positive, the trigonometric ratios sine, cosine and tangent will all be positive in the first quadrant. (Note: $0A$ is always positive since it is the radius of a circle.)

Let $0A$ be further rotated so that θ_2 is any angle in the second quadrant and let AC be constructed to form the right-angled triangle $0AC$. Then

$$\sin\theta_2 = \frac{+}{+} = +\quad \cos\theta_2 = \frac{-}{+} = -$$

$$\tan\theta_2 = \frac{+}{-} = -$$

Let $0A$ be further rotated so that θ_3 is any angle in the third quadrant and let AD be constructed to form the right-angled triangle $0AD$. Then

$$\sin\theta_3 = \frac{-}{+} = -\quad \cos\theta_3 = \frac{-}{+} = -$$

$$\tan\theta_3 = \frac{-}{-} = +$$

Let $0A$ be further rotated so that θ_4 is any angle in the fourth quadrant and let AE be constructed to form the right-angled triangle $0AE$. Then

$$\sin\theta_4 = \frac{-}{+} = -\quad \cos\theta_4 = \frac{+}{+} = +$$

$$\tan\theta_4 = \frac{-}{+} = -$$

The above results are summarized in Fig. 23.4. The letters underlined spell the word CAST when starting in the fourth quadrant and moving in an anticlockwise direction.

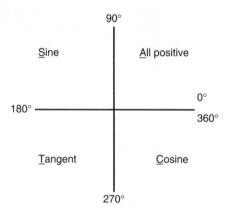

Figure 23.4

In the first quadrant of Fig. 23.1 all of the curves have positive values; in the second only sine is positive; in the third only tangent is positive; in the fourth only cosine is positive — exactly as summarized in Fig. 23.4. A knowledge of angles of any magnitude is needed when finding, for example, all the angles between $0°$ and $360°$ whose sine is, say, 0.3261. If 0.3261 is entered into a calculator and then the inverse sine key pressed (or $\sin^{-1}$ key) the answer $19.03°$ appears. However, there is a second angle between $0°$ and $360°$ which the calculator does not give. Sine is also positive in the second quadrant [either from CAST or from Fig. 23.1(a)]. The other angle is shown in Fig. 23.5 as angle θ where $\theta = 180° - 19.03° = 160.97°$. Thus $19.03°$ **and** $160.97°$ are the angles between $0°$ and $360°$ whose sine is 0.3261 (check that $\sin 160.97° = 0.3261$ on your calculator).

Be careful! Your calculator only gives you one of these answers. The second answer needs to be deduced from a knowledge of angles of any magnitude, as shown in the following worked problems.

Problem 1. Determine all the angles between $0°$ and $360°$ whose sine is -0.4638

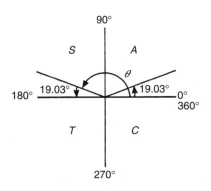

Figure 23.5

The angles whose sine is −0.4638 occurs in the third and fourth quadrants since sine is negative in these quadrants — see Fig. 23.6.

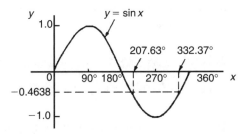

Figure 23.6

From Fig. 23.7, $\theta = \sin^{-1} 0.4638 = 27.63°$. Measured from 0°, the two angles between 0° and 360° whose sine is −0.4638 are $180° + 27.63°$, i.e. **207.63°** and $360° − 27.63°$, i.e. **332.37°**
(Note that a calculator only gives one answer, i.e. −27.632588°)

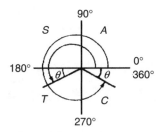

Figure 23.7

Problem 2. Determine all the angles between 0° and 360° whose tangent is 1.7629

A tangent is positive in the first and third quadrants — see Fig. 23.8.
From Fig. 23.9, $\theta = \tan^{-1} 1.7629 = 60.44°$ Measured from 0°, the two angles between 0° and 360°

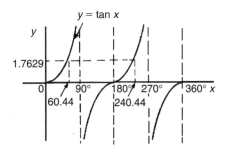

Figure 23.8

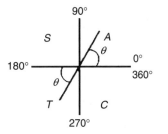

Figure 23.9

whose tangent is 1.7629 are **60.44°** and $180° + 60.44°$, i.e. **240.44°**

Problem 3. Solve the equation $\cos^{-1} (−0.2348) = \alpha$ for angles of α between 0° and 360°

Cosine is positive in the first and fourth quadrants and thus negative in the second and third quadrants — from Fig. 23.5 or from Fig. 23.1(b).
In Fig. 23.10, angle $\theta = \cos^{-1} (0.2348) = 76.42°$

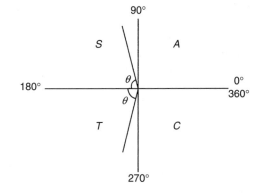

Figure 23.10

Measured from 0° , the two angles whose cosine is −0.2348 are $\alpha = 180° − 76.42°$ i.e. **103.58°** and $\alpha = 180° + 76.42°$, i.e. **256.42°**

Now try the following exercise

Exercise 90   Further problems on angles of any magnitude

1. Determine all of the angles between $0°$ and $360°$ whose sine is:

   (a) 0.6792   (b) −0.1483

   $$\left[\begin{array}{l}\text{(a) } 42.78° \text{ and } 137.22° \\ \text{(b) } 188.53° \text{ and } 351.47°\end{array}\right]$$

2. Solve the following equations for values of x between $0°$ and $360°$:

   (a) $x = \cos^{-1} 0.8739$
   (b) $x = \cos^{-1}(-0.5572)$

   $$\left[\begin{array}{l}\text{(a) } 29.08° \text{ and } 330.92° \\ \text{(b) } 123.86° \text{ and } 236.14°\end{array}\right]$$

3. Find the angles between $0°$ to $360°$ whose tangent is:

   (a) 0.9728   (b) −2.3418

   $$\left[\begin{array}{l}\text{(a) } 44.21° \text{ and } 224.21° \\ \text{(b) } 113.12° \text{ and } 293.12°\end{array}\right]$$

23.3   The production of a sine and cosine wave

In Fig. 23.11, let OR be a vector 1 unit long and free to rotate anticlockwise about O. In one revolution a circle is produced and is shown with $15°$ sectors. Each radius arm has a vertical and a horizontal component. For example, at $30°$, the vertical component is TS and the horizontal component is OS.

From trigonometric ratios,

$$\sin 30° = \frac{TS}{TO} = \frac{TS}{1}, \quad \text{i.e.} \quad TS = \sin 30°$$

and

$$\cos 30° = \frac{OS}{TO} = \frac{OS}{1}, \quad \text{i.e.} \quad OS = \cos 30°$$

The vertical component TS may be projected across to $T'S'$, which is the corresponding value of $30°$ on the graph of y against angle $x°$. If all such vertical components as TS are projected on to the graph, then a **sine wave** is produced as shown in Fig. 23.11.

If all horizontal components such as OS are projected on to a graph of y against angle $x°$, then a **cosine wave** is produced. It is easier to visualize these projections by redrawing the circle with the radius arm OR initially in a vertical position as shown in Fig. 23.12.

From Figs. 23.11 and 23.12 it is seen that a cosine curve is of the same form as the sine curve but is displaced by $90°$ (or $\pi/2$ radians).

23.4   Sine and cosine curves

Graphs of sine and cosine waveforms

(i) A graph of $y = \sin A$ is shown by the broken line in Fig. 23.13 and is obtained by drawing up a table of values as in Section 23.1. A similar table may be produced for $y = \sin 2A$.

A°	0	30	45	60	90	120
2A	0	60	90	120	180	240
sin 2A	0	0.866	1.0	0.866	0	−0.866

A°	135	150	180	210	225	240
2A	270	300	360	420	450	480
sin 2A	−1.0	−0.866	0	0.866	1.0	0.866

A°	270	300	315	330	360
2A	540	600	630	660	720
sin 2A	0	−0.866	−1.0	−0.866	0

A graph of $y = \sin 2A$ is shown in Fig. 23.13.

(ii) A graph of $y = \sin \frac{1}{2}A$ is shown in Fig. 23.14 using the following table of values.

A°	0	30	60	90	120	150	180
$\frac{1}{2}A$	0	15	30	45	60	75	90
sin $\frac{1}{2}A$	0	0.259	0.500	0.707	0.866	0.966	1.00

A°	210	240	270	300	330	360
$\frac{1}{2}A$	105	120	135	150	165	180
sin $\frac{1}{2}A$	0.966	0.866	0.707	0.500	0.259	0

(iii) A graph of $y = \cos A$ is shown by the broken line in Fig. 23.15 and is obtained by drawing up a table of values. A similar table may be produced for $y = \cos 2A$ with the result as shown.

(iv) A graph of $y = \cos \frac{1}{2}A$ is shown in Fig. 23.16 which may be produced by drawing up a table of values, similar to above.

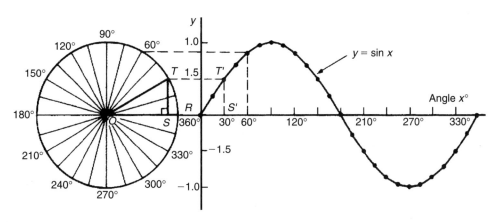

Figure 23.11

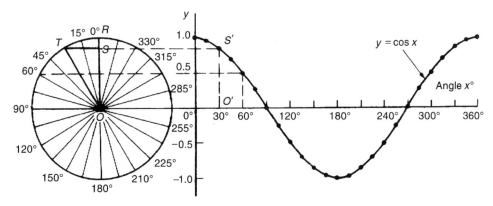

Figure 23.12

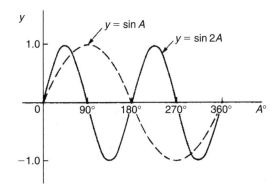

Figure 23.13

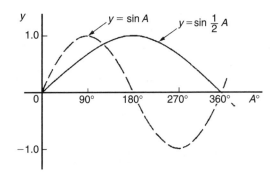

Figure 23.14

Periodic time and period

(i) Each of the graphs shown in Figs. 23.13 to 23.16 will repeat themselves as angle A increases and are thus called **periodic functions**.

(ii) $y = \sin A$ and $y = \cos A$ repeat themselves every $360°$ (or 2π radians); thus $360°$ is called the **period** of these waveforms. $y = \sin 2A$ and $y = \cos 2A$ repeat themselves every $180°$ (or π radians); thus $180°$ is the period of these waveforms.

(iii) In general, if $y = \sin pA$ or $y = \cos pA$ (where p is a constant) then the period of the waveform is $360°/p$ (or $2\pi/p$ rad). Hence if $y = \sin 3A$ then the period is $360/3$, i.e. $120°$, and if $y = \cos 4A$ then the period is $360/4$, i.e. $90°$

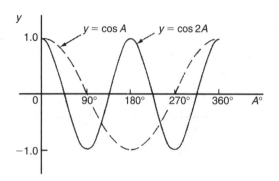

Figure 23.15

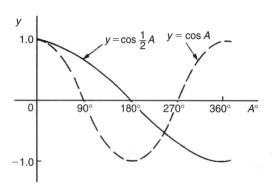

Figure 23.16

Amplitude

Amplitude is the name given to the maximum or peak value of a sine wave. Each of the graphs shown in Figs. 23.13 to 23.16 has an amplitude of $+1$ (i.e. they oscillate between $+1$ and -1). However, if $y = 4\sin A$, each of the values in the table is multiplied by 4 and the maximum value, and thus amplitude, is 4. Similarly, if $y = 5\cos 2A$, the amplitude is 5 and the period is $360°/2$, i.e. $180°$

Problem 4. Sketch $y = \sin 3A$ between $A = 0°$ and $A = 360°$

Amplitude $= 1$ and period $= 360°/3 = 120°$

A sketch of $y = \sin 3A$ is shown in Fig. 23.17.

Problem 5. Sketch $y = 3\sin 2A$ from $A = 0$ to $A = 2\pi$ radians

Amplitude $= 3$ and period $= 2\pi/2 = \pi$ rads (or $180°$)

A sketch of $y = 3\sin 2A$ is shown in Fig. 23.18.

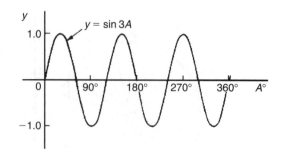

Figure 23.17

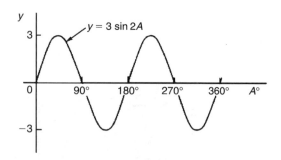

Figure 23.18

Problem 6. Sketch $y = 4\cos 2x$ form $x = 0°$ to $x = 360°$

Amplitude $= 4$ and period $= 360°/2 = 180°$.

A sketch of $y = 4\cos 2x$ is shown in Fig. 23.19.

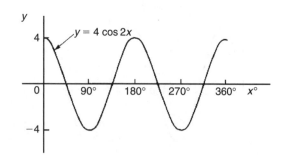

Figure 23.19

Problem 7. Sketch $y = 2\sin \frac{3}{5}A$ over one cycle

Amplitude $= 2$; period $= \dfrac{360°}{\frac{3}{5}} = \dfrac{360° \times 5}{3} = 600°$

A sketch of $y = 2\sin \frac{3}{5}A$ is shown in Fig. 23.20.

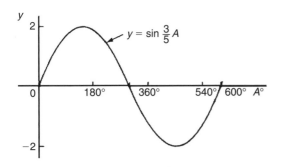

Figure 23.20

Lagging and leading angles

(i) A sine or cosine curve may not always start at $0°$. To show this a periodic function is represented by $y = \sin(A \pm \alpha)$ or $y = \cos(A \pm \alpha)$ where α is a phase displacement compared with $y = \sin A$ or $y = \cos A$.

(ii) By drawing up a table of values, a graph of $y = \sin(A - 60°)$ may be plotted as shown in Fig. 23.21. If $y = \sin A$ is assumed to start at $0°$ then $y = \sin(A - 60°)$ starts $60°$ later (i.e. has a zero value $60°$ later). Thus $y = \sin(A - 60°)$ is said to **lag** $y = \sin A$ by $60°$

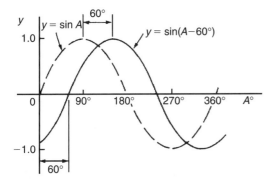

Figure 23.21

(iii) By drawing up a table of values, a graph of $y = \cos(A + 45°)$ may be plotted as shown in Fig. 23.22. If $y = \cos A$ is assumed to start at $0°$ then $y = \cos(A + 45°)$ starts $45°$ earlier (i.e. has a maximum value $45°$ earlier). Thus $y = \cos(A + 45°)$ is said to **lead** $y = \cos A$ by $45°$

(iv) Generally, a graph of $y = \sin(A - \alpha)$ lags $y = \sin A$ by angle α, and a graph of $y = \sin(A + \alpha)$ leads $y = \sin A$ by angle α.

(v) A cosine curve is the same shape as a sine curve but starts $90°$ earlier, i.e. leads by $90°$. Hence

$$\cos A = \sin(A + 90°)$$

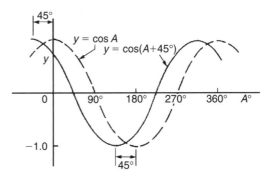

Figure 23.22

Problem 8. Sketch $y = 5 \sin(A + 30°)$ from $A = 0°$ to $A = 360°$

Amplitude $= 5$ and period $= 360°/1 = 360°$.

$5 \sin(A + 30°)$ leads $5 \sin A$ by $30°$ (i.e. starts $30°$ earlier).

A sketch of $y = 5 \sin(A + 30°)$ is shown in Fig. 23.23.

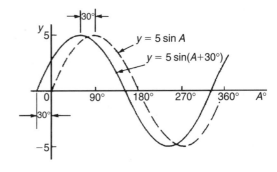

Figure 23.23

Problem 9. Sketch $y = 7 \sin(2A - \pi/3)$ in the range $0 \le A \le 360°$

Amplitude $= 7$ and period $= 2\pi/2 = \pi$ radians.

In general, $y = \sin(pt - \alpha)$ **lags** $y = \sin pt$ by α/p, hence $7 \sin(2A - \pi/3)$ lags $7 \sin 2A$ by $(\pi/3)/2$, i.e. $\pi/6$ rad or $30°$

A sketch of $y = 7 \sin(2A - \pi/3)$ is shown in Fig. 23.24.

Problem 10. Sketch $y = 2 \cos(\omega t - 3\pi/10)$ over one cycle

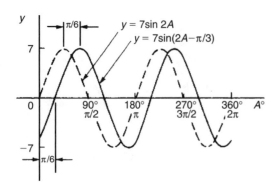

Figure 23.24

Amplitude $= 2$ and period $= 2\pi/\omega$ rad.

$2\cos(\omega t - 3\pi/10)$ lags $2\cos\omega t$ by $3\pi/10\omega$ seconds.

A sketch of $y = 2\cos(\omega t - 3\pi/10)$ is shown in Fig. 23.25.

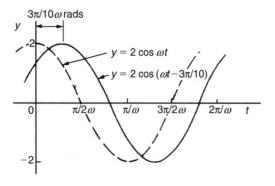

Figure 23.25

Now try the following exercise

Exercise 91   Further problems on sine and cosine curves

In Problems 1 to 7 state the amplitude and period of the waveform and sketch the curve between $0°$ and $360°$.

1. $y = \cos 3A$ [1, 120°]

2. $y = 2\sin\dfrac{5x}{2}$ [2, 144°]

3. $y = 3\sin 4t$ [3, 90°]

4. $y = 3\cos\dfrac{\theta}{2}$ [3, 720°]

5. $y = \dfrac{7}{2}\sin\dfrac{3x}{8}$ $\left[\dfrac{7}{2}, 960°\right]$

6. $y = 6\sin(t - 45°)$ [6, 360°]

7. $y = \cos(2\theta + 30°)$ [4, 180°]

23.5   Sinusoidal form $A\sin(\omega t \pm \alpha)$

In Fig. 23.26, let OR represent a vector that is free to rotate anticlockwise about O at a velocity of ω rad/s. A rotating vector is called a **phasor**. After a time t second OR will have turned through an angle ωt radians (shown as angle TOR in Fig 23.26). If ST is constructed perpendicular to OR, then $\sin\omega t = ST/OT$, i.e. $ST = OT\sin\omega t$.

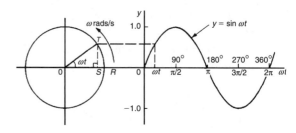

Figure 23.26

If all such vertical components are projected on to a graph of y against ωt, a sine wave results of amplitude OR (as shown in Section 23.3).

If phasor OR makes one revolution (i.e. 2π radians) in T seconds, then the angular velocity, $\omega = 2\pi/T$ rad/s,

from which, $T = 2\pi/\omega$ **seconds**

T is known as the **periodic time**.

The number of complete cycles occurring per second is called the **frequency, f**

$$\text{Frequency} = \frac{\text{number of cycles}}{\text{second}}$$

$$= \frac{1}{T} = \frac{\omega}{2\pi}\text{Hz}$$

i.e. $f = \dfrac{\omega}{2\pi}\text{Hz}$

Hence   **angular velocity,** $\omega = 2\pi f$ **rad/s**

Amplitude is the name given to the maximum or peak value of a sine wave, as explained in Section 23.4. The amplitude of the sine wave shown in Fig. 23.26 has an amplitude of 1.

A sine or cosine wave may not always start at 0°. To show this a periodic function is represented by $y = \sin(\omega t \pm \alpha)$ or $y = \cos(\omega t \pm \alpha)$, where α is a phase displacement compared with $y = \sin A$ or $y = \cos A$. A graph of $y = \sin(\omega t - \alpha)$ **lags** $y = \sin \omega t$ by angle α, and a graph of $y = \sin(\omega t + \alpha)$ **leads** $y = \sin \omega t$ by angle α.

The angle ωt is measured in **radians**

$$\left[\text{i.e.} \left(\omega \frac{\text{rad}}{\text{s}} \right) (t\, s) = \omega t \text{ radians} \right]$$

hence angle α should also be in radians.
The relationship between degrees and radians is:

$$360° = 2\pi \text{ radians} \quad \text{or} \quad \mathbf{180° = \pi \text{ radians}}$$

Hence $1 \text{ rad} = \dfrac{180}{\pi} = 57.30°$ and, for example,

$$71° = 71 \times \frac{\pi}{180} = 1.239 \text{ rad}$$

Summarising, given a general sinusoidal function $\mathbf{y = A\sin(\omega t \pm \alpha)}$, then:

(i) $A = $ amplitude
(ii) $\omega = $ angular velocity $= 2\pi f$ rad/s
(iii) $\dfrac{2\pi}{\omega} = $ periodic time T seconds
(iv) $\dfrac{\omega}{2\pi} = $ frequency, f hertz
(v) $\alpha = $ angle of lead or lag (compared with $y = A \sin \omega t$)

Problem 11.   An alternating current is given by $i = 30 \sin (100\pi t + 0.27)$ amperes. Find the amplitude, periodic time, frequency and phase angle (in degrees and minutes)

$i = 30 \sin(100\pi t + 0.27)$A, hence **amplitude $= 30$ A**. Angular velocity $\omega = 100\pi$, hence

$$\textbf{periodic time, } T = \frac{2\pi}{\omega} = \frac{2\pi}{100\pi} = \frac{1}{50}$$

$$= \mathbf{0.02\, s} \quad \text{or} \quad \mathbf{20\, ms}$$

Frequency, $f = \dfrac{1}{T} = \dfrac{1}{0.02} = \mathbf{50\, Hz}$

Phase angle, $\alpha = 0.27$ rad $= \left(0.27 \times \dfrac{180}{\pi} \right)^{\circ}$

$$= \mathbf{15.47°} \text{ or } \mathbf{15°28'} \textbf{ leading}$$

$$i = 30 \sin(100\pi t)$$

Problem 12.   An oscillating mechanism has a maximum displacement of 2.5 m and a frequency of 60 Hz. At time $t = 0$ the displacement is 90 cm. Express the displacement in the general form $A \sin (\omega t \pm \alpha)$

Amplitude $=$ maximum displacement $= 2.5$ m

Angular velocity, $\omega = 2\pi f = 2\pi(60) = 120\pi$ rad/s

Hence displacement $= 2.5 \sin(120\pi t + \alpha)$ m

When $t = 0$, displacement $= 90$ cm $= 0.90$ m

Hence, $0.90 = 2.5\sin(0 + \alpha)$

i.e. $\sin \alpha = \dfrac{0.90}{2.5} = 0.36$

Hence $\alpha = \sin^{-1} 0.36 = 21.10°$

$$= 21°6' = 0.368 \text{ rad}$$

Thus, **displacement $= 2.5 \sin (120\pi t + 0.368)$ m**

Problem 13.   The instantaneous value of voltage in an a.c. circuit at any time t seconds is given by $v = 340 \sin (50\pi t - 0.541)$ volts. Determine the:

(a) amplitude, periodic time, frequency and phase angle (in degrees)

(b) value of the voltage when $t = 0$

(c) value of the voltage when $t = 10$ ms

(d) time when the voltage first reaches 200 V, and

(e) time when the voltage is a maximum

Sketch one cycle of the waveform

(a) **Amplitude = 340 V**

Angular velocity, $\omega = 50\pi$

Hence **periodic time,** $T = \dfrac{2\pi}{\omega} = \dfrac{2\pi}{50\pi} = \dfrac{1}{25}$

$$= \textbf{0.04 s} \text{ or } \textbf{40 ms}$$

Frequency $f = \dfrac{1}{T} = \dfrac{1}{0.04} = \textbf{25 Hz}$

Phase angle $= 0.541 \text{ rad} = \left(0.541 \times \dfrac{180}{\pi} \right)$

$$= \textbf{31° lagging } v = 340\sin(50\pi t)$$

(b) **When** $t = 0$,

$$v = 340\sin(0 - 0.541)$$

$$= 340\sin(-31°) = \textbf{−175.1 V}$$

(c) **When** $t = 10$ **ms,**

then $v = 340\sin\left(50\pi \dfrac{10}{10^3} - 0.541 \right)$

$$= 340\sin(1.0298)$$

$$= 340\sin 59° = \textbf{291.4 volts}$$

(d) When $v = 200$ volts,

then $200 = 340\sin(50\pi t - 0.541)$

$$\dfrac{200}{340} = \sin(50\pi t - 0.541)$$

Hence $(50\pi t - 0.541) = \sin^{-1}\dfrac{200}{340}$

$$= 36.03° \text{ or } 0.6288 \text{ rad}$$

$$50\pi t = 0.6288 + 0.541$$

$$= 1.1698$$

Hence when $v = 200$ V,

time, $t = \dfrac{1.1698}{50\pi} = \textbf{7.447 ms}$

(e) When the voltage is a maximum, $v = 340$ V

Hence $340 = 340\sin(50\pi t - 0.541)$

$$1 = \sin(50\pi t - 0.541)$$

$$50\pi t - 0.541 = \sin^{-1} 1 = 90° \text{ or } 1.5708 \text{ rad}$$

$$50\pi t = 1.5708 + 0.541 = 2.1118$$

Hence time, $t = \dfrac{2.1118}{50\pi} = \textbf{13.44 ms}$

A sketch of $v = 340\sin(50\pi t - 0.541)$ volts is shown in Fig. 23.27.

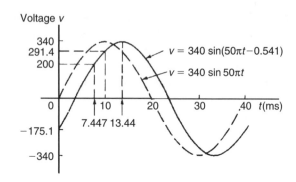

Figure 23.27

Now try the following exercise

Exercise 92   Further problems on the sinusoidal form $A\sin(\omega t \pm \alpha)$

In Problems 1 to 3 find the amplitude, periodic time, frequency and phase angle (stating whether it is leading or lagging $\sin\omega t$) of the alternating quantities given.

1. $i = 40\sin(50\pi t + 0.29)$ mA

$$\left[\begin{array}{l} 40, 0.04 \text{ s}, 25 \text{ Hz}, 0.29 \text{ rad} \\ \text{(or } 16°37') \text{ leading } 40\sin 5\pi t \end{array} \right]$$

2. $y = 75\sin(40t - 0.54)$ cm

$$\left[\begin{array}{l} 75 \text{ cm}, 0.157 \text{ s}, 6.37 \text{ Hz}, 0.54 \text{ rad} \\ \text{(or } 30°56') \text{ lagging } 75\sin 40t \end{array} \right]$$

3. $v = 300\sin(200\pi t - 0.412)$ V

$$\left[\begin{array}{l} 300 \text{ V}, 0.01 \text{ s}, 100 \text{ Hz}, 0.412 \text{ rad} \\ \text{(or } 23°36') \text{ lagging } 300\sin 200\pi t \end{array} \right]$$

4. A sinusoidal voltage has a maximum value of 120 V and a frequency of 50 Hz. At time $t = 0$, the voltage is (a) zero, and (b) 50 V. Express the instantaneous voltage v in the form $v = A\sin(\omega t \pm \alpha)$.

$$\left[\begin{array}{l} \text{(a) } v = 120\sin 100\pi t \text{ volts} \\ \text{(b) } v = 120\sin(100\pi t + 0.43) \text{volts} \end{array} \right]$$

5. An alternating current has a periodic time of 25 ms and a maximum value of 20 A. When time $t = 0$, current $i = -10$ amperes. Express the current i in the form $i = A \sin(\omega t \pm \alpha)$.

$$\left[i = 20 \sin\left(80\pi t - \frac{\pi}{6}\right) \text{amperes}\right]$$

6. An oscillating mechanism has a maximum displacement of 3.2 m and a frequency of 50 Hz. At time $t = 0$ the displacement is 150 cm. Express the displacement in the general form $A \sin(\omega t \pm \alpha)$.

$$[3.2 \sin(100\pi t + 0.488)\text{m}]$$

7. The current in an a.c. circuit at any time t seconds is given by:

$$i = 5 \sin(100\pi t - 0.432) \text{ amperes}$$

Determine (a) the amplitude, periodic time, frequency and phase angle (in degrees) (b) the value of current at $t = 0$, (c) the value of current at $t = 8$ ms, (d) the time when the current is first a maximum, (e) the time when the current first reaches 3A.

Sketch one cycle of the waveform showing relevant points.

$$\begin{bmatrix} \text{(a) 5 A, 20 ms, 50 Hz,} \\ 24°45' \text{ lagging} \\ \text{(b) } -2.093\,\text{A} \\ \text{(c) } 4.363\,\text{A} \\ \text{(d) } 6.375\,\text{ms} \\ \text{(e) } 3.423\,\text{ms} \end{bmatrix}$$

23.6   Waveform harmonics

Let an instantaneous voltage v be represented by $v = V_m \sin 2\pi f\,t$ volts. This is a waveform which varies sinusoidally with time t, has a frequency f, and a maximum value V_m. Alternating voltages are usually assumed to have wave-shapes which are sinusoidal where only one frequency is present. If the waveform is not sinusoidal it is called a **complex wave**, and, whatever its shape, it may be split up mathematically into components called the **fundamental** and a number of **harmonics**. This process is called **harmonic analysis**. The fundamental (or first harmonic) is sinusoidal and has the supply frequency, f; the other harmonics are also sine waves having frequencies which are integer

multiples of f. Thus, if the supply frequency is 50 Hz, then the thid harmonic frequency is 150 Hz, the fifth 250 Hz, and so on.

A complex waveform comprising the sum of the fundamental and a third harmonic of about half the amplitude of the fundamental is shown in Fig. 23.28(a), both waveforms being initially in phase with each other. If further odd harmonic waveforms of the appropriate amplitudes are added, a good approximation to a square wave results. In Fig. 23.28(b), the third harmonic is shown having an initial phase displacement from the fundamental. The positive and negative half cycles of each of the complex waveforms shown in Figs. 23.28(a) and (b) are identical in shape, and this is a feature of waveforms containing the fundamental and only odd harmonics.

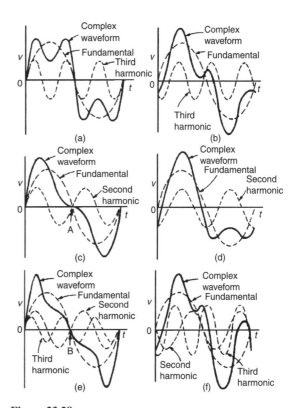

Figure 23.28

A complex waveform comprising the sum of the fundamental and a second harmonic of about half the amplitude of the fundamental is shown in Fig. 23.28(c), each waveform being initially in phase with each other. If further even harmonics of appropriate amplitudes are added a good approximation to a triangular wave results. In Fig. 23.28(c), the negative cycle, if reversed, appears as a mirror image of the positive cycle about point A.

In Fig. 23.28(d) the second harmonic is shown with an initial phase displacement from the fundamental and the positive and negative half cycles are dissimilar.

A complex waveform comprising the sum of the fundamental, a second harmonic and a third harmonic is shown in Fig. 23.28(e), each waveform being initially 'in-phase'. The negative half cycle, if reversed, appears as a mirror image of the positive cycle about point B. In Fig. 23.28(f), a complex waveform comprising the sum of the fundamental, a second harmonic and a third harmonic are shown with initial phase displacement. The positive and negative half cycles are seen to be dissimilar.

The features mentioned relative to Figs. 23.28(a) to (f) make it possible to recognise the harmonics present in a complex waveform.

Chapter 24

Cartesian and polar co-ordinates

24.1 Introduction

There are two ways in which the position of a point in a plane can be represented. These are

(a) by **Cartesian co-ordinates**, i.e. (x, y), and

(b) by **polar co-ordinates**, i.e. (r, θ), where r is a 'radius' from a fixed point and θ is an angle from a fixed point.

24.2 Changing from Cartesian into polar co-ordinates

In Fig. 24.1, if lengths x and y are known, then the length of r can be obtained from Pythagoras' theorem (see Chapter 22) since OPQ is a right-angled triangle.

Hence $\qquad r^2 = (x^2 + y^2)$

from, which $\qquad r = \sqrt{x^2 + y^2}$

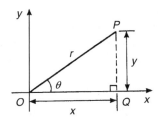

Figure 24.1

From trigonometric ratios (see Chapter 22),

$$\tan \theta = \frac{y}{x}$$

from which $\qquad \theta = \tan^{-1} \frac{y}{x}$

$r = \sqrt{x^2 + y^2}$ and $\theta = \tan^{-1} \frac{y}{x}$ are the two formulae we need to change from Cartesian to polar co-ordinates. The angle θ, which may be expressed in degrees or radians, must **always** be measured from the positive x-axis, i.e. measured from the line OQ in Fig. 24.1. It is suggested that when changing from Cartesian to polar co-ordinates a diagram should always be sketched.

Problem 1. Change the Cartesian co-ordinates (3, 4) into polar co-ordinates

A diagram representing the point (3, 4) is shown in Fig. 24.2.

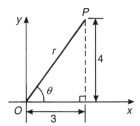

Figure 24.2

From Pythagoras' theorem, $r = \sqrt{3^2 + 4^2} = 5$ (note that -5 has no meaning in this context). By trigonometric ratios, $\theta = \tan^{-1} \frac{4}{3} = 53.13°$ or 0.927 rad

[note that $53.13° = 53.13 \times (\pi/180)$ rad $= 0.927$ rad.]

Hence (3, 4) in Cartesian co-ordinates corresponds to (5, 53.13°) or (5, 0.927 rad) in polar co-ordinates.

Problem 2. Express in polar co-ordinates the position $(-4, 3)$

A diagram representing the point using the Cartesian co-ordinates (−4, 3) is shown in Fig. 24.3.

From Pythagoras' theorem $r = \sqrt{4^2 + 3^2} = 5$

By trigonometric ratios, $\alpha = \tan^{-1}\frac{3}{4} = 36.87°$ or 0.644 rad.

Hence $\theta = 180° − 36.87° = 143.13°$

or $\theta = \pi − 0.644 = 2.498$ rad.

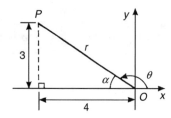

Figure 24.3

Hence the position of point P in polar co-ordinate form is (5, 143.13°) or (5, 2.498 rad).

Problem 3. Express (−5, −12) in polar co-ordinates

A sketch showing the position (−5, −12) is shown in Fig. 24.4.

$$r = \sqrt{5^2 + 12^2} = 13$$

and $\alpha = \tan^{-1}\frac{12}{5} = 67.38°$ or 1.176 rad.

Hence $\theta = 180° + 67.38° = 247.38°$

or $\theta = \pi + 1.176 = 4.318$ rad.

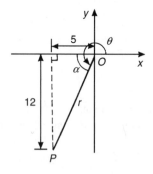

Figure 24.4

Thus (−5, −12) in Cartesian co-ordinates corresponds to (13, 247.38°) or (13, 4.318 rad) in polar co-ordinates.

Problem 4. Express (2, −5) in polar co-ordinates.

A sketch showing the position (2, −5) is shown in Fig. 24.5.

$$r = \sqrt{2^2 + 5^2} = \sqrt{29} = 5.385$$
correct to 3 decimal places

$$\alpha = \tan^{-1}\frac{5}{2} = 68.20° \quad \text{or} \quad 1.190 \text{ rad}$$

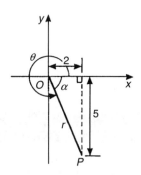

Figure 24.5

Hence $\theta = 360° − 68.20° = 291.80°$

or $\theta = 2\pi − 1.190 = 5.093$ rad

Thus (2, −5) in Cartesian co-ordinates corresponds to (5.385, 291.80°) or (5.385, 5.093 rad) in polar co-ordinates.

Now try the following exercise

Exercise 93   Further problems on changing form Cartesian into polar co-ordinates

In Problems 1 to 8, express the given Cartesian co-ordinates as polar co-ordinates, correct to 2 decimal places, in both degrees and in radians.

1. (3, 5)
   [(5.83, 59.04°) or (5.83, 1.03 rad)]

2. (6.18, 2.35)
   [(6.61, 20.82°) or (6.61, 0.36 rad)]

3. (−2, 4)
   [(4.47, 116.57°) or (4.47, 2.03 rad)]

4. (−5.4, 3.7)
   [(6.55, 145.58°) or (6.55, 2.54 rad)]

5. (−7, −3)
   [(7.62, 203.20°) or (7.62, 3.55 rad)]

6. (−2.4, −3.6)
   [(4.33, 236.31°) or (4.33, 4.12 rad)]

7. (5, −3)
   [(5.83, 329.04°) or (5.83, 5.74 rad)]

8. (9.6, −12.4)
   [(15.68, 307.75°) or (15.68, 5.37 rad)]

24.3 Changing from polar into Cartesian co-ordinates

From the right-angled triangle *OPQ* in Fig. 24.6.

$$\cos\theta = \frac{x}{r} \text{ and } \sin\theta = \frac{y}{r}$$
from trigonometric ratios

Hence $x = r\cos\theta$ and $y = r\sin\theta$

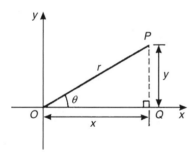

Figure 24.6

If length *r* and angle θ are known then $x = r\cos\theta$ and $y = r\sin\theta$ are the two formulae we need to change from polar to Cartesian co-ordinates.

Problem 5. Change (4, 32°) into Cartesian co-ordinates.

A sketch showing the position (4, 32°) is shown in Fig. 24.7.

Now $x = r\cos\theta = 4\cos 32° = 3.39$

and $y = r\sin\theta = 4\sin 32° = 2.12$

Hence (4, 32°) in polar co-ordinates corresponds to (3.39, 2.12) in Cartesian co-ordinates.

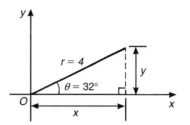

Figure 24.7

Problem 6. Express (6, 137°) in Cartesian co-ordinates

A sketch showing the position (6, 137°) is shown in Fig. 24.8.

$$x = r\cos\theta = 6\cos 137° = -4.388$$

which corresponds to length *OA* in Fig. 24.8.

$$y = r\sin\theta = 6\sin 137° = 4.092$$

which corresponds to length *AB* in Fig. 24.8.

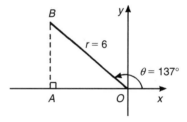

Figure 24.8

Thus (6, 137°) in polar co-ordinates corresponds to (−4.388, 4.092) in Cartesian co-ordinates.

(Note that when changing from polar to Cartesian co-ordinates it is not quite so essential to draw a sketch. Use of $x = r\cos\theta$ and $y = r\sin\theta$ automatically produces the correct signs.)

Problem 7. Express (4.5, 5.16 rad) in Cartesian co-ordinates.

A sketch showing the position (4.5, 5.16 rad) is shown in Fig. 24.9.

$$x = r\cos\theta = 4.5\cos 5.16 = 1.948$$

which corresponds to length *OA* in Fig. 24.9.

$$y = r\sin\theta = 4.5\sin 5.16 = -4.057$$

Section 3

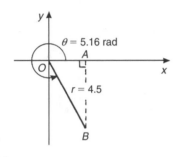

Figure 24.9

which corresponds to length AB in Fig. 24.9.

Thus (1.948, −4.057) in Cartesian co-ordinates corresponds to (4.5, 5.16 rad) in polar co-ordinates.

24.4 Use of $R \to P$ and $P \to R$ functions on calculators

Another name for Cartesian co-ordinates is **rectangular** co-ordinates. Many scientific notation calculators possess $R \to P$ and $P \to R$ functions. The R is the first letter of the word rectangular and the P is the first letter of the word polar. Check the operation manual for your particular calculator to determine how to use these two functions. They make changing from Cartesian to polar co-ordinates, and vice-versa, so much quicker and easier.

Now try the following exercise

Exercise 94 Further problems on changing polar into Cartesian co-ordinates

In Problems 1 to 8, express the given polar co-ordinates as Cartesian co-ordinates, correct to 3 decimal places.

1. (5, 75°)   [(1.294, 4.830)]

2. (4.4, 1.12 rad)   [(1.917, 3.960)]

3. (7, 140°)   [(−5.362, 4.500)]

4. (3.6, 2.5 rad)   [(−2.884, 2.154)]

5. (10.8, 210°)   [(−9.353, −5.400)]

6. (4, 4 rad)   [(−2.615, −3.207)]

7. (1.5, 300°)   [(0.750, −1.299)]

8. (6, 5.5 rad)   [(4.252, −4.233)]

9. Figure 24.10 shows 5 equally spaced holes on an 80 mm pitch circle diameter. Calculate their co-ordinates relative to axes Ox and Oy in (a) polar form, (b) Cartesian form. Calculate also the shortest distance between the centres of two adjacent holes.

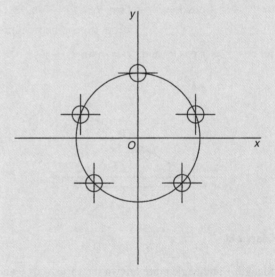

Figure 24.10

[(a) (40, 18°), (40, 90°), (40, 162°), (40, 234°) and (40, 306°)
(b) (38.04, 12.36), (0, 40), (−38.04, 12.36), (−23.51, −32.36), (23.51, −32.36)
(c) 47.02 mm]

Revision Test 6

This Revision test covers the material contained in Chapter 22 to 24. *The marks for each question are shown in brackets at the end of each question.*

1. Fig. R6.1 shows a plan view of a kite design. Calculate the lengths of the dimensions shown as a and b. (4)

2. In Fig. R6.1, evaluate (a) angle θ (b) angle α (5)

3. Determine the area of the plan view of a kite shown in Fig. R6.1. (4)

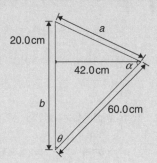

Figure R6.1

4. If the angle of elevation of the top of a 25 m perpendicular building from point A is measured as $27°$, determine the distance to the building. Calculate also the angle of elevation at a point B, 20 m closer to the building than point A. (5)

5. Evaluate, each correct to 4 significant figures: (a) $\sin 231.78°$ (b) $\cos 151°16'$ (c) $\tan \dfrac{3\pi}{8}$ (3)

6. Sketch the following curves labelling relevant points: (a) $y = 4\cos(\theta + 45°)$ (b) $y = 5\sin(2t - 60°)$ (6)

7. Solve the following equations in the range $0°$ to $360°$ (a) $\sin^{-1}(-0.4161) = x$ (b) $\cot^{-1}(2.4198) = \theta$ (6)

8. The current in an alternating current circuit at any time t seconds is given by: $i = 120\sin(100\pi t + 0.274)$ amperes. Determine

   (a) the amplitude, periodic time, frequency and phase angle (with reference to $120\sin 100\pi t$)
   (b) the value of current when $t = 0$
   (c) the value of current when $t = 6$ ms
   (d) the time when the current first reaches 80 A

   Sketch one cycle of the oscillation. (17)

9. Change the following Cartesian coordinates into polar co-ordinates, correct to 2 decimal places, in both degrees and in radians: (a) $(-2.3, 5.4)$ (b) $(7.6, -9.2)$ (6)

10. Change the following polar co-ordinates into Cartesian co-ordinates, correct to 3 decimal places: (a) $(6.5, 132°)$ (b) $(3, 3\,\text{rad})$ (4)

Section 3

Chapter 25

Triangles and some practical applications

25.1 Sine and cosine rules

To '**solve a triangle**' means 'to find the values of unknown sides and angles'. If a triangle is **right-angled**, trigonometric ratios and the theorem of Pythagoras may be used for its solution, as shown in Chapter 22. However, for a **non-right-angled triangle**, trigonometric ratios and Pythagoras' theorem **cannot** be used. Instead, two rules, called the **sine rule** and **cosine rule**, are used.

Sine rule

With reference to triangle ABC of Fig. 25.1, the **sine rule** states:

$$\frac{a}{\sin A} = \frac{b}{\sin B} = \frac{c}{\sin C}$$

Figure 25.1

The rule may be used only when:

(i)   1 side and any 2 angles are initially given, or

(ii) 2 sides and an angle (not the included angle) are initially given.

Cosine rule

With reference to triangle ABC of Fig. 25.1, the **cosine rule** states:

$$a^2 = b^2 + c^2 - 2bc \cos A$$
$$\text{or} \quad b^2 = a^2 + c^2 - 2ac \cos B$$
$$\text{or} \quad c^2 = a^2 + b^2 - 2ab \cos C$$

The rule may be used only when:

(i)   2 sides and the included angle are initially given, or

(ii) 3 sides are initially given.

25.2 Area of any triangle

The **area of any triangle** such as ABC of Fig. 25.1 is given by:

(i) $\frac{1}{2} \times$ **base** $\times$ **perpendicular height, or**

(ii) $\frac{1}{2}ab \sin C$ or $\frac{1}{2}ac \sin B$ or $\frac{1}{2}bc \sin A$, or

(iii) $\sqrt{s(s-a)(s-b)(s-c)}$

where $s = \dfrac{a+b+c}{2}$

25.3 Worked problems on the solution of triangles and their areas

Problem 1.   In a triangle XYZ, $\angle X = 51°$, $\angle Y = 67°$ and $YZ = 15.2$ cm. Solve the triangle and find its area

The triangle XYZ is shown in Fig. 25.2. Since the angles in a triangle add up to 180°, then $z = 180° - 51° - 67° = \mathbf{62°}$.

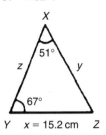

Figure 25.2

Applying the sine rule:

$$\frac{15.2}{\sin 51°} = \frac{y}{\sin 67°} = \frac{z}{\sin 62°}$$

Using $\dfrac{15.2}{\sin 51°} = \dfrac{y}{\sin 67°}$ and transposing gives:

$$y = \frac{15.2 \sin 67°}{\sin 51°} = \mathbf{18.00\,cm} = XZ$$

Using $\dfrac{15.2}{\sin 51°} = \dfrac{z}{\sin 62°}$ and transposing gives:

$$z = \frac{15.2 \sin 62°}{\sin 51°} = \mathbf{17.27\,cm} = XY$$

Area of triangle $XYZ = \frac{1}{2}xy \sin Z$

$$= \tfrac{1}{2}(15.2)(18.00) \sin 62° = \mathbf{120.8\,cm^2}$$

(or area $= \tfrac{1}{2}xz \sin Y = \tfrac{1}{2}(15.2)(17.27) \sin 67°$

$$= \mathbf{120.8\,cm^2})$$

It is always worth checking with triangle problems that the longest side is opposite the largest angle, and vice-versa. In this problem, Y is the largest angle and XZ is the longest of the three sides.

Problem 2. Solve the triangle ABC given $B = 78°51'$, $AC = 22.31$ mm and $AB = 17.92$ mm. Find also its area

Triangle ABC is shown in Fig. 25.3.

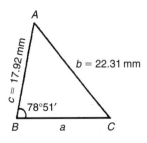

Figure 25.3

Applying the sine rule:

$$\frac{22.31}{\sin 78°51'} = \frac{17.92}{\sin C}$$

from which, $\sin C = \dfrac{17.92 \sin 78°51'}{22.31} = 0.7881$

Hence $C = \sin^{-1} 0.7881 = 52°0'$ or $128°0'$ (see Chapters 22 and 23).

Since $B = 78°51'$, C cannot be $128°0'$, since $128°0' + 78°51'$ is greater than 180°.

Thus only $C = 52°0'$ is valid.

Angle $A = 180° - 78°51' - 52°0' = 49°9'$

Applying the sine rule:

$$\frac{a}{\sin 49°9'} = \frac{22.31}{\sin 78°51'}$$

from which, $a = \dfrac{22.31 \sin 49°9'}{\sin 78°51'} = 17.20$ mm

Hence $A = \mathbf{49°9'}$, $C = \mathbf{52°0'}$ and $BC = \mathbf{17.20\,mm}$.

Area of triangle $ABC = \frac{1}{2}ac \sin B$

$$= \tfrac{1}{2}(17.20)(17.92) \sin 78°51' = \mathbf{151.2\,mm^2}$$

Problem 3. Solve the triangle PQR and find its area given that $QR = 36.5$ mm, $PR = 26.6$ mm and $\angle Q = 36°$

Triangle PQR is shown in Fig. 25.4.

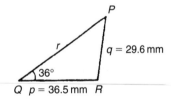

Figure 25.4

Applying the sine rule:

$$\frac{29.6}{\sin 36°} = \frac{36.5}{\sin P}$$

from which, $\sin P = \dfrac{36.5 \sin 36°}{29.6} = 0.7248$

Hence $P = \sin^{-1} 0.7248 = 46°27'$ or $133°33'$

When $P = 46°27'$ and $Q = 36°$

then $R = 180° - 46°27' - 36° = 97°33'$

When $P = 133°33'$ and $Q = 36°$

then $R = 180° - 133°33' - 36° = 10°27'$

Thus, in this problem, there are **two** separate sets of results and both are feasible solutions. Such a situation is called the **ambiguous case**.

Case 1. $P = 46°27'$, $Q = 36°$, $R = 97°33'$, $p = 36.5$ mm and $q = 29.6$ mm

From the sine rule:

$$\frac{r}{\sin 97°33'} = \frac{29.6}{\sin 36°}$$

from which, $r = \dfrac{29.6 \sin 97°33'}{\sin 36°} = \mathbf{49.92\ mm}$

Area $= \frac{1}{2} pq \sin R = \frac{1}{2}(36.5)(29.6)\sin 97°33'$

$$= \mathbf{535.5\ mm^2}$$

Case 2. $P = 133°33'$, $Q = 36°$, $R = 10°27'$, $p = 36.5$ mm and $q = 29.6$ mm

From the sine rule:

$$\frac{r}{\sin 10°27'} = \frac{29.6}{\sin 36°}$$

from which, $r = \dfrac{29.6 \sin 10°2'}{\sin 36°} = \mathbf{9.134\ mm}$

Area $= \frac{1}{2} pq \sin R = \frac{1}{2}(36.5)(29.6)\sin 10°27'$

$$= \mathbf{97.98\ mm^2}$$

Triangle PQR for case 2 is shown in Fig. 25.5.

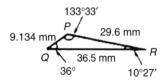

Figure 25.5

Now try the following exercise

Exercise 95 Further problems on the solution of triangles and their areas

In Problems 1 and 2, use the sine rule to solve the triangles ABC and find their areas.

1. $A = 29°$, $B = 68°$, $b = 27$ mm.
    $$\begin{bmatrix} C = 83°, a = 14.4\ \text{mm}, c = 28.9\ \text{mm}, \\ \text{area} = 189\ \text{mm}^2 \end{bmatrix}$$

2. $B = 71°26'$, $C = 56°32'$, $b = 8.60$ cm.
    $$\begin{bmatrix} A = 52°2', c = 7.568\ \text{cm}, \\ a = 7.152\ \text{cm}, \text{area} = 25.65\ \text{cm}^2 \end{bmatrix}$$

In Problems 3 and 4, use the sine rule to solve the triangles DEF and find their areas.

3. $d = 17$ cm, $f = 22$ cm, $F = 26°$
    $$\begin{bmatrix} D = 19°48', E = 134°12', \\ e = 36.0\ \text{cm}, \text{area} = 134\ \text{cm}^2 \end{bmatrix}$$

4. $d = 32.6$ mm, $e = 25.4$ mm, $D = 104°22'$
    $$\begin{bmatrix} E = 49°0', F = 26°38', \\ f = 15.09\ \text{mm}, \text{area} = 185.6\ \text{mm}^2 \end{bmatrix}$$

In Problems 5 and 6, use the sine rule to solve the triangles JKL and find their areas.

5. $j = 3.85$ cm, $k = 3.23$ cm, $K = 36°$
    $$\begin{bmatrix} J = 44°29', L = 99°31', l = 5.420\ \text{cm}, \\ \text{area} = 6.133\ \text{cm}^2 \\ \text{OR } J = 135°31', L = 8°29', \\ l = 0.811\ \text{cm}, \text{area} = 0.917\ \text{cm}^2 \end{bmatrix}$$

6. $k = 46$ mm, $l = 36$ mm, $L = 35°$
    $$\begin{bmatrix} K = 47°8', J = 97°52', \\ j = 62.2\ \text{mm}, \text{area} = 820.2\ \text{mm}^2 \\ \text{OR } K = 132°52', J = 12°8', \\ j = 13.19\ \text{mm}, \text{area} = 174.0\ \text{mm}^2 \end{bmatrix}$$

25.4 Further worked problems on the solution of triangles and their areas

Problem 4. Solve triangle DEF and find its area given that $EF = 35.0$ mm, $DE = 25.0$ mm and $\angle E = 64°$

Triangle DEF is shown in Fig. 25.6.

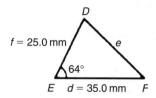

Figure 25.6

Applying the cosine rule:

$$e^2 = d^2 + f^2 - 2df \cos E$$

i.e. $e^2 = (35.0)^2 + (25.0)^2 - [2(35.0)(25.0)\cos 64°]$

$$= 1225 + 625 - 767.1 = 1083$$

from which, $e = \sqrt{1083} = \mathbf{32.91\,mm}$

Applying the sine rule:

$$\frac{32.91}{\sin 64°} = \frac{25.0}{\sin F}$$

from which, $\sin F = \dfrac{25.0 \sin 64°}{32.91} = 0.6828$

Thus $\angle F = \sin^{-1} 0.6828$

$$= 43°4' \text{ or } 136°56'$$

$F = 136°56'$ is not possible in this case since $136°56' + 64°$ is greater than $180°$. Thus only $F = \mathbf{43°4'}$ is valid.

$$\angle D = 180° - 64° - 43°4' = \mathbf{72°56'}$$

Area of triangle $DEF = \frac{1}{2}df \sin E$

$$= \tfrac{1}{2}(35.0)(25.0)\sin 64° = \mathbf{393.2\,mm^2}$$

Problem 5.   A triangle ABC has sides $a = 9.0$ cm, $b = 7.5$ cm and $c = 6.5$ cm. Determine its three angles and its area

---

Triangle ABC is shown in Fig. 25.7. It is usual first to calculate the largest angle to determine whether the triangle is acute or obtuse. In this case the largest angle is A (i.e. opposite the longest side).

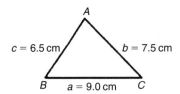

Figure 25.7

Applying the cosine rule:

$$a^2 = b^2 + c^2 - 2bc \cos A$$

from which,

$$2bc \cos A = b^2 + c^2 - a^2$$

and $\cos A = \dfrac{b^2 + c^2 - a^2}{2bc}$

$$= \dfrac{7.5^2 + 6.5^2 - 9.0^2}{2(7.5)(6.5)} = 0.1795$$

Hence $A = \cos^{-1} 0.1795 = \mathbf{79.66°}$ (or 280.33°, which is obviously impossible). The triangle is thus acute angled since $\cos A$ is positive. (If $\cos A$ had been

negative, angle A would be obtuse, i.e. lie between 90° and 180°).

Applying the sine rule:

$$\frac{9.0}{\sin 79.66°} = \frac{7.5}{\sin B}$$

from which, $\sin B = \dfrac{7.5 \sin 79.66°}{9.0} = 0.8198$

Hence $\boldsymbol{B} = \sin^{-1} 0.8198 = \mathbf{55.06°}$

and $\boldsymbol{C} = 180° - 79.66° - 55.06° = \mathbf{45.28°}$

Area $= \sqrt{s(s-a)(s-b)(s-c)}$, where

$$s = \frac{a+b+c}{2} = \frac{9.0 + 7.5 + 6.5}{2} = 11.5\,\text{cm}$$

Hence

$$\mathbf{area} = \sqrt{11.5(11.5 - 9.0)(11.5 - 7.5)(11.5 - 6.5)}$$

$$= \sqrt{11.5(2.5)(4.0)(5.0)} = \mathbf{23.98\,cm^2}$$

Alternatively,

$$\text{area} = \tfrac{1}{2}ab \sin C = \tfrac{1}{2}(9.0)(7.5)\sin 45.28°$$

$$= \mathbf{23.98\,cm^2}$$

Problem 6.   Solve triangle XYZ, shown in Fig. 25.8, and find its area given that $Y = 128°$, $XY = 7.2$ cm and $YZ = 4.5$ cm

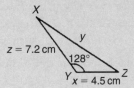

Figure 25.8

Applying the cosine rule:

$$y^2 = x^2 + z^2 - 2xz \cos Y$$

$$= 4.5^2 + 7.2^2 - [2(4.5)(7.2)\cos 128°]$$

$$= 20.25 + 51.84 - [-39.89]$$

$$= 20.25 + 51.84 + 39.89 = 112.0$$

$$y = \sqrt{112.0} = \mathbf{10.58\,cm}$$

Applying the sine rule:

$$\frac{10.58}{\sin 128°} = \frac{7.2}{\sin Z}$$

from which, $\sin Z = \dfrac{7.2 \sin 128°}{10.58} = 0.5363$

Hence $Z = \sin^{-1} 0.5363 = \mathbf{32.43°}$ (or 147.57° which, here, is impossible).

$X = 180° - 128° - 32.43° = \mathbf{19.57°}$

$\textbf{Area} = \frac{1}{2}xz \sin Y = \frac{1}{2}(4.5)(7.2) \sin 128°$

$= \mathbf{12.77\, cm^2}$

Now try the following exercise

Exercise 96   Further problems on the solution of triangles and their areas

In Problems 1 and 2, use the cosine and sine rules to solve the triangles PQR and find their areas.

1. $q = 12$ cm, $r = 16$ cm, $P = 54°$
$$\begin{bmatrix} p = 13.2\,\text{cm}, Q = 47.35°, \\ R = 78.65°, \text{ area} = 77.7\,\text{cm}^2 \end{bmatrix}$$

2. $q = 3.25$ m, $r = 4.42$ m, $P = 105°$
$$\begin{bmatrix} p = 6.127\,\text{m}, Q = 30.83°, \\ R = 44.17°, \text{ area} = 6.938\,\text{m}^2 \end{bmatrix}$$

In Problems 3 and 4, use the cosine and sine rules to solve the triangles XYZ and find their areas.

3. $x = 10.0$ cm, $y = 8.0$ cm, $z = 7.0$ cm.
$$\begin{bmatrix} X = 83.33°, Y = 52.62°, Z = 44.05°, \\ \text{area} = 27.8\,\text{cm}^2 \end{bmatrix}$$

4. $x = 21$ mm, $y = 34$ mm, $z = 42$ mm.
$$\begin{bmatrix} Z = 29.77°, Y = 53.50°, Z = 96.73°, \\ \text{area} = 355\,\text{mm}^2 \end{bmatrix}$$

25.5   Practical situations involving trigonometry

There are a number of **practical situations** where the use of trigonometry is needed to find unknown sides and angles of triangles. This is demonstrated in the following worked problems.

Problem 7.   A room 8.0 m wide has a span roof which slopes at 33° on one side and 40° on the other. Find the length of the roof slopes, correct to the nearest centimetre

A section of the roof is shown in Fig. 25.9.

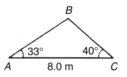

Figure 25.9

Angle at ridge, $B = 180° - 33° - 40° = 107°$

From the sine rule:

$$\frac{8.0}{\sin 107°} = \frac{a}{\sin 33°}$$

from which, $a = \dfrac{8.0 \sin 33°}{\sin 107°} = 4.556\,\text{m}$

Also from the sine rule:

$$\frac{8.0}{\sin 107°} = \frac{c}{\sin 40°}$$

from which, $c = \dfrac{8.0 \sin 40°}{\sin 107°} = 5.377\,\text{m}$

Hence the roof slopes are 4.56 m and 5.38 m, correct to the nearest centimetre.

Problem 8.   A man leaves a point walking at 6.5 km/h in a direction E 20° N (i.e. a bearing of 70°). A cyclist leaves the same point at the same time in a direction E 40° S (i.e. a bearing of 130°) travelling at a constant speed. Find the average speed of the cyclist if the walker and cyclist are 80 km apart after 5 hours

After 5 hours the walker has travelled $5 \times 6.5 = 32.5$ km (shown as AB in Fig. 25.10). If AC is the distance the cyclist travels in 5 hours then $BC = 80$ km.

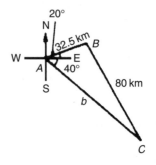

Figure 25.10

Applying the sine rule:

$$\frac{80}{\sin 60°} = \frac{32.5}{\sin C}$$

from which, $\sin C = \dfrac{32.5 \sin 60°}{80} = 0.3518$

Hence $C = \sin^{-1} 0.3518 = 20.60°$ (or $159.40°$, which is impossible in this case).

$B = 180° - 60° - 20.60° = 99.40°$.

Applying the sine rule again:

$$\frac{80}{\sin 60°} = \frac{b}{\sin 99.40°}$$

from which, $b = \dfrac{80 \sin 99.40°}{\sin 60°} = 91.14\,\text{km}$

Since the cyclist travels 91.14 km in 5 hours then

$$\textbf{average speed} = \frac{\text{distance}}{\text{time}} = \frac{91.14}{5} = \textbf{18.23\,km/h}$$

Problem 9. Two voltage phasors are shown in Fig. 25.11. If $V_1 = 40\,\text{V}$ and $V_2 = 100\,\text{V}$ determine the value of their resultant (i.e. length OA) and the angle the resultant makes with V_1

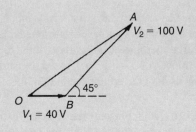

Figure 25.11

Angle $OBA = 180° - 45° = 135°$
Applying the cosine rule:

$$\begin{aligned}
OA^2 &= V_1^2 + V_2^2 - 2V_1 V_2 \cos OBA \\
&= 40^2 + 100^2 - \{2(40)(100) \cos 135°\} \\
&= 1600 + 10\,000 - \{-5657\} \\
&= 1600 + 10\,000 + 5657 = 17\,257
\end{aligned}$$

The resultant

$$OA = \sqrt{17\,257} = 131.4\,\text{V}$$

Applying the sine rule:

$$\frac{131.4}{\sin 135°} = \frac{100}{\sin AOB}$$

from which, $\sin AOB = \dfrac{100 \sin 135°}{131.4} = 0.5381$

Hence angle $AOB = \sin^{-1} 0.5381 = 32.55°$ (or $147.45°$, which is impossible in this case).

Hence the resultant voltage is 131.4 volts at 32.55° to V_1

Problem 10. In Fig. 25.12, PR represents the inclined jib of a crane and is 10.0 m long. PQ is 4.0 m long. Determine the inclination of the jib to the vertical and the length of tie QR

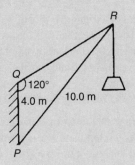

Figure 25.12

Applying the sine rule:

$$\frac{PR}{\sin 120°} = \frac{PQ}{\sin R}$$

from which, $\sin R = \dfrac{PQ \sin 120°}{PR}$

$$= \frac{(4.0) \sin 120°}{10.0} = 0.3464$$

Hence $\angle R = \sin^{-1} 0.3464 = 20.27°$ (or $159.73°$, which is impossible in this case).

$\angle P = 180° - 120° - 20.27° = \textbf{39.73°},$ **which is the inclination of the jib to the vertical**.

Applying the sine rule:

$$\frac{10.0}{\sin 120°} = \frac{QR}{\sin 39.73°}$$

from which, **length of tie,** $QR = \dfrac{10.0 \sin 39.73°}{\sin 120°}$

$$= \textbf{7.38\,m}$$

Now try the following exercise

Exercise 97 Further problems on practical situations involving trigonometry

1. A ship P sails at a steady speed of 45 km/h in a direction of W 32° N (i.e. a bearing of 302°) from a port. At the same time another ship Q

Section 3

leaves the port at a steady speed of 35 km/h in a direction N 15° E (i.e. a bearing of 015°). Determine their distance apart after 4 hours.

[193 km]

2. Two sides of a triangular plot of land are 52.0 m and 34.0 m, respectively. If the area of the plot is 620 m² find (a) the length of fencing required to enclose the plot and (b) the angles of the triangular plot.

[(a) 122.6 m (b) 94.80°, 40.66°, 44.54°]

3. A jib crane is shown in Fig. 25.13. If the tie rod PR is 8.0 long and PQ is 4.5 m long determine (a) the length of jib RQ and (b) the angle between the jib and the tie rod.

[(a) 11.4 m (b) 17.55°]

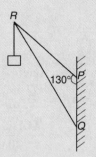

Figure 25.13

4. A building site is in the form of a quadrilateral as shown in Fig. 25.14, and its area is 1510 m². Determine the length of the perimeter of the site.

[163.4 m]

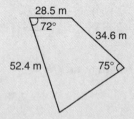

Figure 25.14

5. Determine the length of members BF and EB in the roof truss shown in Fig. 25.15.

[$BF = 3.9$ m, $EB = 4.0$ m]

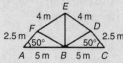

Figure 25.15

6. A laboratory 9.0 m wide has a span roof that slopes at 36° on one side and 44° on the other. Determine the lengths of the roof slopes.

[6.35 m, 5.37 m]

7. PQ and QR are the phasors representing the alternating currents in two branches of a circuit. Phasor PQ is 20.0 A and is horizontal. Phasor QR (which is joined to the end of PQ to form triangle PQR) is 14.0 A and is at an angle of 35° to the horizontal. Determine the resultant phasor PR and the angle it makes with phasor PQ

[32.48 A, 14.31°]

25.6 Further practical situations involving trigonometry

Problem 11. A vertical aerial stands on horizontal ground. A surveyor positioned due east of the aerial measures the elevation of the top as 48°. He moves due south 30.0 m and measures the elevation as 44°. Determine the height of the aerial

In Fig. 25.16, DC represents the aerial, A is the initial position of the surveyor and B his final position.

From triangle ACD, $\tan 48° = \dfrac{DC}{AC}$, from which $AC = \dfrac{DC}{\tan 48°}$

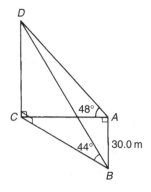

Figure 25.16

Similarly, from triangle BCD, $BC = \dfrac{DC}{\tan 44°}$

For triangle ABC, using Pythagoras' theorem:

$$BC^2 = AB^2 + AC^2$$
$$\left(\frac{DC}{\tan 44°}\right)^2 = (30.0)^2 + \left(\frac{DC}{\tan 48°}\right)^2$$

$$DC^2\left(\frac{1}{\tan^2 44°} - \frac{1}{\tan^2 48°}\right) = 30.0^2$$

$$DC^2(1.072323 - 0.810727) = 30.0^2$$

$$DC^2 = \frac{30.0^2}{0.261596} = 3440.4$$

Hence, height of aerial, $DC = \sqrt{3340.4}$

$$= \mathbf{58.65\,m}.$$

Problem 12. A crank mechanism of a petrol engine is shown in Fig. 25.17. Arm *OA* is 10.0 cm long and rotates clockwise about 0. The connecting rod *AB* is 30.0 cm long and end *B* is constrained to move horizontally

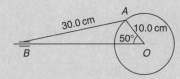

Figure 25.17

(a) For the position shown in Fig. 25.17 determine the angle between the connecting rod *AB* and the horizontal and the length of *OB*.

(b) How far does *B* move when angle *AOB* changes from 50° to 120°?

(a) Applying the sine rule:

$$\frac{AB}{\sin 50°} = \frac{AO}{\sin B}$$

from which, $\sin B = \dfrac{AO\sin 50°}{AB}$

$$= \frac{10.0\sin 50°}{30.0} = 0.2553$$

Hence $B = \sin^{-1} 0.2553 = 14.78°$ (or 165.22°, which is impossible in this case).

Hence the connecting rod *AB* makes an angle of 14.78° with the horizontal.

Angle $OAB = 180° - 50° - 14.78° = 115.22°$

Applying the sine rule:

$$\frac{30.0}{\sin 50°} = \frac{OB}{\sin 115.22°}$$

from which, $\mathbf{OB} = \dfrac{30.0\sin 115.22°}{\sin 50°}$

$$= \mathbf{35.43\,cm}$$

(b) Figure 25.18 shows the initial and final positions of the crank mechanism. In triangle $OA'B'$, applying the sine rule:

$$\frac{30.0}{\sin 120°} = \frac{10.0}{\sin A'B'O}$$

from which, $\sin A'B'O = \dfrac{10.0\sin 120°}{30.0}$

$$= 0.2887$$

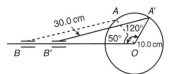

Figure 25.19

Hence $A'B'O = \sin^{-1} 0.2887 = 16.78°$ (or 163.22° which is impossible in this case).

Angle $OA'B' = 180° - 120° - 16.78° = 43.22°$

Applying the sine rule:

$$\frac{30.0}{\sin 120°} = \frac{OB'}{\sin 43.22°}$$

from which, $OB' = \dfrac{30.0\sin 43.22°}{\sin 120°}$

$$= 23.72\,cm$$

Since $OB = 35.43$ cm and $OB' = 23.72$ cm then

$$BB' = 35.43 - 23.72 = 11.71\,cm$$

Hence *B* moves 11.71 cm when angle *AOB* changes from 50° to 120°

Problem 13. The area of a field is in the form of a quadrilateral *ABCD* as shown in Fig. 25.19. Determine its area

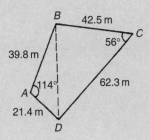

Figure 25.19

A diagonal drawn from B to D divides the quadrilateral into two triangles.

Area of quadrilateral $ABCD$

$= $ area of triangle ABD + area of triangle BCD

$= \frac{1}{2}(39.8)(21.4)\sin 114° + \frac{1}{2}(42.5)(62.3)\sin 56°$

$= 389.04 + 1097.5 = \mathbf{1487\,m^2}$

Now try the following exercise

Exercise 98 Further problems on practical situations involving trigonometry

1. Three forces acting on a fixed point are represented by the sides of a triangle of dimensions 7.2 cm, 9.6 cm and 11.0 cm. Determine the angles between the lines of action and the three forces. [80.42°, 59.38°, 40.20°]

2. A vertical aerial AB, 9.60 m high, stands on ground which is inclined 12° to the horizontal. A stay connects the top of the aerial A to a point C on the ground 10.0 m downhill from B, the foot of the aerial. Determine (a) the length of the stay, and (b) the angle the stay makes with the ground. [(a) 15.23 m (b) 38.07°]

3. A reciprocating engine mechanism is shown in Fig. 25.20. The crank AB is 12.0 cm long and the connecting rod BC is 32.0 cm long. For the position shown determine the length of AC and the angle between the crank and the connecting rod. [40.25 cm, 126.05°]

4. From Fig. 25.20, determine how far C moves, correct to the nearest millimetre when angle CAB changes from 40° to 160°, B moving in an anticlockwise direction. [19.8 cm]

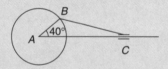

Figure 25.20

5. A surveyor, standing W 25° S of a tower measures the angle of elevation of the top of the tower as 46°30′. From a position E 23°S from the tower the elevation of the top is 37°15′. Determine the height of the tower if the distance between the two observations is 75 m. [36.2 m]

6. Calculate, correct to 3 significant figures, the co-ordinates x and y to locate the hole centre at P shown in Fig. 25.21. [$x = 69.3$ mm, $y = 142$ mm]

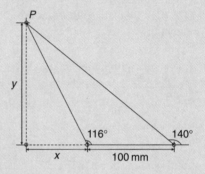

Figure 25.21

7. An idler gear, 30 mm in diameter, has to be fitted between a 70 mm diameter driving gear and a 90 mm diameter driven gear as shown in Fig. 25.22. Determine the value of angle θ between the centre lines. [130°]

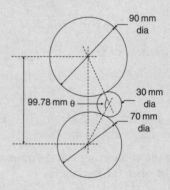

Figure 25.22

8. 16 holes are equally spaced on a pitch circle of 70 mm diameter. Determine the length of the chord joining the centres of two adjacent holes. [13.66 mm]

Chapter 26

Trigonometric identities and equations

26.1 Trigonometric identities

A trigonometric identity is a relationship that is true for all values of the unknown variable.

$$\tan \theta = \frac{\sin \theta}{\cos \theta} \quad \cot \theta = \frac{\cos \theta}{\sin \theta} \quad \sec \theta = \frac{1}{\cos \theta}$$

$$\text{cosec } \theta = \frac{1}{\sin \theta} \quad \text{and} \quad \cot \theta = \frac{1}{\tan \theta}$$

are examples of trigonometric identities from Chapter 22. Applying Pythagoras' theorem to the right-angled triangle shown in Fig. 26.1 gives:

$$a^2 + b^2 = c^2 \tag{1}$$

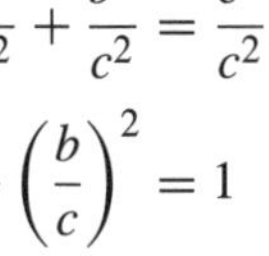

Figure 26.1

Dividing each term of equation (1) by c^2 gives:

$$\frac{a^2}{c^2} + \frac{b^2}{c^2} = \frac{c^2}{c^2}$$

i.e. $$\left(\frac{a}{c}\right)^2 + \left(\frac{b}{c}\right)^2 = 1$$

$$(\cos \theta)^2 + (\sin \theta)^2 = 1$$

Hence $$\cos^2 \theta + \sin^2 \theta = 1 \tag{2}$$

Dividing each term of equation (1) by a^2 gives:

$$\frac{a^2}{a^2} + \frac{b^2}{a^2} = \frac{c^2}{a^2}$$

i.e. $$1 + \left(\frac{b}{a}\right)^2 = \left(\frac{c}{a}\right)^2$$

Hence $$1 + \tan^2 \theta = \sec^2 \theta \tag{3}$$

Dividing each term of equation (1) by b^2 gives:

$$\frac{a^2}{b^2} + \frac{b^2}{b^2} = \frac{c^2}{b^2} \tag{4}$$

i.e. $$\left(\frac{a}{b}\right)^2 + 1 = \left(\frac{c}{b}\right)^2$$

Hence $$\cot^2 \theta + 1 = \text{cosec}^2 \theta \tag{5}$$

Equations (2), (3) and (4) are three further examples of trigonometric identities.

26.2 Worked problems on trigonometric identities

Problem 1. Prove the identity
$$\sin^2 \theta \cot \theta \sec \theta = \sin \theta$$

With trigonometric identities it is necessary to start with the left-hand side (LHS) and attempt to make it equal to the right-hand side (RHS) or vice-versa. It is often useful to change all of the trigonometric ratios into sines and cosines where possible. Thus

$$\text{LHS} = \sin^2 \theta \cot \theta \sec \theta$$

$$= \sin^2 \theta \left(\frac{\cos \theta}{\sin \theta}\right)\left(\frac{1}{\cos \theta}\right)$$

$$= \sin \theta \text{ (by cancelling)} = \text{RHS}$$

Problem 2. Prove that:

$$\frac{\tan x + \sec x}{\sec x \left(1 + \dfrac{\tan x}{\sec x}\right)} = 1$$

$$\begin{aligned}
\text{LHS} &= \frac{\tan x + \sec x}{\sec x \left(1 + \dfrac{\tan x}{\sec x}\right)} \\[2ex]
&= \frac{\dfrac{\sin x}{\cos x} + \dfrac{1}{\cos x}}{\left(\dfrac{1}{\cos x}\right)\left(1 + \dfrac{\dfrac{\sin x}{\cos x}}{\dfrac{1}{\cos x}}\right)} \\[2ex]
&= \frac{\dfrac{\sin x + 1}{\cos x}}{\left(\dfrac{1}{\cos x}\right)\left[1 + \left(\dfrac{\sin x}{\cos x}\right)\left(\dfrac{\cos x}{1}\right)\right]} \\[2ex]
&= \frac{\dfrac{\sin x + 1}{\cos x}}{\left(\dfrac{1}{\cos x}\right)[1 + \sin x]} \\[2ex]
&= \left(\frac{\sin x + 1}{\cos x}\right)\left(\frac{\cos x}{1 + \sin x}\right) \\[2ex]
&= 1 \text{ (by cancelling)} = \text{RHS}
\end{aligned}$$

Problem 3. Prove that: $\dfrac{1 + \cot\theta}{1 + \tan\theta} = \cot\theta$

$$\begin{aligned}
\text{LHS} &= \frac{1 + \cot\theta}{1 + \tan\theta} = \frac{1 + \dfrac{\cos\theta}{\sin\theta}}{1 + \dfrac{\sin\theta}{\cos\theta}} = \frac{\dfrac{\sin\theta + \cos\theta}{\sin\theta}}{\dfrac{\cos\theta + \sin\theta}{\cos\theta}} \\[2ex]
&= \left(\frac{\sin\theta + \cos\theta}{\sin\theta}\right)\left(\frac{\cos\theta}{\cos\theta + \sin\theta}\right) \\[2ex]
&= \frac{\cos\theta}{\sin\theta} = \cot\theta = \text{RHS}
\end{aligned}$$

Problem 4. Show that:

$$\cos^2\theta - \sin^2\theta = 1 - 2\sin^2\theta$$

From equation (2), $\cos^2\theta + \sin^2\theta = 1$, from which, $\cos^2\theta = 1 - \sin^2\theta$

$$\begin{aligned}
\text{Hence,} \quad \text{LHS} &= \cos^2\theta - \sin^2\theta \\
&= (1 - \sin^2\theta) - \sin^2\theta \\
&= 1 - \sin^2\theta - \sin^2\theta \\
&= 1 - 2\sin^2\theta = \text{RHS}
\end{aligned}$$

Problem 5. Prove that:

$$\sqrt{\frac{1 - \sin x}{1 + \sin x}} = \sec x - \tan x$$

$$\begin{aligned}
\text{LHS} &= \sqrt{\frac{1 - \sin x}{1 + \sin x}} \\[2ex]
&= \sqrt{\frac{(1 - \sin x)(1 - \sin x)}{(1 + \sin x)(1 - \sin x)}} \\[2ex]
&= \sqrt{\frac{(1 - \sin x)^2}{(1 - \sin^2 x)}}
\end{aligned}$$

Since $\cos^2 x + \sin^2 x = 1$ then $1 - \sin^2 x = \cos^2 x$

$$\begin{aligned}
\text{LHS} &= \sqrt{\frac{(1 - \sin x)^2}{(1 - \sin^2 x)}} = \sqrt{\frac{(1 - \sin x)^2}{\cos^2 x}} \\[2ex]
&= \frac{1 - \sin x}{\cos x} = \frac{1}{\cos x} - \frac{\sin x}{\cos x} \\[2ex]
&= \sec x - \tan x = \text{RHS}
\end{aligned}$$

Now try the following exercise

Exercise 99 Further problems on trigonometric identities

Prove the following trigonometric identities:

1. $\sin x \cot x = \cos x$

2. $\dfrac{1}{\sqrt{1 - \cos^2\theta}} = \operatorname{cosec}\theta$

3. $2\cos^2 A - 1 = \cos^2 A - \sin^2 A$

4. $\dfrac{\cos x - \cos^3 x}{\sin x} = \sin x \cos x$

5. $(1 + \cot\theta)^2 + (1 - \cot\theta)^2 = 2\operatorname{cosec}^2\theta$

6. $\dfrac{\sin^2 x(\sec x + \operatorname{cosec} x)}{\cos x \tan x} = 1 + \tan x$

26.3 Trigonometric equations

Equations which contain trigonometric ratios are called **trigonometric equations**. There are usually an infinite number of solutions to such equations; however, solutions are often restricted to those between 0° and 360°. A knowledge of angles of any magnitude is essential in the solution of trigonometric equations and calculators cannot be relied upon to give all the solutions (as shown in Chapter 23). Figure 26.2 shows a summary for angles of any magnitude.

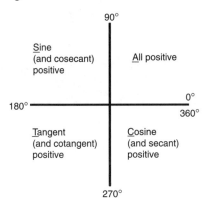

Figure 26.2

Equations of the type $a \sin^2 A + b \sin A + c = 0$

(i) **When $a = 0$,** $b \sin A + c = 0$, hence $\sin A = -\dfrac{c}{b}$
  and $A = \sin^{-1}\left(-\dfrac{c}{b}\right)$
  There are two values of A between $0°$ and $360°$ that satisfy such an equation, provided $-1 \le \dfrac{c}{b} \le 1$ (see Problems 6 to 8).

(ii) **When $b = 0$,** $a \sin^2 A + c = 0$, hence $\sin^2 A = -\dfrac{c}{a}$,
  $\sin A = \sqrt{-\dfrac{c}{a}}$ and $A = \sin^{-1}\sqrt{-\dfrac{c}{a}}$
  If either a or c is a negative number, then the value within the square root sign is positive. Since when a square root is taken there is a positive and negative answer there are four values of A between $0°$ and $360°$ which satisfy such an equation, provided $-1 \le \dfrac{c}{b} \le 1$ (see Problems 9 and 10).

(iii) **When a, b and c are all non-zero:**
  $a \sin^2 A + b \sin A + c = 0$ is a quadratic equation in which the unknown is $\sin A$. The solution of a quadratic equation is obtained either by factorising (if possible) or by using the quadratic formula:
  $$\sin A = \frac{-b \pm \sqrt{b^2 - 4ac}}{2a}$$
  (see Problems 11 and 12).

(iv) Often the trigonometric identities
  $\cos^2 A + \sin^2 A = 1$, $1 + \tan^2 A = \sec^2 A$ and $\cot^2 A + 1 = \operatorname{cosec}^2 A$ need to be used to reduce equations to one of the above forms (see Problems 13 to 15).

26.4 Worked problems (i) on trigonometric equations

Problem 6.   Solve the trigonometric equation: $5 \sin \theta + 3 = 0$ for values of θ from $0°$ to $360°$

$5 \sin \theta + 3 = 0$, from which
$\sin \theta = -3/5 = -0.6000$

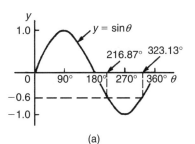

(a)

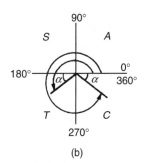

(b)

Figure 26.3

Hence $\theta = \sin^{-1}(-0.6000)$. Sine is negative in the third and fourth quadrants (see Fig. 26.3). The acute angle $\sin^{-1}(0.6000) = 36.87°$ (shown as α in Fig. 26.3(b)). Hence $\theta = 180° + 36.87°$, i.e. **216.87°** or $\theta = 360° - 36.87°$, i.e. **323.13°**

Problem 7.   Solve: $1.5 \tan x - 1.8 = 0$ for $0° \le x \le 360°$

$1.5 \tan x - 1.8 = 0$, from which
$\tan x = \dfrac{1.8}{1.5} = 1.2000$

  Hence $x = \tan^{-1} 1.2000$

Tangent is positive in the first and third quadrants (see Fig. 26.4).

The acute angle $\tan^{-1} 1.2000 = 50.19°$.

Hence, $x = \mathbf{50.19°}$ or $180° + 50.19° = \mathbf{230.19°}$

Section 3

Section 3

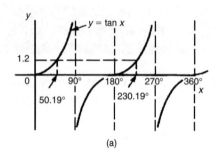

(a)

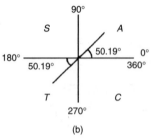

(b)

Figure 26.4

Problem 8. Solve: $4 \sec t = 5$ for values of t between $0°$ and $360°$

$4 \sec t = 5$, from which $\sec t = \frac{5}{4} = 1.2500$

Hence $t = \sec^{-1} 1.2500$

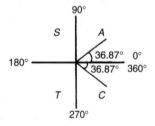

Figure 26.5

Secant ($=1/\text{cosine}$) is positive in the first and fourth quadrants (see Fig. 26.5). The acute angle $\sec^{-1} 1.2500 = 36.87°$. Hence

$$t = \mathbf{36.87°} \text{ or } 360° - 36.87° = \mathbf{323.13°}$$

Now try the following exercise

Exercise 100   Further problems on trigonometric equations

Solve the following equations for angles between $0°$ and $360°$

1. $4 - 7 \sin \theta = 0$ $[\theta = 34.85° \text{ or } 145.15°]$

2. $3 \operatorname{cosec} A + 5.5 = 0$

$[A = 213.06° \text{ or } 326.94°]$

3. $4(2.32 - 5.4 \cot t) = 0$

$[t = 66.75° \text{ or } 246.75°]$

26.5   Worked problems (ii) on trigonometric equations

Problem 9. Solve: $2 - 4 \cos^2 A = 0$ for values of A in the range $0° < A < 360°$

$2 - 4 \cos^2 A = 0$, from which $\cos^2 A = \dfrac{2}{4} = 0.5000$

Hence $\cos A = \sqrt{0.5000} = \pm 0.7071$ and $A = \cos^{-1}(\pm 0.7071)$

Cosine is positive in quadrant one and four and negative in quadrants two and three. Thus in this case there are four solutions, one in each quadrant (see Fig. 26.6). The acute angle $\cos^{-1} 0.7071 = 45°$.

Hence $A = \mathbf{45°, 135°, 225°}$ **or** $\mathbf{315°}$

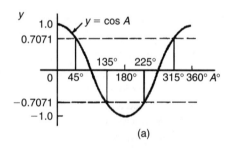

(a)

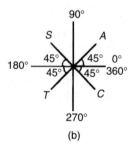

(b)

Figure 26.6

Problem 10. Solve: $\frac{1}{2} \cot^2 y = 1.3$ for $0° < y < 360°$

$\frac{1}{2} \cot^2 y = 1.3$, from which, $\cot^2 y = 2(1.3) = 2.6$

Hence $\cot y = \sqrt{2.6} = \pm 1.6125$, and $y = \cot^{-1}(\pm 1.6125)$. There are four solutions, one in each quadrant. The acute angle $\cot^{-1} 1.6125 = 31.81°$.

Hence **$y = 31.81°, 148.19°, 211.81°$ or $328.19°$**

Now try the following exercise

Exercise 101 Further problems on trigonometric equations

Solve the following equations for angles between $0°$ and $360°$

1. $5 \sin^2 y = 3$ $\begin{bmatrix} y = 50.77°, 129.23°, \\ 230.77° \text{ or } 309.23° \end{bmatrix}$

2. $5 + 3 \operatorname{cosec}^2 D = 8$ $[D = 90° \text{ or } 270°]$

3. $2 \cot^2 \theta = 5$ $[\theta = 32.32°, 147.68°, \\ 212.32° \text{ or } 327.68°]$

26.6 Worked problems (iii) on trigonometric equations

Problem 11. Solve the equation: $8 \sin^2 \theta + 2 \sin \theta - 1 = 0$, for all values of θ between $0°$ and $360°$

Factorising $8 \sin^2 \theta + 2 \sin \theta - 1 = 0$ gives $(4 \sin \theta - 1)(2 \sin \theta + 1) = 0$

Hence $4 \sin \theta - 1 = 0$, from which,

$\sin \theta = \frac{1}{4} = 0.2500$, or $2 \sin \theta + 1 = 0$, from which, $\sin \theta = -\frac{1}{2} = -0.5000$
(Instead of factorising, the quadratic formula can, of course, be used). $\theta = \sin^{-1} 0.2500 = 14.48°$ or $165.52°$, since sine is positive in the first and second quadrants, or $\theta = \sin^{-1}(-0.5000) = 210°$ or $330°$, since sine is negative in the third and fourth quadrants. Hence

$$\theta = 14.48°, 165.52°, 210° \text{ or } 330°$$

Problem 12. Solve: $6 \cos^2 \theta + 5 \cos \theta - 6 = 0$ for values of θ from $0°$ to $360°$

Factorising $6 \cos^2 \theta + 5 \cos \theta - 6 = 0$ gives $(3 \cos \theta - 2)(2 \cos \theta + 3) = 0$.

Hence $3 \cos \theta - 2 = 0$, from which,

$\cos \theta = \frac{2}{3} = 0.6667$, or $2 \cos \theta + 3 = 0$, from which, $\cos = -\frac{3}{2} = -1.5000$

The minimum value of a cosine is -1, hence the latter expression has no solution and is thus neglected. Hence

$$\theta = \cos^{-1} 0.6667 = \textbf{48.18° or 311.82°}$$

since cosine is positive in the first and fourth quadrants.

Now try the following exercise

Exercise 102 Further problems on trigonometric equations

Solve the following equations for angles between $0°$ and $360°$

1. $15 \sin^2 A + \sin A - 2 = 0$
$\begin{bmatrix} A = 19.47°, 160.53°, \\ 203.58° \text{ or } 336.42° \end{bmatrix}$

2. $8 \tan^2 \theta + 2 \tan \theta = 15$
$\begin{bmatrix} \theta = 51.34°, 123.69°, \\ 231.34° \text{ or } 303.69° \end{bmatrix}$

3. $2 \operatorname{cosec}^2 t - 5 \operatorname{cosec} t = 12$
$[t = 14.48°, 165.52°, 221.81° \text{ or } 318.19°]$

26.7 Worked problems (iv) on trigonometric equations

Problem 13. Solve: $5 \cos^2 t + 3 \sin t - 3 = 0$ for values of t from $0°$ to $360°$

Since $\cos^2 t + \sin^2 t = 1$, $\cos^2 t = 1 - \sin^2 t$. Substituting for $\cos^2 t$ in $5 \cos^2 t + 3 \sin t - 3 = 0$ gives

$$5(1 - \sin^2 t) + 3 \sin t - 3 = 0$$
$$5 - 5 \sin^2 t + 3 \sin t - 3 = 0$$
$$-5 \sin^2 t + 3 \sin t + 2 = 0$$
$$5 \sin^2 t - 3 \sin t - 2 = 0$$

Factorising gives $(5 \sin t + 2)(\sin t - 1) = 0$. Hence $5 \sin t + 2 = 0$, from which, $\sin t = -\frac{2}{5} = -0.4000$, or $\sin t - 1 = 0$, from which, $\sin t = 1$.
$t = \sin^{-1}(-0.4000) = 203.58°$ or $336.42°$, since sine is negative in the third and fourth quadrants, or

$t = \sin^{-1} 1 = 90°$. Hence

$$t = \mathbf{90°}, \mathbf{203.58°} \text{ or } \mathbf{336.42°}$$

as shown in Fig. 26.7.

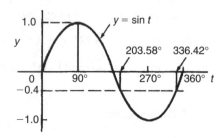

Figure 26.7

Problem 14. Solve: $18 \sec^2 A - 3 \tan A = 21$ for values of A between $0°$ and $360°$

$1 + \tan^2 A = \sec^2 A$. Substituting for $\sec^2 A$ in $18 \sec^2 A - 3 \tan A = 21$ gives

$$18(1 + \tan^2 A) - 3 \tan A = 21$$

i.e. $\quad 18 + 18 \tan^2 A - 3 \tan A - 21 = 0$

$$18 \tan^2 A - 3 \tan A - 3 = 0$$

Factorising gives $(6 \tan A - 3)(3 \tan A + 1) = 0$

Hence $6 \tan A - 3 = 0$, from which, $\tan A = \frac{3}{6} = 0.5000$ or $3 \tan A + 1 = 0$, from which, $\tan A = -\frac{1}{3} = -0.3333$. Thus $A = \tan^{-1}(0.5000) = 26.57°$ or $206.57°$, since tangent is positive in the first and third quadrants, or $A = \tan^{-1}(-0.3333) = 161.57°$ or $341.57°$, since tangent is negative in the second and fourth quadrants. Hence

$$A = \mathbf{26.57°}, \mathbf{161.57°}, \mathbf{206.57°} \text{ or } \mathbf{341.57°}$$

Problem 15. Solve: $3 \,\mathrm{cosec}^2\theta - 5 = 4 \cot \theta$ in the range $0 < \theta < 360°$

$\cot^2 \theta + 1 = \mathrm{cosec}^2 \theta$, Substituting for $\mathrm{cosec}^2 \theta$ in $3 \,\mathrm{cosec}^2 \theta - 5 = 4 \cot \theta$ gives:

$$3(\cot^2 \theta + 1) - 5 = 4 \cot \theta$$

$$3 \cot^2 \theta + 3 - 5 = 4 \cot \theta$$

$$3 \cot^2 \theta - 4 \cot \theta - 2 = 0$$

Since the left-hand side does not factorise the quadratic formula is used.

Thus, $\quad \cot \theta = \dfrac{-(-4) \pm \sqrt{(-4)^2 - 4(3)(-2)}}{2(3)}$

$$= \frac{4 \pm \sqrt{16 + 24}}{6} = \frac{4 \pm \sqrt{40}}{6}$$

$$= \frac{10.3246}{6} \text{ or } -\frac{2.3246}{6}$$

Hence $\cot \theta = 1.7208$ or -0.3874
$\theta = \cot^{-1} 1.7208 = 30.17°$ or $210.17°$, since cotangent is positive in the first and third quadrants, or $\theta = \cot^{-1}(-0.3874) = 111.18°$ or $291.18°$, since cotangent is negative in the second and fourth quadrants.

Hence, $\theta = \mathbf{30.17°}, \mathbf{111.18°}, \mathbf{210.17°} \text{ or } \mathbf{291.18°}$

Now try the following exercise

Exercise 103 Further problems on trigonometric equations

Solve the following equations for angles between $0°$ and $360°$

1. $12 \sin^2 \theta - 6 = \cos \theta$
$$\left[\begin{array}{l} \theta = 48.19°, 138.59°, \\ 221.41° \text{ or } 311.81° \end{array}\right]$$

2. $16 \sec x - 2 = 14 \tan^2 x$
$$[x = 52.93° \text{ or } 307.07°]$$

3. $4 \cot^2 A - 6 \,\mathrm{cosec}\, A + 6 = 0 \qquad [A = 90°]$

4. $5 \sec t + 2 \tan^2 t = 3$
$$[t = 107.83° \text{ or } 252.17°]$$

5. $2.9 \cos^2 a - 7 \sin a + 1 = 0$
$$[a = 27.83° \text{ or } 152.17°]$$

6. $3 \,\mathrm{cosec}^2 \beta = 8 - 7 \cot \beta$
$$\left[\begin{array}{l} \beta = 60.17°, 161.02°, \\ 240.17° \text{ or } 341.02° \end{array}\right]$$

Chapter 27

Compound angles

27.1 Compound angle formulae

An electric current i may be expressed as $i = 5\sin(\omega t - 0.33)$ amperes. Similarly, the displacement x of a body from a fixed point can be expressed as $x = 10\sin(2t + 0.67)$ metres. The angles $(\omega t - 0.33)$ and $(2t + 0.67)$ are called **compound angles** because they are the sum or difference of two angles.

The **compound angle formulae** for sines and cosines of the sum and difference of two angles A and B are:

$$\sin(A + B) = \sin A \cos B + \cos A \sin B$$

$$\sin(A - B) = \sin A \cos B - \cos A \sin B$$

$$\cos(A + B) = \cos A \cos B - \sin A \sin B$$

$$\cos(A - B) = \cos A \cos B + \sin A \sin B$$

(Note, $\sin(A + B)$ is **not** equal to $(\sin A + \sin B)$, and so on.)

The formulae stated above may be used to derive two further compound angle formulae:

$$\tan(A + B) = \frac{\tan A + \tan B}{1 - \tan A \tan B}$$

$$\tan(A - B) = \frac{\tan A - \tan B}{1 + \tan A \tan B}$$

The compound-angle formulae are true for all values of A and B, and by substituting values of A and B into the formulae they may be shown to be true.

Problem 1. Expand and simplify the following expressions:

(a) $\sin(\pi + \alpha)$ (b) $-\cos(90° + \beta)$

(c) $\sin(A - B) - \sin(A + B)$

(a) $\sin(\pi + \alpha) = \sin\pi\cos\alpha + \cos\pi\sin\alpha$ (from the formula for $\sin(A + B)$)

$$= (0)(\cos\alpha) + (-1)\sin\alpha$$

$$= -\sin\alpha$$

(b) $-\cos(90° + \beta)$

$$= -[\cos 90°\cos\beta - \sin 90°\sin\beta]$$

$$= -[(0)(\cos\beta) - (1)\sin\beta] = \sin\beta$$

(c) $\sin(A - B) - \sin(A + B)$

$$= [\sin A\cos B - \cos A\sin B]$$

$$\quad - [\sin A\cos B + \cos A\sin B]$$

$$= -2\cos A\,\sin B$$

Problem 2. Prove that:

$$\cos(y - \pi) + \sin\left(y + \frac{\pi}{2}\right) = 0$$

$$\cos(y - \pi) = \cos y\cos\pi + \sin y\sin\pi$$

$$= (\cos y)(-1) + (\sin y)(0)$$

$$= -\cos y$$

$$\sin\left(y + \frac{\pi}{2}\right) = \sin y\cos\frac{\pi}{2} + \cos y\sin\frac{\pi}{2}$$

$$= (\sin y)(0) + (\cos y)(1) = \cos y$$

Hence $\quad \cos(y - \pi) + \sin\left(y + \frac{\pi}{2}\right)$

$$= (-\cos y) + (\cos y) = 0$$

Problem 3. Show that

$$\tan\left(x + \frac{\pi}{4}\right)\tan\left(x - \frac{\pi}{4}\right) = -1$$

$$\tan\left(x+\frac{\pi}{4}\right)=\frac{\tan x+\tan\frac{\pi}{4}}{1-\tan x\tan\frac{\pi}{4}}$$

(from the formula for $\tan(A+B)$)

$$=\frac{\tan x+1}{1-(\tan x)(1)}=\left(\frac{1+\tan x}{1-\tan x}\right)$$

since $\tan\frac{\pi}{4}=1$

$$\tan\left(x-\frac{\pi}{4}\right)=\frac{\tan x-\tan\frac{\pi}{4}}{1+\tan x\tan\frac{\pi}{4}}=\left(\frac{\tan x-1}{1+\tan x}\right)$$

Hence, $\tan\left(x+\frac{\pi}{4}\right)\tan\left(x-\frac{\pi}{4}\right)$

$$=\left(\frac{1+\tan x}{1-\tan x}\right)\left(\frac{\tan x-1}{1+\tan x}\right)$$

$$=\frac{\tan x-1}{1-\tan x}=\frac{-(1-\tan x)}{1-\tan x}=-1$$

Problem 4. If $\sin P=0.8142$ and $\cos Q=0.4432$ evaluate, correct to 3 decimal places: (a) $\sin(P-Q)$, (b) $\cos(P+Q)$ and (c) $\tan(P+Q)$, using the compound angle formulae

Since $\sin P=0.8142$ then
$P=\sin^{-1}0.8142=54.51°$

Thus $\cos P=\cos 54.51°=0.5806$ and
$\tan P=\tan 54.51°=1.4025$

Since $\cos Q=0.4432$, $Q=\cos^{-1}0.4432=63.69°$

Thus $\sin Q=\sin 63.69°=0.8964$ and
$\tan Q=\tan 63.69°=2.0225$

(a) $\sin(P-Q)$

$=\sin P\cos Q-\cos P\sin Q$

$=(0.8142)(0.4432)-(0.5806)(0.8964)$

$=0.3609-0.5204=\mathbf{-0.160}$

(b) $\cos(P+Q)$

$=\cos P\cos Q-\sin P\sin Q$

$=(0.5806)(0.4432)-(0.8142)(0.8964)$

$=0.2573-0.7298=\mathbf{-0.473}$

(c) $\tan(P+Q)$

$$=\frac{\tan P+\tan Q}{1-\tan P\tan Q}=\frac{(1.4025)+(2.0225)}{1-(1.4025)(2.0225)}$$

$$=\frac{3.4250}{-1.8366}=\mathbf{-1.865}$$

Problem 5. Solve the equation:
$4\sin(x-20°)=5\cos x$ for values of x between $0°$ and $90°$

$4\sin(x-20°)=4[\sin x\cos 20°-\cos x\sin 20°]$

from the formula for $\sin(A-B)$

$=4[\sin x(0.9397)-\cos x(0.3420)]$

$=3.7588\sin x-1.3680\cos x$

Since $4\sin(x-20°)=5\cos x$ then
$3.7588\sin x-1.3680\cos x=5\cos x$
Rearranging gives:

$3.7588\sin x=5\cos x+1.3680\cos x$

$=6.3680\cos x$

and $\frac{\sin x}{\cos x}=\frac{6.3680}{3.7588}=1.6942$

i.e. $\tan x=1.6942$, and
$x=\tan^{-1}1.6942=59.449°$ or $\mathbf{59°27'}$

[Check: LHS $=4\sin(59.449°-20°)$
$=4\sin 39.449°=2.542$
RHS $=5\cos x=5\cos 59.449°=2.542$]

Now try the following exercise

Exercise 104   Further problems on compound angle formulae

1. Reduce the following to the sine of one angle:
   (a) $\sin 37°\cos 21°+\cos 37°\sin 21°$
   (b) $\sin 7t\cos 3t-\cos 7t\sin 3t$
   [(a) $\sin 58°$ (b) $\sin 4t$]

2. Reduce the following to the cosine of one angle:
   (a) $\cos 71°\cos 33°-\sin 71°\sin 33°$

(b) $\cos \dfrac{\pi}{3} \cos \dfrac{\pi}{4} + \sin \dfrac{\pi}{3} \sin \dfrac{\pi}{4}$

$$\left[\begin{array}{l} \text{(a) } \cos 104° \equiv -\cos 76° \\ \text{(b) } \cos \dfrac{\pi}{12} \end{array} \right]$$

3. Show that:

(a) $\sin\left(x + \dfrac{\pi}{3}\right) + \sin\left(x + \dfrac{2\pi}{3}\right) =$

$$\sqrt{3}\cos x$$

(b) $-\sin\left(\dfrac{3\pi}{2} - \phi\right) = \cos \phi$

4. Prove that:

(a) $\sin\left(\theta + \dfrac{\pi}{4}\right) - \sin\left(\theta - \dfrac{3\pi}{4}\right) =$

$$\sqrt{2}(\sin\theta + \cos\theta)$$

(b) $\dfrac{\cos(270° + \theta)}{\cos(360° - \theta)} = \tan\theta$

5. Given $\cos A = 0.42$ and $\sin B = 0.73$ evaluate
(a) $\sin(A - B)$, (b) $\cos(A - B)$, (c) $\tan(A + B)$,
correct to 4 decimal places.

        [(a) 0.3136   (b) 0.9495   (c) −2.4687]

In Problems 6 and 7, solve the equations for values
of θ between 0° and 360°

6. $3\sin(\theta + 30°) = 7\cos\theta$

        [64.72° or 244.72°]

7. $4\sin(\theta - 40°) = 2\sin\theta$

        [67.52° or 247.52°]

27.2 Conversion of $a\,\sin\omega t + b\cos\omega t$ into $R\sin(\omega t + \alpha)$

(i) $R\sin(\omega t + \alpha)$ represents a sine wave of maximum
value R, periodic time $2\pi/\omega$, frequency $\omega/2\pi$ and
leading $R\sin\omega t$ by angle α. (See Chapter 23).

(ii) $R\sin(\omega t + \alpha)$ may be expanded using the
compound-angle formula for $\sin(A + B)$, where
$A = \omega t$ and $B = \alpha$. Hence

$R\sin(\omega t + \alpha)$

$$= R[\sin\omega t\cos\alpha + \cos\omega t\sin\alpha]$$

$$= R\sin\omega t\cos\alpha + R\cos\omega t\sin\alpha$$

$$= (R\cos\alpha)\sin\omega t + (R\sin\alpha)\cos\omega t$$

(iii) If $a = R\cos\alpha$ and $b = R\sin\alpha$, where a and b are
constants, then
$R\sin(\omega t + \alpha) = a\,\sin\omega t + b\cos\omega t$, i.e. a sine
and cosine function of the same frequency when
added produce a sine wave of the same frequency
(which is further demonstrated in Chapter 34).

(iv) Since $a = R\cos\alpha$, then $\cos\alpha = \dfrac{a}{R}$ and since

$b = R\sin\alpha$, then $\sin\alpha = \dfrac{b}{R}$

If the values of a and b are known then the values of
R and α may be calculated. The relationship between
constants a, b, R and α are shown in Fig. 27.1.

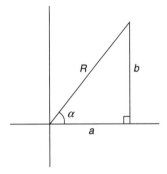

Figure 27.1

From Fig. 27.1, by Pythagoras' theorem:

$$R = \sqrt{a^2 + b^2}$$

and from trigonometric ratios:

$$\alpha = \tan^{-1}\dfrac{b}{a}$$

> **Problem 6.** Find an expression for
> $3\sin\omega t + 4\cos\omega t$ in the form $R\sin(\omega t + \alpha)$ and
> sketch graphs of $3\sin\omega t$, $4\cos\omega t$ and
> $R\sin(\omega t + \alpha)$ on the same axes

Let $3\sin\omega t + 4\cos\omega t = R\sin(\omega t + \alpha)$

then $3\sin\omega t + 4\cos\omega t$

$$= R[\sin\omega t\cos\alpha + \cos\omega t\sin\alpha]$$

$$= (R\cos\alpha)\sin\omega t + (R\sin\alpha)\cos\omega t$$

Equating coefficients of $\sin\omega t$ gives:

$$3 = R\cos\alpha, \text{ from which, } \cos\alpha = \dfrac{3}{R}$$

Equating coefficients of $\cos\omega t$ gives:

$$4 = R\sin\alpha, \text{ from which, } \sin\alpha = \dfrac{4}{R}$$

There is only one quadrant where both sin α **and** cos α are positive, and this is the first, as shown in Fig. 27.2. From Fig. 27.2, by Pythagoras' theorem:

$$R = \sqrt{3^2 + 4^2} = 5$$

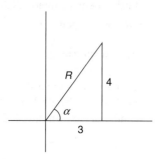

Figure 27.2

From trigonometric ratios:

$$a = \tan^{-1}\frac{4}{3} = 53.13° \quad \text{or} \quad 0.927 \text{ radians}$$

Hence, **3 sin ωt + 4 cos ωt = 5 sin(ωt + 0.927)**

A sketch of 3 sin ωt, 4 cos ωt and 5 sin(ωt + 0.927) is shown in Fig. 27.3.

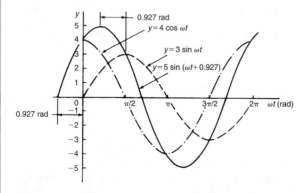

Figure 27.3

Two periodic functions of the same frequency may be combined by

(a) plotting the functions graphically and combining ordinates at intervals, or
(b) by resolution of phasors by drawing or calculation.

Problem 6, together with Problems 7 and 8 following, demonstrate a third method of combining waveforms.

Problem 7. Express: 4.6 sin ωt − 7.3 cos ωt in the form $R \sin(\omega t + \alpha)$

Let   4.6 sin ωt − 7.3 cos ωt = $R \sin(\omega t + \alpha)$

then   4.6 sin ωt − 7.3 cos ωt
$$= R[\sin \omega t \cos \alpha + \cos \omega t \sin \alpha]$$
$$= (R \cos \alpha) \sin \omega t + (R \sin \alpha) \cos \omega t$$

Equating coefficients of sin ωt gives:

$$4.6 = R \cos \alpha, \text{ from which, } \cos \alpha = \frac{4.6}{R}$$

Equating coefficients of cos ωt gives:

$$-7.3 = R \sin \alpha, \text{ from which } \sin \alpha = \frac{-7.3}{R}$$

There is only one quadrant where cosine is positive **and** sine is negative, i.e. the fourth quadrant, as shown in Fig. 27.4. By Pythagoras' theorem:

$$R = \sqrt{4.6^2 + (-7.3)^2} = 8.628$$

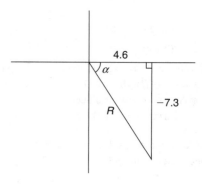

Figure 27.4

By trigonometric ratios:

$$\alpha = \tan^{-1}\left(\frac{-7.3}{4.6}\right)$$
$$= -57.78° \text{ or } -1.008 \text{ radians.}$$

Hence,
   4.6 sin ωt −7.3 cos ωt = 8.628 sin(ωt − 1.008)

Problem 8. Express: −2.7 sin ωt − 4.1 cos ωt in the form $R \sin(\omega t + \alpha)$

Let   −2.7 sin ωt − 4.1 cos ωt = $R \sin(\omega t + \alpha)$
$$= R[\sin \omega t \cos \alpha + \cos \omega t \sin \alpha]$$
$$= (R \cos \alpha) \sin \omega t + (R \sin \alpha) \cos \omega t$$

Equating coefficients gives:

$$-2.7 = R \cos \alpha, \text{ from which, } \cos \alpha = \frac{-2.7}{R}$$

$$\text{and } -4.1 = R \sin \alpha, \text{ from which, } \sin \alpha = \frac{-4.1}{R}$$

There is only one quadrant in which both cosine **and** sine are negative, i.e. the third quadrant, as shown in Fig. 27.5.

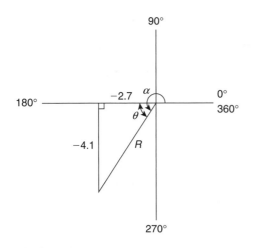

Figure 27.5

From Fig. 27.5,

$$R = \sqrt{(-2.7)^2 + (-4.1)^2} = 4.909$$

and $\qquad \theta = \tan^{-1}\dfrac{4.1}{2.7} = 56.63°$

Hence $\alpha = 180° + 56.63° = 236.63°$ or 4.130 radians. Thus,

$$-2.7 \sin \omega t - 4.1 \cos \omega t = 4.909 \sin(\omega t - 4.130)$$

An angle of $236.63°$ is the same as $-123.37°$ or -2.153 radians.
Hence $-2.7 \sin \omega t - 4.1 \cos \omega t$ may be expressed also as **4.909 $\sin(\omega t - 2.153)$**, which is preferred since it is the **principal value** (i.e. $-\pi \le \alpha \le \pi$).

Problem 9. Express: $3 \sin \theta + 5 \cos \theta$ in the form $R \sin(\theta + \alpha)$, and hence solve the equation $3 \sin \theta + 5 \cos \theta = 4$, for values of θ between $0°$ and $360°$

Let $\quad 3 \sin \theta + 5 \cos \theta = R \sin(\theta + \alpha)$
$$= R[\sin \theta \cos \alpha + \cos \theta \sin \alpha]$$
$$= (R \cos \alpha) \sin \theta + (R \sin \alpha) \cos \theta$$

Equating coefficients gives:

$$3 = R \cos \alpha, \text{ from which}, \cos \alpha = \frac{3}{R}$$

and $\quad 5 = R \sin \alpha, \text{ from which}, \sin \alpha = \frac{5}{R}$

Since both $\sin \alpha$ and $\cos \alpha$ are positive, R lies in the first quadrant, as shown in Fig. 27.6.

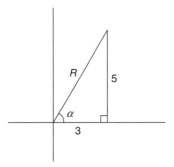

Figure 27.6

From Fig. 27.6, $R = \sqrt{3^2 + 5^2} = 5.831$ and

$$\alpha = \tan^{-1}\frac{5}{3} = 59.03°$$

Hence $\qquad 3 \sin \theta + 5 \cos \theta = 5.831 \sin(\theta + 59.03°)$

However $\quad 3 \sin \theta + 5 \cos \theta = 4$

Thus $\quad 5.831 \sin(\theta + 59.03°) = 4,$

from which $\quad (\theta + 59.03°) = \sin^{-1}\left(\dfrac{4}{5.831}\right)$

i.e. $\qquad \theta + 59.03° = 43.32°$ or $\quad 136.68°$

Hence $\quad \theta = 43.32° - 59.03° = -15.71°$ or

$$\theta = 136.68° - 59.03° = 77.65°$$

Since $-15.71°$ is the same as $-15.71° + 360°$, i.e. $344.29°$, then the solutions are $\theta = $ **77.65° or 344.29°**, which may be checked by substituting into the original equation.

Problem 10. Solve the equation:
$3.5 \cos A - 5.8 \sin A = 6.5$ for $0° \le A \le 360°$

Let $\quad 3.5 \cos A - 5.8 \sin A = R \sin(A + \alpha)$
$$= R[\sin A \cos \alpha + \cos A \sin \alpha]$$
$$= (R \cos \alpha) \sin A + (R \sin \alpha) \cos A$$

Equating coefficients gives:

$$3.5 = R \sin \alpha, \text{ from which}, \sin \alpha = \frac{3.5}{R}$$

and $-5.8 = R \cos \alpha, \text{ from which}, \cos \alpha = \dfrac{-5.8}{R}$

There is only one quadrant in which both sine is positive **and** cosine is negative, i.e. the second, as shown in Fig. 27.7.

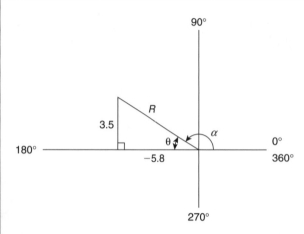

Figure 27.7

From Fig. 27.7, $R = \sqrt{3.5^2 + (-5.8)^2} = 6.774$ and
$\theta = \tan^{-1}\dfrac{3.5}{5.8} = 31.12°$

Hence $\alpha = 180° - 31.12° = 148.88°$

Thus $3.5\cos A - 5.8\sin A$
$$= 6.774\sin(A + 148.88°) = 6.5$$

Hence, $\sin(A + 148.88°) = \dfrac{6.5}{6.774}$

from which, $(A + 148.88°) = \sin^{-1}\dfrac{6.5}{6.774}$
$$= 73.65° \text{ or } 106.35°$$

Thus, $A = 73.65° - 148.88° = -75.23°$
$$\equiv (-75.23° + 360°) = 284.77°$$

or $A = 106.35° - 148.88° = -42.53°$
$$\equiv (-42.53° + 360°) = 317.47°$$

The solutions are thus $A = \textbf{284.77° or 317.47°}$, which may be checked in the original equation.

Now try the following exercise

Exercise 105 Further problems on the conversion of $a\sin\omega t + b\cos\omega t$ into $R\sin(\omega t + \alpha)$

In Problems 1 to 4, change the functions into the form $R\sin(\omega t \pm \alpha)$

1. $5\sin\omega t + 8\cos\omega t$
$$[9.434\sin(\omega t + 1.012)]$$

2. $4\sin\omega t - 3\cos\omega t$ $\qquad [5\sin(\omega t - 0.644)]$

3. $-7\sin\omega t + 4\cos\omega t$
$$[8.062\sin(\omega t + 2.622)]$$

4. $-3\sin\omega t - 6\cos\omega t$
$$[6.708\sin(\omega t - 2.034)]$$

5. Solve the following equations for values of θ between $0°$ and $360°$:
   (a) $2\sin\theta + 4\cos\theta = 3$
   (b) $12\sin\theta - 9\cos\theta = 7$
$$\begin{bmatrix}\text{(a) } 74.44° \text{ or } 338.70°\\ \text{(b) } 64.69° \text{ or } 189.05°\end{bmatrix}$$

6. Solve the following equations for $0° < A < 360°$:
   (a) $3\cos A + 2\sin A = 2.8$
   (b) $12\cos A - 4\sin A = 11$
$$\begin{bmatrix}\text{(a) } 72.73° \text{ or } 354.63°\\ \text{(b) } 11.15° \text{ or } 311.98°\end{bmatrix}$$

7. The third harmonic of a wave motion is given by $4.3\cos 3\theta - 6.9\sin 3\theta$.
   Express this in the form $R\sin(3\theta \pm \alpha)$
$$[8.13\sin(3\theta + 2.584)]$$

8. The displacement x metres of a mass from a fixed point about which it is oscillating is given by $x = 2.4\sin\omega t + 3.2\cos\omega t$, where t is the time in seconds. Express x in the form $R\sin(\omega t + \alpha)$.
$$[x = 4.0\sin(\omega t + 0.927)\text{m}]$$

9. Two voltages, $v_1 = 5\cos\omega t$ and $v_2 = -8\sin\omega t$ are inputs to an analogue circuit. Determine an expression for the output voltage if this is given by $(v_1 + v_2)$.
$$[9.434\sin(\omega t + 2.583)]$$

27.3 Double angles

(i) If, in the compound-angle formula for $\sin(A + B)$, we let $B = A$ then
$$\sin 2A = 2\sin A\cos A$$
   Also, for example, $\sin 4A = 2\sin 2A\cos 2A$ and $\sin 8A = 2\sin 4A\cos 4A$, and so on.

(ii) If, in the compound-angle formula for $\cos(A + B)$, we let $B = A$ then
$$\cos 2A = \cos^2 A - \sin^2 A$$

Since $\cos^2 A + \sin^2 A = 1$, then
$\cos^2 A = 1 - \sin^2 A$, and $\sin^2 A = 1 - \cos^2 A$,
and two further formula for $\cos 2A$ can be produced.

Thus $\cos 2A = \cos^2 A - \sin^2 A$
$= (1 - \sin^2 A) - \sin^2 A$

i.e. $\cos 2A = 1 - 2\sin^2 A$

and $\cos 2A = \cos^2 A - \sin^2 A$
$= \cos^2 A - (1 - \cos^2 A)$

i.e. $\cos 2A = 2\cos^2 A - 1$

Also, for example,
$\cos 4A = \cos^2 2A - \sin^2 2A$ or $1 - 2\sin^2 2A$ or
$2\cos^2 2A - 1$ and $\cos 6A = \cos^2 3A - \sin^2 3A$ or
$1 - 2\sin^2 3A$ or $2\cos^2 3A - 1$ and so on.

(iii) If, in the compound-angle formula for $\tan(A+B)$, we let $B = A$ then

$$\tan 2A = \frac{2\tan A}{1 - \tan^2 A}$$

Also, for example, $\tan 4A = \dfrac{2\tan 2A}{1 - \tan^2 2A}$

and $\tan 5A = \dfrac{2\tan \frac{5}{2}A}{1 - \tan^2 \frac{5}{2}A}$ and so on.

Problem 11. $I_3 \sin 3\theta$ is the third harmonic of a waveform. Express the third harmonic in terms of the first harmonic $\sin\theta$, when $I_3 = 1$

When $I_3 = 1$,

$I_3 \sin 3\theta = \sin 3\theta = \sin(2\theta + \theta)$
$= \sin 2\theta \cos\theta + \cos 2\theta \sin\theta$

from the $\sin(A+B)$ formula

$= (2\sin\theta\cos\theta)\cos\theta + (1 - 2\sin^2\theta)\sin\theta,$

from the double angle expansions

$= 2\sin\theta\cos^2\theta + \sin\theta - 2\sin^3\theta$
$= 2\sin\theta(1 - \sin^2\theta) + \sin\theta - 2\sin^3\theta,$

$(\text{since } \cos^2\theta = 1 - \sin^2\theta)$

$= 2\sin\theta - 2\sin^3\theta + \sin\theta - 2\sin^3\theta$

i.e. $\sin 3\theta = 3\sin\theta - 4\sin^3\theta$

Problem 12. Prove that: $\dfrac{1 - \cos 2\theta}{\sin 2\theta} = \tan\theta$

$\text{LHS} = \dfrac{1 - \cos 2\theta}{\sin 2\theta} = \dfrac{1 - (1 - 2\sin^2\theta)}{2\sin\theta\cos\theta}$

$= \dfrac{2\sin^2\theta}{2\sin\theta\cos\theta} = \dfrac{\sin\theta}{\cos\theta} = \tan\theta = \text{RHS}$

Problem 13. Prove that:
$$\cot 2x + \csc 2x = \cot x$$

$\text{LHS} = \cot 2x + \csc 2x$

$= \dfrac{\cos 2x}{\sin 2x} + \dfrac{1}{\sin 2x} = \dfrac{\cos 2x + 1}{\sin 2x}$

$= \dfrac{(2\cos^2 x - 1) + 1}{\sin 2x} = \dfrac{2\cos^2 x}{\sin 2x}$

$= \dfrac{2\cos^2 x}{2\sin x\cos x} = \dfrac{\cos x}{\sin x} = \cot x = \text{RHS}$

Now try the following exercise

Exercise 106 Further problems on double angles

1. The power p in an electrical circuit is given by $p = \dfrac{v^2}{R}$. Determine the power in terms of V, R and $\cos 2t$ when $v = V\cos t$.
$$\left[\dfrac{V^2}{2R}(1 + \cos 2t)\right]$$

2. Prove the following identities:
   (a) $1 - \dfrac{\cos 2\phi}{\cos^2\phi} = \tan^2\phi$
   (b) $\dfrac{1 + \cos 2t}{\sin^2 t} = 2\cot^2 t$
   (c) $\dfrac{(\tan 2x)(1 + \tan x)}{\tan x} = \dfrac{2}{1 - \tan x}$
   (d) $2\csc 2\theta\cos 2\theta = \cot\theta - \tan\theta$

3. If the third harmonic of a waveform is given by $V_3\cos 3\theta$, express the third harmonic in terms of the first harmonic $\cos\theta$, when $V_3 = 1$.
   $[\cos 3\theta = 4\cos^3\theta - 3\cos\theta]$

27.4   Changing products of sines and cosines into sums or differences

(i) $\sin(A+B)+\sin(A-B)=2\sin A\cos B$ (from the formulae in Section 27.1), i.e.

$$\sin A\,\cos B=\frac{1}{2}[\sin(A+B)+\sin(A-B)] \quad (1)$$

(ii) $\sin(A+B)-\sin(A-B)=2\cos A\sin B$ \quad i.e.

$$\cos A\,\sin B=\frac{1}{2}[\sin(A+B)-\sin(A-B)] \quad (2)$$

(iii) $\cos(A+B)+\cos(A-B)=2\cos A\cos B$ \quad i.e.

$$\cos A\,\cos B=\frac{1}{2}[\cos(A+B)+\cos(A-B)] \quad (3)$$

(iv) $\cos(A+B)-\cos(A-B)=-2\sin A\sin B$ \quad i.e.

$$\sin A\,\sin B=-\frac{1}{2}[\cos(A+B)-\cos(A-B)] \quad (4)$$

Problem 14.   Express: $\sin 4x\cos 3x$ as a sum or difference of sines and cosines

From equation (1),

$$\sin 4x\cos 3x=\frac{1}{2}[\sin(4x+3x)+\sin(4x-3x)]$$
$$=\frac{1}{2}(\sin 7x+\sin x)$$

Problem 15.   Express: $2\cos 5\theta\sin 2\theta$ as a sum or difference of sines or cosines

From equation (2),

$$2\cos 5\theta\sin 2\theta$$
$$=2\left\{\frac{1}{2}[\sin(5\theta+2\theta)-\sin(5\theta-2\theta)]\right\}$$
$$=\sin 7\theta-\sin 3\theta$$

Problem 16.   Express: $3\cos 4t\cos t$ as a sum or difference of sines or cosines

From equation (3),

$$3\cos 4t\cos t=3\left\{\frac{1}{2}[\cos(4t+t)+\cos(4t-t)]\right\}$$
$$=\frac{3}{2}(\cos 5t+\cos 3t)$$

Thus, if the integral $\int 3\cos 4t\cos t\,dt$ was required, then

$$\int 3\cos 4t\cos t\,dt=\int\frac{3}{2}(\cos 5t+\cos 3t)dt$$
$$=\frac{3}{2}\left[\frac{\sin 5t}{5}+\frac{\sin 3t}{3}\right]+c$$

Problem 17.   In an alternating current circuit, voltage $v=5\sin\omega t$ and current $i=10\sin(\omega t-\pi/6)$. Find an expression for the instantaneous power p at time t given that $p=vi$, expressing the answer as a sum or difference of sines and cosines

$$p=vi=(5\sin\omega t)[10\sin(\omega t-\pi/6)]$$
$$=50\sin\omega t\sin(\omega t-\pi/6).$$

From equation (4),

$$50\sin\omega t\sin(\omega t-\pi/6)$$
$$=(50)\left[-\frac{1}{2}\{\cos(\omega t+\omega t-\pi/6)\right.$$
$$\left.-\cos[\omega t-(\omega t-\pi/6)]\}\right]$$
$$=-25[\cos(2\omega t-\pi/6)-\cos\pi/6]$$

i.e. instantaneous power,

$$p=25[\cos\pi/6-\cos(2\omega t-\pi/6)]$$

Now try the following exercise

Exercise 107   Further problems on changing products of sines and cosines into sums or differences

In Problems 1 to 5, express as sums or differences:

1. $\sin 7t\cos 2t$ \qquad $\left[\frac{1}{2}(\sin 9t+\sin 5t)\right]$

2. $\cos 8x \sin 2x$ $\qquad \left[\dfrac{1}{2}(\sin 10x - \sin 6x)\right]$

3. $2 \sin 7t \sin 3t$ $\qquad [\cos 4t - \cos 10t]$

4. $4 \cos 3\theta \cos \theta$ $\qquad [2(\cos 4\theta + \cos 2\theta)]$

5. $3 \sin \dfrac{\pi}{3} \cos \dfrac{\pi}{6}$ $\qquad \left[\dfrac{3}{2}\left(\sin \dfrac{\pi}{2} + \sin \dfrac{\pi}{6}\right)\right]$

6. Determine $\int 2 \sin 3t \cos t \, dt$
$$\left[-\dfrac{\cos 4t}{4} - \dfrac{\cos 2t}{2} + c\right]$$

7. Evaluate $\int_0^{\pi/2} 4 \cos 5x \cos 2x \, dx$ $\qquad \left[-\dfrac{20}{21}\right]$

8. Solve the equation: $2 \sin 2\phi \sin \phi = \cos \phi$ in the range $\phi = 0$ to $\phi = 180°$
$$[30°, 90° \text{ or } 150°]$$

27.5 Changing sums or differences of sines and cosines into products

In the compound-angle formula let $(A+B) = X$ and $(A-B) = Y$
Solving the simultaneous equations gives
$$A = \frac{X+Y}{2} \text{ and } B = \frac{X-Y}{2}$$
Thus $\sin(A+B) + \sin(A-B) = 2 \sin A \cos B$ becomes

$$\sin X + \sin Y = 2 \sin\left(\frac{X+Y}{2}\right)\cos\left(\frac{X-Y}{2}\right) \quad (5)$$

Similarly,

$$\sin X - \sin Y = 2 \cos\left(\frac{X+Y}{2}\right)\sin\left(\frac{X-Y}{2}\right) \quad (6)$$

$$\cos X + \cos Y = 2 \cos\left(\frac{X+Y}{2}\right)\cos\left(\frac{X-Y}{2}\right) \quad (7)$$

$$\cos X - \cos Y = -2 \sin\left(\frac{X+Y}{2}\right)\sin\left(\frac{X-Y}{2}\right) \quad (8)$$

Problem 18. Express: $\sin 5\theta + \sin 3\theta$ as a product

From equation (5),
$$\sin 5\theta + \sin 3\theta = 2 \sin\left(\frac{5\theta + 3\theta}{2}\right)\cos\left(\frac{5\theta - 3\theta}{2}\right)$$
$$= 2 \sin 4\theta \cos \theta$$

Problem 19. Express: $\sin 7x - \sin x$ as a product

From equation (6),
$$\sin 7x - \sin x = 2 \cos\left(\frac{7x + x}{2}\right)\sin\left(\frac{7x - x}{2}\right)$$
$$= 2 \cos 4x \sin 3x$$

Problem 20. Express: $\cos 2t - \cos 5t$ as a product

From equation (8),
$$\cos 2t - \cos 5t = -2 \sin\left(\frac{2t + 5t}{2}\right)\sin\left(\frac{2t - 5t}{2}\right)$$
$$= -2 \sin \frac{7}{2}t \sin\left(-\frac{3}{2}t\right)$$
$$= 2 \sin \frac{7}{2}t \sin \frac{3}{2}t$$
$$\left[\text{since } \sin\left(-\frac{3}{2}t\right) = -\sin\frac{3}{2}t\right]$$

Problem 21. Show that $\dfrac{\cos 6x + \cos 2x}{\sin 6x + \sin 2x} = \cot 4x$

From equation (7), $\cos 6x + \cos 2x = 2 \cos 4x \cos 2x$
From equation (5), $\sin 6x + \sin 2x = 2 \sin 4x \cos 2x$

Hence $\dfrac{\cos 6x + \cos 2x}{\sin 6x + \sin 2x} = \dfrac{2 \cos 4x \cos 2x}{2 \sin 4x \cos 2x}$
$$= \dfrac{\cos 4x}{\sin 4x} = \cot 4x$$

Now try the following exercise

> **Exercise 108** **Further problems on changing sums or differences of sines and cosines into products**
>
> In Problems 1 to 5, express as products:
>
> 1. $\sin 3x + \sin x$ $\qquad$ $[2 \sin 2x \cos x]$
>
> 2. $\dfrac{1}{2}(\sin 9\theta - \sin 7\theta)$ $\qquad$ $[\cos 8\theta \sin \theta]$

> 3. $\cos 5t + \cos 3t$ $\qquad$ $[2 \cos 4t \cos t]$
>
> 4. $\dfrac{1}{8}(\cos 5t - \cos t)$ $\qquad$ $\left[-\dfrac{1}{4} \sin 3t \sin 2t\right]$
>
> 5. $\dfrac{1}{2}\left(\cos \dfrac{\pi}{3} + \cos \dfrac{\pi}{4}\right)$ $\qquad$ $\left[\cos \dfrac{7\pi}{24} \cos \dfrac{\pi}{24}\right]$
>
> 6. Show that: (a) $\dfrac{\sin 4x - \sin 2x}{\cos 4x + \cos 2x} = \tan x$
>
>   (b) $\dfrac{1}{2}[\sin (5x - \alpha) - \sin (x + \alpha)]$
> $\qquad\qquad = \cos 3x \sin (2x - \alpha)$

This Revision test covers the material contained in Chapter 25 to 27. *The marks for each question are shown in brackets at the end of each question.*

1. A triangular plot of land ABC is shown in Fig. R7.1. Solve the triangle and determine its area. (10)

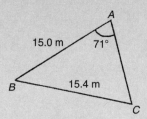

Figure R7.1

2. Figure R7.2 shows a roof truss PQR with rafter $PQ = 3$ m. Calculate the length of (a) the roof rise PP', (b) rafter PR, and (c) the roof span QR.

   Find also (d) the cross-sectional area of the roof truss. (11)

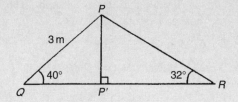

Figure R7.2

3. Prove the following identities:

   (a) $\sqrt{\dfrac{1 - \cos^2 \theta}{\cos^2 \theta}} = \tan \theta$

   (b) $\cos\left(\dfrac{3\pi}{2} + \phi\right) = \sin \phi$ (6)

4. Solve the following trigonometric equations in the range $0° \leq x \leq 360°$:

   (a) $4 \cos x + 1 = 0$ (b) $3.25 \operatorname{cosec} x = 5.25$
   (c) $5 \sin^2 x + 3 \sin x = 4$ (13)

5. Solve the equation $5 \sin(\theta - \pi/6) = 8 \cos \theta$ for values $0 \leq \theta \leq 2\pi$ (8)

6. Express $5.3 \cos t - 7.2 \sin t$ in the form $R \sin(t + \alpha)$. Hence solve the equation $5.3 \cos t - 7.2 \sin t = 4.5$ in the range $0 \leq t \leq 2\pi$ (12)

Multiple choice questions on Chapters 18–27

All questions have only one correct answer (answers on page 570).

1. In the right-angled triangle *ABC* shown in Figure M2.1, sine *A* is given by:

    (a) b/a (b) c/b (c) b/c (d) a/b

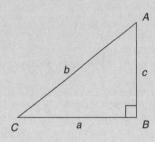

Figure M2.1

2. In the right-angled triangle *ABC* shown in Figure M2.1, cosine *C* is given by:

    (a) a/b (b) c/b (c) a/c (d) b/a

3. In the right-angled triangle shown in Figure M2.1, tangent *A* is given by:

    (a) b/c (b) a/c (c) a/b (d) c/a

4. $\dfrac{3\pi}{4}$ radians is equivalent to:

    (a) $135°$ (b) $270°$ (c) $45°$ (d) $67.5°$

5. In the triangular template *ABC* shown in Figure M2.2, the length *AC* is:

    (a) 6.17 cm   (b) 11.17 cm
    (c) 9.22 cm   (d) 12.40 cm

6. $(-4, 3)$ in polar co-ordinates is:

    (a) (5, 2.498 rad)   (b) (7, 36.87°)
    (c) (5, 36.87°)   (d) (5, 323.13°)

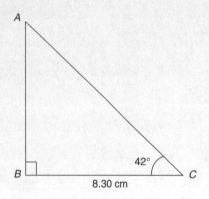

Figure M2.2

7. Correct to 3 decimal places, $\sin(-2.6 \text{ rad})$ is:

    (a) 0.516   (b) −0.045   (c) −0.516   (d) 0.045

8. For the right-angled triangle *PQR* shown in Figure M2.3, angle *R* is equal to:

    (a) $41.41°$ (b) $48.59°$ (c) $36.87°$ (d) $53.13°$

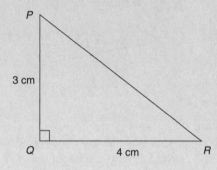

Figure M2.3

9. A hollow shaft has an outside diameter of 6.0 cm and an inside diameter of 4.0 cm. The cross-sectional area of the shaft is:

    (a) 6283 mm^2 (b) 1257 mm^2
    (c) 1571 mm^2 (d) 628 mm^2

10. If $\cos A = \dfrac{12}{13}$, then $\sin A$ is equal to:

    (a) $\dfrac{5}{13}$ (b) $\dfrac{13}{12}$ (c) $\dfrac{5}{12}$ (d) $\dfrac{12}{5}$

11. The area of triangle *XYZ* in Figure M2.4 is:

    (a) 24.22 cm^2 (b) 19.35 cm^2
    (c) 38.72 cm^2 (d) 32.16 cm^2

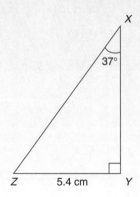

Figure M2.4

12. The value, correct to 3 decimal places, of $\cos\left(\dfrac{-3\pi}{4}\right)$ is:

   (a) 0.999 (b) 0.707 (c) −0.999 (d) −0.707

13. The speed of a car at 1 second intervals is given in the following table:

Time t(s)	0	1	2	3	4	5	6
Speed v(m/s)	0	2.5	5.9	9.0	15.0	22.0	30.0

   The distance travelled in 6 s (i.e. the area under the v/t graph) using the trapezoidal rule is:

   (a) 83.5 m (b) 68 m (c) 68.5 m (d) 204 m

14. A triangle has sides $a = 9.0$ cm, $b = 8.0$ cm and $c = 6.0$ cm. Angle A is equal to:

   (a) 82.42° (b) 56.49°
   (c) 78.58° (d) 79.87°

15. An arc of a circle of length 5.0 cm subtends an angle of 2 radians. The circumference of the circle is:

   (a) 2.5 cm (b) 10.0 cm
   (c) 5.0 cm (d) 15.7 cm

16. In the right-angled triangle ABC shown in Figure M2.5, secant C is given by:

   (a) $\dfrac{a}{b}$ (b) $\dfrac{a}{c}$ (c) $\dfrac{b}{c}$ (d) $\dfrac{b}{a}$

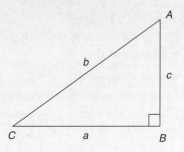

Figure M2.5

17. In the right-angled triangle ABC shown in Figure M2.5, cotangent C is given by:

   (a) $\dfrac{a}{b}$ (b) $\dfrac{b}{c}$ (c) $\dfrac{c}{b}$ (d) $\dfrac{a}{c}$

18. In the right-angled triangle ABC shown in Figure M2.5, cosecant A is given by:

   (a) $\dfrac{c}{a}$ (b) $\dfrac{b}{a}$ (c) $\dfrac{a}{b}$ (d) $\dfrac{b}{c}$

19. The mean value of a sine wave over half a cycle is:

   (a) 0.318 × maximum value
   (b) 0.707 × maximum value
   (c) the peak value
   (d) 0.637 × maximum value

20. Tan 60° is equivalent to:

   (a) $\dfrac{1}{\sqrt{3}}$ (b) $\dfrac{\sqrt{3}}{2}$ (c) $\dfrac{1}{2}$ (d) $\sqrt{3}$

21. An alternating current is given by:
   $i = 15\sin(100\pi t - 0.25)$ amperes. When time $t = 5$ ms, the current i has a value of:

   (a) 0.35 A   (b) 14.53 A
   (c) 15 A   (d) 0.41 A

22. The area of the path shown shaded in Figure M2.6 is:

   (a) 300 m² (b) 234 m² (c) 124 m² (d) 66 m²

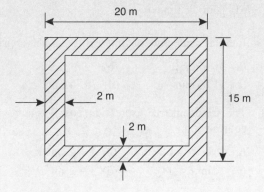

Figure M2.6

23. Correct to 4 significant figures, the value of sec 161° is:

   (a) −1.058   (b) 0.3256
   (c) 3.072   (d) −0.9455

24. Which of the following trigonometrical identities if true?

   (a) $\operatorname{cosec}\theta = \dfrac{1}{\cos\theta}$ (b) $\cot\theta = \dfrac{1}{\sin\theta}$

   (c) $\dfrac{\sin\theta}{\cos\theta} = \tan\theta$ (d) $\sec\theta = \dfrac{1}{\sin\theta}$

25. The displacement x metres of a mass from a fixed point about which it is oscillating is given by $x = 3\cos\omega t - 4\sin\omega t$, where t is the time in seconds. x may be expressed as:

   (a) $5\sin(\omega t + 2.50)$ metres
   (b) $7\sin(\omega t - 36.87°)$ metres
   (c) $5\sin\omega t$ metres
   (d) $-\sin(\omega t - 2.50)$ metres

26. The solutions of the equation $2\tan x - 7 = 0$ for $0° \le x \le 360°$ are:

   (a) $105.95°$ and $254.05°$
   (b) $74.05°$ and $254.05°$
   (c) $74.05°$ and $285.95°$
   (d) $254.05°$ and $285.95°$

27. A sinusoidal current is given by: $i = R\sin(\omega t + \alpha)$. Which of the following statements is incorrect?

   (a) R is the average value of the current
   (b) frequency $= \dfrac{\omega}{2\pi}$ Hz
   (c) ω = angular velocity
   (d) periodic time $= \dfrac{2\pi}{\omega}$ s

28. If the circumference of a circle is 100 mm its area is:

   (a) $314.2\,\text{cm}^2$ (b) $7.96\,\text{cm}^2$
   (c) $31.83\,\text{mm}^2$ (d) $78.54\,\text{cm}^2$

29. The trigonometric expression $\cos^2\theta - \sin^2\theta$ is equivalent to;

   (a) $2\sin^2\theta - 1$ (b) $1 + 2\sin^2\theta$
   (c) $2\sin^2\theta + 1$ (d) $1 - 2\sin^2\theta$

30. A vehicle has a mass of 2000 kg. A model of the vehicle is made to a scale of 1 to 100. If the vehicle and model are made of the same material, the mass of the model is:

   (a) 2 g   (b) 20 kg   (c) 200 g   (d) 20 g

31. A vertical tower stands on level ground. At a point 100 m from the foot of the tower the angle of elevation of the top is $20°$. The height of the tower is:

   (a) 274.7 m   (b) 36.4 m
   (c) 34.3 m   (d) 94.0 m

32. $(7, 141°)$ in Cartesian co-ordinates is:

   (a) $(5.44, -4.41)$ (b) $(-5.44, -4.41)$
   (c) $(5.44, 4.41)$ (d) $(-5.44, 4.41)$

33. If $\tan A = 1.4276$, $\sec A$ is equal to:

   (a) 0.8190 (b) 0.5737 (c) 0.7005 (d) 1.743

34. An indicator diagram for a steam engine is as shown in Figure M2.7. The base has been divided into 6 equally spaced intervals and the lengths of the 7 ordinates measured, with the results shown in centimetres. Using Simpson's rule the area of the indicator diagram is:

   (a) $32\,\text{cm}^2$ (b) $17.9\,\text{cm}^2$
   (c) $16\,\text{cm}^2$ (d) $96\,\text{cm}^2$

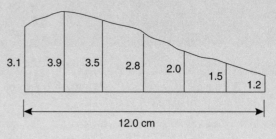

Figure M2.7

35. The acute angle $\cot^{-1}2.562$ is equal to:

   (a) $67.03°$ (b) $21.32°$
   (c) $22.97°$ (d) $68.68°$

36. Correct to 4 significant figures, the value of $\operatorname{cosec}(-125°)$ is:

   (a) -1.221 (b) -1.743
   (c) -0.8192 (d) -0.5736

37. The equation of a circle is $x^2 + y^2 - 2x + 4y - 4 = 0$. Which of the following statements is correct?

   (a) The circle has centre $(1, -2)$ and radius 4
   (b) The circle has centre $(-1, 2)$ and radius 2
   (c) The circle has centre $(-1, -2)$ and radius 4
   (d) The circle has centre $(1, -2)$ and radius 3

38. Cos 30° is equivalent to:

   (a) $\dfrac{1}{2}$ (b) $\dfrac{2}{\sqrt{3}}$ (c) $\dfrac{\sqrt{3}}{2}$ (d) $\dfrac{1}{\sqrt{3}}$

39. The angles between $0°$ and $360°$ whose tangent is -1.7624 are:

   (a) $60.43°$ and $240.43°$
   (b) $119.57°$ and $299.57°$
   (c) $119.57°$ and $240.43°$
   (d) $150.43°$ and $299.57°$

40. The surface are of a sphere of diameter 40 mm is:

   (a) $201.06 \, \text{cm}^2$ (b) $33.51 \, \text{cm}^2$

   (c) $268.08 \, \text{cm}^2$ (d) $50.27 \, \text{cm}^2$

41. In the triangular template DEF show in Figure M2.8, angle F is equal to:

   (a) $43.5°$ (b) $28.6°$ (c) $116.4°$ (d) $101.5°$

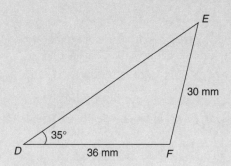

Figure M2.8

42. The area of the triangular template DEF shown in Figure M2.8 is:

   (a) $529.2 \, \text{mm}^2$ (b) $258.5 \, \text{mm}^2$

   (c) $483.7 \, \text{mm}^2$ (d) $371.7 \, \text{mm}^2$

43. A water tank is in the shape of a rectangular prism having length 1.5 m, breadth 60 cm and height 300 mm. If 1 litre $= 1000 \, \text{cm}^3$, the capacity of the tank is:

   (a) 27 litre (b) 2.7 litre

   (c) 2700 litre (d) 270 litre

44. A pendulum of length 1.2 m swings through an angle of $12°$ in a single swing. The length of arc traced by the pendulum bob is:

   (a) 14.40 cm (b) 25.13 cm

   (d) 10.00 cm (d) 45.24 cm

45. In the range $0° \leq \theta \leq 360°$ the solutions of the trigonometrical equation
$9 \tan^2\theta - 12 \tan\theta + 4 = 0$ are:

   (a) $33.69°, 146.31°, 213.69°$ and $326.31°$

   (b) $33.69°,$ and $213.69°$

   (c) $146.31°$ and $213.69°$

   (d) $146.69°$ and $326.31°$

46. A wheel on a car has a diameter of 800 mm. If the car travels 5 miles, the number of complete revolutions the wheel makes (given $1 \, \text{km} = \frac{5}{8}$ mile) is:

   (a) 1989 (b) 1591 (c) 3183 (d) 10 000

47. A rectangular building is shown on a building plan having dimensions 20 mm by 10 mm. If the plan is drawn to a scale of 1 to 300, the true area of the building in m^2 is:

   (a) $60\,000 \, \text{m}^2$ (b) $18 \, \text{m}^2$

   (c) $0.06 \, \text{m}^2$ (d) $1800 \, \text{m}^2$

48. An alternating voltage v is given by
$v = 100 \sin\left(100\pi t + \frac{\pi}{4}\right)$ volts. When $v = 50$ volts, the time t is equal to:

   (a) $0.093 \, \text{s}$ (b) $-0.908 \, \text{ms}$

   (c) $-0.833 \, \text{ms}$ (d) $-0.162 \, \text{s}$

49. Using the theorem of Pappus, the position of the centroid of a semicircle of radius r lies on the axis of symmetry at a distance from the diameter of:

   (a) $\dfrac{3\pi}{4r}$ (b) $\dfrac{3r}{4\pi}$ (c) $\dfrac{4r}{3\pi}$ (d) $\dfrac{4\pi}{3r}$

50. The acute angle $\operatorname{cosec}^{-1} 1.429$ is equal to:

   (a) $55.02°$ (b) $45.59°$

   (c) $44.41°$ (d) $34.98°$

51. The area of triangle PQR is given by:

   (a) $\dfrac{1}{2} pr \cos Q$

   (b) $\sqrt{(s-p)(s-q)(s-r)}$ where $s = \dfrac{p+q+r}{2}$

   (c) $\dfrac{1}{2} rq \sin P$ (d) $\dfrac{1}{2} pq \sin Q$

52. The values of θ that are true for the equation $5 \sin\theta + 2 = 0$ in the range $\theta = 0°$ to $\theta = 360°$ are:

   (a) $23.58°$ and $336.42°$

   (b) $23.58°$ and $203.58°$

   (c) $156.42°$ and $336.42°$

   (d) $203.58°$ and $336.42°$

53. $(-3, -7)$ in polar co-ordinates is:

   (a) $(-7.62, -113.20°)$ (b) $(7.62, 246.80°)$

   (c) $(7.62, 23.20°)$ (d) $(7.62, 203.20°)$

Section 3

54. In triangle ABC in Figure M2.9, length AC is:

   (a) 14.90 cm   (b) 18.15 cm

   (c) 13.16 cm   (d) 14.04 cm

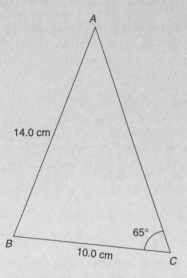

Figure M2.9

55. The total surface area of a cylinder of length 20 cm and diameter 6 cm is:

   (a) 56.55 cm^2 (b) 433.54 cm^2
   (c) 980.18 cm^2 (d) 226.19 cm^2

56. The acute angle $\sec^{-1} 2.4178$ is equal to:

   (a) 24.43°   (b) 22.47°
   (c) 0.426 rad   (d) 65.57°

57. The solution of the equation $3 - 5\cos^2 A = 0$ for values of A in the range $0° \leq A \leq 360°$ are:

   (a) 39.23° and 320.77°
   (b) 39.23°, 140.77°, 219.23° and 320.77°
   (c) 140.77° and 219.23°
   (d) 53.13°, 126.87°, 233.13° and 306.87°

58. An alternating current i has the following values at equal intervals of 2 ms:

Time t (ms)	0	2.0	4.0	6.0
Current I (A)	0	4.6	7.4	10.8
Time t (ms)	8.0	10.0	12.0	
Current I (A)	8.5	3.7	0	

Charge q (in millicoulombs) is given by $q = \int_0^{12.0} i\,dt$. Using the trapezoidal rule, the approximate charge in the 12 ms period is:

   (a) 70 mC   (b) 72.1 mC
   (c) 35 mC   (d) 216.4 mC

59. In triangle ABC in Figure M2.10, the length AC is:

   (a) 18.79 cm   (b) 70.89 cm
   (c) 22.89 cm   (d) 16.10 cm

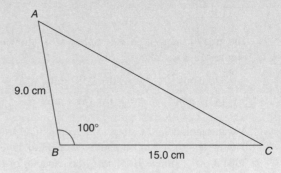

Figure M2.10

60. The total surface area of a solid hemisphere of diameter 6.0 cm is:

   (a) 56.55 cm^2 (b) 339.3 cm^2
   (c) 226.2 cm^2 (d) 84.82 cm^2

Graphs

Graphs

Chapter 28

Straight line graphs

28.1 Introduction to graphs

A **graph** is a pictorial representation of information showing how one quantity varies with another related quantity.

The most common method of showing a relationship between two sets of data is to use **Cartesian** or **rectangular axes** as shown in Fig. 28.1.

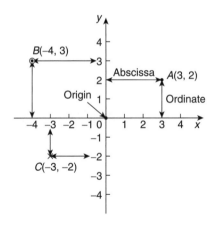

Figure 28.1

The points on a graph are called **co-ordinates**. Point A in Fig. 28.1 has the co-ordinates $(3, 2)$, i.e. 3 units in the x direction and 2 units in the y direction. Similarly, point B has co-ordinates $(-4, 3)$ and C has co-ordinates $(-3, -2)$. The origin has co-ordinates $(0, 0)$.

The horizontal distance of a point from the vertical axis is called the **abscissa** and the vertical distance from the horizontal axis is called the **ordinate**.

28.2 The straight line graph

Let a relationship between two variables x and y be $y = 3x + 2$

When $x = 0$, $y = 3(0) + 2 = 2$.
When $x = 1$, $y = 3(1) + 2 = 5$.
When $x = 2$, $y = 3(2) + 2 = 8$, and so on.

Thus co-ordinates $(0, 2)$, $(1, 5)$ and $(2, 8)$ have been produced from the equation by selecting arbitrary values of x, and are shown plotted in Fig. 28.2. When the points are joined together, a **straight line graph** results.

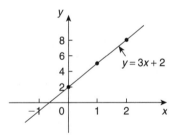

Figure 28.2

The **gradient** or **slope** of a straight line is the ratio of the change in the value of y to the change in the value of x between any two points on the line. If, as x increases, $(\rightarrow)$, y also increases $(\uparrow)$, then the gradient is positive. In Fig. 28.3(a),

$$\text{the gradient of } AC = \frac{\text{change in } y}{\text{change in } x} = \frac{CB}{BA}$$

$$= \frac{7 - 3}{3 - 1} = \frac{4}{2} = 2$$

If as x increases $(\rightarrow)$, y decreases $(\downarrow)$, then the gradient is negative.

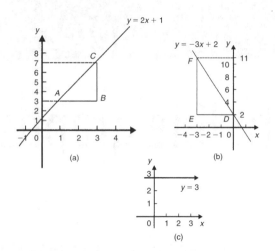

Figure 28.3

In Fig. 28.3(b),

$$\text{the gradient of } DF = \frac{\text{change in } y}{\text{change in } x} = \frac{FE}{ED}$$

$$= \frac{11 - 2}{-3 - 0} = \frac{9}{-3} = -3$$

Figure 28.3(c) shows a straight line graph $y = 3$. Since the straight line is horizontal the gradient is zero.

The value of y when $x = 0$ is called the **y-axis intercept**. In Fig. 28.3(a) the y-axis intercept is 1 and in Fig. 28.3(b) is 2.

If the equation of a graph is of the form $y = mx + c$, where m and c are constants, **the graph will always be a straight line**, m representing the gradient and c the y-axis intercept.

Thus $y = 5x + 2$ represents a straight line of gradient 5 and y-axis intercept 2. Similarly, $y = -3x - 4$ represents a straight line of gradient -3 and y-axis intercept -4.

Summary of general rules to be applied when drawing graphs

(i)   Give the graph a title clearly explaining what is being illustrated.

(ii) Choose scales such that the graph occupies as much space as possible on the graph paper being used.

(iii) Choose scales so that interpolation is made as easy as possible. Usually scales such as 1 cm = 1 unit, or 1 cm = 2 units, or 1 cm = 10 units are used. Awkward scales such as 1 cm = 3 units or 1 cm = 7 units should not be used.

(iv) The scales need not start at zero, particularly when starting at zero produces an accumulation of points within a small area of the graph paper.

(v)   The co-ordinates, or points, should be clearly marked. This may be done either by a cross, or a dot and circle, or just by a dot (see Fig. 28.1).

(vi) A statement should be made next to each axis explaining the numbers represented with their appropriate units.

(vii) Sufficient numbers should be written next to each axis without cramping.

Problem 1. Plot the graph $y = 4x + 3$ in the range $x = -3$ to $x = +4$. From the graph, find (a) the value of y when $x = 2.2$, and (b) the value of x when $y = -3$

Whenever an equation is given and a graph is required, a table giving corresponding values of the variable is necessary. The table is achieved as follows:

When $\quad x = -3, \ y = 4x + 3 = 4(-3) + 3$

$$= -12 + 3 = -9$$

When $\quad x = -2, \ y = 4(-2) + 3$

$$= -8 + 3 = -5, \text{ and so on.}$$

Such a table is shown below:

x	-3	-2	-1	0	1	2	3	4
y	-9	-5	-1	3	7	11	15	19

The co-ordinates $(-3, -9)$, $(-2, -5)$, $(-1, -1)$, and so on, are plotted and joined together to produce the straight line shown in Fig. 28.4. (Note that the scales used on the x and y axes do not have to be the same.) From the graph:

(a) when $x = 2.2, y = \mathbf{11.8}$, and

(b) when $y = -3, x = \mathbf{-1.5}$

Problem 2. Plot the following graphs on the same axes between the range $x = -4$ to $x = +4$, and determine the gradient of each.

(a) $y = x$ (b) $y = x + 2$
(c) $y = x + 5$ (d) $y = x - 3$

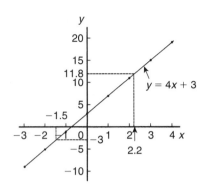

Figure 28.4

A table of co-ordinates is produced for each graph.

(a) $y = x$

x	−4	−3	−2	−1	0	1	2	3	4
y	−4	−3	−2	−1	0	1	2	3	4

(b) $y = x + 2$

x	−4	−3	−2	−1	0	1	2	3	4
y	−2	−1	0	1	2	3	4	5	6

(c) $y = x + 5$

x	−4	−3	−2	−1	0	1	2	3	4
y	1	2	3	4	5	6	7	8	9

(d) $y = x - 3$

x	−4	−3	−2	−1	0	1	2	3	4
y	−7	−6	−5	−4	−3	−2	−1	0	1

The co-ordinates are plotted and joined for each graph. The results are shown in Fig. 28.5. Each of the straight lines produced are parallel to each other, i.e. the slope or gradient is the same for each.

To find the gradient of any straight line, say, $y = x - 3$ a horizontal and vertical component needs to be constructed. In Fig. 28.5, AB is constructed vertically at $x = 4$ and BC constructed horizontally at $y = -3$.

The gradient of $AC = \dfrac{AB}{BC} = \dfrac{1 - (-3)}{4 - 0}$

$$= \frac{4}{4} = 1$$

i.e. the gradient of the straight line $y = x - 3$ is 1. The actual positioning of AB and BC is unimportant for the

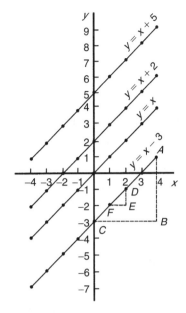

Figure 28.5

gradient is also given by, for example,

$$\frac{DE}{EF} = \frac{-1 - (-2)}{2 - 1} = \frac{1}{1} = 1$$

The slope or gradient of each of the straight lines in Fig. 28.5 is thus 1 since they are all parallel to each other.

Problem 3.   Plot the following graphs on the same axes between the values $x = -3$ to $x = +3$ and determine the gradient and y-axis intercept of each.
(a) $y = 3x$ (b) $y = 3x + 7$
(c) $y = -4x + 4$ (d) $y = -4x - 5$

A table of co-ordinates is drawn up for each equation.

(a) $y = 3x$

x	−3	−2	−1	0	1	2	3
y	−9	−6	−3	0	3	6	9

(b) $y = 3x + 7$

x	−3	−2	−1	0	1	2	3
y	−2	1	4	7	10	13	16

(c) $y = -4x + 4$

x	−3	−2	−1	0	1	2	3
y	16	12	8	4	0	−4	−8

(d) $y = -4x - 5$

x	−3	−2	−1	0	1	2	3
y	7	3	−1	−5	−9	−13	−17

Each of the graphs is plotted as shown in Fig. 28.6, and each is a straight line. $y = 3x$ and $y = 3x + 7$ are parallel to each other and thus have the same gradient. The gradient of AC is given by:

$$\frac{CB}{BA} = \frac{16 - 7}{3 - 0} = \frac{9}{3} = 3$$

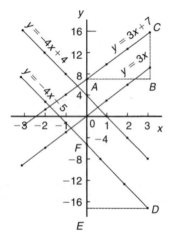

Figure 28.6

Hence the gradient of both $y = 3x$ and $y = 3x + 7$ is 3. $y = -4x + 4$ and $y = -4x - 5$ are parallel to each other and thus have the same gradient. The gradient of DF is given by:

$$\frac{FE}{ED} = \frac{-5 - (-17)}{0 - 3} = \frac{12}{-3} = -4$$

Hence the gradient of both $y = -4x + 4$ and $y = -4x - 5$ is −4.
The y-axis intercept means the value of y where the straight line cuts the y-axis. From Fig. 28.6,

$y = 3x$ cuts the y-axis at $y = 0$

$y = 3x + 7$ cuts the y-axis at $y = +7$

$y = -4x + 4$ cuts the y-axis at $y = +4$

and $y = -4x - 5$ cuts the y-axis at $y = -5$

Some general conclusions can be drawn from the graphs shown in Figs. 28.4, 28.5 and 28.6.
When an equation is of the form $y = mx + c$, where m and c are constants, then

(i) a graph of y against x produces a straight line,

(ii) m represents the slope or gradient of the line, and

(iii) c represents the y-axis intercept.

Thus, given an equation such as $y = 3x + 7$, it may be deduced 'on sight' that its gradient is +3 and its y-axis intercept is +7, as shown in Fig. 28.6. Similarly, if $y = -4x - 5$, then the gradient is −4 and the y-axis intercept is −5, as shown in Fig. 28.6.
When plotting a graph of the form $y = mx + c$, only two co-ordinates need be determined. When the co-ordinates are plotted a straight line is drawn between the two points. Normally, three co-ordinates are determined, the third one acting as a check.

Problem 4. The following equations represent straight lines. Determine, without plotting graphs, the gradient and y-axis intercept for each.

(a) $y = 3$ (b) $y = 2x$

(c) $y = 5x - 1$ (d) $2x + 3y = 3$

(a) $y = 3$ (which is of the form $y = 0x + 3$) represents a horizontal straight line intercepting the y-axis at **3**. Since the line is horizontal its **gradient is zero**.

(b) $y = 2x$ is of the form $y = mx + c$, where c is zero. Hence **gradient = 2** and **y-axis intercept = 0** (i.e. the origin).

(c) $y = 5x - 1$ is of the form $y = mx + c$. Hence **gradient = 5** and **y-axis intercept = −1**.

(d) $2x + 3y = 3$ is not in the form $y = mx + c$ as it stands. Transposing to make y the subject gives $3y = 3 - 2x$, i.e.

$$y = \frac{3 - 2x}{3} = \frac{3}{3} - \frac{2x}{3}$$

i.e. $y = -\dfrac{2x}{3} + 1$

which is of the form $y = mx + c$

Hence **gradient** $= -\dfrac{2}{3}$ and **y-axis intercept** $= +1$

Problem 5. Without plotting graphs, determine the gradient and y-axis intercept values of the following equations:

(a) $y = 7x - 3$

(b) $3y = -6x + 2$

(c) $y - 2 = 4x + 9$

(d) $\dfrac{y}{3} = \dfrac{x}{3} - \dfrac{1}{5}$

(e) $2x + 9y + 1 = 0$

(a) $y = 7x - 3$ is of the form $y = mx + c$, hence **gradient, $m = 7$ and y-axis intercept, $c = -3$**

(b) Rearranging $3y = -6x + 2$ gives

$$y = -\frac{6x}{3} + \frac{2}{3}$$

i.e. $y = -2x + \dfrac{2}{3}$

which is of the form $y = mx + c$. Hence **gradient $m = -2$ and y-axis intercept, $c = \dfrac{2}{3}$**

(c) Rearranging $y - 2 = 4x + 9$ gives $y = 4x + 11$, hence **gradient $= 4$ and y-axis intercept $= 11$**

(d) Rearranging $\dfrac{y}{3} = \dfrac{x}{2} - \dfrac{1}{5}$ gives

$$y = 3\left(\frac{x}{2} - \frac{1}{5}\right) = \frac{3}{2}x - \frac{3}{5}$$

Hence **gradient $= \dfrac{3}{2}$ and y-axis intercept $= -\dfrac{3}{5}$**

(e) Rearranging $2x + 9y + 1 = 0$ gives

$$9y = -2x - 1$$

i.e. $y = -\dfrac{2}{9}x - \dfrac{1}{9}$

Hence **gradient $= -\dfrac{2}{9}$ and y-axis intercept $= -\dfrac{1}{9}$**

Problem 6. Determine the gradient of the straight line graph passing through the co-ordinates
(a) $(-2, 5)$ and $(3, 4)$ (b) $(-2, -3)$ and $(-1, 3)$

A straight line graph passing through co-ordinates (x_1, y_1) and (x_2, y_2) has a gradient given by:

$$m = \frac{y_2 - y_1}{x_2 - x_1} \quad \text{(see Fig. 28.7)}$$

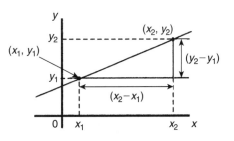

Figure 28.7

(a) A straight line passes through $(-2, 5)$ and $(3, 4)$, hence $x_1 = -2$, $y_1 = 5$, $x_2 = 3$ and $y_2 = 4$, hence gradient

$$m = \frac{y_2 - y_1}{x_2 - x_1} = \frac{4 - 5}{3 - (-2)} = -\frac{1}{5}$$

(b) A straight line passes through $(-2, -3)$ and $(-1, 3)$, hence $x_1 = -2$, $y_1 = -3$, $x_2 = -1$ and $y_2 = 3$, hence gradient,

$$m = \frac{y_2 - y_1}{x_2 - x_1} = \frac{3 - (-3)}{-1 - (-2)}$$

$$= \frac{3 + 3}{-1 + 2} = \frac{6}{1} = 6$$

Problem 7. Plot the graph $3x + y + 1 = 0$ and $2y - 5 = x$ on the same axes and find their point of intersection

Rearranging $3x + y + 1 = 0$ gives: $y = -3x - 1$

Rearranging $2y - 5 = x$ gives: $2y = x + 5$ and $y = \frac{1}{2}x + 2\frac{1}{2}$

Since both equations are of the form $y = mx + c$ both are straight lines. Knowing an equation is a straight line means that only two co-ordinates need to be plotted and a straight line drawn through them. A third co-ordinate is usually determined to act as a check. A table of values is produced for each equation as shown below.

x	1	0	−1
$-3x - 1$	−4	−1	2

x	2	0	−3
$\frac{1}{2}x + 2\frac{1}{2}$	$3\frac{1}{2}$	$2\frac{1}{2}$	1

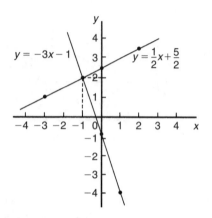

Figure 28.8

The graphs are plotted as shown in Fig. 28.8.

The two straight lines are seen to intersect at $(-1, 2)$

Now try the following exercise

Exercise 109   Further problems on straight line graphs

1. Corresponding values obtained experimentally for two quantities are:

x	-2.0	-0.5	0	1.0	2.5	3.0	5.0
y	-13.0	-5.5	-3.0	2.0	9.5	12.0	22.0

   Use a horizontal scale for x of $1\,\text{cm} = \frac{1}{2}$ unit and a vertical scale for y of $1\,\text{cm} = 2$ units and draw a graph of x against y. Label the graph and each of its axes. By interpolation, find from the graph the value of y when x is 3.5   [14.5]

2. The equation of a line is $4y = 2x + 5$. A table of corresponding values is produced and is shown below. Complete the table and plot a graph of y against x. Find the gradient of the graph.

x	-4	-3	-2	-1	0	1	2	3	4
y		-0.25			1.25				3.25

$\left[\frac{1}{2}\right]$

3. Determine the gradient and intercept on the y-axis for each of the following equations:

   (a) $y = 4x - 2$ (b) $y = -x$

   (c) $y = -3x - 4$ (d) $y = 4$

$$\begin{bmatrix} \text{(a) } 4, -2 & \text{(b) } -1, 0 \\ \text{(c) } -3, -4 & \text{(d) } 0, 4 \end{bmatrix}$$

4. Find the gradient and intercept on the y-axis for each of the following equations:

   (a) $2y - 1 = 4x$ (b) $6x - 2y = 5$

   (c) $3(2y - 1) = \dfrac{x}{4}$

$$\left[\text{(a) } 2, \frac{1}{2} \text{ (b) } 3, -2\frac{1}{2} \text{ (c) } \frac{1}{24}, \frac{1}{2}\right]$$

5. Determine the gradient and y-axis intercept for each of the following equations and sketch the graphs:

   (a) $y = 6x - 3$ (b) $y = 3x$ (c) $y = 7$

   (d) $2x + 3y + 5 = 0$

$$\begin{bmatrix} \text{(a) } 6, -3 & \text{(b) } 3, 0 \\ \text{(c) } 0, 7 & \text{(d) } -\frac{2}{3}, -1\frac{2}{3} \end{bmatrix}$$

6. Determine the gradient of the straight line graphs passing through the co-ordinates:

   (a) $(2, 7)$ and $(-3, 4)$

   (b) $(-4, -1)$ and $(-5, 3)$

   (c) $\left(\dfrac{1}{4}, -\dfrac{3}{4}\right)$ and $\left(-\dfrac{1}{2}, \dfrac{5}{8}\right)$

$$\left[\text{(a) } \frac{3}{5} \text{ (b) } -4 \text{ (c) } -1\frac{5}{6}\right]$$

7. State which of the following equations will produce graphs which are parallel to one another:

   (a) $y - 4 = 2x$ (b) $4x = -(y + 1)$

   (c) $x = \dfrac{1}{2}(y + 5)$ (d) $1 + \dfrac{1}{2}y = \dfrac{3}{2}x$

   (e) $2x = \dfrac{1}{2}(7 - y)$

   [(a) and (c), (b) and (e)]

8. Draw a graph of $y - 3x + 5 = 0$ over a range of $x = -3$ to $x = 4$. Hence determine (a) the value of y when $x = 1.3$ and (b) the value of x when $y = -9.2$ [(a) -1.1 (b) -1.4]

9. Draw on the same axes the graphs of $y = 3x - 5$ and $3y + 2x = 7$. Find the co-ordinates of the point of intersection. Check the result obtained by solving the two simultaneous equations algebraically.
          [(2, 1)]

10. Plot the graphs $y = 2x + 3$ and $2y = 15 - 2x$ on the same axes and determine their point of intersection.   [(1.5, 6)]

28.3 Practical problems involving straight line graphs

When a set of co-ordinate values are given or are obtained experimentally and it is believed that they follow a law of the form $y = mx + c$, then if a straight line can be drawn reasonably close to most of the co-ordinate values when plotted, this verifies that a law of the form $y = mx + c$ exists. From the graph, constants m (i.e. gradient) and c (i.e. y-axis intercept) can be determined. This technique is called **determination of law** (see also Chapter 29).

Problem 8. The temperature in degrees Celsius and the corresponding values in degrees Fahrenheit are shown in the table below. Construct rectangular axes, choose a suitable scale and plot a graph of degrees Celsius (on the horizontal axis) against degrees Fahrenheit (on the vertical scale).

°C	10	20	40	60	80	100
°F	50	68	104	140	176	212

From the graph find (a) the temperature in degrees Fahrenheit at 55°C, (b) the temperature in degrees Celsius at 167°F, (c) the Fahrenheit temperature at 0°C, and (d) the Celsius temperature at 230°F

The co-ordinates (10, 50), (20, 68), (40, 104), and so on are plotted as shown in Fig. 28.9. When the co-ordinates are joined, a straight line is produced. Since a straight line results there is a linear relationship between degrees Celsius and degrees Fahrenheit.

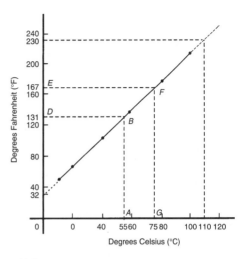

Figure 28.9

(a) To find the Fahrenheit temperature at 55°C a vertical line AB is constructed from the horizontal axis to meet the straight line at B. The point where the horizontal line BD meets the vertical axis indicates the equivalent Fahrenheit temperature.

Hence 55°C is equivalent to 131°F

This process of finding an equivalent value in between the given information in the above table is called **interpolation**.

(b) To find the Celsius temperature at 167°F, a horizontal line EF is constructed as shown in Fig. 28.9. The point where the vertical line FG cuts the horizontal axis indicates the equivalent Celsius temperature.

Hence 167°F is equivalent to 75°C

(c) If the graph is assumed to be linear even outside of the given data, then the graph may be extended at both ends (shown by broken line in Fig. 28.9).

From Fig. 28.9, **0°C corresponds to 32°F**

(d) **230°F is seen to correspond to 110°C.**

The process of finding equivalent values outside of the given range is called **extrapolation**.

Problem 9. In an experiment on Charles's law, the value of the volume of gas, V m^3, was measured for various temperatures T °C. Results are shown below.

V m^3	25.0	25.8	26.6	27.4	28.2	29.0
T °C	60	65	70	75	80	85

Plot a graph of volume (vertical) against temperature (horizontal) and from it find (a) the temperature when the volume is 28.6 m^3, and (b) the volume when the temperature is 67°C

If a graph is plotted with both the scales starting at zero then the result is as shown in Fig. 28.10. All of the points lie in the top right-hand corner of the graph, making interpolation difficult. A more accurate graph is obtained if the temperature axis starts at 55°C and the volume axis starts at 24.5 m^3. The axes corresponding to these values is shown by the broken lines in Fig. 28.10 and are called **false axes**, since the origin is not now at zero. A magnified version of this relevant part of the graph is shown in Fig. 28.11. From the graph:

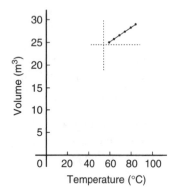

Figure 28.10

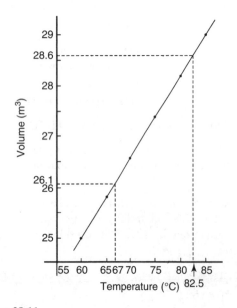

Figure 28.11

(a) when the volume is 28.6 m^3, the equivalent temperature is **82.5°C**, and

(b) when the temperature is 67°C, the equivalent volume is **26.1 m^3**

Problem 10. In an experiment demonstrating Hooke's law, the strain in an aluminium wire was measured for various stresses. The results were:

Stress N/mm^2	4.9	8.7	15.0
Strain	0.00007	0.00013	0.00021

Stress N/mm^2	18.4	24.2	27.3
Strain	0.00027	0.00034	0.00039

Plot a graph of stress (vertically) against strain (horizontally). Find:

(a) Young's Modulus of Elasticity for aluminium which is given by the gradient of the graph,

(b) the value of the strain at a stress of 20 N/mm^2, and

(c) the value of the stress when the strain is 0.00020

The co-ordinates (0.00007, 4.9), (0.00013, 8.7), and so on, are plotted as shown in Fig. 28.12. The graph produced is the best straight line which can be drawn corresponding to these points. (With experimental results it is unlikely that all the points will lie exactly on a straight line.) The graph, and each of its axes, are labelled. Since the straight line passes through the origin, then stress is directly proportional to strain for the given range of values.

(a) The gradient of the straight line AC is given by

$$\frac{AB}{BC} = \frac{28 - 7}{0.00040 - 0.00010} = \frac{21}{0.00030}$$

$$= \frac{21}{3 \times 10^{-4}} = \frac{7}{10^{-4}}$$

$$= 7 \times 10^4 = 70\,000 \text{ N/mm}^2$$

Thus Young's Modulus of Elasticity for aluminium is 70 000 N/mm^2.

Since $1 \text{ m}^2 = 10^6 \text{ mm}^2$, 70 000 N/mm^2 is equivalent to $70\,000 \times 10^6$ N/m^2, i.e. **70 × 10^9 N/m^2 (or Pascals)**.

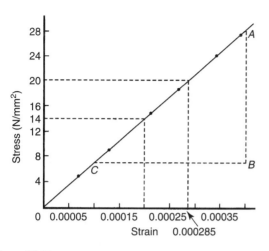

Figure 28.12

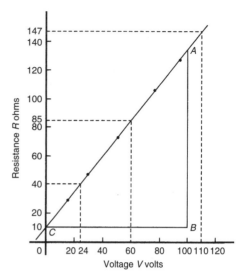

Figure 28.13

From Fig. 28.12:

(b) the value of the strain at a stress of 20 N/mm² is **0.000285**, and

(c) the value of the stress when the strain is 0.00020 is **14 N/mm²**.

Problem 11. The following values of resistance R ohms and corresponding voltage V volts are obtained from a test on a filament lamp.

R ohms	30	48.5	73	107	128
V volts	16	29	52	76	94

Choose suitable scales and plot a graph with R representing the vertical axis and V the horizontal axis. Determine (a) the gradient of the graph, (b) the R axis intercept value, (c) the equation of the graph, (d) the value of resistance when the voltage is 60 V, and (e) the value of the voltage when the resistance is 40 ohms. (f) If the graph were to continue in the same manner, what value of resistance would be obtained at 110 V?

The co-ordinates (16, 30), (29, 48.5), and so on, are shown plotted in Fig. 28.13 where the best straight line is drawn through the points.

(a) The slope or gradient of the straight line AC is given by:

$$\frac{AB}{BC} = \frac{135 - 10}{100 - 0} = \frac{125}{100} = \mathbf{1.25}$$

(Note that the vertical line AB and the horizontal line BC may be constructed anywhere along the length of the straight line. However, calculations are made easier if the horizontal line BC is carefully chosen, in this case, 100.)

(b) The R-axis intercept is at $R = \mathbf{10\ ohms}$ (by extrapolation).

(c) The equation of a straight line is $y = mx + c$, when y is plotted on the vertical axis and x on the horizontal axis. m represents the gradient and c the y-axis intercept. In this case, R corresponds to y, V corresponds to x, $m = 1.25$ and $c = 10$. Hence the equation of the graph is $R = \mathbf{(1.25V + 10)\ \Omega}$

From Fig. 28.13,

(d) when the voltage is 60 V, the resistance is **85 Ω**

(e) when the resistance is 40 ohms, the voltage is **24 V**, and

(f) by extrapolation, when the voltage is 110 V, the resistance is **147 Ω**.

Problem 12. Experimental tests to determine the breaking stress σ of rolled copper at various temperatures t gave the following results.

Stress σ N/cm²	8.46	8.04	7.78
Temperature t °C	70	200	280

Stress σ N/cm^2	7.37	7.08	6.63
Temperature t °C	410	500	640

Show that the values obey the law $\sigma = at + b$, where a and b are constants and determine approximate values for a and b. Use the law to determine the stress at 250°C and the temperature when the stress is 7.54 N/cm^2

The co-ordinates (70, 8.46), (200, 8.04), and so on, are plotted as shown in Fig. 28.14. Since the graph is a straight line then the values obey the law $\sigma = at + b$, and the gradient of the straight line is:

$$a = \frac{AB}{BC} = \frac{8.36 - 6.76}{100 - 600} = \frac{1.60}{-500} = \mathbf{-0.0032}$$

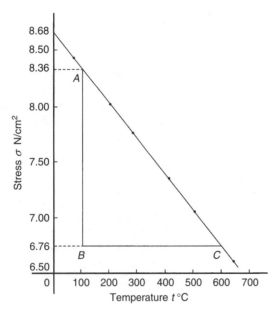

Figure 28.14

Vertical axis intercept, $b = \mathbf{8.68}$

Hence the law of the graph is: $\sigma = \mathbf{0.0032}t + \mathbf{8.68}$

When the temperature is 250°C, stress σ is given by:

$$\sigma = -0.0032(250) + 8.68$$

$$= \mathbf{7.88\ N/cm^2}$$

Rearranging $\sigma = -0.0032t + 8.68$ gives:

$$0.0032t = 8.68 - \sigma,$$

i.e. $\qquad t = \dfrac{8.68 - \sigma}{0.0032}$

Hence when the stress $\sigma = 7.54$ N/cm^2, temperature

$$t = \frac{8.68 - 7.54}{0.0032} = \mathbf{356.3°C}$$

Now try the following exercise

Exercise 110 Further practical problems involving straight line graphs

1. The resistance R ohms of a copper winding is measured at various temperatures t °C and the results are as follows:

R ohms	112	120	126	131	134
t °C	20	36	48	58	64

   Plot a graph of R (vertically) against t (horizontally) and find from it (a) the temperature when the resistance is 122 Ω and (b) the resistance when the temperature is 52°C
   [(a) 40°C (b) 128 Ω]

2. The speed of a motor varies with armature voltage as shown by the following experimental results:

n (rev/min)	285	517	615	750	917	1050
V volts	60	95	110	130	155	175

   Plot a graph of speed (horizontally) against voltage (vertically) and draw the best straight line through the points. Find from the graph: (a) the speed at a voltage of 145 V, and (b) the voltage at a speed of 400 rev/min.
   [(a) 850 rev/min (b) 77.5 V]

3. The following table gives the force F newtons which, when applied to a lifting machine, overcomes a corresponding load of L newtons.

Force F newtons	25	47	64	120	149	187
Load L newtons	50	140	210	430	550	700

Choose suitable scales and plot a graph of F (vertically) against L (horizontally). Draw the best straight line through the points. Determine from the graph
(a) the gradient, (b) the F-axis intercept, (c) the equation of the graph, (d) the force applied when the load is 310 N, and (e) the load that a force of 160 N will overcome. (f) If the graph were to continue in the same manner, what value of force will be needed to overcome a 800 N load?

$$\begin{bmatrix} \text{(a) } 0.25 & \text{(b) } 12 \\ \text{(c) } F = 0.25L + 12 & \text{(d) } 89.5 \text{ N} \\ \text{(e) } 592 \text{ N} & \text{(f) } 212 \text{ N} \end{bmatrix}$$

4. The following table gives the results of tests carried out to determine the breaking stress σ of rolled copper at various temperature, t:

Stress σ (N/cm^2)	8.51	8.07	7.80
Temperature t (°C)	75	220	310

Stress σ (N/cm^2)	7.47	7.23	6.78
Temperature t (°C)	420	500	650

Plot a graph of stress (vertically) against temperature (horizontally). Draw the best straight line through the plotted co-ordinates. Determine the slope of the graph and the vertical axis intercept.

$$[-0.003, 8.73]$$

5. The velocity v of a body after varying time intervals t was measured as follows:

t (seconds)	2	5	8	11	15	18
v (m/s)	16.9	19.0	21.1	23.2	26.0	28.1

Plot v vertically and t horizontally and draw a graph of velocity against time. Determine from the graph (a) the velocity after 10 s, (b) the time at 20 m/s and (c) the equation of the graph.

$$\begin{bmatrix} \text{(a) } 22.5 \text{ m/s} & \text{(b) } 6.5 \text{ s} \\ \text{(c) } v = 0.7t + 15.5 \end{bmatrix}$$

6. The mass m of a steel joint varies with length L as follows:

mass, m (kg)	80	100	120	140	160
length, L (m)	3.00	3.74	4.48	5.23	5.97

Plot a graph of mass (vertically) against length (horizontally). Determine the equation of the graph.

$$[m = 26.9L - 0.63]$$

7. The crushing strength of mortar varies with the percentage of water used in its preparation, as shown below.

Crushing strength, F (tonnes)	1.64	1.36	1.07	0.78	0.50	0.22
% of water used, w%	6	9	12	15	18	21

Plot a graph of F (vertically) against w (horizontally).

(a) Interpolate and determine the crushing strength when 10% of water is used.

(b) Assuming the graph continues in the same manner extrapolate and determine the percentage of water used when the crushing strength is 0.15 tonnes.

(c) What is the equation of the graph?

$$\begin{bmatrix} \text{(a) } 1.25t & \text{(b) } 21.65\% \\ \text{(c) } F = -0.095w + 2.25 \end{bmatrix}$$

8. In an experiment demonstrating Hooke's law, the strain in a copper wire was measured for various stresses. The results were:

Stress (Pascals)	10.6×10^6	18.2×10^6	24.0×10^6
Strain	0.00011	0.00019	0.00025

Section 4

Stress (Pascals)	30.7×10^6	39.4×10^6
Strain	0.00032	0.00041

Plot a graph of stress (vertically) against strain (horizontally). Determine (a) Young's Modulus of Elasticity for copper, which is given by the gradient of the graph, (b) the value of strain at a stress of 21×10^6 Pa, (c) the value of stress when the strain is 0.00030

$$\begin{bmatrix} \text{(a) } 96 \times 10^9 \text{ Pa} \quad \text{(b) } 0.00022 \\ \text{(c) } 29 \times 10^6 \text{ Pa} \end{bmatrix}$$

9. An experiment with a set of pulley blocks gave the following results:

Effort, E (newtons)	9.0	11.0	13.6	17.4	20.8	23.6
Load, L (newtons)	15	25	38	57	74	88

Plot a graph of effort (vertically) against load (horizontally) and determine (a) the gradient, (b) the vertical axis intercept, (c) the law of the graph, (d) the effort when the load is 30 N and (e) the load when the effort is 19 N.

$$\begin{bmatrix} \text{(a) } 0.2 \quad \text{(b) } 6 \quad \text{(c) } E = 0.2L + 6 \\ \text{(d) } 12 \text{ N} \quad \text{(e) } 65 \text{ N} \end{bmatrix}$$

10. The variation of pressure p in a vessel with temperature T is believed to follow a law of the form $p = aT + b$, where a and b are constants. Verify this law for the results given below and determine the approximate values of a and b. Hence determine the pressures at temperatures of 285 K and 310 K and the temperature at a pressure of 250 kPa.

Pressure, p kPa	244	247	252	258	262	267
Temperature, T K	273	277	282	289	294	300

$$\begin{bmatrix} a = 0.85, \quad b = 12, \quad 254.3 \text{ kPa}, \\ 275.5 \text{ kPa}, \quad 280 \text{ K} \end{bmatrix}$$

Chapter 29

Reduction of non-linear laws to linear form

29.1 Determination of law

Frequently, the relationship between two variables, say x and y, is not a linear one, i.e. when x is plotted against y a curve results. In such cases the non-linear equation may be modified to the linear form, $y = mx + c$, so that the constants, and thus the law relating the variables can be determined. This technique is called '**determination of law**'.

Some examples of the reduction of equations to linear form include:

(i) $y = ax^2 + b$ compares with $Y = mX + c$, where $m = a$, $c = b$ and $X = x^2$.
  Hence y is plotted vertically against x^2 horizontally to produce a straight line graph of gradient 'a' and y-axis intercept 'b'

(ii) $y = \dfrac{a}{x} + b$
  y is plotted vertically against $\dfrac{1}{x}$ horizontally to produce a straight line graph of gradient 'a' and y-axis intercept 'b'

(iii) $y = ax^2 + bx$
  Dividing both sides by x gives $\dfrac{y}{x} = ax + b$.
  Comparing with $Y = mX + c$ shows that $\dfrac{y}{x}$ is plotted vertically against x horizontally to produce a straight line graph of gradient 'a' and $\dfrac{y}{x}$ axis intercept 'b'

Problem 1. Experimental values of x and y, shown below, are believed to be related by the law $y = ax^2 + b$. By plotting a suitable graph verify this law and determine approximate values of a and b

x	1	2	3	4	5
y	9.8	15.2	24.2	36.5	53.0

If y is plotted against x a curve results and it is not possible to determine the values of constants a and b from the curve. Comparing $y = ax^2 + b$ with $Y = mX + c$ shows that y is to be plotted vertically against x^2 horizontally. A table of values is drawn up as shown below.

x	1	2	3	4	5
x^2	1	4	9	16	25
y	9.8	15.2	24.2	36.5	53.0

A graph of y against x^2 is shown in Fig. 29.1, with the best straight line drawn through the points. Since a straight line graph results, the law is verified.

From the graph, gradient

$$a = \frac{AB}{BC} = \frac{53 - 17}{25 - 5} = \frac{36}{20} = \mathbf{1.8}$$

and the y-axis intercept,

$$b = \mathbf{8.0}$$

Hence the law of the graph is:

$$y = \mathbf{1.8}x^2 + \mathbf{8.0}$$

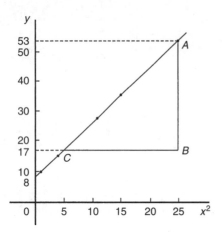

Figure 29.1

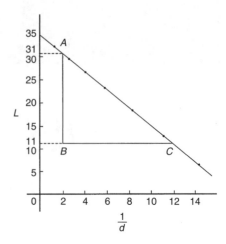

Figure 29.2

Problem 2. Values of load L newtons and distance d metres obtained experimentally are shown in the following table

Load, L N	32.3	29.6	27.0	23.2
distance, d m	0.75	0.37	0.24	0.17

Load, L N	18.3	12.8	10.0	6.4
distance, d m	0.12	0.09	0.08	0.07

Verify that load and distance are related by a law of the form $L = \dfrac{a}{d} + b$ and determine approximate values of a and b. Hence calculate the load when the distance is 0.20 m and the distance when the load is 20 N.

Comparing $L = \dfrac{a}{d} + b$ i.e. $L = a\left(\dfrac{1}{d}\right) + b$ with $Y = mX + c$ shows that L is to be plotted vertically against $\dfrac{1}{d}$ horizontally. Another table of values is drawn up as shown below.

L	32.3	29.6	27.0	23.2	18.3	12.8	10.0	6.4
d	0.75	0.37	0.24	0.17	0.12	0.09	0.08	0.07
$\dfrac{1}{d}$	1.33	2.70	4.17	5.88	8.33	11.11	12.50	14.29

A graph of L against $\dfrac{1}{d}$ is shown in Fig. 29.2. A straight line can be drawn through the points, which verifies that load and distance are related by a law of the form $L = \dfrac{a}{d} + b$

Gradient of straight line,

$$a = \frac{AB}{BC} = \frac{31 - 11}{2 - 12} = \frac{20}{-10} = -2$$

L-axis intercept,

$$b = 35$$

Hence the law of the graph is

$$L = -\frac{2}{d} + 35$$

When the distance $d = 0.20$ m, load

$$L = \frac{-2}{0.20} + 35 = 25.0\,\text{N}$$

Rearranging $L = -\dfrac{2}{d} + 35$ gives:

$$\frac{2}{d} = 35 - L \quad \text{and} \quad d = \frac{2}{35 - L}$$

Hence when the load $L = 20$ N, distance

$$d = \frac{2}{35 - 20} = \frac{2}{15} = 0.13\,\text{m}$$

Problem 3. The solubility s of potassium chlorate is shown by the following table:

$t\,°C$	10	20	30	40	50	60	80	100
s	4.9	7.6	11.1	15.4	20.4	26.4	40.6	58.0

The relationship between s and t is thought to be of the form $s = 3 + at + bt^2$. Plot a graph to test the supposition and use the graph to find approximate values of a and b. Hence calculate the solubility of potassium chlorate at 70°C

Rearranging $s = 3 + at + bt^2$ gives $s - 3 = at + bt^2$ and $\dfrac{s-3}{t} = a + bt$ or $\dfrac{s-3}{t} = bt + a$ which is of the form $Y = mX + c$, showing that $\dfrac{s-3}{t}$ is to be plotted vertically and t horizontally. Another table of values is drawn up as shown below.

t	10	20	30	40	50	60	80	100
s	4.9	7.6	11.1	15.4	20.4	26.4	40.6	58.0
$\dfrac{s-3}{t}$	0.19	0.23	0.27	0.31	0.35	0.39	0.47	0.55

A graph of $\dfrac{s-3}{t}$ against t is shown plotted in Fig. 29.3.

A straight line fits the points, which shows that s and t are related by

$$s = 3 + at + bt^2$$

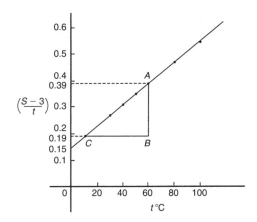

Figure 29.3

Gradient of straight line,

$$b = \frac{AB}{BC} = \frac{0.39 - 0.19}{60 - 10} = \frac{0.20}{50} = \mathbf{0.004}$$

Vertical axis intercept,

$$a = \mathbf{0.15}$$

Hence the law of the graph is:

$$s = 3 + 0.15t + 0.004t^2$$

The solubility of potassium chlorate at 70°C is given by

$$s = 3 + 0.15(70) + 0.004(70)^2$$
$$= 3 + 10.5 + 19.6 = \mathbf{33.1}$$

Now try the following exercise

Exercise 111 Further problems on reducing non-linear laws to linear form

In Problems 1 to 5, x and y are two related variables and all other letters denote constants. For the stated laws to be verified it is necessary to plot graphs of the variables in a modified form. State for each (a) what should be plotted on the vertical axis, (b) what should be plotted on the horizontal axis, (c) the gradient and (d) the vertical axis intercept.

1. $y = d + cx^2$ [(a) y (b) x^2 (c) c (d) d]

2. $y - a = b\sqrt{x}$ [(a) y (b) $\sqrt{x}$ (c) b (d) a]

3. $y - e = \dfrac{f}{x}$ $\left[\text{(a) } y \quad \text{(b) } \dfrac{1}{x} \quad \text{(c) } f \quad \text{(d) } e \right]$

4. $y - cx = bx^2$ $\left[\text{(a) } \dfrac{y}{x} \quad \text{(b) } x \quad \text{(c) } b \quad \text{(d) } c \right]$

5. $y = \dfrac{a}{x} + bx$ $\left[\text{(a) } \dfrac{y}{x} \quad \text{(b) } \dfrac{1}{x^2} \quad \text{(c) } a \quad \text{(d) } b \right]$

6. In an experiment the resistance of wire is measured for wires of different diameters with the following results:

R ohms	1.64	1.14	0.89	0.76	0.63
d mm	1.10	1.42	1.75	2.04	2.56

It is thought that R is related to d by the law $R = (a/d^2) + b$, where a and b are constants. Verify this and find the approximate values for a and b. Determine the cross-sectional area needed for a resistance reading of 0.50 ohms.
[$a = 1.5$, $b = 0.4$, 11.78 mm^2]

7. Corresponding experimental values of two quantities x and y are given below.

x	1.5	3.0	4.5	6.0	7.5	9.0
y	11.5	25.0	47.5	79.0	119.5	169.0

By plotting a suitable graph verify that y and x are connected by a law of the form $y = kx^2 + c$, where k and c are constants. Determine the law of the graph and hence find the value of x when y is 60.0
$\qquad$ $[y = 2x^2 + 7, 5.15]$

8. Experimental results of the safe load L kN, applied to girders of varying spans, d m, are show below:

Span, d m	2.0	2.8	3.6	4.2	4.8
Load, L kN	475	339	264	226	198

It is believed that the relationship between load and span is $L = c/d$, where c is a constant. Determine (a) the value of constant c and (b) the safe load for a span of 3.0 m.
$\qquad$ $[(a)\ 950 \quad (b)\ 317\,\text{kN}]$

9. The following results give corresponding values of two quantities x and y which are believed to be related by a law of the form $y = ax^2 + bx$ where a and b are constants.

x	33.86	55.54	72.80	84.10	111.4	168.1
y	3.4	5.2	6.5	7.3	9.1	12.4

Verify the law and determine approximate values of a and b.

Hence determine (i) the value of y when x is 8.0 and (ii) the value of x when y is 146.5
$\qquad$ $[a = 0.4, b = 8.6 \quad (i)\ 94.4 \quad (ii)\ 11.2]$

29.2 Determination of law involving logarithms

Examples of reduction of equations to linear form involving logarithms include:

(i) $y = ax^n$

Taking logarithms to a base of 10 of both sides gives:

$$\lg y = \lg(ax^n) = \lg a + \lg x^n$$

i.e. $\quad \lg y = n \lg x + \lg a$

by the laws of logarithms which compares with

$$Y = mX + c$$

and shows that $\lg y$ is plotted vertically against $\lg x$ horizontally to produce a straight line graph of gradient n and $\lg y$-axis intercept $\lg a$

(ii) $y = ab^x$

Taking logarithms to a base of 10 of the both sides gives:

$$\lg y = \lg(ab^x)$$

i.e. $\quad \lg y = \lg a + \lg b^x$

i.e. $\quad \lg y = x \lg b + \lg a$

$\qquad$ by the laws of logarithms

or $\quad \lg y = (\lg b)x + \lg a$

which compares with

$$Y = mX + c$$

and shows that $\lg y$ is plotted vertically against x horizontally to produce a straight line graph of gradient $\lg b$ and $\lg y$-axis intercept $\lg a$

(iii) $y = ae^{bx}$

Taking logarithms to a base of e of both sides gives:

$$\ln y = \ln(ae^{bx})$$

i.e. $\quad \ln y = \ln a + \ln e^{bx}$

i.e. $\quad \ln y = \ln a + bx \ln e$

i.e. $\quad \ln y = bx + \ln a$

(since $\ln e = 1$), which compares with

$$Y = mX + c$$

and shows that $\ln y$ is plotted vertically against x horizontally to produce a straight line graph of gradient b and $\ln y$-axis intercept $\ln a$

Problem 4. The current flowing in, and the power dissipated by, a resistor are measured experimentally for various values and the results are as shown below.

Current, I amperes	2.2	3.6	4.1	5.6	6.8
Power, P watts	116	311	403	753	1110

Show that the law relating current and power is of the form $P = RI^n$, where R and n are constants, and determine the law

Taking logarithms to a base of 10 of both sides of $P = RI^n$ gives:

$$\lg P = \lg(RI^n) = \lg R + \lg I^n = \lg R + n \lg I$$

by the laws of logarithms

i.e. $\lg P = n \lg I + \lg R$

which is of the form

$$Y = mX + c$$

showing that $\lg P$ is to be plotted vertically against $\lg I$ horizontally.

A table of values for $\lg I$ and $\lg P$ is drawn up as shown below:

I	2.2	3.6	4.1	5.6	6.8
$\lg I$	0.342	0.556	0.613	0.748	0.833
P	116	311	403	753	1110
$\lg P$	2.064	2.493	2.605	2.877	3.045

A graph of $\lg P$ against $\lg I$ is shown in Fig. 29.4 and since a straight line results the law $P = RI^n$ is verified.

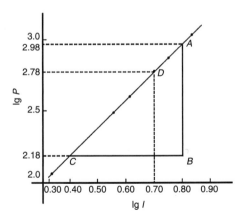

Figure 29.4

Gradient of straight line,

$$n = \frac{AB}{BC} = \frac{2.98 - 2.18}{0.8 - 0.4} = \frac{0.80}{0.4} = 2$$

It is not possible to determine the vertical axis intercept on sight since the horizontal axis scale does not start at zero. Selecting any point from the graph, say point D, where $\lg I = 0.70$ and $\lg P = 2.78$, and substituting values into

$$\lg P = n \lg I + \lg R$$

gives: $2.78 = (2)(0.70) + \lg R$

from which $\lg R = 2.78 - 1.40 = 1.38$

Hence $R = $ antilog $1.38 \; (= 10^{1.38})$

$$= \mathbf{24.0}$$

Hence the law of the graph is $P = 24.0I^2$

Problem 5. The periodic time, T, of oscillation of a pendulum is believed to be related to it length, l, by a law of the form $T = kl^n$, where k and n are constants. Values of T were measured for various lengths of the pendulum and the results are as shown below.

Periodic time, T s	1.0	1.3	1.5	1.8	2.0	2.3
Length, l m	0.25	0.42	0.56	0.81	1.0	1.32

Show that the law is true and determine the approximate values of k and n. Hence find the periodic time when the length of the pendulum is $0.75\,\text{m}$

From para (i), if $T = kl^n$ then

$$\lg T = n \lg l + \lg k$$

and comparing with

$$Y = mX + c$$

shows that $\lg T$ is plotted vertically against $\lg l$ horizontally. A table of values for $\lg T$ and $\lg l$ is drawn up as shown below.

T	1.0	1.3	1.5	1.8	2.0	2.3
$\lg T$	0	0.114	0.176	0.255	0.301	0.362
l	0.25	0.42	0.56	0.81	1.0	1.32
$\lg l$	−0.602	−0.377	−0.252	−0.092	0	0.121

A graph of $\lg T$ against $\lg l$ is shown in Fig. 29.5 and the law $T = kl^n$ is true since a straight line results.

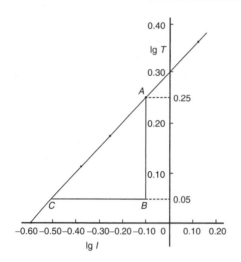

Figure 29.5

From the graph, gradient of straight line,

$$n = \frac{AB}{BC} = \frac{0.25 - 0.05}{-0.10 - (-0.50)} = \frac{0.20}{0.40} = \frac{1}{2}$$

Vertical axis intercept, $\lg k = 0.30$. Hence

$$k = \text{antilog } 0.30 \, (=10^{0.30}) = \mathbf{2.0}$$

Hence the law of the graph is:

$$T = 2.0l^{1/2} \quad \text{or} \quad T = 2.0\sqrt{l}$$

When length $l = 0.75$ m then

$$T = 2.0\sqrt{0.75} = \mathbf{1.73 \, s}$$

Problem 6. Quantities x and y are believed to be related by a law of the form $y = ab^x$, where a and b are constants. Values of x and corresponding values of y are:

x	0	0.6	1.2	1.8	2.4	3.0
y	5.0	9.67	18.7	36.1	69.8	135.0

Verify the law and determine the approximate values of a and b. Hence determine (a) the value of y when x is 2.1 and (b) the value of x when y is 100

From para (ii), if $y = ab^x$ then

$$\lg y = (\lg b)x + \lg a$$

and comparing with

$$Y = mX + c$$

shows that $\lg y$ is plotted vertically and x horizontally. Another table is drawn up as shown below.

x	0	0.6	1.2	1.8	2.4	3.0
y	5.0	9.67	18.7	36.1	69.8	135.0
$\lg y$	0.70	0.99	1.27	1.56	1.84	2.13

A graph of $\lg y$ against x is shown in Fig. 29.6 and since a straight line results, the law $y = ab^x$ is verified.

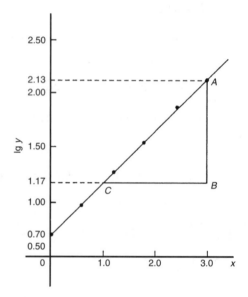

Figure 29.6

Gradient of straight line,

$$\lg b = \frac{AB}{BC} = \frac{2.13 - 1.17}{3.0 - 1.0} = \frac{0.96}{2.0} = 0.48$$

Hence $b = \text{antilog } 0.48 \, (=10^{0.48}) = \mathbf{3.0}$, correct to 2 significant figures.
Vertical axis intercept,

$$\lg a = 0.70, \text{ from which}$$

$$a = \text{antilog } 0.70 \, (=10^{0.70})$$

$$= \mathbf{5.0}, \text{ correct to 2 significant figures.}$$

Hence the law of the graph is $y = 5.0(3.0)^x$

(a)   When $x = 2.1$, $y = 5.0(3.0)^{2.1} = \mathbf{50.2}$
(b)   When $y = 100$, $100 = 5.0(3.0)x$

from which $100/5.0 = (3.0)^x$

i.e. $20 = (3.0)^x$

Taking logarithms of both sides gives

$$lg\, 20 = lg(3.0)^x = x\, lg\, 3.0$$

Hence $x = \dfrac{lg\, 20}{lg\, 3.0} = \dfrac{1.3010}{0.4771} = \mathbf{2.73}$

Problem 7.   The current i mA flowing in a capacitor which is being discharged varies with time t ms as shown below:

i mA	203	61.14	22.49	6.13	2.49	0.615
t ms	100	160	210	275	320	390

Show that these results are related by a law of the form $i = Ie^{t/T}$, where I and T are constants. Determine the approximate values of I and T

Taking Napierian logarithms of both sides of $i = Ie^{t/T}$ gives

$$\ln i = \ln(Ie^{t/T}) = \ln I + \ln e^{t/T} = \ln I + \frac{t}{T}\ln e$$

i.e. $\ln i = \ln I + \dfrac{t}{T}$ (since $\ln e = 1$)

or $\ln i = \left(\dfrac{1}{T}\right)t + \ln I$

which compares with $y = mx + c$, showing that $\ln i$ is plotted vertically against t horizontally. (For methods of evaluating Napierian logarithms see Chapter 14.) Another table of values is drawn up as shown below

t	100	160	210	275	320	390
i	203	61.14	22.49	6.13	2.49	0.615
$\ln i$	5.31	4.11	3.11	1.81	0.91	−0.49

A graph of $\ln i$ against t is shown in Fig. 29.7 and since a straight line results the law $i = Ie^{t/T}$ is verified.

Gradient of straight line,

$$\frac{1}{T} = \frac{AB}{BC} = \frac{5.30 - 1.30}{100 - 300} = \frac{4.0}{-200} = -0.02$$

Hence $T = \dfrac{1}{-0.02} = \mathbf{-50}$

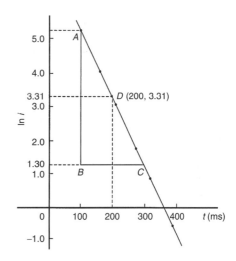

Figure 29.7

Selecting any point on the graph, say point D, where $t = 200$ and $\ln i = 3.31$, and substituting into

$$\ln i = \left(\frac{1}{T}\right)t + \ln I$$

gives: $3.31 = -\dfrac{1}{50}(200) + \ln I$

from which, $\ln I = 3.31 + 4.0 = 7.31$

and $I = \text{antilog}\, 7.31 (= e^{7.31}) = 1495$

or **1500** correct to 3 significant figures.

Hence the law of the graph is, $i = 1500\, e^{-t/50}$

Now try the following exercise

Exercise 112   Further problems on reducing non-linear laws to linear form

In Problem 1 to 3, x and y are two related variables and all other letters denote constants. For the stated laws to be verified it is necessary to plot graphs of the variables in a modified form. State for each (a) what should be plotted on the vertical axis, (b) what should be plotted on the horizontal axis, (c) the gradient and (d) the vertical axis intercept.

1. $y = ba^x$ [(a) $lg\, y$ (b) x (c) $lg\, a$ (d) $lg\, b$]

2. $y = kx^l$ [(a) $lg\, y$ (b) $lg\, x$ (c) l (d) $lg\, k$]

3. $\dfrac{y}{m} = e^{nx}$ [(a) $\ln y$ (b) x (c) n (d) $\ln m$]

Section 4

4. The luminosity I of a lamp varies with the applied voltage V and the relationship between I and V is thought to be $I = kV^n$. Experimental results obtained are:

I candelas	1.92	4.32	9.72
V volts	40	60	90

I candelas	15.87	23.52	30.72
V volts	115	140	160

Verify that the law is true and determine the law of the graph. Determine also the luminosity when 75 V is applied cross the lamp.
$$[I = 0.0012\,V^2,\ 6.75 \text{ candelas}]$$

5. The head of pressure h and the flow velocity v are measured and are believed to be connected by the law $v = ah^b$, where a and b are constants. The results are as shown below:

h	10.6	13.4	17.2	24.6	29.3
v	9.77	11.0	12.44	14.88	16.24

Verify that the law is true and determine values of a and b. $\qquad [a = 3.0, b = 0.5]$

6. Experimental values of x and y are measured as follows:

x	0.4	0.9	1.2	2.3	3.8
y	8.35	13.47	17.94	51.32	215.20

The law relating x and y is believed to be of the form $y = ab^x$, where a and b are constants. Determine the approximate values of a and b. Hence find the value of y when x is 2.0 and the value of x when y is 100.
$$[a = 5.6, b = 2.6, 37.86, 3.0]$$

7. The activity of a mixture of radioactive isotope is believed to vary according to the law $R = R_0 t^{-c}$, where R_0 and c are constants. Experimental results are shown below.

R	9.72	2.65	1.15	0.47	0.32	0.23
t	2	5	9	17	22	28

Verify that the law is true and determine approximate values of R_0 and c.
$$[R_0 = 25.1, c = 1.42]$$

8. Determine the law of the form $y = ae^{kx}$ which relates the following values.

y	0.0306	0.285	0.841	5.21	173.2	1181
x	−4.0	5.3	9.8	17.4	32.0	40.0

$$[y = 0.08e^{0.24x}]$$

9. The tension T in a belt passing round a pulley wheel and in contact with the pulley over an angle of θ radius is given by $T = T_0 e^{\mu\theta}$, where T_0 and μ are constants. Experimental results obtained are:

T newtons	47.9	52.8	60.3	70.1	80.9
θ radians	1.12	1.48	1.97	2.53	3.06

Determine approximate values of T_0 and μ. Hence find the tension when θ is 2.25 radians and the value of θ when the tension is 50.0 newtons.
$$\left[\begin{array}{l} T_0 = 35.3\,\text{N}, \mu = 0.27, \\ 64.8\,\text{N}, 1.29 \text{ radians} \end{array} \right]$$

Chapter 30

Graphs with logarithmic scales

30.1 Logarithmic scales

Graph paper is available where the scale markings along the horizontal and vertical axes are proportional to the logarithms of the numbers. Such graph paper is called **log–log graph paper**.

Figure 30.1

A **logarithmic scale** is shown in Fig. 30.1 where distance between, say 1 and 2, is proportional to $\lg 2 - \lg 1$, i.e. 0.3010 of the total distance from 1 to 10. Similarly, the distance between 7 and 8 is proportional to $\lg 8 - \lg 7$, i.e. 0.05799 of the total distance from 1 to 10. Thus the distance between markings progressively decreases as the numbers increase from 1 to 10.

With log–log graph paper the scale markings are from 1 to 9, and this pattern can be repeated several times. The number of times the pattern of markings is repeated on an axis signifies the number of **cycles**. When the vertical axis has, say, 3 sets of values from 1 to 9, and the horizontal axis has, say, 2 sets of values from 1 to 9, then this log–log graph paper is called 'log 3 cycle × 2 cycle' (see Fig. 30.2). Many different arrangements, are available ranging from 'log 1 cycle × 1 cycle' through to 'log 5 cycle × 5 cycle'.

To depict a set of values, say, from 0.4 to 161, on an axis of log–log graph paper, 4 cycles are required, from 0.1 to 1, 1 to 10, 10 to 100 and 100 to 1000.

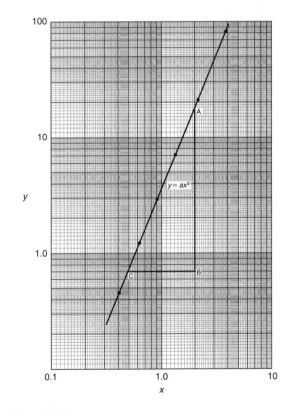

Figure 30.2

30.2 Graphs of the form $y = ax^n$

Taking logarithms to a base of 10 of both sides of $y = ax^n$ gives:

$$\lg y = \lg (ax^n)$$
$$= \lg a + \lg x^n$$

i.e. $\lg y = n \lg x + \lg a$

which compares with $Y = mX + c$

Thus, by plotting $\lg y$ vertically against $\lg x$ horizontally, a straight line results, i.e. the equation $y = ax^n$ is reduced to linear form. With log–log graph paper available x and y may be plotted directly, without having first to determine their logarithms, as shown in Chapter 29.

Problem 1. Experimental values of two related quantities x and y are shown below:

x	0.41	0.63	0.92	1.36	2.17	3.95
y	0.45	1.21	2.89	7.10	20.79	82.46

The law relating x and y is believed to be $y = ax^b$, where a and b are constants. Verify that this law is true and determine the approximate values of a and b

---

If $y = ax^b$ then $\lg y = b \lg x + \lg a$, from above, which is of the form $Y = mX + c$, showing that to produce a straight line graph $\lg y$ is plotted vertically against $\lg x$ horizontally. x and y may be plotted directly on to log–log graph paper as shown in Fig 30.2. The values of y range from 0.45 to 82.46 and 3 cycles are needed (i.e. 0.1 to 1, 1 to 10 and 10 to 100). The values of x range from 0.41 to 3.95 and 2 cycles are needed (i.e. 0.1 to 1 and 1 to 10). Hence 'log 3 cycle × 2 cycle' is used as shown in Fig. 30.2 where the axes are marked and the points plotted. Since the points lie on a straight line the law $y = ax^b$ is verified.

To evaluate constants a and b:

Method 1. Any two points on the straight line, say points A and C, are selected, and AB and BC are measure (say in centimetres).

Then, gradient, $b = \dfrac{AB}{BC} = \dfrac{11.5 \text{ units}}{5 \text{ units}} = \mathbf{2.3}$

Since $\lg y = b \lg x + \lg a$, when $x = 1$, $\lg x = 0$ and $\lg y = \lg a$.

The straight line crosses the ordinate $x = 1.0$ at $y = 3.5$. Hence $\lg a = \lg 3.5$, i.e. $a = \mathbf{3.5}$

Method 2. Any two points on the straight line, say points A and C, are selected. A has coordinates (2, 17.25) and C has coordinates (0.5, 0.7).

Since $y = ax^b$ then $17.25 = a(2)^b$ (1)

and $0.7 = a(0.5)^b$ (2)

i.e. two simultaneous equations are produced and may be solved for a and b.

Dividing equation (1) by equation (2) to eliminate a gives:

$$\frac{17.25}{0.7} = \frac{(2)^b}{(0.5)^b} = \left(\frac{2}{0.5}\right)^b$$

i.e. $24.643 = (4)^b$

Taking logarithms of both sides gives

$$\lg 24.643 = b \lg 4$$

and $b = \dfrac{\lg 24.643}{\lg 4}$

$= 2.3$, correct to 2 significant figures.

Substituting $b = 2.3$ in equation (1) gives:

$17.25 = a(2)^{2.3}$

and $a = \dfrac{17.25}{(2)^{2.3}} = \dfrac{17.25}{4.925}$

$= 3.5$, correct to 2 significant figures.

Hence the law of the graph is: $y = 3.5x^{2.3}$

Problem 2. The power dissipated by a resistor was measured for varying values of current flowing in the resistor and the results are as shown:

Current, I amperes	1.4	4.7	6.8	9.1	11.2	13.1
Power, P watts	49	552	1156	2070	3136	4290

Prove that the law relating current and power is of the form $P = RI^n$, where R and n are constants, and determine the law. Hence calculate the power when the current is 12 amperes and the current when the power is 1000 watts

---

Since $P = RI^n$ then $\lg P = n \lg I + \lg R$, which is of the form $Y = mX + c$, showing that to produce a straight line graph $\lg P$ is plotted vertically against $\lg I$ horizontally. Power values range from 49 to 4290, hence 3 cycles of log–log graph paper are needed (10 to 100, 100 to 1000 and 1000 to 10 000). Current values range from 1.4 to 11.2, hence 2 cycles of log–log graph paper are needed (1 to 10 and 10 to 100). Thus 'log 3 cycles × 2 cycles' is used as shown in Fig. 30.3 (or, if not available, graph paper having a larger number of cycles per axis can be used). The co-ordinates are plotted and a straight line results which proves that the law relating current and power is of the form $P = RI^n$. Gradient of straight line,

$$n = \frac{AB}{BC} = \frac{14 \text{ units}}{7 \text{ units}} = 2$$

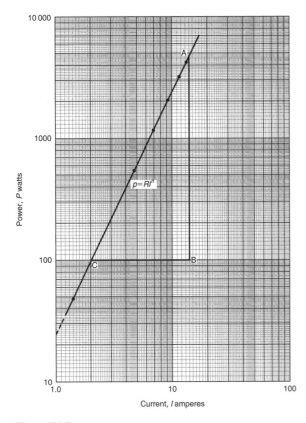

Figure 30.3

At point C, $I = 2$ and $P = 100$. Substituting these values into $P = RI^n$ gives: $100 = R(2)^2$. Hence $R = 100/(2)^2 = 25$ which may have been found from the intercept on the $I = 1.0$ axis in Fig. 30.3.

Hence the law of the graph is $P = 25I^2$

When current $I = 12$, power $P = 25(12)^2 = $ **3600 watts** (which may be read from the graph).

When power $P = 1000$, $1000 = 25I^2$.

Hence

$$I^2 = \frac{1000}{25} = 40,$$

from which, $I = \sqrt{40} = \mathbf{6.32\,A}$

Problem 3. The pressure p and volume v of a gas are believed to be related by a law of the form $p = cv^n$, where c and n are constants. Experimental

values of p and corresponding values of v obtained in a laboratory are:

p pascals	2.28×10^5	8.04×10^5	20.3×10^6
v m^3	3.2×10^{-2}	1.3×10^{-2}	6.7×10^{-3}

p pascals	5.05×10^6	1.82×10^7
v m^3	3.5×10^{-3}	1.4×10^{-3}

Verify that the law is true and determine approximate values of c and n

Since $p = cv^n$, then $\lg p = n \lg v + \lg c$, which is of the form $Y = mX + c$, showing that to produce a straight line graph $\lg p$ is plotted vertically against $\lg v$ horizontally. The co-ordinates are plotted on 'log 3 cycle $\times$ 2 cycle' graph paper as shown in Fig. 30.4. With the data expressed in standard form, the axes are marked in standard form also. Since a straight line results the law $p = cv^n$ is verified.

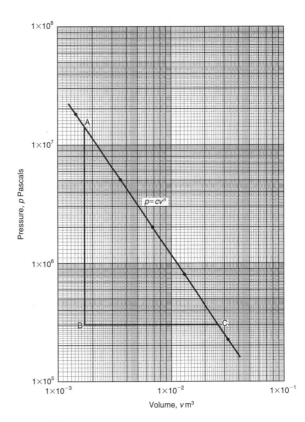

Figure 30.4

The straight line has a negative gradient and the value of the gradient is given by:

$$\frac{AB}{BC} = \frac{14 \text{ units}}{10 \text{ units}} = 1.4,$$

hence $n = -\mathbf{1.4}$

Selecting any point on the straight line, say point C, having co-ordinates $(2.63 \times 10^{-2}, 3 \times 10^5)$, and substituting these values in $p = cv^n$ gives:

$$3 \times 10^5 = c(2.63 \times 10^{-2})^{-1.4}$$

Hence $\qquad c = \dfrac{3 \times 10^5}{(2.63 \times 10^{-2})^{-1.4}} = \dfrac{3 \times 10^5}{(0.0263)^{-1.4}}$

$$= \frac{3 \times 10^5}{1.63 \times 10^2}$$

$= \mathbf{1840}$, correct to 3 significant figures.

Hence the law of the graph is:

$$p = \mathbf{1840}v^{-\mathbf{1.4}} \text{ or } p\, v^{\mathbf{1.4}} = \mathbf{1840}$$

Now try the following exercise

Exercise 113 Further problems on graphs of the form $y = ax^n$

1. Quantities x and y are believed to be related by a law of the form $y = ax^n$, where a and n are constants. Experimental values of x and corresponding values of y are:

x	0.8	2.3	5.4	11.5	21.6	42.9
y	8	54	250	974	3028	10 410

   Show that the law is true and determine the values of a and n. Hence determine the value of y when x is 7.5 and the value of x when y is 5000. $\qquad [a = 12, n = 1.8, 451, 28.5]$

2. Show from the following results of voltage V and admittance Y of an electrical circuit that the law connecting the quantities is of the form $V = kY^n$, and determine the values of k and n.

Voltage, V volts	2.88	2.05	1.60	1.22	0.96
Admittance, Y siemens	0.52	0.73	0.94	1.23	1.57

   $\qquad\qquad\qquad\qquad [k = 1.5, n = -1]$

3. Quantities x and y are believed to be related by a law of the form $y = mx^n$. The values of x and corresponding values of y are:

x	0.5	1.0	1.5	2.0	2.5	3.0
y	0.53	3.0	8.27	16.97	29.65	46.77

   Verify the law and find the values of m and n. $\qquad [m = 3, n = 2.5]$

30.3 Graphs of the form $y = ab^x$

Taking logarithms to a base of 10 of both sides of $y = ab^x$ gives:

$$\lg y = \lg(ab^x) = \lg a + \lg b^x = \lg a + x \lg b$$

i.e. $\qquad\qquad \mathbf{\lg y = (\lg b)x + \lg a}$

which compares with $\quad Y = mX + c$

Thus, by plotting $\lg y$ vertically against x horizontally a straight line results, i.e. the graph $y = ab^x$ is reduced to linear form. In this case, graph paper having a linear horizontal scale and a logarithmic vertical scale may be used. This type of graph paper is called **log–linear graph paper**, and is specified by the number of cycles of the logarithmic scale. For example, graph paper having 3 cycles on the logarithmic scale is called 'log 3 cycle × linear' graph paper.

Problem 4. Experimental values of quantities x and y are believed to be related by a law of the form $y = ab^x$, where a and b are constants. The values of x and corresponding values of y are:

x	0.7	1.4	2.1	2.9	3.7	4.3
y	18.4	45.1	111	308	858	1850

Verify the law and determine the approximate values of a and b. Hence evaluate (i) the value of y when x is 2.5, and (ii) the value of x when y is 1200

Since $y = ab^x$ then $\lg y = (\lg b)x + \lg a$ (from above), which is of the form $Y = mX + c$, showing that to produce a straight line graph $\lg y$ is plotted vertically against x horizontally. Using log-linear graph paper, values of x are marked on the horizontal scale to cover the range 0.7 to 4.3. Values of y range from 18.4 to 1850 and 3 cycles are needed (i.e. 10 to 100, 100 to 1000 and 1000 to 10 000). Thus 'log 3 cycles × linear' graph paper is used as shown in Fig. 30.5. A straight line is drawn through the co-ordinates, hence the law $y = ab^x$ is verified.

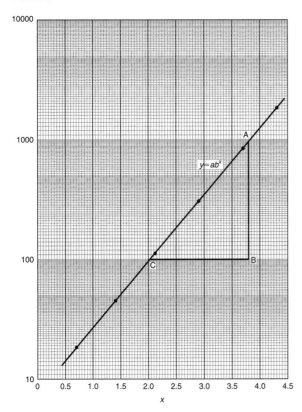

Figure 30.5

Gradient of straight line, $\lg b = AB/BC$. Direct measurement (say in centimeters) is not made with log-linear graph paper since the vertical scale is logarithmic and the horizontal scale is scale is linear. Hence

$$\frac{AB}{BC} = \frac{\lg 1000 - \lg 100}{3.82 - 2.02} = \frac{3 - 2}{1.80}$$

$$= \frac{1}{1.80} = 0.5556$$

Hence $b = \text{antilog } 0.5556 (= 10^{0.5556}) = \textbf{3.6}$, correct to 2 significant figures.
Point A has coordinates (3.82, 1000).
Substituting these values into $y = ab^x$ gives:

$$1000 = a(3.6)^{3.82}$$

i.e. $a = \dfrac{1000}{(3.6)^{3.82}}$

$$= \textbf{7.5,} \text{ correct to 2 significant figures.}$$

Hence the law of the graph is: $y = 7.5(3.6)^x$

(i) When $x = 2.5$, $y = 7.5(3.6)^{2.5} = \textbf{184}$
(ii) When $y = 1200$, $1200 = 7.5(3.6)^x$, hence

$$(3.6)^x = \frac{1200}{7.5} = 160$$

Taking logarithms gives: $x \lg 3.6 = \lg 160$

i.e. $x = \dfrac{\lg 160}{\lg 3.6} = \dfrac{2.2041}{0.5563}$

$$= \textbf{3.96}$$

Now try the following exercise

Exercise 114   Further problem on graphs of the form $y = ab^x$

1. Experimental values of p and corresponding values of q are shown below.

p	−13.2	−27.9	−62.2	−383.2	−1581	−2931
q	0.30	0.75	1.23	2.32	3.17	3.54

Show that the law relating p and q is $p = ab^q$, where a and b are constants. Determine (i) values of a and b, and state the law, (ii) the value of p when q is 2.0, and (iii) the value of q when p is −2000.

$$\begin{bmatrix} \text{(i)} \ \ a = -8, b = 5.3, p = -8(5.3)^q \\ \text{(ii)} - 224.7 \ \ \text{(iii)} \ 3.31 \end{bmatrix}$$

30.4   Graphs of the form $y = ae^{kx}$

Taking logarithms to a base of e of both sides of $y = ae^{kx}$ gives:

$$\ln y = \ln(ae^{kx}) = \ln a + \ln e^{kx} = \ln a + kx \ln e$$

i.e. $\ln y = kx + \ln a$ (since $\ln e = 1$) which compares

with $Y = mX + c$

Thus, by plotting $\ln y$ vertically against x horizontally, a straight line results, i.e. the equation $y = ae^{kx}$ is reduced to linear form. In this case, graph paper having a linear horizontal scale and a logarithmic vertical scale may be used.

Problem 5. The data given below is believed to be related by a law of the form $y = ae^{kx}$, where a

and b are constants. Verify that the law is true and determine approximate values of a and b. Also determine the value of y when x is 3.8 and the value of x when y is 85.

x	−1.2	0.38	1.2	2.5	3.4	4.2	5.3
y	9.3	22.2	34.8	71.2	117	181	332

Since $y = ae^{kx}$ then $\ln y = kx + \ln a$ (from above), which is of the form $Y = mX + c$, showing that to produce a straight line graph $\ln y$ is plotted vertically against x horizontally. The value of y range from 9.3 to 332 hence 'log 3 cycle × linear' graph paper is used. The plotted co-ordinates are shown in Fig. 30.6 and since a straight line passes through the points the law $y = ae^{kx}$ is verified.

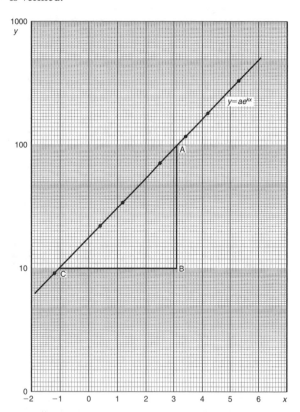

Figure 30.6

Gradient of straight line,

$$k = \frac{AB}{BC} = \frac{\ln 100 - \ln 10}{3.12 - (-1.08)} = \frac{2.3026}{4.20}$$

$$= 0.55, \text{ correct to 2 significant figures.}$$

Since $\ln y = kx + \ln a$, when $x = 0$, $\ln y = \ln a$, i.e. $y = a$.

The vertical axis intercept value at $x = 0$ is 18, hence $a = 18$.

The law of the graph is thus: $y = 18e^{0.55x}$

When x is 3.8,

$y = 18e^{0.55(3.8)} = 18e^{2.09} = 18(8.0849) = \mathbf{146}$

When y is 85, $85 = 18e^{0.55x}$

Hence, $\qquad e^{0.55x} = \dfrac{85}{18} = 4.7222$

and $\qquad 0.55x = \ln 4.7222 = 1.5523$

Hence $\qquad x = \dfrac{1.5523}{0.55} = \mathbf{2.82}$

Problem 6. The voltage, v volts, across an inductor is believed to be related to time, t ms, by the law $v = Ve^{t/T}$, where V and T are constants. Experimental results obtained are:

v volts	883	347	90	55.5	18.6	5.2
t ms	10.4	21.6	37.8	43.6	56.7	72.0

Show that the law relating voltage and time is as stated and determine the approximate values of V and T. Find also the value of voltage after 25 ms and the time when the voltage is 30.0 V

Since $v = Ve^{t/T}$ then $\ln v = \dfrac{1}{T}t + \ln V$

which is of the form $Y = mX + c$.

Using 'log 3 cycle × linear' graph paper, the points are plotted as shown in Fig. 30.7.
Since the points are joined by a straight line the law $v = Ve^{t/T}$ is verified.
Gradient of straight line,

$$\frac{1}{T} = \frac{AB}{BC} = \frac{\ln 100 - \ln 10}{36.5 - 64.2} = \frac{2.3026}{-27.7}$$

Hence $T = \dfrac{-27.7}{2.3026} = \mathbf{-12.0}$, correct to 3 significant figures.

Since the straight line does not cross the vertical axis at $t = 0$ in Fig. 30.7, the value of V is determined by selecting any point, say A, having coordinates $(36.5, 100)$ and substituting these values into $v = Ve^{t/T}$. Thus

$$100 = Ve^{36.5/-12.0}$$

i.e. $\quad V = \dfrac{100}{e^{-36.5/12.0}} = \mathbf{2090 \text{ volts}}$,

correct to 3 significant figures.

Hence the law of the graph is: $v = 2090e^{-t/12.0}$

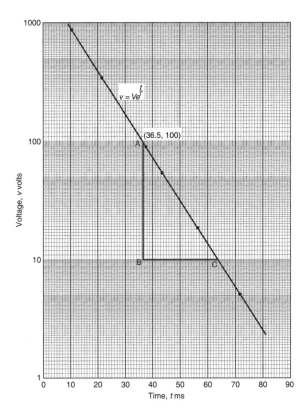

Figure 30.7

When time $t = 25$ ms, voltage $v = 2090e^{-25/12.0}$
$$= \mathbf{260\ V}.$$

When the voltage is 30.0 volts, $30.0 = 2090e^{-t/12.0}$ hence

$$e^{-t/12.0} = \frac{30.0}{2090} \quad \text{and} \quad e^{t/12.0} = \frac{2090}{30.0} = 69.67$$

Taking Napierian logarithms gives:

$$\frac{t}{12.0} = \ln 69.67 = 4.2438$$

from which, time $t = (12.0)(4.2438) = \mathbf{50.9\ ms}$.

Now try the following exercise

Exercise 115 Further problems on reducing exponential laws to linear form

1. Atmospheric pressure p is measured at varying altitudes h and the results are as shown below:

Altitude, h m	500	1500	3000	5000	8000
pressure, p cm	73.39	68.42	61.60	53.56	43.41

Show that the quantities are related by the law $p = ae^{kh}$, where a and k are constants. Determine, the values of a and k and state the law. Find also the atmospheric pressure at 10 000 m.

$$\left[\begin{array}{l} a = 76,\ k = -7 \times 10^{-5}, \\ p = 76e^{-7 \times 10^{-5}\,h},\ 37.74\ \text{cm} \end{array}\right]$$

2. At particular times, t minutes, measurements are made of the temperature, $\theta°C$, of a cooling liquid and the following results are obtained:

Temperature $\theta°C$	92.2	55.9	33.9	20.6	12.5
Time t minutes	10	20	30	40	50

Prove that the quantities follow a law of the form $\theta = \theta_0 e^{kt}$, where θ_0 and k are constants, and determine the approximate value of θ_0 and k. $[\theta_0 = 152,\ k = -0.05]$

Section 4

Chapter 31

Graphical solution of equations

31.1 Graphical solution of simultaneous equations

Linear simultaneous equations in two unknowns may be solved graphically by:

(i) plotting the two straight lines on the same axes, and
(ii) noting their point of intersection.

The co-ordinates of the point of intersection give the required solution.

Problem 1. Solve graphically the simultaneous equations:

$$2x - y = 4$$
$$x + y = 5$$

Rearranging each equation into $y = mx + c$ form gives:

$$y = 2x - 4 \tag{1}$$
$$y = -x + 5 \tag{2}$$

Only three co-ordinates need be calculated for each graph since both are straight lines.

x	0	1	2
$y = 2x - 4$	-4	-2	0

x	0	1	2
$y = -x + 5$	5	4	3

Each of the graphs is plotted as shown in Fig. 31.1. The point of intersection is at (3, 2) and since this is the only

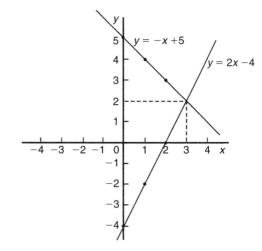

Figure 31.1

point which lies simultaneously on both lines then $x = 3$, $y = 2$ is the solution of the simultaneous equations.

Problem 2. Solve graphically the equations:

$$1.20x + y = 1.80$$
$$x - 5.0y = 8.50$$

Rearranging each equation into $y = mx + c$ form gives:

$$y = -1.20x + 1.80 \tag{1}$$

$$y = \frac{x}{5.0} - \frac{8.5}{5.0}$$

i.e.
$$y = 0.20x - 1.70 \tag{2}$$

Three co-ordinates are calculated for each equation as shown below:

x	0	1	2
$y = -1.20x + 1.80$	1.80	0.60	-0.60

x	0	1	2
$y = 0.20x - 1.70$	-1.70	-1.50	-1.30

The two lines are plotted as shown in Fig. 31.2. The point of intersection is (2.50, -1.20). Hence the solution of the simultaneous equation is $x = \mathbf{2.50}$, $y = \mathbf{-1.20}$

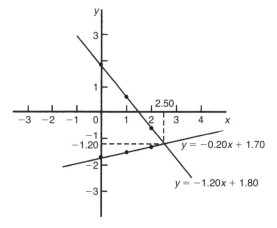

Figure 31.2

(It is sometimes useful initially to sketch the two straight lines to determine the region where the point of intersection is. Then, for greater accuracy, a graph having a smaller range of values can be drawn to 'magnify' the point of intersection).

Now try the following exercise

Exercise 116   Further problems on the graphical solution of simultaneous equations

In Problems 1 to 5, solve the simultaneous equations graphically.

1. $x + y = 2$

    $3y - 2x = 1$ $\qquad\qquad\qquad$ $[x = 1, y = 1]$

2. $y = 5 - x$

    $x - y = 2$ $\qquad\qquad\qquad$ $[x = 3\frac{1}{2}, y = 1\frac{1}{2}]$

3. $3x + 4y = 5$

    $2x - 5y + 12 = 0$ $\qquad\qquad$ $[x = -1, y = 2]$

4. $1.4x - 7.06 = 3.2y$

    $2.1x - 6.7y = 12.87$ $\qquad\quad$ $[x = 2.3, y = -1.2]$

5. $3x - 2y = 0$

    $4x + y + 11 = 0$ $\qquad\qquad$ $[x = -2, y = -3]$

6. The friction force F Newton's and load L Newton's are connected by a law of the form $F = aL + b$, where a and b are constants. When $F = 4$ Newton's, $L = 6$ Newton's and when $F = 2.4$ Newton's, $L = 2$ Newton's. Determine graphically the values of a and b.

    $[a = 0.4, b = 1.6]$

31.2   Graphical solution of quadratic equations

A general **quadratic equation** is of the form $y = ax^2 + bx + c$, where a, b and c are constants and a is not equal to zero.

A graph of a quadratic equation always produces a shape called a **parabola**. The gradient of the curve between 0 and A and between B and C in Fig. 31.3 is positive, whilst the gradient between A and B is negative. Points such as A and B are called **turning points**. At A the gradient is zero and, as x increases, the gradient of the curve changes from positive just before A to negative just after. Such a point is called a **maximum value**. At B the gradient is also zero, and, as x increases, the gradient of the curve changes from negative just before B to positive just after. Such a point is called a **minimum value**.

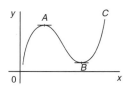

Figure 31.3

Quadratic graphs

(i) $y = ax^2$

Graphs of $y = x^2$, $y = 3x^2$ and $y = \frac{1}{2}x^2$ are shown in Fig. 31.4.

All have minimum values at the origin (0, 0).

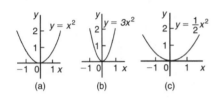

(a)   (b)   (c)

Figure 31.4

Graphs of $y = -x^2$, $y = -3x^2$ and $y = -\frac{1}{2}x^2$ are shown in Fig. 31.5.

All have maximum values at the origin (0, 0).

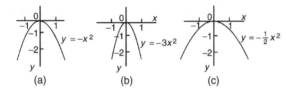

(a)   (b)   (c)

Figure 31.5

When $y = ax^2$,
(a)   curves are symmetrical about the y-axis,
(b)   the magnitude of 'a' affects the gradient of the curve, and
(c)   the sign of 'a' determines whether it has a maximum or minimum value.

(ii) $y = ax^2 + c$

Graphs of $y = x^2 + 3$, $y = x^2 - 2$, $y = -x^2 + 2$ and $y = -2x^2 - 1$ are shown in Fig. 31.6.

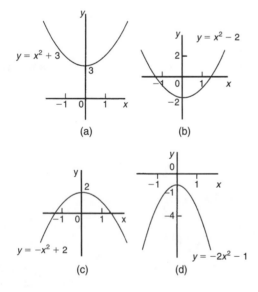

(a)   (b)

(c)   (d)

Figure 31.6

When $y = ax^2 + c$,
(a)   curves are symmetrical about the y-axis,
(b)   the magnitude of 'a' affects the gradient of the curve, and
(c)   the constant 'c' is the y-axis intercept.

(iii) $y = ax^2 + bx + c$

Whenever 'b' has a value other than zero the curve is displaced to the right or left of the y-axis. When b/a is positive, the curve is displaced $b/2a$ to the left of the y-axis, as shown in Fig. 31.7(a). When b/a is negative the curve is displaced $b/2a$ to the right of the y-axis, as shown in Fig. 31.7(b).

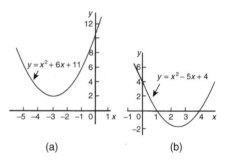

(a)   (b)

Figure 31.7

Quadratic equations of the form $ax^2 + bx + c = 0$ may be solved graphically by:

(i)   plotting the graph $y = ax^2 + bx + c$, and
(ii)   noting the points of intersection on the x-axis (i.e. where $y = 0$).

The x values of the points of intersection give the required solutions since at these points both $y = 0$ and $ax^2 + bx + c = 0$. The number of solutions, or roots of a quadratic equation, depends on how many times the curve cuts the x-axis and there can be no real roots (as in Fig. 31.7(a)) or one root (as in Figs 31.4 and 31.5) or two roots (as in Fig. 31.7(b)).

Problem 3.   Solve the quadratic equation $4x^2 + 4x - 15 = 0$ graphically given that the solutions lie in the range $x = -3$ to $x = 2$. Determine also the co-ordinates and nature of the turning point of the curve

Let $y = 4x^2 + 4x - 15$. A table of values is drawn up as shown below:

x	-3	-2	-1	0	1	2
$4x^2$	36	16	4	0	4	16
$4x$	-12	-8	-4	0	4	8
-15	-15	-15	-15	-15	-15	-15
$y = 4x^2 + 4x - 15$	9	-7	-15	-15	-7	9

A graph of $y = 4x^2 + 4x - 15$ is shown in Fig. 31.8. The only points where $y = 4x^2 + 4x - 15$ and $y = 0$ are the points marked A and B. This occurs at $x = -2.5$ and $x = 1.5$ and these are the solutions of the quadratic equation $4x^2 + 4x - 15 = 0$. (By substituting $x = -2.5$ and $x = 1.5$ into the original equation the solutions may be checked). The curve has a turning point at $(-0.5, -16)$ and the nature of the point is a **minimum**.

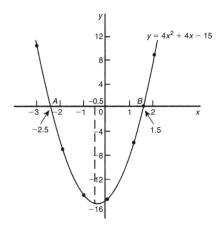

Figure 31.8

An alternative graphical method of solving $4x^2 + 4x - 15 = 0$ is to rearrange the equation as $4x^2 = -4x + 15$ and then plot two separate graphs— in this case $y = 4x^2$ and $y = -4x + 15$. Their points of intersection give the roots of equation $4x^2 = -4x + 15$, i.e. $4x^2 + 4x - 15 = 0$. This is shown in Fig. 31.9, where the roots are $x = -2.5$ and $x = 1.5$ as before.

Problem 4. Solve graphically the quadratic equation $-5x^2 + 9x + 7.2 = 0$ given that the solutions lie between $x = -1$ and $x = 3$. Determine also the co-ordinates of the turning point and state its nature

Let $y = -5x^2 + 9x + 7.2$. A table of values is drawn up as shown to the right. A graph of $y = -5x^2 + 9x + 7.2$

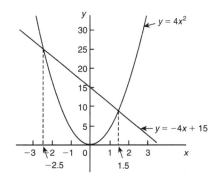

Figure 31.9

is shown plotted in Fig. 31.10. The graph crosses the x-axis (i.e. where $y = 0$) at $x = -0.6$ and $x = 2.4$ and these are the solutions of the quadratic equation $-5x^2 + 9x + 7.2 = 0$. The turning point is a **maximum** having co-ordinates (**0.9, 11.25**).

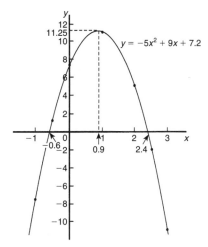

Figure 31.10

x	-1	-0.5	0	1
$-5x^2$	-5	-1.25	0	-5
$+9x$	-9	-4.5	0	9
$+7.2$	7.2	7.2	7.2	7.2
$y = -5x^2 + 9x + 7.2$	-6.8	1.45	7.2	11.2

x	2	2.5	3
$-5x^2$	-20	-31.25	-45
$+9x$	18	22.5	27
$+7.2$	7.2	7.2	7.2
$y = -5x^2 + 9x + 7.2$	5.2	-1.55	-10.8

Section 4

Problem 5. Plot a graph of: $y = 2x^2$ and hence solve the equations: (a) $2x^2 - 8 = 0$ and (b) $2x^2 - x - 3 = 0$

A graph of $y = 2x^2$ is shown in Fig. 31.11.

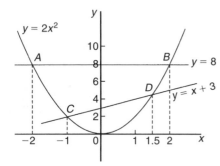

Figure 31.11

(a) Rearranging $2x^2 - 8 = 0$ gives $2x^2 = 8$ and the solution of this equation is obtained from the points of intersection of $y = 2x^2$ and $y = 8$, i.e. at co-ordinates $(-2, 8)$ and $(2, 8)$, shown as A and B, respectively, in Fig. 31.11. Hence the solutions of $2x^2 - 8 = 0$ and $x = -2$ and $x = +2$

(b) Rearranging $2x^2 - x - 3 = 0$ gives $2x^2 = x + 3$ and the solution of this equation is obtained from the points of intersection of $y = 2x^2$ and $y = x + 3$, i.e. at C and D in Fig. 31.11. Hence the solutions of $2x^2 - x - 3 = 0$ are $x = -1$ and $x = 1.5$

Problem 6. Plot the graph of $y = -2x^2 + 3x + 6$ for values of x from $x = -2$ to $x = 4$. Use the graph to find the roots of the following equations:

(a) $-2x^2 + 3x + 6 = 0$
(b) $-2x^2 + 3x + 2 = 0$
(c) $-2x^2 + 3x + 9 = 0$
(d) $-2x^2 + x + 5 = 0$

A table of values is drawn up as shown below.

x	-2	-1	0	1	2	3	4
$-2x^2$	-8	-2	0	-2	-8	-18	-32
$+3x$	-6	-3	0	3	6	9	12
$+6$	6	6	6	6	6	6	6
y	-8	1	6	7	4	-3	-14

A graph of $y = -2x^2 + 3x + 6$ is shown in Fig. 31.12.

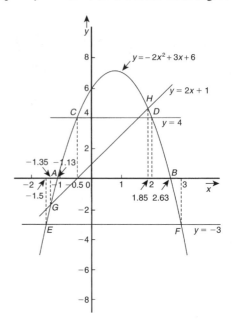

Figure 31.12

(a) The parabola $y = -2x^2 + 3x + 6$ and the straight line $y = 0$ intersect at A and B, where $x = -1.13$ and $x = 2.63$ and these are the roots of the equation $-2x^2 + 3x + 6 = 0$

(b) Comparing

$$y = -2x^2 + 3x + 6 \qquad (1)$$

with $\qquad 0 = -2x^2 + 3x + 2 \qquad (2)$

shows that if 4 is added to both sides of equation (2), the right-hand side of both equations will be the same.

Hence $4 = -2x^2 + 3x + 6$. The solution of this equation is found from the points of intersection of the line $y = 4$ and the parabola $y = -2x^2 + 3x + 6$, i.e. points C and D in Fig. 31.12. Hence the roots of $-2x^2 + 3x + 2 = 0$ are $x = -0.5$ and $x = 2$

(c) $-2x^2 + 3x + 9 = 0$ may be rearranged as $-2x^2 + 3x + 6 = -3$, and the solution of this equation is obtained from the points of intersection of the line $y = -3$ and the parabola $y = -2x^2 + 3x + 6$, i.e. at points E and F in Fig. 31.12. Hence the roots of $-2x^2 + 3x + 9 = 0$ are $x = -1.5$ and $x = 3$

(d) Comparing

$$y = -2x^2 + 3x + 6 \qquad (3)$$

with $\qquad 0 = -2x^2 + x + 5 \qquad (4)$

shows that if $2x + 1$ is added to both sides of equation (4) the right-hand side of both equations will be the same. Hence equation (4) may be written as $2x + 1 = -2x^2 + 3x + 6$. The solution of this equation is found from the points of intersection of the line $y = 2x + 1$ and the parabola $y = -2x^2 + 3x + 6$, i.e. points G and H in Fig. 31.12. Hence the roots of $-2x^2 + x + 5 = 0$ are $x = -1.35$ and $x = 1.85$

Now try the following exercise

Exercise 117 Further problems on solving quadratic equations graphically

1. Sketch the following graphs and state the nature and co-ordinates of their turning points:

   (a) $y = 4x^2$ (b) $y = 2x^2 - 1$
   (c) $y = -x^2 + 3$ (d) $y = -\frac{1}{2}x^2 - 1$

   $$\begin{bmatrix} \text{(a) Minimum } (0, 0) \\ \text{(b) Minimum } (0, -1) \\ \text{(c) Maximum } (0, 3) \\ \text{(d) Maximum } (0, -1) \end{bmatrix}$$

Solve graphically the quadratic equations in Problems 2 to 5 by plotting the curves between the given limits. Give answers correct to 1 decimal place.

2. $4x^2 - x - 1 = 0$; $x = -1$ to $x = 1$
$$[-0.4 \text{ or } 0.6]$$

3. $x^2 - 3x = 27$; $x = -5$ to $x = 8$
$$[-3.9 \text{ or } 6.9]$$

4. $2x^2 - 6x - 9 = 0$; $x = -2$ to $x = 5$
$$[-1.1 \text{ or } 4.1]$$

5. $2x(5x - 2) = 39.6$; $x = -2$ to $x = 3$
$$[-1.8 \text{ or } 2.2]$$

6. Solve the quadratic equation $2x^2 + 7x + 6 = 0$ graphically, given that the solutions lie in the range $x = -3$ to $x = 1$. Determine also the nature and co-ordinates of its turning point.
$$\begin{bmatrix} x = -1.5 \text{ or } -2, \\ \text{Minimum at } (-1.75, -0.1) \end{bmatrix}$$

7. Solve graphically the quadratic equation $10x^2 - 9x - 11.2 = 0$, given that the roots lie between $x = -1$ and $x = 2$.
$$[x = -0.7 \text{ or } 1.6]$$

8. Plot a graph of $y = 3x^2$ and hence solve the equations (a) $3x^2 - 8 = 0$ and (b) $3x^2 - 2x - 1 = 0$
$$[\text{(a) } \pm 1.63 \text{ (b) } 1 \text{ or } -0.3]$$

9. Plot the graphs $y = 2x^2$ and $y = 3 - 4x$ on the same axes and find the co-ordinates of the points of intersection. Hence determine the roots of the equation $2x^2 + 4x - 3 = 0$
$$\begin{bmatrix} (-2.6, 13.2), (0.6, 0.8); \\ x = -2.6 \text{ or } 0.6 \end{bmatrix}$$

10. Plot a graph of $y = 10x^2 - 13x - 30$ for values of x between $x = -2$ and $x = 3$. Solve the equation $10x^2 - 13x - 30 = 0$ and from the graph determine: (a) the value of y when x is 1.3, (b) the values of x when y is 10 and (c) the roots of the equation $10x^2 - 15x - 18 = 0$
$$\begin{bmatrix} x = -1.2 \text{ or } 2.5 & \text{(a) } -30 & \text{(b) } 2.75 \\ \text{and } -1.50 & \text{(c) } 2.3 \text{ or } -0.8 \end{bmatrix}$$

31.3 Graphical solution of linear and quadratic equations simultaneously

The solution of **linear and quadratic equations simultaneously** may be achieved graphically by: (i) plotting the straight line and parabola on the same axes, and (ii) noting the points of intersection. The co-ordinates of the points of intersection give the required solutions.

Problem 7. Determine graphically the values of x and y which simultaneously satisfy the equations: $y = 2x^2 - 3x - 4$ and $y = 2 - 4x$

$y = 2x^2 - 3x - 4$ is a parabola and a table of values is drawn up as shown below:

x	-2	-1	0	1	2	3
$2x^2$	8	2	0	2	8	18
$-3x$	6	3	0	-3	-6	-9
-4	-4	-4	-4	-4	-4	-4
y	10	1	-4	-5	-2	5

$y = 2 - 4x$ is a straight line and only three co-ordinates need be calculated:

x	0	1	2
y	2	−2	−6

The two graphs are plotted in Fig. 31.13 and the points of intersection, shown as A and B, are at co-ordinates $(-2, 10)$ and $(1.5, -4)$. Hence the simultaneous solutions occur when $x = -2, y = 10$ and when $x = 1.5, y = -4$.

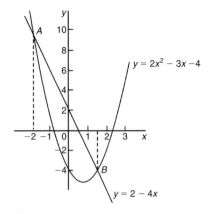

Figure 31.13

(These solutions may be checked by substituting into each of the original equations).

Now try the following exercise

Exercise 118 Further problems on solving linear and quadratic equations simultaneously

1. Determine graphically the values of x and y which simultaneously satisfy the equations $y = 2(x^2 - 2x - 4)$ and $y + 4 = 3x$.
 $\left[x = 4, y = 8 \text{ and } x = -\frac{1}{2}, y = -5\frac{1}{2}\right]$

2. Plot the graph of $y = 4x^2 - 8x - 21$ for values of x from -2 to $+4$. Use the graph to find the roots of the following equations:

 (a) $4x^2 - 8x - 21 = 0$ (b) $4x^2 - 8x - 16 = 0$
 (c) $4x^2 - 6x - 18 = 0$

 $\left[\begin{array}{l}\text{(a) } x = -1.5 \text{ or } 3.5 \quad \text{(b) } x = -1.24 \\ \text{or } 3.24 \quad \text{(c) } x = -1.5 \text{ or } 3.0\end{array}\right]$

31.4 Graphical solution of cubic equations

A **cubic equation** of the form $ax^3 + bx^2 + cx + d = 0$ may be solved graphically by: (i) plotting the graph $y = ax^3 + bx^2 + cx + d$, and (ii) noting the points of intersection on the x-axis (i.e. where $y = 0$). The x-values of the points of intersection give the required solution since at these points both $y = 0$ and $ax^3 + bx^2 + cx + d = 0$.

The number of solutions, or roots of a cubic equation depends on how many times the curve cuts the x-axis and there can be one, two or three possible roots, as shown in Fig. 31.14.

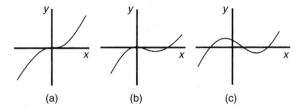

(a)   (b)   (c)

Figure 31.14

Problem 8. Solve graphically the cubic equation $4x^3 - 8x^2 - 15x + 9 = 0$ given that the roots lie between $x = -2$ and $x = 3$. Determine also the co-ordinates of the turning points and distinguish between them

Let $y = 4x^3 - 8x^2 - 15x + 9$. A table of values is drawn up as shown below:

x	−2	−1	0	1	2	3
$4x^3$	−32	−4	0	4	32	108
$-8x^2$	−32	−8	0	−8	−32	−72
$-15x$	30	15	0	−15	−30	−45
$+9$	9	9	9	9	9	9
y	−25	12	9	−10	−21	0

A graph of $y = 4x^3 - 8x^2 - 15x + 9$ is shown in Fig. 31.15.

The graph crosses the x-axis (where $y = 0$) at $x = -1.5$, $x = 0.5$ and $x = 3$ and these are the solutions to the cubic equation $4x^3 - 8x^2 - 15x + 9 = 0$.

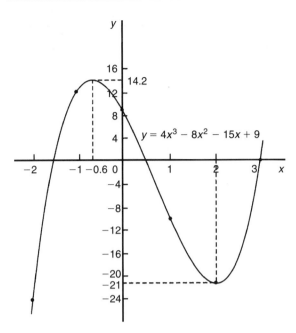

Figure 31.15

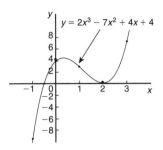

Figure 31.16

The turning points occur at $(-0.6, 14.2)$, which is a **maximum**, and $(2, -21)$, which is a **minimum**.

> **Problem 9.** Plot the graph of $y = 2x^3 - 7x^2 + 4x + 4$ for values of x between $x = -1$ and $x = 3$. Hence determine the roots of the equation:
>
> $$2x^3 - 7x^2 + 4x + 4 = 0$$

A table of values is drawn up as shown below.

x	-1	0	1	2	3
$2x^3$	-2	0	2	16	54
$-7x^2$	-7	0	-7	-28	-63
$+4x$	-4	0	4	8	12
$+4$	4	4	4	4	4
y	-9	4	3	0	7

A graph of $y = 2x^3 - 7x^2 + 4x + 4$ is shown in Fig. 31.16. The graph crosses the x-axis at $x = -0.5$ and touches the x-axis at $x = 2$. Hence the solutions of the equation $2x^3 - 7x^2 + 4x + 4 = 0$ are $x = -0.5$ and $x = 2$.

Now try the following exercise

> **Exercise 119   Further problems on solving cubic equations**
>
> 1. Plot the graph $y = 4x^3 + 4x^2 - 11x - 6$ between $x = -3$ and $x = 2$ and use the graph to solve the cubic equation $4x^3 + 4x^2 - 11x - 6 = 0$
> $$[x = -2.0, -0.5 \text{ or } 1.5]$$
>
> 2. By plotting a graph of $y = x^3 - 2x^2 - 5x + 6$ between $x = -3$ and $x = 4$ solve the equation $x^3 - 2x^2 - 5x + 6 = 0$. Determine also the co-ordinates of the turning points and distinguish between them.
> $$\left[\begin{array}{l} x = -2, 1 \text{ or } 3, \text{ Minimum at} \\ (2.1, -4.1), \text{ Maximum at} \\ (-0.8, 8.2) \end{array} \right]$$
>
> In Problems 3 to 6, solve graphically the cubic equations given, each correct to 2 significant figures.
>
> 3. $x^3 - 1 = 0$ $\qquad\qquad\qquad [x = 1]$
>
> 4. $x^3 - x^2 - 5x + 2 = 0$
> $$[x = -2.0, 0.4 \text{ or } 2.6]$$
>
> 5. $x^3 - 2x^2 = 2x - 2$ $\qquad\quad [x = 0.7 \text{ or } 2.5]$
>
> 6. $2x^3 - x^2 - 9.08x + 8.28 = 0$
> $$[x = -2.3, 1.0 \text{ or } 1.8]$$
>
> 7. Show that the cubic equation $8x^3 + 36x^2 + 54x + 27 = 0$ has only one real root and determine its value.
> $$[x = -1.5]$$

Chapter 32

Functions and their curves

32.1 Standard curves

When a mathematical equation is known, coordinates may be calculated for a limited range of values, and the equation may be represented pictorially as a graph, within this range of calculated values. Sometimes it is useful to show all the characteristic features of an equation, and in this case a sketch depicting the equation can be drawn, in which all the important features are shown, but the accurate plotting of points is less important. This technique is called 'curve sketching' and can involve the use of differential calculus, with, for example, calculations involving turning points.

If, say, y depends on, say, x, then y is said to be a function of x and the relationship is expressed as $y = f(x)$; x is called the independent variable and y is the dependent variable.

In engineering and science, corresponding values are obtained as a result of tests or experiments.

Here is a brief resumé of standard curves, some of which have been met earlier in this text.

(i) Straight line (see Chapter 28, page 249)

The general equation of a straight line is $y = mx + c$, where m is the gradient and c is the y-axis intercept.

Two examples are shown in Fig. 32.1.

(ii) Quadratic graphs (see Chapter 31, page 277)

The general equation of a quadratic graph is $y = ax^2 + bx + c$, and its shape is that of a parabola.

The simplest example of a quadratic graph, $y = x^2$, is shown in Fig. 32.2.

(iii) Cubic equations (see Chapter 31, page 282)

The general equation of a cubic graph is $y = ax^3 + bx^2 + cx + d$.

The simplest example of a cubic graph, $y = x^3$, is shown in Fig. 32.3.

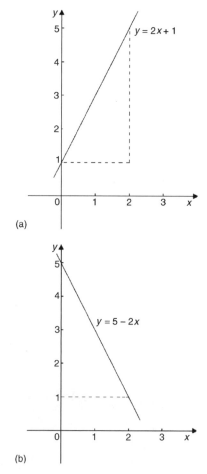

(a)

(b)

Figure 32.1

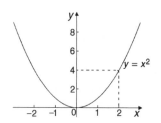

Figure 32.2

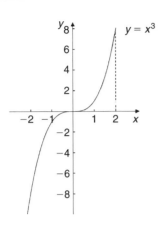

Figure 32.3

(iv) **Trigonometric functions** (see Chapter 23 page 199)

Graphs of $y = \sin\theta$, $y = \cos\theta$ and $y = \tan\theta$ are shown in Fig. 32.4

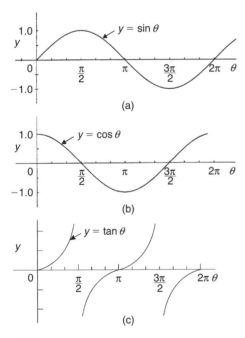

Figure 32.4

(v) **Circle** (see Chapter 18, page 155)

The simplest equation of a circle is $x^2 + y^2 = r^2$, with centre at the origin and radius r, as shown in Fig. 32.5.

More generally, the equation of a circle, centre (a, b), radius r, is given by:

$$(x - a)^2 + (y - b)^2 = r^2 \qquad (1)$$

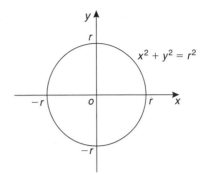

Figure 32.5

Figure 32.6 shows a circle

$$(x - 2)^2 + (y - 3)^2 = 4$$

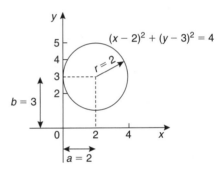

Figure 32.6

(vi) **Ellipse**

The equation of an ellipse is:

$$\frac{x^2}{a^2} + \frac{y^2}{b^2} = 1$$

and the general shape is as shown in Fig. 32.7.

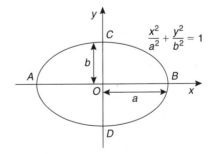

Figure 32.7

The length AB is called the **major axis** and CD the **minor axis**. In the above equation, 'a' is the semi-major axis and 'b' is the semi-minor axis.

Section 4

(Note that if $b = a$, the equation becomes

$$\frac{x^2}{a^2} + \frac{y^2}{a^2} = 1,$$

i.e. $x^2 + y^2 = a^2$, which is a circle of radius a).

(vii) Hyperbola

The equation of a hyperbola is $\dfrac{x^2}{a^2} - \dfrac{y^2}{b^2} = 1$ and the general shape is shown in Fig. 32.8.

The curve is seen to be symmetrical about both the x- and y-axes. The distance AB in Fig. 32.8 is given by $2a$.

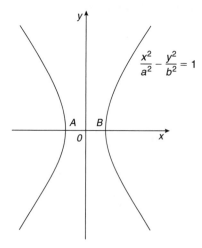

Figure 32.8

(viii) Rectangular hyperbola

The equation of a rectangular hyperbola is $xy = c$ or $y = \dfrac{c}{x}$ and the general shape is shown in Fig. 32.9.

(ix) Logarithmic function (see Chapter 13, page 101)

$y = \ln x$ and $y = \lg x$ are both of the general shape shown in Fig. 32.10.

(x) Exponential functions (see Chapter 14, page 106)

$y = e^x$ is of the general shape shown in Fig. 32.11.

(xi) Polar curves

The equation of a polar curve is of the form $r = f(\theta)$. An example of a polar curve, $r = a \sin \theta$, is shown in Fig. 32.12.

32.2 Simple transformations

From the graph of $y = f(x)$ it is possible to deduce the graphs of other functions which are transformations of $y = f(x)$. For example, knowing the graph of $y = f(x)$,

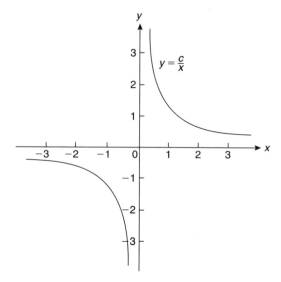

Figure 32.9

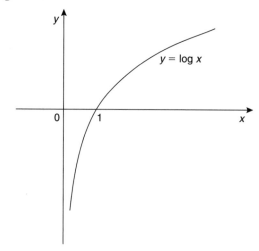

Figure 32.10

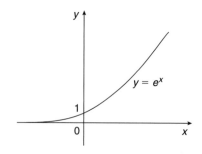

Figure 32.11

can help us draw the graphs of $y = af(x)$, $y = f(x) + a$, $y = f(x + a)$, $y = f(ax)$, $y = -f(x)$ and $y = f(-x)$.

(i) $y = a f(x)$

For each point (x_1, y_1) on the graph of $y = f(x)$ there exists a point (x_1, ay_1) on the graph of

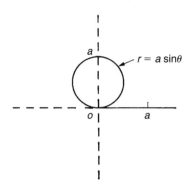

Figure 32.12

$y = a f(x)$. Thus the graph of $y = a f(x)$ can be obtained by stretching $y = f(x)$ parallel to the y-axis by a scale factor 'a'. Graphs of $y = x + 1$ and $y = 3(x + 1)$ are shown in Fig. 32.13(a) and graphs of $y = \sin \theta$ and $y = 2 \sin \theta$ are shown in Fig. 32.13(b).

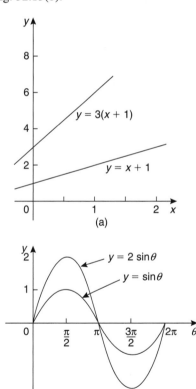

Figure 32.13

(ii) $y = f(x) + a$

The graph of $y = f(x)$ is translated by 'a' units parallel to the y-axis to obtain $y = f(x) + a$. For example, if $f(x) = x$, $y = f(x) + 3$ becomes $y = x + 3$, as shown in Fig. 32.14(a). Similarly, if $f(\theta) = \cos \theta$, then $y = f(\theta) + 2$ becomes

$y = \cos \theta + 2$, as shown in Fig. 32.14(b). Also, if $f(x) = x^2$, then $y = f(x) + 3$ becomes $y = x^2 + 3$, as shown in Fig. 32.14(c).

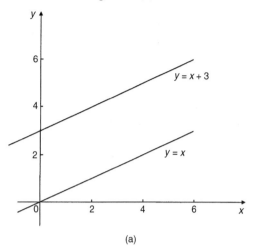

(a)

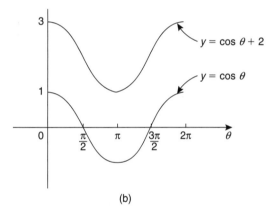

(b)

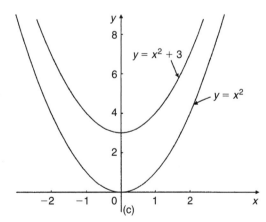

(c)

Figure 32.14

(iii) $y = f(x + a)$

The graph of $y = f(x)$ is translated by 'a' units parallel to the x-axis to obtain $y = f(x + a)$. If 'a' > 0 it moves $y = f(x)$ in the negative direction on the x-axis (i.e. to the left), and if 'a' < 0 it moves $y = f(x)$ in the positive direction on the x-axis (i.e. to the right). For example, if $f(x) = \sin x$, $y = f\left(x - \dfrac{\pi}{3}\right)$ becomes $y = \sin\left(x - \dfrac{\pi}{3}\right)$ as shown in Fig. 32.15(a) and $y = \sin\left(x + \dfrac{\pi}{4}\right)$ is shown in Fig. 32.15(b).

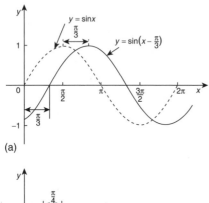

(a)

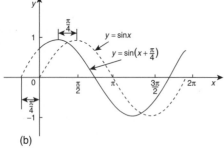

(b)

Figure 32.15

Similarly graphs of $y = x^2$, $y = (x - 1)^2$ and $y = (x + 2)^2$ are shown in Fig. 32.16.

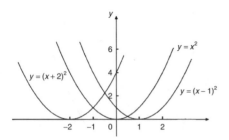

Figure 32.16

(iv) $y = f(ax)$

For each point (x_1, y_1) on the graph of $y = f(x)$, there exists a point $\left(\dfrac{x_1}{a}, y_1\right)$ on the graph of $y = f(ax)$. Thus the graph of $y = f(ax)$ can be obtained by stretching $y = f(x)$ parallel to the x-axis by a scale factor $\dfrac{1}{a}$.

For example, if $f(x) = (x - 1)^2$, and $a = \dfrac{1}{2}$, then $f(ax) = \left(\dfrac{x}{2} - 1\right)^2$.

Both of these curves are shown in Fig. 32.17(a). Similarly, $y = \cos x$ and $y = \cos 2x$ are shown in Fig. 32.17(b).

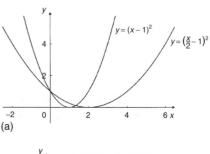

(a)

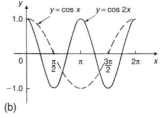

(b)

Figure 32.17

(v) $y = -f(x)$

The graph of $y = -f(x)$ is obtained by reflecting $y = f(x)$ in the x-axis. For example, graphs of $y = e^x$ and $y = -e^x$ are shown in Fig. 32.18(a), and graphs of $y = x^2 + 2$ and $y = -(x^2 + 2)$ are shown in Fig. 32.18(b).

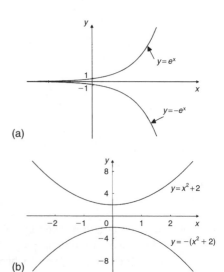

(a)

(b)

Figure 32.18

(vi) $y = f(-x)$

The graph of $y = f(-x)$ is obtained by reflecting $y = f(x)$ in the y-axis. For example, graphs of $y = x^3$ and $y = (-x)^3 = -x^3$ are shown in Fig. 32.19(a) and graphs of $y = \ln x$ and $y = -\ln x$ are shown in Fig. 32.19(b).

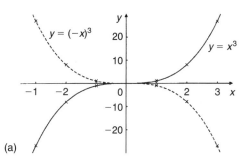

(a)

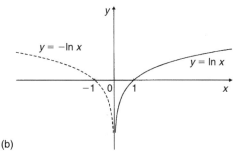

(b)

Figure 32.19

Problem 1. Sketch the following graphs, showing relevant points:

(a) $y = (x - 4)^2$ (b) $y = x^3 - 8$

(a) In Fig. 32.20 a graph of $y = x^2$ is shown by the broken line. The graph of $y = (x - 4)^2$ is of the form $y = f(x + a)$. Since $a = -4$, then $y = (x - 4)^2$ is translated 4 units to the right of $y = x^2$, parallel to the x-axis.
(See section (iii) above).

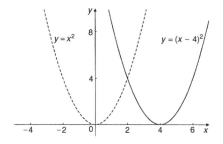

Figure 32.20

(b) In Fig. 32.21 a graph of $y = x^3$ is shown by the broken line. The graph of $y = x^3 - 8$ is of the form $y = f(x) + a$. Since $a = -8$, then $y = x^3 - 8$ is translated 8 units down from $y = x^3$, parallel to the y-axis.
(See section (ii) above)

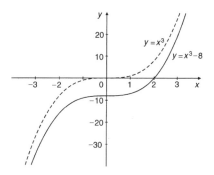

Figure 32.21

Problem 2. Sketch the following graphs, showing relevant points:

(a) $y = 5 - (x + 2)^3$ (b) $y = 1 + 3 \sin 2x$

(a) Figure 32.22(a) shows a graph of $y = x^3$. Figure 32.22(b) shows a graph of $y = (x + 2)^3$ (see $f(x + a)$, section (iii) above). Figure 32.22(c) shows a graph of $y = -(x + 2)^3$ (see $-f(x)$, section (v) above). Figure 32.22(d) shows the graph of $y = 5 - (x + 2)^3$ (see $f(x) + a$, section (ii) above).

(b) Figure 32.23(a) shows a graph of $y = \sin x$. Figure 32.23(b) shows a graph of $y = \sin 2x$ (see $f(ax)$, section (iv) above)

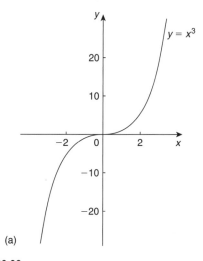

(a)

Figure 32.22

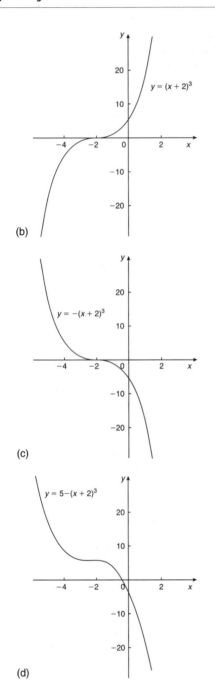

(b)

$y = (x+2)^3$

(c)

$y = -(x+2)^3$

(d)

$y = 5-(x+2)^3$

Figure 32.22 (*Continued*)

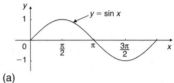

$y = \sin x$

(a)

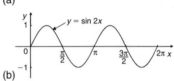

$y = \sin 2x$

(b)

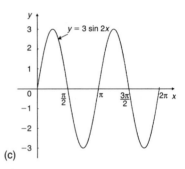

$y = 3 \sin 2x$

(c)

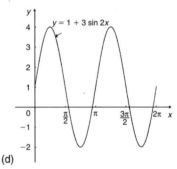

$y = 1 + 3 \sin 2x$

(d)

Figure 32.23

Figure 32.23(c) shows a graph of $y = 3 \sin 2x$ (see $af(x)$, section (i) above). Figure 32.23(d) shows a graph of $y = 1 + 3 \sin 2x$ (see $f(x) + a$, section (ii) above).

Now try the following exercise

Exercise 120   Further problems on simple transformations with curve sketching

Sketch the following graphs, showing relevant points:
(Answers on page 295, Fig. 32.33)

1. $y = 3x - 5$ 2. $y = -3x + 4$
3. $y = x^2 + 3$ 4. $y = (x-3)^2$
5. $y = (x-4)^2 + 2$ 6. $y = x - x^2$

7. $y = x^3 + 2$ 8. $y = 1 + 2\cos 3x$

9. $y = 3 - 2\sin\left(x + \dfrac{\pi}{4}\right)$

10. $y = 2\ln x$

32.3 Periodic functions

A function $f(x)$ is said to be **periodic** if $f(x+T) = f(x)$ for all values of x, where T is some positive number. T is the interval between two successive repetitions and is called the period of the function $f(x)$. For example, $y = \sin x$ is periodic in x with period 2π since $\sin x = \sin(x + 2\pi) = \sin(x + 4\pi)$, and so on. Similarly, $y = \cos x$ is a periodic function with period 2π since $\cos x = \cos(x + 2\pi) = \cos(x + 4\pi)$, and so on. In general, if $y = \sin \omega t$ or $y = \cos \omega t$ then the period of the waveform is $2\pi/\omega$. The function shown in Fig. 32.24 is also periodic of period 2π and is defined by:

$$f(x) = \begin{cases} -1, & \text{when } -\pi \le x \le 0 \\ 1, & \text{when } 0 \le x \le \pi \end{cases}$$

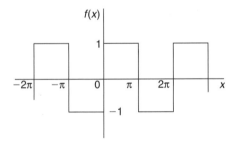

Figure 32.24

32.4 Continuous and discontinuous functions

If a graph of a function has no sudden jumps or breaks it is called a **continuous function**, examples being the graphs of sine and cosine functions. However, other graphs make finite jumps at a point or points in the interval. The square wave shown in Fig. 32.24 has **finite discontinuities** as $x = \pi, 2\pi, 3\pi$, and so on, and is therefore a **discontinuous function**. $y = \tan x$ is another example of a discontinuous function.

32.5 Even and odd functions

Even functions

A function $y = f(x)$ is said to be even if $f(-x) = f(x)$ for all values of x. Graphs of even functions are always symmetrical about the y-axis (i.e. is a mirror image). Two examples of even functions are $y = x^2$ and $y = \cos x$ as shown in Fig. 32.25.

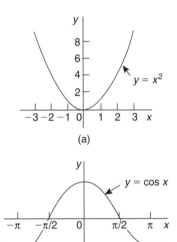

(a)

(b)

Figure 32.25

Odd functions

A function $y = f(x)$ is said to be odd if $f(-x) = -f(x)$ for all values of x. Graphs of odd functions are always symmetrical about the origin. Two examples of odd functions are $y = x^3$ and $y = \sin x$ as shown in Fig. 32.26.

Many functions are neither even nor odd, two such examples being shown in Fig. 32.27.

> **Problem 3.** Sketch the following functions and state whether they are even or odd functions:
>
> (a) $y = \tan x$
>
> (b) $f(x) = \begin{cases} 2, & \text{when } 0 \le x \le \dfrac{\pi}{2} \\ -2, & \text{when } \dfrac{\pi}{2} \le x \le \dfrac{3\pi}{2} \\ 2, & \text{when } \dfrac{3\pi}{2} \le x \le 2\pi \end{cases}$
>
> and is periodic of period 2π

Section 4

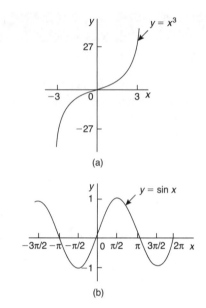

(a)

(b)

Figure 32.26

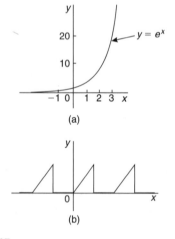

(a)

(b)

Figure 32.27

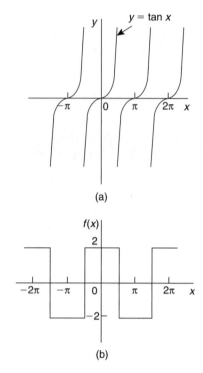

(a)

(b)

Figure 32.28

(a) A graph of $y = \ln x$ is shown in Fig. 32.29(a) and the curve is neither symmetrical about the y-axis nor symmetrical about the origin and is thus **neither even nor odd**.

(a) A graph of $y = \tan x$ is shown in Figure 32.28(a) and is symmetrical about the origin and is thus an **odd function** (i.e. $\tan(-x) = -\tan x$).

(b) A graph of $f(x)$ is shown in Fig. 32.28(b) and is symmetrical about the $f(x)$ axis hence the function is an **even** one, ($f(-x) = f(x)$).

Problem 4. Sketch the following graphs and state whether the functions are even, odd or neither even nor odd:

(a) $y = \ln x$

(b) $f(x) = x$ in the range $-\pi$ to π and is periodic of period 2π

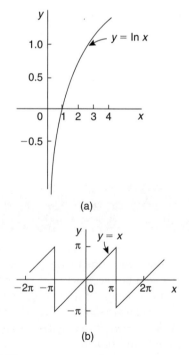

(a)

(b)

Figure 32.29

(b) A graph of $y = x$ in the range $-\pi$ to π is shown in Fig. 32.29(b) and is symmetrical about the origin and is thus an **odd function**.

Now try the following exercise

Exercise 121   Further problems on even and odd functions

In Problems 1 and 2 determine whether the given functions are even, odd or neither even nor odd.

1. (a) x^4 (b) $\tan 3x$ (c) $2e^{3t}$ (d) $\sin^2 x$

$$\begin{bmatrix} \text{(a) even} & \text{(b) odd} \\ \text{(c) neither} & \text{(d) even} \end{bmatrix}$$

2. (a) $5t^3$ (b) $e^x + e^{-x}$ (c) $\dfrac{\cos\theta}{\theta}$ (d) e^x

$$\begin{bmatrix} \text{(a) odd} & \text{(b) even} \\ \text{(c) odd} & \text{(d) neither} \end{bmatrix}$$

3. State whether the following functions, which are periodic of period 2π, are even or odd:

(a) $f(\theta) = \begin{cases} \theta, & \text{when } -\pi \le \theta \le 0 \\ -\theta, & \text{when } 0 \le \theta \le \pi \end{cases}$

(b) $f(x) = \begin{cases} x, & \text{when } -\dfrac{\pi}{2} \le x \le \dfrac{\pi}{2} \\ 0, & \text{when } \dfrac{\pi}{2} \le x \le \dfrac{3\pi}{2} \end{cases}$

$$[\text{(a) even}\quad \text{(b) odd}]$$

32.6   Inverse functions

If y is a function of x, the graph of y against x can be used to find x when any value of y is given. Thus the graph also expresses that x is a function of y. Two such functions are called **inverse functions**.

In general, given a function $y = f(x)$, its inverse may be obtained by inter-changing the roles of x and y and then transposing for y. The inverse function is denoted by $y = f^{-1}(x)$.

For example, if $y = 2x + 1$, the inverse is obtained by

(i)   transposing for x, i.e. $x = \dfrac{y-1}{2} = \dfrac{y}{2} - \dfrac{1}{2}$

and (ii)   interchanging x and y, giving the inverse as
$$y = \frac{x}{2} - \frac{1}{2}$$

Thus if $f(x) = 2x + 1$, then $f^{-1}(x) = \dfrac{x}{2} - \dfrac{1}{2}$

A graph of $f(x) = 2x + 1$ and its inverse $f^{-1}(x) = \dfrac{x}{2} - \dfrac{1}{2}$ is shown in Fig. 32.30 and $f^{-1}(x)$ is seen to be a reflection of $f(x)$ in the line $y = x$.

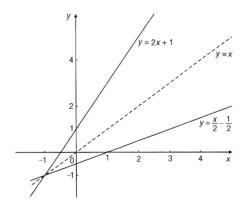

Figure 32.30

Similarly, if $y = x^2$, the inverse is obtained by

(i)   transposing for x, i.e. $x = \pm\sqrt{y}$

and (ii)   interchanging x and y, giving the inverse $y = \pm\sqrt{x}$

Hence the inverse has two values for every value of x. Thus $f(x) = x^2$ does not have a single inverse. In such a case the domain of the original function may be restricted to $y = x^2$ for $x > 0$. Thus the inverse is then $y = +\sqrt{x}$. A graph of $f(x) = x^2$ and its inverse $f^{-1}(x) = \sqrt{x}$ for $x > 0$ is shown in Fig. 32.31 and, again, $f^{-1}(x)$ is seen to be a reflection of $f(x)$ in the line $y = x$.

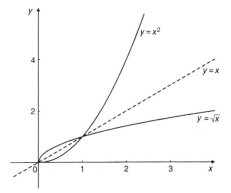

Figure 32.31

It is noted from the latter example, that not all functions have a single inverse. An inverse, however, can be determined if the range is restricted.

Section 4

Problem 5. Determine the inverse for each of the following functions:
(a) $f(x) = x - 1$ (b) $f(x) = x^2 - 4 \ (x > 0)$
(c) $f(x) = x^2 + 1$

(a) If $y = f(x)$, then $y = x - 1$
Transposing for x gives $x = y + 1$
Interchanging x and y gives $y = x + 1$
Hence if $f(x) = x - 1$, then $f^{-1}(x) = x + 1$

(b) If $y = f(x)$, then $y = x^2 - 4 \ (x > 0)$
Transposing for x gives $x = \sqrt{y + 4}$
Interchanging x and y gives $y = \sqrt{x + 4}$
Hence if $f(x) = x^2 - 4 \ (x > 0)$ then
$f^{-1}(x) = \sqrt{x + 4}$ **if $x > -4$**

(c) If $y = f(x)$, then $y = x^2 + 1$
Transposing for x gives $x = \sqrt{y - 1}$
Interchanging x and y gives $y = \sqrt{x - 1}$, which has two values. **Hence there is no single inverse of $f(x) = x^2 + 1$**, since the domain of $f(x)$ is not restricted.

Inverse trigonometric functions

If $y = \sin x$, then x is the angle whose sine is y. Inverse trigonometrical functions are denoted either by prefixing the function with 'arc' or by using $^{-1}$. Hence transposing $y = \sin x$ for x gives $x = \arcsin y$ or $\sin^{-1} y$. Interchanging x and y gives the inverse $y = \arcsin x$ or $\sin^{-1} x$.

Similarly, $y = \cos^{-1} x$, $y = \tan^{-1} x$, $y = \sec^{-1} x$, $y = \csc^{-1} x$ and $y = \cot^{-1} x$ are all inverse trigonometric functions. The angle is always expressed in radians.

Inverse trigonometric functions are periodic so it is necessary to specify the smallest or principal value of the angle. For $y = \sin^{-1} x$, $\tan^{-1} x$, $\csc^{-1} x$ and $\cot^{-1} x$, the principal value is in the range $-\dfrac{\pi}{2} < y < \dfrac{\pi}{2}$. For $y = \cos^{-1} x$ and $\sec^{-1} x$ the principal value is in the range $0 < y < \pi$

Graphs of the six inverse trigonometric functions are shown in Fig. 32.32.

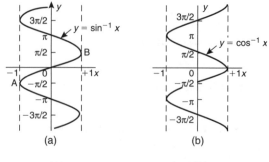

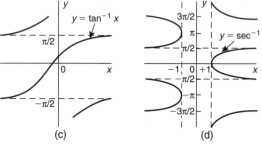

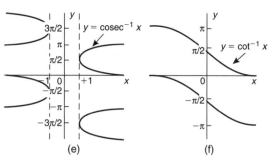

(a)   (b)

(c)   (d)

(e)   (f)

Figure 32.32

(b) $\arctan(-1) \equiv \tan^{-1}(-1) = -45°$
$$= -\frac{\pi}{4} \text{ rad or } -0.7854 \text{ rad}$$

(c) $\arccos\left(-\dfrac{\sqrt{3}}{2}\right) \equiv \cos^{-1}\left(-\dfrac{\sqrt{3}}{2}\right) = 150°$
$$= \frac{5\pi}{6} \text{ rad or } 2.6180 \text{ rad}$$

(d) $\text{arccosec}(\sqrt{2}) = \arcsin\left(\dfrac{1}{\sqrt{2}}\right) \equiv \sin^{-1}\left(\dfrac{1}{\sqrt{2}}\right)$
$$= 45° = \frac{\pi}{4} \text{ rad or } 0.7854 \text{ rad}$$

Problem 6. Determine the principal values of
(a) $\arcsin 0.5$ (b) $\arctan(-1)$
(c) $\arccos\left(-\dfrac{\sqrt{3}}{2}\right)$ (d) $\text{arccosec}(\sqrt{2})$

Using a calculator,

(a) $\arcsin 0.5 \equiv \sin^{-1} 0.5 = 30° = \dfrac{\pi}{6} \text{ rad or } \mathbf{0.5236 \text{ rad}}$

Problem 7. Evaluate (in radians), correct to 3 decimal places: $\sin^{-1} 0.30 + \cos^{-1} 0.65$

$$\sin^{-1} 0.30 = 17.4576° = 0.3047 \text{ rad}$$
$$\cos^{-1} 0.65 = 49.4584° = 0.8632 \text{ rad}$$

Hence $\sin^{-1} 0.30 + \cos^{-1} 0.65 = 0.3047 + 0.8632$
$$= \mathbf{1.168}, \text{ correct to 3 decimal places}$$

Now try the following exercise

Exercise 122   Further problems on inverse functions

Determine the inverse of the functions given in Problems 1 to 4.

1. $f(x) = x + 1$ $[f^{-1}(x) = x - 1]$

2. $f(x) = 5x - 1$ $\left[f^{-1}(x) = \dfrac{1}{5}(x+1)\right]$

3. $f(x) = x^3 + 1$ $[f^{-1}(x) = \sqrt[3]{x - 1}]$

4. $f(x) = \dfrac{1}{x} + 2$ $\left[f^{-1}(x) = \dfrac{1}{x - 2}\right]$

Determine the principal value of the inverse functions in Problems 5 to 11.

5. $\sin^{-1}(-1)$ $\left[-\dfrac{\pi}{2} \text{ or } -1.5708 \text{ rad}\right]$

6. $\cos^{-1} 0.5$ $\left[\dfrac{\pi}{3} \text{ or } 1.0472 \text{ rad}\right]$

7. $\tan^{-1} 1$ $\left[\dfrac{\pi}{4} \text{ or } 0.7854 \text{ rad}\right]$

8. $\cot^{-1} 2$ $[0.4636 \text{ rad}]$

9. $\operatorname{cosec}^{-1} 2.5$ $[0.4115 \text{ rad}]$

10. $\sec^{-1} 1.5$ $[0.8411 \text{ rad}]$

11. $\sin^{-1}\left(\dfrac{1}{\sqrt{2}}\right)$ $\left[\dfrac{\pi}{4} \text{ or } 0.7854 \text{ rad}\right]$

12. Evaluate x, correct to 3 decimal places:
$$x = \sin^{-1}\dfrac{1}{3} + \cos^{-1}\dfrac{4}{5} - \tan^{-1}\dfrac{8}{9}$$
                                     $[0.257]$

13. Evaluate y, correct to 4 significant figures:
$$y = 3\sec^{-1}\sqrt{2} - 4\operatorname{cosec}^{-1}\sqrt{2} + 5\cot^{-1} 2$$
                                     $[1.533]$

Answers to Exercise 120

1.

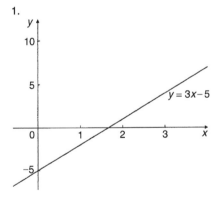

2.

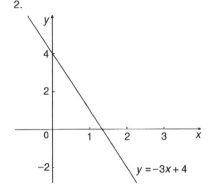

3.

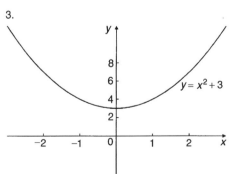

4.

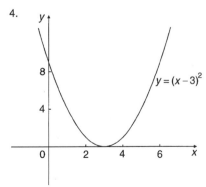

Figure 32.33   Graphical solutions to Exercise 120, page 290.

5.

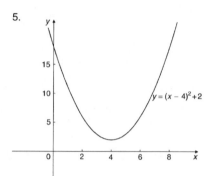

$y = (x - 4)^2 + 2$

6.

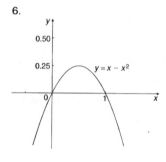

$y = x - x^2$

7.

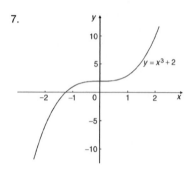

$y = x^3 + 2$

8.

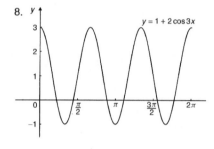

$y = 1 + 2\cos 3x$

9.

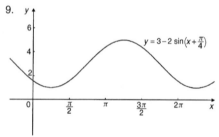

$y = 3 - 2\sin\left(x + \frac{\pi}{4}\right)$

10.

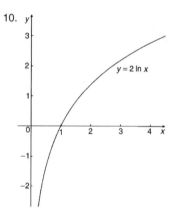

$y = 2\ln x$

Figure 32.33 (*Continued*)

This Revision test covers the material contained in Chapters 28 to 32. *The marks for each question are shown in brackets at the end of each question.*

1. Determine the gradient and intercept on the y-axis for the following equations:

   (a) $y = -5x + 2$ (b) $3x + 2y + 1 = 0$ (5)

2. The equation of a line is $2y = 4x + 7$. A table of corresponding values is produced and is as shown below. Complete the table and plot a graph of y against x. Determine the gradient of the graph. (6)

x	-3	-2	-1	0	1	2	3
y	-2.5					7.5	

3. Plot the graphs $y = 3x + 2$ and $\dfrac{y}{2} + x = 6$ on the same axes and determine the co-ordinates of their point of intersection. (7)

4. The velocity v of a body over varying time intervals t was measured as follows:

t seconds	2	5	7
v m/s	15.5	17.3	18.5

t seconds	10	14	17
v m/s	20.3	22.7	24.5

   Plot a graph with velocity vertical and time horizontal. Determine from the graph (a) the gradient, (b) the vertical axis intercept, (c) the equation of the graph, (d) the velocity after 12.5 s, and (e) the time when the velocity is 18 m/s. (9)

5. The following experimental values of x and y are believed to be related by the law $y = ax^2 + b$, where a and b are constants. By plotting a suitable graph verify this law and find the approximate values of a and b.

x	2.5	4.2	6.0	8.4	9.8	11.4
y	15.4	32.5	60.2	111.8	150.1	200.9

   (9)

6. Determine the law of the form $y = ae^{kx}$ which relates the following values:

y	0.0306	0.285	0.841
x	-4.0	5.3	9.8

y	5.21	173.2	1181
x	17.4	32.0	40.0

   (9)

7. State the minimum number of cycles on logarithmic graph paper needed to plot a set of values ranging from 0.073 to 490. (2)

8. Plot a graph of $y = 2x^2$ from $x = -3$ to $x = +3$ and hence solve the equations:

   (a) $2x^2 - 8 = 0$ (b) $2x^2 - 4x - 6 = 0$ (9)

9. Plot the graph of $y = x^3 + 4x^2 + x - 6$ for values of x between $x = -4$ and $x = 2$. Hence determine the roots of the equation $x^3 + 4x^2 + x - 6 = 0$. (7)

10. Sketch the following graphs, showing the relevant points:

    (a) $y = (x - 2)^2$ (b) $y = 3 - \cos 2x$

    (c) $f(x) = \begin{cases} -1 & -\pi \leq x \leq -\dfrac{\pi}{2} \\ x & -\dfrac{\pi}{2} \leq x \leq \dfrac{\pi}{2} \\ 1 & \dfrac{\pi}{2} \leq x \leq \pi \end{cases}$

    (10)

11. Determine the inverse of $f(x) = 3x + 1$ (3)

12. Evaluate, correct to 3 decimal places:

    $2\tan^{-1}1.64 + \sec^{-1}2.43 - 3\cosec^{-1}3.85$ (4)

Section 5

Vectors

Chapter 33

Vectors

33.1 Introduction

Some physical quantities are entirely defined by a numerical value and are called **scalar quantities** or **scalars**. Examples of scalars include time, mass, temperature, energy and volume. Other physical quantities are defined by both a numerical value and a direction in space and these are called **vector quantities** or **vectors**. Examples of vectors include force, velocity, moment and displacement.

33.2 Vector addition

A vector may be represented by a straight line, the length of line being directly proportional to the magnitude of the quantity and the direction of the line being in the same direction as the line of action of the quantity. An arrow is used to denote the sense of the vector, that is, for a horizontal vector, say, whether it acts from left to right or vice-versa. The arrow is positioned at the end of the vector and this position is called the 'nose' of the vector. Figure 33.1 shows a velocity of 20 m/s at an angle of 45° to the horizontal and may be depicted by $oa = 20$ m/s at 45° to the horizontal.

Figure 33.1

To distinguish between vector and scalar quantities, various ways are used. These include:

(i) **bold print,**

(ii) two capital letters with an arrow above them to denote the sense of direction, e.g. $\overrightarrow{AB}$, where A is the starting point and B the end point of the vector,

(iii) a line over the top of letters, e.g. $\overline{AB}$ or $\bar{a}$

(iv) letters with an arrow above, e.g. $\vec{a}, \vec{A}$

(v) underlined letters, e.g. $\underline{a}$

(vi) $xi + jy$, where i and j are axes at right-angles to each other; for example, $3i + 4j$ means 3 units in the i direction and 4 units in the j direction, as shown in Fig. 33.2.

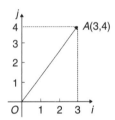

Figure 33.2

(vii) a column matrix $\begin{pmatrix} a \\ b \end{pmatrix}$; for example, the vector OA shown in Fig. 32.2 could be represented by $\begin{pmatrix} 3 \\ 4 \end{pmatrix}$

Thus, in Fig. 33.2,

$$OA \equiv \overrightarrow{OA} \equiv \overline{OA} \equiv 3i + 4j \equiv \begin{pmatrix} 3 \\ 4 \end{pmatrix}$$

The one adopted in this text is to denote vector quantities in **bold print**. Thus, oa represents a vector quantity, but oa is the magnitude of the vector oa. Also, positive angles are measured in an anticlock-wise direction from a horizontal, right facing line and negative angles in a clockwise direction from this line—as with graphical work. Thus 90° is a line vertically upwards and −90° is a line vertically downwards.

The resultant of adding two vectors together, say V_1 at an angle θ_1 and V_2 at angle $(-\theta_2)$, as shown in Fig. 33.3(a), can be obtained by drawing oa to represent

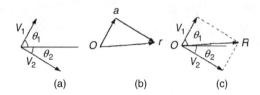

Figure 33.3

V_1 and then drawing *ar* to represent V_2. The resultant of $V_1 + V_2$ is given by *or*. This is shown in Fig. 33.3(b), the vector equation being $oa + ar = or$. This is called the **'nose-to-tail' method** of vector addition.

Alternatively, by drawing lines parallel to V_1 and V_2 from the noses of V_2 and V_1, respectively, and letting the point of intersection of these parallel lines be R, gives **OR** as the magnitude and direction of the resultant of adding V_1 and V_2, as shown in Fig. 33.3(c). This is called the **'parallelogram' method** of vector addition.

Problem 1. A force of 4 N is included at an angle of 45° to a second force of 7 N, both forces acting at a point. Find the magnitude of the resultant of these two forces and the direction of the resultant with respect to the 7 N force by both the 'triangle' and the 'parallelogram' methods

---

The forces are shown in Fig. 33.4(a). Although the 7 N force is shown as a horizontal line, it could have been drawn in any direction.

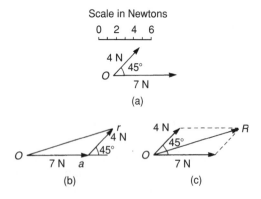

Figure 33.4

Using the **'nose-to-tail' method**, a line 7 units long is drawn horizontally to give vector *oa* in Fig. 33.4(b). To the nose of this vector *ar* is drawn 4 units long at an angle of 45° to *oa*. The resultant of vector addition is *or* and by measurement is **10.2 units long and at an angle of 16° to the 7 N force**.

Figure 33.4(c) uses the **'parallelogram' method** in which lines are drawn parallel to the 7 N and 4 N forces from the noses of the 4 N and 7 N forces, respectively. These intersect at R. Vector **OR** gives the magnitude and direction of the resultant of vector addition and as obtained by the 'nose-to-tail' method is **10.2 units long at an angle of 16° to the 7 N force**.

Problem 2. Use a graphical method to determine the magnitude and direction of the resultant of the three velocities shown in Fig. 33.5.

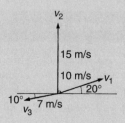

Figure 33.5

Often it is easier to use the 'nose-to-tail' method when more than two vectors are being added. The order in which the vectors are added is immaterial. In this case the order taken is v_1, then v_2, then v_3 but just the same result would have been obtained if the order had been, say, v_1, v_3 and finally v_2. v_1 is drawn 10 units long at an angle of 20° to the horizontal, shown by *oa* in Fig. 33.6. v_2 is added to v_1 by drawing a line 15 units long vertically upwards from a, shown as *ab*. Finally, v_3 is added to $v_1 + v_2$ by drawing a line 7 units long at an angle at 190° from b, shown as *br*. The resultant of vector addition is *or* and by measurement is 17.5 units long at an angle of 82° to the horizontal.

Thus $v_1 + v_2 + v_3 = 17.5\,\text{m/s}$ at **82°** to the **horizontal**.

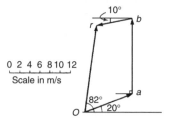

Figure 33.6

33.3 Resolution of vectors

A vector can be resolved into two component parts such that the vector addition of the component parts is equal to the original vector. The two components usually taken are a horizontal component and a vertical component. For the vector shown as F in Fig. 33.7, the horizontal component is $F \cos \theta$ and the vertical component is $F \sin \theta$.

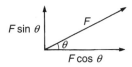

Figure 33.7

For the vectors F_1 and F_2 shown in Fig. 33.8, the horizontal component of vector addition is:

$$H = F_1 \cos \theta_1 + F_2 \cos \theta_2$$

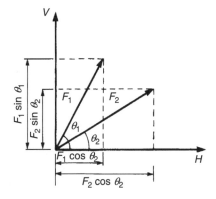

Figure 33.8

and the vertical component of vector addition is:

$$V = F_1 \sin \theta_1 + F_2 \sin \theta_2$$

Having obtained H and V, the magnitude of the resultant vector R is given by: $\sqrt{H^2 + V^2}$ and its angle to the horizontal is given by $\tan^{-1}\dfrac{V}{H}$

Problem 3. Resolve the acceleration vector of $17\,\text{m/s}^2$ at an angle of $120°$ to the horizontal into a horizontal and a vertical component

For a vector A at angle θ to the horizontal, the horizontal component is given by $A \cos \theta$ and the vertical component by $A \sin \theta$. Any convention of signs may be adopted, in this case horizontally from left to right is taken as positive and vertically upwards is taken as positive.

Horizontal component $H = 17 \cos 120° = -8.50\,\text{m/s}^2$, acting from left to right.

Vertical component $V = 17 \sin 120° = 14.72\,\text{m/s}^2$, acting vertically upwards.

These component vectors are shown in Fig. 33.9.

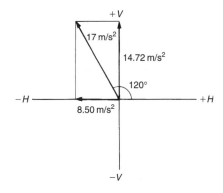

Figure 33.9

Problem 4. Calculate the resultant force of the two forces given in Problem 1

With reference to Fig. 33.4(a):

Horizontal component of force,

$$H = 7 \cos 0° + 4 \cos 45°$$
$$= 7 + 2.828 = 9.828\,\text{N}$$

Vertical component of force,

$$V = 7 \sin 0° + 4 \sin 45°$$
$$= 0 + 2.828 = 2.828\,\text{N}$$

The magnitude of the resultant of vector addition

$$= \sqrt{H^2 + V^2} = \sqrt{9.828^2 + 2.828^2}$$
$$= \sqrt{104.59} = 10.23\,\text{N}$$

The direction of the resultant of vector addition

$$= \tan^{-1}\left(\frac{V}{H}\right) = \tan^{-1}\left(\frac{2.828}{9.828}\right) = 16.05°$$

Thus, the resultant of the two forces is a single vector of 10.23 N at 16.05° to the 7 N vector.

Section 5

Problem 5. Calculate the resultant velocity of the three velocities given in Problem 2

With reference to Fig. 33.5:

Horizontal component of the velocity,

$$H = 10\cos 20° + 15\cos 90° + 7\cos 190°$$

$$= 9.397 + 0 + (-6.894) = \textbf{2.503 m/s}$$

Vertical component of the velocity,

$$V = 10\sin 20° + 15\sin 90° + 7\sin 190°$$

$$= 3.420 + 15 + (-1.216) = \textbf{17.204 m/s}$$

Magnitude of the resultant of vector addition

$$= \sqrt{H^2 + V^2} = \sqrt{2.503^2 + 17.204^2}$$

$$= \sqrt{302.24} = \textbf{17.39 m/s}$$

Direction of the resultant of vector addition

$$= \tan^{-1}\left(\frac{V}{H}\right) = \tan^{-1}\left(\frac{17.204}{2.503}\right)$$

$$= \tan^{-1} 6.8734 = 81.72°$$

Thus, the resultant of the three velocities is a single vector of 17.39 m/s at 81.72° to the horizontal.

Now try the following exercise

Exercise 123   Further problems on vector addition and resolution

1. Forces of 23 N and 41 N act at a point and are inclined at 90° to each other. Find, by drawing, the resultant force and its direction relative to the 41 N force.   [47 N at 29°]

2. Forces A, B and C are coplanar and act at a point. Force A is 12 kN at 90°, B is 5 kN at 180° and C is 13 kN at 293°. Determine graphically the resultant force.   [Zero]

3. Calculate the magnitude and direction of velocities of 3 m/s at 18° and 7 m/s at 115° when acting simultaneously on a point.   [7.27 m/s at 90.8°]

4. Three forces of 2 N, 3 N and 4 N act as shown in Fig. 33.10. Calculate the magnitude of the resultant force and its direction relative to the 2 N force.   [6.24 N at 76.10°]

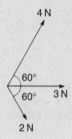

Figure 33.10

5. A load of 5.89 N is lifted by two strings, making angles of 20° and 35° with the vertical. Calculate the tensions in the strings. [For a system such as this, the vectors representing the forces form a closed triangle when the system is in equilibrium].

    [2.46 N, 4.12 N]

6. The acceleration of a body is due to four component, coplanar accelerations. These are 2 m/s^2 due north, 3 m/s^2 due east, 4 m/s^2 to the south-west and 5 m/s^2 to the south-east. Calculate the resultant acceleration and its direction.

    [5.73 m/s^2 at 310.3°]

7. A current phasor i_1 is 5 A and horizontal. A second phasor i_2 is 8 A and is at 50° to the horizontal. Determine the resultant of the two phasors, $i_1 + i_2$, and the angle the resultant makes with current i_1.

    [11.85 A at 31.14°]

8. An object is acted upon by two forces of magnitude 10 N and 8 N at an angle of 60° to each other. Determine the resultant force on the object.

    [15.62 N at 26.33° to the 10 N force]

9. A ship heads in a direction of E 20° S at a speed of 20 knots while the current is 4 knots in a direction of N 30° E. Determine the speed and actual direction of the ship.

    [21.07 knots, E 9.22° S]

33.4 Vector subtraction

In Fig. 33.11, a force vector F is represented by oa. The vector $(-oa)$ can be obtained by drawing a vector from o in the opposite sense to oa but having the same magnitude, shown as ob in Fig. 33.11, i.e. $ob = (-oa)$.

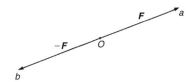

Figure 33.11

For two vectors acting at a point, as shown in Fig. 33.12(a), the resultant of vector addition is $os = oa + ob$. Fig. 33.12(b) shows vectors $ob + (-oa)$, that is $ob - oa$ and the vector equation is $ob - oa = od$. Comparing od in Fig. 33.12(b) with the broken line ab in Fig. 33.12(a) shows that the second diagonal of the 'parallelogram' method of vector addition gives the magnitude and direction of vector subtraction of oa from ob.

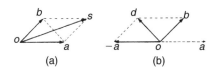

Figure 33.12

> **Problem 6.** Accelerations of $a_1 = 1.5$ m/s² at 90° and $a_2 = 2.6$ m/s² at 145° act at a point. Find $a_1 + a_2$ and $a_1 - a_2$ by:
>
> (i) drawing a scale vector diagram and
> (ii) by calculation

(i) The scale vector diagram is shown in Fig. 33.13. By measurement,

$$a_1 + a_2 = 3.7 \text{ m/s}^2 \text{ at } 126°$$

$$a_1 - a_2 = 2.1 \text{ m/s}^2 \text{ at } 0°$$

(ii) Resolving horizontally and vertically gives:
Horizontal component of $a_1 + a_2$,

$$H = 1.5 \cos 90° + 2.6 \cos 145° = -2.13$$

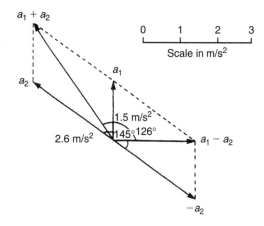

Figure 33.13

Vertical component of $a_1 + a_2$,

$$V = 1.5 \sin 90° + 2.6 \sin 145° = 2.99$$

Magnitude of

$$a_1 + a_2 = \sqrt{(-2.13)^2 + 2.99^2}$$
$$= 3.67 \text{ m/s}^2$$

Direction of $a_1 + a_2 = \tan^{-1}\left(\dfrac{2.99}{-2.13}\right)$ and must lie in the second quadrant since H is negative and V is positive.

$\tan^{-1}\left(\dfrac{2.99}{-2.13}\right) = -54.53°$, and for this to be in the second quadrant, the true angle is 180° displaced, i.e. 180° − 54.53° or 125.47°

Thus $a_1 + a_2 = 3.67$ m/s² at 125.47°.

Horizontal component of $a_1 - a_2$, that is, $a_1 + (-a_2)$

$$= 1.5 \cos 90° + 2.6 \cos(145° - 180°)$$
$$= 2.6 \cos(-35°) = 2.13$$

Vertical component of $a_1 - a_2$, that is, $a_1 + (-a_2)$

$$= 1.5 \sin 90° + 2.6 \sin(-35°) = 0$$
Magnitude of $a_1 - a_2 = \sqrt{2.13^2 + 0^2}$
$$= 2.13 \text{ m/s}^2$$

Direction of $a_1 - a_2 = \tan^{-1}\left(\dfrac{0}{2.13}\right) = 0°$

Thus $a_1 - a_2 = 2.13$ m/s² at 0°

Section 5

Problem 7. Calculate the resultant of
(i) $v_1 - v_2 + v_3$ and (ii) $v_2 - v_1 - v_3$ when
$v_1 = 22$ units at $140°$, $v_2 = 40$ units at $190°$
and $v_3 = 15$ units at $290°$

(i) The vectors are shown in Fig. 33.14.

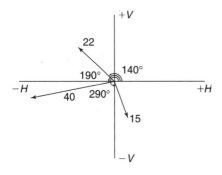

Figure 33.14

The horizontal component of $v_1 - v_2 + v_3$

$$= (22 \cos 140°) - (40 \cos 190°)$$
$$+ (15 \cos 290°)$$
$$= (-16.85) - (-39.39) + (5.13)$$
$$= \textbf{27.67 units}$$

The vertical component of $v_1 - v_2 + v_3$

$$= (22 \sin 140°) - (40 \sin 190°)$$
$$+ (15 \sin 290°)$$
$$= (14.14) - (-6.95) + (-14.10)$$
$$= \textbf{6.99 units}$$

The magnitude of the resultant, R, which can be represented by the mathematical symbol for 'the **modulus** of' as $|v_1 - v_2 + v_3|$ is given by:

$$|R| = \sqrt{27.67^2 + 6.99^2} = \textbf{28.54 units}$$

The direction of the resultant, R, which can be represented by the mathematical symbol for 'the **argument** of' as arg $(v_1 - v_2 + v_3)$ is given by:

$$\textbf{arg } R = \tan^{-1}\left(\frac{6.99}{27.67}\right) = 14.18°$$

Thus $v_1 - v_2 + v_3 = \textbf{28.54 units at 14.18°}$

(ii) The horizontal component of $v_2 - v_1 - v_3$

$$= (40 \cos 190°) - (22 \cos 140°)$$
$$- (15 \cos 290°)$$
$$= (-39.39) - (-16.85) - (5.13)$$
$$= \textbf{-27.67 units}$$

The vertical component of $v_2 - v_1 - v_3$

$$= (40 \sin 190°) - (22 \sin 140°) - (15 \sin 290°)$$
$$= (-6.95) - (14.14) - (-14.10)$$
$$= \textbf{-6.99 units}$$

Let $R = v_2 - v_1 - v_3$ then

$$|R| = \sqrt{(-27.67)^2 + (-6.99)^2} = 28.54 \text{ units}$$

and $\textbf{arg } R = \tan^{-1}\left(\frac{-6.99}{-27.67}\right)$

and must lie in the third quadrant since both H and V are negative quantities.

$$\tan^{-1}\left(\frac{-6.99}{-27.67}\right) = 14.18°,$$

hence the required angle is $180° + 14.18° = 194.18°$

Thus $v_2 - v_1 - v_3 = \textbf{28.54 units at 194.18°}$

This result is as expected, since $v_2 - v_1 - v_3 = -(v_1 - v_2 + v_3)$ and the vector 28.54 units at 194.18° is minus times the vector 28.54 units at 14.18°

Now try the following exercise

Exercise 124 Further problems on vectors subtraction

1. Forces of $F_1 = 40$ N at $45°$ and $F_2 = 30$ N at $125°$ act a point. Determine by drawing and by calculation (a) $F_1 + F_2$ (b) $F_1 - F_2$

$$\begin{bmatrix}(a) \; 54.0 \text{ N at } 78.16° \\ (b) \; 45.64 \text{ N at } 4.66°\end{bmatrix}$$

2. Calculate the resultant of (a) $v_1 + v_2 - v_3$ (b) $v_3 - v_2 + v_1$ when $v_1 = 15$ m/s at $85°$, $v_2 = 25$ m/s at $175°$ and $v_3 = 12$ m/s at $235°$

$$\begin{bmatrix}(a) \; 31.71 \text{ m/s at } 121.81° \\ (b) \; 19.55 \text{ m/s at } 8.63°\end{bmatrix}$$

Section 5

Combination of waveforms

34.1 Combination of two periodic functions

There are a number of instances in engineering and science where waveforms combine and where it is required to determine the single phasor (called the resultant) that could replace two or more separate phasors. (A phasor is a rotating vector). Uses are found in electrical alternating current theory, in mechanical vibrations, in the addition of forces and with sound waves. There are several methods of determining the resultant and two such methods — plotting/measuring, the resolution of phasors by calculation — are explained in this chapter.

34.2 Plotting periodic functions

This may be achieved by sketching the separate functions on the same axes and then adding (or subtracting) ordinates at regular intervals. This is demonstrated in worked problems 1 to 3.

Problem 1. Plot the graph of $y_1 = 3 \sin A$ from $A = 0°$ to $A = 360°$. On the same axes plot $y_2 = 2 \cos A$. By adding ordinates plot $y_R = 3 \sin A + 2 \cos A$ and obtain a sinusoidal expression for this resultant waveform

$y_1 = 3 \sin A$ and $y_2 = 2 \cos A$ are shown plotted in Fig. 34.1. Ordinates may be added at, say, 15° intervals. For example,

at 0°, $\quad y_1 + y_2 = 0 + 2 = 2$

at 15°, $\quad y_1 + y_2 = 0.78 + 1.93 = 2.71$

at 120°, $\quad y_1 + y_2 = 2.60 + -1 = 1.6$

at 210°, $\quad y_1 + y_2 = -1.50 - 1.73$

$\qquad\qquad = -3.23$, and so on

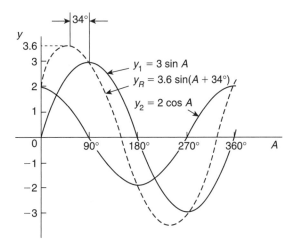

Figure 34.1

The resultant waveform, shown by the broken line, has the same period, i.e. 360° , and thus the same frequency as the single phasors. The maximum value, or amplitude, of the resultant is 3.6. The resultant waveform **leads** $y_1 = 3 \sin A$ by 34° or 0.593 rad. The sinusoidal expression for the resultant waveform is:

$$y_R = 3.6 \sin(A + 34°) \text{ or } y_R = 3.6 \sin(A + 0.593)$$

Problem 2. Plot the graphs of $y_1 = 4 \sin \omega t$ and $y_2 = 3 \sin(\omega t - \pi/3)$ on the same axes, over one cycle. By adding ordinates at intervals plot $y_R = y_1 + y_2$ and obtain a sinusoidal expression for the resultant waveform

$y_1 = 4 \sin \omega t$ and $y_2 = 3 \sin(\omega t - \pi/3)$ are shown plotted in Fig. 34.2. Ordinates are added at 15° intervals and the resultant is shown by the broken line. The amplitude of the resultant is 6.1 and it **lags** y_1 by 25° or 0.436 rad. Hence the sinusoidal expression for the resultant waveform is:

$$y_R = 6.1 \sin(\omega t - 0.436)$$

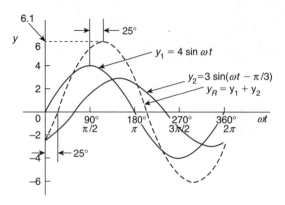

Figure 34.2

Problem 3. Determine a sinusoidal expression for $y_1 - y_2$ when $y_1 = 4 \sin \omega t$ and $y_2 = 3 \sin(\omega t - \pi/3)$

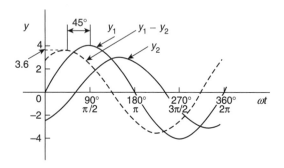

Figure 34.3

y_1 and y_2 are shown plotted in Fig. 34.3. At 15° intervals y_2 is subtracted from y_1. For example:

at 0°, $\quad y_1 - y_2 = 0 - (-2.6) = +2.6$

at 30°, $\quad y_1 - y_2 = 2 - (-1.5) = +3.5$

at 150°, $\quad y_1 - y_2 = 2 - 3 = -1$, and so on.

The amplitude, or peak value of the resultant (shown by the broken line), is 3.6 and it leads y_1 by 45° or 0.79 rad. Hence

$$y_1 - y_2 = 3.6 \sin(\omega t + 0.79)$$

Now try the following exercise

Exercise 125 Further problems on plotting periodic functions

1. Plot the graph of $y = 2 \sin A$ from $A = 0°$ to $A = 360°$. On the same axes plot $y = 4 \cos A$.

By adding ordinates at intervals plot $y = 2 \sin A + 4 \cos A$ and obtain a sinusoidal expression for the waveform.

$$[4.5 \sin(A + 63.5°)]$$

2. Two alternating voltages are given by $v_1 = 10 \sin \omega t$ volts and $v_2 = 14 \sin (\omega t + \pi/3)$ volts. By plotting v_1 and v_2 on the same axes over one cycle obtain a sinusoidal expression for (a) $v_1 + v_2$ (b) $v_1 - v_2$

$$\begin{bmatrix} \text{(a)} & 20.9 \sin(\omega t + 063) \text{ volts} \\ \text{(b)} & 12.5 \sin(\omega t - 1.36) \text{ volts} \end{bmatrix}$$

3. Express $12 \sin \omega t + 5 \cos \omega t$ in the form $A \sin(\omega t \pm \alpha)$ by drawing and measurement.

$$[13 \sin(\omega t + 0.395)]$$

34.3 Determining resultant phasors by calculation

The resultant of two periodic functions may be found from their relative positions when the time is zero. For example, if $y_1 = \sin \omega t$ and $y_2 = 3 \sin(\omega t - \pi/3)$ then each may be represented as phasors as shown in Fig. 34.4, y_1 being 4 units long and drawn horizontally and y_2 being 3 units long, lagging y_1 by $\pi/3$ radians or 60° . To determine the resultant of $y_1 + y_2$, y_1 is drawn horizontally as shown in Fig. 34.5 and y_2 is joined to the end of y_1 at 60° to the horizontal. The resultant is given by y_R. This is the same as the diagonal of a parallelogram that is shown completed in Fig. 34.6.

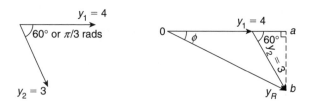

Figure 34.4   **Figure 34.5**

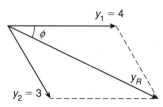

Figure 34.6

Resultant y_R, in Figs. 34.5 and 34.6, is determined either by:

(a) use of the cosine rule (and then sine rule to calculate angle ϕ), or
(b) determining horizontal and vertical components of lengths oa and ob in Fig. 34.5, and then using Pythagoras' theorem to calculate ob.

In the above example, by calculation, $y_R = 6.083$ and angle $\phi = 25.28°$ or 0.441 rad. Thus the resultant may be expressed in sinusoidal form as $y_R = 6.083 \sin(\omega t - 0.441)$. If the resultant phasor, $y_R = y_1 - y_2$ is required, then y_2 is still 3 units long but is drawn in the opposite direction, as shown in Fig. 34.7, and y_R is determined by calculation.

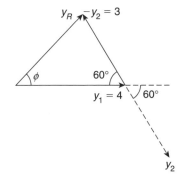

Figure 34.7

Resolution of phasors by calculation is demonstrated in worked problems 4 to 6.

Problem 4. Given $y_1 = 2 \sin \omega t$ and $y_2 = 3 \sin(\omega t + \pi/4)$, obtain an expression for the resultant $y_R = y_1 + y_2$, (a) by drawing, and (b) by calculation

---

(a) When time $t = 0$ the position of phasors y_1 and y_2 are shown in Fig. 34.8(a). To obtain the resultant, y_1 is drawn horizontally, 2 units long, y_2 is drawn 3 units long at an angle of $\pi/4$ rads or 45° and joined to the end of y_1 as shown in Fig. 34.8(b). y_R is measured as 4.6 units long and angle ϕ is measured as 27° or 0.47 rad. Alternativaley, y_R is the diagonal of the parallelogram formed as shown in Fig. 34.8(c).

Hence, by drawing, $y_R = \mathbf{4.6 \sin(\omega t + 0.47)}$

(b) From Fig. 34.8(b), and using the cosine rule:

$$y_R^2 = 2^2 + 3^2 - [2(2)(3)\cos 135°]$$
$$= 4 + 9 - [-8.485] = 21.49$$

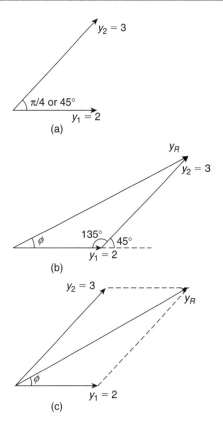

Figure 34.8

Hence $y_R = \sqrt{21.49} = 4.64$

Using the sine rule: $\dfrac{3}{\sin \phi} = \dfrac{4.64}{\sin 135°}$ from

which $\sin \phi = \dfrac{3 \sin 135°}{4.64} = 0.4572$

Hence $\phi = \sin^{-1} 0.4572 = 27.21°$ or 0.475 rad.

By calculation, $y_R = \mathbf{4.64 \sin(\omega t + 0.475)}$

Problem 5. Two alternating voltages are given by $v_1 = 15 \sin \omega t$ volts and $v_2 = 25 \sin(\omega t - \pi/6)$ volts. Determine a sinusoidal expression for the resultant $v_R = v_1 + v_2$ by finding horizontal and vertical components

---

The relative positions of v_1 and v_2 at time $t = 0$ are shown in Fig. 34.9(a) and the phasor diagram is shown in Fig. 34.9(b).

The horizontal component of v_R,

$$H = 15 \cos 0° + 25 \cos(-30°)$$
$$= oa + ab = 36.65\,\text{V}$$

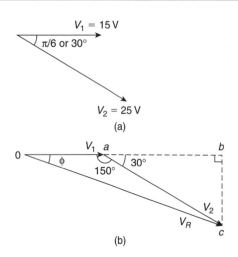

$V_1 = 15$ V

$\pi/6$ or $30°$

$V_2 = 25$ V

(a)

V_1 a b

0

ϕ

$150°$ $30°$

V_2

V_R

c

(b)

Figure 34.9

The vertical component of v_R,

$$V = 15 \sin 0° + 25 \sin (-30°)$$

$$= bc = -12.50 \text{ V}$$

Hence $\qquad v_R(= oc) = \sqrt{36.65^2 + (-12.50)^2}$

by Pythagoras' theorem

$$= \textbf{38.72 volts}$$

$$\tan \phi = \frac{V}{H}\left(= \frac{bc}{ob}\right) = \frac{-12.50}{36.65}$$

$$= -0.3411$$

from which, $\qquad \phi = \tan^{-1}(-0.3411)$

$$= -18.83° \text{ or } -0.329$$

radians.

Hence $\qquad\qquad v_R = v_1 + v_2$

$$= \textbf{38.72} \sin(\omega t - \textbf{0.329}) \textbf{ V}$$

Problem 6. For the voltages in Problem 5, determine the resultant $v_R = v_1 - v_2$

To find the resultant $v_R = v_1 - v_2$, the phasor v_2 of Fig. 34.9(b) is reversed in direction as shown in Fig. 34.10. Using the cosine rule:

$$v_R^2 = 15^2 + 25^2 - 2(15)(25) \cos 30°$$

$$= 225 + 625 - 649.5 = 200.5$$

$$v_R = \sqrt{200.5} = \textbf{14.16 volts}$$

Using the sine rule: $\dfrac{25}{\sin \phi} = \dfrac{14.16}{\sin 30°}$

from which, $\qquad \sin \phi = \dfrac{25 \sin 30°}{14.16} = 0.8828$

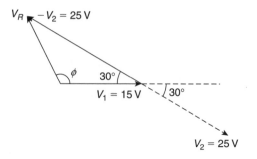

V_R $-V_2 = 25$ V

ϕ $30°$

$V_1 = 15$ V $30°$

$V_2 = 25$ V

Figure 34.10

Hence $\phi = \sin^{-1} 0.8828 = 61.98°$ or $118.02°$. From Fig. 34.10, ϕ is obtuse,

hence $\phi = 118.02°$ or 2.06 radians.

Hence $v_R = v_1 - v_2 = \textbf{14.16} \sin(\omega t + \textbf{2.06}) \textbf{ V}$

Now try the following exercise

Exercise 126   **Further problems on the determination of resultant phasors by calculation**

In Problems 1 to 5, express the combination of periodic functions in the form $A \sin(\omega t \pm \alpha)$ by calculation.

1. $7 \sin \omega t + 5 \sin\left(\omega t + \dfrac{\pi}{4}\right)$

$\qquad\qquad\qquad [11.11 \sin(\omega t + 0.324)]$

2. $6 \sin \omega t + 3 \sin\left(\omega t - \dfrac{\pi}{6}\right)$

$\qquad\qquad\qquad [8.73 \sin(\omega t - 0.173)]$

3. $i = 25 \sin \omega t - 15 \sin\left(\omega t + \dfrac{\pi}{3}\right)$

$\qquad\qquad\qquad [i = 21.79 \sin(\omega t - 0.639)]$

4. $v = 8 \sin \omega t - 5 \sin\left(\omega t - \dfrac{\pi}{4}\right)$

$\qquad\qquad\qquad [v = 5.695 \sin(\omega t + 0.670)]$

5. $x = 9 \sin\left(\omega t + \dfrac{\pi}{3}\right) - 7 \sin\left(\omega t - \dfrac{3\pi}{8}\right)$

$\qquad\qquad\qquad [x = 14.38 \sin(\omega t + 1.444)]$

6. The currents in two parallel branches of an electrical circuit are:

$i_1 = 5 \sin \omega t$ mA and $i_2 = 12 \sin\left(\omega t + \dfrac{\pi}{2}\right)$ mA.

Determine the total current, i_T, given that $i_T = i_1 + i_2$ $\qquad [13 \sin(\omega t + 1.176) \text{ mA}]$

Section 6

Complex Numbers

Chapter 35

Complex numbers

35.1 Cartesian complex numbers

(i) If the quadratic equation $x^2 + 2x + 5 = 0$ is solved using the quadratic formula then:

$$x = \frac{-2 \pm \sqrt{(2)^2 - (4)(1)(5)}}{2(1)}$$

$$= \frac{-2 \pm \sqrt{-16}}{2} = \frac{-2 \pm \sqrt{(16)(-1)}}{2}$$

$$= \frac{-2 \pm \sqrt{16}\sqrt{-1}}{2} = \frac{-2 \pm 4\sqrt{-1}}{2}$$

$$= -1 \pm 2\sqrt{-1}$$

It is not possible to evaluate $\sqrt{-1}$ in real terms. However, if an operator j is defined as $j = \sqrt{-1}$ then the solution may be expressed as $x = -1 \pm j2$.

(ii) $-1 + j2$ and $-1 - j2$ are known as **complex numbers**. Both solutions are of the form $a + jb$, 'a' being termed the **real part** and jb the **imaginary part**. A complex number of the form $a + jb$ is called a **Cartesian complex number**.

(iii) In pure mathematics the symbol i is used to indicate $\sqrt{-1}$ (i being the first letter of the word imaginary). However i is the symbol of electric current in engineering, and to avoid possible confusion the next letter in the alphabet, j, is used to represent $\sqrt{-1}$

Since $x^2 + 4 = 0$ then $x^2 = -4$ and $x = \sqrt{-4}$

i.e., $\quad x = \sqrt{(-1)(4)} = \sqrt{-1}\sqrt{4} = j(\pm 2)$

$$= \pm j2, \text{ (since } j = \sqrt{-1})$$

(Note that $\pm j2$ may also be written as $\pm 2j$).

Problem 2. Solve the quadratic equation:
$$2x^2 + 3x + 5 = 0$$

Using the quadratic formula,

$$x = \frac{-3 \pm \sqrt{(3)^2 - 4(2)(5)}}{2(2)}$$

$$= \frac{-3 \pm \sqrt{-31}}{4} = \frac{-3 \pm \sqrt{-1}\sqrt{31}}{4}$$

$$= \frac{-3 \pm j\sqrt{31}}{4}$$

Hence $\quad x = -\dfrac{3}{4} + j\dfrac{\sqrt{31}}{4}$ or $-0.750 \pm j1.392$,

correct to 3 decimal places.

(Note, a graph of $y = 2x^2 + 3x + 5$ does not cross the x-axis and hence $2x^2 + 3x + 5 = 0$ has no real roots).

Problem 1. Solve the quadratic equation:
$$x^2 + 4 = 0$$

Problem 3. Evaluate

(a) j^3 (b) j^4 (c) j^{23} (d) $\dfrac{-4}{j^9}$

(a) $j^3 = j^2 \times j = (-1) \times j = -j$, since $j^2 = -1$

(b) $j^4 = j^2 \times j^2 = (-1) \times (-1) = 1$

(c) $j^{23} = j \times j^{22} = j \times (j^2)^{11} = j \times (-1)^{11}$

$$= j \times (-1) = -j$$

(d) $j^9 = j \times j^8 = j \times (j^2)^4 = j \times (-1)^4$

$$= j \times 1 = j$$

Hence $\dfrac{-4}{j^9} = \dfrac{-4}{j} = \dfrac{-4}{j} \times \dfrac{-j}{-j} = \dfrac{4j}{-j^2}$

$$= \dfrac{4j}{-(-1)} = 4j \text{ or } j4$$

Now try the following exercise

Exercise 127   Further problems on the introduction to Cartesian complex numbers

In Problems 1 to 3, solve the quadratic equations.

1. $x^2 + 25 = 0$ $[\pm j5]$

2. $2x^2 + 3x + 4 = 0$

$$\left[-\dfrac{3}{4} \pm j\dfrac{\sqrt{23}}{4} \text{ or } -0.750 \pm j1.199 \right]$$

3. $4t^2 - 5t + 7 = 0$

$$\left[\dfrac{5}{8} \pm j\dfrac{\sqrt{87}}{8} \text{ or } 0.625 \pm j1.166 \right]$$

4. Evaluate (a) j^8 (b) $-\dfrac{1}{j^7}$ (c) $\dfrac{4}{2j^{13}}$

$$[\text{(a) } 1 \quad \text{(b) } -j \quad \text{(c) } -j2]$$

35.2   The Argand diagram

A complex number may be represented pictorially on rectangular or Cartesian axes. The horizontal (or x) axis is used to represent the real axis and the vertical (or y) axis is used to represent the imaginary axis. Such a diagram is called an **Argand diagram**. In Fig. 35.1, the point A represents the complex number $(3 + j2)$ and is obtained by plotting the co-ordinates $(3, j2)$ as in graphical work. Figure 35.1 also shows the Argand points B, C and D representing the complex numbers $(-2 + j4)$, $(-3 - j5)$ and $(1 - j3)$ respectively.

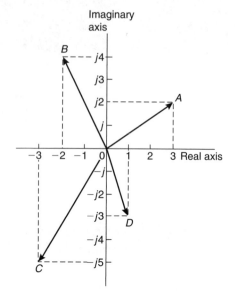

Figure 35.1

35.3   Addition and subtraction of complex numbers

Two complex numbers are added/subtracted by adding/subtracting separately the two real parts and the two imaginary parts.

For example, if $Z_1 = a + jb$ and $Z_2 = c + jd$,

then $\quad Z_1 + Z_2 = (a + jb) + (c + jd)$

$$= (a + c) + j(b + d)$$

and $\quad Z_1 - Z_2 = (a + jb) - (c + jd)$

$$= (a - c) + j(b - d)$$

Thus, for example,

$$(2 + j3) + (3 - j4) = 2 + j3 + 3 - j4$$

$$= 5 - j1$$

and $\quad (2 + j3) - (3 - j4) = 2 + j3 - 3 + j4$

$$= -1 + j7$$

The addition and subtraction of complex numbers may be achieved graphically as shown in the Argand diagram of Fig. 35.2. $(2 + j3)$ is represented by vector OP and $(3 - j4)$ by vector OQ. In Fig. 35.2(a), by vector addition, (i.e. the diagonal of the parallelogram), $OP + OQ = OR$. R is the point $(5, -j1)$.

Hence $(2 + j3) + (3 - j4) = 5 - j1$

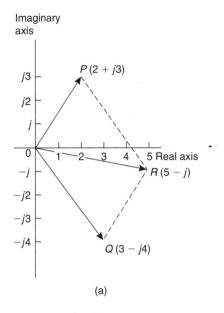

Imaginary axis

P (2 + *j*3)

R (5 − *j*)

Q (3 − *j*4)

(a)

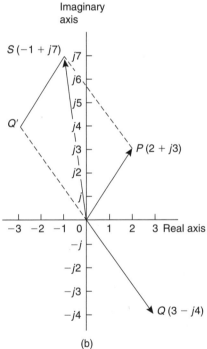

Imaginary axis

S (−1 + *j*7)

Q′

P (2 + *j*3)

Q (3 − *j*4)

(b)

Figure 35.2

In Fig. 35.2(b), vector *OQ* is reversed (shown as *OQ′*) since it is being subtracted. (Note *OQ* = 3 − *j*4 and *OQ′* = −(3 − *j*4) = −3 + *j*4).

OP − *OQ* = *OP* + *OQ′* = *OS* is found to be the Argand point (−1, *j*7).

Hence $(2+j3)-(3-j4)=\mathbf{-1+j7}$

Problem 4. Given $Z_1 = 2+j4$ and $Z_2 = 3-j$ determine (a) $Z_1 + Z_2$, (b) $Z_1 - Z_2$, (c) $Z_2 - Z_1$ and show the results on an Argand diagram

(a) $Z_1 + Z_2 = (2+j4) + (3-j)$
$$= (2+3) + j(4-1) = \mathbf{5+j3}$$

(b) $Z_1 - Z_2 = (2+j4) - (3-j)$
$$= (2-3) + j(4-(-1)) = \mathbf{-1+j5}$$

(c) $Z_2 - Z_1 = (3-j) - (2+j4)$
$$= (3-2) + j(-1-4) = \mathbf{1-j5}$$

Each result is shown in the Argand diagram of Fig. 35.3.

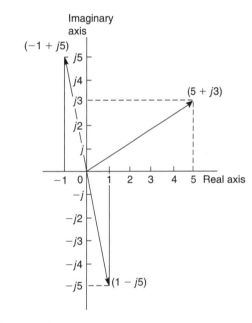

Imaginary axis

(−1 + *j*5)

(5 + *j*3)

(1 − *j*5)

Figure 35.3

35.4 Multiplication and division of complex numbers

(i) **Multiplication of complex numbers** is achieved by assuming all quantities involved are real and then using $j^2 = -1$ to simplify.

Hence $(a+jb)(c+jd)$
$$= ac + a(jd) + (jb)c + (jb)(jd)$$
$$= ac + jad + jbc + j^2bd$$
$$= (ac - bd) + j(ad + bc),$$
$$\text{since } j^2 = -1$$

Thus $(3+j2)(4-j5)$

$$= 12 - j15 + j8 - j^2 10$$
$$= (12-(-10)) + j(-15+8)$$
$$= \mathbf{22 - j7}$$

(ii) The **complex conjugate** of a complex number is obtained by changing the sign of the imaginary part. Hence the complex conjugate of $a+jb$ is $a-jb$. The product of a complex number and its complex conjugate is always a real number.

For example,

$$(3+j4)(3-j4) = 9 - j12 + j12 - j^2 16$$
$$= 9 + 16 = 25$$

$[(a+jb)(a-jb)$ may be evaluated 'on sight' as $a^2+b^2]$

(iii) **Division of complex numbers** is achieved by multiplying both numerator and denominator by the complex conjugate of the denominator.

For example,

$$\frac{2-j5}{3+j4} = \frac{2-j5}{3+j4} \times \frac{(3-j4)}{(3-j4)}$$
$$= \frac{6-j8-j15+j^2 20}{3^2+4^2}$$
$$= \frac{-14-j23}{25} = \frac{-14}{25} - j\frac{23}{25}$$

or $-0.56 - j0.92$

Problem 5. If $Z_1 = 1-j3$, $Z_2 = -2+j5$ and $Z_3 = -3-j4$, determine in $a+jb$ form:

(a) $Z_1 Z_2$ (b) $\dfrac{Z_1}{Z_3}$

(c) $\dfrac{Z_1 Z_2}{Z_1 + Z_2}$ (d) $Z_1 Z_2 Z_3$

(a) $Z_1 Z_2 = (1-j3)(-2+j5)$
$$= -2 + j5 + j6 - j^2 15$$
$$= (-2+15) + j(5+6), \text{ since } j^2 = -1,$$
$$= \mathbf{13 + j11}$$

(b) $\dfrac{Z_1}{Z_3} = \dfrac{1-j3}{-3-j4} = \dfrac{1-j3}{-3-j4} \times \dfrac{-3+j4}{-3+j4}$
$$= \frac{-3+j4+j9-j^2 12}{3^2+4^2}$$
$$= \frac{9+j13}{25} = \frac{9}{25} + j\frac{13}{25}$$
or $\mathbf{0.36 + j0.52}$

(c) $\dfrac{Z_1 Z_2}{Z_1 + Z_2} = \dfrac{(1-j3)(-2+j5)}{(1-j3)+(-2+j5)}$
$$= \frac{13+j11}{-1+j2}, \text{ from part (a),}$$
$$= \frac{13+j11}{-1+j2} \times \frac{-1-j2}{-1-j2}$$
$$= \frac{-13-j26-j11-j^2 22}{1^2+2^2}$$
$$= \frac{9-j37}{5} = \frac{9}{5} - j\frac{37}{5} \text{ or } \mathbf{1.8 - j7.4}$$

(d) $Z_1 Z_2 Z_3 = (13+j11)(-3-j4)$, since
$$Z_1 Z_2 = 13+j11, \text{ from part (a)}$$
$$= -39 - j52 - j33 - j^2 44$$
$$= (-39+44) - j(52+33) = \mathbf{5 - j85}$$

Problem 6. Evaluate:

(a) $\dfrac{2}{(1+j)^4}$ (b) $j\left(\dfrac{1+j3}{1-j2}\right)^2$

(a) $(1+j)^2 = (1+j)(1+j) = 1+j+j+j^2$
$$= 1+j+j-1 = j2$$
$(1+j)^4 = [(1+j)^2]^2 = (j2)^2 = j^2 4 = -4$

Hence $\dfrac{2}{(1+j)^4} = \dfrac{2}{-4} = -\dfrac{1}{2}$

(b) $\dfrac{1+j3}{1-j2} = \dfrac{1+j3}{1-j2} \times \dfrac{1+j2}{1+j2}$
$$= \frac{1+j2+j3+j^2 6}{1^2+2^2} = \frac{-5+j5}{5}$$
$$= -1 + j1 = -1 + j$$

$$\left(\frac{1+j3}{1-j2}\right)^2 = (-1+j)^2 = (-1+j)(-1+j)$$

$$= 1 - j - j + j^2 = -j2$$

Hence $j\left(\dfrac{1+j3}{1-j2}\right)^2 = j(-j2) = -j^2 2 = \mathbf{2},$

$$\text{since } j^2 = -1$$

Now try the following exercise

> **Exercise 128   Further problems on operations involving Cartesian complex numbers**
>
> 1. Evaluate   (a) $(3+j2)+(5-j)$ and (b) $(-2+j6)-(3-j2)$ and show the results on an Argand diagram.
>   [(a) $8+j$ (b) $-5+j8$]
>
> 2. Write down the complex conjugates of (a) $3+j4$, (b) $2-j$ [(a) $3-j4$ (b) $2+j$]
>
> In Problems 3 to 7 evaluate in $a+jb$ form given $Z_1 = 1+j2$, $Z_2 = 4-j3$, $Z_3 = -2+j3$ and $Z_4 = -5-j$.
>
> 3. (a) $Z_1 + Z_2 - Z_3$ (b) $Z_2 - Z_1 + Z_4$
>   [(a) $7-j4$ (b) $-2-j6$]
>
> 4. (a) $Z_1 Z_2$ (b) $Z_3 Z_4$
>   [(a) $10+j5$ (b) $13-j13$]
>
> 5. (a) $Z_1 Z_3 + Z_4$ (b) $Z_1 Z_2 Z_3$
>   [(a) $-13-j2$ (b) $-35+j20$]
>
> 6. (a) $\dfrac{Z_1}{Z_2}$ (b) $\dfrac{Z_1 + Z_3}{Z_2 - Z_4}$
> $\left[\text{(a) } \dfrac{-2}{25}+j\dfrac{11}{25} \quad \text{(b) } \dfrac{-19}{85}+j\dfrac{43}{85}\right]$
>
> 7. (a) $\dfrac{Z_1 Z_3}{Z_1 + Z_3}$ (b) $Z_2 + \dfrac{Z_1}{Z_4} + Z_3$
> $\left[\text{(a) } \dfrac{3}{26}+j\dfrac{41}{26} \quad \text{(b) } \dfrac{45}{26}-j\dfrac{9}{26}\right]$
>
> 8. Evaluate (a) $\dfrac{1-j}{1+j}$ (b) $\dfrac{1}{1+j}$
> $\left[\text{(a) } -j \quad \text{(b) } \dfrac{1}{2}-j\dfrac{1}{2}\right]$
>
> 9. Show that: $\dfrac{-25}{2}\left(\dfrac{1+j2}{3+j4} - \dfrac{2-j5}{-j}\right)$
> $= 57 + j24$

35.5   Complex equations

If two complex numbers are equal, then their real parts are equal and their imaginary parts are equal. Hence **if $a+jb = c+jd$, then $a = c$ and $b = d$**

Problem 7.   Solve the complex equations:

(a) $2(x+jy) = 6-j3$

(b) $(1+j2)(-2-j3) = a+jb$

(a) $2(x+jy) = 6 - j3$ hence $2x + j2y = 6 - j3$

Equating the real parts gives:

$$2x = 6, \text{ i.e. } \mathbf{\mathit{x} = 3}$$

Equating the imaginary parts gives:

$$2y = -3, \text{ i.e. } \mathbf{\mathit{y} = -\frac{3}{2}}$$

(b) $(1+j2)(-2-j3) = a+jb$

$$-2 - j3 - j4 - j^2 6 = a + jb$$

Hence $4 - j7 = a + jb$

Equating real and imaginary terms gives:

$$\mathbf{\mathit{a} = 4} \text{ and } \mathbf{\mathit{b} = -7}$$

Problem 8.   Solve the equations:

(a) $(2-j3) = \sqrt{a+jb}$

(b) $(x - j2y) + (y - j3x) = 2 + j3$

(a) $(2 - j3) = \sqrt{a + jb}$

Hence $(2-j3)^2 = a + jb$

i.e. $(2-j3)(2-j3) = a + jb$

Hence $4 - j6 - j6 + j^2 9 = a + jb$

and $-5 - j12 = a + jb$

Thus $\mathbf{\mathit{a} = -5}$ and $\mathbf{\mathit{b} = -12}$

(b) $(x - j2y) + (y - j3x) = 2 + j3$

Hence $(x + y) + j(-2y - 3x) = 2 + j3$

Equating real and imaginary parts gives:

$$x + y = 2 \qquad (1)$$

and $\quad -3x - 2y = 3 \qquad (2)$

i.e. two stimulaneous equations to solve

Multiplying equation (1) by 2 gives:

$$2x + 2y = 4 \qquad (3)$$

Adding equations (2) and (3) gives:

$$-x = 7, \text{ i.e. } x = -7$$

From equation (1), $y = 9$, which may be checked in equation (2)

Now try the following exercise

Exercise 129 Further problems on complex equations

In Problems 1 to 4 solve the complex equations.

1. $(2 + j)(3 - j2) = a + jb \qquad [a = 8, \quad b = -1]$

2. $\dfrac{2 + j}{1 - j} = j(x + jy) \qquad \left[x = \dfrac{3}{2}, \quad y = -\dfrac{1}{2} \right]$

3. $(2 - j3) = \sqrt{a + jb} \qquad [a = -5, \quad b = -12]$

4. $(x - j2y) - (y - jx) = 2 + j \qquad [x = 3, \quad y = 1]$

5. If $Z = R + j\omega L + 1/j\omega C$, express Z in $(a + jb)$ form when $R = 10$, $L = 5$, $C = 0.04$ and $\omega = 4$
$$[z = 10 + j13.75]$$

35.6 The polar form of a complex number

(i) Let a complex number Z be $x + jy$ as shown in the Argand diagram of Fig. 35.4. Let distance OZ be r and the angle OZ makes with the positive real axis be θ.

From trigonometry, $\quad x = r\cos\theta$ and
$$y = r\sin\theta$$

Hence $\quad Z = x + jy = r\cos\theta + jr\sin\theta$
$$= r(\cos\theta + j\sin\theta)$$

$Z = r(\cos\theta + j\sin\theta)$ is usually abbreviated to $Z = r\angle\theta$ which is known as the **polar form** of a complex number.

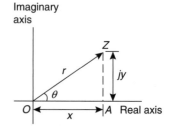

Figure 35.4

(ii) r is called the **modulus** (or magnitude) of Z and is written as mod Z or $|Z|$.
r is determined using Pythagoras' theorem on triangle OAZ in Fig. 35.4,

i.e. $\qquad r = \sqrt{x^2 + y^2}$

(iii) θ is called the **argument** (or amplitude) of Z and is written as arg Z.

By trigonometry on triangle OAZ,

$$\arg Z = \theta = \tan^{-1}\frac{y}{x}$$

(iv) Whenever changing from Cartesian form to polar form, or vice-versa, a sketch is invaluable for determining the quadrant in which the complex number occurs

Problem 9. Determine the modulus and argument of the complex number $Z = 2 + j3$, and express Z in polar form

$Z = 2 + j3$ lies in the first quadrant as shown in Fig. 35.5.
Modulus, $|Z| = r = \sqrt{2^2 + 3^2} = \sqrt{13}$ or **3.606**, correct to 3 decimal places.

Argument, $\quad \arg Z = \theta = \tan^{-1}\dfrac{3}{2}$

$$= 56.31° \text{ or } 56°19'$$

In polar form, $2 + j3$ is written as **3.606** $\angle$ **56.31°** or **3.606** $\angle$ **56°19'**

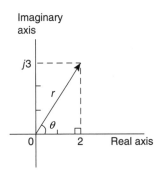

Figure 35.5

Argument $= 180° - 53.13° = 126.87°$ (i.e. the argument must be measured from the positive real axis).

Hence $-3 + j4 = 5\angle\mathbf{126.87°}$

(c) $-3 - j4$ is shown in Fig. 35.6 and lies in the third quadrant.

Modulus, $r = 5$ and $\alpha = 53.13°$, as above.

Hence the argument $= 180° + 53.13° = 233.13°$, which is the same as $-126.87°$

Hence $(-3 - j4) = 5\angle\mathbf{233.13°}$ or $5\angle\mathbf{-126.87°}$

(By convention the **principal value** is normally used, i.e. the numerically least value, such that $-\pi < \theta < \pi$).

(d) $3 - j4$ is shown in Fig. 35.6 and lies in the fourth quadrant.

Modulus, $r = 5$ and angle $\alpha = 53.13°$, as above.

Hence $(3 - j4) = 5\angle\mathbf{-53.13°}$

> **Problem 10.** Express the following complex numbers in polar form:
>
> (a) $3 + j4$ (b) $-3 + j4$
>
> (c) $-3 - j4$ (d) $3 - j4$

(a) $3 + j4$ is shown in Fig. 35.6 and lies in the first quadrant.

Modulus, $r = \sqrt{3^2 + 4^2} = 5$ and argument

$\theta = \tan^{-1}\dfrac{4}{3} = 53.13°$ or $53°8'$

Hence $\mathbf{3 + j4 = 5\angle 53.13°}$

(b) $-3 + j4$ is shown in Fig. 35.6 and lies in the second quadrant.

Modulus, $r = 5$ and angle $\alpha = 53.13°$, from part (a).

> **Problem 11.** Convert (a) $4\angle 30°$ (b) $7\angle -145°$ into $a + jb$ form, correct to 4 significant figures

(a) $4\angle 30°$ is shown in Fig. 35.7(a) and lies in the first quadrant.

Using trigonometric ratios,

$x = 4\cos 30° = 3.464$ and $y = 4\sin 30° = 2.000$

Hence $4\angle 30° = 3.464 + j2.000$

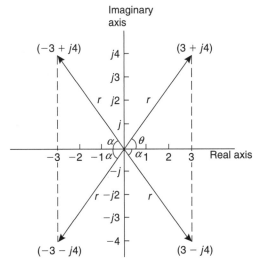

Figure 35.6

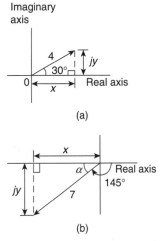

Figure 35.7

Section 6

(b) $7\angle-145°$ is shown in Fig. 35.7(b) and lies in the third quadrant.

Angle $\alpha = 180° - 145° = 35°$

Hence $x = 7\cos 35° = 5.734$

and $y = 7\sin 35° = 4.015$

Hence $7\angle-145° = -5.734 - j4.015$

Alternatively

$$7\angle-145° = 7\cos(-145°) + j7\sin(-145°)$$
$$= -5.734 - j4.015$$

35.7 Multiplication and division in polar form

If $Z_1 = r_1\angle\theta_1$ and $Z_2 = r_2\angle\theta_2$ then:

(i) $Z_1 Z_2 = r_1 r_2\angle(\theta_1 + \theta_2)$ and

(ii) $\dfrac{Z_1}{Z_2} = \dfrac{r_1}{r_2}\angle(\theta_1 - \theta_2)$

Problem 12. Determine, in polar form:

(a) $8\angle25° \times 4\angle60°$

(b) $3\angle16° \times 5\angle-44° \times 2\angle80°$

(a) $8\angle25° \times 4\angle60° = (8\times4)\angle(25°+60°) = \mathbf{32\angle85°}$

(b) $3\angle16° \times 5\angle-44° \times 2\angle80°$
$= (3\times5\times2)\angle[16° + (-44°) + 80°] = \mathbf{30\angle52°}$

Problem 13. Evaluate in polar form:

(a) $\dfrac{16\angle75°}{2\angle15°}$ (b) $\dfrac{10\angle\frac{\pi}{4} \times 12\angle\frac{\pi}{2}}{6\angle-\frac{\pi}{3}}$

(a) $\dfrac{16\angle75°}{2\angle15°} = \dfrac{16}{2}\angle(75° - 15°) = \mathbf{8\angle60°}$

(b) $\dfrac{10\angle\frac{\pi}{4} \times 12\angle\frac{\pi}{2}}{6\angle-\frac{\pi}{3}} = \dfrac{10\times2}{6}\angle\left(\frac{\pi}{4} + \frac{\pi}{2} - \left(-\frac{\pi}{3}\right)\right)$

$$= 20\angle\frac{13\pi}{12} \text{ or } 20\angle-\frac{11\pi}{12} \text{ or}$$
$$\mathbf{20\angle195° \text{ or } 20\angle-165°}$$

Problem 14. Evaluate, in polar form:
$$2\angle30° + 5\angle-45° - 4\angle120°$$

Addition and subtraction in polar form is not possible directly. Each complex number has to be converted into Cartesian form first.

$$2\angle30° = 2(\cos 30° + j\sin 30°)$$
$$= 2\cos 30° + j2\sin 30° = 1.732 + j1.000$$
$$5\angle-45° = 5(\cos(-45°) + j\sin(-45°))$$
$$= 5\cos(-45°) + j5\sin(-45°)$$
$$= 3.536 - j3.536$$
$$4\angle120° = 4(\cos 120° + j\sin 120°)$$
$$= 4\cos 120° + j4\sin 120°$$
$$= -2.000 + j3.464$$

Hence $2\angle30° + 5\angle-45° - 4\angle120°$

$$= (1.732 + j1.000) + (3.536 - j3.536)$$
$$- (-2.000 + j3.464)$$
$$= 7.268 - j6.000, \text{ which lies in the}$$
$$\text{fourth quadrant}$$
$$= \sqrt{7.268^2 + 6.000^2}\angle\tan^{-1}\left(\frac{-6.000}{7.268}\right)$$
$$= \mathbf{9.425\angle-39.54° \text{ or } 9.425\angle-39°32'}$$

Now try the following exercise

Exercise 130 Further problems on polar form

1. Determine the modulus and argument of
    (a) $2 + j4$ (b) $-5 - j2$ (c) $j(2-j)$.

$$\begin{bmatrix} \text{(a) } 4.472, 63.43° & \text{(b) } 5.385, -158.20° \\ \text{(c) } 2.236, 63.43° \end{bmatrix}$$

In Problems 2 and 3 express the given Cartesian complex numbers in polar form, leaving answers in surd form.

2. (a) $2+j3$ (b) -4 (c) $-6+j$

$$\begin{bmatrix} \text{(a) } \sqrt{13}\angle 56.31° & \text{(b) } 4\angle 180° \\ \text{(c) } \sqrt{37}\angle 170.54° \end{bmatrix}$$

3. (a) $-j3$ (b) $(-2+j)^3$ (c) $j^3(1-j)$

$$\begin{bmatrix} \text{(a) } 3\angle -90° & \text{(b) } \sqrt{125}\angle 100.30° \\ \text{(c) } \sqrt{2}\angle -135° \end{bmatrix}$$

In Problems 4 and 5 convert the given polar complex numbers into $(a+jb)$ form giving answers correct to 4 significant figures.

4. (a) $5\angle 30°$ (b) $3\angle 60°$ (c) $7\angle 45°$

$$\begin{bmatrix} \text{(a) } 4.330+j2.500 \\ \text{(b) } 1.500+j2.598 \\ \text{(c) } 4.950+j4.950 \end{bmatrix}$$

5. (a) $6\angle 125°$ (b) $4\angle \pi$ (c) $3.5\angle -120°$

$$\begin{bmatrix} \text{(a) } -3.441+j4.915 \\ \text{(b) } -4.000+j0 \\ \text{(c) } -1.750-j3.031 \end{bmatrix}$$

In Problems 6 to 8, evaluate in polar form.

6. (a) $3\angle 20° \times 15\angle 45°$

(b) $2.4\angle 65° \times 4.4\angle -21°$

$$[\text{(a) } 45\angle 65° \quad \text{(b) } 10.56\angle 44°]$$

7. (a) $6.4\angle 27° \div 2\angle -15°$

(b) $5\angle 30° \times 4\angle 80° \div 10\angle -40°$

$$[\text{(a) } 3.2\angle 4.2° \quad \text{(b) } 2\angle 150°]$$

8. (a) $4\angle\dfrac{\pi}{6} + 3\angle\dfrac{\pi}{8}$

(b) $2\angle 120° + 5.2\angle 58° - 1.6\angle -40°$

$$[\text{(a) } 6.986\angle 26.79° \quad \text{(b) } 7.190\angle 85.77°]$$

35.8 Applications of complex numbers

There are several applications of complex numbers in science and engineering, in particular in electrical alternating current theory and in mechanical vector analysis.

The effect of multiplying a phasor by j is to rotate it in a positive direction (i.e. anticlockwise) on an Argand diagram through $90°$ without altering its length. Similarly, multiplying a phasor by $-j$ rotates the phasor through $-90°$. These facts are used in a.c. theory since certain quantities in the phasor diagrams lie at $90°$ to each other. For example, in the R–L series circuit shown in Fig. 35.8(a), V_L leads I by $90°$ (i.e. I lags V_L by $90°$) and may be written as jV_L, the vertical axis being regarded as the imaginary axis of an Argand diagram. Thus $V_R + jV_L = V$ and since $V_R = IR$, $V = IX_L$ (where X_L is the inductive reactance, $2\pi fL$ ohms) and $V = IZ$ (where Z is the impedance) then $R + jX_L = Z$.

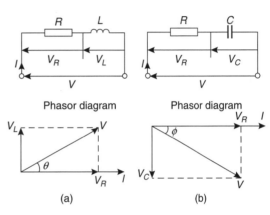

Figure 35.8

Similarly, for the R–C circuit shown in Figure 35.8(b), V_C lags I by $90°$ (i.e. I leads V_C by $90°$) and $V_R - jV_C = V$, from which $R - jX_C = Z$ (where X_C is the capacitive reactance $\dfrac{1}{2\pi f C}$ ohms).

> **Problem 15.** Determine the resistance and series inductance (or capacitance) for each of the following impedances, assuming a frequency of 50 Hz:
>
> (a) $(4.0 + j7.0)\ \Omega$ (b) $-j20\ \Omega$
> (c) $15\angle -60°\ \Omega$

(a) Impedance, $Z = (4.0 + j7.0)\ \Omega$ hence,

resistance $= 4.0\ \Omega$ and reactance $= 7.0\ \Omega$.

Since the imaginary part is positive, the reactance is inductive,

i.e. $X_L = 7.0\ \Omega$

Since $X_L = 2\pi f L$ then **inductance**,

$$L = \frac{X_L}{2\pi f} = \frac{7.0}{2\pi(50)} = 0.0223 \text{ H or } 22.3 \text{ mH}$$

(b) Impedance, $Z = -j20$, i.e. $Z = (0 - j20)\ \Omega$ hence **resistance** $= 0$ and reactance $= 20\ \Omega$. Since the imaginary part is negative, the reactance is capacitive, i.e. $X_C = 20\ \Omega$ and since $X_C = \frac{1}{2\pi f C}$ then:

$$\text{capacitance, } C = \frac{1}{2\pi f X_C} = \frac{1}{2\pi(50)(20)}\ \text{F}$$

$$= \frac{10^6}{2\pi(50)(20)}\mu\text{F} = 159.2\ \mu\text{F}$$

(c) Impedance, Z

$$= 15\angle -60° = 15[\cos(-60°) + j\sin(-60°)]$$

$$= 7.50 - j12.99\ \Omega.$$

Hence **resistance** $= 7.50\ \Omega$ and capacitive reactance, $X_C = 12.99\ \Omega$

Since $X_C = \frac{1}{2\pi f C}$ then **capacitance**,

$$C = \frac{1}{2\pi f X_C} = \frac{10^6}{2\pi(50)(12.99)}\ \mu\text{F}$$

$$= 245\ \mu\text{F}$$

Problem 16. An alternating voltage of 240 V, 50 Hz is connected across an impedance of $(60 - j100)\ \Omega$. Determine (a) the resistance (b) the capacitance (c) the magnitude of the impedance and its phase angle and (d) the current flowing

---

(a) Impedance $Z = (60 - j100)\ \Omega$.

Hence **resistance** $= 60\Omega$

(b) Capacitive reactance $X_C = 100\ \Omega$ and since $X_C = \frac{1}{2\pi f C}$ then

$$\text{capacitance, } C = \frac{1}{2\pi f X_C} = \frac{1}{2\pi(50)(100)}$$

$$= \frac{10^6}{2\pi(50)(100)}\ \mu\text{F}$$

$$= 31.83\ \mu\text{F}$$

(c) Magnitude of impedance,

$$|Z| = \sqrt{60^2 + (-100)^2} = 116.6\ \Omega$$

Phase angle, $\arg Z = \tan^{-1}\left(\frac{-100}{60}\right) = -59.04°$

(d) Current flowing, $I = \dfrac{V}{Z} = \dfrac{240° \angle 0°}{116.6 \angle -59.04°}$

$$= 2.058 \angle 59.04°\text{A}$$

The circuit and phasor diagrams are as shown in Fig. 35.8(b).

Problem 17. For the parallel circuit shown in Fig. 35.9, determine the value of current I, and its phase relative to the 240 V supply, using complex numbers

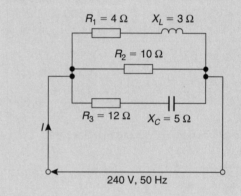

Figure 35.9

Current $I = \dfrac{V}{Z}$. Impedance Z for the three-branch parallel circuit is given by:

$$\frac{1}{Z} = \frac{1}{Z_1} + \frac{1}{Z_2} + \frac{1}{Z_3}$$

where $Z_1 = 4 + j3$, $Z_2 = 10$ and $Z_3 = 12 - j5$

Admittance, $Y_1 = \dfrac{1}{Z_1} = \dfrac{1}{4 + j3}$

$$= \frac{1}{4 + j3} \times \frac{4 - j3}{4 - j3} = \frac{4 - j3}{4^2 + 3^2}$$

$$= 0.160 - j0.120 \text{ siemens}$$

Admittance, $Y_2 = \dfrac{1}{Z_2} = \dfrac{1}{10} = 0.10 \text{ siemens}$

Admittance, $Y_3 = \dfrac{1}{Z_3} = \dfrac{1}{12 - j5}$

$$= \dfrac{1}{12 - j5} \times \dfrac{12 + j5}{12 + j5} = \dfrac{12 + j5}{12^2 + 5^2}$$

$$= 0.0710 + j0.0296 \text{ siemens}$$

Total admittance, $Y = Y_1 + Y_2 + Y_3$

$$= (0.160 - j0.120) + (0.10)$$

$$+ (0.0710 + j0.0296)$$

$$= 0.331 - j0.0904$$

$$= 0.343\angle{-15.28°} \text{ siemens}$$

Current $I = \dfrac{V}{Z} = VY$

$$= (240\angle0°)(0.343\angle{-15.28°})$$

$$= \mathbf{82.32\angle{-15.28°}\ A}$$

Problem 18. Determine the magnitude and direction of the resultant of the three coplanar forces given below, when they act at a point:

Force A, 10 N acting at 45° from the positive horizontal axis,

Force B, 8 N acting at 120° from the positive horizontal axis,

Force C, 15 N acting at 210° from the positive horizontal axis.

The space diagram is shown in Fig. 35.10. The forces may be written as complex numbers.

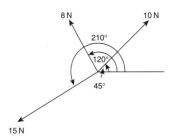

Figure 35.10

Thus force A, $f_A = 10\angle45°$, force B, $f_B = 8\angle120°$ and force C, $f_C = 15\angle210°$.

The resultant force

$$= f_A + f_B + f_C$$

$$= 10\angle45° + 8\angle120° + 15\angle210°$$

$$= 10(\cos 45° + j \sin 45°)$$

$$+ 8(\cos 120° + j \sin 120°)$$

$$+ 15(\cos 210° + j \sin 210°)$$

$$= (7.071 + j7.071) + (-4.00 + j6.928)$$

$$+ (-12.99 - j7.50)$$

$$= -9.919 + j6.499$$

Magnitude of resultant force

$$= \sqrt{(-9.919)^2 + 6.499^2} = \mathbf{11.86\ N}$$

Direction of resultant force

$$= \tan^{-1}\left(\dfrac{6.499}{-9.919}\right) = \mathbf{146.77°}$$

(since $-9.919 + j6.499$ lies in the second quadrant).

Now try the following exercise

Exercise 131   Further problems on applications of complex numbers

1. Determine the resistance R and series inductance L (or capacitance C) for each of the following impedances assuming the frequency to be 50 Hz.

   (a) $(3 + j8)\,\Omega$ (b) $(2 - j3)\,\Omega$
   (c) $j14\,\Omega$ (d) $8\angle{-60°}\,\Omega$

$$\begin{bmatrix} \text{(a) } R = 3\,\Omega, L = 25.5\,\text{mH} \\ \text{(b) } R = 2\,\Omega, C = 1061\,\mu\text{F} \\ \text{(c) } R = 0, L = 44.56\,\text{mH} \\ \text{(d) } R = 4\,\Omega, C = 459.5\,\mu\text{F} \end{bmatrix}$$

2. Two impedances, $Z_1 = (3 + j6)\,\Omega$ and $Z_2 = (4 - j3)\,\Omega$ are connected in series to a supply voltage of 120 V. Determine the magnitude of the current and its phase angle relative to the voltage.
   [15.76 A, 23.20° lagging]

3. If the two impedances in Problem 2 are connected in parallel determine the current flowing and its phase relative to the 120 V supply voltage.
   [27.25 A, 3.37° lagging]

4. A series circuit consists of a 12 Ω resistor, a coil of inductance 0.10 H and a capacitance of

160 μF. Calculate the current flowing and its phase relative to the supply voltage of 240 V, 50 Hz. Determine also the power factor of the circuit.

[14.42 A, 43.85° lagging, 0.721]

5. For the circuit shown in Fig. 35.11, determine the current I flowing and its phase relative to the applied voltage.

[14.58 A, 2.51° leading]

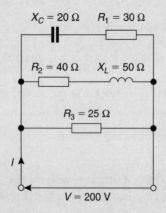

$X_C = 20\,\Omega$ $R_1 = 30\,\Omega$

$R_2 = 40\,\Omega$ $X_L = 50\,\Omega$

$R_3 = 25\,\Omega$

I

$V = 200$ V

Figure 35.11

6. Determine, using complex numbers, the magnitude and direction of the resultant of the coplanar forces given below, which are acting at a point. Force A, 5 N acting horizontally, Force B, 9 N acting at an angle of 135° to force A, Force C, 12 N acting at an angle of 240° to force A.   [8.394 N, 208.68° from force A]

7. A delta-connected impedance Z_A is given by:

$$Z_A = \frac{Z_1 Z_2 + Z_2 Z_3 + Z_3 Z_1}{Z_2}$$

Determine Z_A in both Cartesian and polar form given $Z_1 = (10 + j0)\,\Omega$, $Z_2 = (0 - j10)\,\Omega$ and $Z_3 = (10 + j10)\,\Omega$.

[$(10 + j20)\,\Omega$, $22.36\angle 63.43°\,\Omega$]

8. In the hydrogen atom, the angular momentum, p, of the de Broglie wave is given by:

$$p\psi = -\left(\frac{jh}{2\pi}\right)(\pm jm\psi).$$

Determine an expression for p.

$$\left[\pm \frac{mh}{2\pi}\right]$$

9. An aircraft P flying at a constant height has a velocity of $(400 + j300)$ km/h. Another aircraft Q at the same height has a velocity of $(200 - j600)$ km/h. Determine (a) the velocity of P relative to Q, and (b) the velocity of Q relative to P. Express the answers in polar form, correct to the nearest km/h.

$$\left[\begin{array}{l}\text{(a)} \ 922\,\text{km/h at } 77.47° \\ \text{(b)} \ 922\,\text{km/h at } -102.53°\end{array}\right]$$

10. Three vectors are represented by P, $2\angle 30°$, Q, $3\angle 90°$ and R, $4\angle -60°$. Determine in polar form the vectors represented by (a) $P + Q + R$, (b) $P - Q - R$.

[(a) $3.770\angle 8.17°$ (b) $1.488\angle 100.37°$]

11. In a Schering bridge circuit, $Z_x = (R_X - jX_{C_X})$, $Z_2 = -jX_{C_2}$,

$$Z_3 = \frac{(R_3)(-jX_{C_3})}{(R_3 - jX_{C_3})} \quad \text{and} \quad Z_4 = R_4 \quad \text{where}$$

$$X_C = \frac{1}{2\pi f C}$$

At balance: $(Z_X)(Z_3) = (Z_2)(Z_4)$.

Show that at balance $R_X = \dfrac{C_3 R_4}{C_2}$ and

$$C_X = \frac{C_2 R_3}{R_4}$$

Chapter 36

De Moivre's theorem

36.1 Introduction

From multiplication of complex numbers in polar form,

$$(r\angle\theta) \times (r\angle\theta) = r^2\angle 2\theta$$

Similarly, $(r\angle\theta) \times (r\angle\theta) \times (r\angle\theta) = r^3\angle 3\theta$, and so on.

In general, **de Moivre's theorem** states:

$$[r\angle\theta]^n = r^n\angle n\theta$$

The theorem is true for all positive, negative and fractional values of n. The theorem is used to determine powers and roots of complex numbers.

36.2 Powers of complex numbers

For example, $[3\angle 20°]^4 = 3^4\angle(4 \times 20°) = 81\angle 80°$ by de Moivre's theorem.

Problem 1. Determine, in polar form:

(a) $[2\angle 35°]^5$ (b) $(-2+j3)^6$

(a) $\qquad [2\angle 35°]^5 = 2^5\angle(5 \times 35°),$

$\qquad\qquad$ from De Moivre's theorem

$\qquad\qquad = \mathbf{32\angle 175°}$

(b) $\qquad (-2+j3) = \sqrt{(-2)^2 + (3)^2}\angle\tan^{-1}\left(\dfrac{3}{-2}\right)$

$\qquad\qquad = \sqrt{13}\angle 123.69°$, since $-2+j3$

$\qquad\qquad\qquad$ lies in the second quadrant

$\qquad (-2+j3)^6 = [\sqrt{13}\angle 123.69°]^6$

$\qquad\qquad = \sqrt{13^6}\angle(6 \times 123.69°),$

$\qquad\qquad\qquad$ by De Moivre's theorem

$$= 2197\angle 742.14°$$

$$= 2197\angle 382.14°$$

$$(\text{since } 742.14 \equiv 742.14° - 360° = 382.14°)$$

$$= \mathbf{2197\angle 22.14°}$$

$$(\text{since } 382.14° \equiv 382.14° - 360° = 22.14°)$$

Problem 2. Determine the value of $(-7+j5)^4$, expressing the result in polar and rectangular forms

$$(-7+j5) = \sqrt{(-7)^2 + 5^2}\angle\tan^{-1}\left(\frac{5}{-7}\right)$$

$$= \sqrt{74}\angle 144.46°$$

(Note, by considering the Argand diagram, $-7+j5$ must represent an angle in the second quadrant and **not** in the fourth quadrant).

Applying de Moivre's theorem:

$$(-7+j5)^4 = [\sqrt{74}\angle 144.46°]^4$$

$$= \sqrt{74^4}\angle 4 \times 144.46°$$

$$= 5476\angle 577.84°$$

$$= \mathbf{5476\angle 217.84°} \text{ or }$$

$$\mathbf{5476\angle 217°15'} \text{ in polar form.}$$

Since $r\angle\theta = r\cos\theta + jr\sin\theta$,

$$5476\angle 217.84° = 5476\cos 217.84°$$

$$+ j5476\sin 217.84°$$

$$= -4325 - j3359$$

i.e. $\quad \mathbf{(-7+j5)^4 = -4325 - j3359}$

$\qquad\qquad\qquad\qquad\qquad$ in rectangular form.

Now try the following exercise

Exercise 132 **Further problems on powers of complex numbers**

1. Determine in polar form (a) $[1.5\angle15°]^5$ (b) $(1+j2)^6$

   [(a) $7.594\angle75°$ (b) $125\angle20.61°$]

2. Determine in polar and Cartesian forms (a) $[3\angle41°]^4$ (b) $(-2-j)^5$

   $$\begin{bmatrix}\text{(a) } 81\angle164°, -77.86+j22.33\\ \text{(b) } 55.90\angle-47.18°, 38-j41\end{bmatrix}$$

3. Convert $(3-j)$ into polar form and hence evaluate $(3-j)^7$, giving the answer in polar form.

   $[\sqrt{10}\angle-18.43°, \quad 3162\angle-129°]$

In Problems 4 to 7, express in both polar and rectangular forms:

4. $(6+j5)^3$

   $[476.4\angle119.42°, -234+j415]$

5. $(3-j8)^5$

   $[45\,530\angle12.78°, \; 44\,400+j10\,070]$

6. $(-2+j7)^4$

   $[2809\angle63.78°, \; 1241+j2520]$

7. $(-16-j9)^6$

   $$\begin{bmatrix}(38.27\times10^6)\angle176.15°,\\ 10^6(-38.18+j2.570)\end{bmatrix}$$

36.3 Roots of complex numbers

The **square root** of a complex number is determined by letting $n=\frac{1}{2}$ in De Moivre's theorem,

i.e. $\quad \sqrt{r\angle\theta}=[r\angle\theta]^{1/2}=r^{1/2}\angle\frac{1}{2}\theta=\sqrt{r}\angle\frac{\theta}{2}$

There are two square roots of a real number, equal in size but opposite in sign.

Problem 3. Determine the two square roots of the complex number $(5+j12)$ in polar and Cartesian forms and show the roots on an Argand diagram

$$(5+112)=\sqrt{5^2+12^2}\angle\tan^{-1}\left(\frac{12}{5}\right)=13\angle67.38°$$

When determining square roots two solutions result. To obtain the second solution one way is to express $13\angle67.38°$ also as $13\angle(67.38°+360°)$, i.e. $13\angle427.38°$. When the angle is divided by 2 an angle less than 360° is obtained.

Hence

$$\sqrt{5^2+12^2}=\sqrt{13}\angle67.38° \text{ and } \sqrt{13}\angle427.38°$$

$$=[13\angle67.38°]^{1/2} \text{ and } [13\angle427.38°]^{1/2}$$

$$=13^{1/2}\angle\left(\tfrac{1}{2}\times67.38°\right) \text{ and}$$

$$13^{1/2}\angle\left(\tfrac{1}{2}\times427.38°\right)$$

$$=\sqrt{13}\angle33.69° \text{ and } \sqrt{13}\angle213.69°$$

$$=3.61\angle33.69° \text{ and } 3.61\angle213.69°$$

Thus, in polar form, the two roots are: $3.61\angle33.69°$ and $3.61\angle-146.31°$

$$\sqrt{13}\angle33.69°=\sqrt{13}(\cos33.69°+j\sin33.69°)$$

$$=3.0+j2.0$$

$$\sqrt{13}\angle213.69°=\sqrt{13}(\cos213.69°+j\sin213.69°)$$

$$=-3.0-j2.0$$

Thus, in Cartesian form the two roots are: $\pm(3.0+j2.0)$

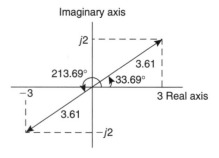

Figure 36.1

From the Argand diagram shown in Fig. 36.1 the two roots are seen to be 180° apart, which is always true when finding square roots of complex numbers.

In general, **when finding the nth root of complex number, there are n solutions.** For example, there are three solutions to a cube root, five solutions to a fifth root, and so on. In the solutions to the roots of a complex number, the modulus, r, is always the same, but the arguments, θ, are different. It is shown in Problem 3 that arguments are symmetrically spaced on an Argand diagram and are $\dfrac{360°}{n}$ apart, where n is the number of the roots required. Thus if one of the solutions to the cube roots of a complex number is, say, $5\angle 20°$, the other two roots are symmetrically spaced $\dfrac{360°}{3}$, i.e. $120°$ from this root, and the three roots are $5\angle 20°$, $5\angle 140°$ and $5\angle 260°$.

Problem 4. Find the roots of $(5+j3)]^{1/2}$ in rectangular form, correct to 4 significant figures

$$(5+j3) = \sqrt{34}\angle 30.96°$$

Applying de Moivre's theorem:

$$(5+j3)^{1/2} = \sqrt{34}^{1/2}\angle\tfrac{1}{2}\times 30.96°$$

$$= 2.415\angle 15.48° \quad \text{or} \quad 2.415\angle 15°29'$$

The second root may be obtained as shown above, i.e. having the same modulus but displaced $\dfrac{360°}{2}$ from the first root.

Thus, $\quad (5+j3)^{1/2} = 2.415\angle(15.48° + 180°)$

$$= 2.415\angle 195.48°$$

In rectangular form:

$$2.415\angle 15.48° = 2.415\cos 15.48°$$
$$+ j2.415\sin 15.48°$$
$$= 2.327 + j0.6446$$

and $\quad 2.415\angle 195.48° = 2.415\cos 195.48°$
$$+ j2.415\sin 195.48°$$
$$= -2.327 - j0.6446$$

Hence $\quad (5+j3)]^{1/2} = \mathbf{2.415\angle 15.48°}$ **and**

$$\mathbf{2.415\angle 195.48°} \text{ or}$$

$$\mathbf{\pm(2.327 + j0.6446)}$$

Problem 5. Express the roots of $(-14+j3)^{-2/5}$ in polar form

$$(-14+j3) = \sqrt{205}\angle 167.905°$$

$$(-14+j3)^{-2/5} = \sqrt{205}^{-2/5}\angle\left[\left(-\tfrac{2}{5}\right)\times 167.905°\right]$$

$$= 0.3449\angle -67.164° \text{ or}$$

$$0.3449\angle -67°10'$$

There are five roots to this complex number,

$$\left(x^{-2/5} = \frac{1}{x^{2/5}} = \frac{1}{\sqrt[5]{x^2}}\right)$$

The roots are symmetrically displaced from one another $\dfrac{360°}{5}$, i.e. $72°$ apart round an Argand diagram.

Thus the required roots are $\mathbf{0.3449\angle -67°10'}$, $\mathbf{0.3449\angle 4°50'}$, $\mathbf{0.3449\angle 76°50'}$, $\mathbf{0.3449\angle 148°50'}$ and $\mathbf{0.3449\angle 220°50'}$.

Now try the following exercise

Exercise 133   Further problems on the roots of complex numbers

In Problems 1 to 3 determine the two square roots of the given complex numbers in Cartesian form and show the results on an Argand diagram.

1. (a) $1+j$ (b) j

$$\begin{bmatrix}\text{(a) } \pm(1.099 + j0.455)\\ \text{(b) } \pm(0.707 + j0.707)\end{bmatrix}$$

2. (a) $3-j4$ (b) $-1-j2$

$$\begin{bmatrix}\text{(a) } \pm(2-j)\\ \text{(b) } \pm(0.786 - j1.272)\end{bmatrix}$$

3. (a) $7\angle 60°$ (b) $12\angle\dfrac{3\pi}{2}$

$$\begin{bmatrix}\text{(a) } \pm(2.291 + j1.323)\\ \text{(b) } \pm(-2.449 + j2.449)\end{bmatrix}$$

Section 6

In Problems 4 to 7, determine the moduli and arguments of the complex roots.

4. $(3+j4)^{1/3}$

$$\begin{bmatrix} \text{Moduli } 1.710, \text{ arguments } 17.71°, \\ 137.71° \text{ and } 257.71° \end{bmatrix}$$

5. $(-2+j)^{1/4}$

$$\begin{bmatrix} \text{Moduli } 1.223, \text{ arguments } 38.36°, \\ 128.36°, 218.36° \text{ and } 308.36° \end{bmatrix}$$

6. $(-6-j5)^{1/2}$

$$\begin{bmatrix} \text{Moduli } 2.795, \text{ arguments} \\ 109.90°, 289.90° \end{bmatrix}$$

7. $(4-j3)^{-2/3}$

$$\begin{bmatrix} \text{Moduli } 0.3420, \text{ arguments } 24.58°, \\ 144.58° \text{ and } 264.58° \end{bmatrix}$$

8. For a transmission line, the characteristic impedance Z_0 and the propagation coefficient γ are given by:

$$Z_0 = \sqrt{\frac{R+j\omega L}{G+j\omega C}} \quad \text{and}$$

$$\gamma = \sqrt{(R+j\omega L)(G+j\omega C)}$$

Given $R = 25\ \Omega$, $L = 5 \times 10^{-3}$ H, $G = 80 \times 10^{-6}$ S, $C = 0.04 \times 10^{-6}$ F and $\omega = 2000\pi$ rad/s, determine, in polar form, Z_0 and γ.

$$\begin{bmatrix} Z_0 = 390.2\angle -10.43°\ \Omega, \\ \gamma = 0.1029\angle 61.92° \end{bmatrix}$$

Section 6

Revision Test 9

This Revision test covers the material contained in Chapters 33 to 36. *The marks for each question are shown in brackets at the end of each question.*

1. Four coplanar forces act at a point A as shown in Fig. R9.1. Determine the value and direction of the resultant force by (a) drawing (b) by calculation. (11)

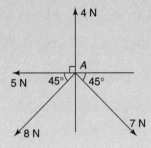

Figure R9.1

2. The instantaneous values of two alternating voltages are given by:

$$v_1 = 150 \sin\left(\omega t + \frac{\pi}{3}\right) \text{ volts and}$$

$$v_2 = 90 \sin\left(\omega t - \frac{\pi}{6}\right) \text{ volts}$$

Plot the two voltages on the same axes to scales of $1 \text{ cm} = 50$ volts and $1 \text{ cm} = \dfrac{\pi}{6}$ rad. Obtain a sinusoidal expression for the resultant $v_1 + v_2$ in the form $R \sin(\omega t + \alpha)$: (a) by adding ordinates at intervals and (b) by calculation. (13)

3. Solve the quadratic equation $x^2 - 2x + 5 = 0$ and show the roots on an Argand diagram. (8)

4. If $Z_1 = 2 + j5$, $Z_2 = 1 - j3$ and $Z_3 = 4 - j$ determine, in both Cartesian and polar forms, the value of: $\dfrac{Z_1 Z_2}{Z_1 + Z_2} + Z_3$, correct to 2 decimal places. (8)

5. Determine in both polar and rectangular forms:
   (a) $[3.2 - j4.8]^5$ (b) $\sqrt{-1 - j3}$ (10)

Section 7

Statistics

Chapter 37

Presentation of statistical data

37.1 Some statistical terminology

Data are obtained largely by two methods:

(a) by counting — for example, the number of stamps sold by a post office in equal periods of time, and

(b) by measurement — for example, the heights of a group of people.

When data are obtained by counting and only whole numbers are possible, the data are called **discrete**. Measured data can have any value within certain limits and are called **continuous** (see Problem 1).

A **set** is a group of data and an individual value within the set is called a **member** of the set. Thus, if the masses of five people are measured correct to the nearest 0.1 kilogram and are found to be 53.1 kg, 59.4 kg, 62.1 kg, 77.8 kg and 64.4 kg, then the set of masses in kilograms for these five people is:

$$\{53.1, 59.4, 62.1, 77.8, 64.4\}$$

and one of the members of the set is 59.4

A set containing all the members is called a **population**. Some member selected at random from a population are called a **sample**. Thus all car registration numbers form a population, but the registration numbers of, say, 20 cars taken at random throughout the country are a sample drawn from that population.

The number of times that the value of a member occurs in a set is called the **frequency** of that member. Thus in the set: $\{2, 3, 4, 5, 4, 2, 4, 7, 9\}$, member 4 has a frequency of three, member 2 has a frequency of 2 and the other members have a frequency of one.

The **relative frequency** with which any member of a set occurs is given by the ratio:

$$\frac{\text{frequency of member}}{\text{total frequency of all members}}$$

For the set: $\{2, 3, 5, 4, 7, 5, 6, 2, 8\}$, the relative frequency of member 5 is $\frac{2}{9}$.

Often, relative frequency is expressed as a percentage and the **percentage relative frequency** is:

(relative frequency $\times$ 100)%

Problem 1. Data are obtained on the topics given below. State whether they are discrete or continuous data.

(a) The number of days on which rain falls in a month for each month of the year.
(b) The mileage travelled by each of a number of salesmen.
(c) The time that each of a batch of similar batteries lasts.
(d) The amount of money spent by each of several families on food.

(a) The number of days on which rain falls in a given month must be an integer value and is obtained by **counting** the number of days. Hence, these data are **discrete**.

(b) A salesman can travel any number of miles (and parts of a mile) between certain limits and these data are **measured**. Hence the data are **continuous**.

(c) The time that a battery lasts is **measured** and can have any value between certain limits. Hence these data are **continuous**.

(d) The amount of money spent on food can only be expressed correct to the nearest pence, the amount being **counted**. Hence, these data are **discrete**.

Now try the following exercise

Exercise 134   Further problems on discrete and continuous data

In Problems 1 and 2, state whether data relating to the topics given are discrete or continuous.

1. (a) The amount of petrol produced daily, for each of 31 days, by a refinery.
   (b) The amount of coal produced daily by each of 15 miners.
   (c) The number of bottles of milk delivered daily by each of 20 milkmen.
   (d) The size of 10 samples of rivets produced by a machine.

   [(a) continuous   (b) continuous]
   [(c) discrete   (d) continuous]

2. (a) The number of people visiting an exhibition on each of 5 days.
   (b) The time taken by each of 12 athletes to run 100 metres.
   (c) The value of stamps sold in a day by each of 20 post offices.
   (d) The number of defective items produced in each of 10 one-hour periods by a machine.

   [(a) discrete   (b) continuous]
   [(c) discrete   (d) discrete]

37.2 Presentation of ungrouped data

Ungrouped data can be presented diagrammatically in several ways and these include:

(a) **pictograms**, in which pictorial symbols are used to represent quantities (see Problem 2),
(b) **horizontal bar charts**, having data represented by equally spaced horizontal rectangles (see Problem 3), and
(c) **vertical bar charts**, in which data are represented by equally spaced vertical rectangles (see Problem 4).

Trends in ungrouped data over equal periods of time can be presented diagrammatically by a **percentage component bar chart**. In such a chart, equally spaced rectangles of any width, but whose height corresponds to 100%, are constructed. The rectangles are then sub-divided into values corresponding to the percentage relative frequencies of the members (see Problem 5).

A **pie diagram** is used to show diagrammatically the parts making up the whole. In a pie diagram, the area of a circle represents the whole, and the areas of the sectors of the circle are made proportional to the parts which make up the whole (see Problem 6).

Problem 2.   The number of television sets repaired in a workshop by a technician in six, one-month periods is as shown below. Present these data as a pictogram.

Month	January	February	March
Number repaired	11	6	15

Month	April	May	June
Number repaired	9	13	8

Each symbol shown in Fig. 37.1 represents two television sets repaired. Thus, in January, $5\frac{1}{2}$ symbols are used to represents the 11 sets repaired, in February, 3 symbols are used to represent the 6 sets repaired, and so on.

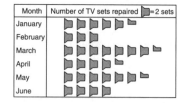

Figure 37.1

Problem 3.   The distance in miles travelled by four salesmen in a week are as shown below.

Salesmen	P	Q	R	S
Distance travelled (miles)	413	264	597	143

Use a horizontal bar chart to represent these data diagrammatically

Equally spaced horizontal rectangles of any width, but whose length is proportional to the distance travelled, are used. Thus, the length of the rectangle for salesman P is proportional to 413 miles, and so on. The horizontal bar chart depicting these data is shown in Fig. 37.2.

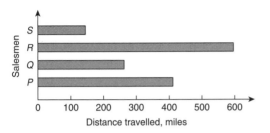

Figure 37.2

> **Problem 4.** The number of issues of tools or materials from a store in a factory is observed for seven, one-hour periods in a day, and the results of the survey are as follows:
>
> | Period | 1 | 2 | 3 | 4 | 5 | 6 | 7 |
> |---|---|---|---|---|---|---|---|
> | Number of issues | 34 | 17 | 9 | 5 | 27 | 13 | 6 |
>
> Present these data on a vertical bar chart.

In a vertical bar chart, equally spaced vertical rectangles of any width, but whose height is proportional to the quantity being represented, are used. Thus the height of the rectangle for period 1 is proportional to 34 units, and so on. The vertical bar chart depicting these data is shown in Fig. 37.3.

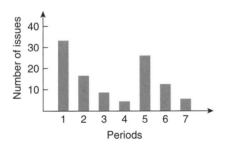

Figure 37.3

> **Problem 5.** The number of various types of dwellings sold by a company annually over a three-year period are as shown below. Draw

percentage component bar charts to present these data.

	Year 1	Year 2	Year 3
4-roomed bungalows	24	17	7
5-roomed bungalows	38	71	118
4-roomed houses	44	50	53
5-roomed houses	64	82	147
6-roomed houses	30	30	25

A table of percentage relative frequency values, correct to the nearest 1%, is the first requirement. Since,

$$\text{percentage relative frequency}$$
$$= \frac{\text{frequency of member} \times 100}{\text{total frequency}}$$

then for 4-roomed bungalows in year 1:

$$\text{percentage relative frequency}$$
$$= \frac{24 \times 100}{24 + 38 + 44 + 64 + 30} = 12\%$$

The percentage relative frequencies of the other types of dwellings for each of the three years are similarly calculated and the results are as shown in the table below.

	Year 1	Year 2	Year 3
4-roomed bungalows	12%	7%	2%
5-roomed bungalows	19%	28%	34%
4-roomed houses	22%	20%	15%
5-roomed houses	32%	33%	42%
6-roomed houses	15%	12%	7%

The percentage component bar chart is produced by constructing three equally spaced rectangles of any width, corresponding to the three years. The heights of the rectangles correspond to 100% relative frequency, and are subdivided into the values in the table of percentages shown above. A key is used (different types of shading or different colour schemes) to indicate corresponding percentage values in the rows of the table of percentages. The percentage component bar chart is shown in Fig. 37.4.

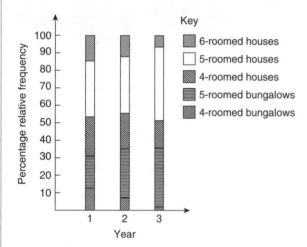

Figure 37.4

Problem 6. The retail price of a product costing £2 is made up as follows: materials 10 p, labour 20 p, research and development 40 p, overheads 70 p, profit 60 p. Present these data on a pie diagram

A circle of any radius is drawn, and the area of the circle represents the whole, which in this case is £2. The circle is subdivided into sectors so that the areas of the sectors are proportional to the parts, i.e. the parts which make up the total retail price. For the area of a sector to be proportional to a part, the angle at the centre of the circle must be proportional to that part. The whole, £2 or 200 p, corresponds to 360°. Therefore,

$$10 \text{ p corresponds to } 360 \times \frac{10}{200} \text{ degrees, i.e. } 18°$$

$$20 \text{ p corresponds to } 360 \times \frac{20}{200} \text{ degrees, i.e. } 36°$$

and so on, giving the angles at the centre of the circle for the parts of the retail price as: 18°, 36°, 72°, 126° and 108°, respectively.
The pie diagram is shown in Fig. 37.5.

Problem 7.

(a) Using the data given in Fig. 37.2 only, calculate the amount of money paid to each salesman for travelling expenses, if they are paid an allowance of 37 p per mile.

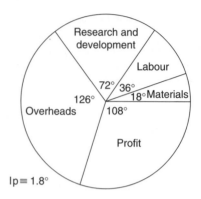

$1p \equiv 1.8°$

Figure 37.5

(b) Using the data presented in Fig. 37.4, comment on the housing trends over the three-year period.
(c) Determine the profit made by selling 700 units of the product shown in Fig. 37.5.

(a) By measuring the length of rectangle P the mileage covered by salesman P is equivalent to 413 miles. Hence salesman P receives a travelling allowance of

$$\frac{£413 \times 37}{100} \text{ i.e. } £152.81$$

Similarly, for salesman Q, the miles travelled are 264 and this allowance is

$$\frac{£264 \times 37}{100} \text{ i.e. } £97.68$$

Salesman R travels 597 miles and he receives

$$\frac{£597 \times 37}{100} \text{ i.e. } £220.89$$

Finally, salesman S receives

$$\frac{£143 \times 37}{100} \text{ i.e. } £52.91$$

(b) An analysis of Fig. 37.4 shows that 5-roomed bungalows and 5-roomed houses are becoming more popular, the greatest change in the three years being a 15% increase in the sales of 5-roomed bungalows.
(c) Since 1.8° corresponds to 1 p and the profit occupies 108° of the pie diagram, then the profit per unit is $\frac{108 \times 1}{1.8}$, that is, 60 p
The profit when selling 700 units of the product is $£\frac{700 \times 60}{100}$, that is, **£420**

Now try the following exercise

Exercise 135   Further problems on presentation of ungrouped data

1. The number of vehicles passing a stationary observer on a road in six ten-minute intervals is as shown. Draw a pictogram to represent these data.

Period of Time	1	2	3	4	5	6
Number of Vehicles	35	44	62	68	49	41

> [If one symbol is used to represent 10 vehicles, working correct to the nearest 5 vehicles, gives 3.5, 4.5, 6, 7, 5 and 4 symbols respectively.]

2. The number of components produced by a factory in a week is as shown below:

Day	Mon	Tues	Wed
Number of Components	1580	2190	1840

Day	Thurs	Fri
Number of Components	2385	1280

Show these data on a pictogram.

> [If one symbol represents 200 components, working correct to the nearest 100 components gives: Mon 8, Tues 11, Wed 9, Thurs 12 and Fri 6.5]

3. For the data given in Problem 1 above, draw a horizontal bar chart.

> [6 equally spaced horizontal rectangles, whose lengths are proportional to 35, 44, 62, 68, 49 and 41, respectively.]

4. Present the data given in Problem 2 above on a horizontal bar chart.

> [5 equally spaced horizontal rectangles, whose lengths are proportional to 1580, 2190, 1840, 2385 and 1280 units, respectively.]

5. For the data given in Problem 1 above, construct a vertical bar chart.

> [6 equally spaced vertical rectangles, whose heights are proportional to 35, 44, 62, 68, 49 and 41 units, respectively.]

6. Depict the data given in Problem 2 above on a vertical bar chart.

> [5 equally spaced vertical rectangles, whose heights are proportional to 1580, 2190, 1840, 2385 and 1280 units, respectively.]

7. A factory produces three different types of components. The percentages of each of these components produced for three, one-month periods are as shown below. Show this information on percentage component bar charts and comment on the changing trend in the percentages of the types of component produced.

Month	1	2	3
Component P	20	35	40
Component Q	45	40	35
Component R	35	25	25

> [Three rectangles of equal height, subdivided in the percentages shown in the columns above. P increases by 20% at the expense of Q and R]

8. A company has five distribution centres and the mass of goods in tonnes sent to each

centre during four, one-week periods, is as shown.

Week	1	2	3	4
Centre A	147	160	174	158
Centre B	54	63	77	69
Centre C	283	251	237	211
Centre D	97	104	117	144
Centre E	224	218	203	194

Use a percentage component bar chart to present these data and comment on any trends.

> Four rectangles of equal heights, subdivided as follows: week 1: 18%, 7%, 35%, 12%, 28% week 2: 20%, 8%, 32%, 13%, 27% week 3: 22%, 10%, 29%, 14%, 25% week 4: 20%, 9%, 27%, 19%, 25%. Little change in centres A and B, a reduction of about 8% in C, an increase of about 7% in D and a reduction of about 3% in E.

9. The employees in a company can be split into the following categories: managerial 3, supervisory 9, craftsmen 21, semi-skilled 67, others 44. Shown these data on a pie diagram.

> A circle of any radius, subdivided into sectors having angles of 7.5°, 22.5°, 52.5°, 167.5° and 110°, respectively.

10. The way in which an apprentice spent his time over a one-month period is a follows:

    drawing office 44 hours, production 64 hours, training 12 hours, at college 28 hours.

    Use a pie diagram to depict this information.

> A circle of any radius, subdivided into sectors having angles of 107°, 156°, 29° and 68°, respectively.

11. (a) With reference to Fig. 37.5, determine the amount spent on labour and materials to produce 1650 units of the product.
    (b) If in year 2 of Fig. 37.4, 1% corresponds to 2.5 dwellings, how many bungalows are sold in that year.

    [(a) £495, (b) 88]

12. (a) If the company sell 23 500 units per annum of the product depicted in Fig. 37.5, determine the cost of their overheads per annum.
    (b) If 1% of the dwellings represented in year 1 of Fig. 37.4 corresponds to 2 dwellings, find the total number of houses sold in that year.

    [(a) £16 450, (b) 138]

37.3 Presentation of grouped data

When the number of members in a set is small, say ten or less, the data can be represented diagrammatically without further analysis, by means of pictograms, bar charts, percentage components bar charts or pie diagrams (as shown in Section 37.2).

For sets having more than ten members, those members having similar values are grouped together in **classes** to form a **frequency distribution**. To assist in accurately counting members in the various classes, a **tally diagram** is used (see Problems 8 and 12).

A frequency distribution is merely a table showing classes and their corresponding frequencies (see Problems 8 and 12).

The new set of values obtained by forming a frequency distribution is called **grouped data**.

The terms used in connection with grouped data are shown in Fig. 37.6(a). The size or range of a class is given by the **upper class boundary value** minus the **lower class boundary value**, and in Fig. 37.6 is $7.65 - 7.35$, i.e. 0.30. The **class interval** for the class shown in Fig. 37.6(b) is 7.4 to 7.6 and the class mid-point value is given by:

$$\frac{\left(\begin{array}{c}\text{upper class}\\\text{boundary value}\end{array}\right) + \left(\begin{array}{c}\text{lower class}\\\text{boundary value}\end{array}\right)}{2}$$

and in Fig. 37.6 is $\dfrac{7.65 + 7.35}{2}$, i.e. 7.5

One of the principal ways of presenting grouped data diagrammatically is by using a **histogram**, in which

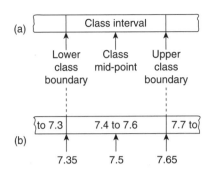

Figure 37.6

the **areas** of vertical, adjacent rectangles are made proportional to frequencies of the classes (see Problem 9). When class intervals are equal, the heights of the rectangles of a histogram are equal to the frequencies of the classes. For histograms having unequal class intervals, the area must be proportional to the frequency. Hence, if the class interval of class A is twice the class interval of class B, then for equal frequencies, the height of the rectangle representing A is half that of B (see Problem 11).

Another method of presenting grouped data diagrammatically is by using a **frequency polygon**, which is the graph produced by plotting frequency against class mid-point values and joining the coordinates with straight lines (see Problem 12).

A **cumulative frequency distribution** is a table showing the cumulative frequency for each value of upper class boundary. The cumulative frequency for a particular value of upper class boundary is obtained by adding the frequency of the class to the sum of the previous frequencies. A cumulative frequency distribution is formed in Problem 13.

The curve obtained by joining the co-ordinates of cumulative frequency (vertically) against upper class boundary (horizontally) is called an **ogive** or a **cumulative frequency distribution curve** (see Problem 13).

Problem 8. The data given below refer to the gain of each of a batch of 40 transistors, expressed correct to the nearest whole number. Form a frequency distribution for these data having seven classes

81	83	87	74	76	89	82	84
86	76	77	71	86	85	87	88
84	81	80	81	73	89	82	79
81	79	78	80	85	77	84	78
83	79	80	83	82	79	80	77

The **range** of the data is the value obtained by taking the value of the smallest member from that of the largest member. Inspection of the set of data shows that, range $= 89 - 71 = 18$. The size of each class is given approximately by range divided by the number of classes. Since 7 classes are required, the size of each class is 18/7, that is, approximately 3. To achieve seven equal classes spanning a range of values from 71 to 89, the class intervals are selected as: 70–72, 73–75, and so on.

To assist with accurately determining the number in each class, a **tally diagram** is produced, as shown in Table 37.1(a). This is obtained by listing the classes in the left-hand column, and then inspecting each of the 40 members of the set in turn and allocating them to the appropriate classes by putting '1s' in the appropriate rows. Every fifth '1' allocated to a particular row is shown as an oblique line crossing the four previous '1s', to help with final counting.

Table 37.1(a)

Class	Tally
70–72	1
73–75	11
76–78	⟋⟋⟋⟋ 11
79–81	⟋⟋⟋⟋ ⟋⟋⟋⟋ 11
82–84	⟋⟋⟋⟋ 1111
85–87	⟋⟋⟋⟋ 1
88–90	111

Table 37.1(b)

Class	Class mid-point	Frequency
70–72	71	1
73–75	74	2
76–78	77	7
79–81	80	12
82–84	83	9
85–87	86	6
88–90	89	3

A **frequency distribution** for the data is shown in Table 37.1(b) and lists classes and their corresponding frequencies, obtained from the tally diagram. (Class mid-point values are also shown in the table, since they are used for constructing the histogram for these data (see Problem 9)).

> **Problem 9.** Construct a histogram for the data given in Table 37.1(b)

The histogram is shown in Fig. 37.7. The width of the rectangles correspond to the upper class boundary values minus the lower class boundary values and the heights of the rectangles correspond to the class frequencies. The easiest way to draw a histogram is to mark the class mid-point values on the horizontal scale and draw the rectangles symmetrically about the appropriate class mid-point values and touching one another.

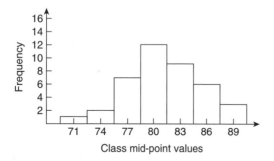

Figure 37.7

> **Problem 10.** The amount of money earned weekly by 40 people working part-time in a factory, correct to the nearest £10, is shown below. Form a frequency distribution having 6 classes for these data.

80	90	70	110	90	160	110	80
140	30	90	50	100	110	60	100
80	90	110	80	100	90	120	70
130	170	80	120	100	110	40	110
50	100	110	90	100	70	110	80

Inspection of the set given shows that the majority of the members of the set lie between £80 and £110 and

Table 37.2

Class	Frequency
20–40	2
50–70	6
80–90	12
100–110	14
120–140	4
150–170	2

that there are a much smaller number of extreme values ranging from £30 to £170. If equal class intervals are selected, the frequency distribution obtained does not give as much information as one with unequal class intervals. Since the majority of members are between £80 and £100, the class intervals in this range are selected to be smaller than those outside of this range. There is no unique solution and one possible solution is shown in Table 37.2.

> **Problem 11.** Draw a histogram for the data given in Table 37.2

When dealing with unequal class intervals, the histogram must be drawn so that the areas, (and not the heights), of the rectangles are proportional to the frequencies of the classes. The data given are shown in columns 1 and 2 of Table 37.3. Columns 3 and 4 give the upper and lower class boundaries, respectively. In column 5, the class ranges (i.e. upper class boundary minus lower class boundary values) are listed. The heights of the rectangles are proportional to the ratio $\frac{\text{frequency}}{\text{class range}}$, as shown in column 6. The histogram is shown in Fig. 37.8.

> **Problem 12.** The masses of 50 ingots in kilograms are measured correct to the nearest 0.1 kg and the results are as shown below. Produce a frequency distribution having about 7 classes for

Table 37.3

1 Class	2 Frequency	3 Upper class boundary	4 Lower class boundary	5 Class range	6 Height of rectangle
20–40	2	45	15	30	$\frac{2}{30} = \frac{1}{15}$
50–70	6	75	45	30	$\frac{6}{30} = \frac{3}{15}$
80–90	12	95	75	20	$\frac{12}{20} = \frac{9}{15}$
100–110	14	115	95	20	$\frac{14}{20} = \frac{10\frac{1}{2}}{15}$
120–140	4	145	115	30	$\frac{4}{30} = \frac{2}{15}$
150–170	2	175	145	30	$\frac{2}{30} = \frac{1}{15}$

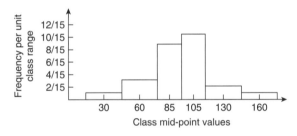

Figure 37.8

these data and then present the grouped data as (a) a frequency polygon and (b) histogram.

8.0	8.6	8.2	7.5	8.0	9.1	8.5	7.6	8.2	7.8
8.3	7.1	8.1	8.3	8.7	7.8	8.7	8.5	8.4	8.5
7.7	8.4	7.9	8.8	7.2	8.1	7.8	8.2	7.7	7.5
8.1	7.4	8.8	8.0	8.4	8.5	8.1	7.3	9.0	8.6
7.4	8.2	8.4	7.7	8.3	8.2	7.9	8.5	7.9	8.0

The **range** of the data is the member having the largest value minus the member having the smallest value. Inspection of the set of data shows that:

$$\text{range} = 9.1 - 7.1 = 2.0$$

The size of each class is given approximately by

$$\frac{\text{range}}{\text{number of classes}}$$

Since about seven classes are required, the size of each class is 2.0/7, that is approximately 0.3, and thus the **class limits** are selected as 7.1 to 7.3, 7.4 to 7.6, 7.7 to 7.9, and so on.

The **class mid-point** for the 7.1 to 7.3 class is $\frac{7.35 + 7.05}{2}$, i.e. 7.2, for the 7.4 to 7.6 class is $\frac{7.65 + 7.35}{2}$, i.e. 7.5, and so on.

To assist with accurately determining the number in each class, a **tally diagram** is produced as shown in Table 37.4. This is obtained by listing the classes in the left-hand column and then inspecting each of the 50 members of the set of data in turn and allocating it to the appropriate class by putting a '1' in the appropriate row. Each fifth '1' allocated to a particular row is marked as an oblique line to help with final counting.

A **frequency distribution** for the data is shown in Table 37.5 and lists classes and their corresponding frequencies. Class mid-points are also shown in this table, since they are used when constructing the frequency polygon and histogram.

A **frequency polygon** is shown in Fig. 37.9, the co-ordinates corresponding to the class mid-point/ frequency values, given in Table 37.5. The co-ordinates are joined by straight lines and the polygon is 'anchored-down' at each end by joining to the next class mid-point value and zero frequency.

Table 37.4

Class	Tally
7.1 to 7.3	111
7.4 to 7.6	ɪɪɪɪ
7.7 to 7.9	ɪɪɪɪ 1111
8.0 to 8.2	ɪɪɪɪ ɪɪɪɪ 1111
8.3 to 8.5	ɪɪɪɪ ɪɪɪɪ 1
8.6 to 8.8	ɪɪɪɪ 1
8.9 to 9.1	11

Table 37.5

Class	Class mid-point	Frequency
7.1 to 7.3	7.2	3
7.4 to 7.6	7.5	5
7.5 to 7.9	7.8	9
8.0 to 8.2	8.1	14
8.3 to 8.5	8.4	11
8.6 to 8.8	8.7	6
8.9 to 9.1	9.0	2

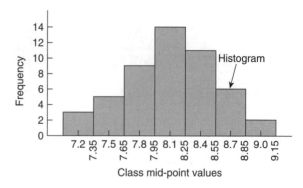

Figure 37.10

value — lower class boundary value) and height corresponding to the class frequency. The easiest way to draw a histogram is to mark class mid-point values on the horizontal scale and to draw the rectangles symmetrically about the appropriate class mid-point values and touching one another. A histogram for the data given in Table 37.5 is shown in Fig. 37.10.

Problem 13. The frequency distribution for the masses in kilograms of 50 ingots is:

7.1 to 7.3   3,   7.4 to 7.6   5,   7.7 to 7.9   9,
8.0 to 8.2   14,   8.3 to 8.5   11,   8.6 to 8.8,   6,
8.9 to 9.1   2,

Form a cumulative frequency distribution for these data and draw the corresponding ogive

A **cumulative frequency distribution** is a table giving values of cumulative frequency for the values of upper class boundaries, and is shown in Table 37.6. Columns 1 and 2 show the classes and their frequencies. Column 3 lists the upper class boundary values for the classes given in column 1. Column 4 gives the cumulative frequency values for all frequencies less than the upper class boundary values given in column 3. Thus, for example, for the 7.7 to 7.9 class shown in row 3, the cumulative frequency value is the sum of all frequencies having values of less than 7.95, i.e. $3 + 5 + 9 = 17$, and so on. The **ogive** for the cumulative frequency distribution given in Table 37.6 is shown in Fig. 37.11. The co-ordinates corresponding to each upper class boundary/cumulative frequency value are plotted and the co-ordinates are joined by straight lines (— not the best curve drawn through the co-ordinates as in experimental work). The ogive is 'anchored' at its start by adding the co-ordinate (7.05, 0).

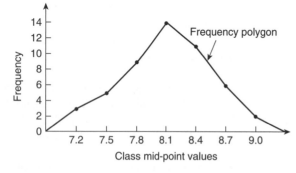

Figure 37.9

A **histogram** is shown in Fig. 37.10, the width of a rectangle corresponding to (upper class boundary

Table 37.6

1 Class	2 Frequency	3 Upper Class boundary	4 Cumulative frequency
		Less than	
7.1–7.3	3	7.35	3
7.4–7.6	5	7.65	8
7.7–7.9	9	7.95	17
8.0–8.2	14	8.25	31
8.3–8.5	11	8.55	42
8.6–8.8	6	8.85	48
8.9–9.1	2	9.15	50

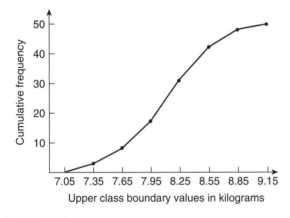

Figure 37.11

Now try the following exercise

Exercise 136   Further problems on presentation of grouped data

1. The mass in kilograms, correct to the nearest one-tenth of a kilogram, of 60 bars of metal are as shown. Form a frequency distribution of about 8 classes for these data.

39.8	40.3	40.6	40.0	39.6
39.6	40.2	40.3	40.4	39.8
40.2	40.3	39.9	39.9	40.0
40.1	40.0	40.1	40.1	40.2
39.7	40.4	39.9	40.1	39.9
39.5	40.0	39.8	39.5	39.9
40.1	40.0	39.7	40.4	39.3
40.7	39.9	40.2	39.9	40.0
40.1	39.7	40.5	40.5	39.9
40.8	40.0	40.2	40.0	39.9
39.8	39.7	39.5	40.1	40.2
40.6	40.1	39.7	40.2	40.3

> There is no unique solution,
> but one solution is:
> 39.3–39.4   1;   39.5–39.6   5;
> 39.7–39.8   9;   39.9–40.0   17;
> 40.1–40.2   15;   40.3–40.4   7;
> 40.5–40.6   4;   40.7–40.8   2

2. Draw a histogram for the frequency distribution given in the solution of Problem 1.

> Rectangles, touching one another,
> having mid-points of 39.35,
> 39.55, 39.75, 39.95, ... and
> heights of 1, 5, 9, 17, ...

3. The information given below refers to the value of resistance in ohms of a batch of 48 resistors of similar value. Form a frequency distribution for the data, having about 6 classes and draw a frequency polygon and histogram to represent these data diagrammatically.

21.0	22.4	22.8	21.5	22.6	21.1	21.6	22.3
22.9	20.5	21.8	22.2	21.0	21.7	22.5	20.7
23.2	22.9	21.7	21.4	22.1	22.2	22.3	21.3
22.1	21.8	22.0	22.7	21.7	21.9	21.1	22.6
21.4	22.4	22.3	20.9	22.8	21.2	22.7	21.6
22.2	21.6	21.3	22.1	21.5	22.0	23.4	21.2

> There is no unique solution,
> but one solution is:
> 20.5–20.9   3;   21.0–21.4   10;
> 21.5–21.9   11;   22.0–22.4   13;
> 22.5–22.9   9;   23.0–23.4   2

4. The time taken in hours to the failure of 50 specimens of a metal subjected to fatigue failure tests are as shown. Form a frequency distribution, having about 8 classes and unequal class intervals, for these data.

28	22	23	20	12	24	37	28	21	25
21	14	30	23	27	13	23	7	26	19
24	22	26	3	21	24	28	40	27	24
20	25	23	26	47	21	29	26	22	33
27	9	13	35	20	16	20	25	18	22

$$\begin{bmatrix} \text{There is no unique solution,} \\ \text{but one solution is: } 1\text{--}10 \qquad 3; \\ 11\text{--}19 \quad 7; \quad 20\text{--}22 \quad 12; \quad 23\text{--}25 \quad 11; \\ 26\text{--}28 \quad 10; \quad 29\text{--}38 \quad 5; \quad 39\text{--}48 \quad 2 \end{bmatrix}$$

5. Form a cumulative frequency distribution and hence draw the ogive for the frequency distribution given in the solution to Problem 3.

$$\begin{bmatrix} 20.95 \quad 3; \quad 21.45 \quad 13; \quad 21.95 \quad 24; \\ 22.45 \quad 37; \quad 22.95 \quad 46; \quad 23.45 \quad 48 \end{bmatrix}$$

6. Draw a histogram for the frequency distribution given in the solution to Problem 4.

$$\begin{bmatrix} \text{Rectangles, touching one another,} \\ \text{having mid-points of 5.5, 15,} \\ \text{21, 24, 27, 33.5 and 43.5. The} \\ \text{heights of the rectangles (frequency} \\ \text{per unit class range) are 0.3,} \\ \text{0.78, 4. 4.67, 2.33, 0.5 and 0.2} \end{bmatrix}$$

7. The frequency distribution for a batch of 50 capacitors of similar value, measured in microfarads, is:

10.5–10.9   2,   11.0–11.4   7,
11.5–11.9   10,   12.0–12.4   12,
12.5–12.9   11,   13.0–13.4   8

Form a cumulative frequency distribution for these data.

$$\begin{bmatrix} (10.95 \quad 2), \quad (11.45 \quad 9), \quad (11.95 \quad 11), \\ (12.45 \quad 31), \quad (12.95 \quad 42), \quad (13.45 \quad 50) \end{bmatrix}$$

8. Draw an ogive for the data given in the solution of Problem 7.

9. The diameter in millimetres of a reel of wire is measured in 48 places and the results are as shown.

2.10	2.29	2.32	2.21	2.14	2.22
2.28	2.18	2.17	2.20	2.23	2.13
2.26	2.10	2.21	2.17	2.28	2.15
2.16	2.25	2.23	2.11	2.27	2.34
2.24	2.05	2.29	2.18	2.24	2.16
2.15	2.22	2.14	2.27	2.09	2.21
2.11	2.17	2.22	2.19	2.12	2.20
2.23	2.07	2.13	2.26	2.16	2.12

(a) Form a frequency distribution of diameters having about 6 classes.

(b) Draw a histogram depicting the data.
(c) Form a cumulative frequency distribution.
(d) Draw an ogive for the the data.

$$\begin{bmatrix} \text{(a) There is unique solution,} \\ \text{but one solution is:} \\ 2.05\text{--}2.09 \quad 3; \quad 2.10\text{--}2.14 \quad 10; \\ 2.15\text{--}2.19 \quad 11; \quad 2.20\text{--}2.24 \quad 13; \\ 2.25\text{--}2.29 \quad 9; \quad 2.30\text{--}2.34 \quad 2 \\ \text{(b) Rectangles, touching one} \\ \text{another, having mid-points of} \\ \text{2.07, 2.12 ... and heights of} \\ \text{3, 10, ...} \\ \text{(c) Using the frequency} \\ \text{distribution given in the} \\ \text{solution to part (a) gives:} \\ 2.095 \quad 3; \quad 2.145 \quad 13; \quad 2.195 \quad 24; \\ 2.245 \quad 37; \quad 2.295 \quad 46; \quad 2.345 \quad 48 \\ \text{(d) A graph of cumulative} \\ \text{frequency against upper} \\ \text{class boundary having} \\ \text{the coordinates given} \\ \text{in part (c).} \end{bmatrix}$$

Chapter 38

Measures of central tendency and dispersion

38.1 Measures of central tendency

A single value, which is representative of a set of values, may be used to give an indication of the general size of the members in a set, the word **'average'** often being used to indicate the single value.

The statistical term used for 'average' is the arithmetic mean or just the **mean**. Other measures of central tendency may be used and these include the **median** and the **modal** values.

38.2 Mean, median and mode for discrete data

Mean

The **arithmetic mean value** is found by adding together the values of the members of a set and dividing by the number of members in the set. Thus, the mean of the set of numbers: $\{4, 5, 6, 9\}$ is:

$$\frac{4+5+6+9}{4} \quad \text{i.e.} \quad 6$$

In general, the mean of the set: $\{x_1, x_2, x_3, \ldots, x_n\}$ is

$$\bar{x} = \frac{x_1 + x_2 + x_3 + \cdots + x_n}{n}, \text{written as } \frac{\sum x}{n}$$

where Σ is the Greek letter 'sigma' and means 'the sum of', and $\bar{x}$ (called x-bar) is used to signify a mean value.

Median

The **median value** often gives a better indication of the general size of a set containing extreme values. The

set: $\{7, 5, 74, 10\}$ has a mean value of 24, which is not really representative of any of the values of the members of the set. The median value is obtained by:

(a) **ranking** the set in ascending order of magnitude, and

(b) selecting the value of the **middle member** for sets containing an odd number of members, or finding the value of the mean of the two middle members for sets containing an even number of members.

For example, the set: $\{7, 5, 74, 10\}$ is ranked as $\{5, 7, 10, 74\}$, and since it contains an even number of members (four in this case), the mean of 7 and 10 is taken, giving a median value of 8.5. Similarly, the set: $\{3, 81, 15, 7, 14\}$ is ranked as $\{3, 7, 14, 15, 81\}$ and the median value is the value of the middle member, i.e. 14.

Mode

The **modal value**, or **mode**, is the most commonly occurring value in a set. If two values occur with the same frequency, the set is 'bi-modal'. The set: $\{5, 6, 8, 2, 5, 4, 6, 5, 3\}$ has a modal value of 5, since the member having a value of 5 occurs three times.

Problem 1. Determine the mean, median and mode for the set:

$$\{2, 3, 7, 5, 5, 13, 1, 7, 4, 8, 3, 4, 3\}$$

The mean value is obtained by adding together the values of the members of the set and dividing by the number of members in the set.

Thus, **mean value**,

$$\bar{x} = \frac{\begin{array}{c}2+3+7+5+5+13+1\\+7+4+8+3+4+3\end{array}}{13} = \frac{65}{13} = 5$$

To obtain the median value the set is ranked, that is, placed in ascending order of magnitude, and since the set contains an odd number of members the value of the middle member is the median value. Ranking the set gives:

$$\{1, 2, 3, 3, 3, 4, 4, 5, 5, 7, 7, 8, 13\}$$

The middle term is the seventh member, i.e. 4, thus the **median value is 4**.

The **modal value** is the value of the most commonly occurring member and is **3**, which occurs three times, all other members only occurring once or twice.

Problem 2. The following set of data refers to the amount of money in £s taken by a news vendor for 6 days. Determine the mean, median and modal values of the set:

$$\{27.90, 34.70, 54.40, 18.92, 47.60, 39.68\}$$

$$\textbf{Mean value} = \frac{\begin{array}{c}27.90 + 34.70 + 54.40\\+18.92 + 47.60 + 39.68\end{array}}{6} = \textbf{£37.20}$$

The ranked set is:

$$\{18.92, 27.90, 34.70, 39.68, 47.60, 54.40\}$$

Since the set has an even number of members, the mean of the middle two members is taken to give the median value, i.e.

$$\textbf{median value} = \frac{34.70 + 39.68}{2} = \textbf{£37.19}$$

Since no two members have the same value, this set has **no mode**.

Now try the following exercise

Exercise 137 Further problems on mean, median and mode for discrete data

In Problems 1 to 4, determine the mean, median and modal values for the sets given.

1. $\{3, 8, 10, 7, 5, 14, 2, 9, 8\}$
   [mean 7.33, median 8, mode 8]

2. $\{26, 31, 21, 29, 32, 26, 25, 28\}$
   [mean 27.25, median 27, mode 26]

3. $\{4.72, 4.71, 4.74, 4.73, 4.72, 4.71, 4.73, 4.72\}$
   [mean 4.7225, median 4.72, mode 4.72]

4. $\{73.8, 126.4, 40.7, 141.7, 28.5, 237.4, 157.9\}$
   [mean 115.2, median 126.4, no mode]

38.3 Mean, median and mode for grouped data

The mean value for a set of grouped data is found by determining the sum of the (frequency × class mid-point values) and dividing by the sum of the frequencies,

i.e. mean value $\quad \bar{x} = \dfrac{f_1 x_1 + f_2 x_2 + \cdots + f_n x_n}{f_1 + f_2 + \cdots + f_n}$

$$= \frac{\sum (fx)}{\sum f}$$

where f is the frequency of the class having a mid-point value of x, and so on.

Problem 3. The frequency distribution for the value of resistance in ohms of 48 resistors is as shown. Determine the mean value of resistance.

20.5–20.9 3, 21.0–21.4 10, 21.5–21.9 11,
22.0–22.4 13, 22.5–22.9 9, 23.0–23.4 2

The class mid-point/frequency values are:

20.7 3, 21.2 10, 21.7 11, 22.2 13,
22.7 9 and 23.2 2

For grouped data, the mean value is given by:

$$\bar{x} = \frac{\sum (fx)}{\sum f}$$

where f is the class frequency and x is the class mid-point value. Hence mean value,

$$\bar{x} = \frac{\begin{array}{c}(3 \times 20.7) + (10 \times 21.2) + (11 \times 21.7)\\+(13 \times 22.2) + (9 \times 22.7) + (2 \times 23.2)\end{array}}{48}$$

$$= \frac{1052.1}{48} = 21.919\ldots$$

i.e. **the mean value is 21.9 ohms**, correct to 3 significant figures.

Histogram

The mean, median and modal values for grouped data may be determined from a **histogram**. In a histogram, frequency values are represented vertically and variable values horizontally. The mean value is given by the value of the variable corresponding to a vertical line drawn through the centroid of the histogram. The median value is obtained by selecting a variable value such that the area of the histogram to the left of a vertical line drawn through the selected variable value is equal to the area of the histogram on the right of the line. The modal value is the variable value obtained by dividing the width of the highest rectangle in the histogram in proportion to the heights of the adjacent rectangles. The method of determining the mean, median and modal values from a histogram is shown in Problem 4.

Problem 4. The time taken in minutes to assemble a device is measured 50 times and the results are as shown. Draw a histogram depicting this data and hence determine the mean, median and modal values of the distribution.

14.5–15.5 5,   16.5–17.5 8,   18.5–19.5 16,
20.5–21.5 12,   22.5–23.5 6,   24.5–25.5 3

The histogram is shown in Fig. 38.1. The mean value lies at the centroid of the histogram. With reference to any arbitrary axis, say YY shown at a time of 14 minutes, the position of the horizontal value of the centroid can be obtained from the relationship $AM = \Sigma(am)$, where A is the area of the histogram, M is the horizontal distance of the centroid from the axis YY, a is the area of a rectangle of the histogram and m is the distance of the centroid of the rectangle from YY. The areas of the individual rectangles are shown circled on the histogram giving a total area of 100 square units. The positions, m, of the centroids of the individual rectangles are 1, 3, 5, …units from YY. Thus

$$100\,M = (10 \times 1) + (16 \times 3) + (32 \times 5)$$
$$+ (24 \times 7) + (12 \times 9) + (6 \times 11)$$

i.e. $M = \dfrac{560}{100} = 5.6$ units from YY

Thus the position of the **mean** with reference to the time scale is $14 + 5.6$, i.e. **19.6 minutes**.

The median is the value of time corresponding to a vertical line dividing the total area of the histogram into two equal parts. The total area is 100 square units, hence the vertical line must be drawn to give 50 units of area on each side. To achieve this with reference to Fig. 38.1, rectangle $ABFE$ must be split so that $50 - (10 + 16)$ units of area lie on one side and $50 - (24 + 12 + 6)$ units of area lie on the other. This shows that the area of $ABFE$ is split so that 24 units of area lie to the left of the line and 8 units of area lie to the right, i.e. the vertical line must pass through 19.5 minutes. Thus the **median value** of the distribution is **19.5 minutes**.

The mode is obtained by dividing the line AB, which is the height of the highest rectangle, proportionally to the heights of the adjacent rectangles. With reference to Fig. 38.1, this is done by joining AC and BD and drawing a vertical line through the point of intersection of these two lines. This gives the **mode** of the distribution and is **19.3 minutes**.

Now try the following exercise

Exercise 138   Further problems on mean, median and mode for grouped data

1. 21 bricks have a mean mass of 24.2 kg, and 29 similar bricks have a mass of 23.6 kg. Determine the mean mass of the 50 bricks.
    [23.85 kg]

2. The frequency distribution given below refers to the heights in centimetres of 100 people. Determine the mean value of the distribution, correct to the nearest millimetre.

    150–156   5,   157–163 18,   164–170   20
    171–177 27,   178–184 22,   185–191   8
    [171.7 cm]

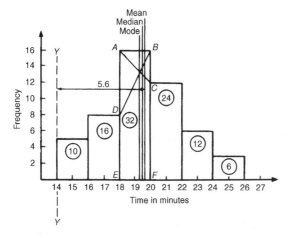

Figure 38.1

3. The gain of 90 similar transistors is measured and the results are as shown.

83.5–85.5 6, 86.5–88.5 39, 89.5–91.5 27, 92.5–94.5 15, 95.5–97.5 3

By drawing a histogram of this frequency distribution, determine the mean, median and modal values of the distribution.
[mean 89.5, median 89, mode 88.2]

4. The diameters, in centimetres, of 60 holes bored in engine castings are measured and the results are as shown. Draw a histogram depicting these results and hence determine the mean, median and modal values of the distribution.

2.011–2.014   7,   2.016–2.019   16, 2.021–2.024   23,   2.026–2.029   9, 2.031–2.034   5

$$\begin{bmatrix} \text{mean } 2.02158\,\text{cm}, \text{median } 2.02152\,\text{cm}, \\ \text{mode } 2.02167\,\text{cm} \end{bmatrix}$$

38.4   Standard deviation

(a) Discrete data

The standard deviation of a set of data gives an indication of the amount of dispersion, or the scatter, of members of the set from the measure of central tendency. Its value is the root-mean-square value of the members of the set and for discrete data is obtained as follows:

(a) determine the measure of central tendency, usually the mean value, (occasionally the median or modal values are specified),

(b) calculate the deviation of each member of the set from the mean, giving

$$(x_1 - \bar{x}), (x_2 - \bar{x}), (x_3 - \bar{x}), \ldots,$$

(c) determine the squares of these deviations, i.e.

$$(x_1 - \bar{x})^2, (x_2 - \bar{x})^2, (x_3 - \bar{x})^2, \ldots,$$

(d) find the sum of the squares of the deviations, that is

$$(x_1 - \bar{x})^2 + (x_2 - \bar{x})^2 + (x_3 - \bar{x})^2, \ldots,$$

(e) divide by the number of members in the set, n, giving

$$\frac{(x_1 - \bar{x})^2 + (x_2 - \bar{x})^2 + (x_3 - \bar{x})^2 + \cdots}{n}$$

(f) determine the square root of (e).

The standard deviation is indicated by σ (the Greek letter small 'sigma') and is written mathematically as:

$$\textbf{standard deviation}, \sigma = \sqrt{\frac{\sum(x - \bar{x})^2}{n}}$$

where x is a member of the set, $\bar{x}$ is the mean value of the set and n is the number of members in the set. The value of standard deviation gives an indication of the distance of the members of a set from the mean value. The set: $\{1, 4, 7, 10, 13\}$ has a mean value of 7 and a standard deviation of about 4.2. The set $\{5, 6, 7, 8, 9\}$ also has a mean value of 7, but the standard deviation is about 1.4. This shows that the members of the second set are mainly much closer to the mean value than the members of the first set. The method of determining the standard deviation for a set of discrete data is shown in Problem 5.

Problem 5. Determine the standard deviation from the mean of the set of numbers: $\{5, 6, 8, 4, 10, 3\}$, correct to 4 significant figures.

The arithmetic mean, $\bar{x} = \dfrac{\sum x}{n}$

$$= \frac{\begin{array}{c}5 + 6 + 8 + 4\\ + 10 + 3\end{array}}{6} = 6$$

Standard deviation, $\sigma = \sqrt{\dfrac{\sum(x - \bar{x})^2}{n}}$

The $(x - \bar{x})^2$ values are: $(5-6)^2$, $(6-6)^2$, $(8-6)^2$, $(4-6)^2$, $(10-6)^2$ and $(3-6)^2$.

The sum of the $(x - \bar{x})^2$ values,

i.e. $\sum(x - \bar{x})^2 = 1 + 0 + 4 + 4 + 16 + 9 = 34$

and $\dfrac{\sum(x - \bar{x})^2}{n} = \dfrac{34}{6} = 5.\dot{6}$

since there are 6 members in the set.
Hence, **standard deviation**,

$$\sigma = \sqrt{\frac{\sum(x - \bar{x}^2)}{n}} = \sqrt{5.\dot{6}} = \textbf{2.380}$$

correct to 4 significant figures

(b) Grouped data

For **grouped data, standard deviation**

$$\sigma = \sqrt{\frac{\sum \{f(x - \bar{x})^2\}}{\sum f}}$$

where f is the class frequency value, x is the class mid-point value and $\bar{x}$ is the mean value of the grouped data. The method of determining the standard deviation for a set of grouped data is shown in Problem 6.

> **Problem 6.** The frequency distribution for the values of resistance in ohms of 48 resistors is as shown. Calculate the standard deviation from the mean of the resistors, correct to 3 significant figures.
>
> 20.5–20.9 3, 21.0–21.4 10, 21.5–21.9 11,
> 22.0–22.4 13, 22.5–22.9 9, 23.0–23.4 2

The standard deviation for grouped data is given by:

$$\sigma = \sqrt{\frac{\sum \{f(x - \bar{x})^2\}}{\sum f}}$$

From Problem 3, the distribution mean value, $\bar{x} = 21.92$, correct to 4 significant figures.

The 'x-values' are the class mid-point values, i.e. 20.7, 21.2, 21.7, ...,

Thus the $(x - \bar{x})^2$ values are $(20.7 - 21.92)^2$, $(21.2 - 21.92)^2$, $(21.7 - 21.92)^2$, ...

and the $f(x - \bar{x})^2$ values are $3(20.7 - 21.92)^2$, $10(21.2 - 21.92)^2$, $11(21.7 - 21.92)^2$, ...

The $\sum f(x - \bar{x})^2$ values are

$$4.4652 + 5.1840 + 0.5324 + 1.0192$$
$$+ 5.4756 + 3.2768 = 19.9532$$

$$\frac{\sum \{f(x - \bar{x})^2\}}{\sum f} = \frac{19.9532}{48} = 0.41569$$

and **standard deviation**,

$$\sigma = \sqrt{\frac{\sum \{f(x - \bar{x})^2\}}{\sum f}}$$

$$= \sqrt{0.41569} = \mathbf{0.645},$$

correct to 3 significant figures

Now try the following exercise

> **Exercise 139 Further problems on standard deviation**
>
> 1. Determine the standard deviation from the mean of the set of numbers:
>
> $$\{35, 22, 25, 23, 28, 33, 30\}$$
>
>   correct to 3 significant figures.   [4.60]
>
> 2. The values of capacitances, in microfarads, of ten capacitors selected at random from a large batch of similar capacitors are:
>
> $$34.3, 25.0, 30.4, 34.6, 29.6, 28.7,$$
> $$33.4, 32.7, 29.0 \text{ and } 31.3$$
>
>   Determine the standard deviation from the mean for these capacitors, correct to 3 significant figures.   [2.83 µF]
>
> 3. The tensile strength in megapascals for 15 samples of tin were determined and found to be:
>
> $$34.61, 34.57, 34.40, 34.63, 34.63, 34.51,$$
> $$34.49, 34.61, 34.52, 34.55, 34.58, 34.53,$$
> $$34.44, 34.48 \text{ and } 34.40$$
>
>   Calculate the mean and standard deviation from the mean for these 15 values, correct to 4 significant figures.
>   [mean 34.53 MPa, standard deviation 0.07474 MPa]
>
> 4. Calculate the standard deviation from the mean for the mass of the 50 bricks given in Problem 1 of Exercise 138, page 347, correct to 3 significant figures.   [0.296 kg]
>
> 5. Determine the standard deviation from the mean, correct to 4 significant figures, for the heights of the 100 people given in Problem 2 of Exercise 138, page 347.   [9.394 cm]
>
> 6. Calculate the standard deviation from the mean for the data given in Problem 4 of Exercise 138, page 348, correct to 3 significant figures.   [0.00544 cm]

Section 7

38.5 Quartiles, deciles and percentiles

Other measures of dispersion, which are sometimes used, are the quartile, decile and percentile values. The **quartile values** of a set of discrete data are obtained by selecting the values of members that divide the set into four equal parts. Thus for the set: {2, 3, 4, 5, 5, 7, 9, 11, 13, 14, 17} there are 11 members and the values of the members dividing the set into four equal parts are 4, 7, and 13. These values are signified by Q_1, Q_2 and Q_3 and called the first, second and third quartile values, respectively. It can be seen that the second quartile value, Q_2, is the value of the middle member and hence is the median value of the set.

For grouped data the ogive may be used to determine the quartile values. In this case, points are selected on the vertical cumulative frequency values of the ogive, such that they divide the total value of cumulative frequency into four equal parts. Horizontal lines are drawn from these values to cut the ogive. The values of the variable corresponding to these cutting points on the ogive give the quartile values (see Problem 7).

When a set contains a large number of members, the set can be split into ten parts, each containing an equal number of members. These ten parts are then called **deciles**. For sets containing a very large number of members, the set may be split into one hundred parts, each containing an equal number of members. One of these parts is called a **percentile**.

> **Problem 7.** The frequency distribution given below refers to the overtime worked by a group of craftsmen during each of 48 working weeks in a year.
>
> 25–29 5, 30–34 4, 35–39 7, 40–44 11,
> 45–49 12, 50–54 8, 55–59 1
>
> Draw an ogive for this data and hence determine the quartile values.

The cumulative frequency distribution (i.e. upper class boundary/cumulative frequency values) is:

    29.5   5,   34.5   9,   39.5   16,
    44.5   27,   49.5   39,   54.5   47,
    59.5   48

The ogive is formed by plotting these values on a graph, as shown in Fig. 38.2. The total frequency is divided into four equal parts, each having a range of 48/4, i.e. 12. This gives cumulative frequency values of 0 to 12

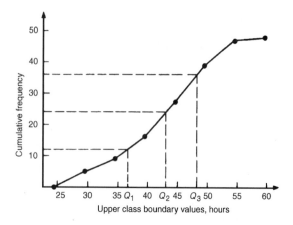

Figure 38.2

corresponding to the first quartile, 12 to 24 corresponding to the second quartile, 24 to 36 corresponding to the third quartile and 36 to 48 corresponding to the fourth quartile of the distribution, i.e. the distribution is divided into four equal parts. The quartile values are those of the variable corresponding to cumulative frequency values of 12, 24 and 36, marked Q_1, Q_2 and Q_3 in Fig. 38.2. These values, correct to the nearest hour, are **37 hours, 43 hours and 48 hours**, respectively. The Q_2 value is also equal to the median value of the distribution. One measure of the dispersion of a distribution is called the **semi-interquartile range** and is given by: $(Q_3 - Q_1)/2$, and is $(48 - 37)/2$ in this case, i.e. $5\frac{1}{2}$ **hours**.

> **Problem 8.** Determine the numbers contained in the (a) 41st to 50th percentile group, and (b) 8th decile group of the set of numbers shown below:
>
> 14 22 17 21 30 28 37 7 23 32
> 24 17 20 22 27 19 26 21 15 29

The set is ranked, giving:

    7   14   15   17   17   19   20   21   21   22
    22   23   24   26   27   28   29   30   32   37

(a) There are 20 numbers in the set, hence the first 10% will be the two numbers 7 and 14, the second 10% will be 15 and 17, and so on. Thus the 41st to 50th percentile group will be the numbers **21 and 22**

(b) The first decile group is obtained by splitting the ranked set into 10 equal groups and selecting the first group, i.e. the numbers 7 and 14. The second decile group are the numbers 15 and 17, and so on. Thus the 8th decile group contains the numbers **27 and 28**

Now try the following exercise

Exercise 140   Further problems on quartiles, deciles and percentiles

1. The number of working days lost due to accidents for each of 12 one-monthly periods are as shown. Determine the median and first and third quartile values for this data.

    27   37   40   28   23   30
    35   24   30   32   31   28
    [30, 25.5, 33.5 days]

2. The number of faults occurring on a production line in a nine-week period are as shown below. Determine the median and quartile values for the data.

    30   27   25   24   27
    37   31   27   35
    [27, 26, 33 faults]

3. Determine the quartile values and semi-interquartile range for the frequency distribution given in Problem 2 of Exercise 138, page 347.
    $$\left[\begin{array}{l} Q_1 = 164.5\,\text{cm}, Q_2 = 172.5\,\text{cm}, \\ Q_3 = 179\,\text{cm}, 7.25\,\text{cm} \end{array} \right]$$

4. Determine the numbers contained in the 5th decile group and in the 61st to 70th percentile groups for the set of numbers:

    40   46   28   32   37   42   50   31   48   45
    32   38   27   33   40   35   25   42   38   41
    [37 and 38; 40 and 41]

5. Determine the numbers in the 6th decile group and in the 81st to 90th percentile group for the set of numbers:

    43 47 30 25 15 51 17 21 37 33 44 56 40 49 22
    36 44 33 17 35 58 51 35 44 40 31 41 55 50 16
    [40, 40, 41; 50, 51, 51]

Chapter 39

Probability

39.1 Introduction to probability

The **probability** of something happening is the likelihood or chance of it happening. Values of probability lie between 0 and 1, where 0 represents an absolute impossibility and 1 represents an absolute certainty. The probability of an event happening usually lies somewhere between these two extreme values and is expressed either as a proper or decimal fraction. Examples of probability are:

that a length of copper wire
    has zero resistance at 100°C   0

that a fair, six-sided dice
    will stop with a 3 upwards $\frac{1}{6}$ or 0.1667

that a fair coin will land
    with a head upwards $\frac{1}{2}$ or 0.5

that a length of copper wire
    has some resistance at 100°C   1

If p is the probability of an event happening and q is the probability of the same event not happening, then the total probability is $p + q$ and is equal to unity, since it is an absolute certainty that the event either does or does not occur, i.e. $p + q = 1$

Expectation

The **expectation**, E, of an event happening is defined in general terms as the product of the probability p of an event happening and the number of attempts made, n, i.e. $E = pn$.

Thus, since the probability of obtaining a 3 upwards when rolling a fair dice is $\frac{1}{6}$, the expectation of getting a 3 upwards on four throws of the dice is $\frac{1}{6} \times 4$, i.e. $\frac{2}{3}$

Thus expectation is the average occurrence of an event.

Dependent event

A **dependent event** is one in which the probability of an event happening affects the probability of another event happening. Let 5 transistors be taken at random from a batch of 100 transistors for test purposes, and the probability of there being a defective transistor, p_1, be determined. At some later time, let another 5 transistors be taken at random from the 95 remaining transistors in the batch and the probability of there being a defective transistor, p_2, be determined. The value of p_2 is different from p_1 since batch size has effectively altered from 100 to 95, i.e. probability p_2 is dependent on probability p_1. Since transistors are drawn, and then another 5 transistors drawn without replacing the first 5, the second random selection is said to be **without replacement**.

Independent event

An independent event is one in which the probability of an event happening does not affect the probability of another event happening. If 5 transistors are taken at random from a batch of transistors and the probability of a defective transistor p_1 is determined and the process is repeated after the original 5 have been replaced in the batch to give p_2, then p_1 is equal to p_2. Since the 5 transistors are replaced between draws, the second selection is said to be **with replacement**.

Conditional probability

Conditional probability is concerned with the probability of say event B occurring, given that event A has already taken place. If A and B are independent events, then the fact that event A has already occurred will not affect the probability of event B. If A and B are dependent events, then event A having occurred will effect the probability of event B.

39.2 Laws of probability

The addition law of probability

The addition law of probability is recognized by the word **'or'** joining the probabilities. If p_A is the probability of event A happening and p_B is the probability of event B happening, the probability of **event A or event B** happening is given by $p_A + p_B$ (provided events A and B are **mutually exclusive**, i.e. A and B are events which cannot occur together). Similarly, the probability of events **A or B or C or $\ldots$ N** happening is given by

$$p_A + p_B + p_C + \cdots + p_N$$

The multiplication law of probability

The multiplication law of probability is recognized by the word **'and'** joining the probabilities. If p_A is the probability of event A happening and p_B is the probability of event B happening, the probability of **event A and event B** happening is given by $p_A \times p_B$. Similarly, the probability of events **A and B and C and $\ldots$ N** happening is given by:

$$p_A \times p_B \times p_C \times \cdots \times p_N$$

39.3 Worked problems on probability

Problem 1. Determine the probabilities of selecting at random (a) a man, and (b) a woman from a crowd containing 20 men and 33 women

(a) The probability of selecting at random a man, p, is given by the ratio

$$\frac{\text{number of men}}{\text{number in crowd}}$$

i.e. $p = \dfrac{20}{20+33} = \dfrac{20}{53}$ or **0.3774**

(b) The probability of selecting at random a woman, q, is given by the ratio

$$\frac{\text{number of women}}{\text{number in crowd}}$$

i.e. $q = \dfrac{33}{20+33} = \dfrac{33}{53}$ or **0.6226**

(Check: the total probability should be equal to 1;

$$p = \frac{20}{53} \text{ and } q = \frac{33}{53},$$

thus the total probability,

$$p + q = \frac{20}{53} + \frac{33}{53} = 1$$

hence no obvious error has been made.)

Problem 2. Find the expectation of obtaining a 4 upwards with 3 throws of a fair dice

Expectation is the average occurrence of an event and is defined as the probability times the number of attempts. The probability, p, of obtaining a 4 upwards for one throw of the dice, is $\frac{1}{6}$.

Also, 3 attempts are made, hence $n = 3$ and the expectation, E, is pn, i.e.

$$E = \tfrac{1}{6} \times 3 = \frac{1}{2} \text{ or } \mathbf{0.50}$$

Problem 3. Calculate the probabilities of selecting at random:

(a) the winning horse in a race in which 10 horses are running

(b) the winning horses in both the first and second races if there are 10 horses in each race

(a) Since only one of the ten horses can win, the probability of selecting at random the winning horse is

$$\frac{\text{number of winners}}{\text{number of horses}} \text{ i.e. } \frac{1}{10} \text{ or } \mathbf{0.10}$$

(b) The probability of selecting the winning horse in the first race is $\dfrac{1}{10}$. The probability of selecting the winning horse in the second race is $\dfrac{1}{10}$. The probability of selecting the winning horses in the first **and** second race is given by the multiplication law of probability,

i.e. **probability** $= \dfrac{1}{10} \times \dfrac{1}{10}$

$$= \frac{1}{100} \text{ or } \mathbf{0.01}$$

Problem 4. The probability of a component failing in one year due to excessive temperature is $\frac{1}{20}$, due to excessive vibration is $\frac{1}{25}$ and due to excessive humidity is $\frac{1}{50}$. Determine the probabilities that during a one-year period a component: (a) fails due to excessive (b) fails due to excessive vibration or excessive humidity, and (c) will not fail because of both excessive temperature and excessive humidity

Let p_A be the probability of failure due to excessive temperature, then

$$p_A = \frac{1}{20} \quad \text{and} \quad \overline{p_A} = \frac{19}{20}$$

(where $\overline{p_A}$ is the probability of not failing.)

Let p_B be the probability of failure due to excessive vibration, then

$$p_B = \frac{1}{25} \quad \text{and} \quad \overline{p_B} = \frac{24}{25}$$

Let p_C be the probability of failure due to excessive humidity, then

$$p_C = \frac{1}{50} \quad \text{and} \quad \overline{p_C} = \frac{49}{50}$$

(a) The probability of a component failing due to excessive temperature **and** excessive vibration is given by:

$$p_A \times p_B = \frac{1}{20} \times \frac{1}{25} = \frac{1}{500} \quad \text{or} \quad \mathbf{0.002}$$

(b) The probability of a component failing due to excessive vibration **or** excessive humidity is:

$$p_B + p_C = \frac{1}{25} + \frac{1}{50} = \frac{3}{50} \quad \text{or} \quad \mathbf{0.06}$$

(c) The probability that a component will not fail due to excessive temperature **and** will not fail due to excess humidity is:

$$\overline{p_A} \times \overline{p_C} = \frac{19}{20} \times \frac{49}{50} = \frac{931}{1000} \quad \text{or} \quad \mathbf{0.931}$$

Problem 5. A batch of 100 capacitors contains 73 that are within the required tolerance values, 17 which are below the required tolerance values, and the remainder are above the required tolerance values. Determine the probabilities that when randomly selecting a capacitor and then a second capacitor: (a) both are within the required tolerance values when selecting with replacement, and (b) the first one drawn is below and the second one drawn is above the required tolerance value, when selection is without replacement

(a) The probability of selecting a capacitor within the required tolerance values is $\frac{73}{100}$. The first capacitor drawn is now replaced and a second one is drawn from the batch of 100. The probability of this capacitor being within the required tolerance values is also $\frac{73}{100}$.

Thus, the probability of selecting a capacitor within the required tolerance values for both the first **and** the second draw is:

$$\frac{73}{100} \times \frac{73}{100} = \frac{5329}{10\,000} \quad \text{or} \quad \mathbf{0.5329}$$

(b) The probability of obtaining a capacitor below the required tolerance values on the first draw is $\frac{17}{100}$. There are now only 99 capacitors left in the batch, since the first capacitor is not replaced. The probability of drawing a capacitor above the required tolerance values on the second draw is $\frac{10}{99}$, since there are $(100 - 73 - 17)$, i.e. 10 capacitors above the required tolerance value. Thus, the probability of randomly selecting a capacitor below the required tolerance values and followed by randomly selecting a capacitor above the tolerance values is

$$\frac{17}{100} \times \frac{10}{99} = \frac{170}{9900} = \frac{17}{990} \quad \text{or} \quad \mathbf{0.0172}$$

Now try the following exercise

Exercise 141 Further problems on probability

1. In a batch of 45 lamps there are 10 faulty lamps. If one lamp is drawn at random, find the probability of it being (a) faulty and (b) satisfactory.

$$\left[(a) \frac{2}{9} \quad \text{or} \quad 0.2222 \right.$$
$$\left. (b) \frac{7}{9} \quad \text{or} \quad 0.7778 \right]$$

2. A box of fuses are all of the same shape and size and comprises 23 2 A fuses, 47 5 A fuses and 69 13 A fuses. Determine the probability of selecting at random (a) a 2 A fuse, (b) a 5 A fuse and (c) a 13 A fuse.

$$\left[\begin{array}{l} \text{(a) } \dfrac{23}{139} \quad \text{or} \quad 0.1655 \\[2mm] \text{(b) } \dfrac{47}{139} \quad \text{or} \quad 0.3381 \\[2mm] \text{(c) } \dfrac{69}{139} \quad \text{or} \quad 0.4964 \end{array} \right]$$

3. (a) Find the probability of having a 2 upwards when throwing a fair 6-sided dice. (b) Find the probability of having a 5 upwards when throwing a fair 6-sided dice. (c) Determine the probability of having a 2 and then a 5 on two successive throws of a fair 6-sided dice.

$$\left[\text{(a) } \dfrac{1}{6} \quad \text{(b) } \dfrac{1}{6} \quad \text{(c) } \dfrac{1}{36} \right]$$

4. Determine the probability that the total score is 8 when two like dice are thrown.

$$\left[\dfrac{5}{36} \right]$$

5. The probability of event A happening is $\frac{3}{5}$ and the probability of event B happening is $\frac{2}{3}$. Calculate the probabilities of (a) both A and B happening, (b) only event A happening, i.e. event A happening and event B not happening, (c) only event B happening, and (d) either A, or B, or A and B happening.

$$\left[\text{(a) } \dfrac{2}{5} \quad \text{(b) } \dfrac{1}{5} \quad \text{(c) } \dfrac{4}{15} \quad \text{(d) } \dfrac{13}{15} \right]$$

6. When testing 1000 soldered joints, 4 failed during a vibration test and 5 failed due to having a high resistance. Determine the probability of a joint failing due to (a) vibration, (b) high resistance, (c) vibration or high resistance and (d) vibration and high resistance.

$$\left[\begin{array}{ll} \text{(a) } \dfrac{1}{250} & \text{(b) } \dfrac{1}{200} \\[2mm] \text{(c) } \dfrac{9}{1000} & \text{(d) } \dfrac{1}{50\,000} \end{array} \right]$$

39.4 Further worked problems on probability

Problem 6. A batch of 40 components contains 5 which are defective. A component is drawn at random from the batch and tested and then a second component is drawn. Determine the probability that neither of the components is defective when drawn (a) with replacement, and (b) without replacement.

(a) With replacement

The probability that the component selected on the first draw is satisfactory is $\dfrac{35}{40}$, i.e. $\dfrac{7}{8}$. The component is now replaced and a second draw is made. The probability that this component is also satisfactory is $\dfrac{7}{8}$. Hence, the probability that both the first component drawn **and** the second component drawn are satisfactory is:

$$\dfrac{7}{8} \times \dfrac{7}{8} = \dfrac{49}{64} \quad \text{or} \quad \mathbf{0.7656}$$

(b) Without replacement

The probability that the first component drawn is satisfactory is $\dfrac{7}{8}$. There are now only 34 satisfactory components left in the batch and the batch number is 39. Hence, the probability of drawing a satisfactory component on the second draw is $\dfrac{34}{39}$. Thus the probability that the first component drawn **and** the second component drawn are satisfactory, i.e. neither is defective, is:

$$\dfrac{7}{8} \times \dfrac{34}{39} = \dfrac{238}{312} \quad \text{or} \quad \mathbf{0.7628}$$

Problem 7. A batch of 40 components contains 5 that are defective. If a component is drawn at random from the batch and tested and then a second component is drawn at random, calculate the probability of having one defective component, both with and without replacement

The probability of having one defective component can be achieved in two ways. If p is the probability of drawing a defective component and q is the probability of drawing a satisfactory component, then the probability of having one defective component is given by drawing a satisfactory component and then a defective component **or** by drawing a defective component and then a satisfactory one, i.e. by $q \times p + p \times q$

With replacement:

$$p = \frac{5}{40} = \frac{1}{8} \quad \text{and} \quad q = \frac{35}{40} = \frac{7}{8}$$

Hence, probability of having one defective component is:

$$\frac{1}{8} \times \frac{7}{8} + \frac{7}{8} \times \frac{1}{8}$$

i.e. $\quad \dfrac{7}{64} + \dfrac{7}{64} = \dfrac{7}{32} \quad$ or $\quad \mathbf{0.2188}$

Without replacement:

$p_1 = \dfrac{1}{8}$ and $q_1 = \dfrac{7}{8}$ on the first of the two draws. The batch number is now 39 for the second draw, thus,

$$p_2 = \frac{5}{39} \quad \text{and} \quad q_2 = \frac{35}{39}$$

$$p_1 q_2 + q_1 p_2 = \frac{1}{8} \times \frac{35}{39} + \frac{7}{8} \times \frac{5}{39}$$

$$= \frac{35 + 35}{312}$$

$$= \frac{70}{312} \quad \text{or} \quad \mathbf{0.2244}$$

Problem 8. A box contains 74 brass washer, 86 steel washers and 40 aluminium washers. Three washers are drawn at random from the box without replacement. Determine the probability that all three are steel washers

Assume, for clarity of explanation, that a washer is drawn at random, then a second, then a third (although this assumption does not affect the results obtained). The total number of washers is $74 + 86 + 40$, i.e. 200.

The probability of randomly selecting a steel washer on the first draw is $\dfrac{86}{200}$. There are now 85 steel washers in a batch of 199. The probability of randomly selecting a steel washer on the second draw is $\dfrac{85}{199}$. There are now 84 steel washers in a batch of 198. The probability of randomly selecting a steel washer on the third draw is $\dfrac{84}{198}$. Hence the probability of selecting a steel washer

on the first draw **and** the second draw **and** the third draw is:

$$\frac{86}{200} \times \frac{85}{199} \times \frac{84}{198} = \frac{614\,040}{7\,880\,400}$$

$$= \mathbf{0.0779}$$

Problem 9. For the box of washers given in Problem 8 above, determine the probability that there are no aluminium washers drawn, when three washers are drawn at random from the box without replacement

The probability of not drawing an aluminium washer on the first draw is $1 - \left(\dfrac{40}{200}\right)$, i.e. $\dfrac{160}{200}$. There are now 199 washers in the batch of which 159 are not aluminium washers. Hence, the probability of not drawing an aluminium washer on the second draw is $\dfrac{159}{199}$. Similarly, the probability of not drawing an aluminium washer on the third draw is $\dfrac{158}{198}$. Hence the probability of not drawing an aluminium washer on the first **and** second **and** third draw is

$$\frac{160}{200} \times \frac{159}{199} \times \frac{158}{198} = \frac{4\,019\,520}{7\,880\,400}$$

$$= \mathbf{0.5101}$$

Problem 10. For the box of washers in Problem 8 above, find the probability that there are two brass washers and either a steel or an aluminium washer when three are drawn at random, without replacement

Two brass washers (A) and one steel washer (B) can be obtained in any of the following ways:

1st draw	2nd draw	3rd draw
A	A	B
A	B	A
B	A	A

Two brass washers and one aluminium washer (C) can also be obtained in any of the following ways:

1st draw	2nd draw	3rd draw
A	A	C
A	C	A
C	A	A

Thus there are six possible ways of achieving the combinations specified. If A represents a brass washer, B a steel washer and C an aluminium washer, then the combinations and their probabilities are as shown:

First	Draw Second	Third	Probability
A	A	B	$\frac{74}{200} \times \frac{73}{199} \times \frac{86}{198} = 0.0590$
A	B	A	$\frac{74}{200} \times \frac{86}{199} \times \frac{73}{198} = 0.0590$
B	A	A	$\frac{86}{200} \times \frac{74}{199} \times \frac{73}{198} = 0.0590$
A	A	C	$\frac{74}{200} \times \frac{73}{199} \times \frac{40}{198} = 0.0274$
A	C	A	$\frac{74}{200} \times \frac{40}{199} \times \frac{73}{198} = 0.0274$
C	A	A	$\frac{40}{200} \times \frac{74}{199} \times \frac{73}{198} = 0.0274$

The probability of having the first combination **or** the second, **or** the third, and so on, is given by the sum of the probabilities,

i.e. by $3 \times 0.0590 + 3 \times 0.0274$, that is **0.2592**

Now try the following exercise

Exercise 142   Further problems on probability

1. The probability that component A will operate satisfactorily for 5 years is 0.8 and that B will operate satisfactorily over that same period of time is 0.75. Find the probabilities that in a 5 year period: (a) both components operate satisfactorily, (b) only component A will operate satisfactorily, and (c) only component B will operate satisfactorily.
[(a) 0.6 (b) 0.2 (c) 0.15]

2. In a particular street, 80% of the houses have telephones. If two houses selected at random are visited, calculate the probabilities that (a) they both have a telephone and (b) one has a telephone but the other does not have a telephone. [(a) 0.64 (b) 0.32]

3. Veroboard pins are packed in packets of 20 by a machine. In a thousand packets, 40 have less than 20 pins. Find the probability that if 2 packets are chosen at random, one will contain less than 20 pins and the other will contain 20 pins or more. [0.0768]

4. A batch of 1 kW fire elements contains 16 which are within a power tolerance and 4 which are not. If 3 elements are selected at random from the batch, calculate the probabilities that (a) all three are within the power tolerance and (b) two are within but one is not within the power tolerance.
[(a) 0.4912 (b) 0.4211]

5. An amplifier is made up of three transistors, A, B and C. The probabilities of A, B or C being defective are $\frac{1}{20}$, $\frac{1}{25}$ and $\frac{1}{50}$, respectively. Calculate the percentage of amplifiers produced (a) which work satisfactorily and (b) which have just one defective transistor.
[(a) 89.38% (b) 10.25%]

6. A box contains 14 40 W lamps, 28 60 W lamps and 58 25 W lamps, all the lamps being of the same shape and size. Three lamps are drawn at random from the box, first one, then a second, then a third. Determine the probabilities of: (a) getting one 25 W, one 40 W and one 60 W lamp, with replacement, (b) getting one 25 W, one 40 W and one 60 W lamp without replacement, and (c) getting either one 25 W and two 40 W or one 60 W and two 40 W lamps with replacement.
[(a) 0.0227 (b) 0.0234 (c) 0.0169]

39.5   Permutations and combinations

Permutations

If n different objects are available, they can be arranged in different orders of selection. Each different ordered arrangement is called a **permutation**. For example, permutations of the three letters X, Y and Z taken together are:

$$XYZ, XZY, YXZ, YZX, ZXY \text{ and } ZYX$$

This can be expressed as $^3P_3 = 6$, the upper 3 denoting the number of items from which the arrangements are made, and the lower 3 indicating the number of items used in each arrangement.

Section 7

If we take the same three letters XYZ two at a time the permutations

$$XY, YZ, XZ, ZX, YZ, ZY$$

can be found, and denoted by $^3P_2 = 6$
(Note that the order of the letters matter in permutations, i.e. YX is a different permutation from XY). In general, $^nP_r = n(n-1)(n-2)\ldots(n-r+1)$ or

$$^nP_r = \frac{n!}{(n-r)!} \text{ as stated in Chapter 15}$$

For example, $^5P_4 = 5(4)(3)(2) = 120$ or

$$^5P_4 = \frac{5!}{(5-4)!} = \frac{5!}{1!} = (5)(4)(3)(2) = 120$$

Also, $^3P_3 = 6$ from above; using $^nP_r = \frac{n!}{(n-r)!}$ gives

$^3P_3 = \frac{3!}{(3-3)!} = \frac{6}{0!}$. Since this must equal 6, then $0! = 1$ (check this with your calculator).

Combinations

If selections of the three letters X, Y, Z are made without regard to the order of the letters in each group, i.e. XY is now the same as YX for example, then each group is called a **combination**. The number of possible combinations is denoted by ^nC_r, where n is the total number of items and r is the number in each selection. In general,

$$^nC_r = \frac{n!}{r!(n-r)!}$$

For example,

$$^5C_4 = \frac{5!}{4!(5-4)!} = \frac{5!}{4!}$$

$$= \frac{5 \times 4 \times 3 \times 2 \times 1}{4 \times 3 \times 2 \times 1} = 5$$

Problem 11. Calculate the number of permutations there are of: (a) 5 distinct objects taken 2 at a time, (b) 4 distinct objects taken 2 at a time

(a) $\quad ^5P_2 = \frac{5!}{(5-2)!} = \frac{5!}{3!} = \frac{5 \times 4 \times 3 \times 2}{3 \times 2} = 20$

(b) $\quad ^4P_2 = \frac{4!}{(4-2)!} = \frac{4!}{2!} = 12$

Problem 12. Calculate the number of combinations there are of: (a) 5 distinct objects taken 2 at a time, (b) 4 distinct objects taken 2 at a time

(a) $\quad ^5C_2 = \frac{5!}{2!(5-2)!} = \frac{5!}{2!3!}$

$$= \frac{5 \times 4 \times 3 \times 2 \times 1}{(2 \times 1)(3 \times 2 \times 1)} = 10$$

(b) $\quad ^4C_2 = \frac{4!}{2!(4-2)!} = \frac{4!}{2!2!} = 6$

Problem 13. A class has 24 students. 4 can represent the class at an exam board. How many combinations are possible when choosing this group

Number of combinations possible,

$$^nC_r = \frac{n!}{r!(n-r!)}$$

i.e. $\quad ^{24}C_4 = \frac{24!}{4!(24-4)!} = \frac{24!}{4!20!} = \mathbf{10\,626}$

Problem 14. In how many ways can a team of eleven be picked from sixteen possible players?

Number of ways $= {}^nC_r = {}^{16}C_{11}$

$$= \frac{16!}{11!(16-11)!} = \frac{16!}{11!5!} = \mathbf{4368}$$

Now try the following exercise

Exercise 143 Further problems on permutations and combinations

1. Calculate the number of permutations there are of: (a) 15 distinct objects taken 2 at a time, (b) 9 distinct objects taken 4 at a time.
   [(a) 210 (b) 3024]

2. Calculate the number of combinations there are of: (a) 12 distinct objects taken 5 at a time, (b) 6 distinct objects taken 4 at a time.
   [(a) 792 (b) 15]

3. In how many ways can a team of six be picked from ten possible players? [210]

4. 15 boxes can each hold one object. In how many ways can 10 identical objects be placed in the boxes? [3003]

This Revision test covers the material in Chapters 37 to 39. *The marks for each question are shown in brackets at the end of each question.*

1. A company produces five products in the following proportions:
   Product A 24 Product B 16 Product C 15 Product D 11 Product E 6
   Present these data visually by drawing (a) a vertical bar chart (b) a percentage bar chart (c) a pie diagram. (13)

2. The following lists the diameters of 40 components produced by a machine, each measured correct to the nearest hundredth of a centimetre:

1.39	1.36	1.38	1.31	1.33	1.40	1.28
1.40	1.24	1.28	1.42	1.34	1.43	1.35
1.36	1.36	1.35	1.45	1.29	1.39	1.38
1.38	1.35	1.42	1.30	1.26	1.37	1.33
1.37	1.34	1.34	1.32	1.33	1.30	1.38
1.41	1.35	1.38	1.27	1.37		

   (a) Using 8 classes form a frequency distribution and a cumulative frequency distribution.

   (b) For the above data draw a histogram, a frequency polygon and an ogive. (21)

3. Determine for the 10 measurements of lengths shown below:

   (a) the arithmetic mean, (b) the median, (c) the mode, and (d) the standard deviation.

   28 m, 20 m, 32 m, 44 m, 28 m, 30 m, 30 m, 26 m, 28 m and 34 m (9)

4. The heights of 100 people are measured correct to the nearest centimetre with the following results:

150–157 cm	5	158–165 cm	18
166–173 cm	42	174–181 cm	27
182–189 cm	8		

   Determine for the data (a) the mean height and (b) the standard deviation. (10)

5. Determine the probabilities of:
   (a) drawing a white ball from a bag containing 6 black and 14 white balls
   (b) winning a prize in a raffle by buying 6 tickets when a total of 480 tickets are sold
   (c) selecting at random a female from a group of 12 boys and 28 girls
   (d) winning a prize in a raffle by buying 8 tickets when there are 5 prizes and a total of 800 tickets are sold. (8)

6. In a box containing 120 similar transistors 70 are satisfactory, 37 give too high a gain under normal operating conditions and the remainder give too low a gain.

   Calculate the probability that when drawing two transistors in turn, at random, **with replacement**, of having (a) two satisfactory, (b) none with low gain, (c) one with high gain and one satisfactory, (d) one with low gain and none satisfactory.

   Determine the probabilities in (a), (b) and (c) above if the transistors are drawn **without replacement**. (14)

Chapter 40

The binomial and Poisson distribution

40.1 The binomial distribution

The binomial distribution deals with two numbers only, these being the probability that an event will happen, p, and the probability that an event will not happen, q. Thus, when a coin is tossed, if p is the probability of the coin landing with a head upwards, q is the probability of the coin landing with a tail upwards. $p+q$ must always be equal to unity. A binomial distribution can be used for finding, say, the probability of getting three heads in seven tosses of the coin, or in industry for determining defect rates as a result of sampling. One way of defining a binomial distribution is as follows:

'if p is the probability that an event will happen and q is the probability that the event will not happen, then the probabilities that the event will happen 0, 1, 2, 3, ..., n times in n trials are given by the successive terms of the expansion of $(q+p)^n$ taken from left to right'.

The binomial expansion of $(q+p)^n$ is:

$$q^n + nq^{n-1}p + \frac{n(n-1)}{2!}q^{n-2}p^2$$
$$+ \frac{n(n-1)(n-2)}{3!}q^{n-3}p^3 + \cdots$$

from Chapter 16

This concept of a binomial distribution is used in Problems 1 and 2.

Problem 1. Determine the probabilities of having (a) at least 1 girl and (b) at least 1 girl and 1 boy in a family of 4 children, assuming equal probability of male and female birth

The probability of a girl being born, p, is 0.5 and the probability of a girl not being born (male birth), q, is also 0.5. The number in the family, n, is 4. From above, the probabilities of 0, 1, 2, 3, 4 girls in a family of 4 are given by the successive terms of the expansion of $(q+p)^4$ taken from left to right.

From the binomial expansion:

$$(q+p)^4 = q^4 + 4q^3p + 6q^2p^2 + 4qp^3 + p^4$$

Hence the probability of no girls is q^4,

i.e. $\qquad 0.5^4 = 0.0625$

the probability of 1 girl is $4q^3p$,

i.e. $\qquad 4 \times 0.5^3 \times 0.5 = 0.2500$

the probability of 2 girls is $6q^2p^2$,

i.e. $\qquad 6 \times 0.5^2 \times 0.5^2 = 0.3750$

the probability of 3 girls is $4qp^3$,

i.e. $\qquad 4 \times 0.5 \times 0.5^3 = 0.2500$

the probability of 4 girls is p^4,

i.e. $\qquad 0.5^4 = 0.0625$

$\qquad$ Total probability, $(q+p)^4 = 1.0000$

(a) The probability of having at least one girl is the sum of the probabilities of having 1, 2, 3 and 4 girls, i.e.

$$0.2500 + 0.3750 + 0.2500 + 0.0625 = \mathbf{0.9375}$$

(Alternatively, the probability of having at least 1 girl is: 1 − (the probability of having no girls), i.e. 1 − 0.0625, giving **0.9375**, as obtained previously).

(b) The probability of having at least 1 girl and 1 boy is given by the sum of the probabilities of having: 1 girl and 3 boys, 2 girls and 2 boys and 3 girls and 2 boys, i.e.

$$0.2500 + 0.3750 + 0.2500 = \mathbf{0.8750}$$

(Alternatively, this is also the probability of having 1 − (probability of having no girls + probability of having no boys), i.e. $1 - 2 \times 0.0625 = \mathbf{0.8750}$, as obtained previously).

Problem 2. A dice is rolled 9 times. Find the probabilities of having a 4 upwards (a) 3 times and (b) less than 4 times

---

Let p be the probability of having a 4 upwards. Then $p = 1/6$, since dice have six sides.

Let q be the probability of not having a 4 upwards. Then $q = 5/6$. The probabilities of having a 4 upwards 0, 1, 2...n times are given by the successive terms of the expansion of $(q + p)^n$, taken from left to right. From the binomial expansion:

$$(q + q)^9 = q^9 + 9q^8p + 36q^7p^2 + 84q^6p^3 + \ldots$$

The probability of having a 4 upwards no times is

$$q^9 = (5/6)^9 = 0.1938$$

The probability of having a 4 upwards once is

$$9q^8p = 9(5/6)^8(1/6) = 0.3489$$

The probability of having a 4 upwards twice is

$$36q^7p^2 = 36(5/6)^7(1/6)^2 = 0.2791$$

The probability of having a 4 upwards 3 times is

$$84q^6p^3 = 84(5/6)^6(1/6)^3 = 0.1302$$

(a) The probability of having a 4 upwards 3 times is **0.1302**

(b) The probability of having a 4 upwards less than 4 times is the sum of the probabilities of having a 4 upwards 0, 1, 2, and 3 times, i.e.

$$0.1938 + 0.3489 + 0.2791 + 0.1302 = \mathbf{0.9520}$$

Industrial inspection

In industrial inspection, p is often taken as the probability that a component is defective and q is the probability that the component is satisfactory. In this case, a binomial distribution may be defined as:

'the probabilities that 0, 1, 2, 3, …, n components are defective in a sample of n components, drawn at random from a large batch of components, are given by the successive terms of the expansion of $(q + p)^n$, taken from left to right'.

This definition is used in Problems 3 and 4.

Problem 3. A machine is producing a large number of bolts automatically. In a box of these bolts, 95% are within the allowable tolerance values with respect to diameter, the remainder being outside of the diameter tolerance values. Seven bolts are drawn at random from the box. Determine the probabilities that (a) two and (b) more than two of the seven bolts are outside of the diameter tolerance values

---

Let p be the probability that a bolt is outside of the allowable tolerance values, i.e. is defective, and let q be the probability that a bolt is within the tolerance values, i.e. is satisfactory. Then $p = 5\%$, i.e. 0.05 per unit and $q = 95\%$, i.e. 0.95 per unit. The sample number is 7.

The probabilities of drawing 0, 1, 2, . . . , n defective bolts are given by the successive terms of the expansion of $(q + p)^n$, taken from left to right. In this problem

$$\begin{aligned}(q + p)^n &= (0.95 + 0.05)^7 \\ &= 0.95^7 + 7 \times 0.95^6 \times 0.05 \\ &\quad + 21 \times 0.95^5 \times 0.05^2 + \ldots\end{aligned}$$

Thus the probability of no defective bolts is:

$$0.95^7 = 0.6983$$

The probability of 1 defective bolt is:

$$7 \times 0.95^6 \times 0.05 = 0.2573$$

The probability of 2 defective bolts is:

$$21 \times 0.95^5 \times 0.05^2 = 0.0406, \text{ and so on.}$$

(a) The probability that two bolts are outside of the diameter tolerance values is **0.0406**

(b) To determine the probability that more than two bolts are defective, the sum of the probabilities

Section 7

of 3 bolts, 4 bolts, 5 bolts, 6 bolts and 7 bolts being defective can be determined. An easier way to find this sum is to find $1 - $ (sum of 0 bolts, 1 bolt and 2 bolts being defective), since the sum of all the terms is unity. Thus, the probability of there being more than two bolts outside of the tolerance values is:

$$1 - (0.6983 + 0.2573 + 0.0406), \text{ i.e. } \mathbf{0.0038}$$

Problem 4. A package contains 50 similar components and inspection shows that four have been damaged during transit. If six components are drawn at random from the contents of the package determine the probabilities that in this sample (a) one and (b) less than three are damaged

The probability of a component being damaged, p, is 4 in 50, i.e. 0.08 per unit. Thus, the probability of a component not being damaged, q, is $1 - 0.08$, i.e. 0.92 The probability of there being 0, 1, 2, ..., 6 damaged components is given by the successive terms of $(q+p)^6$, taken from left to right.

$$(q + p)^6 = q^6 + 6q^5p + 15q^4p^2 + 20q^3p^3 + \cdots$$

(a) The probability of one damaged component is

$$6q^5p = 6 \times 0.92^5 \times 0.08 = \mathbf{0.3164}$$

(b) The probability of less than three damaged components is given by the sum of the probabilities of 0, 1 and 2 damaged components.

$$q^6 + 6q^5p + 15q^4p^2$$
$$= 0.92^6 + 6 \times 0.92^5 \times 0.08$$
$$+ 15 \times 0.92^4 \times 0.08^2$$
$$= 0.6064 + 0.3164 + 0.0688 = \mathbf{0.9916}$$

Histogram of probabilities

The terms of a binomial distribution may be represented pictorially by drawing a histogram, as shown in Problem 5.

Problem 5. The probability of a student successfully completing a course of study in three years is 0.45. Draw a histogram showing the probabilities of 0, 1, 2, ..., 10 students successfully completing the course in three years

Let p be the probability of a student successfully completing a course of study in three years and q be the probability of not doing so. Then $p = 0.45$ and $q = 0.55$. The number of students, n, is 10.

The probabilities of 0, 1, 2, ..., 10 students successfully completing the course are given by the successive terms of the expansion of $(q+p)^{10}$, taken from left to right.

$$(q + p)^{10} = q^{10} + 10q^9p + 45q^8p^2 + 120q^7p^3$$
$$+ 210q^6p^4 + 252q^5p^5 + 210q^4p^6$$
$$+ 120q^3p^7 + 45q^2p^8 + 10qp^9 + p^{10}$$

Substituting $q = 0.55$ and $p = 0.45$ in this expansion gives the values of the success terms as:

0.0025, 0.0207, 0.0763, 0.1665, 0.2384, 0.2340, 0.1596, 0.0746, 0.0229, 0.0042 and 0.0003. The histogram depicting these probabilities is shown in Fig. 40.1.

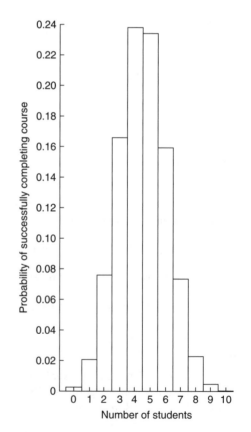

Figure 40.1

Now try the following exercise

Exercise 144   Further problems on the binomial distribution

1. Concrete blocks are tested and it is found that, on average, 7% fail to meet the required specification. For a batch of 9 blocks, determine the probabilities that (a) three blocks and (b) less than four blocks will fail to meet the specification.

   [(a) 0.0186 (b) 0.9976]

2. If the failure rate of the blocks in Problem 1 rises to 15%, find the probabilities that (a) no blocks and (b) more than two blocks will fail to meet the specification in a batch of 9 blocks.

   [(a) 0.2316 (b) 0.1408]

3. The average number of employees absent from a firm each day is 4%. An office within the firm has seven employees. Determine the probabilities that (a) no employee and (b) three employees will be absent on a particular day.

   [(a) 0.7514 (b) 0.0019]

4. A manufacturer estimates that 3% of his output of a small item is defective. Find the probabilities that in a sample of 10 items (a) less than two and (b) more than two items will be defective.

   [(a) 0.9655 (b) 0.0028]

5. Five coins are tossed simultaneously. Determine the probabilities of having 0, 1, 2, 3, 4 and 5 heads upwards, and draw a histogram depicting the results.

   ⎡ Vertical adjacent rectangles,
   whose heights are proportional to
   0.0313, 0.1563, 0.3125, 0.3125,
   0.1563 and 0.0313 ⎤

6. If the probability of rain falling during a particular period is 2/5, find the probabilities of having 0, 1, 2, 3, 4, 5, 6 and 7 wet days in a week. Show these results on a histogram.

   ⎡ Vertical adjacent rectangles,
   whose heights are proportional
   to 0.0280, 0.1306, 0.2613,
   0.2903, 0.1935, 0.0774,
   0.0172 and 0.0016 ⎤

7. An automatic machine produces, on average, 10% of its components outside of the tolerance required. In a sample of 10 components from this machine, determine the probability of having three components outside of the tolerance required by assuming a binomial distribution.

   [0.0574]

40.2   The Poisson distribution

When the number of trials, n, in a binomial distribution becomes large (usually taken as larger than 10), the calculations associated with determining the values of the terms become laborious. If n is large and p is small, and the product np is less than 5, a very good approximation to a binomial distribution is given by the corresponding Poisson distribution, in which calculations are usually simpler.

The Poisson approximation to a binomial distribution may be defined as follows:

'the probabilities that an event will happen 0, 1, 2, 3, ..., n times in n trials are given by the successive terms of the expression

$$e^{-\lambda}\left(1 + \lambda + \frac{\lambda^2}{2!} + \frac{\lambda^3}{3!} + \cdots\right)$$

taken from left to right'

The symbol λ is the expectation of an event happening and is equal to np.

Problem 6. If 3% of the gearwheels produced by a company are defective, determine the probabilities that in a sample of 80 gearwheels (a) two and (b) more than two will be defective.

The sample number, n, is large, the probability of a defective gearwheel, p, is small and the product np is 80×0.03, i.e. 2.4, which is less than 5. Hence a Poisson approximation to a binomial distribution may be used. The expectation of a defective gearwheel, $\lambda = np = 2.4$

The probabilities of 0, 1, 2, ... defective gearwheels are given by the successive terms of the expression

$$e^{-\lambda}\left(1 + \lambda + \frac{\lambda^2}{2!} + \frac{\lambda^3}{3!} + \cdots\right)$$

Section 7

taken from left to right, i.e. by

$$e^{-\lambda}, \lambda e^{-\lambda}, \frac{\lambda^2 e^{-\lambda}}{2!}, \ldots \quad \text{Thus:}$$

probability of no defective gearwheels is

$$e^{-\lambda} = e^{-2.4} = 0.0907$$

probability of 1 defective gearwheel is

$$\lambda e^{-\lambda} = 2.4 e^{-2.4} = 0.2177$$

probability of 2 defective gearwheels is

$$\frac{\lambda^2 e^{-\lambda}}{2!} = \frac{2.4^2 e^{-2.4}}{2 \times 1} = 0.2613$$

(a) The probability of having 2 defective gearwheels is **0.2613**

(b) The probability of having more than 2 defective gearwheels is 1 − (the sum of the probabilities of having 0, 1, and 2 defective gearwheels), i.e.

$$1 - (0.0907 + 0.2177 + 0.2613),$$

that is, **0.4303**

The principal use of a Poisson distribution is to determine the theoretical probabilities when p, the probability of an event happening, is known, but q, the probability of the event not happening is unknown. For example, the average number of goals scored per match by a football team can be calculated, but it is not possible to quantify the number of goals that were not scored. In this type of problem, a Poisson distribution may be defined as follows:

'the probabilities of an event occurring 0, 1, 2, 3…times are given by the successive terms of the expression

$$e^{-\lambda}\left(1 + \lambda + \frac{\lambda^2}{2!} + \frac{\lambda^3}{3!} + \cdots\right),$$

taken from left to right'

The symbol λ is the value of the average occurrence of the event.

Problem 7. A production department has 35 similar milling machines. The number of breakdowns on each machine averages 0.06 per week. Determine the probabilities of having (a) one, and (b) less than three machines breaking down in any week

Since the average occurrence of a breakdown is known but the number of times when a machine did not break down is unknown, a Poisson distribution must be used.

The expectation of a breakdown for 35 machines is 35×0.06, i.e. 2.1 breakdowns per week. The probabilities of a breakdown occurring 0, 1, 2, … times are given by the successive terms of the expression

$$e^{-\lambda}\left(1 + \lambda + \frac{\lambda^2}{2!} + \frac{\lambda^3}{3!} + \cdots\right),$$

taken from left to right. Hence:

probability of no breakdowns

$$e^{-\lambda} = e^{-2.1} = 0.1225$$

probability of 1 breakdown is

$$\lambda e^{-\lambda} = 2.1 e^{-2.1} = 0.2572$$

probability of 2 breakdowns is

$$\frac{\lambda^2 e^{-\lambda}}{2!} = \frac{2.1^2 e^{-2.1}}{2 \times 1} = 0.2700$$

(a) The probability of 1 breakdown per week is **0.2572**

(b) The probability of less than 3 breakdowns per week is the sum of the probabilities of 0, 1 and 2 breakdowns per week,

i.e. $0.1225 + 0.2572 + 0.2700 = \mathbf{0.6497}$

Histogram of probabilities

The terms of a Poisson distribution may be represented pictorially by drawing a histogram, as shown in Problem 8.

Problem 8. The probability of a person having an accident in a certain period of time is 0.0003. For a population of 7500 people, draw a histogram showing the probabilities of 0, 1, 2, 3, 4, 5 and 6 people having an accident in this period.

The probabilities of 0, 1, 2, … people having an accident are given by the terms of the expression

$$e^{-\lambda}\left(1 + \lambda + \frac{\lambda^2}{2!} + \frac{\lambda^3}{3!} + \cdots\right),$$

taken from left to right.

The average occurrence of the event, λ, is 7500×0.0003, i.e. 2.25

The probability of no people having an accident is

$$e^{-\lambda} = e^{-2.25} = 0.1054$$

The probability of 1 person having an accident is

$$\lambda e^{-\lambda} = 2.25 e^{-2.25} = 0.2371$$

The probability of 2 people having an accident is

$$\frac{\lambda^2 e^{-\lambda}}{2!} = \frac{2.25^2 e^{-2.25}}{2!} = 0.2668$$

and so on, giving probabilities of 0.2001, 0.1126, 0.0506 and 0.0190 for 3, 4, 5 and 6 respectively having an accident. The histogram for these probabilities is shown in Fig. 40.2.

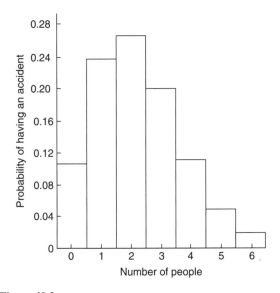

Figure 40.2

Now try the following exercise

Exercise 145   Further problems on the Poisson distribution

1. In problem 7 of Exercise 144, page 363, determine the probability of having three components outside of the required tolerance using the Poisson distribution.   [0.0613]

2. The probability that an employee will go to hospital in a certain period of time is 0.0015. Use a Poisson distribution to determine the probability of more than two employees going to hospital during this period of time if there are 2000 employees on the payroll.   [0.5768]

3. When packaging a product, a manufacturer finds that one packet in twenty is underweight. Determine the probabilities that in a box of 72 packets (a) two and (b) less than four will be underweight.   [(a) 0.1771   (b) 0.5153]

4. A manufacturer estimates that 0.25% of his output of a component are defective. The components are marketed in packets of 200. Determine the probability of a packet containing less than three defective components.   [0.9856]

5. The demand for a particular tool from a store is, on average, five times a day and the demand follows a Poisson distribution. How many of these tools should be kept in the stores so that the probability of there being one available when required is greater than 10%?

    ⎡ The probabilities of the demand
    for 0, 1, 2, ... tools are
    0.0067, 0.0337, 0.0842, 0.1404,
    0.1755, 0.1755, 0.1462, 0.1044,
    0.0653, ... This shows that the
    probability of wanting a tool
    8 times a day is 0.0653, i.e.
    less than 10%. Hence 7 should
    be kept in the store ⎦

6. Failure of a group of particular machine tools follows a Poisson distribution with a mean value of 0.7. Determine the probabilities of 0, 1, 2, 3, 4 and 5 failures in a week and present these results on a histogram.

    ⎡ Vertical adjacent rectangles
    having heights proportional
    to 0.4966, 0.3476, 0.1217,
    0.0284, 0.0050 and 0.0007 ⎦

Chapter 41

The normal distribution

41.1 Introduction to the normal distribution

When data is obtained, it can frequently be considered to be a sample (i.e. a few members) drawn at random from a large population (i.e. a set having many members). If the sample number is large, it is theoretically possible to choose class intervals which are very small, but which still have a number of members falling within each class. A frequency polygon of this data then has a large number of small line segments and approximates to a continuous curve. Such a curve is called a **frequency or a distribution curve**.

An extremely important symmetrical distribution curve is called the **normal curve** and is as shown in Fig. 41.1. This curve can be described by a mathematical equation and is the basis of much of the work done in more advanced statistics. Many natural occurrences such as the heights or weights of a group of people, the sizes of components produced by a particular machine and the life length of certain components approximate to a normal distribution.

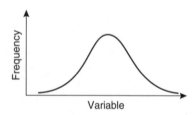

Figure 41.1

Normal distribution curves can differ from one another in the following four ways:

(a) by having different mean values
(b) by having different values of standard deviations
(c) the variables having different values and different units

and (d) by having different areas between the curve and the horizontal axis.

A normal distribution curve is **standardised** as follows:

(a) The mean value of the unstandardised curve is made the origin, thus making the mean value, $\bar{x}$, zero.
(b) The horizontal axis is scaled in standard deviations. This is done by letting $z = \dfrac{x - \bar{x}}{\sigma}$, where z is called the **normal standard variate**, x is the value of the variable, $\bar{x}$ is the mean value of the distribution and σ is the standard deviation of the distribution.
(c) The area between the normal curve and the horizontal axis is made equal to unity.

When a normal distribution curve has been standardised, the normal curve is called a **standardised normal curve** or a **normal probability curve**, and any normally distributed data may be represented by the **same** normal probability curve.

The area under part of a normal probability curve is directly proportional to probability and the value of the shaded area shown in Fig. 41.2 can be determined by evaluating:

$$\int \frac{1}{\sqrt{2\pi}} e^{(z^2/2)} dz, \quad \text{where } z = \frac{x - \bar{x}}{\sigma}$$

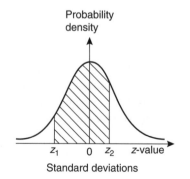

Figure 41.2

To save repeatedly determining the values of this function, tables of partial areas under the standardised normal curve are available in many mathematical formulae books, and such a table is shown in Table 41.1.

Problem 1. The mean height of 500 people is 170 cm and the standard deviation is 9 cm.

Assuming the heights are normally distributed, determine the number of people likely to have heights between 150 cm and 195 cm

The mean value, $\bar{x}$, is 170 cm and corresponds to a normal standard variate value, z, of zero on the standardised normal curve. A height of 150 cm has a z-value given

Table 41.1 Partial areas under the standardised normal curve

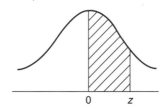

$z = \dfrac{x - \bar{x}}{\sigma}$	0	1	2	3	4	5	6	7	8	9
0.0	0.0000	0.0040	0.0080	0.0120	0.0159	0.0199	0.0239	0.0279	0.0319	0.0359
0.1	0.0398	0.0438	0.0478	0.0517	0.0557	0.0596	0.0636	0.0678	0.0714	0.0753
0.2	0.0793	0.0832	0.0871	0.0910	0.0948	0.0987	0.1026	0.1064	0.1103	0.1141
0.3	0.1179	0.1217	0.1255	0.1293	0.1331	0.1388	0.1406	0.1443	0.1480	0.1517
0.4	0.1554	0.1591	0.1628	0.1664	0.1700	0.1736	0.1772	0.1808	0.1844	0.1879
0.5	0.1915	0.1950	0.1985	0.2019	0.2054	0.2086	0.2123	0.2157	0.2190	0.2224
0.6	0.2257	0.2291	0.2324	0.2357	0.2389	0.2422	0.2454	0.2486	0.2517	0.2549
0.7	0.2580	0.2611	0.2642	0.2673	0.2704	0.2734	0.2760	0.2794	0.2823	0.2852
0.8	0.2881	0.2910	0.2939	0.2967	0.2995	0.3023	0.3051	0.3078	0.3106	0.3133
0.9	0.3159	0.3186	0.3212	0.3238	0.3264	0.3289	0.3315	0.3340	0.3365	0.3389
1.0	0.3413	0.3438	0.3451	0.3485	0.3508	0.3531	0.3554	0.3577	0.3599	0.3621
1.1	0.3643	0.3665	0.3686	0.3708	0.3729	0.3749	0.3770	0.3790	0.3810	0.3830
1.2	0.3849	0.3869	0.3888	0.3907	0.3925	0.3944	0.3962	0.3980	0.3997	0.4015
1.3	0.4032	0.4049	0.4066	0.4082	0.4099	0.4115	0.4131	0.4147	0.4162	0.4177
1.4	0.4192	0.4207	0.4222	0.4236	0.4251	0.4265	0.4279	0.4292	0.4306	0.4319
1.5	0.4332	0.4345	0.4357	0.4370	0.4382	0.4394	0.4406	0.4418	0.4430	0.4441

(Continued)

Section 7

Table 41.1 (*Continued*)

$z = \dfrac{x - \bar{x}}{\sigma}$	0	1	2	3	4	5	6	7	8	9
1.6	0.4452	0.4463	0.4474	0.4484	0.4495	0.4505	0.4515	0.4525	0.4535	0.4545
1.7	0.4554	0.4564	0.4573	0.4582	0.4591	0.4599	0.4608	0.4616	0.4625	0.4633
1.8	0.4641	0.4649	0.4656	0.4664	0.4671	0.4678	0.4686	0.4693	0.4699	0.4706
1.9	0.4713	0.4719	0.4726	0.4732	0.4738	0.4744	0.4750	0.4756	0.4762	0.4767
2.0	0.4772	0.4778	0.4783	0.4785	0.4793	0.4798	0.4803	0.4808	0.4812	0.4817
2.1	0.4821	0.4826	0.4830	0.4834	0.4838	0.4842	0.4846	0.4850	0.4854	0.4857
2.2	0.4861	0.4864	0.4868	0.4871	0.4875	0.4878	0.4881	0.4884	0.4887	0.4890
2.3	0.4893	0.4896	0.4898	0.4901	0.4904	0.4906	0.4909	0.4911	0.4913	0.4916
2.4	0.4918	0.4920	0.4922	0.4925	0.4927	0.4929	0.4931	0.4932	0.4934	0.4936
2.5	0.4938	0.4940	0.4941	0.4943	0.4945	0.4946	0.4948	0.4949	0.4951	0.4952
2.6	0.4953	0.4955	0.4956	0.4957	0.4959	0.4960	0.4961	0.4962	0.4963	0.4964
2.7	0.4965	0.4966	0.4967	0.4968	0.4969	0.4970	0.4971	0.4972	0.4973	0.4974
2.8	0.4974	0.4975	0.4976	0.4977	0.4977	0.4978	0.4979	0.4980	0.4980	0.4981
2.9	0.4981	0.4982	0.4982	0.4983	0.4984	0.4984	0.4985	0.4985	0.4986	0.4986
3.0	0.4987	0.4987	0.4987	0.4988	0.4988	0.4989	0.4989	0.4989	0.4990	0.4990
3.1	0.4990	0.4991	0.4991	0.4991	0.4992	0.4992	0.4992	0.4992	0.4993	0.4993
3.2	0.4993	0.4993	0.4994	0.4994	0.4994	0.4994	0.4994	0.4995	0.4995	0.4995
3.3	0.4995	0.4995	0.4995	0.4996	0.4996	0.4996	0.4996	0.4996	0.4996	0.4997
3.4	0.4997	0.4997	0.4997	0.4997	0.4997	0.4997	0.4997	0.4997	0.4997	0.4998
3.5	0.4998	0.4998	0.4998	0.4998	0.4998	0.4998	0.4998	0.4998	0.4998	0.4998
3.6	0.4998	0.4998	0.4999	0.4999	0.4999	0.4999	0.4999	0.4999	0.4999	0.4999
3.7	0.4999	0.4999	0.4999	0.4999	0.4999	0.4999	0.4999	0.4999	0.4999	0.4999
3.8	0.4999	0.4999	0.4999	0.4999	0.4999	0.4999	0.4999	0.4999	0.4999	0.4999
3.9	0.5000	0.5000	0.5000	0.5000	0.5000	0.5000	05000	0.5000	0.5000	0.5000

by $z = \dfrac{x - \bar{x}}{\sigma}$ standard deviations, i.e. $\dfrac{150 - 170}{9}$ or -2.22 standard deviations. Using a table of partial areas beneath the standardised normal curve (see Table 41.1), a z-value of -2.22 corresponds to an area of 0.4868 between the mean value and the ordinate $z = -2.22$.

The negative z-value shows that it lies to the left of the $z = 0$ ordinate.

This area is shown shaded in Fig. 41.3(a). Similarly, 195 cm has a z-value of $\dfrac{195 - 170}{9}$ that is 2.78 standard deviations. From Table 41.1, this value of z corresponds

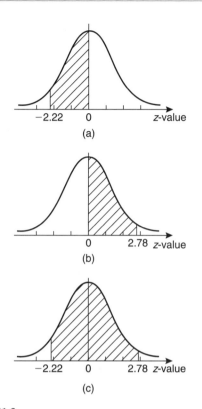

Figure 41.3

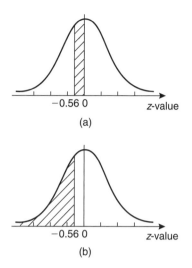

Figure 41.4

to an area of 0.4973, the positive value of z showing that it lies to the right of the $z = 0$ ordinate. This area is show shaded in Fig. 41.3(b). The total area shaded in Fig. 41.3(a) and (b) is shown in Fig. 41.3(c) and is $0.4868 + 0.4973$, i.e. 0.9841 of the total area beneath the curve.

However, the area is directly proportional to probability. Thus, the probability that a person will have a height of between 150 and 195 cm is 0.9841. For a group of 500 people, 500×0.9841, i.e. **492 people are likely to have heights in this range**. The value of 500×0.9841 is 492.05, but since answers based on a normal probability distribution can only be approximate, results are usually given correct to the nearest whole number.

> **Problem 2.** For the group of people given in Problem 1, find the number of people likely to have heights of less than 165 cm

A height of 165 cm corresponds to $\dfrac{165 - 170}{9}$, i.e. -0.56 standard deviations. The area between $z = 0$ and $z = -0.56$ (from Table 41.1) is 0.2123, shown shaded in Fig. 41.4(a). The total area under the standardised normal curve is unity and since the curve is symmetrical, it follows that the total area to the left of the $z = 0$ ordinate is 0.5000. Thus the area to the left of the $z = -0.56$ ordinate ('left' means 'less than', 'right' means 'more than') is $0.5000 - 0.2123$, i.e. 0.2877 of the total area, which is shown shaded in Fig. 41.4(b). The area is directly proportional to probability and since the total area beneath the standardised normal curve is unity, the probability of a person's height being less than 165 cm is 0.2877. For a group of 500 people, 500×0.2877, i.e. **144 people are likely to have heights of less than 165 cm**.

> **Problem 3.** For the group of people given in Problem 1 find how many people are likely to have heights of more than 194 cm

194 cm correspond to a z-value of $\dfrac{194 - 170}{9}$ that is, 2.67 standard deviations. From Table 41.1, the area between $z = 0$, $z = 2.67$ and the standardised normal curve is 0.4962, shown shaded in Fig. 41.5(a). Since the standardised normal curve is symmetrical, the total area to the right of the $z = 0$ ordinate is 0.5000, hence the shaded area shown in Fig. 41.5(b) is $0.5000 - 0.4962$, i.e. 0.0038. This area represents the probability of a person having a height of more than 194 cm, and for 500 people, the number of people likely to have a height of more than 194 cm is 0.0038×500, i.e. **2 people**.

> **Problem 4.** A batch of 1500 lemonade bottles have an average contents of 753 ml and the standard deviation of the contents is 1.8 ml. If the volumes of the content are normally distributed, find the

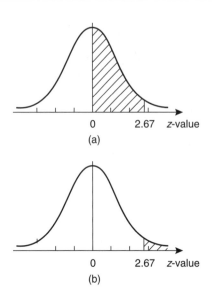

Figure 41.5

number of bottles likely to contain: (a) less than 750 ml, (b) between 751 and 754 ml, (c) more than 757 ml, and (d) between 750 and 751 ml

(a) The z-value corresponding to 750 ml is given by $\dfrac{x-\bar{x}}{\sigma}$ i.e. $\dfrac{750-753}{1.8}=-1.67$ standard deviations. From Table 41.1, the area between $z=0$ and $z=-1.67$ is 0.4525. Thus the area to the left of the $z=-1.67$ ordinate is $0.5000-0.4525$ (see Problem 2), i.e. 0.0475. This is the probability of a bottle containing less than 750 ml. Thus, for a batch of 1500 bottles, it is likely that 1500×0.0475, i.e. **71 bottles will contain less than 750 ml**.

(b) The z-value corresponding to 751 and 754 ml are $\dfrac{751-753}{1.8}$ and $\dfrac{754-753}{1.8}$ i.e. -1.11 and 0.56 respectively. From Table 41.1, the areas corresponding to these values are 0.3665 and 0.2123 respectively. Thus the probability of a bottle containing between 751 and 754 ml is $0.3665+0.2123$ (see Problem 1), i.e. 0.5788. For 1500 bottles, it is likely that 1500×0.5788, i.e. **868 bottles will contain between 751 and 754 ml**.

(c) The z-value corresponding to 757 ml is $\dfrac{757-753}{1.8}$, i.e. 2.22 standard deviations. From Table 41.1, the area corresponding to a z-value of 2.22 is 0.4868. The area to the right of the $z=2.22$ ordinate is

0.5000 − 0.4868 (see Problem 3), i.e. 0.0132 Thus, for 1500 bottles, it is likely that 1500×0.0132, i.e. **20 bottles will have contents of more than 750 ml**.

(d) The z-value corresponding to 750 ml is −1.67 (see part (a)), and the z-value corresponding to 751 ml is −1.11 (see part (b)). The areas corresponding to these z-values area 0.4525 and 0.3665 respectively, and both these areas lie on the left of the $z=0$ ordinate. The area between $z=-1.67$ and $z=-1.11$ is $0.4525-0.3665$, i.e. 0.0860 and this is the probability of a bottle having contents between 750 and 751 ml. For 1500 bottles, it is likely that 1500×0.0860, i.e. **129 bottles will be in this range**.

Now try the following exercise

Exercise 146 Further problems on the introduction to the normal distribution

1. A component is classed as defective if it has a diameter of less than 69 mm. In a batch of 350 components, the mean diameter is 75 mm and the standard deviation is 2.8 mm. Assuming the diameters are normally distributed, determine how many are likely to be classed as defective [6]

2. The masses of 800 people are normally distributed, having a mean value of 64.7 kg, and a standard deviation of 5.4 kg. Find how many people are likely to have masses of less than 54.4 kg. [22]

3. 500 tins of paint have a mean content of 1010 ml and the standard deviation of the contents is 8.7 ml. Assuming the volumes of the contents are normally distributed, calculate the number of tins likely to have contents whose volumes are less than (a) 1025 ml (b) 1000 ml and (c) 995 ml.
[(a) 479 (b) 63 (c) 21]

4. For the 350 components in Problem 1, if those having a diameter of more than 81.5 mm are rejected, find, correct to the nearest component, the number likely to be rejected due to being oversized. [4]

5. For the 800 people in Problem 2, determine how many are likely to have masses of more than (a) 70 kg, and (b) 62 kg.

[(a) 131 (b) 553]

6. The mean diameter of holes produced by a drilling machine bit is 4.05 mm and the standard deviation of the diameters is 0.0028 mm. For twenty holes drilled using this machine, determine, correct to the nearest whole number, how many are likely to have diameters of between (a) 4.048 and 4.0553 mm, and (b) 4.052 and 4.056 mm, assuming the diameters are normally distributed.

[(a) 15 (b) 4]

7. The intelligence quotients of 400 children have a mean value of 100 and a standard deviation of 14. Assuming that I.Q.'s are normally distributed, determine the number of children likely to have I.Q.'s of between (a) 80 and 90, (b) 90 and 110, and (c) 110 and 130.

[(a) 65 (b) 209 (c) 89]

8. The mean mass of active material in tablets produced by a manufacturer is 5.00 g and the standard deviation of the masses is 0.036 g. In a bottle containing 100 tablets, find how many tablets are likely to have masses of (a) between 4.88 and 4.92 g, (b) between 4.92 and 5.04 g, and (c) more than 5.04 g.

[(a) 1 (b) 85 (c) 13]

41.2 Testing for a normal distribution

It should never be assumed that because data is continuous it automatically follows that it is normally distributed. One way of checking that data is normally distributed is by using **normal probability paper**, often just called **probability paper**. This is special graph paper which has linear markings on one axis and percentage probability values from 0.01 to 99.99 on the other axis (see Figs. 41.6 and 41.7). The divisions on the probability axis are such that a straight line graph results for normally distributed data when percentage cumulative frequency values are plotted against upper class boundary values. If the points do not lie in a reasonably straight line, then the data is not normally distributed. The method used to test the normality of a distribution

is shown in Problems 5 and 6. The mean value and standard deviation of normally distributed data may be determined using normal probability paper. For normally distributed data, the area beneath the standardised normal curve and a z-value of unity (i.e. one standard deviation) may be obtained from Table 41.1. For one standard deviation, this area is 0.3413, i.e. 34.13%. An area of ± 1 standard deviation is symmetrically placed on either side of the $z = 0$ value, i.e. is symmetrically placed on either side of the 50 per cent cumulative frequency value. Thus an area corresponding to ± 1 standard deviation extends from percentage cumulative frequency values of $(50 + 34.13)\%$ to $(50 - 34.13)\%$, i.e. from 84.13% to 15.87%. For most purposes, these values area taken as 84% and 16%. Thus, when using normal probability paper, the standard deviation of the distribution is given by:

$$\frac{\text{(variable value for 84\% cumulative frequency)} - \text{(variable value for 16\% cumulative frequency)}}{2}$$

Problem 5. Use normal probability paper to determine whether the data given below, which refers to the masses of 50 copper ingots, is approximately normally distributed. If the data is normally distributed, determine the mean and standard deviation of the data from the graph drawn

Class mid-point value (kg)	29.5	30.5	31.5	32.5	33.5
Frequency	2	4	6	8	9

Class mid-point value (kg)	34.5	35.5	36.5	37.5	38.5
Frequency	8	6	4	2	1

To test the normality of a distribution, the upper class boundary/percentage cumulative frequency values are plotted on normal probability paper. The upper class boundary values are: 30, 31, 32, ..., 38, 39. The corresponding cumulative frequency values (for 'less than' the upper class boundary values) are: 2, $(4+2)=6$, $(6+4+2)=12$, 20, 29, 37, 43, 47, 49 and 50. The corresponding percentage cumulative frequency values are $\frac{2}{50} \times 100 = 4$, $\frac{6}{50} \times 100 = 12$, 24, 40, 58, 74, 86, 94, 98 and 100%.

The co-ordinates of upper class boundary/percentage cumulative frequency values are plotted as shown in

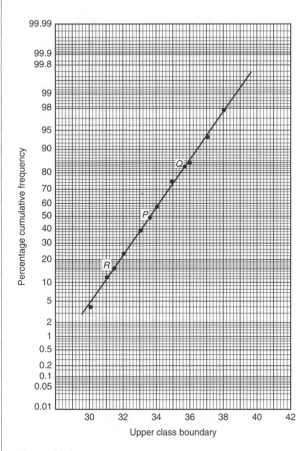

Figure 41.6

Fig. 41.6. When plotting these values, it will always be found that the co-ordinate for the 100% cumulative frequency value cannot be plotted, since the maximum value on the probability scale is 99.99. **Since the points plotted in Fig. 41.6 lie very nearly in a straight line, the data is approximately normally distributed.**

The mean value and standard deviation can be determined from Fig. 41.6. Since a normal curve is symmetrical, the mean value is the value of the variable corresponding to a 50% cumulative frequency value, shown as point P on the graph. This shows that **the mean value is 33.6 kg**. The standard deviation is determined using the 84% and 16% cumulative frequency values, shown as Q and R in Fig. 41.6. The variable values for Q and R are 35.7 and 31.4 respectively; thus two standard deviations correspond to $35.7 - 31.4$, i.e. 4.3, showing that the standard deviation of the distribution is approximately $\dfrac{4.3}{2}$ i.e. **2.15 standard deviations**.

The mean value and standard deviation of the distribution can be calculated using mean, $\bar{x} = \dfrac{\sum fx}{\sum f}$ and

standard deviation, $\sigma = \sqrt{\dfrac{\sum [f(x - \bar{x})^2]}{\sum f}}$ where f is the frequency of a class and x is the class mid-point value. Using these formulae gives a mean value of the distribution of 33.6 (as obtained graphically) and a standard deviation of 2.12, showing that the graphical method of determining the mean and standard deviation give quite realistic results.

Problem 6. Use normal probability paper to determine whether the data given below is normally distributed. Use the graph and assume a normal distribution whether this is so or not, to find approximate values of the mean and standard deviation of the distribution.

Class mid-point Values	5	15	25	35	45
Frequency	1	2	3	6	9

Class mid-point Values	55	65	75	85	95
Frequency	6	2	2	1	1

To test the normality of a distribution, the upper class boundary/percentage cumulative frequency values are plotted on normal probability paper. The upper class boundary values are: 10, 20, 30, ..., 90 and 100. The corresponding cumulative frequency values are 1, $1 + 2 = 3$, $1 + 2 + 3 = 6$, 12, 21, 27, 29, 31, 32 and 33. The percentage cumulative frequency values are $\dfrac{1}{33} \times 100 = 3$, $\dfrac{3}{33} \times 100 = 9$, 18, 36, 64, 82, 88, 94, 97 and 100.

The co-ordinates of upper class boundary values/ percentage cumulative frequency values are plotted as shown in Fig. 41.7. Although six of the points lie approximately in a straight line, three points corresponding to upper class boundary values of 50, 60 and 70 are not close to the line and indicate that **the distribution is not normally** distributed. However, if a normal distribution is assumed, the **mean value** corresponds to the variable value at a cumulative frequency of 50% and, from Fig. 41.7, point A is **48**. The value of the standard deviation of the distribution can be obtained from the variable values corresponding to the 84% and 16% cumulative frequency values, shown as B and C in Fig. 41.7 and give: $2\sigma = 69 - 28$, i.e. **the standard**

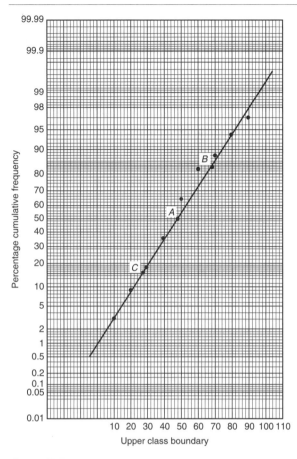

Figure 41.7

Class mid-point value and frequency table (top right):

Class mid-point value	27.2	27.4	27.6
Frequency	36	25	12

Use normal probability paper to show that this data approximates to a normal distribution and hence determine the approximate values of the mean and standard deviation of the distribution. Use the formula for mean and standard deviation to verify the results obtained.

$$\left[\begin{array}{l} \text{Graphically, } \bar{x} = 27.1, \sigma = 0.3; \text{ by} \\ \text{calculation, } \bar{x} = 27.079, \sigma = 0.3001 \end{array} \right]$$

2. A frequency distribution of the class mid-point values of the breaking loads for 275 similar fibres is as shown below:

Load (kN)	17	19	21	23
Frequency	9	23	55	78

Load (kN)	25	27	29	31
Frequency	64	28	14	4

Use normal probability paper to show that this distribution is approximately normally distributed and determine the mean and standard deviation of the distribution (a) from the graph and (b) by calculation.

$$\left[\begin{array}{l} \text{(a) } \bar{x} = 23.5 \text{ kN}, \sigma = 2.9 \text{ kN} \\ \text{(b) } \bar{x} = 23.364 \text{ kN}, \sigma = 2.917 \text{ kN} \end{array} \right]$$

deviation $\sigma = 20.5$. The calculated values of the mean and standard deviation of the distribution are 45.9 and 19.4 respectively, showing that errors are introduced if the graphical method of determining these values is used for data that is not normally distributed.

Now try the following exercise

Exercise 147 Further problems on testing for a normal distribution

1. A frequency distribution of 150 measurements is as shown:

Class mid-point value	26.4	26.6	26.8	27.0
Frequency	5	12	24	36

Section 7

This Revision test covers the material contained in Chapters 40 and 41. *The marks for each question are shown in brackets at the end of each question.*

1. A machine produces 15% defective components. In a sample of 5, drawn at random, calculate, using the binomial distribution, the probability that:

    (a) there will be 4 defective items

    (b) there will be not more than 3 defective items

    (c) all the items will be non-defective   (14)

2. 2% of the light bulbs produced by a company are defective. Determine, using the Poisson distribution, the probability that in a sample of 80 bulbs:

    (a) 3 bulbs will be defective, (b) not more than 3 bulbs will be defective, (c) at least 2 bulbs will be defective.   (13)

3. Some engineering components have a mean length of 20 mm and a standard deviation of 0.25 mm. Assume that the data on the lengths of the components is normally distributed.

    In a batch of 500 components, determine the number of components likely to:

    (a) have a length of less than 19.95 mm

    (b) be between 19.95 mm and 20.15 mm

    (c) be longer than 20.54 mm.   (12)

4. In a factory, cans are packed with an average of 1.0 kg of a compound and the masses are normally distributed about the average value. The standard deviation of a sample of the contents of the cans is 12 g. Determine the percentage of cans containing (a) less than 985 g, (b) more than 1030 g, (c) between 985 g and 1030 g.   (11)

All questions have only one correct answer (answers on page 570).

1. A graph of resistance against voltage for an electrical circuit is shown in Figure M3.1. The equation relating resistance R and voltage V is:

   (a) $R = 1.45\,V + 40$ (b) $R = 0.8\,V + 20$
   (c) $R = 1.45\,V + 20$ (d) $R = 1.25\,V + 20$

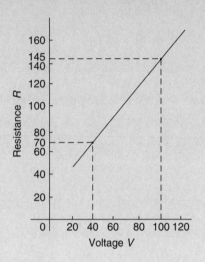

Figure M3.1

2. $\dfrac{5}{j^6}$ is equivalent to:

   (a) $j5$ (b) -5 (c) $-j5$ (d) 5

3. Two voltage phasors are shown in Figure M3.2. If $V_1 = 40$ volts and $V_2 = 100$ volts, the resultant (i.e. length OA) is:

   (a) 131.4 volts at 32.55° to V_1
   (b) 105.0 volts at 32.55° to V_1
   (c) 131.4 volts at 68.30° to V_1
   (d) 105.0 volts at 42.31° to V_1

4. Which of the straight lines shown in Figure M3.3 has the equation $y + 4 = 2x$?

   (a) (i)   (b) (ii)   (c) (iii)   (d) (iv)

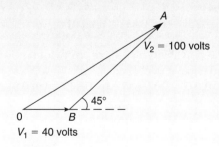

Figure M3.2

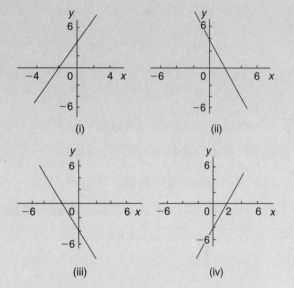

Figure M3.3

5. A pie diagram is shown in Figure M3.4 where P, Q, R and S represent the salaries of four employees of a firm. P earns £24 000 p.a. Employee S earns:

   (a) £40 000   (b) £36 000
   (c) £20 000   (d) £24 000

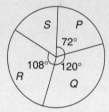

Figure M3.4

6. A force of 4 N is inclined at an angle of 45° to a second force of 7 N, both forces acting at a point, as shown in Figure M3.5. The magnitude of the resultant of these two forces and the direction of the resultant with respect to the 7 N force is:

   (a) 3 N at 45°   (b) 5 N at 146°
   (c) 11 N at 135°   (d) 10.2 N at 16°

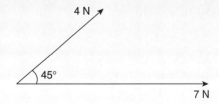

Figure M3.5

Questions 7 to 10 relate to the following information:

The capacitance (in pF) of 6 capacitors is as follows: {5, 6, 8, 5, 10, 2}

7. The median value is:
   (a) 36 pF   (b) 6 pF   (c) 5.5 pF   (d) 5 pF

8. The modal value is:
   (a) 36 pF   (b) 6 pF   (c) 5.5 pF   (d) 5 pF

9. The mean value is:
   (a) 36 pF   (b) 6 pF   (c) 5.5 pF   (d) 5 pF

10. The standard deviation is:
    (a) 2.66 pF   (b) 2.52 pF
    (c) 2.45 pF   (d) 6.33 pF

11. A graph of y against x, two engineering quantities, produces a straight line.
    A table of values is shown below:

x	2	-1	p
y	9	3	5

    The value of p is:
    (a) $-\dfrac{1}{2}$ (b) -2 (c) 3   (d) 0

Questions 12 and 13 relate to the following information. The voltage phasors V_1 and V_2 are shown in Figure M3.6.

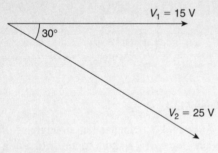

Figure M3.6

12. The resultant $V_1 + V_2$ is given by:
    (a) 38.72 V at $-19°$ to V_1
    (b) 14.16 V at $62°$ to V_1
    (c) 38.72 V at $161°$ to V_1
    (d) 14.16 V at $118°$ to V_1

13. The resultant $V_1 - V_2$ is given by:
    (a) 38.72 V at $-19°$ to V_1
    (b) 14.16 V at $62°$ to V_1
    (c) 38.72 V at $161°$ to V_1
    (d) 14.16 V at $118°$ to V_1

14. The curve obtained by joining the co-ordinates of cumulative frequency against upper class boundary values is called;
    (a) a historgram   (b) a frequency polygon
    (c) a tally diagram   (d) an ogive

15. A graph relating effort E (plotted vertically) against load L (plotted horizontally) for a set of pulleys is given by $L + 30 = 6E$. The gradient of the graph is:
    (a) $\dfrac{1}{6}$ (b) 5   (c) 6   (d) $\dfrac{1}{5}$

Questions 16 to 19 relate to the following information:

x and y are two related engineering variables and p and q are constants.

For the law $y - p = \dfrac{q}{x}$ to be verified it is necessary to plot a graph of the variables.

16. On the vertical axis is plotted:
    (a) y   (b) p   (c) q   (d) x

17. On the horizontal axis is plotted:
    (a) x   (b) $\dfrac{q}{x}$ (c) $\dfrac{1}{x}$ (d) p

18. The gradient of the graph is:
    (a) y   (b) p   (c) q   (d) x

19. The vertical axis intercept is:
    (a) y   (b) p   (c) q   (d) x

Questions 20 to 22 relate to the following information:

A box contains 35 brass washers, 40 steel washers and 25 aluminium washers. 3 washers are drawn at random from the box without replacement.

20. The probability that all three are steel washers is:
    (a) 0.0611   (b) 1.200   (c) 0.0640   (d) 1.182

21. The probability that there are no aluminium washers is:
    (a) 2.250   (b) 0.418   (c) 0.014   (d) 0.422

22. The probability that there are two brass washers and either a steel or an aluminium washer is:
    (a) 0.071   (b) 0.687   (c) 0.239   (d) 0.343

23. $(-4 - j3)$ in polar form is :
    (a) $5\angle -143.13°$ (b) $5\angle 126.87°$
    (c) $5\angle 143.13°$ (d) $5\angle -126.87°$

24. The magnitude of the resultant of velocities of 3 m/s at 20° and 7 m/s at 120° when acting simultaneously at a point is:
    (a) 21 m/s   (b) 10 m/s
    (c) 7.12 m/s   (d) 4 m/s

25. Here are four equations in x and y. When x is plotted against y, in each case a straight line results.

    (i) $y + 3 = 3x$ (ii) $y + 3x = 3$

    (iii) $\dfrac{y}{2} - \dfrac{3}{2} = x$ (iv) $\dfrac{y}{3} = x + \dfrac{2}{3}$

    Which of these equations are parallel to each other?
    (a) (i) and (ii)   (b) (i) and (iv)
    (c) (ii) and (iii)   (d) (ii) and (iv)

26. The relationship between two related engineering variables x and y is $y - cx = bx^2$ where b and c are constants. To produce a straight line graph it is necessary to plot:
    (a) x vertically against y horizontally
    (b) y vertically against x^2 horizontally
    (c) $\dfrac{y}{x}$ vertically against x horizontally
    (d) y vertically against x horizontally

27. The number of faults occurring on a production line in a 9-week period are as shown:
    32   29   27   26   29   39   33   29   37
    The third quartile value is:
    (a) 29   (b) 35   (c) 31   (d) 28

28. $(1 + j)^4$ is equivalent to:
    (a) 4   (b) $-j4$ (c) $j4$ (d) -4

29. 2% of the components produced by a manufacturer are defective. Using the Poisson distribution the percentage probability that more than two will be defective in a sample of 100 components is:
    (a) 13.5%   (b) 32.3%
    (c) 27.1%   (d) 59.4%

30. The equation of the graph shown in Figure M3.7 is:
    (a) $x(x + 1) = \dfrac{15}{4}$ (b) $4x^2 - 4x - 15 = 0$
    (c) $x^2 - 4x - 5 = 0$ (d) $4x^2 + 4x - 15 = 0$

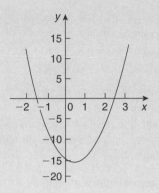

Figure M3.7

31. In an experiment demonstrating Hooke's law, the strain in a copper wire was measured for various stresses. The results included

Stress (megapascals)	18.24	24.00	39.36
Strain	0.00019	0.00025	0.00041

    When stress is plotted vertically against strain horizontally a straight line graph results.
    Young's modulus of elasticity for copper, which is given by the gradient of the graph, is:
    (a) 96×10^9 Pa   (b) 1.04×10^{-11} Pa
    (c) 96 Pa   (d) 96 000 Pa

Questions 32 and 33 relate to the following information:
The frequency distribution for the values of resistance in ohms of 40 transistors is as follows:
    15.5–15.9   3   16.0–16.4   10
    16.5–16.9   13   17.0–17.4   8
    17.5–17.9   6

32. The mean value of the resistance is:
    (a) 16.75 Ω   (b) 1.0 Ω
    (c) 15.85 Ω   (d) 16.95 Ω

33. The standard deviation is:
    (a) 0.335 Ω   (b) 0.251 Ω
    (c) 0.682 Ω   (d) 0.579 Ω

34. The depict a set of values from 0.05 to 275, the minimum number of cycles required on logarithmic graph paper is:
    (a) 2   (b) 3   (c) 4   (d) 5

35. A manufacturer estimates that 4% of components produced are defective. Using the binomial distribution, the percentage probability that less than two

components will be defective in a sample of 10 components is:

(a) 0.40%   (b) 5.19%

(c) 0.63%   (d) 99.4%

Questions 36 to 39 relate to the following information.

A straight line graph is plotted for the equation $y = ax^n$, where y and x are the variables and a and n are constants.

36. On the vertical axis is plotted:

(a) y (b) x (c) $\ln y$ (d) a

37. On the horizontal axis is plotted:

(a) $\ln x$ (b) x (c) x^n (d) a

38. The gradient of the graph is given by:

(a) y (b) a (c) x (d) n

39. The vertical axis intercept is given by:

(a) n (b) $\ln a$ (c) x (d) $\ln y$

Questions 40 to 42 relate to the following information.

The probability of a component failing in one year due to excessive temperature is $\dfrac{1}{16}$, due to excessive vibration is $\dfrac{1}{20}$ and due to excessive humidity is $\dfrac{1}{40}$.

40. The probability that a component fails due to excessive temperature and excessive vibration is:

(a) $\dfrac{285}{320}$ (b) $\dfrac{1}{320}$ (c) $\dfrac{9}{80}$ (d) $\dfrac{1}{800}$

41. The probability that a component fails due to excessive vibration or excessive humidity is:

(a) 0.00125   (b) 0.00257

(c) 0.0750   (d) 0.1125

42. The probability that a component will not fail because of both excessive temperature and excessive humidity is:

(a) 0.914   (b) 1.913

(c) 0.00156   (d) 0.0875

43. Three forces of 2 N, 3 N and 4 N act as shown in Figure M3.8. The magnitude of the resultant force is:

(a) 8.08 N   (b) 9 N   (c) 7.17 N   (d) 1 N

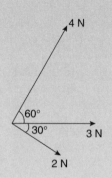

Figure M3.8

44. $2\angle\dfrac{\pi}{3} + 3\angle\dfrac{\pi}{6}$ in polar form is:

(a) $5\angle\dfrac{\pi}{2}$ (b) $4.84\angle0.84$

(c) $6\angle0.55$ (d) $4.84\angle0.73$

Questions 45 and 46 relate to the following information.

Two alternating voltages are given by:

$v_1 = 2 \sin \omega t$ and $v_2 = 3 \sin \left(\omega t + \dfrac{\pi}{4}\right)$ volts.

45. Which of the phasor diagrams shown in Figure M3.9 represents $v_R = v_1 + v_2$?

(a) (i)   (b) (ii)   (c) (iii)   (d) (iv)

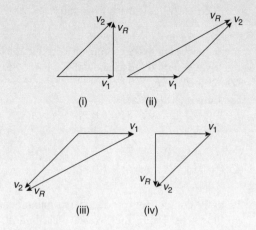

Figure M3.9

46. Which of the phasor diagrams shown represents $v_R = v_1 - v_2$?

(a) (i)   (b) (ii)   (c) (iii)   (d) (iv)

47. The two square roots of $(-3 + j4)$ are:

(a) $\pm(1 + j2)$ (b) $\pm(0.71 + j2.12)$

(c) $\pm(1 - j2)$ (d) $\pm(0.71 - j2.12)$

Questions 48 and 49 relate to the following information.

A set of measurements (in mm) is as follows:
$\{4, 5, 2, 11, 7, 6, 5, 1, 5, 8, 12, 6\}$

48. The median is:
    (a) 6 mm   (b) 5 mm   (c) 72 mm   (d) 5.5 mm

49. The mean is:
    (a) 6 mm   (b) 5 mm   (c) 72 mm   (d) 5.5 mm

50. The graph of $y = 2 \tan 3\theta$ is:
    (a) a continuous, periodic, even function
    (b) a discontinuous, non-periodic, odd function
    (c) a discontinuous, periodic, odd function
    (d) a continuous, non-periodic, even function

Questions 51 to 53 relate to the following information.

The mean height of 400 people is 170 cm and the standard deviation is 8 cm. Assume a normal distribution. (See Table 41.1 on pages 367/368)

51. The number of people likely to have heights of between 154 cm and 186 cm is:
    (a) 390   (b) 380   (c) 190   (d) 185

52. The number of people likely to have heights less than 162 cm is:
    (a) 133   (b) 380   (c) 67   (d) 185

53. The number of people likely to have a height of more than 186 cm is:
    (a) 10   (b) 67   (c) 137   (d) 20

54. $[2\angle 30°]^4$ in Cartesian form is:
    (a) $(0.50 + j0.06)$ (b) $(-8 + j13.86)$
    (c) $(-4 + j6.93)$ (d) $(13.86 + j8)$

Section 8

Differential Calculus

Chapter 42

Introduction to differentiation

42.1 Introduction to calculus

Calculus is a branch of mathematics involving or leading to calculations dealing with continuously varying functions.

Calculus is a subject that falls into two parts:

(i) **differential calculus** (or **differentiation**) and
(ii) **integral calculus** (or **integration**).

Differentiation is used in calculations involving velocity and acceleration, rates of change and maximum and minimum values of curves.

42.2 Functional notation

In an equation such as $y = 3x^2 + 2x - 5$, y is said to be a function of x and may be written as $y = f(x)$.

An equation written in the form $f(x) = 3x^2 + 2x - 5$ is termed **functional notation**. The value of $f(x)$ when $x = 0$ is denoted by $f(0)$, and the value of $f(x)$ when $x = 2$ is denoted by $f(2)$ and so on. Thus when $f(x) = 3x^2 + 2x - 5$, then

$$f(0) = 3(0)^2 + 2(0) - 5 = -5$$

and $f(2) = 3(2)^2 + 2(2) - 5 = 11$ and so on.

Problem 1. If $f(x) = 4x^2 - 3x + 2$ find:
$f(0), f(3), f(-1)$ and $f(3) - f(-1)$

$$f(x) = 4x^2 - 3x + 2$$

$$f(0) = 4(0)^2 - 3(0) + 2 = \mathbf{2}$$

$$f(3) = 4(3)^2 - 3(3) + 2$$
$$= 36 - 9 + 2 = \mathbf{29}$$

$$f(-1) = 4(-1)^2 - 3(-1) + 2$$
$$= 4 + 3 + 2 = \mathbf{9}$$

$$f(3) - f(-1) = 29 - 9 = \mathbf{20}$$

Problem 2. Given that $f(x) = 5x^2 + x - 7$ determine:

(i) $f(2) \div f(1)$ (iii) $f(3+a) - f(3)$

(ii) $f(3+a)$ (iv) $\dfrac{f(3+a) - f(3)}{a}$

$$f(x) = 5x^2 + x - 7$$

(i) $f(2) = 5(2)^2 + 2 - 7 = 15$

   $f(1) = 5(1)^2 + 1 - 7 = -1$

   $f(2) \div f(1) = \dfrac{15}{-1} = \mathbf{-15}$

(ii) $f(3+a) = 5(3+a)^2 + (3+a) - 7$

           $= 5(9 + 6a + a^2) + (3+a) - 7$

           $= 45 + 30a + 5a^2 + 3 + a - 7$

           $= \mathbf{41 + 31a + 5a^2}$

(iii) $f(3) = 5(3)^2 + 3 - 7 = 41$

    $f(3+a) - f(3) = (41 + 31a + 5a^2) - (41)$

                $= \mathbf{31a + 5a^2}$

(iv) $\dfrac{f(3+a) - f(3)}{a} = \dfrac{31a + 5a^2}{a} = \mathbf{31 + 5a}$

Now try the following exercise

Exercise 148 Further problems on functional notation

1. If $f(x) = 6x^2 - 2x + 1$ find $f(0)$, $f(1)$, $f(2)$, $f(-1)$ and $f(-3)$.

$$[1, 5, 21, 9, 61]$$

2. If $f(x) = 2x^2 + 5x - 7$ find $f(1)$, $f(2)$, $f(-1)$, $f(2) - f(-1)$.

$$[0, 11, -10, 21]$$

3. Given $f(x) = 3x^3 + 2x^2 - 3x + 2$ prove that $f(1) = \frac{1}{7}f(2)$

4. If $f(x) = -x^2 + 3x + 6$ find $f(2)$, $f(2+a)$, $f(2+a) - f(2)$ and $\dfrac{f(2+a) - f(2)}{a}$

$$[8, -a^2 - a + 8, -a^2 - a, -a - 1]$$

42.3 The gradient of a curve

(a) If a tangent is drawn at a point P on a curve, then the gradient of this tangent is said to be the **gradient of the curve** at P. In Fig. 42.1, the gradient of the curve at P is equal to the gradient of the tangent PQ.

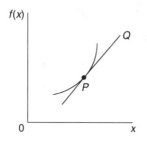

Figure 42.1

(b) For the curve shown in Fig. 42.2, let the points A and B have co-ordinates (x_1, y_1) and (x_2, y_2), respectively. In functional notation, $y_1 = f(x_1)$ and $y_2 = f(x_2)$ as shown.

The gradient of the chord AB

$$= \frac{BC}{AC} = \frac{BD - CD}{ED}$$

$$= \frac{f(x_2) - f(x_1)}{(x_2 - x_1)}$$

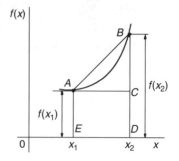

Figure 42.2

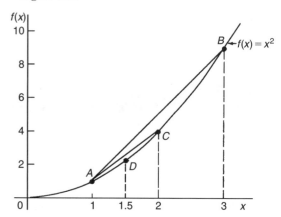

Figure 42.3

(c) For the curve $f(x) = x^2$ shown in Fig. 42.3:

(i) the gradient of chord AB

$$= \frac{f(3) - f(1)}{3 - 1} = \frac{9 - 1}{2} = 4$$

(ii) the gradient of chord AC

$$= \frac{f(2) - f(1)}{2 - 1} = \frac{4 - 1}{1} = 3$$

(iii) the gradient of chord AD

$$= \frac{f(1.5) - f(1)}{1.5 - 1} = \frac{2.25 - 1}{0.5} = 2.5$$

(iv) if E is the point on the curve $(1.1, f(1.1))$ then the gradient of chord AE

$$= \frac{f(1.1) - f(1)}{1.1 - 1}$$

$$= \frac{1.21 - 1}{0.1} = 2.1$$

(v) if F is the point on the curve $(1.01, f(1.01))$ then the gradient of chord AF

$$= \frac{f(1.01) - f(1)}{1.01 - 1}$$

$$= \frac{1.0201 - 1}{0.01} = 2.01$$

Thus as point B moves closer and closer to point A the gradient of the chord approaches nearer and nearer to the value 2. This is called the **limiting value** of the gradient of the chord AB and when B coincides with A the chord becomes the tangent to the curve.

Now try the following exercise

Exercise 149   A further problem on the gradient of a curve

1. Plot the curve $f(x) = 4x^2 - 1$ for values of x from $x = -1$ to $x = +4$. Label the co-ordinates $(3, f(3))$ and $(1, f(1))$ as J and K, respectively. Join points J and K to form the chord JK. Determine the gradient of chord JK. By moving J nearer and nearer to K determine the gradient of the tangent of the curve at K.

[16, 8]

42.4 Differentiation from first principles

(i) In Fig. 42.4, A and B are two points very close together on a curve, δx (delta x) and δy (delta y) representing small increments in the x and y directions, respectively.

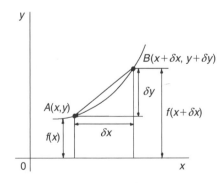

Figure 42.4

Gradient of chord $AB = \dfrac{\delta y}{\delta x}$

However, $\delta y = f(x + \delta x) - f(x)$

Hence $\dfrac{\delta y}{\delta x} = \dfrac{f(x + \delta x) - f(x)}{\delta x}$

As δx approaches zero, $\dfrac{\delta y}{\delta x}$ approaches a limiting value and the gradient of the chord approaches the gradient of the tangent at A.

(ii) When determining the gradient of a tangent to a curve there are two notations used. The gradient of the curve at A in Fig. 42.4 can either be written as:

$$\underset{\delta x \to 0}{\text{limit}} \frac{\delta y}{\delta x} \quad \text{or} \quad \underset{\delta x \to 0}{\text{limit}} \left\{ \frac{f(x + \delta x) - f(x)}{\delta x} \right\}$$

In **Leibniz notation**, $\dfrac{dy}{dx} = \underset{\delta x \to 0}{\text{limit}} \dfrac{\delta y}{\delta x}$

In **functional notation**,

$$f'(x) = \underset{\delta x \to 0}{\text{limit}} \left\{ \frac{f(x + \delta x) - f(x)}{\delta x} \right\}$$

(iii) $\dfrac{dy}{dx}$ is the same as $f'(x)$ and is called the **differential coefficient** or the **derivative**. The process of finding the differential coefficient is called **differentiation**.

Summarising, the differential coefficient,

$$\frac{dy}{dx} = f'(x) = \underset{\delta x \to 0}{\text{limit}} \frac{\delta y}{\delta x}$$

$$= \underset{\delta x \to 0}{\text{limit}} \left\{ \frac{f(x + \delta x) - f(x)}{\delta x} \right\}$$

Problem 3. Differentiate from first principles $f(x) = x^2$ and determine the value of the gradient of the curve at $x = 2$

To 'differentiate from first principles' means 'to find $f'(x)$' by using the expression

$$f'(x) = \underset{\delta x \to 0}{\text{limit}} \left\{ \frac{f(x + \delta x) - f(x)}{\delta x} \right\}$$

$$f(x) = x^2$$

Substituting $(x + \delta x)$ for x gives
$f(x + \delta x) = (x + \delta x)^2 = x^2 + 2x\delta x + \delta x^2$, hence

$$f'(x) = \underset{\delta x \to 0}{\text{limit}} \left\{ \frac{(x^2 + 2x\delta x + \delta x^2) - (x^2)}{\delta x} \right\}$$

$$= \underset{\delta x \to 0}{\text{limit}} \left\{ \frac{2x\delta x + \delta x^2}{\delta x} \right\} = \underset{\delta x \to 0}{\text{limit}} \{2x + \delta x\}$$

Section 8

As $\delta x \to 0$, $[2x + \delta x] \to [2x + 0]$. Thus $f'(x) = 2x$, i.e. the differential coefficient of x^2 is $2x$. At $x = 2$, the gradient of the curve, $f'(x) = 2(2) = \mathbf{4}$

> **Problem 4.** Find the differential coefficient of
> $$y = 5x$$

By definition, $\quad \dfrac{dy}{dx} = f'(x)$

$$= \underset{\delta x \to 0}{\text{limit}} \left\{ \frac{f(x + \delta x) - f(x)}{\delta x} \right\}$$

The function being differentiated is $y = f(x) = 5x$. Substituting $(x + \delta x)$ for x gives:
$f(x + \delta x) = 5(x + \delta x) = 5x + 5\delta x$. Hence

$$\frac{dy}{dx} = f'(x) = \underset{\delta x \to 0}{\text{limit}} \left\{ \frac{(5x + 5\delta x) - (5x)}{\delta x} \right\}$$

$$= \underset{\delta x \to 0}{\text{limit}} \left\{ \frac{5\delta x}{\delta x} \right\} = \underset{\delta x \to 0}{\text{limit}}\{5\}$$

Since the term δx does not appear in [5] the limiting value as $\delta x \to 0$ of [5] is 5. Thus $\dfrac{dy}{dx} = \mathbf{5}$, i.e. the differential coefficient of $5x$ is 5. The equation $y = 5x$ represents a straight line of gradient 5 (see Chapter 28). The 'differential coefficient' (i.e. $\dfrac{dy}{dx}$ or $f'(x)$) means 'the gradient of the curve', and since the slope of the line $y = 5x$ is 5 this result can be obtained by inspection. Hence, in general, if $y = kx$ (where k is a constant), then the gradient of the line is k and $\dfrac{dy}{dx}$ or $f'(x) = k$.

> **Problem 5.** Find the derivative of $y = 8$

$y = f(x) = 8$. Since there are no x-values in the original equation, substituting $(x + \delta x)$ for x still gives $f(x + \delta x) = 8$. Hence

$$\frac{dy}{dx} = f'(x) = \underset{\delta x \to 0}{\text{limit}} \left\{ \frac{f(x + \delta x) - f(x)}{\delta x} \right\}$$

$$= \underset{\delta x \to 0}{\text{limit}} \left\{ \frac{8 - 8}{\delta x} \right\} = 0$$

Thus, when $y = 8$, $\dfrac{dy}{dx} = \mathbf{0}$

The equation $y = 8$ represents a straight horizontal line and the gradient of a horizontal line is zero, hence the

result could have been determined by inspection. 'Finding the derivative' means 'finding the gradient', hence, in general, for any horizontal line if $y = k$ (where k is a constant) then $\dfrac{dy}{dx} = 0$.

> **Problem 6.** Differentiate from first principles
> $$f(x) = 2x^3$$

Substituting $(x + \delta x)$ for x gives

$$f(x + \delta x) = 2(x + \delta x)^3$$
$$= 2(x + \delta x)(x^2 + 2x\delta x + \delta x^2)$$
$$= 2(x^3 + 3x^2\delta x + 3x\delta x^2 + \delta x^3)$$
$$= 2x^3 + 6x^2\delta x + 6x\delta x^2 + 2\delta x^3$$

$$\frac{dy}{dx} = f'(x) = \underset{\delta x \to 0}{\text{limit}} \left\{ \frac{f(x + \delta x) - f(x)}{\delta x} \right\}$$

$$= \underset{\delta x \to 0}{\text{limit}} \left\{ \frac{(2x^3 + 6x^2\delta x + 6x\delta x^2 + 2\delta x^3) - (2x^3)}{\delta x} \right\}$$

$$= \underset{\delta x \to 0}{\text{limit}} \left\{ \frac{6x^2\delta x + 6x\delta x^2 + 2\delta x^3}{\delta x} \right\}$$

$$= \underset{\delta x \to 0}{\text{limit}}\{6x^2 + 6x\delta x + 2\delta x^2\}$$

Hence $f'(x) = \mathbf{6x^2}$, i.e. the differential coefficient of $2x^3$ is $6x^2$.

> **Problem 7.** Find the differential coefficient of $y = 4x^2 + 5x - 3$ and determine the gradient of the curve at $x = -3$

$$y = f(x) = 4x^2 + 5x - 3$$
$$f(x + \delta x) = 4(x + \delta x)^2 + 5(x + \delta x) - 3$$
$$= 4(x^2 + 2x\delta x + \delta x^2) + 5x + 5\delta x - 3$$
$$= 4x^2 + 8x\delta x + 4\delta x^2 + 5x + 5\delta x - 3$$

$$\frac{dy}{dx} = f'(x) = \underset{\delta x \to 0}{\text{limit}} \left\{ \frac{f(x + \delta x) - f(x)}{\delta x} \right\}$$

$$= \underset{\delta x \to 0}{\text{limit}} \left\{ \frac{\begin{array}{c}(4x^2 + 8x\delta x + 4\delta x^2 + 5x + 5\delta x - 3) \\ - (4x^2 + 5x - 3)\end{array}}{\delta x} \right\}$$

$$= \underset{\delta x \to 0}{\text{limit}} \left\{ \frac{8x\delta x + 4\delta x^2 + 5\delta x}{\delta x} \right\}$$

$$= \underset{\delta x \to 0}{\text{limit}} \{8x + 4\delta x + 5\}$$

i.e. $\dfrac{dy}{dx} = f'(x) = 8x + 5$

At $x = -3$, the gradient of the curve

$$= \frac{dy}{dx} = f'(x) = 8(-3) + 5 = -19$$

Now try the following exercise

Exercise 150   Further problems on differentiation from first principles

In Problems 1 to 12, differentiate from first principles.

1. $y = x$ [1]

2. $y = 7x$ [7]

3. $y = 4x^2$ [8x]

4. $y = 5x^3$ [15x^2]

5. $y = -2x^2 + 3x - 12$ [-4x + 3]

6. $y = 23$ [0]

7. $f(x) = 9x$ [9]

8. $f(x) = \dfrac{2x}{3}$ $\left[\dfrac{2}{3}\right]$

9. $f(x) = 9x^2$ [18x]

10. $f(x) = -7x^3$ [-21x^2]

11. $f(x) = x^2 + 15x - 4$ [2x + 15]

12. $f(x) = 4$ [0]

13. Determine $\dfrac{d}{dx}(4x^3)$ from first principles

                            [12x^2]

14. Find $\dfrac{d}{dx}(3x^2 + 5)$ from first principles

                            [6x]

42.5 Differentiation of $y = ax^n$ by the general rule

From differentiation by first principles, a general rule for differentiating ax^n emerges where a and n are any constants. This rule is:

$$\text{if } y = ax^n \text{ then } \frac{dy}{dx} = anx^{n-1}$$

$$\text{or, if } f(x) = ax^n \text{ then } f'(x) = anx^{n-1}$$

(Each of the results obtained in worked problems 3 to 7 may be deduced by using this general rule).

When differentiating, results can be expressed in a number of ways.

For example:

(i) if $y = 3x^2$ then $\dfrac{dy}{dx} = 6x$,

(ii) if $f(x) = 3x^2$ then $f'(x) = 6x$,

(iii) the differential coefficient of $3x^2$ is $6x$,

(iv) the derivative of $3x^2$ is $6x$, and

(v) $\dfrac{d}{dx}(3x^2) = 6x$

Problem 8. Using the general rule, differentiate the following with respect to x:

(a) $y = 5x^7$ (b) $y = 3\sqrt{x}$ (c) $y = \dfrac{4}{x^2}$

(a) Comparing $y = 5x^7$ with $y = ax^n$ shows that $a = 5$ and $n = 7$. Using the general rule,

$$\frac{dy}{dx} = anx^{n-1} = (5)(7)x^{7-1} = 35x^6$$

(b) $y = 3\sqrt{x} = 3x^{\frac{1}{2}}$. Hence $a = 3$ and $n = \dfrac{1}{2}$

$$\frac{dy}{dx} = anx^{n-1} = (3)\frac{1}{2}x^{\frac{1}{2}-1}$$

$$= \frac{3}{2}x^{-\frac{1}{2}} = \frac{3}{2x^{\frac{1}{2}}} = \frac{3}{2\sqrt{x}}$$

(c) $y = \dfrac{4}{x^2} = 4x^{-2}$. Hence $a = 4$ and $n = -2$

$$\frac{dy}{dx} = anx^{n-1} = (4)(-2)x^{-2-1}$$

$$= -8x^{-3} = -\frac{8}{x^3}$$

Section 8

Problem 9. Find the differential coefficient of $y = \frac{2}{5}x^3 - \frac{4}{x^3} + 4\sqrt{x^5} + 7$

$$y = \frac{2}{5}x^3 - \frac{4}{x^3} + 4\sqrt{x^5} + 7$$

i.e. $\quad y = \frac{2}{5}x^3 - 4x^{-3} + 4x^{5/2} + 7$

$$\frac{dy}{dx} = \left(\frac{2}{5}\right)(3)x^{3-1} - (4)(-3)x^{-3-1}$$

$$+ (4)\left(\frac{5}{2}\right)x^{(5/2)-1} + 0$$

$$= \frac{6}{5}x^2 + 12x^{-4} + 10x^{3/2}$$

i.e. $\quad \dfrac{dy}{dx} = \dfrac{6}{5}x^2 + \dfrac{12}{x^4} + 10\sqrt{x^3}$

Problem 10. If $f(t) = 5t + \dfrac{1}{\sqrt{t^3}}$ find $f'(t)$

$$f(t) = 5t + \frac{1}{\sqrt{t^3}} = 5t + \frac{1}{t^{\frac{3}{2}}} = 5t^1 + t^{-\frac{3}{2}}$$

Hence $\quad f'(t) = (5)(1)t^{1-1} + \left(-\frac{3}{2}\right)t^{-\frac{3}{2}-1}$

$$= 5t^0 - \frac{3}{2}t^{-\frac{5}{2}}$$

i.e. $\quad f'(t) = 5 - \dfrac{3}{2t^{\frac{5}{2}}} = \mathbf{5 - \dfrac{3}{2\sqrt{t^5}}}$

Problem 11. Differentiate $y = \dfrac{(x+2)^2}{x}$ with respect to x

$$y = \frac{(x+2)^2}{x} = \frac{x^2 + 4x + 4}{x}$$

$$= \frac{x^2}{x} + \frac{4x}{x} + \frac{4}{x}$$

i.e. $\quad y = x + 4 + 4x^{-1}$

Hence $\quad \dfrac{dy}{dx} = 1 + 0 + (4)(-1)x^{-1-1}$

$$= 1 - 4x^{-2} = \mathbf{1 - \dfrac{4}{x^2}}$$

Now try the following exercise

Exercise 151 Further problems on differentiation of $y = ax^n$ by the general rule

In Problems 1 to 8, determine the differential coefficients with respect to the variable.

1. $y = 7x^4$ $\qquad\qquad\qquad\qquad [28x^3]$

2. $y = \sqrt{x}$ $\qquad\qquad\qquad\quad \left[\dfrac{1}{2\sqrt{x}}\right]$

3. $y = \sqrt{t^3}$ $\qquad\qquad\qquad\quad \left[\dfrac{3}{2}\sqrt{t}\right]$

4. $y = 6 + \dfrac{1}{x^3}$ $\qquad\qquad\quad \left[-\dfrac{3}{x^4}\right]$

5. $y = 3x - \dfrac{1}{\sqrt{x}} + \dfrac{1}{x}$ $\quad \left[3 + \dfrac{1}{2\sqrt{x^3}} - \dfrac{1}{x^2}\right]$

6. $y = \dfrac{5}{x^2} - \dfrac{1}{\sqrt{x^7}} + 2$ $\quad \left[-\dfrac{10}{x^3} + \dfrac{7}{2\sqrt{x^9}}\right]$

7. $y = 3(t-2)^2$ $\qquad\qquad\qquad [6t - 12]$

8. $y = (x+1)^3$ $\qquad\qquad\quad [3x^2 + 6x + 3]$

9. Using the general rule for ax^n check the results of Problems 1 to 12 of Exercise 150, page 387.

10. Differentiate $f(x) = 6x^2 - 3x + 5$ and find the gradient of the curve at (a) $x = -1$, and (b) $x = 2$. $\qquad [12x - 3$ (a) -15 (b) $21]$

11. Find the differential coefficient of $y = 2x^3 + 3x^2 - 4x - 1$ and determine the gradient of the curve at $x = 2$. $\qquad\qquad\qquad [6x^2 + 6x - 4, 32]$

12. Determine the derivative of $y = -2x^3 + 4x + 7$ and determine the gradient of the curve at $x = -1.5$ $\qquad\qquad\qquad [-6x^2 + 4, -9.5]$

42.6 Differentiation of sine and cosine functions

Figure 42.5(a) shows a graph of $y = \sin \theta$. The gradient is continually changing as the curve moves from O to A to B to C to D. The gradient, given by $\dfrac{dy}{d\theta}$, may be

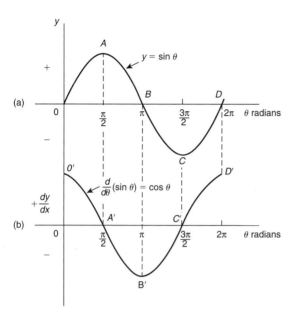

Figure 42.5

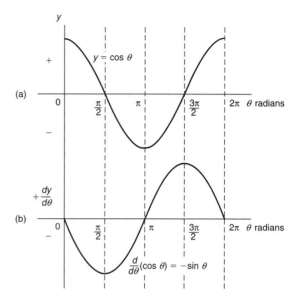

Figure 42.6

plotted in a corresponding position below $y = \sin\theta$, as shown in Fig. 42.5(b).

(i) At 0, the gradient is positive and is at its steepest. Hence $0'$ is the maximum positive value.

(ii) Between 0 and A the gradient is positive but is decreasing in value until at A the gradient is zero, shown as A'.

(iii) Between A and B the gradient is negative but is increasing in value until at B the gradient is at its steepest. Hence B' is a maximum negative value.

(iv) If the gradient of $y = \sin\theta$ is further investigated between B and C and C and D then the resulting graph of $\dfrac{dy}{d\theta}$ is seen to be a cosine wave.

Hence the rate of change of $\sin\theta$ is $\cos\theta$, i.e.

if $y = \sin\theta$ then $\dfrac{dy}{d\theta} = \cos\theta$

It may also be shown that:

if $y = \sin a\theta$, $\dfrac{dy}{d\theta} = a\cos a\theta$

(where a is a constant)

and if $y = \sin(a\theta + \alpha)$, $\dfrac{dy}{d\theta} = a\cos(a\theta + \alpha)$

(where a and α are constants).

If a similar exercise is followed for $y = \cos\theta$ then the graphs of Fig. 42.6 result, showing $\dfrac{dy}{d\theta}$ to be a graph of $\sin\theta$, but displaced by π radians. If each point on the curve $y = \sin\theta$ (as shown in Fig. 42.5(a)) were to be made negative, (i.e. $+\dfrac{\pi}{2}$ is made $-\dfrac{\pi}{2}$, $-\dfrac{3\pi}{2}$ is made

$+\dfrac{3\pi}{2}$, and so on) then the graph shown in Fig. 42.6(b) would result. This latter graph therefore represents the curve of $-\sin\theta$.

Thus, if $y = \cos\theta$, $\dfrac{dy}{d\theta} = -\sin\theta$

It may also be shown that:

if $y = \cos a\theta$, $\dfrac{dy}{d\theta} = -a\sin a\theta$

(where a is a constant)

and if $y = \cos(a\theta + \alpha)$, $\dfrac{dy}{d\theta} = -a\sin(a\theta + \alpha)$

(where a and α are constants).

Problem 12. Differentiate the following with respect to the variable: (a) $y = 2\sin 5\theta$ (b) $f(t) = 3\cos 2t$

(a) $y = 2\sin 5\theta$

$\dfrac{dy}{d\theta} = (2)(5)\cos 5\theta = \mathbf{10\cos 5\theta}$

(b) $f(t) = 3\cos 2t$

$f'(t) = (3)(-2)\sin 2t = \mathbf{-6\sin 2t}$

Problem 13. Find the differential coefficient of $y = 7\sin 2x - 3\cos 4x$

$y = 7\sin 2x - 3\cos 4x$

$\dfrac{dy}{dx} = (7)(2)\cos 2x - (3)(-4)\sin 4x$

$= \mathbf{14\cos 2x + 12\sin 4x}$

Section 8

Problem 14. Differentiate the following with respect to the variable:

(a) $f(\theta) = 5 \sin(100\pi\theta - 0.40)$
(b) $f(t) = 2 \cos(5t + 0.20)$

(a) If $f(\theta) = 5 \sin(100\pi\theta - 0.40)$

$$f'(\theta) = 5[100\pi \cos(100\pi\theta - 0.40)]$$

$$= \mathbf{500\pi \cos(100\pi\theta - 0.40)}$$

(b) If $f(t) = 2 \cos(5t + 0.20)$

$$f'(t) = 2[-5 \sin(5t + 0.20)]$$

$$= \mathbf{-10 \sin(5t + 0.20)}$$

Problem 15. An alternating voltage is given by: $v = 100 \sin 200t$ volts, where t is the time in seconds. Calculate the rate of change of voltage when (a) $t = 0.005$ s and (b) $t = 0.01$ s

$v = 100 \sin 200t$ volts. The rate of change of v is given by $\dfrac{dv}{dt}$.

$$\frac{dv}{dt} = (100)(200) \cos 200t = 20\,000 \cos 200t$$

(a) When $t = 0.005$ s,

$$\frac{dv}{dt} = 20\,000 \cos(200)(0.005) = 20\,000 \cos 1$$

cos 1 means 'the cosine of 1 radian' (make sure your calculator is on radians — not degrees).

Hence $\dfrac{dv}{dt} = \mathbf{10\,806}$ **volts per second**

(b) When $t = 0.01$ s,

$$\frac{dv}{dt} = 20\,000 \cos(200)(0.01) = 20\,000 \cos 2.$$

Hence $\dfrac{dv}{dt} = \mathbf{-8323}$ **volts per second**

Now try the following exercise

Exercise 152   Further problems on the differentiation of sine and cosine functions

1. Differentiate with respect to x: (a) $y = 4 \sin 3x$
   (b) $y = 2 \cos 6x$
   [(a) $12 \cos 3x$ (b) $-12 \sin 6x$]

2. Given $f(\theta) = 2 \sin 3\theta - 5 \cos 2\theta$, find $f'(\theta)$
   [$6 \cos 3\theta + 10 \sin 2\theta$]

3. An alternating current is given by $i = 5 \sin 100t$ amperes, where t is the time in seconds. Determine the rate of change of current when $t = 0.01$ seconds.
   [270.2 A/s]

4. $v = 50 \sin 40t$ volts represents an alternating voltage where t is the time in seconds. At a time of 20×10^{-3} seconds, find the rate of change of voltage.
   [1393.4 V/s]

5. If $f(t) = 3 \sin(4t + 0.12) - 2 \cos(3t - 0.72)$ determine $f'(t)$
   [$12 \cos(4t + 0.12) + 6 \sin(3t - 0.72)$]

42.7   Differentiation of e^{ax} and ln ax

A graph of $y = e^x$ is shown in Fig. 42.7(a). The gradient of the curve at any point is given by $\dfrac{dy}{dx}$ and is continually changing. By drawing tangents to the curve at many points on the curve and measuring the gradient of the tangents, values of $\dfrac{dy}{dx}$ for corresponding values of x may be obtained. These values are shown graphically

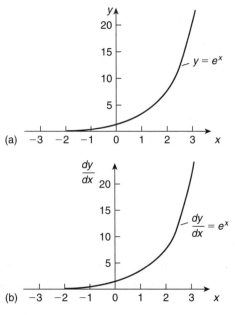

Figure 42.7

in Fig. 42.7(b). The graph of $\dfrac{dy}{dx}$ against x is identical to the original graph of $y = e^x$. It follows that:

$$\text{if } y = e^x, \text{ then } \frac{dy}{dx} = e^x$$

It may also be shown that

$$\text{if } y = e^{ax}, \text{ then } \frac{dy}{dx} = ae^{ax}$$

Therefore if $y = 2e^{6x}$, then $\dfrac{dy}{dx} = (2)(6e^{6x}) = \mathbf{12e^{6x}}$

A graph of $y = \ln x$ is shown in Fig. 42.8(a). The gradient of the curve at any point is given by $\dfrac{dy}{dx}$ and is continually changing. By drawing tangents to the curve at many points on the curve and measuring the gradient of the tangents, values of $\dfrac{dy}{dx}$ for corresponding values of x may be obtained. These values are shown graphically in Fig. 42.8(b). The graph of $\dfrac{dy}{dx}$ against x is the graph of $\dfrac{dy}{dx} = \dfrac{1}{x}$.

It follows that: **if $y = \ln x$, then $\dfrac{dy}{dx} = \dfrac{1}{x}$**

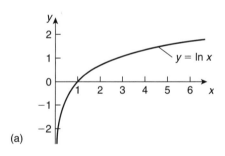

(a)

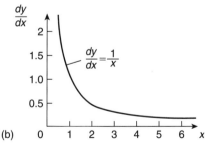

(b)

Figure 42.8

It may also be shown that

$$\text{if } y = \ln ax, \text{ then } \frac{dy}{dx} = \frac{1}{x}$$

(Note that in the latter expression 'a' does not appear in the $\dfrac{dy}{dx}$ term).

Thus if $y = \ln 4x$, then $\dfrac{dy}{dx} = \dfrac{1}{x}$

Problem 16. Differentiate the following with respect to the variable: (a) $y = 3e^{2x}$ (b) $f(t) = \dfrac{4}{3e^{5t}}$

(a) If $y = 3e^{2x}$ then $\dfrac{dy}{dx} = (3)(2e^{2x}) = \mathbf{6e^{2x}}$

(b) If $f(t) = \dfrac{4}{3e^{5t}} = \dfrac{4}{3}e^{-5t}$, then

$$f'(t) = \frac{4}{3}(-5e^{-5t}) = -\frac{20}{3}e^{-5t} = -\frac{20}{3e^{5t}}$$

Problem 17. Differentiate $y = 5 \ln 3x$

If $y = 5 \ln 3x$, then $\dfrac{dy}{dx} = (5)\left(\dfrac{1}{x}\right) = \dfrac{5}{x}$

Now try the following exercise

Exercise 153   Further problems on the differentiation of e^{ax} and ln ax

1. Differentiate with respect to x:

    (a) $y = 5e^{3x}$ (b) $y = \dfrac{2}{7e^{2x}}$

    $$\left[\text{(a) } 15e^{3x} \quad \text{(b) } -\frac{4}{7e^{2x}}\right]$$

2. Given $f(\theta) = 5 \ln 2\theta - 4 \ln 3\theta$, determine $f'(\theta)$

    $$\left[\frac{5}{\theta} - \frac{4}{\theta} = \frac{1}{\theta}\right]$$

3. If $f(t) = 4 \ln t + 2$, evaluate $f'(t)$ when $t = 0.25$

    [16]

4. Evaluate $\dfrac{dy}{dx}$ when $x = 1$, given

    $$y = 3e^{4x} - \frac{5}{2e^{3x}} + 8 \ln 5x.$$ Give the answer correct to 3 significant figures.

    [664]

Section 8

Chapter 43

Methods of differentiation

43.1 Differentiation of common functions

The **standard derivatives** summarised below were derived in Chapter 42 and are true for all real values of x.

y or $f(x)$	$\dfrac{dy}{dx}$ of $f'(x)$
ax^n	ax^{n-1}
$\sin ax$	$a\cos ax$
$\cos ax$	$-a\sin ax$
e^{ax}	ae^{ax}
$\ln ax$	$\dfrac{1}{x}$

The **differential coefficient of a sum or difference** is the sum or difference of the differential coefficients of the separate terms.

Thus, if $f(x) = p(x) + q(x) - r(x)$, (where f, p, q and r are functions), then $f'(x) = p'(x) + q'(x) - r'(x)$

Differentiation of common functions is demonstrated in the following worked problems.

Problem 1. Find the differential coefficients of:

(a) $y = 12x^3$ (b) $y = \dfrac{12}{x^3}$

If $y = ax^n$ then $\dfrac{dy}{dx} = anx^{n-1}$

(a) Since $y = 12x^3$, $a = 12$ and $n = 3$ thus
$\dfrac{dy}{dx} = (12)(3)x^{3-1} = \mathbf{36x^2}$

(b) $y = \dfrac{12}{x^3}$ is rewritten in the standard ax^n form as $y = 12x^{-3}$ and in the general rule $a = 12$ and $n = -3$

Thus $\dfrac{dy}{dx} = (12)(-3)x^{-3-1}$

$= -36x^{-4} = -\dfrac{36}{x^4}$

Problem 2. Differentiate: (a) $y = 6$ (b) $y = 6x$

(a) $y = 6$ may be written as $y = 6x^0$, i.e. in the general rule $a = 6$ and $n = 0$.

Hence $\dfrac{dy}{dx} = (6)(0)x^{0-1} = \mathbf{0}$

In general, **the differential coefficient of a constant is always zero**.

(b) Since $y = 6x$, in the general rule $a = 6$ and $n = 1$

Hence $\dfrac{dy}{dx} = (6)(1)x^{1-1} = 6x^0 = \mathbf{6}$

In general, the differential coefficient of kx, where k is a constant, is always k.

Problem 3. Find the derivatives of:

(a) $y = 3\sqrt{x}$ (b) $y = \dfrac{5}{\sqrt[3]{x^4}}$

(a) $y = 3\sqrt{x}$ is rewritten in the standard differential form as $y = 3x^{1/2}$

In the general rule, $a = 3$ and $n = \dfrac{1}{2}$

Thus $\dfrac{dy}{dx} = (3)\left(\dfrac{1}{2}\right)x^{\frac{1}{2}-1} = \dfrac{3}{2}x^{-\frac{1}{2}}$

$= \dfrac{3}{2x^{1/2}} = \dfrac{\mathbf{3}}{\mathbf{2\sqrt{x}}}$

(b) $y = \dfrac{5}{\sqrt[3]{x^4}} = \dfrac{5}{x^{4/3}} = 5x^{-4/3}$

In the general rule, $a = 5$ and $n = -\dfrac{4}{3}$

Thus $\dfrac{dy}{dx} = (5)\left(-\dfrac{4}{3}\right)x^{(-4/3)-1}$

$= \dfrac{-20}{3}x^{-7/3} = \dfrac{-20}{3x^{7/3}} = \dfrac{-20}{3\sqrt[3]{x^7}}$

Problem 4. Differentiate:
$y = 5x^4 + 4x - \dfrac{1}{2x^2} + \dfrac{1}{\sqrt{x}} - 3$ with respect to x

$y = 5x^4 + 4x - \dfrac{1}{2x^2} + \dfrac{1}{\sqrt{x}} - 3$ is rewritten as

$y = 5x^4 + 4x - \dfrac{1}{2}x^{-2} + x^{-1/2} - 3$

When differentiating a sum, each term is differentiated in turn.

Thus $\dfrac{dy}{dx} = (5)(4)x^{4-1} + (4)(1)x^{1-1} - \dfrac{1}{2}(-2)x^{-2-1}$

$+ (1)\left(-\dfrac{1}{2}\right)x^{(-1/2)-1} - 0$

$= 20x^3 + 4 + x^{-3} - \dfrac{1}{2}x^{-3/2}$

i.e. $\dfrac{dy}{dx} = \mathbf{20x^3 + 4 - \dfrac{1}{x^3} - \dfrac{1}{2\sqrt{x^3}}}$

Problem 5. Find the differential coefficients of:
(a) $y = 3\sin 4x$ (b) $f(t) = 2\cos 3t$ with respect to the variable

(a) When $y = 3\sin 4x$ then $\dfrac{dy}{dx} = (3)(4\cos 4x)$
$= \mathbf{12\cos 4x}$

(b) When $f(t) = 2\cos 3t$ then
$f'(t) = (2)(-3\sin 3t) = \mathbf{-6\sin 3t}$

Problem 6. Determine the derivatives of:
(a) $y = 3e^{5x}$ (b) $f(\theta) = \dfrac{2}{e^{3\theta}}$ (c) $y = 6\ln 2x$

(a) When $y = 3e^{5x}$ then $\dfrac{dy}{dx} = (3)(5)e^{5x} = \mathbf{15e^{5x}}$

(b) $f(\theta) = \dfrac{2}{e^{3\theta}} = 2e^{-3\theta}$, thus
$f'(\theta) = (2)(-3)e^{-3\theta} = -6e^{-3\theta} = \dfrac{\mathbf{-6}}{\mathbf{e^{3\theta}}}$

(c) When $y = 6\ln 2x$ then $\dfrac{dy}{dx} = 6\left(\dfrac{1}{x}\right) = \dfrac{\mathbf{6}}{\mathbf{x}}$

Problem 7. Find the gradient of the curve
$y = 3x^4 - 2x^2 + 5x - 2$ at the points $(0, -2)$ and $(1, 4)$

The gradient of a curve at a given point is given by the corresponding value of the derivative. Thus, since $y = 3x^4 - 2x^2 + 5x - 2$ then the
gradient $= \dfrac{dy}{dx} = 12x^3 - 4x + 5$

At the point $(0, -2)$, $x = 0$.
Thus the gradient $= 12(0)^3 - 4(0) + 5 = \mathbf{5}$

At the point $(1, 4)$, $x = 1$.
Thus the gradient $= 12(1)^3 - 4(1) + 5 = \mathbf{13}$

Problem 8. Determine the co-ordinates of the point on the graph $y = 3x^2 - 7x + 2$ where the gradient is -1

The gradient of the curve is given by the derivative.
When $y = 3x^2 - 7x + 2$ then $\dfrac{dy}{dx} = 6x - 7$
Since the gradient is -1 then $6x - 7 = -1$, from which,
$x = 1$
When $x = 1$, $y = 3(1)^2 - 7(1) + 2 = -2$
Hence the gradient is -1 at the point $(1, -2)$

Now try the following exercise

Exercise 154 Further problems on differentiating common functions

In Problems 1 to 6 find the differential coefficients of the given functions with respect to the variable.

1. (a) $5x^5$ (b) $2.4x^{3.5}$ (c) $\dfrac{1}{x}$

$\left[\text{(a) } 25x^4 \quad \text{(b) } 8.4x^{2.5} \quad \text{(c) } -\dfrac{1}{x^2}\right]$

2. (a) $\dfrac{-4}{x^2}$ (b) 6 (c) $2x$

$\left[\text{(a) } \dfrac{8}{x^3} \quad \text{(b) } 0 \quad \text{(c) } 2\right]$

3. (a) $2\sqrt{x}$ (b) $3\sqrt[3]{x^5}$ (c) $\dfrac{4}{\sqrt{x}}$

$$\left[\text{(a) } \frac{1}{\sqrt{x}} \quad \text{(b) } 5\sqrt[3]{x^2} \quad \text{(c) } -\frac{2}{\sqrt{x^3}}\right]$$

4. (a) $\dfrac{-3}{\sqrt[3]{x}}$ (b) $(x-1)^2$ (c) $2\sin 3x$

$$\left[\text{(a) } \frac{-3}{\sqrt[3]{x^4}} \quad \text{(b) } 2(x-1) \quad \text{(c) } 6\cos 3x\right]$$

5. (a) $-4\cos 2x$ (b) $2e^{6x}$ (c) $\dfrac{3}{e^{5x}}$

$$\left[\text{(a) } 8\sin 2x \quad \text{(b) } 12e^{6x} \quad \text{(c) } \frac{-15}{e^{5x}}\right]$$

6. (a) $4\ln 9x$ (b) $\dfrac{e^x - e^{-x}}{2}$ (c) $\dfrac{1-\sqrt{x}}{x}$

$$\left[\text{(a) } \frac{4}{x} \quad \text{(b) } \frac{e^x + e^{-x}}{2} \quad \text{(c) } \frac{-1}{x^2} + \frac{1}{2\sqrt{x^3}}\right]$$

7. Find the gradient of the curve $y = 2t^4 + 3t^3 - t + 4$ at the points $(0, 4)$ and $(1, 8)$.　　　　$[-1, 16]$

8. Find the co-ordinates of the point on graph $y = 5x^2 - 3x + 1$ where the gradient is 2.
$$\left[\left(\frac{1}{2}, \frac{3}{4}\right)\right]$$

9. (a) Differentiate
$$y = \frac{2}{\theta^2} + 2\ln 2\theta - 2(\cos 5\theta + 3\sin 2\theta) - \frac{2}{e^{3\theta}}$$

    (b) Evaluate $\dfrac{dy}{d\theta}$ when $\theta = \dfrac{\pi}{2}$, correct to 4 significant figures.

$$\left[\begin{array}{l}\text{(a) } \dfrac{-4}{\theta^3} + \dfrac{2}{\theta} + 10\sin 5\theta \\[2mm] - 12\cos 2\theta + \dfrac{6}{e^{3\theta}} \quad \text{(b) } 22.30\end{array}\right]$$

10. Evaluate $\dfrac{ds}{dt}$, correct to 3 significant figures, when $t = \dfrac{\pi}{6}$ given $s = 3\sin t - 3 + \sqrt{t}$
$$[3.29]$$

43.2 Differentiation of a product

When $y = uv$, and u and v are both functions of x,

then $\qquad \dfrac{dy}{dx} = u\dfrac{dv}{dx} + v\dfrac{du}{dx}$

This is known as the **product rule**.

Problem 9. Find the differential coefficient of:
$$y = 3x^2 \sin 2x$$

$3x^2 \sin 2x$ is a product of two terms $3x^2$ and $\sin 2x$. Let $u = 3x^2$ and $v = \sin 2x$

Using the product rule:
$$\frac{dy}{dx} = \underset{\downarrow}{u}\ \ \underset{\downarrow}{\frac{dv}{dx}}\ \ + \ \ \underset{\downarrow}{v}\ \ \underset{\downarrow}{\frac{du}{dx}}$$

gives: $\quad \dfrac{dy}{dx} = (3x^2)(2\cos 2x) + (\sin 2x)(6x)$

i.e. $\quad \dfrac{dy}{dx} = 6x^2 \cos 2x + 6x\sin 2x$

$$= \mathbf{6x(x\cos 2x + \sin 2x)}$$

Note that the differential coefficient of a product is **not** obtained by merely differentiating each term and multiplying the two answers together. The product rule formula **must** be used when differentiating products.

Problem 10. Find the rate of change of y with respect to x given: $y = 3\sqrt{x}\ln 2x$

The rate of change of y with respect to x is given by $\dfrac{dy}{dx}$.

$y = 3\sqrt{x}\ln 2x = 3x^{1/2}\ln 2x$, which is a product.

Let $u = 3x^{1/2}$ and $v = \ln 2x$

Then $\quad \dfrac{dy}{dx} = \underset{\downarrow}{u}\ \ \underset{\downarrow}{\frac{dv}{dx}}\ \ + \ \ \underset{\downarrow}{v}\ \ \underset{\downarrow}{\frac{du}{dx}}$

$$= (3x^{1/2})\left(\frac{1}{x}\right) + (\ln 2x)\left[3\left(\frac{1}{2}\right)x^{(1/2)-1}\right]$$

$$= 3x^{(1/2)-1} + (\ln 2x)\left(\frac{3}{2}\right)x^{-1/2}$$

$$= 3x^{-1/2}\left(1 + \frac{1}{2}\ln 2x\right)$$

i.e. $\quad \dfrac{dy}{dx} = \dfrac{3}{\sqrt{x}}\left(1 + \dfrac{1}{2}\mathbf{ln\ 2x}\right)$

Problem 11. Differentiate: $y = x^3 \cos 3x \ln x$

Let $u = x^3 \cos 3x$ (i.e. a product) and $v = \ln x$

Then $\quad \dfrac{dy}{dx} = u\dfrac{dv}{dx} + v\dfrac{du}{dx}$

where $\quad \dfrac{du}{dx} = (x^3)(-3\sin 3x) + (\cos 3x)(3x^2)$

and $\dfrac{dv}{dx} = \dfrac{1}{x}$

Hence $\dfrac{dy}{dx} = (x^3 \cos 3x)\left(\dfrac{1}{x}\right)$

$$+ (\ln x)[-3x^3 \sin 3x + 3x^2 \cos 3x]$$

$$= x^2 \cos 3x + 3x^2 \ln x(\cos 3x - x \sin 3x)$$

i.e. $\dfrac{dy}{dx} = x^2\{\cos 3x + 3 \ln x(\cos 3x - x \sin 3x)\}$

Problem 12. Determine the rate of change of voltage, given $v = 5t \sin 2t$ volts, when $t = 0.2$ s

Rate of change of voltage

$$= \dfrac{dv}{dt} = (5t)(2 \cos 2t) + (\sin 2t)(5)$$

$$= 10t \cos 2t + 5 \sin 2t$$

When $t = 0.2$,

$$\dfrac{dv}{dt} = 10(0.2) \cos 2(0.2) + 5 \sin 2(0.2)$$

$$= 2 \cos 0.4 + 5 \sin 0.4$$

(where cos 0.4 means the cosine of 0.4 radians = 0.92106)

Hence $\dfrac{dv}{dt} = 2(0.92106) + 5(0.38942)$

$$= 1.8421 + 1.9471 = 3.7892$$

i.e. **the rate of change of voltage when $t = 0.2$ s is 3.79 volts/s, correct to 3 significant figures.**

Now try the following exercise

Exercise 155 Further problems on differentiating products

In Problems 1 to 5 differentiate the given products with respect to the variable.

1. $2x^3 \cos 3x$ $\qquad$ $[6x^2(\cos 3x - x \sin 3x)]$

2. $\sqrt{x^3} \ln 3x$ $\qquad$ $\left[\sqrt{x}\left(1 + \dfrac{3}{2} \ln 3x\right)\right]$

3. $e^{3t} \sin 4t$ $\qquad$ $[e^{3t}(4 \cos 4t + 3 \sin 4t)]$

4. $e^{4\theta} \ln 3\theta$ $\qquad$ $\left[e^{4\theta}\left(\dfrac{1}{\theta} + 4 \ln 3\theta\right)\right]$

5. $e^t \ln t \cos t$

$$\left[e^t\left\{\left(\dfrac{1}{t} + \ln t\right) \cos t - \ln t \sin t\right\}\right]$$

6. Evaluate $\dfrac{di}{dt}$, correct to 4 significant figure, when $t = 0.1$, and $i = 15t \sin 3t$ $\qquad$ [8.732]

7. Evaluate $\dfrac{dz}{dt}$, correct to 4 significant figures, when $t = 0.5$, given that $z = 2e^{3t} \sin 2t$ $\qquad$ [32.31]

43.3 Differentiation of a quotient

When $y = \dfrac{u}{v}$, and u and v are both functions of x

then $\qquad \dfrac{dy}{dx} = \dfrac{v \dfrac{du}{dx} - u \dfrac{dv}{dx}}{v^2}$

This is known as the **quotient rule**.

Problem 13. Find the differential coefficient of:
$$y = \dfrac{4 \sin 5x}{5x^4}$$

$\dfrac{4 \sin 5x}{5x^4}$ is a quotient. Let $u = 4 \sin 5x$ and $v = 5x^4$

(Note that v is **always** the denominator and u the numerator)

$$\dfrac{dy}{dx} = \dfrac{v \dfrac{du}{dx} - u \dfrac{dv}{dx}}{v^2}$$

where $\dfrac{du}{dx} = (4)(5) \cos 5x = 20 \cos 5x$

and $\dfrac{dv}{dx} = (5)(4)x^3 = 20x^3$

Hence $\dfrac{dy}{dx} = \dfrac{(5x^4)(20 \cos 5x) - (4 \sin 5x)(20x^3)}{(5x^4)^2}$

$$= \dfrac{100x^4 \cos 5x - 80x^3 \sin 5x}{25x^8}$$

$$= \dfrac{20x^3[5x \cos 5x - 4 \sin 5x]}{25x^8}$$

i.e. $\dfrac{dy}{dx} = \dfrac{4}{5x^5}(5x \cos 5x - 4 \sin 5x)$

Note that the differential coefficient is **not** obtained by merely differentiating each term in turn and then dividing the numerator by the denominator. The quotient formula **must** be used when differentiating quotients.

Problem 14. Determine the differential coefficient of: $y = \tan ax$

$y = \tan ax = \dfrac{\sin ax}{\cos ax}$. Differentiation of $\tan ax$ is thus

treated as a quotient with $u = \sin ax$ and $v = \cos ax$

$$\frac{dy}{dx} = \frac{v\dfrac{du}{dx} - u\dfrac{dv}{dx}}{v^2}$$

$$= \frac{(\cos ax)(a\cos ax) - (\sin ax)(-a\sin ax)}{(\cos ax)^2}$$

$$= \frac{a\cos^2 ax + a\sin^2 ax}{(\cos ax)^2}$$

$$= \frac{a(\cos^2 ax + \sin^2 ax)}{\cos^2 ax}$$

$$= \frac{a}{\cos^2 ax} \quad \text{since } \cos^2 ax + \sin^2 ax = 1$$

(see Chapter 26)

Hence $\dfrac{dy}{dx} = a\sec^2 ax$ since $\sec^2 ax = \dfrac{1}{\cos^2 ax}$

(see Chapter 22)

Problem 15. Find the derivative of: $y = \sec ax$

$y = \sec ax = \dfrac{1}{\cos ax}$ (i.e. a quotient), Let $u = 1$ and $v = \cos ax$

$$\frac{dy}{dx} = \frac{v\dfrac{du}{dx} - u\dfrac{dv}{dx}}{v^2}$$

$$= \frac{(\cos ax)(0) - (1)(-a\sin ax)}{(\cos ax)^2}$$

$$= \frac{a\sin ax}{\cos^2 ax} = a\left(\frac{1}{\cos ax}\right)\left(\frac{\sin ax}{\cos ax}\right)$$

i.e. $\dfrac{dy}{dx} = a\sec ax \tan ax$

Problem 16. Differentiate: $y = \dfrac{te^{2t}}{2\cos t}$

The function $\dfrac{te^{2t}}{2\cos t}$ is a quotient, whose numerator is a product.

Let $u = te^{2t}$ and $v = 2\cos t$ then

$$\frac{du}{dt} = (t)(2e^{2t}) + (e^{2t})(1) \text{ and } \frac{dv}{dt} = -2\sin t$$

Hence $\dfrac{dy}{dx} = \dfrac{v\dfrac{du}{dx} - u\dfrac{dv}{dx}}{v^2}$

$$= \frac{(2\cos t)[2te^{2t} + e^{2t}] - (te^{2t})(-2\sin t)}{(2\cos t)^2}$$

$$= \frac{4te^{2t}\cos t + 2e^{2t}\cos t + 2te^{2t}\sin t}{4\cos^2 t}$$

$$= \frac{2e^{2t}[2t\cos t + \cos t + t\sin t]}{4\cos^2 t}$$

i.e. $\dfrac{dy}{dx} = \dfrac{e^{2t}}{2\cos^2 t}(2t\cos t + \cos t + t\sin t)$

Problem 17. Determine the gradient of the curve $y = \dfrac{5x}{2x^2 + 4}$ at the point $\left(\sqrt{3}, \dfrac{\sqrt{3}}{2}\right)$

Let $y = 5x$ and $v = 2x^2 + 4$

$$\frac{dy}{dx} = \frac{v\dfrac{du}{dx} - u\dfrac{dv}{dx}}{v^2} = \frac{(2x^2 + 4)(5) - (5x)(4x)}{(2x^2 + 4)^2}$$

$$= \frac{10x^2 + 20 - 20x^2}{(2x^2 + 4)^2} = \frac{20 - 10x^2}{(2x^2 + 4)^2}$$

At the point $\left(\sqrt{3}, \dfrac{\sqrt{3}}{2}\right)$, $x = \sqrt{3}$,

hence the gradient $= \dfrac{dy}{dx} = \dfrac{20 - 10(\sqrt{3})^2}{[2(\sqrt{3})^2 + 4]^2}$

$$= \frac{20 - 30}{100} = -\frac{1}{10}$$

Now try the following exercise

Exercise 156   Further problems on differentiating quotients

In Problems 1 to 5, differentiate the quotients with respect to the variable.

1. $\dfrac{2\cos 3x}{x^3}$ $\qquad \left[\dfrac{-6}{x^4}(x\sin 3x + \cos 3x)\right]$

2. $\dfrac{2x}{x^2+1}$ $\left[\dfrac{2(1-x^2)}{(x^2+1)^2}\right]$

3. $\dfrac{3\sqrt{\theta^3}}{2\sin 2\theta}$ $\left[\dfrac{3\sqrt{\theta}(3\sin 2\theta-4\theta\cos 2\theta)}{4\sin^2 2\theta}\right]$

4. $\dfrac{\ln 2t}{\sqrt{t}}$ $\left[\dfrac{1-\dfrac{1}{2}\ln 2t}{\sqrt{t^3}}\right]$

5. $\dfrac{2xe^{4x}}{\sin x}$ $\left[\dfrac{2e^{4x}}{\sin^2 x}\{(1+4x)\sin x-x\cos x\}\right]$

6. Find the gradient of the curve $y=\dfrac{2x}{x^2-5}$ at the point $(2,-4)$ $[-18]$

7. Evaluate $\dfrac{dy}{dx}$ at $x=2.5$, correct to 3 significant figures, given $y=\dfrac{2x^2+3}{\ln 2x}$ $[3.82]$

43.4 Function of a function

It is often easier to make a substitution before differentiating.

If y is a function of x then $\qquad \dfrac{dy}{dx}=\dfrac{dy}{du}\times\dfrac{du}{dx}$

This is known as the **'function of a function'** rule (or sometimes the **chain rule**).

For example, if $y=(3x-1)^9$ then, by making the substitution $u=(3x-1)$, $y=u^9$, which is of the 'standard' from.

Hence $\dfrac{dy}{du}=9u^8$ and $\dfrac{du}{dx}=3$

Then $\dfrac{dy}{dx}=\dfrac{dy}{du}\times\dfrac{du}{dx}=(9u^8)(3)=27u^8$

Rewriting u as $(3x-1)$ gives: $\dfrac{dy}{dx}=27(3x-1)^8$

Since y is a function of u, and u is a function of x, then y is a function of a function of x.

Problem 18. Differentiate: $y=3\cos(5x^2+2)$

Let $u=5x^2+2$ then $y=3\cos u$

Hence $\dfrac{du}{dx}=10x$ and $\dfrac{dy}{du}=-3\sin u$

Using the function of a function rule,

$$\dfrac{dy}{dx}=\dfrac{dy}{du}\times\dfrac{du}{dx}=(-3\sin u)(10x)=-30x\sin u$$

Rewriting u as $5x^2+2$ gives:

$$\dfrac{dy}{dx}=-30x\sin(5x^2+2)$$

Problem 19. Find the derivative of:
$$y=(4t^3-3t)^6$$

Let $u=4t^3-3t$, then $y=u^6$

Hence $\dfrac{du}{dt}=12t^2-3$ and $\dfrac{dy}{dt}=6u^5$

Using the function of a function rule,

$$\dfrac{dy}{dx}=\dfrac{dy}{du}\times\dfrac{du}{dx}=(6u^5)(12t^2-3)$$

Rewriting u as $(4t^3-3t)$ gives:

$$\dfrac{dy}{dt}=6(4t^3-3t)^5(12t^2-3)$$

$$=18(4t^2-1)(4t^3-3t)^5$$

Problem 20. Determine the differential coefficient of: $y=\sqrt{3x^2+4x-1}$

$y=\sqrt{3x^2+4x-1}=(3x^2+4x-1)^{1/2}$

Let $u=3x^2+4x-1$ then $y=u^{1/2}$

Hence $\dfrac{du}{dx}=6x+4$ and $\dfrac{dy}{du}=\dfrac{1}{2}u^{-1/2}=\dfrac{1}{2\sqrt{u}}$

Using the function of a function rule,

$$\dfrac{dy}{dx}=\dfrac{dy}{du}\times\dfrac{du}{dx}=\left(\dfrac{1}{2\sqrt{u}}\right)(6x+4)=\dfrac{3x+2}{\sqrt{u}}$$

i.e. $\dfrac{dy}{dx}=\dfrac{3x+2}{\sqrt{3x^2+4x-1}}$

Problem 21. Differentiate: $y=3\tan^4 3x$

Let $u=\tan 3x$ then $y=3u^4$

Hence $\dfrac{du}{dx}=3\sec^2 3x$, (from Problem 14),

and $\dfrac{dy}{du}=12u^3$

Then $\dfrac{dy}{dx} = \dfrac{dy}{du} \times \dfrac{du}{dx} = (12u^3)(3\sec^2 3x)$

$$= 12(\tan 3x)^3(3\sec^2 3x)$$

i.e. $\dfrac{dy}{dx} = \mathbf{36\tan^3 3x \sec^2 3x}$

Problem 22. Find the differential coefficient of:

$$y = \dfrac{2}{(2t^3 - 5)^4}$$

$y = \dfrac{2}{(2t^3 - 5)^4} = 2(2t^3 - 5)^{-4}$. Let $u = (2t^3 - 5)$, then $y = 2u^{-4}$

Hence $\dfrac{du}{dt} = 6t^2$ and $\dfrac{dy}{du} = -8u^{-5} = \dfrac{-8}{u^5}$

Then $\dfrac{dy}{dt} = \dfrac{dy}{du} \times \dfrac{du}{dt} = \left(\dfrac{-8}{u^5}\right)(6t^2) = \dfrac{-48t^2}{(2t^3 - 5)^5}$

Now try the following exercise

Exercise 157 Further problems on the function of a function

In Problems 1 to 8, find the differential coefficients with respect to the variable.

1. $(2x^3 - 5x)^5$ $[5(6x^2 - 5)(2x^3 - 5x)^4]$

2. $2\sin(3\theta - 2)$ $[6\cos(3\theta - 2)]$

3. $2\cos^5 \alpha$ $[-10\cos^4 \alpha \sin \alpha]$

4. $\dfrac{1}{(x^3 - 2x + 1)^5}$ $\left[\dfrac{5(2 - 3x^2)}{(x^3 - 2x + 1)^6}\right]$

5. $5e^{2t+1}$ $[10e^{2t+1}]$

6. $2\cot(5t^2 + 3)$ $[-20t \operatorname{cosec}^2(5t^2 + 3)]$

7. $6\tan(3y + 1)$ $[18\sec^2(3y + 1)]$

8. $2e^{\tan \theta}$ $[2\sec^2 \theta\, e^{\tan \theta}]$

9. Differentiate: $\theta \sin\left(\theta - \dfrac{\pi}{3}\right)$ with respect to θ, and evaluate, correct to 3 significant figures, when $\theta = \dfrac{\pi}{2}$ $[1.86]$

43.5 Successive differentiation

When a function $y = f(x)$ is differentiated with respect to x the differential coefficient is written as $\dfrac{dy}{dx}$ or $f'(x)$. If the expression is differentiated again, the second differential coefficient is obtained and is written as $\dfrac{d^2 y}{dx^2}$ (pronounced dee two y by dee x squared) or $f''(x)$ (pronounced f double–dash x). By successive differentiation further higher derivatives such as $\dfrac{d^3 y}{dx^3}$ and $\dfrac{d^4 y}{dx^4}$ may be obtained.

Thus if $y = 3x^4$,

$$\dfrac{dy}{dx} = 12x^3, \quad \dfrac{d^2 y}{dx^2} = 36x^2,$$

$$\dfrac{d^3 y}{dx^3} = 72x, \quad \dfrac{d^4 y}{dx^4} = 72 \text{ and } \dfrac{d^5 y}{dx^5} = 0$$

Problem 23. If $f(x) = 2x^5 - 4x^3 + 3x - 5$, find $f''(x)$

$$f(x) = 2x^5 - 4x^3 + 3x - 5$$

$$f'(x) = 10x^4 - 12x^2 + 3$$

$$f''(x) = 40x^3 - 24x = \mathbf{4x(10x^2 - 6)}$$

Problem 24. If $y = \cos x - \sin x$, evaluate x, in the range $0 \le x \le \dfrac{\pi}{2}$, when $\dfrac{d^2 y}{dx^2}$ is zero

Since $y = \cos x - \sin x$, $\dfrac{dy}{dx} = -\sin x - \cos x$ and $\dfrac{d^2 y}{dx^2} = -\cos x + \sin x$

When $\dfrac{d^2 y}{dx^2}$ is zero, $-\cos x + \sin x = 0$,

i.e. $\sin x = \cos x$ or $\dfrac{\sin x}{\cos x} = 1$

Hence $\tan x = 1$ and $x = \tan^{-1} 1 = \mathbf{45°}$ or $\dfrac{\pi}{4}$ **rads** in the range $0 \le x \le \dfrac{\pi}{2}$

Problem 25. Given $y = 2xe^{-3x}$ show that

$$\dfrac{d^2 y}{dx^2} + 6\dfrac{dy}{dx} + 9y = 0$$

$y = 2xe^{-3x}$ (i.e. a product)

Hence $\dfrac{dy}{dx} = (2x)(-3e^{-3x}) + (e^{-3x})(2)$

$= -6xe^{-3x} + 2e^{-3x}$

$\dfrac{d^2y}{dx^2} = [(-6x)(-3e^{-3x}) + (e^{-3x})(-6)]$

$+ (-6e^{-3x})$

$= 18xe^{-3x} - 6e^{-3x} - 6e^{-3x}$

i.e. $\dfrac{d^2y}{dx^2} = 18xe^{-3x} - 12e^{-3x}$

Substituting values into $\dfrac{d^2y}{dx^2} + 6\dfrac{dy}{dx} + 9y$ gives:

$(18xe^{-3x} - 12e^{-3x}) + 6(-6xe^{-3x} + 2e^{-3x})$

$+ 9(2xe^{-3x})$

$= 18xe^{-3x} - 12e^{-3x} - 36xe^{-3x}$

$+ 12e^{-3x} + 18xe^{-3x} = 0$

Thus when $y = 2xe^{-3x}$, $\dfrac{d^2y}{dx^2} + 6\dfrac{dy}{dx} + 9y = 0$

Problem 26. Evaluate $\dfrac{d^2y}{d\theta^2}$ when $\theta = 0$ given: $y = 4\sec 2\theta$

Since $y = 4\sec 2\theta$, then

$\dfrac{dy}{d\theta} = (4)(2)\sec 2\theta \tan 2\theta$ (from Problem 15)

$= 8\sec 2\theta \tan 2\theta$ (i.e. a product)

$\dfrac{d^2y}{d\theta^2} = (8\sec 2\theta)(2\sec^2 2\theta)$

$+ (\tan 2\theta)[(8)(2)\sec 2\theta \tan 2\theta]$

$= 16\sec^3 2\theta + 16\sec 2\theta \tan^2 2\theta$

When $\theta = 0$,

$\dfrac{d^2y}{d\theta^2} = 16\sec^3 0 + 16\sec 0 \tan^2 0$

$= 16(1) + 16(1)(0) = \mathbf{16}$

Now try the following exercise

Exercise 158   Further problems on successive differentiation

1. If $y = 3x^4 + 2x^3 - 3x + 2$ find

    (a) $\dfrac{d^2y}{dx^2}$ (b) $\dfrac{d^3y}{dx^3}$

    [(a) $36x^2 + 12x$ (b) $72x + 12$]

2. (a) Given $f(t) = \dfrac{2}{5}t^2 - \dfrac{1}{t^3} + \dfrac{3}{t} - \sqrt{t} + 1$
        determine $f''(t)$

    (b) Evaluate $f''(t)$ when $t = 1$.

    $\left[\begin{array}{l}\text{(a) } \dfrac{4}{5} - \dfrac{12}{t^5} + \dfrac{6}{t^3} + \dfrac{1}{4\sqrt{t^3}} \\ \text{(b) } -4.95\end{array}\right]$

In Problems 3 and 4, find the second differential coefficient with respect to the variable.

3. (a) $3\sin 2t + \cos t$ (b) $2\ln 4\theta$

    $\left[\text{(a) } -(12\sin 2t + \cos t)\quad \text{(b) } \dfrac{-2}{\theta^2}\right]$

4. (a) $2\cos^2 x$ (b) $(2x-3)^4$

    [(a) $4(\sin^2 x - \cos^2 x)$ (b) $48(2x-3)^2$]

5. Evaluate $f''(\theta)$ when $\theta = 0$ given $f(\theta) = 2\sec 3\theta$ [18]

6. Show that the differential equation $\dfrac{d^2y}{dx^2} - 4\dfrac{dy}{dx} + 4y = 0$ is satisfied when $y = xe^{2x}$

7. Show that, if P and Q are constants and $y = P\cos(\ln t) + Q\sin(\ln t)$, then

    $$t^2\dfrac{d^2y}{dt^2} + t\dfrac{dy}{dt} + y = 0$$

Section 8

Chapter 44

Some applications of differentiation

44.1 Rates of change

If a quantity y depends on and varies with a quantity x then the rate of change of y with respect to x is $\dfrac{dy}{dx}$.

Thus, for example, the rate of change of pressure p with height h is $\dfrac{dp}{dh}$.

A rate of change with respect to time is usually just called 'the rate of change', the 'with respect to time' being assumed. Thus, for example, a rate of change of current, i, is $\dfrac{di}{dt}$ and a rate of change of temperature, θ, is $\dfrac{d\theta}{dt}$, and so on.

Problem 1. The length l metres of a certain metal rod at temperature $\theta°C$ is given by: $l = 1 + 0.00005\theta + 0.0000004\theta^2$. Determine the rate of change of length, in mm/°C, when the temperature is (a) 100°C and (b) 400°C

The rate of change of length means $\dfrac{dl}{d\theta}$

Since length $l = 1 + 0.00005\theta + 0.0000004\theta^2$,

then $\quad \dfrac{dl}{d\theta} = 0.00005 + 0.0000008\theta$

(a) When $\theta = 100°C$,

$$\frac{dl}{d\theta} = 0.00005 + (0.0000008)(100)$$

$$= 0.00013\,\text{m/°C} = \mathbf{0.13\,mm/°C}$$

(b) When $\theta = 400°C$,

$$\frac{dl}{d\theta} = 0.00005 + (0.0000008)(400)$$

$$= 0.00037\,\text{m/°C} = \mathbf{0.37\,mm/°C}$$

Problem 2. The luminous intensity I candelas of a lamp at varying voltage V is given by: $I = 4 \times 10^{-4}\,V^2$. Determine the voltage at which the light is increasing at a rate of 0.6 candelas per volt

---

The rate of change of light with respect to voltage is given by $\dfrac{dI}{dV}$

Since $I = 4 \times 10^{-4}\,V^2$, $\dfrac{dI}{dV} = (4 \times 10^{-4})(2)V$

$$= 8 \times 10^{-4}\,V$$

When the light is increasing at 0.6 candelas per volt then $+0.6 = 8 \times 10^{-4}\,V$, from which, voltage

$$V = \frac{0.6}{8 \times 10^{-4}} = 0.075 \times 10^{+4} = \mathbf{750\,volts}$$

Problem 3. Newtons law of cooling is given by: $\theta = \theta_0 e^{-kt}$, where the excess of temperature at zero time is $\theta_0°C$ and at time t seconds is $\theta°C$. Determine the rate of change of temperature after 40 s, given that $\theta_0 = 16°C$ and $k = -0.03$

The rate of change of temperture is $\dfrac{d\theta}{dt}$

Since $\theta = \theta_0 e^{-kt}$ then $\dfrac{d\theta}{dt} = (\theta_0)(-k)e^{-kt}$

$$= -k\theta_0 e^{-kt}$$

When $\theta_0 = 16$, $k = -0.03$ and $t = 40$ then

$$\frac{d\theta}{dt} = -(-0.03)(16)e^{-(-0.03)(40)}$$

$$= 0.48e^{1.2} = \mathbf{1.594°C/s}$$

Problem 4. The displacement s cm of the end of a stiff spring at time t seconds is given by: $s = ae^{-kt} \sin 2\pi ft$. Determine the velocity of the end of the spring after 1 s, if $a = 2$, $k = 0.9$ and $f = 5$

Velocity $v = \dfrac{ds}{dt}$ where $s = ae^{-kt} \sin 2\pi ft$ (i.e. a product)

Using the product rule,

$$\frac{ds}{dt} = (ae^{-kt})(2\pi f \cos 2\pi ft)$$

$$+ (\sin 2\pi ft)(-ake^{-kt})$$

When $a = 2$, $k = 0.9$, $f = 5$ and $t = 1$,

$$\mathbf{velocity,}\ v = (2e^{-0.9})(2\pi 5 \cos 2\pi 5)$$

$$+ (\sin 2\pi 5)(-2)(0.9)e^{-0.9}$$

$$= 25.5455 \cos 10\pi - 0.7318 \sin 10\pi$$

$$= 25.5455(1) - 0.7318(0)$$

$$= \mathbf{25.55\ cm/s}$$

(Note that $\cos 10\pi$ means 'the cosine of 10π radians', *not* degrees, and $\cos 10\pi \equiv \cos 2\pi = 1$).

Now try the following exercise

Exercise 159   Further problems on rates of change

1. An alternating current, i amperes, is given by $i = 10 \sin 2\pi ft$, where f is the frequency in hertz and t the time in seconds. Determine the rate of change of current when $t = 20$ ms, given that $f = 150$ Hz.   [3000π A/s]

2. The luminous intensity, I candelas, of a lamp is given by $I = 6 \times 10^{-4} V^2$, where V is the voltage. Find (a) the rate of change of luminous intensity with voltage when $V = 200$ volts, and (b) the voltage at which the light is increasing at a rate of 0.3 candelas per volt.
   [(a) 0.24 cd/V   (b) 250 V]

3. The voltage across the plates of a capacitor at any time t seconds is given by $v = Ve^{-t/CR}$, where V, C and R are constants. Given $V = 300$ volts, $C = 0.12 \times 10^{-6}$ farads and $R = 4 \times 10^6$ ohms find (a) the initial rate of change of voltage, and (b) the rate of change of voltage after 0.5 s.
   [(a) -625 V/s   (b) -220.5 V/s]

4. The pressure p of the atmosphere at height h above ground level is given by $p = p_0 e^{-h/c}$, where p_0 is the pressure at ground level and c is a constant. Determine the rate of change of pressure with height when $p_0 = 1.013 \times 10^5$ Pascals and $c = 6.05 \times 10^4$ at 1450 metres.
   [-1.635 Pa/m]

44.2   Velocity and acceleration

When a car moves a distance x metres in a time t seconds along a straight road, if the **velocity v** is constant then $v = \dfrac{x}{t}$ m/s, i.e. the gradient of the distance/time graph shown in Fig. 44.1 is constant.

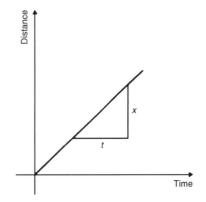

Figure 44.1

If, however, the velocity of the car is not constant then the distance/time graph will not be a straight line. It may be as shown in Fig. 44.2.

The average velocity over a small time δt and distance δx is given by the gradient of the chord AB, i.e. the average velocity over time δt is $\dfrac{\delta x}{\delta t}$. As $\delta t \to 0$, the chord AB becomes a tangent, such that at point A, the velocity is given by: $v = \dfrac{dx}{dt}$

Hence the velocity of the car at any instant is given by the gradient of the distance/time graph. If an expression

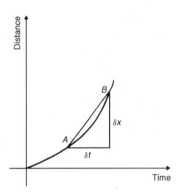

Figure 44.2

for the distance x is known in terms of time t then the velocity is obtained by differentiating the expression.

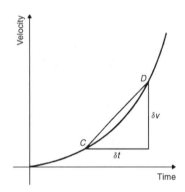

Figure 44.3

The **acceleration a** of the car is defined as the rate of change of velocity. A velocity/time graph is shown in Fig. 44.3. If δv is the change in v and δt the corresponding change in time, then $a = \dfrac{\delta v}{\delta t}$. As $\delta t \to 0$, the chord CD becomes a tangent, such that at point C, the acceleration is given by: $a = \dfrac{dv}{dt}$

Hence the acceleration of the car at any instant is given by the gradient of the velocity/time graph. If an expression for velocity is known in terms of time t then the acceleration is obtained by differentiating the expression.

Acceleration $\quad a = \dfrac{dv}{dt}$

However, $\quad\quad v = \dfrac{dx}{dt}$

Hence $\quad\quad a = \dfrac{d}{dt}\left(\dfrac{dx}{dt}\right) = \dfrac{d^2x}{dx^2}$

The acceleration is given by the second differential coefficient of distance x with respect to time t

Summarising, if a body moves a distance x metres in a time t seconds then:

(i) distance $x = f(t)$

(ii) velocity $v = f'(t)$ or $\dfrac{dx}{dt}$, which is the gradient of the distance/time graph

(iii) acceleration $a = \dfrac{dv}{dt} = f''$ or $\dfrac{d^2x}{dt^2}$, which is the gradient of the velocity/time graph.

Problem 5. The distance x metres moved by a car in a time t seconds is given by: $x = 3t^3 - 2t^2 + 4t - 1$. Determine the velocity and acceleration when (a) $t = 0$, and (b) $t = 1.5$ s

Distance $\quad\quad x = 3t^3 - 2t^2 + 4t - 1$ m.

Velocity $\quad\quad v = \dfrac{dx}{dt} = 9t^2 - 4t + 4$ m/s

Acceleration $\quad a = \dfrac{d^2x}{dx^2} = 18t - 4$ m/s^2

(a) When time $t = 0$,
     velocity $v = 9(0)^2 - 4(0) + 4 = \mathbf{4\ m/s}$
     and acceleration $a = 18(0) - 4 = \mathbf{-4\ m/s^2}$
     $\quad\quad\quad\quad\quad\quad\quad\quad$ (i.e. a deceleration)

(b) When time $t = 1.5$ s,
     velocity $v = 9(1.5)^2 - 4(1.5) + 4 = \mathbf{18.25\ m/s}$
     and acceleration $a = 18(1.5) - 4 = \mathbf{23\ m/s^2}$

Problem 6. Supplies are dropped from a helicopter and the distance fallen in a time t seconds is given by: $x = \frac{1}{2}gt^2$, where $g = 9.8$ m/s^2. Determine the velocity and acceleration of the supplies after it has fallen for 2 seconds

Distance $\quad\quad x = \dfrac{1}{2}gt^2 = \dfrac{1}{2}(9.8)t^2 = 4.9t^2$ m

Velocity $\quad\quad v = \dfrac{dv}{dt} = 9.8t$ m/s

and acceleration $\quad a = \dfrac{d^2x}{dx^2} = 9.8$ m/s^2

When time $t = 2$ s,
velocity $v = (9.8)(2) = \mathbf{19.6\ m/s}$
and **acceleration $a = 9.8$ m/s^2** (which is acceleration due to gravity).

Problem 7. The distance x metres travelled by a vehicle in time t seconds after the brakes are

applied is given by: $x = 20t - \dfrac{5}{3}t^2$. Determine
(a) the speed of the vehicle (in km/h) at the instant the brakes are applied, and (b) the distance the car travels before it stops

(a) Distance, $x = 20t - \dfrac{5}{3}t^2$

Hence velocity $v = \dfrac{dx}{dt} = 20 - \dfrac{10}{3}t$

At the instant the brakes are applied, time $= 0$
Hence

$$\textbf{velocity } v = 20 \text{ m/s} = \dfrac{20 \times 60 \times 60}{1000} \text{ km/h}$$

$$= \textbf{72 km/h}$$

(Note: changing from m/s to km/h merely involves multiplying by 3.6).

(b) When the car finally stops, the velocity is zero, i.e.
$v = 20 - \dfrac{10}{3}t = 0$, from which, $20 = \dfrac{10}{3}t$, giving
$t = 6$ s. Hence the distance travelled before the car stops is given by:

$$x = 20t - \dfrac{5}{3}t^2 = 20(6) - \dfrac{5}{3}(6)^2$$

$$= 120 - 60 = \textbf{60 m}$$

Problem 8. The angular displacement θ radians of a flywheel varies with time t seconds and follows the equation: $\theta = 9t^2 - 2t^3$. Determine (a) the angular velocity and acceleration of the flywheel when time, $t = 1$ s, and (b) the time when the angular acceleration is zero

(a) Angular displacement $\theta = 9t^2 - 2t^3$ rad.

Angular velocity $\omega = \dfrac{d\theta}{dt} = 18t - 6t^2$ rad/s.

When time $t = 1$ s,

$\omega = 18(1) - 6(1)^2 = \textbf{12 rad/s}$.

Angular acceleration $\alpha = \dfrac{d^2\theta}{dt^2} = 18 - 12t$ rad/s.

When time $t = 1$ s, $\alpha = 18 - 12(1)$

$$= \textbf{6 rad/s}^2$$

(b) When the angular acceleration is zero, $18 - 12t = 0$, from which, $18 = 12t$, giving time, $t = \textbf{1.5 s}$

Problem 9. The displacement x cm of the slide valve of an engine is given by: $x = 2.2 \cos 5\pi t + 3.6 \sin 5\pi t$. Evaluate the velocity (in m/s) when time $t = 30$ ms

Displacement $x = 2.2 \cos 5\pi t + 3.6 \sin 5\pi t$

Velocity $v = \dfrac{dx}{dt} = (2.2)(-5\pi) \sin 5\pi t$

$$+ (3.6)(5\pi) \cos 5\pi t$$

$$= -11\pi \sin 5\pi t + 18\pi \cos 5\pi t \text{ cm/s}$$

When time $t = 30$ ms,

velocity $= -11\pi \sin (5\pi \times 30 \times 10^{-3})$

$$+ 18\pi \cos (5\pi \times 30 \times 10^{-3})$$

$$= -11\pi \sin 0.4712 + 18\pi \cos 0.4712$$

$$= -11\pi \sin 27° + 18\pi \cos 27°$$

$$= -15.69 + 50.39$$

$$= 34.7 \text{ cm/s} = \textbf{0.347 m/s}$$

Now try the following exercise

Exercise 160 Further problems on velocity and acceleration

1. A missile fired from ground level rises x metres vertically upwards in t seconds and $x = 100t - \dfrac{25}{2}t^2$. Find (a) the initial velocity of the missile, (b) the time when the height of the missile is a maximum, (c) the maximum height reached, (d) the velocity with which the missile strikes the ground.
$$\begin{bmatrix} \text{(a) } 100 \text{ m/s} & \text{(b) } 4 \text{ s} \\ \text{(c) } 200 \text{ m} & \text{(d) } -100 \text{ m/s} \end{bmatrix}$$

2. The distance s metres travelled by a car in t seconds after the brakes are applied is given by $s = 25t - 2.5t^2$. Find (a) the speed of the car (in km/h) when the brakes are applied, (b) the distance the car travels before it stops.
$$[\text{(a) } 90 \text{ km/h}\quad \text{(b) } 62.5 \text{ m}]$$

3. The equation $\theta = 10\pi + 24t - 3t^2$ gives the angle θ, in radians, through which a wheel turns in t seconds. Determine (a) the time the wheel takes to come to rest, (b) the angle turned through in the last second of movement.
$$[\text{(a) } 4 \text{ s}\quad \text{(b) } 3 \text{ rads}]$$

Section 8

4. At any time t seconds the distance x metres of a particle moving in a straight line from a fixed point is given by: $x = 4t + \ln(1 - t)$. Determine (a) the initial velocity and acceleration, (b) the velocity and acceleration after 1.5 s, and (c) the time when the velocity is zero.

$$\begin{bmatrix} \text{(a) } 3\,\text{m/s}; -1\,\text{m/s}^2 \\ \text{(b) } 6\,\text{m/s}; -4\,\text{m/s}^2 \quad \text{(c) } \tfrac{3}{4}\,\text{s} \end{bmatrix}$$

5. The angular displacement θ of a rotating disc is given by: $\theta = 6 \sin \dfrac{t}{4}$, where t is the time in seconds. Determine (a) the angular velocity of the disc when t is 1.5 s, (b) the angular acceleration when t is 5.5 s, and (c) the first time when the angular velocity is zero.

$$\begin{bmatrix} \text{(a) } \omega = 1.40\,\text{rad/s} \\ \text{(b) } \alpha = -0.37\,\text{rad/s}^2 \\ \text{(c) } t = 6.28\,\text{s} \end{bmatrix}$$

6. $x = \dfrac{20t^3}{3} - \dfrac{23t^2}{2} + 6t + 5$ represents the distance, x metres, moved by a body in t seconds. Determine (a) the velocity and acceleration at the start, (b) the velocity and acceleration when $t = 3$ s, (c) the values of t when the body is at rest, (d) the value of t when the acceleration is 37 m/s^2, and (e) the distance travelled in the third second.

$$\begin{bmatrix} \text{(a) } 6\,\text{m/s}, -23\,\text{m/s}^2 \\ \text{(b) } 117\,\text{m/s}, 97\,\text{m/s}^2 \\ \text{(c) } \tfrac{3}{4}\,\text{s or } \tfrac{2}{5}\,\text{s} \\ \text{(d) } 1\tfrac{1}{2}\,\text{s} \quad \text{(e) } 75\tfrac{1}{6}\,\text{m} \end{bmatrix}$$

44.3 Turning points

In Fig. 44.4, the gradient (or rate of change) of the curve changes from positive between O and P to negative between P and Q, and then positive again between Q and R. At point P, the gradient is zero and, as x increases, the gradient of the curve changes from positive just before P to negative just after. Such a point is called a **maximum point** and appears as the 'crest of a wave'. At point Q, the gradient is also zero and, as x increases, the gradient of the curve changes from negative just before Q to positive just after. Such a point is called a **minimum point**, and appears as the 'bottom of

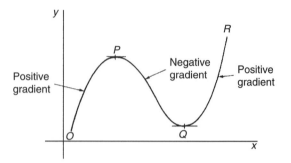

Figure 44.4

a valley'. Points such as P and Q are given the general name of **turning points**.

It is possible to have a turning point, the gradient on either side of which is the same. Such a point is given the special name of a **point of inflexion**, and examples are shown in Fig. 44.5.

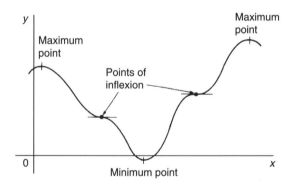

Figure 44.5

Maximum and minimum points and points of inflexion are given the general term of **stationary points**.

Procedure for finding and distinguishing between stationary points.

(i) Given $y = f(x)$, determine $\dfrac{dy}{dx}$ (i.e. $f'(x)$)

(ii) Let $\dfrac{dy}{dx} = 0$ and solve for the values of x

(iii) Substitute the values of x into the original equation, $y = f(x)$, to find the corresponding y-ordinate values. This establishes the co-ordinates of the stationary points.

To determine the nature of the stationary points: Either

(iv) Find $\dfrac{d^2y}{dx^2}$ and substitute into it the values of x found in (ii).

If the result is: (a) positive — the point is a minimum one,

(b) negative — the point is a maximum one,

(c) zero — the point is a point of inflexion

or

(v) Determine the sign of the gradient of the curve just before and just after the stationary points. If the sign change for the gradient of the curve is:

(a) positive to negative — the point is a maximum one

(b) negative to positive — the point is a minimum one

(c) positive to positive or negative to negative — the point is a point of inflexion.

Problem 10. Locate the turning point on the curve $y = 3x^2 - 6x$ and determine its nature by examining the sign of the gradient on either side

Following the above procedure:

(i) Since $y = 3x^2 - 6x$, $\dfrac{dy}{dx} = 6x - 6$

(ii) At a turning point, $\dfrac{dy}{dx} = 0$, hence $6x - 6 = 0$, from which, $x = 1$.

(iii) When $x = 1$, $y = 3(1)^2 - 6(1) = -3$

Hence the co-ordinates of the turning point is (1, −3)

(v) If x is slightly less than 1, say, 0.9, then $\dfrac{dy}{dx} = 6(0.9) - 6 = -0.6$, i.e. negative

If x is slightly greater than 1, say, 1.1, then $\dfrac{dy}{dx} = 6(1.1) - 6 = 0.6$, i.e. positive

Since the gradient of the curve is negative just before the turning point and positive just after (i.e. $-\cup+$), **(1, −3) is a minimum point**

Problem 11. Find the maximum and minimum values of the curve $y = x^3 - 3x + 5$ by (a) examining the gradient on either side of the turning

points, and (b) determining the sign of the second derivative

Since $y = x^3 - 3x + 5$ then $\dfrac{dy}{dx} = 3x^2 - 3$

For a maximum or minimum value $\dfrac{dy}{dx} = 0$

Hence $\qquad 3x^2 - 3 = 0$

from which, $\qquad 3x^2 = 3$

and $\qquad\qquad x = \pm 1$

When $x = 1$, $y = (1)^3 - 3(1) + 5 = 3$

When $x = -1$, $y = (-1)^3 - 3(-1) + 5 = 7$

Hence (1, 3) and (−1, 7) are the co-ordinates of the turning points.

(a) Considering the point (1, 3):

If x is slightly less than 1, say 0.9, then $\dfrac{dy}{dx} = 3(0.9)^2 - 3$, which is negative.

If x is slightly more than 1, say 1.1, then $\dfrac{dy}{dx} = 3(1.1)^2 - 3$, which is positive.

Since the gradient changes from negative to positive, **the point (1, 3) is a minimum point**.

Considering the point (−1, 7):

If x is slightly less than −1, say −1.1, then $\dfrac{dy}{dx} = 3(-1.1)^2 - 3$, which is positive.

If x is slightly more than −1, say −0.9, then $\dfrac{dy}{dx} = 3(-0.9)^2 - 3$, which is negative.

Since the gradient changes from positive to negative, **the point (−1, 7) is a maximum point**.

(b) Since $\dfrac{dy}{dx} = 3x^2 - 3$, then $\dfrac{d^2y}{dx^2} = 6x$

When $x = 1$, $\dfrac{d^2y}{dx^2}$ is positive, hence (1, 3) is a **minimum value**.

When $x = -1$, $\dfrac{d^2y}{dx^2}$ is negative, hence (−1, 7) is a **maximum value**.

Thus the maximum value is 7 and the minimum value is 3.

It can be seen that the second differential method of determining the nature of the turning points is, in this case, quicker than investigating the gradient.

Section 8

Problem 12. Locate the turning point on the following curve and determine whether it is a maximum or minimum point: $y = 4\theta + e^{-\theta}$

Since $y = 4\theta + e^{-\theta}$ then $\dfrac{dy}{d\theta} = 4 - e^{-\theta} = 0$ for a maximum or minimum value.

Hence $4 = e^{-\theta}$ and $\dfrac{1}{4} = e^{\theta}$

giving $\theta = \ln \dfrac{1}{4} = -1.3863$

When $\theta = -1.3863$,

$y = 4(-1.3863) + e^{-(-1.3863)} = 5.5452 + 4.0000$

$= -1.5452$

Thus $(-1.3863, -1.5452)$ are the co-ordinates of the turning point.

$$\frac{d^2 y}{d\theta^2} = e^{-\theta}$$

When $\theta = -1.3863$, $\dfrac{d^2 y}{d\theta^2} = e^{+1.3863} = 4.0$, which is positive, hence
$(-1.3863, -1.5452)$ is a minimum point.

Problem 13. Determine the co-ordinates of the maximum and minimum values of the graph $y = \dfrac{x^3}{3} - \dfrac{x^2}{2} - 6x + \dfrac{5}{3}$ and distinguish between them. Sketch the graph

Following the given procedure:

(i) Since $y = \dfrac{x^3}{3} - \dfrac{x^2}{2} - 6x + \dfrac{5}{3}$ then

$\dfrac{dy}{dx} = x^2 - x - 6$

(ii) At a turning point, $\dfrac{dy}{dx} = 0$.

Hence $\qquad x^2 - x - 6 = 0$

i.e. $\qquad (x + 2)(x - 3) = 0$

from which $\qquad x = -2$ or $x = 3$

(iii) When $x = -2$

$y = \dfrac{(-2)^3}{3} - \dfrac{(-2)^2}{2} - 6(-2) + \dfrac{5}{3} = 9$

When $x = 3, y = \dfrac{(3)^3}{3} - \dfrac{(3)^2}{2} - 6(3) + \dfrac{5}{3}$

$= -11\dfrac{5}{6}$

Thus the co-ordinates of the turning points are $(-2, 9)$ and $\left(3, -11\dfrac{5}{6}\right)$

(iv) Since $\dfrac{dy}{dx} = x^2 - x - 6$ then $\dfrac{d^2 y}{dx^2} = 2x - 1$

When $x = -2$, $\dfrac{d^2 y}{dx^2} = 2(-2) - 1 = -5$, which is negative.

Hence $(-2, 9)$ is a maximum point.

When $x = 3$, $\dfrac{d^2 y}{dx^2} = 2(3) - 1 = 5$, which is positive.

Hence $\left(3, -11\dfrac{5}{6}\right)$ is a minimum point.

Knowing $(-2, 9)$ is a maximum point (i.e. crest of a wave), and $\left(3, -11\dfrac{5}{6}\right)$ is a minimum point (i.e. bottom of a valley) and that when $x = 0, y = \dfrac{5}{3}$, a sketch may be drawn as shown in Fig. 44.6.

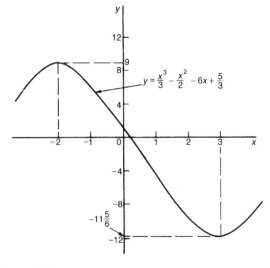

Figure 44.6

Problem 14. Determine the turning points on the curve $y = 4 \sin x - 3 \cos x$ in the range $x = 0$ to $x = 2\pi$ radians, and distinguish between them. Sketch the curve over one cycle

Since $y = 4 \sin x - 3 \cos x$ then

$\dfrac{dy}{dx} = 4 \cos x + 3 \sin x = 0$, for a turning point,

from which, $4 \cos x = -3 \sin x$ and $\dfrac{-4}{3} = \dfrac{\sin x}{\cos x}$

$$= \tan x.$$

Hence $x = \tan^{-1}\left(\dfrac{-4}{3}\right) = 126.87°$ or $306.87°$, since tangent is negative in the second and fourth quadrants.

When $x = 126.87°$,

$\qquad y = 4 \sin 126.87° - 3 \cos 126.87° = 5$

When $x = 306.87°$

$\qquad y = 4 \sin 306.87° - 3 \cos 306.87° = -5$

$126.87° = \left(125.87° \times \dfrac{\pi}{180}\right)$ radians

$\qquad = 2.214\,\text{rad}$

$306.87° = \left(306.87° \times \dfrac{\pi}{180}\right)$ radians

$\qquad = 5.356\,\text{rad}$

Hence (2.214, 5) and (5.356, −5) are the co-ordinates of the turning points.

$$\dfrac{d^2y}{dx^2} = -4 \sin x + 3 \cos x$$

When $x = 2.214$ rad,

$\dfrac{d^2y}{dx^2} = -4 \sin 2.214 + 3 \cos 2.214$, which is negative.

Hence (2.214, 5) is a maximum point.

When $x = 5.356$ rad,

$\dfrac{d^2y}{dx^2} = -4 \sin 5.356 + 3 \cos 5.356$, which is positive.

Hence (5.356, −5) is a minimum point.

A sketch of $y = 4 \sin x - 3 \cos x$ is shown in Fig. 44.7.

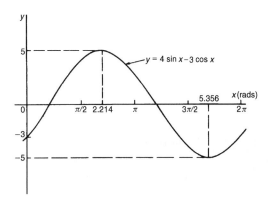

Figure 44.7

Now try the following exercise

Exercise 161   Further problems on turning points

In Problems 1 to 7, find the turning points and distinguish between them.

1. $y = 3x^2 - 4x + 2$ $\left[\text{Minimum at }\left(\dfrac{2}{3}, \dfrac{2}{3}\right)\right]$

2. $x = \theta(6 - \theta)$ [Maximum at (3, 9)]

3. $y = 4x^3 + 3x^2 - 60x - 12$
    $\left[\begin{matrix}\text{Minimum } (2, -88) \\ \text{Maximum } (-2.5, 94.25)\end{matrix}\right]$

4. $y = 5x - 2 \ln x$
    [Minimum at (0.4000, 3.8326)]

5. $y = 2x - e^x$
    [Maximum at (0.6931, −0.6136)]

6. $y = t^3 - \dfrac{t^2}{2} - 2t + 4$
    $\left[\begin{matrix}\text{Minimum at } (1,\ 2.5) \\ \text{Maximum at } \left(-\dfrac{2}{3}, 4\dfrac{22}{27}\right)\end{matrix}\right]$

7. $x = 8t + \dfrac{1}{2t^2}$ [Minimum at (0.5, 6)]

8. Determine the maximum and minimum values on the graph $y = 12 \cos\theta - 5 \sin\theta$ in the range $\theta = 0$ to $\theta = 360°$. Sketch the graph over one cycle showing relevant points.
    $\left[\begin{matrix}\text{Maximum of 13 at } 337.38°, \\ \text{Minimum of } -13 \text{ at } 157.38°\end{matrix}\right]$

Section 8

9. Show that the curve $y = \dfrac{2}{3}(t-1)^3 + 2t(t-2)$ has a maximum value of $\dfrac{2}{3}$ and a minimum value of -2.

44.4 Practical problems involving maximum and minimum values

There are many **practical problems** involving maximum and minimum values which occur in science and engineering. Usually, an equation has to be determined from given data, and rearranged where necessary, so that it contains only one variable. Some examples are demonstrated in Problems 15 to 20.

Problem 15. A rectangular area is formed having a perimeter of 40 cm. Determine the length and breadth of the rectangle if it is to enclose the maximum possible area

Let the dimensions of the rectangle be x and y. Then the perimeter of the rectangle is $(2x + 2y)$. Hence

$$2x + 2y = 40, \quad \text{or} \quad x + y = 20 \qquad (1)$$

Since the rectangle is to enclose the maximum possible area, a formula for area A must be obtained in terms of one variable only.

Area $A = xy$. From equation (1), $x = 20 - y$

Hence, area $A = (20 - y)y = 20y - y^2$

$\dfrac{dA}{dy} = 20 - 2y = 0$ for a turning point, from which, $y = 10$ cm.

$\dfrac{d^2A}{dy^2} = -2$, which is negative, giving a maximum point.

When $y = 10$ cm, $x = 10$ cm, from equation (1).

Hence the length and breadth of the rectangle are each 10 cm, i.e. a square gives the maximum possible area. When the perimeter of a rectangle is 40 cm, the maximum possible area is $10 \times 10 = \mathbf{100\ cm^2}$.

Problem 16. A rectangular sheet of metal having dimensions 20 cm by 12 cm has squares removed from each of the four corners and the sides bent upwards to form an open box. Determine the maximum possible volume of the box

The squares to be removed from each corner are shown in Fig. 44.8, having sides x cm. When the sides are bent upwards the dimensions of the box will be: length $(20 - 2x)$ cm, breadth $(12 - 2x)$ cm and height, x cm.

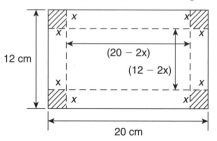

Figure 44.8

Volume of box, $V = (20 - 2x)(12 - 2x)(x)$

$$= 240x - 64x^2 + 4x^3$$

$\dfrac{dV}{dx} = 240 - 128x + 12x^2 = 0$ for a turning point.

Hence $4(60 - 32x + 3x^2) = 0$, i.e. $3x^2 - 32x + 60 = 0$

Using the quadratic formula,

$$x = \frac{32 \pm \sqrt{(-32)^2 - 4(3)(60)}}{2(3)}$$

$$= 8.239\ \text{cm or } 2.427\ \text{cm.}$$

Since the breadth is $(12 - 2x)$ cm then $x = 8.239$ cm is not possible and is neglected.

Hence $x = 2.427$ cm.

$$\frac{d^2V}{dx^2} = -128 + 24x$$

When $x = 2.427$, $\dfrac{d^2V}{dx^2}$ is negative, giving a maximum value.

The dimensions of the box are:
length $= 20 - 2(2.427) = 15.146$ cm,
breadth $= 12 - 2(2.427) = 7.146$ cm, and
height $= 2.427$ cm.

$$\textbf{Maximum volume} = (15.146)(7.146)(2.427)$$

$$= \mathbf{262.7\ cm^3}$$

Problem 17. Determine the height and radius of a cylinder of volume 200 cm³ which has the least surface area

Let the cylinder have radius r and perpendicular height h.

Volume of cylinder, $V = \pi r^2 h = 200$ (1)

Surface area of cylinder, $A = 2\pi rh + 2\pi r^2$

Least surface area means minimum surface area and a formula for the surface area in terms of one variable only is required.

From equation (1), $h = \dfrac{200}{\pi r^2}$ (2)

Hence surface area,

$$A = 2\pi r \left(\frac{200}{\pi r^2}\right) + 2\pi r^2$$

$$= \frac{400}{r} + 2\pi r^2 = 400r^{-1} + 2\pi r^2$$

$\dfrac{dA}{dr} = \dfrac{-400}{r^2} + 4\pi r = 0$, for a turning point.

Hence $4\pi r = \dfrac{400}{r^2}$

and $r^3 = \dfrac{400}{4\pi}$

from which, $r = \sqrt[3]{\dfrac{100}{\pi}} = 3.169\,\text{cm}.$

$$\frac{d^2 A}{dr^2} = \frac{800}{r^3} + 4\pi$$

When $r = 3.169\,\text{cm}$, $\dfrac{d^2 A}{dr^2}$ is positive, giving a minimum value.
From equation (2), when $r = 3.169\,\text{cm}$,

$$h = \frac{200}{\pi(3.169)^2} = 6.339\,\text{cm}.$$

Hence for the least surface area, a cylinder of volume 200 cm³ has a radius of 3.169 cm and height of 6.339 cm.

Problem 18.   Determine the area of the largest piece of rectangular ground that can be enclosed by 100 m of fencing, if part of an existing straight wall is used as one side

Let the dimensions of the rectangle be x and y as shown in Fig. 44.9, where PQ represents the straight wall.

From Fig. 44.9, $x + 2y = 100$ (1)

Area of rectangle, $A = xy$ (2)

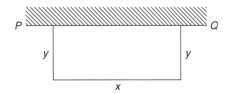

Figure 44.9

Since the maximum area is required, a formula for area A is needed in terms of one variable only.
From equation (1), $x = 100 - 2y$
Hence,   area $A = xy = (100 - 2y)y = 100y - 2y^2$
$\dfrac{dA}{dy} = 100 - 4y = 0$, for a turning point, from which, $y = 25\,\text{m}.$
$\dfrac{d^2 A}{dy^2} = -4$, which is negative, giving a maximum value.
When $y = 25\,\text{m}$, $x = 50\,\text{m}$ from equation (1).
Hence the **maximum possible area**
$$= xy = (50)(25) = \mathbf{1250\,m^2}$$

Problem 19.   An open rectangular box with square ends is fitted with an overlapping lid which covers the top and the front face. Determine the maximum volume of the box if 6 m² of metal are used in its construction

A rectangular box having square ends of side x and length y is shown in Fig. 44.10.

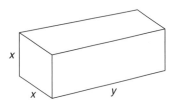

Figure 44.10

Surface area of box, A, consists of two ends and five faces (since the lid also covers the front face).

Hence $A = 2x^2 + 5xy = 6$ (1)

Since it is the maximum volume required, a formula for the volume in terms of one variable only is needed.
Volume of box, $V = x^2 y$
From equation (1),

$$y = \frac{6 - 2x^2}{5x} = \frac{6}{5x} - \frac{2x}{5}$$ (2)

Section 8

Hence volume $V = x^2 y = x^2 \left(\dfrac{6}{5x} - \dfrac{2x}{5} \right)$

$$= \frac{6x}{5} - \frac{2x^3}{5}$$

$\dfrac{dV}{dx} = \dfrac{6}{5} - \dfrac{6x^2}{5} = 0$ for a maximum or minimum value.

Hence $6 = 6x^2$, giving $x = 1$ m ($x = -1$ is not possible, and is thus neglected).

$$\frac{d^2V}{dx^2} = \frac{-12x}{5}$$

When $x = 1$, $\dfrac{d^2V}{dx^2}$ is negative, giving a maximum value.

From equation (2), when $x = 1$, $y = \dfrac{6}{5(1)} - \dfrac{2(1)}{5} = \dfrac{4}{5}$

Hence the maximum volume of the box is given by

$V = x^2 y = (1)^2 \left(\dfrac{4}{5} \right) = \dfrac{4}{5}$ **m³**

Problem 20. Find the diameter and height of a cylinder of maximum volume which can be cut from a sphere of radius 12 cm

---

A cylinder of radius r and height h is shown enclosed in a sphere of radius $R = 12$ cm in Fig. 44.11.

Volume of cylinder, $\quad V = \pi r^2 h \qquad (1)$

Using the right-angled triangle OPQ shown in Fig. 44.11,

$$r^2 + \left(\frac{h}{2} \right)^2 = R^2 \text{ by Pythagoras' theorem,}$$

i.e. $\qquad r^2 + \dfrac{h^2}{4} = 144 \qquad (2)$

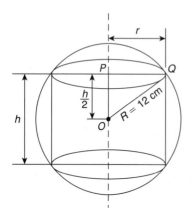

Figure 44.11

Since the maximum volume is required, a formula for the volume V is needed in terms of one variable only.

From equation (2), $r^2 = 144 - \dfrac{h^2}{4}$

Substituting into equation (1) gives:

$$V = \pi \left(144 - \frac{h^2}{4} \right) h = 144\pi h - \frac{\pi h^3}{4}$$

$\dfrac{dV}{dh} = 144\pi - \dfrac{3\pi h^2}{4} = 0$, for a maximum or minimum value.

Hence $144\pi = \dfrac{3\pi h^2}{4}$, from which,

$$h = \sqrt{\frac{(144)(4)}{3}} = 13.86 \text{ cm.}$$

$$\frac{d^2V}{dh^2} = \frac{-6\pi h}{4}$$

When $h = 13.86$, $\dfrac{d^2V}{dh^2}$ is negative, giving a maximum value.

From equation (2),

$r^2 = 144 - \dfrac{h^2}{4} = 144 - \dfrac{13.86^2}{4}$, from which, radius $r = 9.80$ cm

Diameter of cylinder $= 2r = 2(9.80) = 19.60$ cm.

Hence the cylinder having the maximum volume that can be cut from a sphere of radius 12 cm is one in which the diameter is 19.60 cm and the height is 13.86 cm.

Now try the following exercise

Exercise 162 Further problems on practical maximum and minimum problems

1. The speed, v, of a car (in m/s) is related to time t s by the equation $v = 3 + 12t - 3t^2$. Determine the maximum speed of the car in km/h. [54 km/h]

2. Determine the maximum area of a rectangular piece of land that can be enclosed by 1200 m of fencing. [90 000 m²]

3. A shell is fired vertically upwards and its vertical height, x metres, is given

by: $x = 24t - 3t^2$, where t is the time in seconds. Determine the maximum height reached.   [48 m]

4. A lidless box with square ends is to be made from a thin sheet of metal. Determine the least area of the metal for which the volume of the box is $3.5\,\mathrm{m}^3$.   [11.42 m²]

5. A closed cylindrical container has a surface area of $400\,\mathrm{cm}^2$. Determine the dimensions for maximum volume.
[radius $= 4.607$ cm, height $= 9.212$ cm]

6. Calculate the height of a cylinder of maximum volume that can be cut from a cone of height 20 cm and base radius 80 cm.
[6.67 cm]

7. The power developed in a resistor R by a battery of emf E and internal resistance r is given by $P = \dfrac{E^2 R}{(R+r)^2}$. Differentiate P with respect to R and show that the power is a maximum when $R = r$.

8. Find the height and radius of a closed cylinder of volume $125\,\mathrm{cm}^3$ which has the least surface area.   [height $= 5.42$ cm, radius $= 2.71$ cm]

9. Resistance to motion, F, of a moving vehicle, is given by: $F = \dfrac{5}{x} + 100x$. Determine the minimum value of resistance.   [44.72]

10. An electrical voltage E is given by: $E = (15 \sin 50\pi t + 40 \cos 50\pi t)$ volts, where t is the time in seconds. Determine the maximum value of voltage.   [42.72 volts]

11. The fuel economy E of a car, in miles per gallon, is given by:

$$E = 21 + 2.10 \times 10^{-2} v^2 - 3.80 \times 10^{-6} v^4$$

where v is the speed of the car in miles per hour.
Determine, correct to 3 significant figures, the most economical fuel consumption, and the speed at which it is achieved.
[50.0 miles/gallon, 52.6 miles/hour]

44.5   Tangents and normals

Tangents

The equation of the tangent to a curve $y = f(x)$ at the point (x_1, y_1) is given by:

$$y - y_1 = m(x - x_1)$$

where $m = \dfrac{dy}{dx} =$ gradient of the curve at (x_1, y_1).

> **Problem 21.** Find the equation of the tangent to the curve $y = x^2 - x - 2$ at the point $(1, -2)$

Gradient, $m = \dfrac{dy}{dx} = 2x - 1$

At the point $(1, -2)$, $x = 1$ and $m = 2(1) - 1 = 1$
Hence the equation of the tangent is:

$$y - y_1 = m(x - x_1)$$

i.e. $y - -2 = 1(x - 1)$

i.e. $y + 2 = x - 1$

or $\mathbf{y = x - 3}$

The graph of $y = x^2 - x - 2$ is shown in Fig. 46.12. The line AB is the tangent to the curve at the point C, i.e. $(1, -2)$, and the equation of this line is $y = x - 3$.

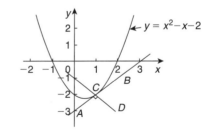

Figure 44.12

Normals

The normal at any point on a curve is the line that passes through the point and is at right angles to the tangent. Hence, in Fig. 44.12, the line CD is the normal.

It may be shown that if two lines are at right angles then the product of their gradients is -1. Thus if m is the gradient of the tangent, then the gradient of the normal is $-\dfrac{1}{m}$

Hence the equation of the normal at the point (x_1, y_1) is given by:

$$y - y_1 = -\frac{1}{m}(x - x_1)$$

Problem 22. Find the equation of the normal to the curve $y = x^2 - x - 2$ at the point $(1, -2)$

$m = 1$ from Problem 21, hence the equation of the normal is $y - y_1 = -\frac{1}{m}(x - x_1)$

i.e. $\quad y - -2 = -\frac{1}{1}(x - 1)$

i.e. $\quad y + 2 = -x + 1 \quad$ or $\quad y = -x - 1$

Thus the line *CD* in Fig. 44.12 has the equation $y = -x - 1$

Problem 23. Determine the equations of the tangent and normal to the curve $y = \dfrac{x^3}{5}$ at the point $\left(-1, -\frac{1}{5}\right)$

Gradient m of curve $y = \dfrac{x^3}{5}$ is given by

$$m = \frac{dy}{dx} = \frac{3x^2}{5}$$

At the point $\left(-1, -\frac{1}{5}\right)$, $x = -1$ and

$$m = \frac{3(-1)^2}{5} = \frac{3}{5}$$

Equation of the tangent is:

$$y - y_1 = m(x - x_1)$$

i.e. $\quad y - -\dfrac{1}{5} = \dfrac{3}{5}(x - -1)$

i.e. $\quad y + \dfrac{1}{5} = \dfrac{3}{5}(x + 1)$

or $\quad 5y + 1 = 3x + 3$

or $\quad \mathbf{5y - 3x = 2}$

Equation of the normal is:

$$y - y_1 = -\frac{1}{m}(x - x_1)$$

i.e. $\quad y - -\dfrac{1}{5} = \dfrac{-1}{\left(\dfrac{3}{5}\right)}(x - -1)$

i.e. $\quad y + \dfrac{1}{5} = -\dfrac{5}{3}(x + 1)$

i.e. $\quad y + \dfrac{1}{5} = -\dfrac{5}{3}x - \dfrac{5}{3}$

Multiplying each term by 15 gives:

$$15y + 3 = -25x - 25$$

Hence **equation of the normal** is:

$$\mathbf{15y + 25x + 28 = 0}$$

Now try the following exercise

Exercise 163   Further problems on tangents and normals

For the following curves, at the points given, find (a) the equation of the tangent, and (b) the equation of the normal

1. $y = 2x^2$ at the point $(1, 2)$
$$[(a)\ y = 4x - 2 \quad (b)\ 4y + x = 9]$$

2. $y = 3x^2 - 2x$ at the point $(2, 8)$
$$[(a)\ y = 10x - 12 \quad (b)\ 10y + x = 82]$$

3. $y = \dfrac{x^3}{2}$ at the point $\left(-1, -\dfrac{1}{2}\right)$
$$\left[(a)\ y = \frac{3}{2}x + 1 \quad (b)\ 6y + 4x + 7 = 0\right]$$

4. $y = 1 + x - x^2$ at the point $(-2, -5)$
$$[(a)\ y = 5x + 5 \quad (b)\ 5y + x + 27 = 0]$$

5. $\theta = \dfrac{1}{t}$ at the point $\left(3, \dfrac{1}{3}\right)$
$$\left[\begin{array}{l}(a)\ 9\theta + t = 6 \\ (b)\ \theta = 9t - 26\dfrac{2}{3} \quad \text{or} \quad 3\theta = 27t - 80\end{array}\right]$$

44.6   Small changes

If y is a function of x, i.e. $y = f(x)$, and the approximate change in y corresponding to a small change δx in x is required, then:

$$\frac{\delta y}{\delta x} \approx \frac{dy}{dx}$$

and $\quad \delta y \approx \dfrac{dy}{dx} \cdot \delta x \quad$ or $\quad \delta y \approx f'(x) \cdot \delta x$

Section 8

Problem 24. Given $y = 4x^2 - x$, determine the approximate change in y if x changes from 1 to 1.02

Since $y = 4x^2 - x$, then $\dfrac{dy}{dx} = 8x - 1$

Approximate change in y,

$$\delta y \approx \frac{dy}{dx} \cdot \delta x \approx (8x - 1)\delta x$$

When $x = 1$ and $\delta x = 0.02$, $\delta y \approx [8(1) - 1](0.02)$

$$\approx \mathbf{0.14}$$

[Obviously, in this case, the exact value of δy may be obtained by evaluating y when $x = 1.02$, i.e. $y = 4(1.02)^2 - 1.02 = 3.1416$ and then subtracting from it the value of y when $x = 1$, i.e. $y = 4(1)^2 - 1 = 3$, giving $\delta y = 3.1416 - 3 = \mathbf{0.1416}$. Using $\delta y = \dfrac{dy}{dx} \cdot \delta x$ above gave 0.14, which shows that the formula gives the approximate change in y for a small change in x].

Problem 25. The time of swing T of a pendulum is given by $T = k\sqrt{l}$, where k is a constant. Determine the percentage change in the time of swing if the length of the pendulum l changes from 32.1 cm to 32.0 cm

If $\qquad T = k\sqrt{l} = kl^{1/2}$

then $\quad \dfrac{dT}{dl} = k\left(\dfrac{1}{2}l^{-1/2}\right) = \dfrac{k}{2\sqrt{l}}$

Approximate change in T,

$$\delta T \approx \frac{dT}{dl}\delta l \approx \left(\frac{k}{2\sqrt{l}}\right)\delta l \approx \left(\frac{k}{2\sqrt{l}}\right)(-0.1)$$

(negative since l decreases)

Percentage error

$$= \left(\frac{\text{approximate change in } T}{\text{original value of } T}\right)100\%$$

$$= \frac{\left(\dfrac{k}{2\sqrt{l}}\right)(-0.1)}{k\sqrt{l}} \times 100\%$$

$$= \left(\frac{-0.1}{2l}\right)100\% = \left(\frac{-0.1}{2(32.1)}\right)100\%$$

$$= \mathbf{-0.156\%}$$

Hence the percentage change in the time of swing is a decrease of **0.156%**

Problem 26. A circular template has a radius of 10 cm (± 0.02). Determine the possible error in calculating the area of the template. Find also the percentage error

Area of circular template, $A = \pi r^2$, hence

$$\frac{dA}{dr} = 2\pi r$$

Approximate change in area,

$$\delta A \approx \frac{dA}{dr} \cdot \delta r \approx (2\pi r)\delta r$$

When $r = 10$ cm and $\delta r = 0.02$,

$$\delta A = (2\pi 10)(0.02) \approx 0.4\pi \text{ cm}^2$$

i.e. **the possible error in calculating the template area is approximately 1.257 cm^2**.

Percentage error $\approx \left(\dfrac{0.4\pi}{\pi(10)^2}\right)100\% = \mathbf{0.40\%}$

Now try the following exercise

Exercise 164 Further problems on small changes

1. Determine the change in y if x changes from 2.50 to 2.51 when (a) $y = 2x - x^2$ (b) $y = \dfrac{5}{x}$

   [(a) -0.03 (b) -0.008]

2. The pressure p and volume v of a mass of gas are related by the equation $pv = 50$. If the pressure increases from 25.0 to 25.4, determine the approximate change in the volume of the gas. Find also the percentage change in the volume of the gas. $[-0.032, -1.6\%]$

3. Determine the approximate increase in (a) the volume, and (b) the surface area of a cube of side x cm if x increases from 20.0 cm to 20.05 cm.   [(a) 60 cm^3 (b) 12 cm^2]

4. The radius of a sphere decreases from 6.0 cm to 5.96 cm. Determine the approximate change in (a) the surface area, and (b) the volume.

   [(a) -6.03 cm^2 (b) -18.10 cm^3]

5. The rate of flow of a liquid through a tube is given by Poiseuilles's equation as: $Q = \dfrac{p\pi r^4}{8\eta L}$ where Q is the rate of flow, p is the pressure difference between the ends of the tube, r is the radius of the tube, L is the length of the tube and η is the coefficient of viscosity of the liquid. η is obtained by measuring Q, p, r and L. If Q can be measured accurate to $\pm 0.5\%$, p accurate to $\pm 3\%$, r accurate to $\pm 2\%$ and L accurate to $\pm 1\%$, calculate the maximum possible percentage error in the value of η.

[12.5%]

Revision Test 12

This Revision test covers the material contained in Chapters 42 to 44. *The marks for each question are shown in brackets at the end of each question.*

1. Differentiate the following with respect to the variable:

   (a) $y = 5 + 2\sqrt{x^3} - \dfrac{1}{x^2}$ (b) $s = 4e^{2\theta}\sin 3\theta$

   (c) $y = \dfrac{3\ln 5t}{\cos 2t}$ (d) $x = \dfrac{2}{\sqrt{t^2 - 3t + 5}}$

   (15)

2. If $f(x) = 2.5x^2 - 6x + 2$ find the co-ordinates at the point at which the gradient is -1. (5)

3. The displacement s cm of the end of a stiff spring at time t seconds is given by: $s = ae^{-kt}\sin 2\pi f t$. Determine the velocity and acceleration of the end of the spring after 2 seconds if $a = 3$, $k = 0.75$ and $f = 20$. (10)

4. Find the co-ordinates of the turning points on the curve $y = 3x^3 + 6x^2 + 3x - 1$ and distinguish between them. (9)

5. The heat capacity C of a gas varies with absolute temperature θ as shown:

   $$C = 26.50 + 7.20 \times 10^{-3}\theta - 1.20 \times 10^{-6}\theta^2$$

   Determine the maximum value of C and the temperature at which it occurs. (7)

6. Determine for the curve $y = 2x^2 - 3x$ at the point $(2, 2)$: (a) the equation of the tangent (b) the equation of the normal. (7)

7. A rectangular block of metal with a square cross-section has a total surface area of $250\,\text{cm}^2$. Find the maximum volume of the block of metal. (7)

Section 8

Chapter 45

Differentiation of parametric equations

45.1 Introduction to parametric equations

Certain mathematical functions can be expressed more simply by expressing, say, x and y separately in terms of a third variable. For example, $y = r \sin \theta$, $x = r \cos \theta$. Then, any value given to θ will produce a pair of values for x and y, which may be plotted to provide a curve of $y = f(x)$.

The third variable, θ, is called a **parameter** and the two expressions for y and x are called **parametric equations**.

The above example of $y = r \sin \theta$ and $x = r \cos \theta$ are the parametric equations for a circle. The equation of any point on a circle, centre at the origin and of radius r is given by: $x^2 + y^2 = r^2$, as shown in Chapter 19.

To show that $y = r \sin \theta$ and $x = r \cos \theta$ are suitable parametric equations for such a circle:

Left hand side of equation
$$= x^2 + y^2$$
$$= (r \cos \theta)^2 + (r \sin \theta)^2$$
$$= r^2 \cos^2 \theta + r^2 \sin^2 \theta$$
$$= r^2 (\cos^2 \theta + \sin^2 \theta)$$
$$= r^2 = \text{right hand side}$$
$$(\text{since } \cos^2 \theta + \sin^2 \theta = 1, \text{ as shown in}$$
Chapter 26)

45.2 Some common parametric equations

The following are some of the most common parametric equations, and Figure 45.1 shows typical shapes of these curves.

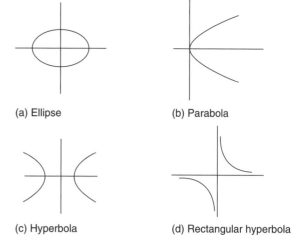

(a) Ellipse   (b) Parabola

(c) Hyperbola   (d) Rectangular hyperbola

(e) Cardioid   (f) Astroid

(g) Cycloid

Figure 45.1

(a) Ellipse $x = a \cos \theta, y = b \sin \theta$
(b) Parabola $x = at^2, y = 2at$
(c) Hyperbola $x = a \sec \theta, y = b \tan \theta$
(d) Rectangular $x = ct, y = \dfrac{c}{t}$
    hyperbola

(e) Cardioid $x = a(2\cos\theta - \cos 2\theta)$,
$y = a(2\sin\theta - \sin 2\theta)$
(f) Astroid $x = a\cos^3\theta, y = a\sin^3\theta$
(g) Cycloid $x = a(\theta - \sin\theta), y = a(1-\cos\theta)$

45.3 Differentiation in parameters

When x and y are given in terms of a parameter say θ, then by the function of a function rule of differentiation (from Chapter 43):

$$\frac{dy}{dx} = \frac{dy}{d\theta} \times \frac{d\theta}{dx}$$

It may be shown that this can be written as:

$$\frac{dy}{dx} = \frac{\frac{dy}{d\theta}}{\frac{dx}{d\theta}} \quad (1)$$

For the second differential,

$$\frac{d^2y}{dx^2} = \frac{d}{dx}\left(\frac{dy}{dx}\right) = \frac{d}{d\theta}\left(\frac{dy}{dx}\right)\cdot\frac{d\theta}{dx}$$

or

$$\frac{d^2y}{dx^2} = \frac{\frac{d}{d\theta}\left(\frac{dy}{dx}\right)}{\frac{dx}{d\theta}} \quad (2)$$

Problem 1. Given $x = 5\theta - 1$ and $y = 2\theta(\theta - 1)$, determine $\dfrac{dy}{dx}$ in terms of θ

$x = 5\theta - 1$, hence $\dfrac{dy}{d\theta} = 5$

$y = 2\theta(\theta - 1) = 2\theta^2 - 2\theta$,

hence $\dfrac{dy}{d\theta} = 4\theta - 2 = 2(2\theta - 1)$

From equation (1),

$$\frac{dy}{dx} = \frac{\frac{dy}{d\theta}}{\frac{dx}{d\theta}} = \frac{2(2\theta-1)}{5} \text{ or } \frac{2}{5}(2\theta-1)$$

Problem 2. The parametric equations of a function are given by $y = 3\cos 2t, x = 2\sin t$.
Determine expressions for (a) $\dfrac{dy}{dx}$ (b) $\dfrac{d^2y}{dx^2}$

(a) $y = 3\cos 2t$, hence $\dfrac{dy}{dt} = -6\sin 2t$

$x = 2\sin t$, hence $\dfrac{dx}{dt} = 2\cos t$

From equation (1),

$$\frac{dy}{dx} = \frac{\frac{dy}{dt}}{\frac{dx}{dt}} = \frac{-6\sin 2t}{2\cos t} = \frac{-6(2\sin t\cos t)}{2\cos t}$$

from double angles, Chapter 27

i.e. $\dfrac{dy}{dx} = -6\sin t$

(b) From equation (2),

$$\frac{d^2y}{dx^2} = \frac{\frac{d}{dt}\left(\frac{dy}{dx}\right)}{\frac{dx}{dt}} = \frac{\frac{d}{dt}(-6\sin t)}{2\cos t} = \frac{-6\cos t}{2\cos t}$$

i.e. $\dfrac{d^2y}{dx^2} = -3$

Problem 3. The equation of a tangent drawn to a curve at point (x_1, y_1) is given by:

$$y - y_1 = \frac{dy_1}{dx_1}(x - x_1)$$

Determine the equation of the tangent drawn to the parabola $x = 2t^2, y = 4t$ at the point t.

At point t, $x_1 = 2t^2$, hence $\dfrac{dx_1}{dt} = 4t$

and $y_1 = 4t$, hence $\dfrac{dy_1}{dt} = 4$

From equation (1),

$$\frac{dy}{dx} = \frac{\frac{dy}{dt}}{\frac{dx}{dt}} = \frac{4}{4t} = \frac{1}{t}$$

Hence, the equation of the tangent is:

$$y - 4t = \frac{1}{t}(x - 2t^2)$$

Problem 4. The parametric equations of a cycloid are $x = 4(\theta - \sin\theta), y = 4(1 - \cos\theta)$.
Determine (a) $\dfrac{dy}{dx}$ (b) $\dfrac{d^2y}{dx^2}$

(a) $x = 4(\theta - \sin\theta)$

hence $\dfrac{dx}{d\theta} = 4 - 4\cos\theta = 4(1 - \cos\theta)$

$y = 4(1 - \cos\theta)$, hence $\dfrac{dy}{d\theta} = 4\sin\theta$

From equation (1),

$$\frac{dy}{dx} = \frac{\dfrac{dy}{d\theta}}{\dfrac{dx}{d\theta}} = \frac{4\sin\theta}{4(1-\cos\theta)} = \frac{\mathbf{\sin\theta}}{\mathbf{(1-\cos\theta)}}$$

(b) From equation (2),

$$\frac{d^2y}{dx^2} = \frac{\dfrac{d}{d\theta}\left(\dfrac{dy}{dx}\right)}{\dfrac{dx}{d\theta}} = \frac{\dfrac{d}{d\theta}\left(\dfrac{\sin\theta}{1-\cos\theta}\right)}{4(1-\cos\theta)}$$

$$= \frac{\dfrac{(1-\cos\theta)(\cos\theta) - (\sin\theta)(\sin\theta)}{(1-\cos\theta)^2}}{4(1-\cos\theta)}$$

$$= \frac{\cos\theta - \cos^2\theta - \sin^2\theta}{4(1-\cos\theta)^3}$$

$$= \frac{\cos\theta - (\cos^2\theta + \sin^2\theta)}{4(1-\cos\theta)^3}$$

$$= \frac{\cos\theta - 1}{4(1-\cos\theta)^3}$$

$$= \frac{-(1-\cos\theta)}{4(1-\cos\theta)^3} = \frac{\mathbf{-1}}{\mathbf{4(1-\cos\theta)^2}}$$

Now try the following exercise

Exercise 165 Further problems on differentiation of parametric equations

1. Given $x = 3t - 1$ and $y = t(t-1)$, determine $\dfrac{dy}{dx}$ in terms of t.
$$\left[\frac{1}{3}(2t-1)\right]$$

2. A parabola has parametric equations: $x = t^2$, $y = 2t$. Evaluate $\dfrac{dy}{dx}$ when $t = 0.5$ [2]

3. The parametric equations for an ellipse are $x = 4\cos\theta$, $y = \sin\theta$. Determine (a) $\dfrac{dy}{dx}$ (b) $\dfrac{d^2y}{dx^2}$
$$\left[(a) -\frac{1}{4}\cot\theta \quad (b) -\frac{1}{16}\cosec^3\theta\right]$$

4. Evaluate $\dfrac{dy}{dx}$ at $\theta = \dfrac{\pi}{6}$ radians for the hyperbola whose parametric equations are $x = 3\sec\theta$, $y = 6\tan\theta$.   [4]

5. The parametric equations for a rectangular hyperbola are $x = 2t$, $y = \dfrac{2}{t}$. Evaluate $\dfrac{dy}{dx}$ when $t = 0.40$
$$[-6.25]$$
The equation of a tangent drawn to a curve at point (x_1, y_1) is given by:
$$y - y_1 = \frac{dy_1}{dx_1}(x - x_1)$$
Use this in Problems 6 and 7.

6. Determine the equation of the tangent drawn to the ellipse $x = 3\cos\theta$, $y = 2\sin\theta$ at $\theta = \dfrac{\pi}{6}$.
$$[y = -1.155x + 4]$$

7. Determine the equation of the tangent drawn to the rectangular hyperbola $x = 5t$, $y = \dfrac{5}{t}$ at $t = 2$.
$$\left[y = -\frac{1}{4}x + 5\right]$$

45.4 Further worked problems on differentiation of parametric equations

Problem 5. The equation of the normal drawn to a curve at point (x_1, y_1) is given by:
$$y - y_1 = -\frac{1}{\dfrac{dy_1}{dx_1}}(x - x_1)$$

Determine the equation of the normal drawn to the astroid $x = 2\cos^3\theta$, $y = 2\sin^3\theta$ at the point $\theta = \dfrac{\pi}{4}$

$x = 2\cos^3 \theta$, hence $\dfrac{dx}{d\theta} = -6\cos^2 \theta \sin \theta$

$y = 2\sin^3 \theta$, hence $\dfrac{dy}{d\theta} = 6\sin^2 \theta \cos \theta$

From equation (1),

$$\dfrac{dy}{dx} = \dfrac{\dfrac{dy}{d\theta}}{\dfrac{dx}{d\theta}} = \dfrac{6\sin^2 \theta \cos \theta}{-6\cos^2 \theta \sin \theta} = -\dfrac{\sin \theta}{\cos \theta} = -\tan \theta$$

When $\theta = \dfrac{\pi}{4}$,　$\dfrac{dy}{dx} = -\tan \dfrac{\pi}{4} = -1$

$x_1 = 2\cos^3 \dfrac{\pi}{4} = 0.7071$ and $y_1 = 2\sin^3 \dfrac{\pi}{4} = 0.7071$

Hence, **the equation of the normal is**:

$$y - 0.7071 = -\dfrac{1}{-1}(x - 0.7071)$$

i.e.　$y - 0.7071 = x - 0.7071$

i.e.　　　　$y = x$

Problem 6.　The parametric equations for a hyperbola are $x = 2\sec \theta$, $y = 4\tan \theta$. Evaluate (a) $\dfrac{dy}{dx}$ (b) $\dfrac{d^2 y}{dx^2}$, correct to 4 significant figures, when $\theta = 1$ radian.

(a)　$x = 2\sec \theta$, hence $\dfrac{dx}{d\theta} = 2\sec \theta \tan \theta$

$y = 4\tan \theta$, hence $\dfrac{dy}{d\theta} = 4\sec^2 \theta$

From equation (1),

$$\dfrac{dy}{dx} = \dfrac{\dfrac{dy}{d\theta}}{\dfrac{dx}{d\theta}} = \dfrac{4\sec^2 \theta}{2\sec \theta \tan \theta} = \dfrac{2\sec \theta}{\tan \theta}$$

$$= \dfrac{2\left(\dfrac{1}{\cos \theta}\right)}{\left(\dfrac{\sin \theta}{\cos \theta}\right)} = \dfrac{2}{\sin \theta} \text{ or } 2\operatorname{cosec} \theta$$

When $\theta = 1$ rad, $\dfrac{dy}{dx} = \dfrac{2}{\sin 1} = \mathbf{2.377}$, correct to 4 significant figures.

(b)　From equation (2),

$$\dfrac{d^2 y}{dx^2} = \dfrac{\dfrac{d}{d\theta}\left(\dfrac{dy}{dx}\right)}{\dfrac{dx}{d\theta}} = \dfrac{\dfrac{d}{d\theta}(2\operatorname{cosec} \theta)}{2\sec \theta \tan \theta}$$

$$= \dfrac{-2\operatorname{cosec} \theta \cot \theta}{2\sec \theta \tan \theta}$$

$$= \dfrac{-\left(\dfrac{1}{\sin \theta}\right)\left(\dfrac{\cos \theta}{\sin \theta}\right)}{\left(\dfrac{1}{\cos \theta}\right)\left(\dfrac{\sin \theta}{\cos \theta}\right)}$$

$$= -\left(\dfrac{\cos \theta}{\sin^2 \theta}\right)\left(\dfrac{\cos^2 \theta}{\sin \theta}\right)$$

$$= -\dfrac{\cos^3 \theta}{\sin^3 \theta} = -\cot^3 \theta$$

When $\theta = 1$ rad, $\dfrac{d^2 y}{dx^2} = -\cot^3 1 = -\dfrac{1}{(\tan 1)^3} = $ $\mathbf{-0.2647}$, correct to 4 significant figures.

Problem 7.　When determining the surface tension of a liquid, the radius of curvature, ρ, of part of the surface is given by:

$$\rho = \dfrac{\sqrt{\left[1 + \left(\dfrac{dy}{dx}\right)^2\right]^3}}{\dfrac{d^2 y}{dx^2}}$$

Find the radius of curvature of the part of the surface having the parametric equations $x = 3t^2$, $y = 6t$ at the point $t = 2$.

$x = 3t^2$, hence $\dfrac{dx}{dt} = 6t$

$y = 6t$, hence $\dfrac{dy}{dt} = 6$

From equation (1), $\dfrac{dy}{dx} = \dfrac{\dfrac{dy}{dt}}{\dfrac{dx}{dt}} = \dfrac{6}{6t} = \dfrac{1}{t}$

From equation (2),

$$\dfrac{d^2 y}{dx^2} = \dfrac{\dfrac{d}{dt}\left(\dfrac{dy}{dx}\right)}{\dfrac{dx}{dt}} = \dfrac{\dfrac{d}{dt}\left(\dfrac{1}{t}\right)}{6t} = \dfrac{-\dfrac{1}{t^2}}{6t} = -\dfrac{1}{6t^3}$$

Hence, radius of curvature, $\rho = \dfrac{\sqrt{\left[1+\left(\dfrac{dy}{dx}\right)^2\right]^3}}{\dfrac{d^2y}{dx^2}}$

$= \dfrac{\sqrt{\left[1+\left(\dfrac{1}{t}\right)^2\right]^3}}{-\dfrac{1}{6t^3}}$

When $t = 2$, $\rho = \dfrac{\sqrt{\left[1+\left(\dfrac{1}{2}\right)^2\right]^3}}{-\dfrac{1}{6(2)^3}} = \dfrac{\sqrt{(1.25)^3}}{-\dfrac{1}{48}}$

$= -48\sqrt{(1.25)^3} = \mathbf{-67.08}$

Now try the following exercise

Exercise 166 Further problems on differentiation of parametric equations

1. A cycloid has parametric equations $x = 2(\theta - \sin\,\theta)$, $y = 2(1 - \cos\,\theta)$. Evaluate, at $\theta = 0.62$ rad, correct to 4 significant figures, (a) $\dfrac{dy}{dx}$ (b) $\dfrac{d^2y}{dx^2}$

[(a) 3.122 (b) −14.43]

The equation of the normal drawn to a curve at point (x_1, y_1) is given by:

$$y - y_1 = -\dfrac{1}{\dfrac{dy_1}{dx_1}}(x - x_1)$$

Use this in Problems 2 and 3.

2. Determine the equation of the normal drawn to the parabola $x = \dfrac{1}{4}t^2$, $y = \dfrac{1}{2}t$ at $t = 2$.

[$y = -2x + 3$]

3. Find the equation of the normal drawn to the cycloid $x = 2(\theta - \sin\,\theta)$, $y = 2(1 - \cos\,\theta)$ at $\theta = \dfrac{\pi}{2}$ rad.

[$y = -x + \pi$]

4. Determine the value of $\dfrac{d^2y}{dx^2}$, correct to 4 significant figures, at $\theta = \dfrac{\pi}{6}$ rad for the cardioid $x = 5(2\theta - \cos\,2\theta)$, $y = 5(2\sin\,\theta - \sin\,2\theta)$.

[0.02975]

5. The radius of curvature, ρ, of part of a surface when determining the surface tension of a liquid is given by:

$$\rho = \dfrac{\left[1+\left(\dfrac{dy}{dx}\right)^2\right]^{3/2}}{\dfrac{d^2y}{dx^2}}$$

Find the radius of curvature (correct to 4 significant figures) of the part of the surface having parametric equations

(a) $x = 3t$, $y = \dfrac{3}{t}$ at the point $t = \dfrac{1}{2}$

(b) $x = 4\cos^3 t$, $y = 4\sin^3 t$ at $t = \dfrac{\pi}{6}$ rad

[(a) 13.14 (b) 5.196]

Chapter 46

Differentiation of implicit functions

46.1 Implicit functions

When an equation can be written in the form $y = f(x)$ it is said to be an **explicit function** of x. Examples of explicit functions include

$$y = 2x^3 - 3x + 4, \quad y = 2x \ln x$$

and $\quad y = \dfrac{3e^x}{\cos x}$

In these examples y may be differentiated with respect to x by using standard derivatives, the product rule and the quotient rule of differentiation respectively.

Sometimes with equations involving, say, y and x, it is impossible to make y the subject of the formula. The equation is then called an **implicit function** and examples of such functions include $y^3 + 2x^2 = y^2 - x$ and $\sin y = x^2 + 2xy$.

46.2 Differentiating implicit functions

It is possible to **differentiate an implicit function** by using the **function of a function rule**, which may be stated as

$$\frac{du}{dx} = \frac{du}{dy} \times \frac{dy}{dx}$$

Thus, to differentiate y^3 with respect to x, the substitution $u = y^3$ is made, from which, $\dfrac{du}{dy} = 3y^2$.

Hence, $\dfrac{d}{dx}(y^3) = (3y^2) \times \dfrac{dy}{dx}$, by the function of a function rule.

A simple rule for differentiating an implicit function is summarised as:

$$\frac{d}{dx}[f(y)] = \frac{d}{dy}[f(y)] \times \frac{dy}{dx} \qquad (1)$$

Problem 1. Differentiate the following functions with respect to x:

(a) $2y^4$ (b) $\sin 3t$

(a) Let $u = 2y^4$, then, by the function of a function rule:

$$\frac{du}{dx} = \frac{du}{dy} \times \frac{dy}{dx} = \frac{d}{dy}(2y^4) \times \frac{dy}{dx}$$

$$= 8y^3 \frac{dy}{dx}$$

(b) Let $u = \sin 3t$, then, by the function of a function rule:

$$\frac{du}{dx} = \frac{du}{dt} \times \frac{dt}{dx} = \frac{d}{dt}(\sin 3t) \times \frac{dt}{dx}$$

$$= 3\cos 3t \frac{dy}{dx}$$

Problem 2. Differentiate the following functions with respect to x:

(a) $4 \ln 5y$ (b) $\dfrac{1}{5}e^{3\theta - 2}$

(a) Let $u = 4 \ln 5y$, then, by the function of a function rule:

$$\frac{du}{dx} = \frac{du}{dy} \times \frac{dy}{dx} = \frac{d}{dy}(4 \ln 5y) \times \frac{dy}{dx}$$

$$= \frac{4}{y}\frac{dy}{dx}$$

(b) Let $u = \frac{1}{5}e^{3\theta - 2}$, then, by the function of a function rule:

$$\frac{du}{dx} = \frac{du}{d\theta} \times \frac{d\theta}{dx} = \frac{d}{d\theta}\left(\frac{1}{5}e^{3\theta-2}\right) \times \frac{d\theta}{dx}$$

$$= \frac{3}{5}e^{3\theta-2}\frac{d\theta}{dx}$$

Now try the following exercise.

Exercise 167 Further problems on differentiating implicit functions

In Problems 1 and 2 differentiate the given functions with respect to x.

1. (a) $3y^5$ (b) $2\cos 4\theta$ (c) $\sqrt{k}$

$$\left[\begin{array}{l} \text{(a) } 15y^4\dfrac{dy}{dx} \quad \text{(b) } -8\sin 4\theta\dfrac{d\theta}{dx} \\[2mm] \text{(c) } \dfrac{1}{2\sqrt{k}}\dfrac{dk}{dx} \end{array}\right]$$

2. (a) $\frac{5}{2}\ln 3t$ (b) $\frac{3}{4}e^{2y+1}$ (c) $2\tan 3y$

$$\left[\begin{array}{l} \text{(a) } \dfrac{5}{2t}\dfrac{dt}{dx} \quad \text{(b) } \dfrac{3}{2}e^{2y+1}\dfrac{dy}{dx} \\[2mm] \text{(c) } 6\sec^2 3y\dfrac{dy}{dx} \end{array}\right]$$

3. Differentiate the following with respect to y:
   (a) $3\sin 2\theta$ (b) $4\sqrt{x^3}$ (c) $\dfrac{2}{e^t}$

$$\left[\begin{array}{l} \text{(a) } 6\cos 2\theta\dfrac{d\theta}{dy} \quad \text{(b) } 6\sqrt{x}\dfrac{dx}{dy} \\[2mm] \text{(c) } \dfrac{-2}{e^t}\dfrac{dt}{dy} \end{array}\right]$$

4. Differentiate the following with respect to u:
   (a) $\dfrac{2}{(3x+1)}$ (b) $3\sec 2\theta$ (c) $\dfrac{2}{\sqrt{y}}$

$$\left[\begin{array}{l} \text{(a) } \dfrac{-6}{(3x+1)^2}\dfrac{dx}{du} \\[2mm] \text{(b) } 6\sec 2\theta\tan 2\theta\dfrac{d\theta}{du} \\[2mm] \text{(c) } \dfrac{-1}{\sqrt{y^3}}\dfrac{dy}{du} \end{array}\right]$$

46.3 Differentiating implicit functions containing products and quotients

The product and quotient rules of differentiation must be applied when differentiating functions containing products and quotients of two variables.

For example, $\dfrac{d}{dx}(x^2 y) = (x^2)\dfrac{d}{dx}(y) + (y)\dfrac{d}{dx}(x^2),$

by the product rule

$$= (x^2)\left(1\dfrac{dy}{dx}\right) + y(2x),$$

by using equation (1)

$$= x^2\dfrac{dy}{dx} + 2xy$$

Problem 3. Determine $\dfrac{d}{dx}(2x^3 y^2)$

In the product rule of differentiation let $u = 2x^3$ and $v = y^2$.

Thus $\dfrac{d}{dx}(2x^3 y^2) = (2x^3)\dfrac{d}{dx}(y^2) + (y^2)\dfrac{d}{dx}(2x^3)$

$$= (2x^3)\left(2y\dfrac{dy}{dx}\right) + (y^2)(6x^2)$$

$$= 4x^3 y\dfrac{dy}{dx} + 6x^2 y^2$$

$$= 2x^2 y\left(2x\dfrac{dy}{dx} + 3y\right)$$

Problem 4. Find $\dfrac{d}{dx}\left(\dfrac{3y}{2x}\right)$

In the quotient rule of differentiation let $u = 3y$ and $v = 2x$.

Thus $\dfrac{d}{dx}\left(\dfrac{3y}{2x}\right) = \dfrac{(2x)\dfrac{d}{dx}(3y) - (3y)\dfrac{d}{dx}(2x)}{(2x)^2}$

$$= \dfrac{(2x)\left(3\dfrac{dy}{dx}\right) - (3y)(2)}{4x^2}$$

$$= \dfrac{6x\dfrac{dy}{dx} - 6y}{4x^2} = \dfrac{3}{2x^2}\left(x\dfrac{dy}{dx} - y\right)$$

Problem 5. Differentiate $z = x^2 + 3x\cos 3y$ with respect to y.

$$\frac{dz}{dy} = \frac{d}{dy}(x^2) + \frac{d}{dy}(3x\cos 3y)$$

$$= 2x\frac{dx}{dy} + \left[(3x)(-3\sin 3y) + (\cos 3y)\left(3\frac{dx}{dy}\right)\right]$$

$$= 2x\frac{dx}{dy} - 9x\sin 3y + 3\cos 3y\frac{dx}{dy}$$

Now try the following exercise

Exercise 168 Further problems on differentiating implicit functions involving products and quotients

1. Determine $\frac{d}{dx}(3x^2y^3)$

$$\left[3xy^2\left(3x\frac{dy}{dx} + 2y\right)\right]$$

2. Find $\frac{d}{dx}\left(\frac{2y}{5x}\right)$

$$\left[\frac{2}{5x^2}\left(x\frac{dy}{dx} - y\right)\right]$$

3. Determine $\frac{d}{du}\left(\frac{3u}{4v}\right)$

$$\left[\frac{3}{4v^2}\left(v - u\frac{dv}{du}\right)\right]$$

4. Given $z = 3\sqrt{y}\cos 3x$ find $\frac{dz}{dx}$

$$\left[3\left(\frac{\cos 3x}{2\sqrt{y}}\right)\frac{dy}{dx} - 9\sqrt{y}\sin 3x\right]$$

5. Determine $\frac{dz}{dy}$ given $z = 2x^3\ln y$

$$\left[2x^2\left(\frac{x}{y} + 3\ln y\frac{dx}{dy}\right)\right]$$

46.4 Further implicit differentiation

An implicit function such as $3x^2 + y^2 - 5x + y = 2$, may be differentiated term by term with respect to x. This gives:

$$\frac{d}{dx}(3x^2) + \frac{d}{dx}(y^2) - \frac{d}{dx}(5x) + \frac{d}{dx}(y) = \frac{d}{dx}(2)$$

i.e. $\quad 6x + 2y\frac{dy}{dx} - 5 + 1\frac{dy}{dx} = 0,$

using equation (1) and standard derivatives.

An expression for the derivative $\frac{dy}{dx}$ in terms of x and y may be obtained by rearranging this latter equation. Thus:

$$(2y + 1)\frac{dy}{dx} = 5 - 6x$$

from which, $\quad \frac{dy}{dx} = \frac{5 - 6x}{2y + 1}$

Problem 6. Given $2y^2 - 5x^4 - 2 - 7y^3 = 0$, determine $\frac{dy}{dx}$

Each term in turn is differentiated with respect to x:

Hence $\quad \frac{d}{dx}(2y^2) - \frac{d}{dx}(5x^4) - \frac{d}{dx}(2) - \frac{d}{dx}(7y^3)$

$$= \frac{d}{dx}(0)$$

i.e. $\quad 4y\frac{dy}{dx} - 20x^3 - 0 - 21y^2\frac{dy}{dx} = 0$

Rearranging gives:

$$(4y - 21y^2)\frac{dy}{dx} = 20x^3$$

i.e. $\quad \frac{dy}{dx} = \frac{20x^3}{(4y - 21y^2)}$

Problem 7. Determine the values of $\frac{dy}{dx}$ when $x = 4$ given that $x^2 + y^2 = 25$.

Differentiating each term in turn with respect to x gives:

$$\frac{d}{dx}(x^2) + \frac{d}{dx}(y^2) = \frac{d}{dx}(25)$$

i.e. $\quad 2x + 2y\frac{dy}{dx} = 0$

Hence $\quad \frac{dy}{dx} = -\frac{2x}{2y} = -\frac{x}{y}$

Since $x^2 + y^2 = 25$, when $x = 4$, $y = \sqrt{(25 - 4^2)} = \pm 3$

Thus when $x = 4$ and $y = \pm 3$, $\frac{dy}{dx} = -\frac{4}{\pm 3} = \pm\frac{4}{3}$

$x^2 + y^2 = 25$ is the equation of a circle, centre at the origin and radius 5, as shown in Fig. 46.1. At $x = 4$, the two gradients are shown.

Above, $x^2 + y^2 = 25$ was differentiated implicitly; actually, the equation could be transposed to

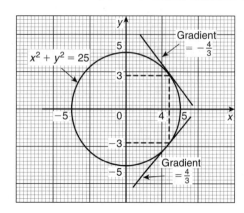

Figure 46.1

$y = \sqrt{(25 - x^2)}$ and differentiated using the function of a function rule. This gives

$$\frac{dy}{dx} = \frac{1}{2}(25 - x^2)^{\frac{-1}{2}}(-2x) = -\frac{x}{\sqrt{(25 - x^2)}}$$

and when $x = 4$, $\dfrac{dy}{dx} = -\dfrac{4}{\sqrt{(25 - 4^2)}} = \pm\dfrac{4}{3}$ as obtained above.

Problem 8.

(a) Find $\dfrac{dy}{dx}$ in terms of x and y given $4x^2 + 2xy^3 - 5y^2 = 0$.

(b) Evalate $\dfrac{dy}{dx}$ when $x = 1$ and $y = 2$.

(a) Differentiating each term in turn with respect to x gives:

$$\frac{d}{dx}(4x^2) + \frac{d}{dx}(2xy^3) - \frac{d}{dx}(5y^2) = \frac{d}{dx}(0)$$

i.e. $8x + \left[(2x)\left(3y^2\dfrac{dy}{dx}\right) + (y^3)(2)\right]$

$$- 10y\frac{dy}{dx} = 0$$

i.e. $8x + 6xy^2\dfrac{dy}{dx} + 2y^3 - 10y\dfrac{dy}{dx} = 0$

Rearranging gives:

$$8x + 2y^3 = (10y - 6xy^2)\frac{dy}{dx}$$

and $\dfrac{dy}{dx} = \dfrac{8x + 2y^3}{10y - 6xy^2} = \dfrac{4x + y^3}{y(5 - 3xy)}$

(b) When $x = 1$ and $y = 2$,

$$\frac{dy}{dx} = \frac{4(1) + (2)^3}{2[5 - (3)(1)(2)]} = \frac{12}{-2} = -6$$

Problem 9. Find the gradients of the tangents drawn to the circle $x^2 + y^2 - 2x - 2y = 3$ at $x = 2$.

The gradient of the tangent is given by $\dfrac{dy}{dx}$

Differentiating each term in turn with respect to x gives:

$$\frac{d}{dx}(x^2) + \frac{d}{dx}(y^2) - \frac{d}{dx}(2x) - \frac{d}{dx}(2y) = \frac{d}{dx}(3)$$

i.e. $2x + 2y\dfrac{dy}{dx} - 2 - 2\dfrac{dy}{dx} = 0$

Hence $(2y - 2)\dfrac{dy}{dx} = 2 - 2x,$

from which $\dfrac{dy}{dx} = \dfrac{2 - 2x}{2y - 2} = \dfrac{1 - x}{y - 1}$

The value of y when $x = 2$ is determined from the original equation

Hence $(2)^2 + y^2 - 2(2) - 2y = 3$

i.e. $4 + y^2 - 4 - 2y = 3$

or $y^2 - 2y - 3 = 0$

Factorising gives: $(y + 1)(y - 3) = 0$, from which $y = -1$ or $y = 3$

When $x = 2$ and $y = -1$,

$$\frac{dy}{dx} = \frac{1 - x}{y - 1} = \frac{1 - 2}{-1 - 1} = \frac{-1}{-2} = \frac{1}{2}$$

When $x = 2$ and $y = 3$,

$$\frac{dy}{dx} = \frac{1 - 2}{3 - 1} = \frac{-1}{2}$$

Hence the gradients of the tangents are $\pm\dfrac{1}{2}$

The circle having the given equation has its centre at $(1, 1)$ and radius $\sqrt{5}$ (see Chapter 19) and is shown in Fig. 46.2 with the two gradients of the tangents.

Problem 10. Pressure p and volume v of a gas are related by the law $pv^\gamma = k$, where γ and k are constants. Show that the rate of change of pressure $\dfrac{dp}{dt} = -\gamma\dfrac{p}{v}\dfrac{dv}{dt}$

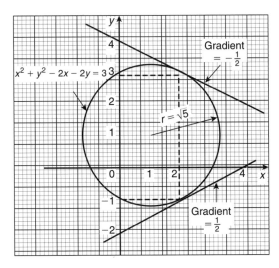

Figure 46.2

Since $pv^\gamma = k$, then $p = \dfrac{k}{v^\gamma} = kv^{-\gamma}$

$$\frac{dp}{dt} = \frac{dp}{dv} \times \frac{dv}{dt}$$

by the function of a function rule

$$\frac{dp}{dv} = \frac{d}{dv}(kv^{-\gamma})$$

$$= -\gamma kv^{-\gamma-1} = \frac{-\gamma k}{v^{\gamma+1}}$$

$$\frac{dp}{dt} = \frac{-\gamma k}{v^{\gamma+1}} \times \frac{dv}{dt}$$

Since $\quad k = pv^\gamma$

$$\frac{dp}{dt} = \frac{-\gamma(pv^\gamma)}{v^{r+1}}\frac{dv}{dt} = \frac{-\gamma p v^\gamma}{v^\gamma v^1}\frac{dv}{dt}$$

i.e. $\quad \dfrac{dp}{dt} = -\gamma\dfrac{p}{v}\dfrac{dv}{dt}$

Now try the following exercise

Exercise 169　Further problems on implicit differentiation

In Problems 1 and 2 determine $\dfrac{dy}{dx}$

1. $x^2 + y^2 + 4x - 3y + 1 = 0$ $\qquad \left[\dfrac{2x+4}{3-2y}\right]$

2. $2y^3 - y + 3x - 2 = 0$ $\qquad \left[\dfrac{3}{1-6y^2}\right]$

3. Given $x^2 + y^2 = 9$ evaluate $\dfrac{dy}{dx}$ when $x = \sqrt{5}$
   and $y = 2$ $\qquad \left[-\dfrac{\sqrt{5}}{2}\right]$

In Problems 4 to 7, determine $\dfrac{dy}{dx}$

4. $x^2 + 2x \sin 4y = 0$ $\qquad \left[\dfrac{-(x + \sin 4y)}{4x \cos 4y}\right]$

5. $3y^2 + 2xy - 4x^2 = 0$ $\qquad \left[\dfrac{4x - y}{3y + x}\right]$

6. $2x^2y + 3x^3 = \sin y$ $\qquad \left[\dfrac{x(4y + 9x)}{\cos y - 2x^2}\right]$

7. $3y + 2x \ln y = y^4 + x$ $\qquad \left[\dfrac{1 - 2\ln y}{3 + (2x/y) - 4y^3}\right]$

8. If $3x^2 + 2x^2y^3 - \dfrac{5}{4}y^2 = 0$ evaluate $\dfrac{dy}{dx}$ when $x = \dfrac{1}{2}$ and $y = 1$ $\qquad [5]$

9. Determine the gradients of the tangents drawn to the circle $x^2 + y^2 = 16$ at the point where $x = 2$. Give the answer correct to 4 significant figures $\qquad [\pm 0.5774]$

10. Find the gradients of the tangents drawn to the ellipse $\dfrac{x^2}{4} + \dfrac{y^2}{9} = 2$ at the point where $x = 2$ $\qquad [\pm 1.5]$

11. Determine the gradient of the curve $3xy + y^2 = -2$ at the point $(1, -2)$ $\qquad [-6]$

Chapter 47

Logarithmic differentiation

47.1 Introduction to logarithmic differentiation

With certain functions containing more complicated products and quotients, differentiation is often made easier if the logarithm of the function is taken before differentiating. This technique, called **'logarithmic differentiation'** is achieved with a knowledge of (i) the laws of logarithms, (ii) the differential coefficients of logarithmic functions, and (iii) the differentiation of implicit functions.

47.2 Laws of logarithms

Three laws of logarithms may be expressed as:

(i) $\log (A \times B) = \log A + \log B$

(ii) $\log \left(\dfrac{A}{B}\right) = \log A - \log B$

(iii) $\log A^n = n \log A$

In calculus, Napierian logarithms (i.e. logarithms to a base of 'e') are invariably used. Thus for two functions $f(x)$ and $g(x)$ the laws of logarithms may be expressed as:

(i) $\ln[f(x) \cdot g(x)] = \ln f(x) + \ln g(x)$

(ii) $\ln \left(\dfrac{f(x)}{g(x)}\right) = \ln f(x) - \ln g(x)$

(iii) $\ln[f(x)]^n = n \ln f(x)$

Taking Napierian logarithms of both sides of the equation $y = \dfrac{f(x) \cdot g(x)}{h(x)}$ gives :

$$\ln y = \ln \left(\frac{f(x) \cdot g(x)}{h(x)}\right)$$

which may be simplified using the above laws of logarithms, giving;

$$\ln y = \ln f(x) + \ln g(x) - \ln h(x)$$

This latter form of the equation is often easier to differentiate.

47.3 Differentiation of logarithmic functions

The differential coefficient of the logarithmic function $\ln x$ is given by:

$$\frac{d}{dx}(\ln x) = \frac{1}{x}$$

More generally, it may be shown that:

$$\frac{d}{dx}[\ln f(x)] = \frac{f'(x)}{f(x)} \qquad (1)$$

For example, if $y = \ln(3x^2 + 2x - 1)$ then,

$$\frac{dy}{dx} = \frac{6x + 2}{3x^2 + 2x - 1}$$

Similarly, if $y = \ln(\sin 3x)$ then

$$\frac{dy}{dx} = \frac{3\cos 3x}{\sin 3x} = 3 \cot 3x.$$

As explained in Chapter 46, by using the function of a function rule:

$$\frac{d}{dx}(\ln y) = \left(\frac{1}{y}\right)\frac{dy}{dx} \qquad (2)$$

Differentiation of an expression such as

$y = \dfrac{(1+x)^2 \sqrt{(x-1)}}{x\sqrt{(x+2)}}$ may be achieved by using

the product and quotient rules of differentiation; however the working would be rather complicated. With logarithmic differentiation the following procedure is adopted:

(i)   Take Napierian logarithms of both sides of the equation.

$$\text{Thus } \ln y = \ln\left\{\frac{(1+x)^2\sqrt{(x-1)}}{x\sqrt{(x+2)}}\right\}$$

$$= \ln\left\{\frac{(1+x)^2(x-1)^{\frac{1}{2}}}{x(x+2)^{\frac{1}{2}}}\right\}$$

(ii) Apply the laws of logarithms.

$$\text{Thus } \ln y = \ln(1+x)^2 + \ln(x-1)^{\frac{1}{2}}$$
$$- \ln x - \ln(x+2)^{\frac{1}{2}}, \text{ by laws (i)}$$
$$\text{and (ii) of Section 47.2}$$

$$\text{i.e. } \ln y = 2\ln(1+x) + \frac{1}{2}\ln(x-1)$$
$$- \ln x - \frac{1}{2}\ln(x+2), \text{ by law (iii)}$$
$$\text{of Section 47.2}$$

(iii) Differentiate each term in turn with respect of x using equations (1) and (2).

$$\text{Thus } \frac{1}{y}\frac{dy}{dx} = \frac{2}{(1+x)} + \frac{\frac{1}{2}}{(x-1)} - \frac{1}{x} - \frac{\frac{1}{2}}{(x+2)}$$

(iv) Rearrange the equation to make $\dfrac{dy}{dx}$ the subject.

$$\text{Thus } \frac{dy}{dx} = y\left\{\frac{2}{(1+x)} + \frac{1}{2(x-1)} - \frac{1}{x}\right.$$
$$\left. - \frac{1}{2(x+2)}\right\}$$

(v)   Substitute for y in terms of x.

$$\text{Thus } \frac{dy}{dx} = \frac{(1+x)^2(\sqrt{(x-1)})}{x\sqrt{(x+2)}}\left\{\frac{2}{(1+x)}\right.$$
$$\left. + \frac{1}{2(x-1)} - \frac{1}{x} - \frac{1}{2(x+2)}\right\}$$

Problem 1.   Use logarithmic differentiation to differentiate $y = \dfrac{(x+1)(x-2)^3}{(x-3)}$

Following the above procedure:

(i)   Since $y = \dfrac{(x+1)(x-2)^3}{(x-3)}$

then $\ln y = \ln\left\{\dfrac{(x+1)(x-2)^3}{(x-3)}\right\}$

(ii) $\ln y = \ln(x+1) + \ln(x-2)^3 - \ln(x-3)$,
       by laws (i) and (ii) of Section 47.2,
       i.e. $\ln y = \ln(x+1) + 3\ln(x-2) - \ln(x-3)$,
       by law (iii) of Section 47.2.

(iii) Differentiating with respect to x gives:

$$\frac{1}{y}\frac{dy}{dx} = \frac{1}{(x+1)} + \frac{3}{(x-2)} - \frac{1}{(x-3)}$$

by using equations (1) and (2)

(iv) Rearranging gives:

$$\frac{dy}{dx} = y\left\{\frac{1}{(x+1)} + \frac{3}{(x-2)} - \frac{1}{(x-3)}\right\}$$

(v)   Substituting for y gives:

$$\frac{dy}{dx} = \frac{(x+1)(x-2)^3}{(x-3)}\left\{\frac{1}{(x+1)} + \frac{3}{(x-2)} - \frac{1}{(x-3)}\right\}$$

Problem 2.   Differentiate $y = \dfrac{\sqrt{(x-2)^3}}{(x+1)^2(2x-1)}$

with respect to x and evalualte $\dfrac{dy}{dx}$ when $x = 3$.

Using logarithmic differentiation and following the above procedure:

(i)   Since $y = \dfrac{\sqrt{(x-2)^3}}{(x+1)^2(2x-1)}$

then $\ln y = \ln\left\{\dfrac{\sqrt{(x-2)^3}}{(x+1)^2(2x-1)}\right\}$

$$= \ln\left\{\frac{(x-2)^{\frac{3}{2}}}{(x+1)^2(2x-1)}\right\}$$

(ii) $\ln y = \ln(x-2)^{\frac{3}{2}} - \ln(x+1)^2 - \ln(2x-1)$
       i.e. $\ln y = \frac{3}{2}\ln(x-2) - 2\ln(x+1) - \ln(2x-1)$

(iii) $\dfrac{1}{y}\dfrac{dy}{dx} = \dfrac{\frac{3}{2}}{(x-2)} - \dfrac{2}{(x+1)} - \dfrac{2}{(2x-1)}$

(iv) $\dfrac{dy}{dx} = y\left\{\dfrac{3}{2(x-2)} - \dfrac{2}{(x+1)} - \dfrac{2}{(2x-1)}\right\}$

(v) $\dfrac{dy}{dx} = \dfrac{\sqrt{(x-2)^3}}{(x+1)^2(2x-1)}\left\{\dfrac{3}{2(x-2)}\right.$
$$\left. - \dfrac{2}{(x+1)} - \dfrac{2}{(2x-1)}\right\}$$

When $x = 3$, $\dfrac{dy}{dx} = \dfrac{\sqrt{(1)^3}}{(4)^2(5)}\left(\dfrac{3}{2} - \dfrac{2}{4} - \dfrac{2}{5}\right)$

$= \pm\dfrac{1}{80}\left(\dfrac{3}{5}\right) = \pm\dfrac{3}{400}$ or ± 0.0075

Problem 3. Given $y = \dfrac{3e^{2\theta}\sec 2\theta}{\sqrt{(\theta - 2)}}$ determine $\dfrac{dy}{d\theta}$

Using logarithmic differentiation and following the procedure:

(i) Since $y = \dfrac{3e^{2\theta}\sec 2\theta}{\sqrt{(\theta - 2)}}$

then $\ln y = \ln\left\{\dfrac{3e^{2\theta}\sec 2\theta}{\sqrt{(\theta - 2)}}\right\}$

$= \ln\left\{\dfrac{3e^{2\theta}\sec 2\theta}{(\theta - 2)^{\frac{1}{2}}}\right\}$

(ii) $\ln y = \ln 3e^{2\theta} + \ln\sec 2\theta - \ln(\theta - 2)^{\frac{1}{2}}$

i.e. $\ln y = \ln 3 + \ln e^{2\theta} + \ln\sec 2\theta$
$\qquad\qquad - \frac{1}{2}\ln(\theta - 2)$

i.e. $\ln y = \ln 3 + 2\theta + \ln\sec 2\theta - \frac{1}{2}\ln(\theta - 2)$

(iii) Differentiating with respect to θ gives:

$\dfrac{1}{y}\dfrac{dy}{d\theta} = 0 + 2 + \dfrac{2\sec 2\theta\tan 2\theta}{\sec 2\theta} - \dfrac{\frac{1}{2}}{(\theta - 2)}$

from equations (1) and (2)

(iv) Rearranging gives:

$\dfrac{dy}{d\theta} = y\left\{2 + 2\tan 2\theta - \dfrac{1}{2(\theta - 2)}\right\}$

(v) Substituting for y gives:

$\dfrac{dy}{d\theta} = \dfrac{3e^{2\theta}\sec 2\theta}{\sqrt{(\theta - 2)}}\left\{2 + 2\tan 2\theta - \dfrac{1}{2(\theta - 2)}\right\}$

Problem 4. Differentiate $y = \dfrac{x^3\ln 2x}{e^x\sin x}$ with respect to x

---

Using logarithmic differentiation and following the procedure gives:

(i) $\ln y = \ln\left\{\dfrac{x^3\ln 2x}{e^x\sin x}\right\}$

(ii) $\ln y = \ln x^3 + \ln(\ln 2x) - \ln(e^x) - \ln(\sin x)$

i.e. $\ln y = 3\ln x + \ln(\ln 2x) - x - \ln(\sin x)$

(iii) $\dfrac{1}{y}\dfrac{dy}{dx} = \dfrac{3}{x} + \dfrac{\frac{1}{x}}{\ln 2x} - 1 - \dfrac{\cos x}{\sin x}$

(iv) $\dfrac{dy}{dx} = y\left\{\dfrac{3}{x} + \dfrac{1}{x\ln 2x} - 1 - \cot x\right\}$

(v) $\dfrac{dy}{dx} = \dfrac{x^3\ln 2x}{e^x\sin x}\left\{\dfrac{3}{x} + \dfrac{1}{x\ln 2x} - 1 - \cot x\right\}$

Now try the following exercise.

Exercise 170 Further problems on differentiating logarithmic functions

In Problems 1 to 6, use logarithmic differentiation to differentiate the given functions with respect to the variable.

1. $y = \dfrac{(x - 2)(x + 1)}{(x - 1)(x + 3)}$

$\left[\dfrac{(x - 2)(x + 1)}{(x - 1)(x + 3)}\left\{\dfrac{1}{(x - 2)} + \dfrac{1}{(x + 1)}\right.\right.$
$\left.\left.- \dfrac{1}{(x - 1)} - \dfrac{1}{(x + 3)}\right\}\right]$

2. $y = \dfrac{(x + 1)(2x + 1)^3}{(x - 3)^2(x + 2)^4}$

$\left[\dfrac{(x + 1)(2x + 1)^3}{(x - 3)^2(x + 2)^4}\left\{\dfrac{1}{(x + 1)} + \dfrac{6}{(2x + 1)}\right.\right.$
$\left.\left.- \dfrac{2}{(x - 3)} - \dfrac{4}{(x + 2)}\right\}\right]$

3. $y = \dfrac{(2x - 1)\sqrt{(x + 2)}}{(x - 3)\sqrt{(x + 1)^3}}$

$\left[\dfrac{(2x - 1)\sqrt{(x + 2)}}{(x - 3)\sqrt{(x + 1)^3}}\left\{\dfrac{2}{(2x - 1)} + \dfrac{1}{2(x + 2)}\right.\right.$
$\left.\left.- \dfrac{1}{(x - 3)} - \dfrac{3}{2(x + 1)}\right\}\right]$

4. $y = \dfrac{e^{2x} \cos 3x}{\sqrt{(x-4)}}$

$$\left[\dfrac{e^{2x} \cos 3x}{\sqrt{(x-4)}} \left\{ 2 - 3 \tan 3x - \dfrac{1}{2(x-4)} \right\} \right]$$

5. $y = 3\theta \sin\theta \cos\theta$

$$\left[3\theta \sin\theta \cos\theta \left\{ \dfrac{1}{\theta} + \cot\theta - \tan\theta \right\} \right]$$

6. $y = \dfrac{2x^4 \tan x}{e^{2x} \ln 2x}$

$$\left[\dfrac{2x^4 \tan x}{e^{2x} \ln 2x} \left\{ \dfrac{4}{x} + \dfrac{1}{\sin x \cos x} - 2 - \dfrac{1}{x \ln 2x} \right\} \right]$$

7. Evaluate $\dfrac{dy}{dx}$ when $x = 1$ given

$$y = \dfrac{(x+1)^2 \sqrt{(2x-1)}}{\sqrt{(x+3)^3}} \qquad \left[\dfrac{13}{16} \right]$$

8. Evaluate $\dfrac{dy}{d\theta}$, correct to 3 significant figures,

when $\theta = \dfrac{\pi}{4}$ given $y = \dfrac{2e^\theta \sin\theta}{\sqrt{\theta^5}}$ $\qquad [-6.71]$

47.4 Differentiation of $[f(x)]^x$

Whenever an expression to be differentiated contains a term raised to a power which is itself a function of the variable, then logarithmic differentiation must be used. For example, the differentiation of expressions such as x^x, $(x+2)^x$, $\sqrt[x]{(x-1)}$ and x^{3x+2} can only be achieved using logarithmic differentiation.

Problem 5. Determine $\dfrac{dy}{dx}$ given $y = x^x$.

Taking Napierian logarithms of both sides of $y = x^x$ gives:

$\ln y = \ln x^x = x \ln x$, by law (iii) of Section 47.2

Differentiating both sides with respect to x gives:

$\dfrac{1}{y} \dfrac{dy}{dx} = (x) \left(\dfrac{1}{x} \right) + (\ln x)(1)$, using the product rule

i.e. $\dfrac{1}{y} \dfrac{dy}{dx} = 1 + \ln x$

from which, $\dfrac{dy}{dx} = y(1 + \ln x)$

i.e. $\dfrac{dy}{dx} = x^x(1 + \ln x)$

Problem 6. Evaluate $\dfrac{dy}{dx}$ when $x = -1$ given $y = (x+2)^x$

Taking Napierian logarithms of both sides of $y = (x+2)^x$ gives:

$\ln y = \ln(x+2)^x = x \ln(x+2)$, by law (iii) of Section 47.2

Differentiating both sides with respect to x gives:

$$\dfrac{1}{y} \dfrac{dy}{dx} = (x) \left(\dfrac{1}{x+2} \right) + [\ln(x+2)](1),$$

by the product rule.

Hence $\dfrac{dy}{dx} = y \left(\dfrac{x}{x+2} + \ln(x+2) \right)$

$$= (x+2)^x \left\{ \dfrac{x}{x+2} + \ln(x+2) \right\}$$

When $x = -1$, $\dfrac{dy}{dx} = (1)^{-1} \left(\dfrac{-1}{1} + \ln 1 \right)$

$$= (+1)(-1) = -1$$

Problem 7. Determine (a) the differential coefficient of $y = \sqrt[x]{(x-1)}$ and (b) evaluate $\dfrac{dy}{dx}$ when $x = 2$.

(a) $y = \sqrt[x]{(x-1)} = (x-1)^{\frac{1}{x}}$, since by the laws of indices $\sqrt[n]{a^m} = a^{\frac{m}{n}}$

Taking Napierian logarithms of both sides gives:

$$\ln y = \ln(x-1)^{\frac{1}{x}} = \dfrac{1}{x} \ln(x-1),$$

by law (iii) of Section 47.2.

Differentiating each side with respect to x gives:

$$\dfrac{1}{y} \dfrac{dy}{dx} = \left(\dfrac{1}{x} \right) \left(\dfrac{1}{x-1} \right) + [\ln(x-1)] \left(\dfrac{-1}{x^2} \right)$$

by the product rule.

Hence $\dfrac{dy}{dx} = y \left\{ \dfrac{1}{x(x-1)} - \dfrac{\ln(x-1)}{x^2} \right\}$

i.e. $\dfrac{dy}{dx} = \sqrt[x]{(x-1)}\left\{\dfrac{1}{x(x-1)} - \dfrac{\ln(x-1)}{x^2}\right\}$

(b) When $x = 2$, $\dfrac{dy}{dx} = \sqrt[2]{(1)}\left\{\dfrac{1}{2(1)} - \dfrac{\ln(1)}{4}\right\}$

$$= \pm 1\left\{\dfrac{1}{2} - 0\right\} = \pm\dfrac{1}{2}$$

Problem 8. Differentiate x^{3x+2} with respect to x.

Let $y = x^{3x+2}$

Taking Napierian logarithms of both sides gives:

$\ln y = \ln x^{3x+2}$

i.e. $\ln y = (3x+2)\ln x$, by law (iii) of Section 47.2

Differentiating each term with respect to x gives:

$$\dfrac{1}{y}\dfrac{dy}{dx} = (3x+2)\left(\dfrac{1}{x}\right) + (\ln x)(3),$$

by the product rule.

Hence $\dfrac{dy}{dx} = y\left\{\dfrac{3x+2}{x} + 3\ln x\right\}$

$$= x^{3x+2}\left\{\dfrac{3x+2}{x} + 3\ln x\right\}$$

$$= x^{3x+2}\left\{3 + \dfrac{2}{x} + 3\ln x\right\}$$

Now try the following exercise

Exercise 171 Further problems on differentiating $[f(x)]^x$ type functions

In Problems 1 to 4, differentiate with respect to x

1. $y = x^{2x}$ $[2x^{2x}(1+\ln x)]$

2. $y = (2x-1)^x$
$$\left[(2x-1)^x\left\{\dfrac{2x}{2x-1} + \ln(2x-1)\right\}\right]$$

3. $y = \sqrt[x]{(x+3)}$
$$\left[\sqrt[x]{(x+3)}\left\{\dfrac{1}{x(x+3)} - \dfrac{\ln(x+3)}{x^2}\right\}\right]$$

4. $y = 3x^{4x+1}$ $\left[3x^{4x+1}\left\{4+\dfrac{1}{x}+4\ln x\right\}\right]$

5. Show that when $y = 2x^x$ and $x = 1$, $\dfrac{dy}{dx} = 2$.

6. Evaluate $\dfrac{d}{dx}\{\sqrt[x]{(x-2)}\}$ when $x = 3$. $\left[\dfrac{1}{3}\right]$

7. Show that if $y = \theta^\theta$ and $\theta = 2$, $\dfrac{dy}{d\theta} = 6.77$, correct to 3 significant figures.

Revision Test 13

This Revision test covers the material contained in Chapters 45 to 47. *The marks for each question are shown in brackets at the end of each question.*

1. A cycloid has parametric equations given by: $x = 5(\theta - \sin\theta)$ and $y = 5(1 - \cos\theta)$. Evaluate
   (a) $\dfrac{dy}{dx}$ (b) $\dfrac{d^2y}{dx^2}$ when $\theta = 1.5$ radians. Give answers correct to 3 decimal places. (8)

2. Determine the equation of (a) the tangent, and (b) the normal, drawn to an ellipse $x = 4\cos\theta$, $y = \sin\theta$ at $\theta = \dfrac{\pi}{3}$ (8)

3. Determine expressions for $\dfrac{dz}{dy}$ for each of the following functions:
   (a) $z = 5y^2\cos x$ (b) $z = x^2 + 4xy - y^2$ (5)

4. If $x^2 + y^2 + 6x + 8y + 1 = 0$, find $\dfrac{dy}{dx}$ in terms of x and y. (4)

5. Determine the gradient of the tangents drawn to the hyperbola $x^2 - y^2 = 8$ at $x = 3$. (4)

6. Use logarithmic differentiation to differentiate
   $$y = \frac{(x+1)^2\sqrt{(x-2)}}{(2x-1)\sqrt[3]{(x-3)^4}}$$ with respect to x. (6)

7. Differentiate $y = \dfrac{3e^{\theta}\sin 2\theta}{\sqrt{\theta^5}}$ and hence evaluate $\dfrac{dy}{d\theta}$, correct to 2 decimal places, when $\theta = \dfrac{\pi}{3}$ (9)

8. Evaluate $\dfrac{d}{dt}[\sqrt{(2t+1)}]$ when $t = 2$, correct to 4 significant figures. (6)

Section 8

Section 9

Integral Calculus

Chapter 48

Standard integration

48.1 The process of integration

The process of integration reverses the process of differentiation. In differentiation, if $f(x) = 2x^2$ then $f'(x) = 4x$. Thus the integral of $4x$ is $2x^2$, i.e. integration is the process of moving from $f'(x)$ to $f(x)$. By similar reasoning, the integral of $2t$ is t^2.

Integration is a process of summation or adding parts together and an elongated S, shown as $\int$, is used to replace the words 'the integral of'. Hence, from above, $\int 4x = 2x^2$ and $\int 2t$ is t^2.

In differentiation, the differential coefficient $\dfrac{dy}{dx}$ indicates that a function of x is being differentiated with respect to x, the dx indicating that it is 'with respect to x'. In integration the variable of integration is shown by adding d(the variable) after the function to be integrated.

Thus $\displaystyle\int 4x\,dx$ means 'the integral of $4x$

with respect to x',

and $\displaystyle\int 2t\,dt$ means 'the integral of $2t$

with respect to t'

As stated above, the differential coefficient of $2x^2$ is $4x$, hence $\int 4x\,dx = 2x^2$. However, the differential coefficient of $2x^2 + 7$ is also $4x$. Hence $\int 4x\,dx$ is also equal to $2x^2 + 7$. To allow for the possible presence of a constant, whenever the process of integration is performed, a constant 'c' is added to the result.

Thus $\displaystyle\int 4x\,dx = 2x^2 + c$ and $\displaystyle\int 2t\,dt = t^2 + c$

'c' is called the **arbitrary constant of integration.**

48.2 The general solution of integrals of the form ax^n

The general solution of integrals of the form $\int ax^n\,dx$, where a and n are constants is given by:

$$\int ax^n\,dx = \frac{ax^{n+1}}{n+1} + c$$

This rule is true when n is fractional, zero, or a positive or negative integer, with the exception of $n = -1$. Using this rule gives:

(i) $\displaystyle\int 3x^4\,dx = \frac{3x^{4+1}}{4+1} + c = \frac{3}{5}x^5 + c$

(ii) $\displaystyle\int \frac{2}{x^2}\,dx = \int 2x^{-2}\,dx = \frac{2x^{-2+1}}{-2+1} + c$

$$= \frac{2x^{-1}}{-1} + c = \frac{-2}{x} + c, \text{ and}$$

(iii) $\displaystyle\int \sqrt{x}\,dx = \int x^{1/2}\,dx = \frac{x^{\frac{1}{2}+1}}{\frac{1}{2}+1} + c$

$$= \frac{x^{\frac{3}{2}}}{\frac{3}{2}} + c = \frac{2}{3}\sqrt{x^3} + c$$

Each of these three results may be checked by differentiation.

(a) The integral of a constant k is $kx + c$. For example,

$$\int 8\,dx = 8x + c$$

(b) When a sum of several terms is integrated the result is the sum of the integrals of the separate terms. For example,

$$\int (3x + 2x^2 - 5)dx$$

$$= \int 3x\, dx + \int 2x^2 dx - \int 5\, dx$$

$$= \frac{3x^2}{2} + \frac{2x^3}{3} - 5x + c$$

48.3 Standard integrals

Since integration is the reverse process of differentiation the **standard integrals** listed in Table 48.1 may be deduced and readily checked by differentiation.

Table 48.1 Standard integrals

(i) $\int ax^n\, dx = \dfrac{ax^{n+1}}{n+1} + c$
(except when $n = -1$)
(ii) $\int \cos ax\, dx = \dfrac{1}{a}\sin ax + c$
(iii) $\int \sin ax\, dx = -\dfrac{1}{a}\cos ax + c$
(iv) $\int \sec^2 ax\, dx = \dfrac{1}{a}\tan ax + c$
(v) $\int \operatorname{cosec}^2 ax\, dx = -\dfrac{1}{a}\cot ax + c$
(vi) $\int \operatorname{cosec} ax \cot ax\, dx = -\dfrac{1}{a}\operatorname{cosec} ax + c$
(vii) $\int \sec ax \tan ax\, dx = \dfrac{1}{a}\sec ax + c$
(viii) $\int e^{ax}\, dx = \dfrac{1}{a}e^{ax} + c$
(ix) $\int \dfrac{1}{x}dx = \ln x + c$

Problem 1. Determine:

(a) $\int 5x^2 dx$ (b) $\int 2t^3 dt$

The standard integral, $\int ax^n\, dx = \dfrac{ax^{n+1}}{n+1} + c$

(a) When $a = 5$ and $n = 2$ then

$$\int 5x^2 dx = \frac{5x^{2+1}}{2+1} + c = \frac{5x^3}{3} + c$$

(b) When $a = 2$ and $n = 3$ then

$$\int 2t^3 dt = \frac{2t^{3+1}}{3+1} + c = \frac{2t^4}{4} + c = \frac{1}{2}t^4 + c$$

Each of these results may be checked by differentiating them.

Problem 2. Determine $\int \left(4 + \frac{3}{7}x - 6x^2\right) dx$

$\int \left(4 + \frac{3}{7}x - 6x^2\right) dx$ may be written as

$$\int 4\, dx + \int \frac{3}{7}x\, dx - \int 6x^2\, dx$$

i.e. each term is integrated separately. (This splitting up of terms only applies, however, for addition and subtraction).

Hence $\int \left(4 + \frac{3}{7}x - 6x^2\right) dx$

$$= 4x + \left(\frac{3}{7}\right)\frac{x^{1+1}}{1+1} - (6)\frac{x^{2+1}}{2+1} + c$$

$$= 4x + \left(\frac{3}{7}\right)\frac{x^2}{2} - (6)\frac{x^3}{3} + c$$

$$= 4x + \frac{3}{14}x^2 - 2x^3 + c$$

Note that when an integral contains more than one term there is no need to have an arbitrary constant for each; just a single constant at the end is sufficient.

Problem 3. Determine

(a) $\int \frac{2x^3 - 3x}{4x}dx$ (b) $\int (1 - t)^2 dt$

(a) Rearranging into standard integral form gives:

$$\int \frac{2x^3 - 3x}{4x}dx = \int \frac{2x^3}{4x} - \frac{3x}{4x}dx$$

$$= \int \frac{x^2}{2} - \frac{3}{4}dx = \left(\frac{1}{2}\right)\frac{x^{2+1}}{2+1} - \frac{3}{4}x + c$$

$$= \left(\frac{1}{2}\right)\frac{x^3}{3} - \frac{3}{4}x + c = \frac{1}{6}x^3 - \frac{3}{4}x + c$$

(b) Rearranging $\int (1-t)^2 dt$ gives:

$$\int (1 - 2t + t^2)dt = t - \frac{2t^{1+1}}{1+1} + \frac{t^{2+1}}{2+1} + c$$

$$= t - \frac{2t^2}{2} + \frac{t^3}{3} + c$$

$$= t - t^2 + \frac{1}{3}t^3 + c$$

This problem shows that functions often have to be rearranged into the standard form of $\int ax^n dx$ before it is possible to integrate them.

Problem 4.   Determine $\int \frac{3}{x^2}dx$

$\int \frac{3}{x^2}dx = \int 3x^{-2}$. Using the standard integral, $\int ax^n dx$ when $a=3$ and $n=-2$ gives:

$$\int 3x^{-2}dx = \frac{3x^{-2+1}}{-2+1} + c = \frac{3x^{-1}}{-1} + c$$

$$= -3x^{-1} + c = \frac{-3}{x} + c$$

Problem 5.   Determine $\int 3\sqrt{x}dx$

For fractional powers it is necessary to appreciate $\sqrt[n]{a^m} = a^{\frac{m}{n}}$

$$\int 3\sqrt{x}dx = \int 3x^{1/2}dx = \frac{3x^{\frac{1}{2}+1}}{\frac{1}{2}+1} + c$$

$$= \frac{3x^{\frac{3}{2}}}{\frac{3}{2}} + c = 2x^{\frac{3}{2}} + c = 2\sqrt{x^3} + c$$

Problem 6.   Determine $\int \frac{-5}{9\sqrt[4]{t^3}}dt$

$$\int \frac{-5}{9\sqrt[4]{t^3}}dt = \int \frac{-5}{9t^{\frac{3}{4}}}dt = \int \left(-\frac{5}{9}\right)t^{-\frac{3}{4}}dt$$

$$= \left(-\frac{5}{9}\right)\frac{t^{-\frac{3}{4}+1}}{-\frac{3}{4}+1} + c$$

$$= \left(-\frac{5}{9}\right)\frac{t^{\frac{1}{4}}}{\frac{1}{4}} + c = \left(-\frac{5}{9}\right)\left(\frac{4}{1}\right)t^{1/4} + c$$

$$= -\frac{20}{9}\sqrt[4]{t} + c$$

Problem 7.   Determine $\int \frac{(1+\theta)^2}{\sqrt{\theta}}d\theta$

$$\int \frac{(1+\theta)^2}{\sqrt{\theta}}d\theta = \int \frac{(1+2\theta+\theta^2)}{\sqrt{\theta}}d\theta$$

$$= \int \left(\frac{1}{\theta^{\frac{1}{2}}} + \frac{2\theta}{\theta^{\frac{1}{2}}} + \frac{\theta^2}{\theta^{\frac{1}{2}}}\right)d\theta$$

$$= \int \left(\theta^{-\frac{1}{2}} + 2\theta^{1-(\frac{1}{2})} + \theta^{2-(\frac{1}{2})}\right)d\theta$$

$$= \int \left(\theta^{-\frac{1}{2}} + 2\theta^{\frac{1}{2}} + \theta^{\frac{3}{2}}\right)d\theta$$

$$= \frac{\theta^{(-\frac{1}{2})+1}}{-\frac{1}{2}+1} + \frac{2\theta^{(\frac{1}{2})+1}}{\frac{1}{2}+1} + \frac{\theta^{(\frac{3}{2})+1}}{\frac{3}{2}+1} + c$$

$$= \frac{\theta^{\frac{1}{2}}}{\frac{1}{2}} + \frac{2\theta^{\frac{3}{2}}}{\frac{3}{2}} + \frac{\theta^{\frac{5}{2}}}{\frac{5}{2}} + c$$

$$= 2\theta^{\frac{1}{2}} + \frac{4}{3}\theta^{\frac{3}{2}} + \frac{2}{5}\theta^{\frac{5}{2}} + c$$

$$= 2\sqrt{\theta} + \frac{4}{3}\sqrt{\theta^3} + \frac{2}{5}\sqrt{\theta^5} + c$$

Problem 8.   Determine

(a) $\int 4\cos 3x\, dx$ (b) $\int 5\sin 2\theta\, d\theta$

(a) From Table 48.1 (ii),

$$\int 4\cos 3x\, dx = (4)\left(\frac{1}{3}\right)\sin 3x + c$$

$$= \frac{4}{3}\sin 3x + c$$

(b) From Table 48.1(iii),

$$\int 5\sin 2\theta\, d\theta = (5)\left(-\frac{1}{2}\right)\cos 2\theta + c$$

$$= -\frac{5}{2}\cos 2\theta + c$$

Section 9

Problem 9. Determine (a) $\int 7\sec^2 4t\, dt$

(b) $3\int \csc^2 2\theta\, d\theta$

(a) From Table 48.1(iv),

$$\int 7\sec^2 4t\, dt = (7)\left(\frac{1}{4}\right)\tan 4t + c$$

$$= \frac{7}{4}\tan 4t + c$$

(b) From Table 48.1(v),

$$3\int \csc^2 2\theta\, d\theta = (3)\left(-\frac{1}{2}\right)\cot 2\theta + c$$

$$= -\frac{3}{2}\cot 2\theta + c$$

Problem 10. Determine (a) $\int 5e^{3x}\, dx$

(b) $\int \frac{2}{3e^{4t}}\, dt$

(a) From Table 48.1(viii),

$$\int 5e^{3x}\, dx = (5)\left(\frac{1}{3}\right)e^{3x} + c = \frac{5}{3}e^{3x} + c$$

(b) $\int \frac{2}{3e^{4t}}\, dt = \int \frac{2}{3}e^{-4t}\, dt$

$$= \left(\frac{2}{3}\right)\left(-\frac{1}{4}\right)e^{-4t} + c$$

$$= -\frac{1}{6}e^{-4t} + c = -\frac{1}{6e^{4t}} + c$$

Problem 11. Determine

(a) $\int \frac{3}{5x}\, dx$ (b) $\int \left(\frac{2m^2 + 1}{m}\right)\, dm$

(a) $\int \frac{3}{5x}\, dx = \int \left(\frac{3}{5}\right)\left(\frac{1}{x}\right)\, dx = \frac{3}{5}\ln x + c$

(from Table 48.1(ix))

(b) $\int \left(\frac{2m^2 + 1}{m}\right)\, dm = \int \left(\frac{2m^2}{m} + \frac{1}{m}\right)\, dm$

$$= \int \left(2m + \frac{1}{m}\right)\, dm$$

$$= \frac{2m^2}{2} + \ln m + c$$

$$= m^2 + \ln m + c$$

Now try the following exercise

Exercise 172   Further problems on standard integrals

Determine the following integrals:

1. (a) $\int 4\, dx$ (b) $\int 7x\, dx$

$$\left[\text{(a) } 4x + c \quad \text{(b) } \frac{7x^2}{2} + c\right]$$

2. (a) $\int \frac{2}{5}x^2\, dx$ (b) $\int \frac{5}{6}x^3\, dx$

$$\left[\text{(a) } \frac{2}{15}x^3 + c \quad \text{(b) } \frac{5}{24}x^4 + c\right]$$

3. (a) $\int \left(\frac{3x^2 - 5x}{x}\right)\, dx$ (b) $\int (2 + \theta)^2\, d\theta$

$$\left[\begin{array}{l}\text{(a) } \dfrac{3x^2}{2} - 5x + c \\[2mm] \text{(b) } 4\theta + 2\theta^2 + \dfrac{\theta^3}{3} + c\end{array}\right]$$

4. (a) $\int \frac{4}{3x^2}\, dx$ (b) $\int \frac{3}{4x^4}\, dx$

$$\left[\text{(a) } \frac{-4}{3x} + c \quad \text{(b) } \frac{-1}{4x^3} + c\right]$$

5. (a) $2\int \sqrt{x^3}\, dx$ (b) $\int \frac{1}{4}\sqrt[4]{x^5}\, dx$

$$\left[\text{(a) } \frac{4}{5}\sqrt{x^5} + c \quad \text{(b) } \frac{1}{9}\sqrt[4]{x^9} + c\right]$$

6. (a) $\int \frac{-5}{\sqrt{t^3}}\, dt$ (b) $\int \frac{3}{7\sqrt[5]{x^4}}\, dx$

$$\left[\text{(a) } \frac{10}{\sqrt{t}} + c \quad \text{(b) } \frac{15}{7}\sqrt[5]{x} + c\right]$$

7. (a) $\int 3\cos 2x\, dx$ (b) $\int 7\sin 3\theta\, d\theta$

$$\left[\begin{array}{l}\text{(a) } \dfrac{3}{2}\sin 2x + c \\[2mm] \text{(b) } -\dfrac{7}{3}\cos 3\theta + c\end{array}\right]$$

8. (a) $\int \frac{3}{4}\sec^2 3x\, dx$ (b) $\int 2\csc^2 4\theta\, d\theta$

$$\left[\text{(a) } \frac{1}{4}\tan 3x + c \quad \text{(b) } -\frac{1}{2}\cot 4\theta + c\right]$$

Section 9

9. (a) $5\displaystyle\int \cot 2t\,\operatorname{cosec} 2t\,dt$

(b) $\displaystyle\int \frac{4}{3}\sec 4t\,\tan 4t\,dt$

$$\left[(a)\ -\frac{5}{2}\operatorname{cosec} 2t + c \quad (b)\ \frac{1}{3}\sec 4t + c \right]$$

10. (a) $\displaystyle\int \frac{3}{4}e^{2x}\,dx$ (b) $\displaystyle\frac{2}{3}\int \frac{dx}{e^{5x}}$

$$\left[(a)\ \frac{3}{8}e^{2x} + c \quad (b)\ \frac{-2}{15e^{5x}} + c \right]$$

11. (a) $\displaystyle\int \frac{2}{3x}\,dx$ (b) $\displaystyle\int \left(\frac{u^2-1}{u}\right)du$

$$\left[(a)\ \frac{2}{3}\ln x + c \quad (b)\ \frac{u^2}{2} - \ln u + c \right]$$

12. (a) $\displaystyle\int \frac{(2+3x)^2}{\sqrt{x}}\,dx$ (b) $\displaystyle\int \left(\frac{1}{t}+2t\right)^2 dt$

$$\left[\begin{array}{l} (a)\ 8\sqrt{x}+8\sqrt{x^3}+\dfrac{18}{5}\sqrt{x^5}+c \\[2mm] (b)\ -\dfrac{1}{t}+4t+\dfrac{4t^3}{3}+c \end{array} \right]$$

48.4 Definite integrals

Integrals containing an arbitrary constant c in their results are called **indefinite integrals** since their precise value cannot be determined without further information. **Definite integrals** are those in which limits are applied. If an expression is written as $[x]_a^b$, 'b' is called the upper limit and 'a' the lower limit.

The operation of applying the limits is defined as: $[x]_a^b = (b) - (a)$

The increase in the value of the integral x^2 as x increases from 1 to 3 is written as $\int_1^3 x^2\,dx$

Applying the limits gives:

$$\int_1^3 x^2\,dx = \left[\frac{x^3}{3}+c\right]_1^3 = \left(\frac{3^3}{3}+c\right) - \left(\frac{1^3}{3}+c\right)$$

$$= (9+c) - \left(\frac{1}{3}+c\right) = 8\frac{2}{3}$$

Note that the 'c' term always cancels out when limits are applied and it need not be shown with definite integrals.

Problem 12. Evaluate (a) $\displaystyle\int_1^2 3x\,dx$

(b) $\displaystyle\int_{-2}^3 (4-x^2)\,dx$

(a) $\displaystyle\int_1^2 3x\,dx = \left[\frac{3x^2}{2}\right]_1^2 = \left\{\frac{3}{2}(2)^2\right\} - \left\{\frac{3}{2}(1)^2\right\}$

$$= 6 - 1\frac{1}{2} = 4\frac{1}{2}$$

(b) $\displaystyle\int_{-2}^3 (4-x^2)\,dx = \left[4x - \frac{x^3}{3}\right]_{-2}^3$

$$= \left\{4(3) - \frac{(3)^3}{3}\right\} - \left\{4(-2) - \frac{(-2)^3}{3}\right\}$$

$$= \{12-9\} - \left\{-8 - \frac{-8}{3}\right\}$$

$$= \{3\} - \left\{-5\frac{1}{3}\right\} = 8\frac{1}{3}$$

Problem 13. Evaluate $\displaystyle\int_1^4 \left(\frac{\theta+2}{\sqrt{\theta}}\right)d\theta$, taking positive square roots only

$$\int_1^4 \left(\frac{\theta+2}{\sqrt{\theta}}\right)d\theta = \int_1^4 \left(\frac{\theta}{\theta^{\frac{1}{2}}} + \frac{2}{\theta^{\frac{1}{2}}}\right)d\theta$$

$$= \int_1^4 \left(\theta^{\frac{1}{2}} + 2\theta^{-\frac{1}{2}}\right)d\theta$$

$$= \left[\frac{\theta^{(\frac{1}{2})+1}}{\frac{1}{2}+1} + \frac{2\theta^{(-\frac{1}{2})+1}}{-\frac{1}{2}+1}\right]_1^4$$

$$= \left[\frac{\theta^{\frac{3}{2}}}{\frac{3}{2}} + \frac{2\theta^{\frac{1}{2}}}{\frac{1}{2}}\right]_1^4 = \left[\frac{2}{3}\sqrt{\theta^3} + 4\sqrt{\theta}\right]_1^4$$

$$= \left\{\frac{2}{3}\sqrt{(4)^3} + 4\sqrt{4}\right\} - \left\{\frac{2}{3}\sqrt{(1)^3} + 4\sqrt{1}\right\}$$

$$= \left\{\frac{16}{3} + 8\right\} - \left\{\frac{2}{3} + 4\right\}$$

$$= 5\frac{1}{3} + 8 - \frac{2}{3} - 4 = 8\frac{2}{3}$$

Section 9

Problem 14. Evaluate: $\int_0^{\pi/2} 3\sin 2x\, dx$

$$\int_0^{\frac{\pi}{2}} 3\sin 2x\, dx$$

$$= \left[(3)\left(-\frac{1}{2}\right)\cos 2x\right]_0^{\frac{\pi}{2}} = \left[-\frac{3}{2}\cos 2x\right]_0^{\frac{\pi}{2}}$$

$$= \left\{-\frac{3}{2}\cos 2\left(\frac{\pi}{2}\right)\right\} - \left\{-\frac{3}{2}\cos 2(0)\right\}$$

$$= \left\{-\frac{3}{2}\cos \pi\right\} - \left\{-\frac{3}{2}\cos 0\right\}$$

$$= \left\{-\frac{3}{2}(-1)\right\} - \left\{-\frac{3}{2}(1)\right\} = \frac{3}{2} + \frac{3}{2} = \mathbf{3}$$

Problem 15. Evaluate $\int_1^2 4\cos 3t\, dt$

$$\int_1^2 4\cos 3t\, dt = \left[(4)\left(\frac{1}{3}\right)\sin 3t\right]_1^2 = \left[\frac{4}{3}\sin 3t\right]_1^2$$

$$= \left\{\frac{4}{3}\sin 6\right\} - \left\{\frac{4}{3}\sin 3\right\}$$

Note that limits of trigonometric functions are always expressed in radians—thus, for example, sin 6 means the sine of 6 radians $= -0.279415\ldots$

Hence $\int_1^2 4\cos 3t\, dt = \left\{\frac{4}{3}(-0.279415\ldots)\right\}$

$$- \left\{\frac{4}{3}(-0.141120\ldots)\right\}$$

$$= (-0.37255) - (0.18816) = \mathbf{-0.5607}$$

Problem 16. Evaluate

(a) $\int_1^2 4e^{2x}\, dx$ (b) $\int_1^4 \frac{3}{4u}\, du$,

each correct to 4 significant figures

(a) $\int_1^2 4e^{2x}\, dx = \left[\frac{4}{2}e^{2x}\right]_1^2$

$$= 2[e^{2x}]_1^2 = 2[e^4 - e^2]$$

$$= 2[54.5982 - 7.3891] = \mathbf{94.42}$$

(b) $\int_1^4 \frac{3}{4u}\, du = \left[\frac{3}{4}\ln u\right]_1^4 = \frac{3}{4}[\ln 4 - \ln 1]$

$$= \frac{3}{4}[1.3863 - 0] = \mathbf{1.040}$$

Now try the following exercise

Exercise 173 Further problems on definite integrals

In Problems 1 to 8, evaluate the definite integrals (where necessary, correct to 4 significant figures).

1. (a) $\int_1^4 5x^2\, dx$ (b) $\int_{-1}^1 -\frac{3}{4}t^2\, dt$

$$\left[\text{(a) } 105 \quad \text{(b) } -\frac{1}{2}\right]$$

2. (a) $\int_{-1}^2 (3 - x^2)\, dx$ (b) $\int_1^3 (x^2 - 4x + 3)\, dx$

$$\left[\text{(a) } 6 \quad \text{(b) } -1\frac{1}{3}\right]$$

3. (a) $\int_0^{\pi} \frac{3}{2}\cos\theta\, d\theta$ (b) $\int_0^{\frac{\pi}{2}} 4\cos\theta\, d\theta$

$$[\text{(a) } 0 \quad \text{(b) } 4]$$

4. (a) $\int_{\frac{\pi}{6}}^{\frac{\pi}{3}} 2\sin 2\theta\, d\theta$ (b) $\int_0^2 3\sin t\, dt$

$$[\text{(a) } 1 \quad \text{(b) } 4.248]$$

5. (a) $\int_0^1 5\cos 3x\, dx$ (b) $\int_0^{\frac{\pi}{6}} 3\sec^2 2x\, dx$

$$[\text{(a) } 0.2352 \quad \text{(b) } 2.598]$$

6. (a) $\int_1^2 \operatorname{cosec}^2 4t\, dt$

(b) $\int_{\frac{\pi}{4}}^{\frac{\pi}{2}} (3\sin 2x - 2\cos 3x)\, dx$

$$[\text{(a) } 0.2572 \quad \text{(b) } 2.638]$$

7. (a) $\int_0^1 3e^{3t}\, dt$ (b) $\int_{-1}^2 \frac{2}{3e^{2x}}\, dx$

$$[\text{(a) } 19.09 \quad \text{(b) } 2.457]$$

8. (a) $\int_2^3 \frac{2}{3x}\, dx$ (b) $\int_1^3 \frac{2x^2 + 1}{x}\, dx$

$$[\text{(a) } 0.2703 \quad \text{(b) } 9.099]$$

9. The entropy change ΔS, for an ideal gas is given by:

$$\Delta S = \int_{T_1}^{T_2} C_v \frac{dT}{T} - R \int_{V_1}^{V_2} \frac{dV}{V}$$

where T is the thermodynamic temperature, V is the volume and $R = 8.314$. Determine the entropy change when a gas expands from 1 litre to 3 litres for a temperature rise from 100 K to 400 K given that:

$$C_v = 45 + 6 \times 10^{-3}T + 8 \times 10^{-6}T^2$$

[55.65]

10. The p.d. between boundaries a and b of an electric field is given by: $V = \int_{a}^{b} \frac{Q}{2\pi r \varepsilon_0 \varepsilon_r} dr$

If $a = 10$, $b = 20$, $Q = 2 \times 10^{-6}$ coulombs, $\varepsilon_0 = 8.85 \times 10^{-12}$ and $\varepsilon_r = 2.77$, show that $V = 9\,\text{kV}$.

11. The average value of a complex voltage waveform is given by:

$$V_{AV} = \frac{1}{\pi} \int_{0}^{\pi} (10 \sin \omega t + 3 \sin 3\omega t + 2 \sin 5\omega t) d(\omega t)$$

Evaluate V_{AV} correct to 2 decimal places.

[7.26]

Chapter 49

Integration using algebraic substitutions

49.1 Introduction

Functions that require integrating are not always in the 'standard form' shown in Chapter 48. However, it is often possible to change a function into a form which can be integrated by using either:

(i) an algebraic substitution (see Section 49.2),

(ii) trigonometric substitutions (see Chapter 50),

(iii) partial fractions (see Chapter 51),

(iv) the $t = \tan\frac{\theta}{2}$ substitution (see Chapter 52), or

(v) integration by parts (see Chapter 53).

49.2 Algebraic substitutions

With **algebraic substitutions**, the substitution usually made is to let u be equal to $f(x)$ such that $f(u)\,du$ is a standard integral. It is found that integrals of the forms:

$$k\int [f(x)]^n f'(x)\,dx \quad \text{and} \quad k\int \frac{f'(x)^n}{[f(x)]}dx$$

(where k and n are constants) can both be integrated by substituting u for $f(x)$.

49.3 Worked problems on integration using algebraic substitutions

Problem 1. Determine $\int \cos(3x + 7)\,dx$

$\int \cos(3x+7)\,dx$ is not a standard integral of the form shown in Table 48.1, page 436, thus an algebraic substitution is made.

Let $u = 3x + 7$ then $\dfrac{du}{dx} = 3$ and rearranging gives $dx = \dfrac{du}{3}$

Hence $\displaystyle\int \cos(3x + 7)\,dx = \int (\cos u)\frac{du}{3}$

$$= \int \frac{1}{3}\cos u\,du,$$

which is a standard integral

$$= \frac{1}{3}\sin u + c$$

Rewriting u as $(3x + 7)$ gives:

$$\int \cos(3x + 7)\,dx = \frac{1}{3}\sin(3x + 7) + c,$$

which may be checked by differentiating it.

Problem 2. Find: $\displaystyle\int (2x - 5)^7 dx$

$(2x - 5)$ may be multiplied by itself 7 times and then each term of the result integrated. However, this would be a lengthy process, and thus an algebraic substitution is made.

Let $u = (2x - 5)$ then $\dfrac{du}{dx} = 2$ and $dx = \dfrac{du}{2}$

Hence

$$\int (2x - 5)^7 dx = \int u^7 \frac{du}{2} = \frac{1}{2}\int u^7 du$$

$$= \frac{1}{2}\left(\frac{u^8}{8}\right) + c = \frac{1}{16}u^8 + c$$

Rewriting u as $(2x - 5)$ gives:

$$\int (2x - 5)^7 dx = \frac{1}{16}(2x - 5)^8 + c$$

Problem 3.   Find: $\int \frac{4}{(5x - 3)} dx$

Let $u = (5x - 3)$ then $\frac{du}{dx} = 5$ and $dx = \frac{du}{5}$

Hence $\int \frac{4}{(5x - 3)} dx = \int \frac{4}{u} \frac{du}{5} = \frac{4}{5} \int \frac{1}{u} du$

$$= \frac{4}{5} \ln u + c$$

$$= \frac{4}{5} \ln(5x - 3) + c$$

Problem 4.   Evaluate $\int_0^1 2e^{6x-1} dx$, correct to 4 significant figures

Let $u = 6x - 1$ then $\frac{du}{dx} = 6$ and $dx = \frac{du}{6}$

Hence $\int 2e^{6x-1} dx = \int 2e^u \frac{du}{6} = \frac{1}{3} \int e^u du$

$$= \frac{1}{3} e^u + c = \frac{1}{3} e^{6x-1} + c$$

Thus $\int_0^1 2e^{6x-1} dx = \frac{1}{3} \left[e^{6x-1} \right]_0^1$

$$= \frac{1}{3} [e^5 - e^{-1}] = \mathbf{49.35},$$

correct to 4 significant figures.

Problem 5.   Determine: $\int 3x(4x^2 + 3)^5 dx$

Let $u = (4x^2 + 3)$ then $\frac{du}{dx} = 8x$ and $dx = \frac{du}{8x}$
Hence

$$\int 3x(4x^2 + 3)^5 dx = \int 3x(u)^5 \frac{du}{8x}$$

$$= \frac{3}{8} \int u^5 du, \text{ by cancelling}$$

The original variable 'x' has been completely removed and the integral is now only in terms of u and is a standard integral.

Hence $\frac{3}{8} \int u^5 du = \frac{3}{8} \left(\frac{u^6}{6} \right) + c = \frac{1}{16} u^6 + c$

$$= \frac{1}{16}(4x^2 + 3)^6 + c$$

Problem 6.   Evaluate: $\int_0^{\pi/6} 24 \sin^5 \theta \cos \theta d\theta$

Let $u = \sin \theta$ then $\frac{du}{d\theta} = \cos \theta$ and $d\theta = \frac{du}{\cos \theta}$

Hence $\int 24 \sin^5 \theta \cos \theta d\theta$

$$= \int 24 u^5 \cos \theta \frac{du}{\cos \theta}$$

$$= 24 \int u^5 du, \text{ by cancelling}$$

$$= 24 \frac{u^6}{6} + c = 4u^6 + c = 4(\sin \theta)^6 + c$$

$$= 4 \sin^6 \theta + c$$

Thus $\int_0^{\pi/6} 24 \sin^5 \theta \cos \theta d\theta$

$$= \left[4 \sin^6 \theta \right]_0^{\pi/6} = 4 \left[\left(\sin \frac{\pi}{6} \right)^6 - (\sin 0)^6 \right]$$

$$= 4 \left[\left(\frac{1}{2} \right)^6 - 0 \right] = \frac{1}{16} \text{ or } \mathbf{0.0625}$$

Now try the following exercise

Exercise 174   Further problems on integration using algebraic substitutions

In Problems 1 to 6, integrate with respect to the variable.

1. $2 \sin(4x + 9)$ $\left[-\frac{1}{2} \cos(4x + 9) + c \right]$

2. $3 \cos(2\theta - 5)$ $\left[\frac{3}{2} \sin(2\theta - 5) + c \right]$

3. $4\sec^2(3t+1)$ $\left[\dfrac{4}{3}\tan(3t+1)+c\right]$

4. $\dfrac{1}{2}(5x-3)^6$ $\left[\dfrac{1}{70}(5x-3)^7+c\right]$

5. $\dfrac{-3}{(2x-1)}$ $\left[-\dfrac{3}{2}\ln(2x-1)+c\right]$

6. $3e^{3\theta+5}$ $[e^{3\theta+5}+c]$

In Problems 7 to 10, evaluate the definite integrals correct to 4 significant figures.

7. $\displaystyle\int_0^1 (3x+1)^5 dx$ $[227.5]$

8. $\displaystyle\int_0^2 x\sqrt{2x^2+1}\,dx$ $[4.333]$

9. $\displaystyle\int_0^{\pi/3} 2\sin\left(3t+\dfrac{\pi}{4}\right) dt$ $[0.9428]$

10. $\displaystyle\int_0^1 3\cos(4x-3)dx$ $[0.7369]$

49.4 Further worked problems on integration using algebraic substitutions

Problem 7. Find: $\displaystyle\int \dfrac{x}{2+3x^2}dx$

Let $u=2+3x^2$ then $\dfrac{du}{dx}=6x$ and $dx=\dfrac{du}{6x}$

Hence $\displaystyle\int \dfrac{x}{2+3x^2}dx$

$= \displaystyle\int \dfrac{x}{u}\dfrac{du}{6x} = \dfrac{1}{6}\int \dfrac{1}{u}\,du$, by cancelling,

$= \dfrac{1}{6}\ln u + x$

$= \dfrac{1}{6}\ln(2+3x^2)+c$

Problem 8. Determine: $\displaystyle\int \dfrac{2x}{\sqrt{4x^2-1}}dx$

Let $u=4x^2-1$ then $\dfrac{du}{dx}=8x$ and $dx=\dfrac{du}{8x}$

Hence $\displaystyle\int \dfrac{2x}{\sqrt{4x^2-1}}dx$

$= \displaystyle\int \dfrac{2x}{\sqrt{u}}\dfrac{du}{8x} = \dfrac{1}{4}\int \dfrac{1}{\sqrt{u}}\,du$, by cancelling

$= \dfrac{1}{4}\displaystyle\int u^{-1/2}du$

$= \dfrac{1}{4}\left[\dfrac{u^{(-1/2)+1}}{-\frac{1}{2}+1}\right]+c = \dfrac{1}{4}\left[\dfrac{u^{1/2}}{\frac{1}{2}}\right]+c$

$= \dfrac{1}{2}\sqrt{u}+c = \dfrac{1}{2}\sqrt{4x^2-1}+c$

Problem 9. Show that:

$$\int \tan\theta\, d\theta = \ln(\sec\theta)+c$$

$$\int \tan\theta\, d\theta = \int \dfrac{\sin\theta}{\cos\theta}d\theta.$$

Let $u=\cos\theta$

then $\dfrac{du}{d\theta}=-\sin\theta$ and $d\theta=\dfrac{-du}{\sin\theta}$

Hence

$$\int \dfrac{\sin\theta}{\cos\theta}d\theta = \int \dfrac{\sin\theta}{u}\left(\dfrac{-du}{\sin\theta}\right)$$

$$= -\int \dfrac{1}{u}du = -\ln u + c$$

$$= -\ln(\cos\theta)+c$$

$$= \ln(\cos\theta)^{-1}+c,$$

by the laws of logarithms

Hence $\displaystyle\int \tan\theta\, d\theta = \ln(\sec\theta)+c,$

since $(\cos\theta)^{-1}=\dfrac{1}{\cos\theta}=\sec\theta$

49.5 Change of limits

When evaluating definite integrals involving substitutions it is sometimes more convenient to **change the limits** of the integral as shown in Problems 10 and 11.

Problem 10. Evaluate: $\int_1^3 5x\sqrt{2x^2+7}\ dx$, taking positive values of square roots only

Let $u = 2x^2 + 7$, then $\dfrac{du}{dx} = 4x$ and $dx = \dfrac{du}{4x}$

It is possible in this case to change the limits of integration. Thus when $x = 3$, $u = 2(3)^2 + 7 = 25$ and when $x = 1$, $u = 2(1)^2 + 7 = 9$

Hence $\displaystyle\int_{x=1}^{x=3} 5x\sqrt{2x^2+7}\ dx$

$$= \int_{u=9}^{u=25} 5x\sqrt{u}\ \frac{du}{4x} = \frac{5}{4}\int_9^{25} \sqrt{u}\ du$$

$$= \frac{5}{4}\int_9^{25} u^{1/2}\,du$$

Thus the limits have been changed, and it is unnecessary to change the integral back in terms of x.

Thus $\displaystyle\int_{x=1}^{x=3} 5x\sqrt{2x^2+7}\ dx$

$$= \frac{5}{4}\left[\frac{u^{3/2}}{3/2}\right]_9^{25} = \frac{5}{6}\left[\sqrt{u^3}\right]_9^{25}$$

$$= \frac{5}{6}[\sqrt{25^3} - \sqrt{9^3}] = \frac{5}{6}(125 - 27) = 81\frac{2}{3}$$

Problem 11. Evaluate: $\displaystyle\int_0^2 \frac{3x}{\sqrt{2x^2+1}}\ dx$, taking positive values of square roots only

Let $u = 2x^2 + 1$ then $\dfrac{du}{dx} = 4x$ and $dx = \dfrac{du}{4x}$

Hence $\displaystyle\int_0^2 \frac{3x}{\sqrt{2x^2+1}}\ dx = \int_{x=0}^{x=2} \frac{3x}{\sqrt{u}}\ \frac{du}{4x}$

$$= \frac{3}{4}\int_{x=0}^{x=2} u^{-1/2}\,du$$

Since $u = 2x^2 + 1$, when $x = 2$, $u = 9$ and when $x = 0$, $u = 1$

Thus $\dfrac{3}{4}\displaystyle\int_{x=0}^{x=2} u^{-1/2}\,du = \dfrac{3}{4}\int_{u=1}^{u=9} u^{-1/2}\,du$,

i.e. the limits have been changed

$$= \frac{3}{4}\left[\frac{u^{1/2}}{\frac{1}{2}}\right]_1^9 = \frac{3}{2}[\sqrt{9} - \sqrt{1}] = 3,$$

taking positive values of square roots only.

Now try the following exercise

Exercise 175 Further problems on integration using algebraic substitutions

In Problems 1 to 7, integrate with respect to the variable.

1. $2x(2x^2 - 3)^5$ $\qquad\left[\dfrac{1}{12}(2x^2 - 3)^6 + c\right]$

2. $5\cos^5 t \sin t$ $\qquad\left[-\dfrac{5}{6}\cos^6 t + c\right]$

3. $3\sec^2 3x \tan 3x$

$\qquad\left[\dfrac{1}{2}\sec^2 3x + c\ \text{or}\ \dfrac{1}{2}\tan^2 3x + c\right]$

4. $2t\sqrt{3t^2 - 1}$ $\qquad\left[\dfrac{2}{9}\sqrt{(3t^2 - 1)^3} + c\right]$

5. $\dfrac{\ln\theta}{\theta}$ $\qquad\left[\dfrac{1}{2}(\ln\theta)^2 + c\right]$

6. $3\tan 2t$ $\qquad\left[\dfrac{3}{2}\ln(\sec 2t) + c\right]$

7. $\dfrac{2e^t}{\sqrt{e^t + 4}}$ $\qquad[4\sqrt{e^t + 4} + c]$

In Problems 8 to 10, evaluate the definite integrals correct to 4 significant figures.

8. $\displaystyle\int_0^1 3xe^{(2x^2-1)}dx$ $\qquad$ [1.763]

9. $\displaystyle\int_0^{\pi/2} 3\sin^4\theta\cos\theta\ d\theta$ $\qquad$ [0.6000]

10. $\displaystyle\int_0^1 \frac{3x}{(4x^2-1)^5}dx$ $\qquad$ [0.09259]

Section 9

11. The electrostatic potential on all parts of a conducting circular disc of radius r is given by the equation:

$$V = 2\pi\sigma \int_0^9 \frac{R}{\sqrt{R^2 + r^2}} dR$$

Solve the equation by determining the integral. $\left[V = 2\pi\sigma \left\{ \sqrt{(9^2 + r^2)} - r \right\} \right]$

12. In the study of a rigid rotor the following integration occurs:

$$Z_r = \int_0^\infty (2J+1)\, e^{\frac{-J(J+1)h^2}{8\pi^2 IkT}}\, dJ$$

Determine Z_r for constant temperature T assuming h, I and k are constants.

$$\left[\frac{8\pi^2 IkT}{h^2} \right]$$

13. In electrostatics,

$$E = \int_0^\pi \left\{ \frac{a^2\sigma \sin\theta}{2\varepsilon \sqrt{(a^2 - x^2 - 2ax\cos\theta)}} d\theta \right\}$$

where a, σ and ε are constants, x is greater than a, and x is independent of θ. Show that

$$E = \frac{a^2\sigma}{\varepsilon x}$$

Chapter 50

Integration using trigonometric substitutions

50.1 Introduction

Table 50.1 gives a summary of the integrals that require the use of **trigonometric substitutions**, and their application is demonstrated in Problems 1 to 19.

50.2 Worked problems on integration of $\sin^2 x$, $\cos^2 x$, $\tan^2 x$ and $\cot^2 x$

Problem 1. Evaluate: $\int_0^{\frac{\pi}{4}} 2\cos^2 4t\, dt$

Since $\cos 2t = 2\cos^2 t - 1$ (from Chapter 27),

then $\cos^2 t = \frac{1}{2}(1 + \cos 2t)$ and

$\cos^2 4t = \frac{1}{2}(1 + \cos 8t)$

Hence $\int_0^{\frac{\pi}{4}} 2\cos^2 4t\, dt$

$= 2\int_0^{\frac{\pi}{4}} \frac{1}{2}(1 + \cos 8t)\, dt$

$= \left[t + \frac{\sin 8t}{8} \right]_0^{\frac{\pi}{4}}$

$= \left[\frac{\pi}{4} + \frac{\sin 8\left(\frac{\pi}{4}\right)}{8} \right] - \left[0 + \frac{\sin 0}{8} \right]$

$= \frac{\pi}{4}$ or **0.7854**

Problem 2. Determine: $\int \sin^2 3x\, dx$

Since $\cos 2x = 1 - 2\sin^2 x$ (from Chapter 27),

then $\sin^2 x = \frac{1}{2}(1 - \cos 2x)$ and

$\sin^2 3x = \frac{1}{2}(1 - \cos 6x)$

Hence $\int \sin^2 3x\, dx = \int \frac{1}{2}(1 - \cos 6x)\, dx$

$= \frac{1}{2}\left(x - \frac{\sin 6x}{6} \right) + c$

Problem 3. Find: $3\int \tan^2 4x\, dx$

Since $1 + \tan^2 x = \sec^2 x$, then $\tan^2 x = \sec^2 x - 1$ and $\tan^2 4x = \sec^2 4x - 1$

Hence $3\int \tan^2 4x\, dx = 3\int (\sec^2 4x - 1)\, dx$

$= 3\left(\frac{\tan 4x}{4} - x \right) + c$

Problem 4. Evaluate $\int_{\frac{\pi}{6}}^{\frac{\pi}{3}} \frac{1}{2}\cot^2 2\theta\, d\theta$

Engineering Mathematics

Table 50.1 Integrals using trigonometric substitutions

$f(x)$	$\int f(x)dx$	Method	See problem
1. $\cos^2 x$	$\dfrac{1}{2}\left(x+\dfrac{\sin 2x}{2}\right)+c$	Use $\cos 2x = 2\cos^2 x - 1$	1
2. $\sin^2 x$	$\dfrac{1}{2}\left(x-\dfrac{\sin 2x}{2}\right)+c$	Use $\cos 2x = 1 - 2\sin^2 x$	2
3. $\tan^2 x$	$\tan x - x + c$	Use $1 + \tan^2 x = \sec^2 x$	3
4. $\cot^2 x$	$-\cot x - x + c$	Use $\cot^2 x + 1 = \operatorname{cosec}^2 x$	4
5. $\cos^m x \sin^n x$	(a) If either m or n is odd (but not both), use $$\cos^2 x + \sin^2 x = 1$$		5, 6
	(b) If both m and n are even, use either $$\cos 2x = 2\cos^2 x - 1 \text{ or } \cos 2x = 1 - 2\sin^2 x$$		7, 8
6. $\sin A \cos B$		Use $\dfrac{1}{2}[\sin(A+B)+\sin(A-B)]$	9
7. $\cos A \sin B$		Use $\dfrac{1}{2}[\sin(A+B)-\sin(A-B)]$	10
8. $\cos A \cos B$		Use $\dfrac{1}{2}[\cos(A+B)+\cos(A-B)]$	11
9. $\sin A \sin B$		Use $-\dfrac{1}{2}[\cos(A+B)-\cos(A-B)]$	12
10. $\dfrac{1}{\sqrt{a^2-x^2}}$	$\sin^{-1}\dfrac{x}{a}+c$	Use $x = a\sin\theta$ substitution	13, 14
11. $\sqrt{a^2-x^2}$	$\dfrac{a^2}{2}\sin^{-1}\dfrac{x}{a}+\dfrac{x}{2}\sqrt{a^2-x^2}+c$		15, 16
12. $\dfrac{1}{a^2+x^2}$	$\dfrac{1}{a}\tan^{-1}\dfrac{x}{a}+c$	Use $x = a\tan\theta$ substitution	17–19

Since $\cot^2\theta + 1 = \operatorname{cosec}^2\theta$, then

$\cot^2\theta = \operatorname{cosec}^2\theta - 1$ and $\cot^2 2\theta = \operatorname{cosec}^2 2\theta - 1$

Hence $\displaystyle\int_{\frac{\pi}{6}}^{\frac{\pi}{3}} \frac{1}{2}\cot^2 2\theta \, d\theta$

$= \dfrac{1}{2}\displaystyle\int_{\frac{\pi}{6}}^{\frac{\pi}{3}} (\operatorname{cosec}^2 2\theta - 1)\, d\theta = \dfrac{1}{2}\left[\dfrac{-\cot 2\theta}{2} - \theta\right]_{\frac{\pi}{6}}^{\frac{\pi}{3}}$

$= \dfrac{1}{2}\left[\left(\dfrac{-\cot 2\left(\frac{\pi}{3}\right)}{2} - \dfrac{\pi}{3}\right) - \left(\dfrac{-\cot 2\left(\frac{\pi}{6}\right)}{2} - \dfrac{\pi}{6}\right)\right]$

$= \dfrac{1}{2}[(--0.2887 - 1.0472) - (-0.2887 - 0.5236)]$

$= \mathbf{0.0269}$

Section 9

Now try the following exercise

Exercise 176 Further problems on integration of $\sin^2 x$, $\cos^2 x$, $\tan^2 x$ and $\cot^2 x$

In Problems 1 to 4, integrate with respect to the variable.

1. $\sin^2 2x$ $\qquad \left[\dfrac{1}{2}\left(x - \dfrac{\sin 4x}{4}\right) + c\right]$

2. $3\cos^2 t$ $\qquad \left[\dfrac{3}{2}\left(t + \dfrac{\sin 2t}{2}\right) + c\right]$

3. $5\tan^2 3\theta$ $\qquad \left[5\left(\dfrac{1}{3}\tan 3\theta - \theta\right) + c\right]$

4. $2\cot^2 2t$ $\qquad [-(\cot 2t + 2t) + c]$

In Problems 5 to 8, evaluate the definite integrals, correct to 4 significant figures.

5. $\displaystyle\int_0^{\pi/3} 3\sin^2 3x\,dx$ $\qquad \left[\dfrac{\pi}{2}\text{ or }1.571\right]$

6. $\displaystyle\int_0^{\pi/4} \cos^2 4x\,dx$ $\qquad \left[\dfrac{\pi}{8}\text{ or }0.3927\right]$

7. $\displaystyle\int_0^{0.5} 2\tan^2 2t\,dt$ $\qquad [0.5574]$

8. $\displaystyle\int_{\pi/6}^{\pi/3} \cot^2 \theta\,d\theta$ $\qquad [0.6311]$

50.3 Worked problems on powers of sines and cosines

Problem 5. Determine: $\displaystyle\int \sin^5 \theta\,d\theta$

Since $\cos^2\theta + \sin^2\theta = 1$ then $\sin^2\theta = (1 - \cos^2\theta)$. Hence $\displaystyle\int \sin^5\theta\,d\theta$

$= \displaystyle\int \sin\theta(\sin^2\theta)^2\,d\theta = \int \sin\theta(1 - \cos^2\theta)^2 d\theta$

$= \displaystyle\int \sin\theta(1 - 2\cos^2\theta + \cos^4\theta)\,d\theta$

$= \displaystyle\int (\sin\theta - 2\sin\theta\cos^2\theta + \sin\theta\cos^4\theta)d\theta$

$= -\cos\theta + \dfrac{2\cos^3\theta}{3} - \dfrac{\cos^5\theta}{5} + c$

[Whenever a power of a cosine is multiplied by a sine of power 1, or vice-versa, the integral may be determined by inspection as shown.

In general, $\displaystyle\int \cos^n\theta\sin\theta\,d\theta = \dfrac{-\cos^{n+1}\theta}{(n+1)} + c$

and $\displaystyle\int \sin^n\theta\cos\theta\,d\theta = \dfrac{\sin^{n+1}\theta}{(n+1)} + c$

Alternatively, an algebraic substitution may be used as shown in Problem 6, chapter 49, page 443].

Problem 6. Evaluate: $\displaystyle\int_0^{\frac{\pi}{2}} \sin^2 x\cos^3 x\,dx$

$\displaystyle\int_0^{\frac{\pi}{2}} \sin^2 x\cos^3 x\,dx = \int_0^{\frac{\pi}{2}} \sin^2 x\cos^2 x\cos x\,dx$

$= \displaystyle\int_0^{\frac{\pi}{2}} (\sin^2 x)(1 - \sin^2 x)(\cos x)\,dx$

$= \displaystyle\int_0^{\frac{\pi}{2}} (\sin^2 x\cos x - \sin^4 x\cos x)\,dx$

$= \left[\dfrac{\sin^3 x}{3} - \dfrac{\sin^5 x}{5}\right]_0^{\frac{\pi}{2}}$

$= \left[\dfrac{\left(\sin\dfrac{\pi}{2}\right)^3}{3} - \dfrac{\left(\sin\dfrac{\pi}{2}\right)^5}{5}\right] - [0 - 0]$

$= \dfrac{1}{3} - \dfrac{1}{5} = \dfrac{2}{15}$ or **0.1333**

Problem 7. Evaluate: $\displaystyle\int_0^{\frac{\pi}{4}} 4\cos^4\theta\,d\theta$, correct to 4 significant figures

$\displaystyle\int_0^{\frac{\pi}{4}} 4\cos^4\theta\,d\theta = 4\int_0^{\frac{\pi}{4}} (\cos^2\theta)^2 d\theta$

$= 4\displaystyle\int_0^{\frac{\pi}{4}} \left[\dfrac{1}{2}(1 + \cos 2\theta)\right]^2 d\theta$

$= \displaystyle\int_0^{\frac{\pi}{4}} (1 + 2\cos 2\theta + \cos^2 2\theta)\,d\theta$

$= \displaystyle\int_0^{\frac{\pi}{4}} \left[1 + 2\cos 2\theta + \dfrac{1}{2}(1 + \cos 4\theta)\right] d\theta$

$$= \int_0^{\frac{\pi}{4}} \left[\frac{3}{2} + 2\cos 2\theta + \frac{1}{2}\cos 4\theta \right] d\theta$$

$$= \left[\frac{3\theta}{2} + \sin 2\theta + \frac{\sin 4\theta}{8} \right]_0^{\frac{\pi}{4}}$$

$$= \left[\frac{3}{2}\left(\frac{\pi}{4}\right) + \sin \frac{2\pi}{4} + \frac{\sin 4(\pi/4)}{8} \right] - [0]$$

$$= \frac{3\pi}{8} + 1$$

$$= \mathbf{2.178}, \text{ correct to 4 significant figures.}$$

Problem 8. Find: $\int \sin^2 t \cos^4 t \, dt$

$$\int \sin^2 t \cos^4 t \, dt = \int \sin^2 t (\cos^2 t)^2 \, dt$$

$$= \int \left(\frac{1 - \cos 2t}{2}\right) \left(\frac{1 + \cos 2t}{2}\right)^2 dt$$

$$= \frac{1}{8} \int (1 - \cos 2t)(1 + 2\cos 2t + \cos^2 2t) \, dt$$

$$= \frac{1}{8} \int (1 + 2\cos 2t + \cos^2 2t - \cos 2t$$
$$- 2\cos^2 2t - \cos^3 2t) \, dt$$

$$= \frac{1}{8} \int (1 + \cos 2t - \cos^2 2t - \cos^3 2t) \, dt$$

$$= \frac{1}{8} \int \left[1 + \cos 2t - \left(\frac{1 + \cos 4t}{2}\right)\right.$$
$$\left. - \cos 2t(1 - \sin^2 2t) \right] dt$$

$$= \frac{1}{8} \int \left(\frac{1}{2} - \frac{\cos 4t}{2} + \cos 2t \sin^2 2t \right) dt$$

$$= \frac{1}{8} \left(\frac{t}{2} - \frac{\sin 4t}{8} + \frac{\sin^3 2t}{6} \right) + c$$

Now try the following exercise

Exercise 177 Further problems on integration of powers of sines and cosines

Integrate the following with respect to the variable:

1. $\sin^3 \theta$ $\left[-\cos\theta + \frac{\cos^3\theta}{3} + c \right]$

2. $2\cos^3 2x$ $\left[\sin 2x - \frac{\sin^3 2x}{3} + c \right]$

3. $2\sin^3 t \cos^2 t$ $\left[\frac{-2}{3}\cos^3 t + \frac{2}{5}\cos^5 t + c \right]$

4. $\sin^3 x \cos^4 x$ $\left[\frac{-\cos^5 x}{5} + \frac{\cos^7 x}{7} + c \right]$

5. $2\sin^4 2\theta$ $\left[\frac{3\theta}{4} - \frac{1}{4}\sin 4\theta + \frac{1}{32}\sin 8\theta + c \right]$

6. $\sin^2 t \cos^2 t$ $\left[\frac{t}{8} - \frac{1}{32}\sin 4t + c \right]$

50.4 Worked problems on integration of products of sines and cosines

Problem 9. Determine: $\int \sin 3t \cos 2t \, dt$

$$\int \sin 3t \cos 2t \, dt$$

$$= \int \frac{1}{2}[\sin(3t + 2t) + \sin(3t - 2t)] \, dt,$$

from 6 of Table 50.1, which follows from Section 27.4, page 238,

$$= \frac{1}{2} \int (\sin 5t + \sin t) \, dt$$

$$= \frac{1}{2} \left(\frac{-\cos 5t}{5} - \cos t \right) + c$$

Problem 10. Find: $\int \frac{1}{3} \cos 5x \sin 2x \, dx$

$$\int \frac{1}{3} \cos 5x \sin 2x \, dx$$

$$= \frac{1}{3} \int \frac{1}{2}[\sin(5x + 2x) - \sin(5x - 2x)] \, dx,$$

from 7 of Table 50.1

$$= \frac{1}{6} \int (\sin 7x - \sin 3x) \, dx$$

$$= \frac{1}{6} \left(\frac{-\cos 7x}{7} + \frac{\cos 3x}{3} \right) + c$$

Problem 11. Evaluate: $\int_0^1 2\cos 6\theta \cos\theta\, d\theta$, correct to 4 decimal places

$$\int_0^1 2\cos 6\theta \cos\theta\, d\theta$$

$$= 2\int_0^1 \frac{1}{2}[\cos(6\theta+\theta)+\cos(6\theta-\theta)]\, d\theta,$$
$$\text{from 8 of Table 50.1}$$

$$= \int_0^1 (\cos 7\theta + \cos 5\theta)\, d\theta = \left[\frac{\sin 7\theta}{7}+\frac{\sin 5\theta}{5}\right]_0^1$$

$$= \left(\frac{\sin 7}{7}+\frac{\sin 5}{5}\right)-\left(\frac{\sin 0}{7}+\frac{\sin 0}{5}\right)$$

'sin 7' means 'the sine of 7 radians' ($\equiv 401.07°$) and $\sin 5 \equiv 286.48°$.

Hence $\int_0^1 2\cos 6\theta \cos\theta\, d\theta$

$$= (0.09386 + -0.19178) - (0)$$
$$= -0.0979, \text{ correct to 4 decimal places}$$

Problem 12. Find: $3\int \sin 5x \sin 3x\, dx$

$$3\int \sin 5x \sin 3x\, dx$$

$$= 3\int -\frac{1}{2}[\cos(5x+3x)-\cos(5x-3x)]\, dx,$$
$$\text{from 9 of Table 50.1}$$

$$= -\frac{3}{2}\int (\cos 8x - \cos 2x)\, dx$$

$$= -\frac{3}{2}\left(\frac{\sin 8x}{8}-\frac{\sin 2x}{2}\right)+c \quad\text{or}$$

$$\frac{3}{16}(4\sin 2x - \sin 8x)+c$$

Now try the following exercise

Exercise 178 Further problems on integration of products of sines and cosines

In Problems 1 to 4, integrate with respect to the variable.

1. $\sin 5t \cos 2t$ $\left[-\frac{1}{2}\left(\frac{\cos 7t}{7}+\frac{\cos 3t}{3}\right)+c\right]$

2. $2\sin 3x \sin x$ $\left[\frac{\sin 2x}{2}-\frac{\sin 4x}{4}+c\right]$

3. $3\cos 6x \cos x$ $\left[\frac{3}{2}\left(\frac{\sin 7x}{7}+\frac{\sin 5x}{5}\right)+c\right]$

4. $\frac{1}{2}\cos 4\theta \sin 2\theta$ $\left[\frac{1}{4}\left(\frac{\cos 2\theta}{2}-\frac{\cos 6\theta}{6}\right)+c\right]$

In Problems 5 to 8, evaluate the definite integrals.

5. $\int_0^{\pi/2} \cos 4x \cos 3x\, dx$ $\left[\frac{3}{7}\text{ or }0.4286\right]$

6. $\int_0^1 2\sin 7t \cos 3t\, dt$ $[0.5973]$

7. $-4\int_0^{\pi/3} \sin 5\theta \sin 2\theta\, d\theta$ $[0.2474]$

8. $\int_1^2 3\cos 8t \sin 3t\, dt$ $[-0.1999]$

50.5 Worked problems on integration using the $\sin\theta$ substitution

Problem 13. Determine: $\int \frac{1}{\sqrt{a^2-x^2}}\, dx$

Let $x = a\sin\theta$, then $\frac{dx}{d\theta} = a\cos\theta$ and $dx = a\cos\theta\, d\theta$.

Hence $\int \frac{1}{\sqrt{a^2-x^2}}\, dx$

$$= \int \frac{1}{\sqrt{a^2-a^2\sin^2\theta}}\, a\cos\theta\, d\theta$$

$$= \int \frac{a\cos\theta\, d\theta}{\sqrt{a^2(1-\sin^2\theta)}}$$

$$= \int \frac{a\cos\theta\, d\theta}{\sqrt{a^2\cos^2\theta}}, \text{ since } \sin^2\theta+\cos^2\theta=1$$

$$= \int \frac{a\cos\theta\, d\theta}{a\cos\theta} = \int d\theta = \theta + c$$

Since $x = a\sin\theta$, then $\sin\theta = \frac{x}{a}$ and $\theta = \sin^{-1}\frac{x}{a}$

Hence $\int \frac{1}{\sqrt{a^2-x^2}}\, dx = \sin^{-1}\frac{x}{a}+c$

Problem 14.   Evaluate $\displaystyle\int_0^3 \frac{1}{\sqrt{9-x^2}}\,dx$

From Problem 13, $\displaystyle\int_0^3 \frac{1}{\sqrt{9-x^2}}\,dx$

$$= \left[\sin^{-1}\frac{x}{3}\right]_0^3 \quad \text{since } a=3$$

$$= (\sin^{-1}1 - \sin^{-1}0) = \frac{\pi}{2} \text{ or } \mathbf{1.5708}$$

Problem 15.   Find: $\displaystyle\int \sqrt{a^2-x^2}\,dx$

Let $x=a\sin\theta$ then $\dfrac{dx}{d\theta}=a\cos\theta$ and $dx=a\cos\theta\,d\theta$

Hence $\displaystyle\int \sqrt{a^2-x^2}\,dx$

$$= \int \sqrt{a^2-a^2\sin^2\theta}\,(a\cos\theta\,d\theta)$$

$$= \int \sqrt{a^2(1-\sin^2\theta)}\,(a\cos\theta\,d\theta)$$

$$= \int \sqrt{a^2\cos^2\theta}\,(a\cos\theta\,d\theta)$$

$$= \int (a\cos\theta)(a\cos\theta\,d\theta)$$

$$= a^2\int \cos^2\theta\,d\theta = a^2\int\left(\frac{1+\cos2\theta}{2}\right)d\theta$$

$$(\text{since } \cos2\theta = 2\cos^2\theta - 1)$$

$$= \frac{a^2}{2}\left(\theta + \frac{\sin2\theta}{2}\right) + c$$

$$= \frac{a^2}{2}\left(\theta + \frac{2\sin\theta\cos\theta}{2}\right) + c$$

since from Chapter 27, $\sin2\theta = 2\sin\theta\cos\theta$

$$= \frac{a^2}{2}[\theta + \sin\theta\cos\theta] + c$$

Since $x=a\sin\theta$, then $\sin\theta=\dfrac{x}{a}$ and $\theta=\sin^{-1}\dfrac{x}{a}$

Also, $\cos^2\theta + \sin^2\theta = 1$, from which,

$$\cos\theta = \sqrt{1-\sin^2\theta} = \sqrt{1-\left(\frac{x}{a}\right)^2}$$

$$= \sqrt{\frac{a^2-x^2}{a^2}} = \frac{\sqrt{a^2-x^2}}{a}$$

Thus $\displaystyle\int \sqrt{a^2-x^2}\,dx = \frac{a^2}{2}[\theta + \sin\theta\cos\theta]$

$$= \frac{a^2}{2}\left[\sin^{-1}\frac{x}{a} + \left(\frac{x}{a}\right)\frac{\sqrt{a^2-x^2}}{a}\right] + c$$

$$= \frac{a^2}{2}\sin^{-1}\frac{x}{a} + \frac{x}{2}\sqrt{a^2-x^2} + c$$

Problem 16.   Evaluate: $\displaystyle\int_0^4 \sqrt{16-x^2}\,dx$

From Problem 15, $\displaystyle\int_0^4 \sqrt{16-x^2}\,dx$

$$= \left[\frac{16}{2}\sin^{-1}\frac{x}{4} + \frac{x}{2}\sqrt{16-x^2}\right]_0^4$$

$$= \left[8\sin^{-1}1 + 2\sqrt{0}\right] - \left[8\sin^{-1}0 + 0\right]$$

$$= 8\sin^{-1}1 = 8\left(\frac{\pi}{2}\right)$$

$$= \mathbf{4\pi} \text{ or } \mathbf{12.57}$$

Now try the following exercise

Exercise 179   Further problems on integration using the sine θ substitution

1.   Determine: $\displaystyle\int \frac{5}{\sqrt{4-t^2}}\,dt$

$$\left[5\sin^{-1}\frac{t}{2} + c\right]$$

2.   Determine: $\displaystyle\int \frac{3}{\sqrt{9-x^2}}\,dx$

$$\left[3\sin^{-1}\frac{x}{3} + c\right]$$

3.   Determine: $\displaystyle\int \sqrt{4-x^2}\,dx$

$$\left[2\sin^{-1}\frac{x}{2} + \frac{x}{2}\sqrt{4-x^2} + c\right]$$

4.   Determine: $\displaystyle\int \sqrt{16-9t^2}\,dt$

$$\left[\frac{8}{3}\sin^{-1}\frac{3t}{4} + \frac{t}{2}\sqrt{16-9t^2} + c\right]$$

5. Evaluate: $\int_0^4 \dfrac{1}{\sqrt{16-x^2}}dx$ $\left[\dfrac{\pi}{2}\text{ or }1.571\right]$

6. Evaluate: $\int_0^1 \sqrt{9-4x^2}\,dx$ [2.760]

50.6 Worked problems on integration using the tan θ substitution

Problem 17. Determine: $\int \dfrac{1}{(a^2+x^2)}dx$

Let $x=a\tan\theta$ then $\dfrac{dx}{d\theta}=a\sec^2\theta$ and $dx=a\sec^2\theta\,d\theta$

Hence $\int \dfrac{1}{(a^2+x^2)}dx$

$=\int \dfrac{1}{(a^2+a^2\tan^2\theta)}(a\sec^2\theta\,d\theta)$

$=\int \dfrac{a\sec^2\theta\,d\theta}{a^2(1+\tan^2\theta)}$

$=\int \dfrac{a\sec^2\theta\,d\theta}{a^2\sec^2\theta}$ since $1+\tan^2\theta=\sec^2\theta$

$=\int \dfrac{1}{a}d\theta = \dfrac{1}{a}(\theta)+c$

Since $x=a\tan\theta$, $\theta=\tan^{-1}\dfrac{x}{a}$

Hence $\int \dfrac{1}{(a^2+x^2)}dx = \dfrac{1}{a}\tan^{-1}\dfrac{x}{a}+c$

Problem 18. Evaluate: $\int_0^2 \dfrac{1}{(4+x^2)}dx$

From Problem 17, $\int_0^2 \dfrac{1}{(4+x^2)}dx$

$=\dfrac{1}{2}\left[\tan^{-1}\dfrac{x}{2}\right]_0^2$ since $a=2$

$=\dfrac{1}{2}(\tan^{-1}1-\tan^{-1}0)=\dfrac{1}{2}\left(\dfrac{\pi}{4}-0\right)$

$=\dfrac{\pi}{8}$ or **0.3927**

Problem 19. Evaluate: $\int_0^1 \dfrac{5}{(3+2x^2)}dx$, correct to 4 decimal places

$\int_0^1 \dfrac{5}{(3+2x^2)}dx = \int_0^1 \dfrac{5}{2[(3/2)+x^2]}dx$

$=\dfrac{5}{2}\int_0^1 \dfrac{1}{[\sqrt{3/2}]^2+x^2}dx$

$=\dfrac{5}{2}\left[\dfrac{1}{\sqrt{3/2}}\tan^{-1}\dfrac{x}{\sqrt{3/2}}\right]_0^1$

$=\dfrac{5}{2}\sqrt{\dfrac{2}{3}}\left[\tan^{-1}\sqrt{\dfrac{2}{3}}-\tan^{-1}0\right]$

$=(2.0412)[0.6847-0]$

$=\mathbf{1.3976}$, correct to 4 decimal places.

Now try the following exercise

Exercise 180 Further problems on integration using the tan θ substitution

1. Determine: $\int \dfrac{3}{4+t^2}dt$ $\left[\dfrac{3}{2}\tan^{-1}\dfrac{x}{2}+c\right]$

2. Determine: $\int \dfrac{5}{16+9\theta^2}d\theta$ $\left[\dfrac{5}{12}\tan^{-1}\dfrac{3\theta}{4}+c\right]$

3. Evaluate: $\int_0^1 \dfrac{3}{1+t^2}dt$ [2.356]

4. Evaluate: $\int_0^3 \dfrac{5}{4+x^2}dx$ [2.457]

This Revision test covers the material contained in Chapters 48 to 50. *The marks for each question are shown in brackets at the end of each question.*

1. Determine:

    (a) $\displaystyle\int 3\sqrt{t^5}\,dt$

    (b) $\displaystyle\int \frac{2}{\sqrt[3]{x^2}}\,dx$

    (c) $\displaystyle\int (2+\theta)^2\,d\theta$ (9)

2. Evaluate the following integrals, each correct to 4 significant figures:

    (a) $\displaystyle\int_0^{\pi/3} 3\sin 2t\,dt$

    (b) $\displaystyle\int_1^2 \left(\frac{2}{x^2} + \frac{1}{x} + \frac{3}{4}\right)dx$ (10)

3. Determine the following integrals:

    (a) $\displaystyle\int 5(6t+5)^7\,dt$

    (b) $\displaystyle\int \frac{3\ln x}{x}\,dx$

    (c) $\displaystyle\int \frac{2}{\sqrt{(2\theta-1)}}\,d\theta$ (9)

4. Evaluate the following definite integrals:

    (a) $\displaystyle\int_0^{\pi/2} 2\sin\left(2t + \frac{\pi}{3}\right)dt$

    (b) $\displaystyle\int_0^1 3xe^{4x^2-3}\,dx$ (10)

5. Determine the following integrals:

    (a) $\displaystyle\int \cos^3 x \sin^2 x\,dx$

    (b) $\displaystyle\int \frac{2}{\sqrt{9-4x^2}}\,dx$ (8)

6. Evaluate the following definite integrals, correct to 4 significant figures:

    (a) $\displaystyle\int_0^{\pi/2} 3\sin^2 t\,dt$

    (b) $\displaystyle\int_0^{\pi/3} 3\cos 5\theta \sin 3\theta\,d\theta$

    (c) $\displaystyle\int_0^2 \frac{5}{4+x^2}\,dx$ (14)

Chapter 51

Integration using partial fractions

51.1 Introduction

The process of expressing a fraction in terms of simpler fractions—called **partial fractions**—is discussed in Chapter 7, with the forms of partial fractions used being summarised in Table 7.1, page 54.

Certain functions have to be resolved into partial fractions before they an be integrated, as demonstrated in the following worked problems.

51.2 Worked problems on integration using partial fractions with linear factors

Problem 1. Determine: $\displaystyle\int \frac{11 - 3x}{x^2 + 2x - 3}\, dx$

As shown in Problem 1, page 54:

$$\frac{11 - 3x}{x^2 + 2x - 3} \equiv \frac{2}{(x - 1)} - \frac{5}{(x + 3)}$$

Hence $\displaystyle\int \frac{11 - 3x}{x^2 + 2x - 3}\, dx$

$$= \int \left\{ \frac{2}{(x - 1)} - \frac{5}{(x + 3)} \right\} dx$$

$$= \mathbf{2\, ln(x - 1) - 5\, ln(x + 3) + c}$$

(by algebraic substitutions—see chapter 49)

$$\text{or } \mathbf{ln} \left\{ \frac{(x - 1)^2}{(x + 3)^5} \right\} + c \text{ by the laws of logarithms}$$

Problem 2. Find: $\displaystyle\int \frac{2x^2 - 9x - 35}{(x + 1)(x - 2)(x + 3)}\, dx$

It was shown in Problem 2, page 55:

$$\frac{2x^2 - 9x - 35}{(x + 1)(x - 2)(x + 3)} \equiv \frac{4}{(x + 1)} - \frac{3}{(x - 2)} + \frac{1}{(x + 3)}$$

Hence $\displaystyle\int \frac{2x^2 - 9x - 35}{(x + 1)(x - 2)(x + 3)}\, dx$

$$\equiv \int \left\{ \frac{4}{(x + 1)} - \frac{3}{(x - 2)} + \frac{1}{(x + 3)} \right\} dx$$

$$= \mathbf{4\, ln(x + 1) - 3\, ln(x - 2) + ln(x + 3) + c}$$

$$\text{or } \mathbf{ln} \left\{ \frac{(x + 1)^4 (x + 3)}{(x - 2)^3} \right\} + c$$

Problem 3. Determine: $\displaystyle\int \frac{x^2 + 1}{x^2 - 3x + 2}\, dx$

By dividing out (since the numerator and denominator are of the same degree) and resolving into partial fractions it was shown in Problem 3, page 55:

$$\frac{x^2 + 1}{x^2 - 3x + 2} \equiv 1 - \frac{2}{(x - 1)} + \frac{5}{(x - 2)}$$

Hence $\displaystyle\int \frac{x^2 + 1}{x^2 - 3x + 2}\, dx$

$$\equiv \int \left\{ 1 - \frac{2}{(x - 1)} + \frac{5}{(x - 2)} \right\} dx$$

$$= x - 2\ln(x - 1) + 5\ln(x - 2) + c$$

$$\text{or } x + \ln\left\{\frac{(x - 2)^5}{(x - 1)^2}\right\} + c$$

Problem 4. Evaluate:

$\displaystyle\int_2^3 \frac{x^3 - 2x^2 - 4x - 4}{x^2 + x - 2}\,dx$, correct to 4 significant figures

By dividing out and resolving into partial fractions, it was shown in Problem 4, page 56:

$$\frac{x^3 - 2x^2 - 4x - 4}{x^2 + x - 2} \equiv x - 3 + \frac{4}{(x + 2)} - \frac{3}{(x - 1)}$$

Hence $\displaystyle\int_2^3 \frac{x^3 - 2x^2 - 4x - 4}{x^2 + x - 2}\,dx$

$$\equiv \int_2^3 \left\{x - 3 + \frac{4}{(x + 2)} - \frac{3}{(x - 1)}\right\}dx$$

$$= \left[\frac{x^2}{2} - 3x + 4\ln(x + 2) - 3\ln(x - 1)\right]_2^3$$

$$= \left(\frac{9}{2} - 9 + 4\ln 5 - 3\ln 2\right)$$

$$\qquad\qquad - (2 - 6 + 4\ln 4 - 3\ln 1)$$

$$= -1.687, \text{ correct to 4 significant figures}$$

Now try the following exercise

Exercise 181 Further problems on integration using partial fractions with linear factors

In Problems 1 to 5, integrate with respect to x

1. $\displaystyle\int \frac{12}{(x^2 - 9)}\,dx$

$$\left[\begin{array}{c} 2\ln(x - 3) - 2\ln(x + 3) + c \\ \text{or } \ln\left\{\dfrac{x - 3}{x + 3}\right\}^2 + c \end{array}\right]$$

2. $\displaystyle\int \frac{4(x - 4)}{(x^2 - 2x - 3)}\,dx$

$$\left[\begin{array}{c} 5\ln(x + 1) - \ln(x - 3) + c \\ \text{or } \ln\left\{\dfrac{(x + 1)^5}{(x - 3)}\right\} + c \end{array}\right]$$

3. $\displaystyle\int \frac{3(2x^2 - 8x - 1)}{(x + 4)(x + 1)(2x - 1)}\,dx$

$$\left[\begin{array}{c} 7\ln(x + 4) - 3\ln(x + 1) - \ln(2x - 1) + c \\ \text{or } \ln\left\{\dfrac{(x + 4)^7}{(x + 1)^3(2x - 1)}\right\} + c \end{array}\right]$$

4. $\displaystyle\int \frac{x^2 + 9x + 8}{x^2 + x - 6}\,dx$

$$\left[\begin{array}{c} x + 2\ln(x + 3) + 6\ln(x - 2) + c \\ \text{or } x + \ln\{(x + 3)^2(x - 2)^6\} + c \end{array}\right]$$

5. $\displaystyle\int \frac{3x^3 - 2x^2 - 16x + 20}{(x - 2)(x + 2)}\,dx$

$$\left[\begin{array}{c} \dfrac{3x^2}{2} - 2x + \ln(x - 2) \\ -5\ln(x + 2) + c \end{array}\right]$$

In Problems 6 and 7, evaluate the definite integrals correct to 4 significant figures.

6. $\displaystyle\int_3^4 \frac{x^2 - 3x + 6}{x(x - 2)(x - 1)}\,dx$ $\qquad$ [0.6275]

7. $\displaystyle\int_4^6 \frac{x^2 - x - 14}{x^2 - 2x - 3}\,dx$ $\qquad$ [0.8122]

51.3 Worked problems on integration using partial fractions with repeated linear factors

Problem 5. Determine: $\displaystyle\int \frac{2x + 3}{(x - 2)^2}\,dx$

It was shown in Problem 5, page 57:

$$\frac{2x + 3}{(x - 2)^2} \equiv \frac{2}{(x - 2)} + \frac{7}{(x - 2)^2}$$

Thus $\displaystyle\int \frac{2x + 3}{(x - 2)^2}\,dx$

$$\equiv \int \left\{\frac{2}{(x - 2)} + \frac{7}{(x - 2)^2}\right\}dx$$

$$= 2\ln(x - 2) - \frac{7}{(x - 2)} + c$$

$$\left[\int \frac{7}{(x-2)^2}\,dx \text{ is determined using the algebraic}\right.$$
$$\left. \text{substitution } u=(x-2), \text{ see Chapter 49}\right]$$

Problem 6. Find: $\displaystyle\int \frac{5x^2-2x-19}{(x+3)(x-1)^2}\,dx$

It was shown in Problem 6, page 57:

$$\frac{5x^2-2x-19}{(x+3)(x-1)^2} \equiv \frac{2}{(x+3)} + \frac{3}{(x-1)} - \frac{4}{(x-1)^2}$$

Hence $\displaystyle\int \frac{5x^2-2x-19}{(x+3)(x-1)^2}\,dx$

$$\equiv \int \left\{\frac{2}{(x+3)} + \frac{3}{(x-1)} - \frac{4}{(x-1)^2}\right\} dx$$

$$= 2\ln(x+3) + 3\ln(x-1) + \frac{4}{(x-1)} + c$$

or $\quad \ln(x+3)^2\,(x-1)^3 + \dfrac{4}{(x-1)} + c$

Problem 7. Evaluate:

$\displaystyle\int_{-2}^{1} \frac{3x^2+16x+15}{(x+3)^3}\,dx$, correct to

4 significant figures

It was shown in Problem 7, page 58:

$$\frac{3x^2+16x+15}{(x+3)^3} \equiv \frac{3}{(x+3)} - \frac{2}{(x+3)^2} - \frac{6}{(x+3)^3}$$

Hence $\displaystyle\int \frac{3x^2+16x+15}{(x+3)^3}\,dx$

$$\equiv \int_{-2}^{1} \left\{\frac{3}{(x+3)} - \frac{2}{(x+3)^2} - \frac{6}{(x+3)^3}\right\} dx$$

$$= \left[3\ln(x+3) + \frac{2}{(x+3)} + \frac{3}{(x+3)^2}\right]_{-2}^{1}$$

$$= \left(3\ln 4 + \frac{2}{4} + \frac{3}{16}\right) - \left(3\ln 1 + \frac{2}{1} + \frac{3}{1}\right)$$

$$= -\mathbf{0.1536}, \text{ correct to 4 significant figures.}$$

Now try the following exercise

Exercise 182 Further problems on integration using partial fractions with repeated linear factors

In Problems 1 and 2, integrate with respect to x.

1. $\displaystyle\int \frac{4x-3}{(x+1)^2}\,dx \qquad \left[4\ln(x+1) + \frac{7}{(x+1)} + c\right]$

2. $\displaystyle\int \frac{5x^2-30x+44}{(x-2)^3}\,dx$

$$\left[5\ln(x-2) + \frac{10}{(x-2)} - \frac{2}{(x-2)^2} + c\right]$$

In Problems 3 and 4, evaluate the definite integrals correct to 4 significant figures.

3. $\displaystyle\int_{1}^{2} \frac{x^2+7x+3}{x^2(x+3)}\,dx \qquad\qquad [1.663]$

4. $\displaystyle\int_{6}^{7} \frac{18+21x-x^2}{(x-5)(x+2)^2}\,dx \qquad\qquad [1.089]$

51.4 Worked problems on integration using partial fractions with quadratic factors

Problem 8. Find: $\displaystyle\int \frac{3+6x+4x^2-2x^3}{x^2(x^2+3)}\,dx$

It was shown in Problem 9, page 59:

$$\frac{3+6x+4x^2-2x^2}{x^2(x^2+3)} \equiv \frac{2}{x} + \frac{1}{x^2} + \frac{3-4x}{(x^2+3)}$$

Thus $\displaystyle\int \frac{3+6x+4x^2-2x^3}{x^2(x^2+3)}\,dx$

$$\equiv \int \left(\frac{2}{x} + \frac{1}{x^2} + \frac{3-4x}{(x^2+3)}\right) dx$$

$$= \int \left\{\frac{2}{x} + \frac{1}{x^2} + \frac{3}{(x^2+3)} - \frac{4x}{(x^2+3)}\right\} dx$$

$$\int \frac{3}{(x^2+3)}\,dx = 3\int \frac{1}{x^2+(\sqrt{3})^2}\,dx$$

$$= \frac{3}{\sqrt{3}}\tan^{-1}\frac{x}{\sqrt{3}}$$

from 12, Table 50.1, page 448.

$\int \dfrac{4x}{x^2+3}\,dx$ is determined using the algebraic substitutions $u=(x^2+3)$.

Hence $\displaystyle\int \left\{\frac{2}{x}+\frac{1}{x^2}+\frac{3}{(x^2+3)}-\frac{4x}{(x^2+3)}\right\}dx$

$$= 2\ln x - \frac{1}{x} + \frac{3}{\sqrt{3}}\tan^{-1}\frac{x}{\sqrt{3}} - 2\ln(x^2+3) + c$$

$$= \ln\left(\frac{x}{x^2+3}\right)^2 - \frac{1}{x} + \sqrt{3}\tan^{-1}\frac{x}{\sqrt{3}} + c$$

Problem 9.   Determine: $\displaystyle\int \frac{1}{(x^2-a^2)}\,dx$

Let $\dfrac{1}{(x^2-a^2)} \equiv \dfrac{A}{(x-a)} + \dfrac{B}{(x+a)}$

$$\equiv \frac{A(x+a)+B(x-a)}{(x+a)(x-a)}$$

Equating the numerators gives:

$$1 \equiv A(x+a)+B(x-a)$$

Let $x=a$, then $A=\dfrac{1}{2a}$

and let $x=-a$,

then $B=-\dfrac{1}{2a}$

Hence $\displaystyle\int \frac{1}{(x^2-a^2)}\,dx \equiv \int \frac{1}{2a}\left[\frac{1}{(x-a)}-\frac{1}{(x+a)}\right]dx$

$$= \frac{1}{2a}[\ln(x-a)-\ln(x+a)] + c$$

$$= \frac{1}{2a}\ln\left(\frac{x-a}{x+a}\right) + c$$

Problem 10.   Evaluate: $\displaystyle\int_3^4 \frac{3}{(x^2-4)}\,dx$, correct to 3 significant figures

From Problem 9,

$$\int_3^4 \frac{3}{(x^2-4)}\,dx = 3\left[\frac{1}{2(2)}\ln\left(\frac{x-2}{x+2}\right)\right]_3^4$$

$$= \frac{3}{4}\left[\ln\frac{2}{6}-\ln\frac{1}{5}\right]$$

$$= \frac{3}{4}\ln\frac{5}{3} = \textbf{0.383}, \text{ correct to 3}$$

significant figures.

Problem 11.   Determine: $\displaystyle\int \frac{1}{(a^2-x^2)}\,dx$

Using partial fractions, let

$$\frac{1}{(a^2-x^2)} \equiv \frac{1}{(a-x)(a+x)} \equiv \frac{A}{(a-x)} + \frac{B}{(a+x)}$$

$$\equiv \frac{A(a+x)+B(a-x)}{(a-x)(a+x)}$$

Then $1 \equiv A(a+x)+B(a-x)$

Let $x=a$ then $A=\dfrac{1}{2a}$. Let $x=-a$ then $B=\dfrac{1}{2a}$

Hence $\displaystyle\int \frac{1}{(a^2-x^2)}\,dx$

$$= \int \frac{1}{2a}\left[\frac{1}{(a-x)}+\frac{1}{(a+x)}\right]dx$$

$$= \frac{1}{2a}[-\ln(a-x)+\ln(a+x)] + c$$

$$= \frac{1}{2a}\ln\left(\frac{a+x}{a-x}\right) + c$$

Problem 12.   Evaluate: $\displaystyle\int_0^2 \frac{5}{(9-x^2)}\,dx$, correct to 4 decimal places

From Problem 11,

$$\int_0^2 \frac{5}{(9-x^2)}\,dx = 5\left[\frac{1}{2(3)}\ln\left(\frac{3+x}{3-x}\right)\right]_0^2$$

$$= \frac{5}{6}\left[\ln\frac{5}{1}-\ln 1\right] = \textbf{1.3412},$$

correct to 4 decimal places

Now try the following exercise

In Problems 2 to 4, evaluate the definite integrals correct to 4 significant figures.

Exercise 183 Further problems on integration using partial fractions with quadratic factors

1. Determine $\displaystyle\int \frac{x^2 - x - 13}{(x^2 + 7)(x - 2)}\, dx$

$$\left[\begin{array}{l} \ln(x^2 + 7) + \dfrac{3}{\sqrt{7}} \tan^{-1} \dfrac{x}{\sqrt{7}} \\ \quad - \ln(x - 2) + c \end{array} \right]$$

2. $\displaystyle\int_5^6 \frac{6x - 5}{(x - 4)(x^2 + 3)}\, dx$ [0.5880]

3. $\displaystyle\int_1^2 \frac{4}{(16 - x^2)}\, dx$ [0.2939]

4. $\displaystyle\int_4^5 \frac{2}{(x^2 - 9)}\, dx$ [0.1865]

Section 9

Chapter 52

The $t = \tan \dfrac{\theta}{2}$ substitution

52.1 Introduction

Integrals of the form $\displaystyle\int \dfrac{1}{a\cos\theta + b\sin\theta + c}\, d\theta$, where a, b and c are constants, may be determined by using the substitution $t = \tan\dfrac{\theta}{2}$. The reason is explained below.

If angle A in the right-angled triangle ABC shown in Fig. 52.1 is made equal to $\dfrac{\theta}{2}$ then, since

$$\text{tangent} = \dfrac{\text{opposite}}{\text{adjacent}}, \quad \text{if} \quad BC = t \quad \text{and} \quad AB = 1, \quad \text{then}$$

$\tan\dfrac{\theta}{2} = t$.

By Pythagoras' theorem, $AC = \sqrt{1 + t^2}$

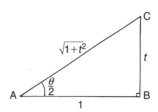

Figure 52.1

Therefore $\sin\dfrac{\theta}{2} = \dfrac{t}{\sqrt{1 + t^2}}$ and $\cos\dfrac{\theta}{2} = \dfrac{1}{\sqrt{1 + t^2}}$

Since $\sin 2x = 2\sin x \cos x$ (from double angle formulae, Chapter 27), then

$$\sin\theta = 2\sin\dfrac{\theta}{2}\cos\dfrac{\theta}{2}$$

$$= 2\left(\dfrac{t}{\sqrt{1 + t^2}}\right)\left(\dfrac{1}{\sqrt{1 + t^2}}\right)$$

i.e. $$\sin\theta = \dfrac{2t}{(1 + t^2)} \qquad (1)$$

Since $\cos 2x = \cos^2\dfrac{\theta}{2} - \sin^2\dfrac{\theta}{2}$

$$= \left(\dfrac{1}{\sqrt{1 + t^2}}\right)^2 - \left(\dfrac{1}{\sqrt{1 + t^2}}\right)^2$$

i.e. $$\cos\theta = \dfrac{1 - t^2}{1 + t^2} \qquad (2)$$

Also, since $t = \tan\dfrac{\theta}{2}$

$\dfrac{dt}{d\theta} = \dfrac{1}{2}\sec^2\dfrac{\theta}{2} = \dfrac{1}{2}\left(1 + \tan^2\dfrac{\theta}{2}\right)$ from trigonometric identities,

i.e. $$\dfrac{dt}{d\theta} = \dfrac{1}{2}(1 + t^2)$$

from which, $$d\theta = \dfrac{2\,dt}{1 + t^2} \qquad (3)$$

Equations (1), (2) and (3) are used to determine integrals of the form $\displaystyle\int \dfrac{1}{a\cos\theta + b\sin\theta + c}\, d\theta$ where a, b or c may be zero.

52.2 Worked problems on the $t = \tan\dfrac{\theta}{2}$ substitution

Problem 1. Determine: $\displaystyle\int \dfrac{d\theta}{\sin\theta}$

If $t = \tan\dfrac{\theta}{2}$ then $\sin\theta = \dfrac{2t}{1 + t^2}$ and $d\theta = \dfrac{2\,dt}{1 + t^2}$ from equations (1) and (3).

Thus $$\int \dfrac{d\theta}{\sin\theta} = \int \dfrac{1}{\sin\theta}\, d\theta$$

$$= \int \frac{1}{\dfrac{2t}{1+t^2}} \left(\frac{2\,dt}{1+t^2} \right)$$

$$= \int \frac{1}{t}\,dt = \ln t + c$$

Hence $\displaystyle \int \frac{d\theta}{\sin \theta} = \ln \left(\tan \frac{\theta}{2} \right) + c$

Problem 2. Determine: $\displaystyle \int \frac{dx}{\cos x}$

If $t = \tan \dfrac{x}{2}$ then $\cos x = \dfrac{1-t^2}{1+t^2}$ and $dx = \dfrac{2\,dt}{1+t^2}$ from equations (2) and (3).

Thus $\displaystyle \int \frac{dx}{\cos x} = \int \frac{1}{\dfrac{1-t^2}{1+t^2}} \left(\frac{2\,dt}{1+t^2} \right)$

$$= \int \frac{2}{1-t^2}\,dt$$

$\dfrac{2}{1-t^2}$ may be resolved into partial fractions (see Chapter 7).

Let $\displaystyle \frac{2}{1-t^2} = \frac{2}{(1-t)(1+t)}$

$$= \frac{A}{(1-t)} + \frac{B}{(1+t)}$$

$$= \frac{A(1+t) + B(1-t)}{(1-t)(1+t)}$$

Hence $\quad 2 = A(1+t) + B(1-t)$

When $t = 1$, $2 = 2A$, from which, $A = 1$

When $t = -1$, $2 = 2B$, from which, $B = 1$

Hence $\displaystyle \int \frac{2dt}{1-t^2} = \int \frac{1}{(1-t)} + \frac{1}{(1+t)}\,dt$

$$= -\ln(1-t) + \ln(1+t) + c$$

$$= \ln \left\{ \frac{(1+t)}{(1-t)} \right\} + c$$

Thus $\displaystyle \int \frac{dx}{\cos x} = \ln \left\{ \frac{1 + \tan \dfrac{x}{2}}{1 - \tan \dfrac{x}{2}} \right\} + c$

Note that since $\tan \dfrac{\pi}{4} = 1$, the above result may be written as:

$$\int \frac{dx}{\cos x} = \ln \left\{ \frac{\tan \dfrac{\pi}{4} + \tan \dfrac{\pi}{2}}{1 - \tan \dfrac{\pi}{4} \tan \dfrac{x}{2}} \right\} + c$$

$$= \ln \left\{ \tan \left(\frac{\pi}{4} + \frac{x}{2} \right) \right\} + c$$

from compound angles, Chapter 27

Problem 3. Determine: $\displaystyle \int \frac{dx}{1 + \cos x}$

If $t = \tan \dfrac{x}{2}$ then $\cos x = \dfrac{1-t^2}{1+t^2}$ and $dx = \dfrac{2dt}{1+t^2}$ from equations (2) and (3).

Thus $\displaystyle \int \frac{dx}{1 + \cos x} = \int \frac{1}{1 + \cos x}\,dx$

$$= \int \frac{1}{1 + \dfrac{1-t^2}{1+t^2}} \left(\frac{2\,dt}{1+t^2} \right)$$

$$= \int \frac{1}{\dfrac{(1+t^2) + (1-t^2)}{1+t^2}} \left(\frac{2\,dt}{1+t^2} \right)$$

$$= \int dt$$

Hence $\displaystyle \int \frac{dx}{1 + \cos x} = t + c = \tan \frac{x}{2} + c$

Problem 4. Determine: $\displaystyle \int \frac{d\theta}{5 + 4\cos \theta}$

If $t = \tan \dfrac{\theta}{2}$ then $\cos \theta = \dfrac{1-t^2}{1+t^2}$ and $dx = \dfrac{2\,dt}{1+t^2}$ from equations (2) and (3).

Thus $\displaystyle \int \frac{d\theta}{5 + 4\cos \theta} = \int \frac{\left(\dfrac{2\,dt}{1+t^2} \right)}{5 + 4 \left(\dfrac{1-t^2}{1+t^2} \right)}$

$$= \int \frac{\left(\dfrac{2\,dt}{1+t^2} \right)}{\dfrac{5(1+t^2) + 4(1-t^2)}{1+t^2}}$$

$$= 2 \int \frac{dt}{t^2 + 9} = 2 \int \frac{dt}{t^2 + 3^2}$$

$$= 2 \left(\frac{1}{3} \tan^{-1} \frac{t}{3} \right) + c,$$

from 12 of Table 50.1, page 448. Hence

$$\int \frac{d\theta}{5+4\cos\theta} = \frac{2}{3}\tan^{-1}\left(\frac{1}{3}\tan\frac{\theta}{2}\right) + c$$

Now try the following exercise

Exercise 184   Further problems on the $t = \tan\dfrac{\theta}{2}$ substitution

Integrate the following with respect to the variable:

1. $\displaystyle\int \frac{d\theta}{1+\sin\theta}$ $\left[\dfrac{-2}{1+\tan\dfrac{\theta}{2}} + c\right]$

2. $\displaystyle\int \frac{dx}{1-\cos x + \sin x}$ $\left[\ln\left\{\dfrac{\tan\dfrac{x}{2}}{1+\tan\dfrac{x}{2}}\right\} + c\right]$

3. $\displaystyle\int \frac{d\alpha}{3+2\cos\alpha}$ $\left[\dfrac{2}{\sqrt{5}}\tan^{-1}\left(\dfrac{1}{\sqrt{5}}\tan\dfrac{\alpha}{2}\right) + c\right]$

4. $\displaystyle\int \frac{dx}{3\sin x - 4\cos x}$ $\left[\dfrac{1}{5}\ln\left\{\dfrac{2\tan\dfrac{x}{2}-1}{\tan\dfrac{x}{2}+2}\right\} + c\right]$

52.3   Further worked problems on the $t = \tan\dfrac{\theta}{2}$ substitution

Problem 5.   Determine: $\displaystyle\int \frac{dx}{\sin x + \cos x}$

If $t = \tan\dfrac{x}{2}$ then $\sin x = \dfrac{2t}{1+t^2}$, $\cos x = \dfrac{1-t^2}{1+t^2}$ and $dx = \dfrac{2\,dt}{1+t^2}$ from equations (1), (2) and (3).

Thus

$$\int \frac{dx}{\sin x + \cos x} = \int \frac{\dfrac{2\,dt}{1+t^2}}{\left(\dfrac{2t}{1+t^2}\right) + \left(\dfrac{1-t^2}{1+t^2}\right)}$$

$$= \int \frac{\dfrac{2\,dt}{1+t^2}}{\dfrac{2t+1-t^2}{1+t^2}} = \int \frac{2\,dt}{1+2t-t^2}$$

$$= \int \frac{-2\,dt}{t^2-2t-1} = \int \frac{-2\,dt}{(t-1)^2 - 2}$$

$$= \int \frac{2\,dt}{(\sqrt{2})^2 - (t-1)^2}$$

$$= 2\left[\frac{1}{2\sqrt{2}}\ln\left\{\frac{\sqrt{2}+(t-1)}{\sqrt{2}-(t-1)}\right\}\right] + c$$

(see problem 11, Chapter 51, page 458),

i.e. $\displaystyle\int \frac{dx}{\sin x + \cos x}$

$$= \frac{1}{\sqrt{2}}\ln\left\{\frac{\sqrt{2}-1+\tan\dfrac{x}{2}}{\sqrt{2}+1-\tan\dfrac{x}{2}}\right\} + c$$

Problem 6.   Determine: $\displaystyle\int \frac{dx}{7-3\sin x + 6\cos x}$

From equations (1) and (3),

$$\int \frac{dx}{7-3\sin x + 6\cos x}$$

$$= \int \frac{\dfrac{2\,dt}{1+t^2}}{7-3\left(\dfrac{2t}{1+t^2}\right) + 6\left(\dfrac{1-t^2}{1+t^2}\right)}$$

$$= \int \frac{\dfrac{2\,dt}{1+t^2}}{\dfrac{7(1+t^2)-3(2t)+6(1-t^2)}{1+t^2}}$$

$$= \int \frac{2\,dt}{7+7t^2-6t+6-6t^2}$$

$$= \int \frac{2\,dt}{t^2-6t+13} = \int \frac{2\,dt}{(t-3)^2 + 2^2}$$

$$= 2\left[\frac{1}{2}\tan^{-1}\left(\frac{t-3}{2}\right)\right] + c$$

from 12, Table 50.1, page 448. Hence

$$\int \frac{dx}{7 - 3\sin x + 6\cos x}$$

$$= \tan^{-1}\left(\frac{\tan\frac{x}{2} - 3}{2}\right) + c$$

Problem 7. Determine: $\int \frac{d\theta}{4\cos\theta + 3\sin\theta}$

From equations (1) to (3),

$$\int \frac{d\theta}{4\cos\theta + 3\sin\theta}$$

$$= \int \frac{\frac{2\,dt}{1+t^2}}{4\left(\frac{1-t^2}{1+t^2}\right) + 3\left(\frac{2t}{1+t^2}\right)}$$

$$= \int \frac{2\,dt}{4 - 4t^2 + 6t} = \int \frac{dt}{2 + 3t - 2t^2}$$

$$= -\frac{1}{2}\int \frac{dt}{t^2 - \frac{3}{2}t - 1}$$

$$= -\frac{1}{2}\int \frac{dt}{\left(t - \frac{3}{4}\right)^2 - \frac{25}{16}}$$

$$= \frac{1}{2}\int \frac{dt}{\left(\frac{5}{4}\right)^2 - \left(t - \frac{3}{4}\right)^2}$$

$$= \frac{1}{2}\left[\frac{1}{2\left(\frac{5}{4}\right)}\ln\left\{\frac{\frac{5}{4} + \left(t - \frac{3}{4}\right)}{\frac{5}{4} - \left(t - \frac{3}{4}\right)}\right\}\right] + c$$

from problem 11, Chapter 51, page 458,

$$= \frac{1}{5}\ln\left\{\frac{\frac{1}{2} + t}{2 - t}\right\} + c$$

Hence $\int \frac{d\theta}{4\cos\theta + 3\sin\theta}$

$$= \frac{1}{5}\ln\left\{\frac{\frac{1}{2} + \tan\frac{\theta}{2}}{2 - \tan\frac{\theta}{2}}\right\} + c$$

or $\quad \frac{1}{5}\ln\left\{\frac{1 + 2\tan\frac{\theta}{2}}{4 - 2\tan\frac{\theta}{2}}\right\} + c$

Now try the following exercise

Exercise 185   Further problems on the $t = \tan\frac{\theta}{2}$ substitution

In Problems 1 to 4, integrate with respect to the variable.

1. $\int \frac{d\theta}{5 + 4\sin\theta}$

$$\left[\frac{2}{3}\tan^{-1}\left(\frac{5\tan\frac{\theta}{2} + 4}{3}\right) + c\right]$$

2. $\int \frac{dx}{1 + 2\sin x}$

$$\left[\frac{1}{\sqrt{3}}\ln\left\{\frac{\tan\frac{x}{2} + 2 - \sqrt{3}}{\tan\frac{x}{2} + 2 + \sqrt{3}}\right\} + c\right]$$

3. $\int \frac{dp}{3 - 4\sin p + 2\cos p}$

$$\left[\frac{1}{\sqrt{11}}\ln\left\{\frac{\tan\frac{p}{2} - 4 - \sqrt{11}}{\tan\frac{p}{2} - 4 + \sqrt{11}}\right\} + c\right]$$

4. $\int \frac{d\theta}{3 - 4\sin\theta}$

$$\left[\frac{1}{\sqrt{7}}\ln\left\{\frac{3\tan\frac{\theta}{2} - 4 - \sqrt{7}}{3\tan\frac{\theta}{2} - 4 + \sqrt{7}}\right\} + c\right]$$

5. Show that

$$\int \frac{dt}{1 + 3\cos t} = \frac{1}{2\sqrt{2}}\ln\left\{\frac{\sqrt{2} + \tan\frac{t}{2}}{\sqrt{2} - \tan\frac{t}{2}}\right\} + c$$

6. Show that $\int_0^{\pi/3} \frac{3\,d\theta}{\cos\theta} = 3.95$, correct to 3 significant figures.

7. Show that $\int_0^{\pi/2} \frac{d\theta}{2 + \cos\theta} = \frac{\pi}{3\sqrt{3}}$

Chapter 53

Integration by parts

53.1 Introduction

From the product rule of differentiation:

$$\frac{d}{dx}(uv) = v\frac{du}{dx} + u\frac{dv}{dx}$$

where u and v are both functions of x.

Rearranging gives: $u\frac{dv}{dx} = \frac{d}{dx}(uv) - v\frac{du}{dx}$

Integrating both sides with respect to x gives:

$$\int u\frac{dv}{dx}dx = \int \frac{d}{dx}(uv)dx - \int v\frac{du}{dx}dx$$

ie $\qquad \int u\frac{dv}{dx}dx = uv - \int v\frac{du}{dx}dx$

or $\qquad \int u\, dv = uv - \int v\, du$

This is known as the **integration by parts formula** and provides a method of integrating such products of simple functions as $\int xe^x dx$, $\int t\sin t\, dt$, $\int e^\theta \cos\theta\, d\theta$ and $\int x\ln x\, dx$.

Given a product of two terms to integrate the initial choice is: 'which part to make equal to u' and 'which part to make equal to dv'. The choice must be such that the 'u part' becomes a constant after successive differentiation and the 'dv part' can be integrated from standard integrals. Invariable, the following rule holds: 'If a product to be integrated contains an algebraic term (such as x, t^2 or 3θ) then this term is chosen as the u part. The one exception to this rule is when a '$\ln x$' term is involved; in this case $\ln x$ is chosen as the 'u part'.

53.2 Worked problems on integration by parts

Problem 1. Determine: $\displaystyle\int x\cos x\, dx$

From the integration by parts formula,

$$\int u\, dv = uv - \int v\, du$$

Let $u = x$, from which $\dfrac{du}{dx} = 1$, i.e. $du = dx$ and let $dv = \cos x\, dx$, from which $v = \int \cos x\, dx = \sin x$.

Expressions for u, du and v are now substituted into the 'by parts' formula as shown below.

$$\int \boxed{u}\ \boxed{dv} = \boxed{u}\ \boxed{v} - \int \boxed{v}\ \boxed{du}$$
$$\int \boxed{x}\ \boxed{\cos x\, dx} = \boxed{(x)}\ \boxed{(\sin x)} - \int \boxed{(\sin x)}\ \boxed{(dx)}$$

i.e. $\displaystyle\int x\cos x\, dx = x\sin x - (-\cos x) + c$

$$= x\sin x + \cos x + c$$

[This result may be checked by differentiating the right hand side,

i.e. $\dfrac{d}{dx}(x\sin x + \cos x + c)$

$$= [(x)(\cos x) + (\sin x)(1)] - \sin x + 0$$

$$\text{using the product rule}$$

$$= x\cos x, \text{ which is the function being integrated}$$

Problem 2. Find: $\int 3te^{2t}\,dt$

Let $u = 3t$, from which, $\dfrac{du}{dt} = 3$, i.e. $du = 3\,dt$ and

let $dv = e^{2t}\,dt$, from which, $v = \int e^{2t}\,dt = \dfrac{1}{2}e^{2t}$

Substituting into $\int u\,dv = uv - \int v\,du$ gives:

$$\int 3te^{2t}\,dt = (3t)\left(\frac{1}{2}e^{2t}\right) - \int\left(\frac{1}{2}e^{2t}\right)(3\,dt)$$

$$= \frac{3}{2}te^{2t} - \frac{3}{2}\int e^{2t}\,dt$$

$$= \frac{3}{2}te^{2t} - \frac{3}{2}\left(\frac{e^{2t}}{2}\right) + c$$

Hence $\int 3te^{2t}\,dt = \dfrac{3}{2}e^{2t}\left(t - \dfrac{1}{2}\right) + c$,

which may be checked by differentiating.

Problem 3. Evaluate $\int_0^{\frac{\pi}{2}} 2\theta\sin\theta\,d\theta$

Let $u = 2\theta$, from which, $\dfrac{du}{d\theta} = 2$, i.e. $du = 2\,d\theta$ and let

$dv = \sin\theta\,d\theta$, from which,

$$v = \int \sin\theta\,d\theta = -\cos\theta$$

Substituting into $\int u\,dv = uv - \int v\,du$ gives:

$$\int 2\theta\sin\theta\,d\theta = (2\theta)(-\cos\theta) - \int(-\cos\theta)(2\,d\theta)$$

$$= -2\theta\cos\theta + 2\int\cos\theta\,d\theta$$

$$= -2\theta\cos\theta + 2\sin\theta + c$$

Hence $\int_0^{\frac{\pi}{2}} 2\theta\sin\theta\,d\theta$

$$= \left[-2\theta\cos\theta + 2\sin\theta\right]_0^{\frac{\pi}{2}}$$

$$= \left[-2\left(\frac{\pi}{2}\right)\cos\frac{\pi}{2} + 2\sin\frac{\pi}{2}\right] - [0 + 2\sin 0]$$

$$= (-0 + 2) - (0 + 0) = 2$$

$$\text{since } \cos\frac{\pi}{2} = 0 \quad\text{and}\quad \sin\frac{\pi}{2} = 1$$

Problem 4. Evaluate: $\int_0^1 5xe^{4x}\,dx$, correct to 3 significant figures

Let $u = 5x$, from which $\dfrac{du}{dx} = 5$, i.e. $du = 5\,dx$ and let

$dv = e^{4x}\,dx$, from which, $v = \int e^{4x}\,dx = \dfrac{1}{4}e^{4x}$

Substituting into $\int u\,dv = uv - \int v\,du$ gives:

$$\int 5xe^{4x}\,dx = (5x)\left(\frac{e^{4x}}{4}\right) - \int\left(\frac{e^{4x}}{4}\right)(5\,dx)$$

$$= \frac{5}{4}xe^{4x} - \frac{5}{4}\int e^{4x}\,dx$$

$$= \frac{5}{4}xe^{4x} - \frac{5}{4}\left(\frac{e^{4x}}{4}\right) + c$$

$$= \frac{5}{4}e^{4x}\left(x - \frac{1}{4}\right) + c$$

Hence $\int_0^1 5xe^{4x}\,dx$

$$= \left[\frac{5}{4}e^{4x}\left(x - \frac{1}{4}\right)\right]_0^1$$

$$= \left[\frac{5}{4}e^4\left(1 - \frac{1}{4}\right)\right] - \left[\frac{5}{4}e^0\left(0 - \frac{1}{4}\right)\right]$$

$$= \left(\frac{15}{16}e^4\right) - \left(-\frac{5}{16}\right)$$

$$= 51.186 + 0.313 = 51.499 = \mathbf{51.5},$$

correct to 3 significant figures.

Problem 5. Determine: $\int x^2 \sin x\,dx$

Let $u = x^2$, from which, $\dfrac{du}{dx} = 2x$, i.e. $du = 2x\,dx$, and

let $dv = \sin x\,dx$, from which, $v = \int \sin x\,dx = -\cos x$

Substituting into $\int u\,dv = uv - \int v\,du$ gives:

$$\int x^2 \sin x\,dx = (x^2)(-\cos x) - \int(-\cos x)(2x\,dx)$$

$$= -x^2\cos x + 2\left[\int x\cos x\,dx\right]$$

The integral, $\int x\cos x\,dx$, is not a 'standard integral' and it can only be determined by using the integration by parts formula again.

From Problem 1, $\int x \cos x \, dx = x \sin x + \cos x$

Hence $\int x^2 \sin x \, dx$

$$= -x^2 \cos x + 2\{x \sin x + \cos x\} + c$$

$$= -x^2 \cos x + 2x \sin x + 2 \cos x + c$$

$$= (2 - x^2)\cos x + 2x \sin x + c$$

In general, if the algebraic term of a product is of power n, then the integration by parts formula is applied n times.

Now try the following exercise

Exercise 186   Further problems on integration by parts

Determine the integrals in Problems 1 to 5 using integration by parts.

1. $\int x e^{2x} \, dx$ $\left[\dfrac{e^{2x}}{2}\left(x - \dfrac{1}{2}\right) + c\right]$

2. $\int \dfrac{4x}{e^{3x}} \, dx$ $\left[-\dfrac{4}{3}e^{-3x}\left(x + \dfrac{1}{3}\right) + c\right]$

3. $\int x \sin x \, dx$ $[-x \cos x + \sin x + c]$

4. $\int 5\theta \cos 2\theta \, d\theta$

$\left[\dfrac{5}{2}\left(\theta \sin 2\theta + \dfrac{1}{2}\cos 2\theta\right) + c\right]$

5. $\int 3t^2 e^{2t} \, dt$ $\left[\dfrac{3}{2}e^{2t}\left(t^2 - t + \dfrac{1}{2}\right) + c\right]$

Evaluate the integrals in Problems 6 to 9, correct to 4 significant figures.

6. $\int_0^2 2x e^x \, dx$ [16.78]

7. $\int_0^{\frac{\pi}{4}} x \sin 2x \, dx$ [0.2500]

8. $\int_0^{\frac{\pi}{2}} t^2 \cos t \, dt$ [0.4674]

9. $\int_1^2 3x^2 e^{\frac{x}{2}} \, dx$ [15.78]

53.3   Further worked problems on integration by parts

Problem 6. Find: $\int x \ln x \, dx$

The logarithmic function is chosen as the 'u part' Thus when $u = \ln x$, then $\dfrac{du}{dx} = \dfrac{1}{x}$ i.e. $du = \dfrac{dx}{x}$

Letting $dv = x \, dx$ gives $v = \int x \, dx = \dfrac{x^2}{2}$

Substituting into $\int u \, dv = uv - \int v \, du$ gives:

$$\int x \ln x \, dx = (\ln x)\left(\frac{x^2}{2}\right) - \int \left(\frac{x^2}{2}\right)\frac{dx}{x}$$

$$= \frac{x^2}{2}\ln x - \frac{1}{2}\int x \, dx$$

$$= \frac{x^2}{2}\ln x - \frac{1}{2}\left(\frac{x^2}{2}\right) + c$$

Hence $\int x \ln x \, dx = \dfrac{x^2}{2}\left(\ln x - \dfrac{1}{2}\right) + c$

or $\dfrac{x^2}{4}(2\ln x - 1) + c$

Problem 7. Determine: $\int \ln x \, dx$

$\int \ln x \, dx$ is the same as $\int (1) \ln x \, dx$

Let $u = \ln x$, from which, $\dfrac{du}{dx} = \dfrac{1}{x}$ i.e. $du = \dfrac{dx}{x}$ and let $dv = 1 \, dx$, from which, $v = \int 1 \, dx = x$

Substituting into $\int u \, dv = uv - \int v \, du$ gives:

$$\int \ln x \, dx = (\ln x)(x) - \int x \frac{dx}{x}$$

$$= x \ln x - \int dx = x \ln x - x + c$$

Hence $\int \ln x \, dx = x(\ln x - 1) + c$

Problem 8. Evaluate: $\int_1^9 \sqrt{x} \ln x \, dx$, correct to 3 significant figures

Let $u = \ln x$, from which $du = \dfrac{dx}{x}$

and let $dv = \sqrt{x}\,dx = x^{\frac{1}{2}}\,dx$, from which,

$$v = \int x^{\frac{1}{2}}\,dx = \frac{2}{3}x^{\frac{3}{2}}$$

Substituting into $\int u\,dv = uv - \int v\,du$ gives:

$$\int \sqrt{x}\ln x\,dx = (\ln x)\left(\frac{2}{3}x^{\frac{3}{2}}\right) - \int \left(\frac{2}{3}x^{\frac{3}{2}}\right)\left(\frac{dx}{x}\right)$$

$$= \frac{2}{3}\sqrt{x^3}\ln x - \frac{2}{3}\int x^{\frac{1}{2}}\,dx$$

$$= \frac{2}{3}\sqrt{x^3}\ln x - \frac{2}{3}\left(\frac{2}{3}x^{\frac{3}{2}}\right) + c$$

$$= \frac{2}{3}\sqrt{x^3}\left[\ln x - \frac{2}{3}\right] + c$$

Hence $\displaystyle\int_1^9 \sqrt{x}\ln x\,dx = \left[\frac{2}{3}\sqrt{x^3}\left(\ln x - \frac{2}{3}\right)\right]_1^9$

$$= \left[\frac{2}{3}\sqrt{9^3}\left(\ln 9 - \frac{2}{3}\right)\right] - \left[\frac{2}{3}\sqrt{1^3}\left(\ln 1 - \frac{2}{3}\right)\right]$$

$$= \left[18\left(\ln 9 - \frac{2}{3}\right)\right] - \left[\frac{2}{3}\left(0 - \frac{2}{3}\right)\right]$$

$$= 27.550 + 0.444 = 27.994 = \mathbf{28.0},$$

correct to 3 significant figures.

Problem 9. Find: $\int e^{ax}\cos bx\,dx$

When integrating a product of an exponential and a sine or cosine function it is immaterial which part is made equal to 'u'.

Let $u = e^{ax}$, from which $\dfrac{du}{dx} = ae^{ax}$, i.e. $du = ae^{ax}\,dx$

and let $dv = \cos bx\,dx$, from which,

$$v = \int \cos bx\,dx = \frac{1}{b}\sin bx$$

Substituting into $\int u\,dv = uv - \int v\,du$ gives:

$$\int e^{ax}\cos bx\,dx$$

$$= (e^{ax})\left(\frac{1}{b}\sin bx\right) - \int \left(\frac{1}{b}\sin bx\right)(ae^{ax}dx)$$

$$= \frac{1}{b}e^{ax}\sin bx - \frac{a}{b}\left[\int e^{ax}\sin bx\,dx\right] \qquad (1)$$

$\int e^{ax}\sin bx\,dx$ is now determined separately using integration by parts again:

Let $u = e^{ax}$ then $du = ae^{ax}\,dx$, and let $dv = \sin bx\,dx$, from which

$$v = \int \sin bx\,dx = -\frac{1}{b}\cos bx$$

Substituting into the integration by parts formula gives:

$$\int e^{ax}\sin bx\,dx = (e^{ax})\left(-\frac{1}{b}\cos bx\right)$$

$$- \int \left(-\frac{1}{b}\cos bx\right)(ae^{ax}\,dx)$$

$$= -\frac{1}{b}e^{ax}\cos bx$$

$$+ \frac{a}{b}\int e^{ax}\cos bx\,dx$$

Substituting this result into equation (1) gives:

$$\int e^{ax}\cos bx\,dx = \frac{1}{b}e^{ax}\sin bx - \frac{a}{b}\left[-\frac{1}{b}e^{ax}\cos bx\right.$$

$$\left. + \frac{a}{b}\int e^{ax}\cos bx\,dx\right]$$

$$= \frac{1}{b}e^{ax}\sin bx + \frac{a}{b^2}e^{ax}\cos bx$$

$$- \frac{a^2}{b^2}\int e^{ax}\cos bx\,dx$$

The integral on the far right of this equation is the same as the integral on the left hand side and thus they may be combined.

$$\int e^{ax}\cos bx\,dx + \frac{a^2}{b^2}\int e^{ax}\cos bx\,dx$$

$$= \frac{1}{b}e^{ax}\sin bx + \frac{a}{b^2}e^{ax}\cos bx$$

i.e. $\left(1 + \dfrac{a^2}{b^2}\right)\displaystyle\int e^{ax}\cos bx\,dx$

$$= \frac{1}{b}e^{ax}\sin bx + \frac{a}{b^2}e^{ax}\cos bx$$

i.e. $\left(\dfrac{b^2 + a^2}{b^2}\right)\displaystyle\int e^{ax}\cos bx\,dx$

$$= \frac{e^{ax}}{b^2}(b\sin bx + a\cos bx)$$

Hence $\int e^{ax} \cos bx\, dx$

$$= \left(\frac{b^2}{b^2+a^2}\right)\left(\frac{e^{ax}}{b^2}\right)(b \sin bx + a \cos bx)$$

$$= \frac{e^{ax}}{a^2+b^2}(b \sin bx + a \cos bx) + c$$

Using a similar method to above, that is, integrating by parts twice, the following result may be proved:

$$\int e^{ax} \sin bx\, dx$$

$$= \frac{e^{ax}}{a^2+b^2}(a \sin bx - b \cos bx) + c \qquad (2)$$

Problem 10. Evaluate $\int_0^{\frac{\pi}{4}} e^t \sin 2t\, dt$, correct to 4 decimal places

Comparing $\int e^t \sin 2t\, dt$ with $\int e^{ax} \sin bx\, dx$ shows that $x=t$, $a=1$ and $b=2$.
Hence, substituting into equation (2) gives:

$$\int_0^{\frac{\pi}{4}} e^t \sin 2t\, dt$$

$$= \left[\frac{e^t}{1^2+2^2}(1 \sin 2t - 2 \cos 2t)\right]_0^{\frac{\pi}{4}}$$

$$= \left[\frac{e^{\frac{\pi}{4}}}{5}\left(\sin 2\left(\frac{\pi}{4}\right) - 2 \cos 2\left(\frac{\pi}{4}\right)\right)\right]$$

$$\quad - \left[\frac{e^0}{5}(\sin 0 - 2 \cos 0)\right]$$

$$= \left[\frac{e^{\frac{\pi}{4}}}{5}(1-0)\right] - \left[\frac{1}{5}(0-2)\right] = \frac{e^{\frac{\pi}{4}}}{5} + \frac{2}{5}$$

$$= \mathbf{0.8387}, \text{ correct to 4 decimal places}$$

Now try the following exercise

Exercise 187   Further problems on integration by parts

Determine the integrals in Problems 1 to 5 using integration by parts.

1. $\int 2x^2 \ln x\, dx$ $\left[\dfrac{2}{3}x^3\left(\ln x - \dfrac{1}{3}\right) + c\right]$

2. $\int 2 \ln 3x\, dx$ $[2x(\ln 3x - 1) + c]$

3. $\int x^2 \sin 3x\, dx$

$$\left[\frac{\cos 3x}{27}(2-9x^2) + \frac{2}{9}x \sin 3x + c\right]$$

4. $\int 2e^{5x} \cos 2x\, dx$

$$\left[\frac{2}{29}e^{5x}(2 \sin 2x + 5 \cos 2x) + c\right]$$

5. $\int 2\theta \sec^2 \theta\, d\theta$ $[2[\theta \tan \theta - \ln(\sec \theta)] + c]$

Evaluate the integrals in Problems 6 to 9, correct to 4 significant figures.

6. $\int_1^2 x \ln x\, dx$ $[0.6363]$

7. $\int_0^1 2e^{3x} \sin 2x\, dx$ $[11.31]$

8. $\int_0^{\frac{\pi}{2}} e^t \cos 3t\, dt$ $[-1.543]$

9. $\int_1^4 \sqrt{x^3} \ln x\, dx$ $[12.78]$

10. In determining a Fourier series to represent $f(x)=x$ in the range $-\pi$ to π, Fourier coefficients are given by:

$$a_n = \frac{1}{\pi}\int_{-\pi}^{\pi} x \cos nx\, dx$$

and $$b_n = \frac{1}{\pi}\int_{-\pi}^{\pi} x \sin nx\, dx$$

where n is a positive integer. Show by using integration by parts that $a_n = 0$ and $b_n = -\dfrac{2}{n} \cos n\pi$

11. The equations:

$$C = \int_0^1 e^{-0.4\theta} \cos 1.2\theta\, d\theta$$

and $$S = \int_0^1 e^{-0.4\theta} \sin 1.2\theta\, d\theta$$

are involved in the study of damped oscillations. Determine the values of C and S.
$$[C = 0.66, S = 0.41]$$

Chapter 54

Numerical integration

54.1 Introduction

Even with advanced methods of integration there are many mathematical functions which cannot be integrated by analytical methods and thus approximate methods have then to be used. Approximate methods of definite integrals may be determined by what is termed **numerical integration**.

It may be shown that determining the value of a definite integral is, in fact, finding the area between a curve, the horizontal axis and the specified ordinates. Three methods of finding approximate areas under curves are the trapezoidal rule, the mid-ordinate rule and Simpson's rule, and these rules are used as a basis for numerical integration.

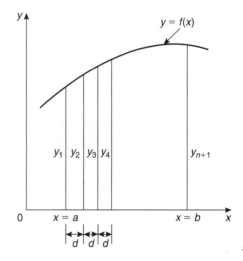

Figure 54.1

54.2 The trapezoidal rule

Let a required definite integral be denoted by $\int_a^b y \, dx$ and be represented by the area under the graph of $y = f(x)$ between the limits $x = a$ and $x = b$ as shown in Fig. 54.1.

Let the range of integration be divided into n equal intervals each of width d, such that $nd = b - a$,

i.e. $d = \dfrac{b-a}{n}$

The ordinates are labelled $y_1, y_2, y_3, \ldots, y_{n+1}$ as shown.

An approximation to the area under the curve may be determined by joining the tops of the ordinates by straight lines. Each interval is thus a trapezium, and since the area of a trapezium is given by:

$$\text{area} = \frac{1}{2}(\text{sum of parallel sides})(\text{perpendicular distance between them})$$

then

$$\int_a^b y \, dx \approx \frac{1}{2}(y_1 + y_2)d + \frac{1}{2}(y_2 + y_3)d$$

$$+ \frac{1}{2}(y_3 + y_4)d + \cdots + \frac{1}{2}(y_n + y_{n+1})d$$

$$\approx d\left[\frac{1}{2}y_1 + y_2 + y_3 + y_4 + \cdots + y_n + \frac{1}{2}y_{n+1}\right]$$

i.e. **the trapezoidal rule states**:

$$\int_a^b y \, dx \approx \left(\begin{array}{c}\text{width of}\\\text{interval}\end{array}\right)\left\{\frac{1}{2}\left(\begin{array}{c}\text{first + last}\\\text{ordinate}\end{array}\right) + \left(\begin{array}{c}\text{sum of}\\\text{remaining}\\\text{ordinates}\end{array}\right)\right\}$$

$$(1)$$

Problem 1. (a) Use integration to evaluate, correct to 3 decimal places, $\displaystyle\int_1^3 \frac{2}{\sqrt{x}} \, dx$
(b) Use the trapezoidal rule with 4 intervals to evaluate the integral in part (a), correct to 3 decimal places

(a) $\int_1^3 \frac{2}{\sqrt{x}} \, dx = \int_1^3 2x^{-\frac{1}{2}} \, dx$

$$= \left[\frac{2x^{(-\frac{1}{2})+1}}{-\frac{1}{2}+1} \right]_1^3 = \left[4x^{\frac{1}{2}} \right]_1^3$$

$$= 4[\sqrt{x}]_1^3 = 4[\sqrt{3} - \sqrt{1}]$$

$$= \mathbf{2.928}, \text{ correct to 3 decimal places.}$$

(b) The range of integration is the difference between the upper and lower limits, i.e. $3 - 1 = 2$. Using the trapezoidal rule with 4 intervals gives an interval width $d = \frac{3-1}{4} = 0.5$ and ordinates situated at 1.0, 1.5, 2.0, 2.5 and 3.0. Corresponding values of $\frac{2}{\sqrt{x}}$ are shown in the table below, each correct to 4 decimal places (which is one more decimal place than required in the problem).

x	$\frac{2}{\sqrt{x}}$
1.0	2.0000
1.5	1.6330
2.0	1.4142
2.5	1.2649
3.0	1.1547

From equation (1):

$$\int_1^3 \frac{2}{\sqrt{x}} \, dx \approx (0.5) \left\{ \frac{1}{2}(2.0000 + 1.1547) \right.$$

$$\left. + 1.6330 + 1.4142 + 1.2649 \right\}$$

$$= \mathbf{2.945}, \text{ correct to 3 decimal places.}$$

This problem demonstrates that even with just 4 intervals a close approximation to the true value of 2.928 (correct to 3 decimal places) is obtained using the trapezoidal rule.

Problem 2. Use the trapezoidal rule with 8 intervals to evaluate $\int_1^3 \frac{2}{\sqrt{x}} \, dx$, correct to 3 decimal places

With 8 intervals, the width of each is $\frac{3-1}{8}$ i.e. 0.25 giving ordinates at 1.00, 1.25, 1.50, 1.75, 2.00, 2.25, 2.50, 2.75 and 3.00. Corresponding values of $\frac{2}{\sqrt{x}}$ are shown in the table below:

x	$\frac{2}{\sqrt{x}}$
1.00	2.000
1.25	1.7889
1.50	1.6330
1.75	1.5119
2.00	1.4142
2.25	1.3333
2.50	1.2649
2.75	1.2060
3.00	1.1547

From equation (1):

$$\int_1^3 \frac{2}{\sqrt{x}} \, dx \approx (0.25) \left\{ \frac{1}{2}(2.000 + 1.1547) + 1.7889 \right.$$

$$+ 1.6330 + 1.5119 + 1.4142$$

$$\left. + 1.3333 + 1.2649 + 1.2060 \right\}$$

$$= \mathbf{2.932}, \text{ correct to 3 decimal places}$$

This problem demonstrates that the greater the number of intervals chosen (i.e. the smaller the interval width) the more accurate will be the value of the definite integral. The exact value is found when the number of intervals is infinite, which is what the process of integration is based upon.

Problem 3. Use the trapezoidal rule to evaluate $\int_0^{\pi/2} \frac{1}{1 + \sin x} \, dx$ using 6 intervals. Give the answer correct to 4 significant figures

With 6 intervals, each will have a width of $\frac{\frac{\pi}{2} - 0}{6}$ i.e. $\frac{\pi}{12}$ rad (or 15°) and the ordinates occur at 0,

$\dfrac{\pi}{12}, \dfrac{\pi}{6}, \dfrac{\pi}{4}, \dfrac{\pi}{3}, \dfrac{5\pi}{12}$ and $\dfrac{\pi}{2}$. Corresponding values of $\dfrac{1}{1+\sin x}$ are shown in the table below:

x	$\dfrac{1}{1+\sin x}$
0	1.0000
$\dfrac{\pi}{12}$ (or 15°)	0.79440
$\dfrac{\pi}{6}$ (or 30°)	0.66667
$\dfrac{\pi}{4}$ (or 45°)	0.58579
$\dfrac{\pi}{3}$ (or 60°)	0.53590
$\dfrac{5\pi}{12}$ (or 75°)	0.50867
$\dfrac{\pi}{2}$ (or 90°)	0.50000

From equation (1):

$$\int_0^{\frac{\pi}{2}} \dfrac{1}{1+\sin x}\,dx \approx \left(\dfrac{\pi}{12}\right)\left\{\dfrac{1}{2}(1.00000 + 0.50000)\right.$$

$$+\, 0.79440 + 0.66667 + 0.58579$$

$$\left. +\, 0.53590 + 0.50867\right\}$$

$$= \mathbf{1.006},\ \text{correct to 4 significant figures}$$

Now try the following exercise

Exercise 188   Further problems on the trapezoidal rule

Evaluate the following definite integrals using the **trapezoidal rule**, giving the answers correct to 3 decimal places:

1. $\displaystyle\int_0^1 \dfrac{2}{1+x^2}\,dx$ (Use 8 intervals) [1.569]

2. $\displaystyle\int_1^3 2\ln 3x\,dx$ (Use 8 intervals) [6.979]

3. $\displaystyle\int_0^{\pi/3} \sqrt{\sin\theta}\,d\theta$ (Use 6 intervals) [0.672]

4. $\displaystyle\int_0^{1.4} e^{-x^2}\,dx$ (Use 7 intervals) [0.843]

54.3   The mid-ordinate rule

Let a required definite integral be denoted again by $\int_a^b y\,dx$ and represented by the area under the graph of $y = f(x)$ between the limits $x = a$ and $x = b$, as shown in Fig. 54.2.

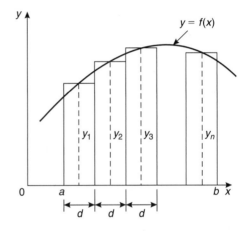

Figure 54.2

With the mid-ordinate rule each interval of width d is assumed to be replaced by a rectangle of height equal to the ordinate at the middle point of each interval, shown as $y_1, y_2, y_3, \ldots, y_n$ in Fig. 54.2.

Thus $\displaystyle\int_a^b y\,dx \approx dy_1 + dy_2 + dy_3 + \cdots + dy_n$

$$\approx d(y_1 + y_2 + y_3 + \cdots + y_n)$$

i.e. **the mid-ordinate rule states**:

$$\int_a^b y\,dx \approx \begin{pmatrix}\text{width of} \\ \text{interval}\end{pmatrix}\begin{pmatrix}\text{sum of} \\ \text{mid-ordinates}\end{pmatrix} \quad (2)$$

Problem 4. Use the mid-ordinate rule with (a) 4 intervals, (b) 8 intervals, to evaluate $\displaystyle\int_1^3 \dfrac{2}{\sqrt{x}}\,dx$, correct to 3 decimal places

Section 9

(a) With 4 intervals, each will have a width of $\dfrac{3-1}{4}$, i.e. 0.5. and the ordinates will occur at 1.0, 1.5, 2.0, 2.5 and 3.0. Hence the mid-ordinates y_1, y_2, y_3 and y_4 occur at 1.25, 1.75, 2.25 and 2.75

Corresponding values of $\dfrac{2}{\sqrt{x}}$ are shown in the following table:

x	$\dfrac{2}{\sqrt{x}}$
1.25	1.7889
1.75	1.5119
2.25	1.3333
2.75	1.2060

From equation (2):

$$\int_1^3 \frac{2}{\sqrt{x}}\,dx \approx (0.5)[1.7889 + 1.5119$$
$$+1.3333 + 1.2060]$$
$$= \mathbf{2.920}, \text{ correct to 3 decimal places}$$

(b) With 8 intervals, each will have a width of 0.25 and the ordinates will occur at 1.00, 1.25, 1.50, 1.75,... and thus mid-ordinates at 1.125, 1.375, 1.625, 1.875.... Corresponding values of $\dfrac{2}{\sqrt{x}}$ are shown in the following table:

x	$\dfrac{2}{\sqrt{x}}$
1.125	1.8856
1.375	1.7056
1.625	1.5689
1.875	1.4606
2.125	1.3720
2.375	1.2978
2.625	1.2344
2.875	1.1795

From equation (2):

$$\int_1^3 \frac{2}{\sqrt{x}}\,dx \approx (0.25)[1.8856 + 1.7056$$
$$+ 1.5689 + 1.4606$$
$$+ 1.3720 + 1.2978$$
$$+ 1.2344 + 1.1795]$$
$$= \mathbf{2.926}, \text{ correct to 3 decimal places}$$

As previously, the greater the number of intervals the nearer the result is to the true value of 2.928, correct to 3 decimal places.

Problem 5. Evaluate $\int_0^{2.4} e^{-x^2/3}\,dx$, correct to 4 significant figures, using the mid-ordinate rule with 6 intervals

With 6 intervals each will have a width of $\dfrac{2.4-0}{6}$, i.e. 0.40 and the ordinates will occur at 0, 0.40, 0.80, 1.20, 1.60, 2.00 and 2.40 and thus mid-ordinates at 0.20, 0.60, 1.00, 1.40, 1.80 and 2.20.

Corresponding values of $e^{-x^2/3}$ are shown in the following table:

x	$e^{-\frac{x^2}{3}}$
0.20	0.98676
0.60	0.88692
1.00	0.71653
1.40	0.52031
1.80	0.33960
2.20	0.19922

From equation (2):

$$\int_0^{2.4} e^{-\frac{x^2}{3}}\,dx \approx (0.40)[0.98676 + 0.88692$$
$$+ 0.71653 + 0.52031$$
$$+ 0.33960 + 0.19922]$$
$$= \mathbf{1.460}, \text{ correct to 4 significant figures.}$$

Now try the following exercise

Exercise 189   Further problems on the mid-ordinate rule

Evaluate the following definite integrals using the **mid-ordinate rule**, giving the answers correct to 3 decimal places.

1. $\displaystyle\int_0^2 \frac{3}{1+t^2}\,dt$ (Use 8 intervals)   [3.323]

2. $\displaystyle\int_0^{\pi/2} \frac{1}{1+\sin\theta}\,d\theta$ (Use 6 intervals)

    [0.997]

3. $\displaystyle\int_1^3 \frac{\ln x}{x}\,dx$ (Use 10 intervals)   [0.605]

4. $\displaystyle\int_0^{\pi/3} \sqrt{\cos^3 x}\,dx$ (Use 6 intervals)

    [0.799]

54.4   Simpson's rule

The approximation made with the trapezoidal rule is to join the top of two successive ordinates by a straight line, i.e. by using a linear approximation of the form $a+bx$. With Simpson's rule, the approximation made is to join the tops of three successive ordinates by a parabola, i.e. by using a quadratic approximation of the form $a+bx+cx^2$.

Figure 54.3 shows a parabola $y=a+bx+cx^2$ with ordinates y_1, y_2 and y_3 at $x=-d$, $x=0$ and $x=d$ respectively.

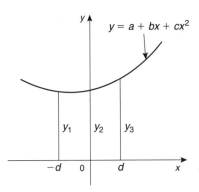

Figure 54.3

Thus the width of each of the two intervals is d. The area enclosed by the parabola, the x-axis and ordinates $x=-d$ and $x=d$ is given by:

$$\int_{-d}^{d} (a+bx+cx^2)\,dx = \left[ax + \frac{bx^2}{2} + \frac{cx^3}{3}\right]_{-d}^{d}$$

$$= \left(ad + \frac{bd^2}{2} + \frac{cd^3}{3}\right)$$

$$- \left(-ad + \frac{bd^2}{2} - \frac{cd^3}{3}\right)$$

$$= 2ad + \frac{2}{3}cd^3$$

$$\text{or } \frac{1}{3}d(6a+2cd^2) \qquad (3)$$

Since $y = a+bx+cx^2$

at $x=-d$, $y_1 = a - bd + cd^2$

at $x=0$, $y_2 = a$

and at $x=d$, $y_3 = a + bd + cd^2$

Hence $y_1 + y_3 = 2a + 2cd^2$

and $y_1 + 4y_2 + y_3 = 6a + 2cd^2 \qquad (4)$

Thus the area under the parabola between $x=-d$ and $x=d$ in Fig. 54.3 may be expressed as $\frac{1}{3}d(y_1 + 4y_2 + y_3)$, from equation (3) and (4), and the result is seen to be independent of the position of the origin.

Let a definite integral be denoted by $\int_a^b y\,dx$ and represented by the area under the graph of $y=f(x)$ between the limits $x=a$ and $x=b$, as shown in Fig. 54.4. The range of integration, $b-a$, is divided into an **even** number of intervals, say $2n$, each of width d.

Since an even number of intervals is specified, an odd number of ordinates, $2n+1$, exists. Let an approximation to the curve over the first two intervals be a parabola of the form $y=a+bx+cx^2$ which passes through the tops of the three ordinates y_1, y_2 and y_3. Similarly, let an approximation to the curve over the next two intervals be the parabola which passes through the tops of the ordinates y_3, y_4 and y_5, and so on. Then

$$\int_a^b y\,dx \approx \frac{1}{3}d(y_1 + 4y_2 + y_3) + \frac{1}{3}d(y_3 + 4y_4 + y_5)$$

$$+ \frac{1}{3}d(y_{2n-1} + 4y_{2n} + y_{2n+1})$$

Section 9

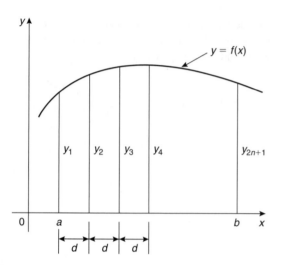

Figure 54.4

$$\approx \frac{1}{3}d[(y_1 + y_{2n+1}) + 4(y_2 + y_4 + \cdots + y_{2n})$$

$$+ 2(y_3 + y_5 + \cdots + y_{2n-1})]$$

i.e. **Simpson's rule states:**

$$\int_a^b y \, dx \approx \frac{1}{3} \left(\begin{array}{c}\textbf{width of}\\\textbf{interval}\end{array}\right) \left\{\left(\begin{array}{c}\textbf{first} + \textbf{last}\\\textbf{ordinate}\end{array}\right)\right.$$

$$+ 4 \left(\begin{array}{c}\textbf{sum of even}\\\textbf{ordinates}\end{array}\right)$$

$$\left. + 2 \left(\begin{array}{c}\textbf{sum of remaining}\\\textbf{odd ordinates}\end{array}\right)\right\}$$

(5)

Note that Simpson's rule can only be applied when an **even** number of intervals is chosen, i.e. an odd number of ordinates.

Problem 6. Use Simpson's rule with (a) 4 intervals, (b) 8 intervals, to evaluate $\int_1^3 \frac{2}{\sqrt{x}} \, dx$, correct to 3 decimal places

---

(a) With 4 intervals, each will have a width of $\frac{3-1}{4}$ i.e. 0.5 and the ordinates will occur at 1.0, 1.5, 2.0, 2.5 and 3.0.

The values of the ordinates are as shown in the table of Problem 1(b), page 470.

Thus, from equation (5):

$$\int_1^3 \frac{2}{\sqrt{x}} \, dx \approx \frac{1}{3}(0.5)[(2.0000 + 1.1547)$$

$$+ 4(1.6330 + 1.2649)$$

$$+ 2(1.4142)]$$

$$= \frac{1}{3}(0.5)[3.1547 + 11.5916$$

$$+ 2.8284]$$

$$= \textbf{2.929}, \text{ correct to 3 decimal places.}$$

(b) With 8 intervals, each will have a width of $\frac{3-1}{8}$ i.e. 0.25 and the ordinates occur at 1.00, 1.25, 1.50, 1.75, …, 3.0.

The values of the ordinates are as shown in the table in Problem 2, page 470.

Thus, from equation (5):

$$\int_1^3 \frac{2}{\sqrt{x}} \, dx \approx \frac{1}{3}(0.25)[(2.0000 + 1.1547)$$

$$+ 4(1.7889 + 1.5119 + 1.3333$$

$$+ 1.2060) + 2(1.6330$$

$$+ 1.4142 + 1.2649)]$$

$$= \frac{1}{3}(0.25)[3.1547 + 23.3604$$

$$+ 8.6242]$$

$$= \textbf{2.928}, \text{ correct to 3 decimal places.}$$

It is noted that the latter answer is exactly the same as that obtained by integration. In general, Simpson's rule is regarded as the most accurate of the three approximate methods used in numerical integration.

Problem 7. Evaluate $\int_0^{\pi/3} \sqrt{1 - \frac{1}{3}\sin^2\theta} \, d\theta$, correct to 3 decimal places, using Simpson's rule with 6 intervals

---

With 6 intervals, each will have a width of $\dfrac{\frac{\pi}{3} - 0}{6}$ i.e. $\frac{\pi}{18}$ rad (or 10°), and the ordinates will occur at 0, $\frac{\pi}{18}, \frac{\pi}{9}, \frac{\pi}{6}, \frac{2\pi}{9}, \frac{5\pi}{18}$ and $\frac{\pi}{3}$

Corresponding values of $\sqrt{1 - \frac{1}{3}\sin^2\theta}$ are shown in the table below:

θ	0	$\frac{\pi}{18}$ (or 10°)	$\frac{\pi}{9}$ (or 20°)	$\frac{\pi}{6}$ (or 30°)
$\sqrt{1 - \frac{1}{3}\sin^2\theta}$	1.0000	0.9950	0.9803	0.9574

θ	$\frac{2\pi}{9}$ (or 40°)	$\frac{5\pi}{18}$ (or 50°)	$\frac{\pi}{3}$ (or 60°)
$\sqrt{1 - \frac{1}{3}\sin^2\theta}$	0.9286	0.8969	0.8660

From equation (5):

$$\int_0^{\frac{\pi}{3}} \sqrt{1 - \frac{1}{3}\sin^2\theta}\, d\theta$$
$$\approx \frac{1}{3}\left(\frac{\pi}{18}\right)[(1.0000 + 0.8660) + 4(0.9950 + 0.9574 + 0.8969) + 2(0.9803 + 0.9286)]$$
$$= \frac{1}{3}\left(\frac{\pi}{18}\right)[1.8660 + 11.3972 + 3.8178]$$
$$= \mathbf{0.994}, \text{ correct to 3 decimal places.}$$

Problem 8. An alternating current i has the following values at equal intervals of 2.0 milliseconds:

Time (ms)	0	2.0	4.0	6.0	8.0	10.0	12.0
Current i (A)	0	3.5	8.2	10.0	7.3	2.0	0

Charge, q, in millicoulombs, is given by $q = \int_0^{12.0} i\, dt$. Use Simpson's rule to determine the approximate charge in the 12 ms period

From equation (5):

Charge, $\quad q = \int_0^{12.0} i\, dt$
$$\approx \frac{1}{3}(2.0)[(0 + 0) + 4(3.5 + 10.0 + 2.0) + 2(8.2 + 7.3)]$$
$$= \mathbf{62\ mC}$$

Now try the following exercise

Exercise 190   Further problems on Simpson's rule

In Problems 1 to 5, evaluate the definite integrals using **Simpson's rule**, giving the answers correct to 3 decimal places.

1. $\int_0^{\pi/2} \sqrt{\sin x}\, dx$ (Use 6 intervals)   [1.187]

2. $\int_0^{1.6} \frac{1}{1+\theta^4}\, d\theta$ (Use 8 intervals)   [1.034]

3. $\int_{0.2}^{1.0} \frac{\sin\theta}{\theta}\, d\theta$ (Use 8 intervals)   [0.747]

4. $\int_0^{\pi/2} x \cos x\, dx$ (Use 6 intervals)   [0.571]

5. $\int_0^{\pi/3} e^{x^2} \sin 2x\, dx$ (Use 10 intervals)   [1.260]

In Problems 6 and 7 evaluate the definite integrals using (a) integration, (b) the trapezoidal rule, (c) the mid-ordinate rule, (d) Simpson's rule. Give answers correct to 3 decimal places.

6. $\int_1^4 \frac{4}{x^3}\, dx$ (Use 6 intervals)

$$\begin{bmatrix} \text{(a) } 1.875 & \text{(b) } 2.107 \\ \text{(c) } 1.765 & \text{(d) } 1.916 \end{bmatrix}$$

7. $\int_2^6 \frac{1}{\sqrt{2x-1}}\, dx$ (Use 8 intervals)

$$\begin{bmatrix} \text{(a) } 1.585 & \text{(b) } 1.588 \\ \text{(c) } 1.583 & \text{(d) } 1.585 \end{bmatrix}$$

In Problems 8 and 9 evaluate the definite integrals using (a) the trapezoidal rule, (b) the mid-ordinate rule, (c) Simpson's rule. Use 6 intervals in each case and give answers correct to 3 decimal places.

8. $\int_0^3 \sqrt{1+x^4}\, dx$
[(a) 10.194 (b) 10.007 (c) 10.070]

9. $\int_{0.1}^{0.7} \frac{1}{\sqrt{1-y^2}}\, dy$
[(a) 0.677 (b) 0.674 (c) 0.675]

10. A vehicle starts from rest and its velocity is measured every second for 8 seconds, with values as follows:

time t (s)	velocity v (ms^{-1})
0	0
1.0	0.4
2.0	1.0
3.0	1.7
4.0	2.9
5.0	4.1
6.0	6.2
7.0	8.0
8.0	9.4

The distance travelled in 8.0 seconds is given by $\int_0^{8.0} v\,dt$.

Estimate this distance using Simpson's rule, giving the answer correct to 3 significant figures.   [28.8 m]

11. A pin moves along a straight guide so that its velocity v (m/s) when it is a distance x (m) from the beginning of the guide at time t (s) is given in the table below:

t (s)	v (m/s)
0	0
0.5	0.052
1.0	0.082
1.5	0.125
2.0	0.162
2.5	0.175
3.0	0.186
3.5	0.160
4.0	0

Use Simpson's rule with 8 intervals to determine the approximate total distance travelled by the pin in the 4.0 second period.   [0.485 m]

This Revision test covers the material contained in Chapters 51 to 54. *The marks for each question are shown in brackets at the end of each question.*

1. Determine: (a) $\int \dfrac{x-11}{x^2-x-2}\,dx$

   (b) $\int \dfrac{3-x}{(x^2+3)(x+3)}\,dx$ (21)

2. Evaluate: $\int_1^2 \dfrac{3}{x^2(x+2)}\,dx$ correct to 4 significant figures.   (11)

3. Determine: $\int \dfrac{dx}{2\sin x + \cos x}$ (7)

4. Determine the following integrals:

   (a) $\int 5xe^{2x}\,dx$ (b) $\int t^2 \sin 2t\,dt$ (12)

5. Evaluate correct to 3 decimal places:

   $\int_1^4 \sqrt{x}\ln x\,dx$ (9)

6. Evaluate: $\int_1^3 \dfrac{5}{x^2}\,dx$ using

   (a) integration

   (b) the trapezoidal rule

   (c) the mid-ordinate rule

   (d) Simpson's rule.

   In each of the approximate methods use 8 intervals and give the answers correct to 3 decimal places.   (16)

7. An alternating current i has the following values at equal intervals of 5 ms:

Time t (ms)	0	5	10	15	20	25	30
Current i (A)	0	4.8	9.1	12.7	8.8	3.5	0

   Charge q, in coulombs, is given by

   $$q = \int_0^{30\times10^{-3}} i\,dt.$$

   Use Simpson's rule to determine the approximate charge in the 30 ms period.   (4)

Chapter 55

Areas under and between curves

55.1 Area under a curve

The area shown shaded in Fig. 55.1 may be determined using approximate methods (such as the trapezoidal rule, the mid-ordinate rule or Simpson's rule) or, more precisely, by using integration.

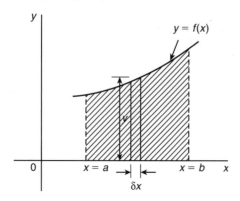

Figure 55.1

(i) Let A be the area shown shaded in Fig. 55.1 and let this area be divided into a number of strips each of width δx. One such strip is shown and let the area of this strip be δA.

Then: $\quad \delta A \approx y\delta x$ (1)

The accuracy of statement (1) increases when the width of each strip is reduced, i.e. area A is divided into a greater number of strips.

(ii) Area A is equal to the sum of all the strips from $x = a$ to $x = b$,

i.e. $\quad A = \displaystyle\lim_{\delta x \to 0} \sum_{x=a}^{x=b} y\,\delta x$ (2)

(iii) From statement (1), $\dfrac{\delta A}{\delta x} \approx y$ (3)

In the limit, as δx approaches zero, $\dfrac{\delta A}{\delta x}$ becomes the differential coefficient $\dfrac{dA}{dx}$

Hence $\displaystyle\lim_{\delta x \to 0}\left(\dfrac{\delta A}{\delta x}\right) = \dfrac{dA}{dx} = y$, from statement (3).

By integration,

$$\int \frac{dA}{dx}dx = \int y\,dx \quad \text{i.e.} \quad A = \int y\,dx$$

The ordinates $x = a$ and $x = b$ limit the area and such ordinate values are shown as limits. Hence

$$A = \int_a^b y\,dx \qquad (4)$$

(iv) Equating statements (2) and (4) gives:

$$\textbf{Area } A = \lim_{\delta x \to 0}\sum_{x=a}^{x=b} y\,\delta x = \int_a^b y\,dx$$

$$= \int_a^b f(x)\,dx$$

(v) If the area between a curve $x = f(y)$, the y-axis and ordinates $y = p$ and $y = q$ is required, then

$$\text{area} = \int_p^q x\,dy$$

Thus, determining the area under a curve by integration merely involves evaluating a definite integral.
There are several instances in engineering and science where the area beneath a curve needs to be accurately

determined. For example, the areas between limits of a:

> **velocity/time graph gives distance travelled,**
> **force/distance graph gives work done,**
> **voltage/current graph gives power, and so on.**

Should a curve drop below the x-axis, then $y\ (=f(x))$ becomes negative and $f(x)\,dx$ is negative. When determining such areas by integration, a negative sign is placed before the integral. For the curve shown in Fig. 55.2, the total shaded area is given by (area E + area F + area G).

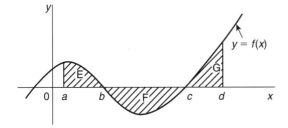

Figure 55.2

By integration, **total shaded area**

$$= \int_a^b f(x)\,dx - \int_b^c f(x)\,dx + \int_c^d f(x)\,dx$$

(Note that this is **not** the same as $\int_a^d f(x)\,dx$.)
It is usually necessary to sketch a curve in order to check whether it crosses the x-axis.

55.2 Worked problems on the area under a curve

Problem 1. Determine the area enclosed by $y = 2x + 3$, the x-axis and ordinates $x = 1$ and $x = 4$

$y = 2x + 3$ is a straight line graph as shown in Fig. 55.3, where the required area is shown shaded.
By integration,

$$\text{shaded area} = \int_1^4 y\,dx$$

$$= \int_1^4 (2x + 3)\,dx$$

$$= \left[\frac{2x^2}{2} + 3x\right]_1^4$$

$$= [(16 + 12) - (1 + 3)]$$

$$= \textbf{24 square units}$$

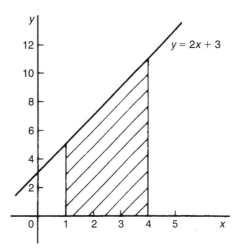

Figure 55.3

[This answer may be checked since the shaded area is a trapezium.

Area of trapezium

$$= \frac{1}{2}\begin{pmatrix}\text{sum of parallel}\\\text{sides}\end{pmatrix}\begin{pmatrix}\text{perpendicular distance}\\\text{between parallel sides}\end{pmatrix}$$

$$= \frac{1}{2}(5 + 11)(3)$$

$$= \textbf{24 square units}]$$

> **Problem 2.** The velocity v of a body t seconds after a certain instant is: $(2t^2 + 5)$ m/s. Find by integration how far it moves in the interval from $t = 0$ to $t = 4$ s

Since $2t^2 + 5$ is a quadratic expression, the curve $v = 2t^2 + 5$ is a parabola cutting the v-axis at $v = 5$, as shown in Fig. 55.4.
The distance travelled is given by the area under the v/t curve (shown shaded in Fig. 55.4).
By integration,

$$\text{shaded area} = \int_0^4 v\,dt$$

$$= \int_0^4 (2t^2 + 5)\,dt$$

$$= \left[\frac{2t^3}{3} + 5t\right]_0^4$$

$$= \left(\frac{2(4^3)}{3} + 5(4)\right) - (0)$$

i.e. **distance travelled = 62.67 m**

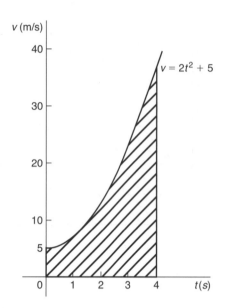

Figure 55.4

x	-3	-2	-1	0	1	2
x^3	-27	-8	-1	0	1	8
$2x^2$	18	8	2	0	2	8
$-5x$	15	10	5	0	-5	-10
-6	-6	-6	-6	-6	-6	-6
y	0	4	0	-6	-8	0

Shaded area $= \displaystyle\int_{-3}^{-1} y\,dx - \int_{-1}^{2} y\,dx$, the minus sign before the second integral being necessary since the enclosed area is below the x-axis.

Hence shaded area

$$= \int_{-3}^{-1} (x^3 + 2x^2 - 5x - 6)\,dx$$

$$- \int_{-1}^{2} (x^3 + 2x^2 - 5x - 6)\,dx$$

$$= \left[\frac{x^4}{4} + \frac{2x^3}{3} - \frac{5x^2}{2} - 6x\right]_{-3}^{-1}$$

$$- \left[\frac{x^4}{4} + \frac{2x^3}{3} - \frac{5x^2}{2} - 6x\right]_{-1}^{2}$$

$$= \left[\left\{\frac{1}{4} - \frac{2}{3} - \frac{5}{2} + 6\right\}\right.$$

$$\left. - \left\{\frac{81}{4} - 18 - \frac{45}{2} + 18\right\}\right]$$

$$- \left[\left\{4 + \frac{16}{3} - 10 - 12\right\}\right.$$

$$\left. - \left\{\frac{1}{4} - \frac{2}{3} - \frac{5}{2} + 6\right\}\right]$$

$$= \left[\left\{3\frac{1}{12}\right\} - \left\{-2\frac{1}{4}\right\}\right]$$

$$- \left[\left\{-12\frac{2}{3}\right\} - \left\{3\frac{1}{12}\right\}\right]$$

$$= \left[5\frac{1}{3}\right] - \left[-15\frac{3}{4}\right]$$

$$= 21\frac{1}{12} \quad \text{or} \quad \textbf{21.08 square units}$$

Problem 3. Sketch the graph $y = x^3 + 2x^2 - 5x - 6$ between $x = -3$ and $x = 2$ and determine the area enclosed by the curve and the x-axis

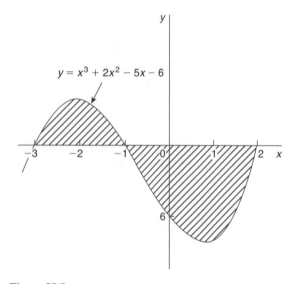

Figure 55.5

A table of values is produced and the graph sketched as shown in Fig. 55.5 where the area enclosed by the curve and the x-axis is shown shaded.

Problem 4. Determine the area enclosed by the curve $y = 3x^2 + 4$, the x-axis and ordinates $x = 1$ and $x = 4$ by (a) the trapezoidal rule, (b) the

mid-ordinate rule, (c) Simpson's rule, and
(d) integration

x	0	1.0	1.5	2.0	2.5	3.0	3.5	4.0
y	4	7	10.75	16	22.75	31	40.75	52

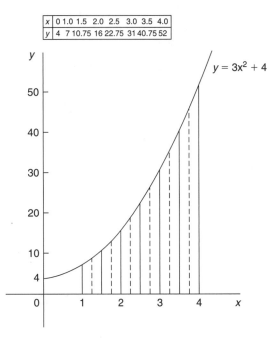

Figure 55.6

The curve $y = 3x^2 + 4$ is shown plotted in Fig. 55.6.

(a) **By the trapezoidal rule**

$$\text{Area} = \begin{pmatrix}\text{width of}\\\text{interval}\end{pmatrix}\left[\frac{1}{2}\begin{pmatrix}\text{first} + \text{last}\\\text{ordinate}\end{pmatrix} + \begin{pmatrix}\text{sum of}\\\text{remaining}\\\text{ordinates}\end{pmatrix}\right]$$

Selecting 6 intervals each of width 0.5 gives:

$$\text{Area} = (0.5)\left[\frac{1}{2}(7 + 52) + 10.75 + 16\right.$$
$$\left. + 22.75 + 31 + 40.75\right]$$
$$= \textbf{75.375 square units}$$

(b) **By the mid-ordinate rule**,
area = (width of interval) (sum of mid-ordinates).
Selecting 6 intervals, each of width 0.5 gives the
mid-ordinates as shown by the broken lines in
Fig. 55.6.

Thus, area $= (0.5)(8.5 + 13 + 19 + 26.5$
$$+ 35.5 + 46)$$
$$= \textbf{74.25 square units}$$

(c) **By Simpson's rule**,

$$\text{area} = \frac{1}{3}\begin{pmatrix}\text{width of}\\\text{interval}\end{pmatrix}\left[\begin{pmatrix}\text{first} + \text{last}\\\text{ordinates}\end{pmatrix}\right.$$

$$+ 4\begin{pmatrix}\text{sum of even}\\\text{ordinates}\end{pmatrix}$$

$$\left. + 2\begin{pmatrix}\text{sum of remaining}\\\text{odd ordinates}\end{pmatrix}\right]$$

Selecting 6 intervals, each of width 0.5, gives:

$$\text{area} = \frac{1}{3}(0.5)[(7 + 52) + 4(10.75 + 22.75$$
$$+ 40.75) + 2(16 + 31)]$$
$$= \textbf{75 square units}$$

(d) **By integration**, shaded area

$$= \int_1^4 y\, dx$$
$$= \int_1^4 (3x^2 + 4)\, dx$$
$$= \left[x^3 + 4x\right]_1^4$$
$$= \textbf{75 square units}$$

Integration gives the precise value for the area
under a curve. In this case Simpson's rule is seen
to be the most accurate of the three approximate
methods.

Problem 5. Find the area enclosed by the curve
$y = \sin 2x$, the x-axis and the ordinates $x = 0$ and
$x = \pi/3$

A sketch of $y = \sin 2x$ is shown in Fig. 55.7.

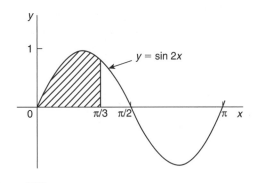

Figure 55.7

(Note that $y = \sin 2x$ has a period of $\dfrac{2\pi}{2}$, i.e. π radians.)

Section 9

Shaded area $= \displaystyle\int_0^{\pi/3} y\,dx$

$= \displaystyle\int_0^{\pi/3} \sin 2x\,dx$

$= \left[-\dfrac{1}{2}\cos 2x \right]_0^{\pi/3}$

$= \left\{ -\dfrac{1}{2}\cos\dfrac{2\pi}{3} \right\} - \left\{ -\dfrac{1}{2}\cos 0 \right\}$

$= \left\{ -\dfrac{1}{2}\left(-\dfrac{1}{2} \right) \right\} - \left\{ -\dfrac{1}{2}(1) \right\}$

$= \dfrac{1}{4} + \dfrac{1}{2} = \dfrac{3}{4}$ **square units**

Now try the following exercise

Exercise 191   Further problems on area under curves

Unless otherwise stated all answers are in square units.

1. Shown by integration that the area of the triangle formed by the line $y = 2x$, the ordinates $x = 0$ and $x = 4$ and the x-axis is 16 square units.

2. Sketch the curve $y = 3x^2 + 1$ between $x = -2$ and $x = 4$. Determine by integration the area enclosed by the curve, the x-axis and ordinates $x = -1$ and $x = 3$. Use an approximate method to find the area and compare your result with that obtained by integration.   [32]

In Problems 3 to 8, find the area enclosed between the given curves, the horizontal axis and the given ordinates.

3. $y = 5x$; $x = 1, x = 4$ [37.5]

4. $y = 2x^2 - x + 1$; $x = -1, x = 2$ [7.5]

5. $y = 2\sin 2\theta$; $\theta = 0, \theta = \dfrac{\pi}{4}$ [1]

6. $\theta = t + e^t$; $t = 0, t = 2$ [8.389]

7. $y = 5\cos 3t$; $t = 0, t = \dfrac{\pi}{6}$ [1.67]

8. $y = (x - 1)(x - 3)$; $x = 0, x = 3$ [2.67]

55.3   Further worked problems on the area under a curve

Problem 6.   A gas expands according to the law $pv = \text{constant}$. When the volume is $3\,\text{m}^3$ the pressure is $150\,\text{kPa}$. Given that work done $= \displaystyle\int_{v_1}^{v_2} p\,dv$, determine the work done as the gas expands from $2\,\text{m}^3$ to a volume of $6\,\text{m}^3$

$pv = \text{constant}$.   When $v = 3\,\text{m}^3$ and $p = 150\,\text{kPa}$ the constant is given by $(3 \times 150) = 450\,\text{kPa m}^3$ or $450\,\text{kJ}$.

Hence $pv = 450$, or $p = \dfrac{450}{v}$

Work done $= \displaystyle\int_2^6 \dfrac{450}{v}\,dv$

$= \left[450\ln v \right]_2^6 = 450[\ln 6 - \ln 2]$

$= 450\ln\dfrac{6}{2} = 450\ln 3 = \mathbf{494.4\,kJ}$

Problem 7.   Determine the area enclosed by the curve $y = 4\cos\left(\dfrac{\theta}{2}\right)$, the θ-axis and ordinates $\theta = 0$ and $\theta = \dfrac{\pi}{2}$

The curve $y = 4\cos(\theta/2)$ is shown in Fig. 55.8.

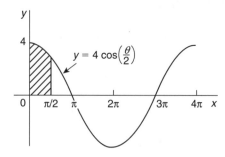

Figure 55.8

(Note that $y = 4\cos\left(\dfrac{\theta}{2}\right)$ has a maximum value of 4 and period $2\pi/(1/2)$, i.e. 4π rads.)

Shaded area $= \displaystyle\int_0^{\pi/2} y\,d\theta = \int_0^{\pi/2} 4\cos\dfrac{\theta}{2}\,d\theta$

$= \left[4\left(\dfrac{1}{\frac{1}{2}} \right)\sin\dfrac{\theta}{2} \right]_0^{\pi/2}$

$$= \left(8\sin\frac{\pi}{4}\right) - (8\sin 0)$$

$$= \textbf{5.657 square units}$$

Problem 8. Determine the area bounded by the curve $y = 3e^{t/4}$, the t-axis and ordinates $t = -1$ and $t = 4$, correct to 4 significant figures

A table of values is produced as shown.

t	-1	0	1	2	3	4
$y = 3e^{t/4}$	2.34	3.0	3.85	4.95	6.35	8.15

Since all the values of y are positive the area required is wholly above the t-axis.

Hence area $= \displaystyle\int_1^4 y\,dt$

$$= \int_1^4 3e^{t/4}dt = \left[\frac{3}{\left(\frac{1}{4}\right)}e^{t/4}\right]_{-1}^4$$

$$= 12\left[e^{t/4}\right]_{-1}^4 = 12(e^1 - e^{-1/4})$$

$$= 12(2.7183 - 0.7788)$$

$$= 12(1.9395) = \textbf{23.27 square units}$$

Problem 9. Sketch the curve $y = x^2 + 5$ between $x = -1$ and $x = 4$. Find the area enclosed by the curve, the x-axis and the ordinates $x = 0$ and $x = 3$. Determine also, by integration, the area enclosed by the curve and the y-axis, between the same limits

A table of values is produced and the curve $y = x^2 + 5$ plotted as shown in Fig. 55.9.

x	-1	0	1	2	3
y	6	5	6	9	14

Shaded area $= \displaystyle\int_0^3 y\,dx = \int_0^3 (x^2 + 5)\,dx$

$$= \left[\frac{x^3}{5} + 5x\right]_0^3$$

$$= \textbf{24 square units}$$

When $x = 3$, $y = 3^2 + 5 = 14$, and when $x = 0$, $y = 5$.

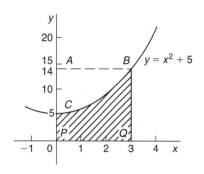

Figure 55.9

Since $y = x^2 + 5$ then $x^2 = y - 5$ and $x = \sqrt{y - 5}$
The area enclosed by the curve $y = x^2 + 5$ (i.e. $x = \sqrt{y - 5}$), the y-axis and the ordinates $y = 5$ and $y = 14$ (i.e. area ABC of Fig. 55.9) is given by:

$$\text{Area} = \int_{y=5}^{y=14} x\,dy = \int_5^{14} \sqrt{y - 5}\,dy$$

$$= \int_5^{14} (y - 5)^{1/2}\,dy$$

Let $u = y - 5$, then $\dfrac{du}{dy} = 1$ and $dy = du$

Hence $\displaystyle\int (y - 5)^{1/2}dy = \int u^{1/2}du = \frac{2}{3}u^{3/2}$
(for algebraic substitutions, see Chapter 49)
Since $u = y - 5$ then

$$\int_5^{14} \sqrt{y - 5}\,dy = \frac{2}{3}\left[(y - 5)^{3/2}\right]_5^{14}$$

$$= \frac{2}{3}[\sqrt{9^3} - 0]$$

$$= \textbf{18 square units}$$

(Check: From Fig. 55.9, area $BCPQ$ + area $ABC = 24 + 18 = 42$ square units, which is the area of rectangle $ABQP$.)

Problem 10. Determine the area between the curve $y = x^3 - 2x^2 - 8x$ and the x-axis

$$y = x^3 - 2x^2 - 8x = x(x^2 - 2x - 8)$$
$$= x(x + 2)(x - 4)$$

When $y = 0$, then $x = 0$ or $(x + 2) = 0$ or $(x - 4) = 0$, i.e. when $y = 0$, $x = 0$ or -2 or 4, which means that the curve crosses the x-axis at 0, -2 and 4. Since the curve is a continuous function, only one other co-ordinate value needs to be calculated before a sketch

Section 9

of the curve can be produced. When $x=1$, $y=-9$, showing that the part of the curve between $x=0$ and $x=4$ is negative. A sketch of $y=x^3-2x^2-8x$ is shown in Fig. 55.10. (Another method of sketching Fig. 55.10 would have been to draw up a table of values.)

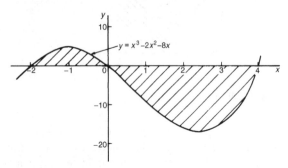

Figure 55.10

$$\text{Shaded area} = \int_{-2}^{0} (x^3 - 2x^2 - 8x)\,dx$$

$$- \int_{0}^{4} (x^3 - 2x^2 - 8x)\,dx$$

$$= \left[\frac{x^4}{4} - \frac{2x^3}{3} - \frac{8x^2}{2}\right]_{-2}^{0}$$

$$- \left[\frac{x^4}{4} - \frac{2x^3}{3} - \frac{8x^2}{2}\right]_{0}^{4}$$

$$= \left(6\frac{2}{3}\right) - \left(-42\frac{2}{3}\right)$$

$$= 49\frac{1}{3} \text{ square units}$$

Now try the following exercise

Exercise 192 Further problems on areas under curves

In Problems 1 and 2, find the area enclosed between the given curves, the horizontal axis and the given ordinates.

1. $y=2x^3$; $x=-2, x=2$ [16 square units]

2. $xy=4$; $x=1, x=4$ [5.545 square units]

3. The force F newtons acting on a body at a distance x metres from a fixed point is given by: $F=3x+2x^2$. If work done $= \int_{x_1}^{x_2} F\,dx$, determine the work done when the body moves from the position where $x=1$ m to that where $x=3$ m.   [29.33 Nm]

4. Find the area between the curve $y=4x-x^2$ and the x-axis.   [10.67 square units]

5. Determine the area enclosed by the curve $y=5x^2+2$, the x-axis and the ordinates $x=0$ and $x=3$. Find also the area enclosed by the curve and the y-axis between the same limits.   [51 sq. units, 90 sq. units]

6. Calculate the area enclosed between $y=x^3-4x^2-5x$ and the x-axis.   [73.83 sq. units]

7. The velocity v of a vehicle t seconds after a certain instant is given by: $v=(3t^2+4)$ m/s. Determine how far it moves in the interval from $t=1$ s to $t=5$ s.   [140 m]

8. A gas expands according to the law $pv=$ constant. When the volume is $2\,\text{m}^3$ the pressure is $250\,\text{kPa}$. Find the work done as the gas expands from $1\,\text{m}^3$ to a volume of $4\,\text{m}^3$ given that work done $= \int_{v_1}^{v_2} p\,dv$ [693.1 kJ]

55.4 The area between curves

The area enclosed between curves $y=f_1(x)$ and $y=f_2(x)$ (shown shaded in Fig. 55.11) is given by:

$$\text{shaded area} = \int_{a}^{b} f_2(x)\,dx - \int_{a}^{b} f_1(x)\,dx$$

$$= \int_{a}^{b} [f_2(x) - f_2(x)]\,dx$$

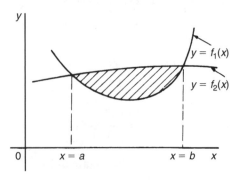

Figure 55.11

Problem 11. Determine the area enclosed between the curves $y = x^2 + 1$ and $y = 7 - x$

At the points of intersection, the curves are equal. Thus, equating the y-values of each curve gives: $x^2 + 1 = 7 - x$, from which $x^2 + x - 6 = 0$. Factorising gives $(x - 2)(x + 3) = 0$, from which, $x = 2$ and $x = -3$. By firstly determining the points of intersection the range of x-values has been found. Tables of values are produced as shown below.

x	-3	-2	-1	0	1	2
$y = x^2 + 1$	10	5	2	1	2	5

x	-3	0	2
$y = 7 - x$	10	7	5

A sketch of the two curves is shown in Fig. 55.12.

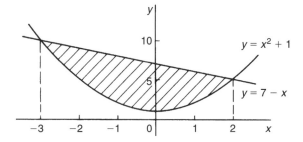

Figure 55.12

$$\text{Shaded area} = \int_{-3}^{2} (7 - x)dx - \int_{-3}^{2} (x^2 + 1)dx$$

$$= \int_{-3}^{2} [(7 - x) - (x^2 + 1)]dx$$

$$= \int_{-3}^{2} (6 - x - x^2)dx$$

$$= \left[6x - \frac{x^2}{2} - \frac{x^3}{3}\right]_{-3}^{2}$$

$$= \left(12 - 2 - \frac{8}{3}\right) - \left(-18 - \frac{9}{2} + 9\right)$$

$$= \left(7\frac{1}{3}\right) - \left(-13\frac{1}{2}\right)$$

$$= 20\frac{5}{6} \text{ square units}$$

Problem 12. (a) Determine the coordinates of the points of intersection of the curves $y = x^2$ and $y^2 = 8x$. (b) Sketch the curves $y = x^2$ and $y^2 = 8x$ on the same axes. (c) Calculate the area enclosed by the two curves

(a) At the points of intersection the coordinates of the curves are equal. When $y = x^2$ then $y^2 = x^4$.

Hence at the points of intersection $x^4 = 8x$, by equating the y^2 values.

Thus $x^4 - 8x = 0$, from which $x(x^3 - 8) = 0$, i.e. $x = 0$ or $(x^3 - 8) = 0$.

Hence at the points of intersection $x = 0$ or $x = 2$.

When $x = 0$, $y = 0$ and when $x = 2$, $y = 2^2 = 4$.

Hence the points of intersection of the curves $y = x^2$ and $y^2 = 8x$ are (0, 0) and (2, 4)

(b) A sketch of $y = x^2$ and $y^2 = 8x$ is shown in Fig. 55.13

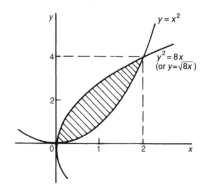

Figure 55.13

(c) **Shaded area** $= \int_{0}^{2} \{\sqrt{8x} - x^2\}dx$

$$= \int_{0}^{2} \{(\sqrt{8})x^{1/2} - x^2\}dx$$

$$= \left[(\sqrt{8})\frac{x^{3/2}}{(\frac{3}{2})} - \frac{x^3}{3}\right]_{0}^{2}$$

$$= \left\{\frac{\sqrt{8}\sqrt{8}}{(\frac{3}{2})} - \frac{8}{3}\right\} - \{0\}$$

$$= \frac{16}{3} - \frac{8}{3} = \frac{8}{3}$$

$$= 2\frac{2}{3} \text{ square units}$$

Problem 13. Determine by integration the area bounded by the three straight lines $y = 4 - x$, $y = 3x$ and $3y = x$

Each of the straight lines is shown sketched in Fig. 55.14.

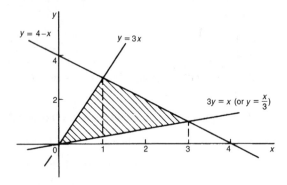

Figure 55.14

Shaded area $= \displaystyle\int_0^1 \left(3x - \frac{x}{3}\right) dx$

$$+ \int_1^3 \left[(4 - x) - \frac{x}{3}\right] dx$$

$$= \left[\frac{3x^2}{2} - \frac{x^2}{6}\right]_0^1 + \left[4x - \frac{x^2}{2} - \frac{x^2}{6}\right]_1^3$$

$$= \left[\left(\frac{3}{2} - \frac{1}{6}\right) - (0)\right]$$

$$+ \left[\left(12 - \frac{9}{2} - \frac{9}{6}\right) - \left(4 - \frac{1}{2} - \frac{1}{6}\right)\right]$$

$$= \left(1\frac{1}{3}\right) + \left(6 - 3\frac{1}{3}\right)$$

$$= \textbf{4 square units}$$

Now try the following exercise

Exercise 193   Further problems on areas between curves

1.   Determine the coordinates of the points of intersection and the area enclosed between the parabolas $y^2 = 3x$ and $x^2 = 3y$.
         [(0, 0) and (3, 3), 3 sq. units]

2.   Sketch the curves $y = x^2 + 3$ and $y = 7 - 3x$ and determine the area enclosed by them.
         [20.83 square units]

3.   Determine the area enclosed by the curves $y = \sin x$ and $y = \cos x$ and the y-axis.
         [0.4142 square units]

4.   Determine the area enclosed by the three straight lines $y = 3x$, $2y = x$ and $y + 2x = 5$
         [2.5 sq. units]

Chapter 56

Mean and root mean square values

56.1 Mean or average values

(i) The mean or average value of the curve shown in Fig. 56.1, between $x = a$ and $x = b$, is given by:

mean or average value,

$$\bar{y} = \frac{\text{area under curve}}{\text{length of base}}$$

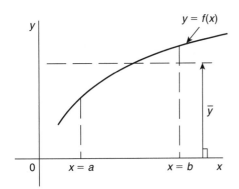

Figure 56.1

(ii) When the area under a curve may be obtained by integration then:

mean or average value,

$$\bar{y} = \frac{\int_a^b y\, dx}{b - a}$$

i.e. $$\bar{y} = \frac{1}{b - a} \int_a^b f(x)\, dx$$

(iii) For a periodic function, such as a sine wave, the mean value is assumed to be 'the mean value over

half a cycle', since the mean value over a complete cycle is zero.

Problem 1. Determine, using integration, the mean value of $y = 5x^2$ between $x = 1$ and $x = 4$

Mean value,

$$\bar{y} = \frac{1}{4 - 1} \int_1^4 y\, dx = \frac{1}{3} \int_1^4 5x^2\, dx$$

$$= \frac{1}{3} \left[\frac{5x^3}{3} \right]_1^4 = \frac{5}{9} [x^3]_1^4 = \frac{5}{9}(64 - 1) = \mathbf{35}$$

Problem 2. A sinusoidal voltage is given by $v = 100 \sin \omega t$ volts. Determine the mean value of the voltage over half a cycle using integration

Half a cycle means the limits are 0 to π radians. Mean value,

$$\bar{v} = \frac{1}{\pi - 0} \int_0^\pi v\, d(\omega t)$$

$$= \frac{1}{\pi} \int_0^\pi 100 \sin \omega t\, d(\omega t) = \frac{100}{\pi} [-\cos \omega t]_0^\pi$$

$$= \frac{100}{\pi} [(-\cos \pi) - (-\cos 0)]$$

$$= \frac{100}{\pi} [(+1) - (-1)] = \frac{200}{\pi}$$

$$= \mathbf{63.66\ volts}$$

[Note that for a sine wave,

$$\mathbf{mean\ value = \frac{2}{\pi} \times maximum\ value}$$

In this case, mean value $= \dfrac{2}{\pi} \times 100 = 63.66\,\text{V}$]

Problem 3. Calculate the mean value of $y = 3x^2 + 2$ in the range $x = 0$ to $x = 3$ by (a) the mid-ordinate rule and (b) integration

(a) A graph of $y = 3x^2$ over the required range is shown in Fig. 56.2 using the following table:

x	0	0.5	1.0	1.5	2.0	2.5	3.0
y	2.0	2.75	5.0	8.75	14.0	20.75	29.0

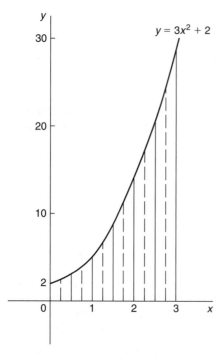

Figure 56.2

Using the mid-ordinate rule, mean value

$$= \frac{\text{area under curve}}{\text{length of base}}$$

$$= \frac{\text{sum of mid-ordinates}}{\text{number of mid-ordinates}}$$

Selecting 6 intervals, each of width 0.5, the mid-ordinates are erected as shown by the broken lines in Fig. 56.2.

$$\text{Mean value} = \frac{\begin{array}{c} 2.2 + 3.7 + 6.7 + 11.2 \\ + 17.2 + 24.7 \end{array}}{6}$$

$$= \frac{65.7}{6} = \mathbf{10.95}$$

(b) By integration, mean value

$$= \frac{1}{3 - 0} \int_0^3 y\, dx = \frac{1}{3} \int_0^3 (3x^2 + 2)\, dx$$

$$= \frac{1}{3}[x^3 + 2x]_0^3 = \frac{1}{3}[(27 + 6) - (0)]$$

$$= \mathbf{11}$$

The answer obtained by integration is exact; greater accuracy may be obtained by the mid-ordinate rule if a larger number of intervals are selected.

Problem 4. The number of atoms, N, remaining in a mass of material during radioactive decay after time t seconds is given by: $N = N_0 e^{-\lambda t}$, where N_0 and λ are constants. Determine the mean number of atoms in the mass of material for the time period $t = 0$ and $t = \dfrac{1}{\lambda}$

Mean number of atoms

$$= \frac{1}{\frac{1}{\lambda} - 0} \int_0^{1/\lambda} N\, dt = \frac{1}{\frac{1}{\lambda}} \int_0^{1/\lambda} N_0 e^{-\lambda t}\, dt$$

$$= \lambda N_0 \int_0^{1/\lambda} e^{-\lambda t}\, dt = \lambda N_0 \left[\frac{e^{-\lambda t}}{-\lambda}\right]_0^{1/\lambda}$$

$$= -N_0 [e^{-\lambda(1/\lambda)} - e^0] = -N_0 [e^{-1} - e^0]$$

$$= +N_0 [e^0 - e^{-1}] = N_0[1 - e^{-1}] = \mathbf{0.632\,N_0}$$

Now try the following exercise

Exercise 194   Further problems on mean or average values

1. Determine the mean value of (a) $y = 3\sqrt{x}$ from $x = 0$ to $x = 4$ (b) $y = \sin 2\theta$ from $\theta = 0$ to $\theta = \dfrac{\pi}{4}$ (c) $y = 4e^t$ from $t = 1$ to $t = 4$

$$\left[\text{(a) } 4 \quad \text{(b) } \frac{2}{\pi} \text{ or } 0.637 \quad \text{(c) } 69.17\right]$$

2. Calculate the mean value of $y = 2x^2 + 5$ in the range $x = 1$ to $x = 4$ by (a) the mid-ordinate rule, and (b) integration. [19]

3. The speed v of a vehicle is given by: $v = (4t + 3)$ m/s, where t is the time in seconds. Determine the average value of the speed from $t = 0$ to $t = 3$ s. [9 m/s]

4. Find the mean value of the curve $y = 6 + x - x^2$ which lies above the x-axis by using an approximate method. Check the result using integration. [4.17]

5. The vertical height h km of a missile varies with the horizontal distance d km, and is given by $h = 4d - d^2$. Determine the mean height of the missile from $d = 0$ to $d = 4$ km. [2.67 km]

6. The velocity v of a piston moving with simple harmonic motion at any time t is given by: $v = c \sin \omega t$, where c is a constant. Determine the mean velocity between $t = 0$ and $t = \dfrac{\pi}{\omega}$
$$\left[\frac{2c}{\pi}\right]$$

56.2 Root mean square values

The **root mean square value** of a quantity is 'the sqaure root of the mean value of the squared values of the quantity' taken over an interval. With reference to Fig. 56.1, the r.m.s. value of $y = f(x)$ over the range $x = a$ to $x = b$ is given by:

$$\text{r.m.s. value} = \sqrt{\frac{1}{b-a} \int_a^b y^2 \, dx}$$

One of the principal applications of r.m.s. values is with alternating currents and voltages. The r.m.s. value of an alternating current is defined as that current which will give the same heating effect as the equivalent direct current.

Problem 5. Determine the r.m.s. value of $y = 2x^2$ between $x = 1$ and $x = 4$

R.m.s. value

$$= \sqrt{\frac{1}{4-1} \int_1^4 y^2 \, dx} = \sqrt{\frac{1}{3} \int_1^4 (2x^2)^2 \, dx}$$

$$= \sqrt{\frac{1}{3} \int_1^4 4x^4 \, dx} = \sqrt{\frac{4}{3} \left[\frac{x^5}{5}\right]_1^4}$$

$$= \sqrt{\frac{4}{15}(1024 - 1)} = \sqrt{272.8} = \mathbf{16.5}$$

Problem 6. A sinusoidal voltage has a maximum value of 100 V. Calculate its r.m.s. value

A sinusoidal voltage v having a maximum value of 100 V may be written as: $v = 100 \sin \theta$. Over the range $\theta = 0$ to $\theta = \pi$, ·

r.m.s. value

$$= \sqrt{\frac{1}{\pi - 0} \int_0^\pi v^2 \, d\theta}$$

$$= \sqrt{\frac{1}{\pi} \int_0^\pi (100 \sin \theta)^2 \, d\theta}$$

$$= \sqrt{\frac{10\,000}{\pi} \int_0^\pi \sin^2 \theta \, d\theta}$$

which is not a 'standard' integral. It is shown in Chapter 27 that $\cos 2A = 1 - 2\sin^2 A$ and this formula is used whenever $\sin^2 A$ needs to be integrated. Rearranging $\cos 2A = 1 - 2\sin^2 A$ gives $\sin^2 A = \frac{1}{2}(1 - \cos 2A)$

Hence $\sqrt{\dfrac{10\,000}{\pi} \displaystyle\int_0^\pi \sin^2 \theta \, d\theta}$

$$= \sqrt{\frac{10\,000}{\pi} \int_0^\pi \frac{1}{2}(1 - \cos 2\theta) \, d\theta}$$

$$= \sqrt{\frac{10\,000}{\pi} \frac{1}{2} \left[\theta - \frac{\sin 2\theta}{2}\right]_0^\pi}$$

$$= \sqrt{\frac{10\,000}{\pi} \frac{1}{2} \left[\left(\pi - \frac{\sin 2\pi}{2}\right) - \left(0 - \frac{\sin 0}{2}\right)\right]}$$

$$= \sqrt{\frac{10\,000}{\pi} \frac{1}{2}[\pi]} = \sqrt{\frac{10\,000}{2}}$$

$$= \frac{100}{\sqrt{2}} = \mathbf{70.71 \text{ volts}}$$

[Note that for a sine wave,
$$\textbf{r.m.s. value} = \frac{1}{\sqrt{2}} \times \textbf{maximum value}.$$

In this case, r.m.s. value $= \dfrac{1}{\sqrt{2}} \times 100 = 70.71$ V]

Problem 7. In a frequency distribution the average distance from the mean, y, is related to the variable, x, by the equation $y = 2x^2 - 1$. Determine, correct to 3 significant figures, the r.m.s. deviation from the mean for values of x from -1 to $+4$

R.m.s. deviation

$$= \sqrt{\frac{1}{4 - -1} \int_{-1}^{4} y^2 \, dx}$$

$$= \sqrt{\frac{1}{5} \int_{-1}^{4} (2x^2 - 1)^2 dx}$$

$$= \sqrt{\frac{1}{5} \int_{-1}^{4} (4x^4 - 4x^2 + 1) dx}$$

$$= \sqrt{\frac{1}{5} \left[\frac{4x^5}{5} - \frac{4x^3}{3} + x \right]_{-1}^{4}}$$

$$= \sqrt{\frac{1}{5} \left[\left(\frac{4}{5}(4)^5 - \frac{4}{3}(4)^3 + 4 \right) - \left(\frac{4}{5}(-1)^5 - \frac{4}{3}(-1)^3 + (-1) \right) \right]}$$

$$= \sqrt{\frac{1}{5}[(737.87) - (-0.467)]}$$

$$= \sqrt{\frac{1}{5}[738.34]}$$

$$= \sqrt{147.67} = 12.152 = \mathbf{12.2},$$

correct to 3 significant figures.

Now try the following exercise

Exercise 195 Further problems on root mean square values

1. Determine the r.m.s. values of:
   (a) $y = 3x$ from $x = 0$ to $x = 4$
   (b) $y = t^2$ from $t = 1$ to $t = 3$
   (c) $y = 25 \sin \theta$ from $\theta = 0$ to $\theta = 2\pi$

$$\left[\text{(a) } 6.928 \quad \text{(b) } 4.919 \quad \text{(c) } \frac{25}{\sqrt{2}} \text{ or } 17.68 \right]$$

2. Calculate the r.m.s. values of:
   (a) $y = \sin 2\theta$ from $\theta = 0$ to $\theta = \frac{\pi}{4}$
   (b) $y = 1 + \sin t$ from $t = 0$ to $t = 2\pi$
   (c) $y = 3 \cos 2x$ from $x = 0$ to $x = \pi$

   (Note that $\cos^2 t = \frac{1}{2}(1 + \cos 2t)$, from Chapter 27).

$$\left[\text{(a) } \frac{1}{\sqrt{2}} \text{ or } 0.707 \quad \text{(b) } 1.225 \quad \text{(c) } 2.121 \right]$$

3. The distance, p, of points from the mean value of a frequency distribution are related to the variable, q, by the equation $p = \frac{1}{q} + q$. Determine the standard deviation (i.e. the r.m.s. value), correct to 3 significant figures, for values from $q = 1$ to $q = 3$. [2.58]

4. A current, $i = 30 \sin 100\pi t$ amperes is applied across an electric circuit. Determine its mean and r.m.s. values, each correct to 4 significant figures, over the range $t = 0$ to $t = 10$ ms. [19.10 A, 21.21 A]

5. A sinusoidal voltage has a peak value of 340 V. Calculate its mean and r.m.s. values, correct to 3 significant figures. [216 V, 240 V]

6. Determine the form factor, correct to 3 significant figures, of a sinusoidal voltage of maximum value 100 volts, given that form factor $= \frac{\text{r.m.s. value}}{\text{average value}}$ [1.11]

7. A wave is defined by the equation:
$$v = E_1 \sin \omega t + E_3 \sin 3\omega t$$
where, E_1, E_3 and ω are constants.
Determine the r.m.s. value of v over the interval $0 \leq t \leq \frac{\pi}{\omega}$
$$\left[\sqrt{\frac{E_1^2 + E_3^2}{2}} \right]$$

Chapter 57

Volumes of solids of revolution

57.1 Introduction

If the area under the curve $y = f(x)$, (shown in Fig. 57.1(a)), between $x = a$ and $x = b$ is rotated $360°$ about the x-axis, then a volume known as a **solid of revolution** is produced as shown in Fig. 57.1(b).

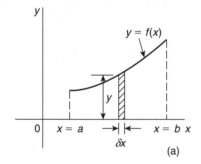

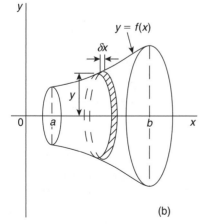

Figure 57.1

The volume of such a solid may be determined precisely using integration.

(i) Let the area shown in Fig. 57.1(a) be divided into a number of strips each of width δx. One such strip is shown shaded.

(ii) When the area is rotated $360°$ about the x-axis, each strip produces a solid of revolution approximating to a circular disc of radius y and thickness δx. Volume of disc = (circular cross-sectional area) (thickness) = $(\pi y^2)(\delta x)$

(iii) Total volume, V, between ordinates $x = a$ and $x = b$ is given by:

$$\textbf{Volume } V = \lim_{\delta x \to 0} \sum_{x=a}^{x=b} \pi y^2 \, \delta x = \int_a^b \pi y^2 \, dx$$

If a curve $x = f(y)$ is rotated about the y-axis $360°$ between the limits $y = c$ and $y = d$, as shown in Fig. 57.2, then the volume generated is given by:

$$\textbf{Volume } V = \lim_{\delta y \to 0} \sum_{y=c}^{y=d} \pi x^2 \, \delta y = \int_c^d \pi x^2 \, dy$$

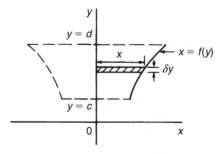

Figure 57.2

57.2 Worked problems on volumes of solids of revolution

Problem 1. Determine the volume of the solid of revolution formed when the curve $y = 2$ is rotated $360°$ about the x-axis between the limits $x = 0$ to $x = 3$

When $y = 2$ is rotated $360°$ about the x-axis between $x = 0$ and $x = 3$ (see Fig. 57.3):

volume generated

$$= \int_0^3 \pi y^2 \, dx = \int_0^3 \pi(2)^2 \, dx$$

$$= \int_0^3 4\pi \, dx = 4\pi[x]_0^3 = \mathbf{12\pi \text{ cubic units}}$$

[Check: The volume generated is a cylinder of radius 2 and height 3.
Volume of cylinder $= \pi r^2 h = \pi(2)^2(3) = \mathbf{12\pi \text{ cubic units}}.$]

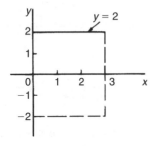

Figure 57.3

Problem 2. Find the volume of the solid of revolution when the cure $y = 2x$ is rotated one revolution about the x-axis between the limits $x = 0$ and $x = 5$

When $y = 2x$ is revolved one revolution about the x-axis between $x = 0$ and $x = 5$ (see Fig. 57.4) then:

volume generated

$$= \int_0^5 \pi y^2 \, dx = \int_0^5 \pi(2x)^2 \, dx$$

$$= \int_0^5 4\pi x^2 \, dx = 4\pi \left[\frac{x^3}{3} \right]_0^5$$

$$= \frac{500\pi}{3} = 166\frac{2}{3}\pi \text{ cubic units}$$

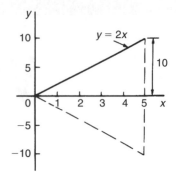

Figure 57.4

[Check: The volume generated is a cone of radius 10 and height 5. Volume of cone

$$= \frac{1}{3}\pi r^2 h = \frac{1}{3}\pi(10)^2 5 = \frac{500\pi}{3}$$

$$= \mathbf{166\frac{2}{3}\pi \text{ cubic units}.}]$$

Problem 3. The curve $y = x^2 + 4$ is rotated one revolution about the x-axis between the limits $x = 1$ and $x = 4$. Determine the volume of the solid of revolution produced

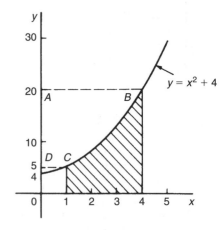

Figure 57.5

Revolving the shaded area shown in Fig. 57.5 about the x-axis $360°$ produces a solid of revolution given by:

$$\text{Volume} = \int_1^4 \pi y^2 \, dx = \int_1^4 \pi(x^2 + 4)^2 \, dx$$

$$= \int_1^4 \pi(x^4 + 8x^2 + 16) \, dx$$

$$= \pi \left[\frac{x^5}{5} + \frac{8x^3}{3} + 16x \right]_1^4$$

$$= \pi[(204.8 + 170.67 + 64) - (0.2 + 2.67 + 16)]$$

$$= \textbf{420.6}\pi \textbf{ cubic units}$$

Problem 4. If the curve in Problem 3 is revolved about the y-axis between the same limits, determine the volume of the solid of revolution produced

The volume produced when the curve $y = x^2 + 4$ is rotated about the y-axis between $y = 5$ (when $x = 1$) and $y = 20$ (when $x = 4$), i.e. rotating area ABCD of Fig. 57.5 about the y-axis is given by:

$$\text{volume} = \int_5^{20} \pi x^2 \, dy$$

Since $y = x^2 + 4$, then $x^2 = y - 4$

$$\text{Hence volume} = \int_5^{20} \pi(y - 4)dy = \pi \left[\frac{y^2}{2} - 4y \right]_5^{20}$$

$$= \pi[(120) - (-7.5)]$$

$$= \textbf{127.5}\pi \textbf{ cubic units}$$

Now try the following exercise

Exercise 196   Further problems on volumes of solids of revolution

(Answers are in cubic units and in terms of π).

In Problems 1 to 5, determine the volume of the solid of revolution formed by revolving the areas enclosed by the given curve, the x-axis and the given ordinates through one revolution about the x-axis.

1. $y = 5x$; $x = 1, x = 4$ $[525\pi]$

2. $y = x^2$; $x = -2, x = 3$ $[55\pi]$

3. $y = 2x^2 + 3$; $x = 0, x = 2$ $[75.6\pi]$

4. $\dfrac{y^2}{4} = x$; $x = 1, x = 5$ $[48\pi]$

5. $xy = 3$; $x = 2, x = 3$ $[1.5\pi]$

In Problems 6 to 8, determine the volume of the solid of revolution formed by revolving the areas enclosed by the given curves, the y-axis and the given ordinates through one revolution about the y-axis.

6. $y = x^2$; $y = 1, y = 3$ $[4\pi]$

7. $y = 3x^2 - 1$; $y = 2, y = 4$ $[2.67\pi]$

8. $y = \dfrac{2}{x}$; $y = 1, y = 3$ $[2.67\pi]$

9. The curve $y = 2x^2 + 3$ is rotated about (a) the x-axis between the limits $x = 0$ and $x = 3$, and (b) the y-axis, between the same limits. Determine the volume generated in each case. $[(a)\ 329.4\pi \quad (b)\ 81\pi]$

57.3   Further worked problems on volumes of solids of revolution

Problem 5. The area enclosed by the curve $y = 3e^{\frac{x}{3}}$, the x-axis and ordinates $x = -1$ and $x = 3$ is rotated $360°$ about the x-axis. Determine the volume generated

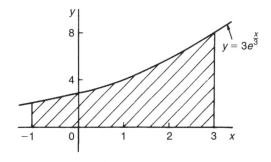

Figure 57.6

A sketch of $y = 3e^{\frac{x}{3}}$ is shown in Fig. 57.6. When the shaded area is rotated $360°$ about the x-axis then:

$$\text{volume generated} = \int_{-1}^{3} \pi y^2 \, dx$$

$$= \int_{-1}^{3} \pi \left(3e^{\frac{x}{3}} \right)^2 dx$$

$$= 9\pi \int_{-1}^{3} e^{\frac{2x}{3}} \, dx$$

$$= 9\pi \left[\frac{e^{\frac{2x}{3}}}{\frac{2}{3}} \right]_{-1}^{3}$$

$$= \frac{27\pi}{2} \left(e^2 - e^{-\frac{2}{3}} \right)$$

$$= \mathbf{92.82\pi} \text{ cubic units}$$

Problem 6. Determine the volume generated when the area above the x-axis bounded by the curve $x^2 + y^2 = 9$ and the ordinates $x = 3$ and $x = -3$ is rotated one revolution about the x-axis

Figure 57.7 shows the part of the curve $x^2 + y^2 = 9$ lying above the x-axis, Since, in general, $x^2 + y^2 = r^2$ represents a circle, centre 0 and radius r, then $x^2 + y^2 = 9$ represents a circle, centre 0 and radius 3. When the semi-circular area of Fig. 57.7 is rotated one revolution about the x-axis then:

$$\text{volume generated} = \int_{-3}^{3} \pi y^2 dx$$

$$= \int_{-3}^{3} \pi (9 - x^2) \, dx$$

$$= \pi \left[9x - \frac{x^3}{3} \right]_{-3}^{3}$$

$$= \pi[(18) - (-18)]$$

$$= \mathbf{36\pi} \text{ cubic units}$$

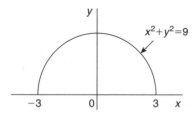

Figure 57.7

(Check: The volume generated is a sphere of radius 3. Volume of sphere $= \frac{4}{3}\pi r^3 = \frac{4}{3}\pi(3)^3 =$ **36π cubic units**.)

Problem 7. Calculate the volume of a frustum of a sphere of radius 4 cm that lies between two parallel planes at 1 cm and 3 cm from the centre and on the same side of it

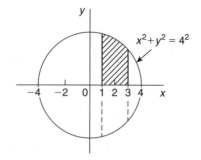

Figure 57.8

The volume of a frustum of a sphere may be determined by integration by rotating the curve $x^2 + y^2 = 4^2$ (i.e. a circle, centre 0, radius 4) one revolution about the x-axis, between the limits $x = 1$ and $x = 3$ (i.e. rotating the shaded area of Fig. 57.8).

$$\text{Volume of frustum} = \int_{1}^{3} \pi y^2 \, dx$$

$$= \int_{1}^{3} \pi (4^2 - x^2) \, dx$$

$$= \pi \left[16x - \frac{x^3}{3} \right]_{1}^{3}$$

$$= \pi \left[(39) - \left(15\frac{2}{3} \right) \right]$$

$$= \mathbf{23\frac{1}{3}\pi} \text{ cubic units}$$

Problem 8. The area enclosed between the two parabolas $y = x^2$ and $y^2 = 8x$ of Problem 12, Chapter 55, page 485, is rotated 360° about the x-axis. Determine the volume of the solid produced

The area enclosed by the two curves is shown in Fig. 55.13, page 485. The volume produced by revolving the shaded area about the x-axis is given by: [(volume produced by revolving $y^2 = 8x$) − (volume produced by revolving $y = x^2$)]

i.e. **volume** $= \int_{0}^{2} \pi(8x) \, dx - \int_{0}^{2} \pi(x^4) \, dx$

$$= \pi \int_{0}^{2} (8x - x^4) \, dx = \pi \left[\frac{8x^2}{2} - \frac{x^5}{5} \right]_{0}^{2}$$

$$= \pi \left[\left(16 - \frac{32}{5} \right) - (0) \right]$$

$$= \mathbf{9.6\pi} \text{ cubic units}$$

Now try the following exercise

Exercise 197   Further problems on volumes of solids of revolution

(Answers to volumes are in cubic units and in terms of π.)

In Problems 1 and 2, determine the volume of the solid of revolution formed by revolving the areas enclosed by the given curve, the x-axis and the given ordinates through one revolution about the x-axis.

1. $y = 4e^x$; $x = 0$; $x = 2$ $[428.8\pi]$

2. $y = \sec x$; $x = 0$, $x = \dfrac{\pi}{4}$ $[\pi]$

In Problems 3 and 4, determine the volume of the solid of revolution formed by revolving the areas enclosed by the given curves, the y-axis and the given ordinates through one revolution about the y-axis.

3. $x^2 + y^2 = 16$; $y = 0$, $y = 4$ $[42.67\pi]$

4. $x\sqrt{y} = 2$; $y = 2$, $y = 3$ $[1.622\pi]$

5. Determine the volume of a plug formed by the frustum of a sphere of radius 6 cm which lies between two parallel planes at 2 cm and 4 cm from the centre and on the same side of it. (The equation of a circle, centre 0, radius r is $x^2 + y^2 = r^2$). $[53.33\pi]$

6. The area enclosed between the two curves $x^2 = 3y$ and $y^2 = 3x$ is rotated about the x-axis. Determine the volume of the solid formed. $[8.1\pi]$

7. The portion of the curve $y = x^2 + \dfrac{1}{x}$ lying between $x = 1$ and $x = 3$ is revolved $360°$ about the x-axis. Determine the volume of the solid formed. $[57.07\pi]$

8. Calculate the volume of the frustum of a sphere of radius 5 cm that lies between two parallel planes at 3 cm and 2 cm from the centre and on opposite sides of it. $[113.33\pi]$

9. Sketch the curves $y = x^2 + 2$ and $y - 12 = 3x$ from $x = -3$ to $x = 6$. Determine (a) the co-ordinates of the points of intersection of the two curves, and (b) the area enclosed by the two curves. (c) If the enclosed area is rotated $360°$ about the x-axis, calculate the volume of the solid produced

$$\left[\begin{array}{l} \text{(a) } (-2, 6) \text{ and } (5, 27) \\ \text{(b) } 57.17 \text{ square units} \\ \text{(c) } 1326\pi \text{ cubic units} \end{array}\right]$$

Chapter 58

Centroids of simple shapes

58.1 Centroids

A **lamina** is a thin flat sheet having uniform thickness. The **centre of gravity** of a lamina is the point where it balances perfectly, i.e. the lamina's **centre of mass**. When dealing with an area (i.e. a lamina of negligible thickness and mass) the term **centre of area** or **centroid** is used for the point where the centre of gravity of a lamina of that shape would lie.

58.2 The first moment of area

The **first moment of area** is defined as the product of the area and the perpendicular distance of its centroid from a given axis in the plane of the area. In Fig. 58.1, the first moment of area A about axis XX is given by (Ay) cubic units.

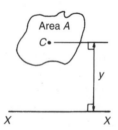

Figure 58.1

58.3 Centroid of area between a curve and the x-axis

(i) Figure 58.2 shows an area $PQRS$ bounded by the curve $y = f(x)$, the x-axis and ordinates $x = a$ and $x = b$. Let this area be divided into a large number of strips, each of width δx. A typical strip is shown shaded drawn at point (x, y) on $f(x)$. The area of the strip is approximately rectangular and is given by $y\delta x$. The centroid, C, has coordinates $\left(x, \dfrac{y}{2}\right)$.

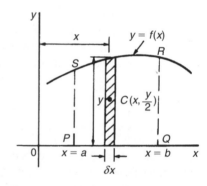

Figure 58.2

(ii) First moment of area of shaded strip about axis $Oy = (y\delta x)(x) = xy\delta x$.

Total first moment of area $PQRS$ about axis $Oy = \lim\limits_{\delta x \to 0} \sum_{x=a}^{x=b} xy\delta x = \int_a^b xy\,dx$

(iii) First moment of area of shaded strip about axis $Ox = (y\delta x)\left(\dfrac{y}{2}\right) = \dfrac{1}{2}y^2x$.

Total first moment of area $PQRS$ about axis $Ox = \lim\limits_{\delta x \to 0} \sum_{x=a}^{x=b} \dfrac{1}{2}y^2\delta x = \dfrac{1}{2}\int_a^b y^2\,dx$

(iv) Area of $PQRS$, $A = \int_a^b y\,dx$ (from Chapter 55)

(v) Let $\bar{x}$ and $\bar{y}$ be the distances of the centroid of area A about Oy and Ox respectively then: $(\bar{x})(A) =$ total first moment of area A about axis $Oy = \int_a^b xy\,dx$

from which, $\qquad \bar{x} = \dfrac{\displaystyle\int_a^b xy\,dx}{\displaystyle\int_a^b y\,dx}$

and $(\bar{y})(A)=$ total moment of area A about axis

$Ox = \dfrac{1}{2}\displaystyle\int_a^b y^2\,dx$

from which, $\bar{y} = \dfrac{\dfrac{1}{2}\displaystyle\int_a^b y^2\,dx}{\displaystyle\int_a^b y\,dx}$

58.4 Centroid of area between a curve and the y-axis

If $\bar{x}$ and $\bar{y}$ are the distances of the centroid of area EFGH in Fig. 58.3 from Oy and Ox respectively, then, by similar reasoning as above:

$(\bar{x})(\text{total area}) = \displaystyle\lim_{\delta y \to 0}\sum_{y=c}^{y=d} x\,\delta y\left(\frac{x}{2}\right) = \frac{1}{2}\int_c^d x^2\,dy$

from which, $\bar{x} = \dfrac{\dfrac{1}{2}\displaystyle\int_c^d x^2\,dy}{\displaystyle\int_c^d x\,dy}$

and $(\bar{y})(\text{total area}) = \displaystyle\lim_{\delta y \to 0}\sum_{y=c}^{y=d}(x\,\delta y)y = \int_c^d xy\,dy$

from which, $\bar{y} = \dfrac{\displaystyle\int_c^d xy\,dy}{\displaystyle\int_c^d x\,dy}$

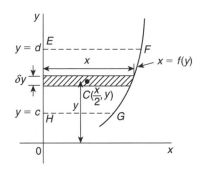

Figure 58.3

58.5 Worked problems on centroids of simple shapes

Problem 1. Show, by integration, that the centroid of a rectangle lies at the intersection of the diagonals

Let a rectangle be formed by the line $y=b$, the x-axis and ordinates $x=0$ and $x=l$ as shown in Fig. 58.4. Let the coordinates of the centroid C of this area be $(\bar{x},\bar{y})$.

By integration, $\bar{x} = \dfrac{\displaystyle\int_0^l xy\,dx}{\displaystyle\int_0^l y\,dx} = \dfrac{\displaystyle\int_0^l (x)(b)\,dx}{\displaystyle\int_0^l b\,dx}$

$= \dfrac{\left[b\dfrac{x^2}{2}\right]_0^l}{[bx]_0^l} = \dfrac{\dfrac{bl^2}{2}}{bl} = \dfrac{1}{2}$

and $\bar{y} = \dfrac{\dfrac{1}{2}\displaystyle\int_0^l y^2\,dx}{\displaystyle\int_0^l y\,dx} = \dfrac{\dfrac{1}{2}\displaystyle\int_0^l b^2\,dx}{bl}$

$= \dfrac{\dfrac{1}{2}[b^2 x]_0^l}{bl} = \dfrac{\dfrac{b^2 l}{2}}{bl} = \dfrac{b}{2}$

Figure 58.4

i.e. **the centroid lies at** $\left(\dfrac{l}{2}, \dfrac{b}{2}\right)$ **which is at the intersection of the diagonals.**

Problem 2. Find the position of the centroid of the area bounded by the curve $y=3x^2$, the x-axis and the ordinates $x=0$ and $x=2$

If, $(\bar{x}, \bar{y})$ are the co-ordinates of the centroid of the given area then:

$\bar{x} = \dfrac{\displaystyle\int_0^2 xy\,dx}{\displaystyle\int_0^2 y\,dx} = \dfrac{\displaystyle\int_0^2 x(3x^2)\,dx}{\displaystyle\int_0^2 3x^2\,dx}$

$= \dfrac{\displaystyle\int_0^2 3x^3\,dx}{\displaystyle\int_0^2 3x^2\,dx} = \dfrac{\left[\dfrac{3x^4}{4}\right]_0^2}{[x^3]_0^2} = \dfrac{12}{8} = \mathbf{1.5}$

$$\bar{y} = \frac{\frac{1}{2}\int_0^2 y^2 dx}{\int_0^2 y dx} = \frac{\frac{1}{2}\int_0^2 (3x^2)^2 dx}{8}$$

$$= \frac{\frac{1}{2}\int_0^2 9x^4 dx}{8} = \frac{\frac{9}{2}\left[\frac{x^5}{5}\right]_0^2}{8} = \frac{\frac{9}{2}\left(\frac{32}{5}\right)}{8}$$

$$= \frac{18}{5} = \mathbf{3.6}$$

Hence the centroid lies at (1.5, 3.6)

> **Problem 3.** Determine by integration the position of the centroid of the area enclosed by the line $y = 4x$, the x-axis and ordinates $x = 0$ and $x = 3$

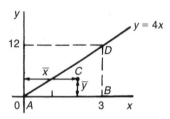

Figure 58.5

Let the coordinates of the area be $(\bar{x}, \bar{y})$ as shown in Fig. 58.5.

$$\text{Then}\quad \bar{x} = \frac{\int_0^3 xy\, dx}{\int_0^3 y\, dx} = \frac{\int_0^3 (x)(4x)dx}{\int_0^3 4x\, dx}$$

$$= \frac{\int_0^3 4x^2 dx}{\int_0^3 4x\, dx} = \frac{\left[\frac{4x^3}{3}\right]_0^3}{[2x^2]_0^3} = \frac{36}{18} = 2$$

$$\bar{y} = \frac{\frac{1}{2}\int_0^3 y^2 dx}{\int_0^3 y\, dx} = \frac{\frac{1}{2}\int_0^3 (4x)^2 dx}{18}$$

$$= \frac{\frac{1}{2}\int_0^3 16x^2 dx}{18} = \frac{\frac{1}{2}\left[\frac{16x^3}{3}\right]_0^3}{18} = \frac{72}{18} = 4$$

Hence the centroid lies at (2, 4).

In Fig. 58.5, ABD is a right-angled triangle. The centroid lies 4 units from AB and 1 unit from BD showing

that the centroid of a triangle lies at one-third of the perpendicular height above any side as base.

Now try the following exercise

> **Exercise 198   Further problems on centroids of simple shapes**
>
> In Problems 1 to 5, find the position of the centroids of the areas bounded by the given curves, the x-axis and the given ordinates.
>
> 1. $y = 2x$; $x = 0, x = 3$ [(2, 2)]
>
> 2. $y = 3x + 2$; $x = 0, x = 4$
>
>   [(2.50, 4.75)]
>
> 3. $y = 5x^2$; $x = 1, x = 4$
>
>   [(3.036, 24.36)]
>
> 4. $y = 2x^3$; $x = 0, x = 2$
>
>   [(1.60, 4.57)]
>
> 5. $y = x(3x + 1)$; $x = -1, x = 0$
>
>   [(−0.833, 0.633)]

58.6   Further worked problems on centroids of simple shapes

> **Problem 4.** Determien the co-ordinates of the centroid of the area lying between the curve $y = 5x - x^2$ and the x-axis

$y = 5x - x^2 = x(5 - x)$. When $y = 0$, $x = 0$ or $x = 5$, Hence the curve cuts the x-axis at 0 and 5 as shown in Fig. 58.6. Let the co-ordinates of the centroid be $(\bar{x}, \bar{y})$ then, by integration,

$$\bar{x} = \frac{\int_0^5 xy\, dx}{\int_0^5 y\, dx} = \frac{\int_0^5 x(5x - x^2)dx}{\int_0^5 (5x - x^2)dx}$$

$$= \frac{\int_0^5 (5x^2 - x^3)dx}{\int_0^5 (5x - x^2)dx} = \frac{\left[\frac{5x^3}{3} - \frac{x^4}{4}\right]_0^5}{\left[\frac{5x^2}{2} - \frac{x^3}{3}\right]_0^5}$$

$$= \frac{\dfrac{625}{3} - \dfrac{625}{4}}{\dfrac{125}{2} - \dfrac{125}{3}} = \frac{\dfrac{625}{12}}{\dfrac{125}{6}}$$

$$= \left(\frac{625}{12}\right)\left(\frac{6}{125}\right) = \frac{5}{2} = \mathbf{2.5}$$

$$\bar{y} = \frac{\dfrac{1}{2}\displaystyle\int_0^5 y^2 dx}{\displaystyle\int_0^5 y\,dx} = \frac{\dfrac{1}{2}\displaystyle\int_0^5 (5x - x^2)^2 dx}{\displaystyle\int_0^5 (5x - x^2)\,dx}$$

$$= \frac{\dfrac{1}{2}\displaystyle\int_0^5 (25x^2 - 10x^3 + x^4)\,dx}{\dfrac{125}{6}}$$

$$= \frac{\dfrac{1}{2}\left[\dfrac{25x^3}{3} - \dfrac{10x^4}{4} + \dfrac{x^5}{5}\right]_0^5}{\dfrac{125}{6}}$$

$$= \frac{\dfrac{1}{2}\left(\dfrac{25(125)}{3} - \dfrac{6250}{4} + 625\right)}{\dfrac{125}{6}} = \mathbf{2.5}$$

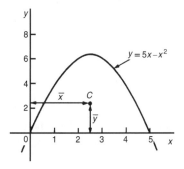

Figure 58.6

Hence the centroid of the area lies at (2.5, 2.5)
(Note from Fig. 58.6 that the curve is symmetrical about $x = 2.5$ and thus $\bar{x}$ could have been determined 'on sight').

Problem 5. Locate the centroid of the area enclosed by the curve $y = 2x^2$, the y-axis and ordinates $y = 1$ and $y = 4$, correct to 3 decimal places

From Section 58.4,

$$\bar{x} = \frac{\dfrac{1}{2}\displaystyle\int_1^4 x^2 dy}{\displaystyle\int_1^4 x\,dy} = \frac{\dfrac{1}{2}\displaystyle\int_1^4 \dfrac{y}{2}\,dy}{\displaystyle\int_1^4 \sqrt{\dfrac{y}{2}}\,dy}$$

$$= \frac{\dfrac{1}{2}\left[\dfrac{y^2}{4}\right]_1^4}{\left[\dfrac{2y^{3/2}}{3\sqrt{2}}\right]_1^4} = \frac{\dfrac{15}{8}}{\dfrac{14}{3\sqrt{2}}} = \mathbf{0.568}$$

and $\quad \bar{y} = \dfrac{\displaystyle\int_1^4 xy\,dy}{\displaystyle\int_1^4 x\,dy} = \dfrac{\displaystyle\int_1^4 \sqrt{\dfrac{y}{2}}(y)\,dy}{\dfrac{14}{3\sqrt{2}}}$

$$= \frac{\displaystyle\int_1^4 \dfrac{y^{3/2}}{\sqrt{2}}\,dy}{\dfrac{14}{3\sqrt{2}}} = \frac{\dfrac{1}{\sqrt{2}}\left[\dfrac{y^{5/2}}{\dfrac{5}{2}}\right]_1^4}{\dfrac{14}{3\sqrt{2}}}$$

$$= \frac{\dfrac{2}{5\sqrt{2}}(31)}{\dfrac{14}{3\sqrt{2}}} = \mathbf{2.657}$$

Hence the position of the centroid is at (0.568, 2.657)

Problem 6. Locate the position of the centroid enclosed by the curves $y = x^2$ and $y^2 = 8x$

Figure 58.7 shows the two curves intersection at (0, 0) and (2, 4). These are the same curves as used in Problem 12, Chapter 55 where the shaded area was

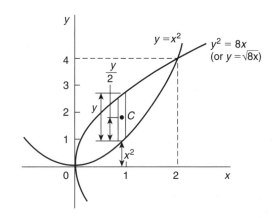

Figure 58.7

calculated as $2\frac{2}{3}$ square units. Let the co-ordinates of centroid C be $\bar{x}$ and $\bar{y}$.

By integration, $\bar{x} = \dfrac{\displaystyle\int_0^2 xy\,dx}{\displaystyle\int_0^2 y\,dx}$

The value of y is given by the height of the typical strip shown in Fig. 58.7, i.e. $y = \sqrt{8x} - x^2$. Hence,

$$\bar{x} = \frac{\displaystyle\int_0^2 x(\sqrt{8x} - x^2)dx}{2\frac{2}{3}} = \frac{\displaystyle\int_0^2 (\sqrt{8}\,x^{3/2} - x^3)}{2\frac{2}{3}}$$

$$= \frac{\left[\sqrt{8}\dfrac{x^{5/2}}{\frac{5}{2}} - \dfrac{x^4}{4}\right]_0^2}{2\frac{2}{3}} = \left(\frac{\sqrt{8}\dfrac{\sqrt{2^5}}{\frac{5}{2}} - 4}{2\frac{2}{3}}\right)$$

$$= \frac{2\frac{2}{5}}{2\frac{2}{3}} = \mathbf{0.9}$$

Care needs to be taken when finding $\bar{y}$ in such examples as this. From Fig. 58.7, $y = \sqrt{8x} - x^2$ and $\dfrac{y}{2} = \dfrac{1}{2}(\sqrt{8x} - x^2)$. The perpendicular distance from centroid C of the strip to Ox is $\dfrac{1}{2}(\sqrt{8x} - x^2) + x^2$. Taking moments about Ox gives:

(total area) $(\bar{y}) = \sum_{x=0}^{x=2}$(area of strip) (perpendicular

distance of centroid of strip to Ox)

Hence (area) $(\bar{y})$

$$= \int \left[\sqrt{8x} - x^2\right]\left[\frac{1}{2}(\sqrt{8x} - x^2) + x^2\right]dx$$

i.e. $\left(2\dfrac{2}{3}\right)(\bar{y}) = \displaystyle\int_0^2 \left[\sqrt{8x} - x^2\right]\left(\dfrac{\sqrt{8x}}{2} + \dfrac{x^2}{2}\right)dx$

$$= \int_0^2 \left(\frac{8x}{2} - \frac{x^4}{2}\right)dx = \left[\frac{8x^2}{4} - \frac{x^5}{10}\right]_0^2$$

$$= \left(8 - 3\frac{1}{5}\right) - (0) = 4\frac{4}{5}$$

Hence $\quad \bar{y} = \dfrac{4\frac{4}{5}}{2\frac{2}{3}} = \mathbf{1.8}$

Thus the position of the centroid of the enclosed area in Fig. 58.7 is at (0.9, 1.8)

Now try the following exercise

Exercise 199 Further problems on centroids of simple shapes

1. Determine the position of the centroid of a sheet of metal formed by the curve $y = 4x - x^2$ which lies above the x-axis. $\quad$ [(2, 1.6)]

2. Find the coordinates of the centroid of the area that lies between curve $\dfrac{y}{x} = x - 2$ and the x-axis. $\quad$ [(1, −0.4)]

3. Determine the coordinates of the centroid of the area formed between the curve $y = 9 - x^2$ and the x-axis. $\quad$ [(0, 3.6)]

4. Determine the centroid of the area lying between $y = 4x^2$, the y-axis and the ordinates $y = 0$ and $y = 4$. $\quad$ [(0.375, 2.40]

5. Find the position of the centroid of the area enclosed by the curve $y = \sqrt{5x}$, the x-axis and the ordinate $x = 5$. $\quad$ [(3.0, 1.875)]

6. Sketch the curve $y^2 = 9x$ between the limits $x = 0$ and $x = 4$. Determine the position of the centroid of this area. $\quad$ [(2.4, 0)]

7. Calculate the points of intersection of the curves $x^2 = 4y$ and $\dfrac{y^2}{4} = x$, and determine the position of the centroid of the area enclosed by them. $\quad$ [(0, 0) and (4, 4), (1.80, 1.80)]

8. Determine the position of the centroid of the sector of a circle of radius 3 cm whose angle subtended at the centre is 40°.
$$\left[\begin{array}{l}\text{On the centre line, 1.96 cm}\\ \text{from the centre}\end{array}\right]$$

9. Sketch the curves $y = 2x^2 + 5$ and $y - 8 = x(x + 2)$ on the same axes and determine their points of intersection. Calculate the coordinates of the centroid of the area enclosed by the two curves.
$\quad$ [(−1, 7) and (3, 23), (1, 10.20)]

58.7 Theorem of Pappus

A theorem of Pappus states:
'If a plane area is rotated about an axis in its own plane but not intersecting it, the volume of the solid formed is given by the product of the area and the distance moved by the centroid of the area'.

With reference to Fig. 58.8, when the curve $y=f(x)$ is rotated one revolution about the x-axis between the limits $x=a$ and $x=b$, the volume V generated is given by:
volume $V=(A)(2\pi\bar{y})$, from which,

$$\bar{y}=\frac{V}{2\pi A}$$

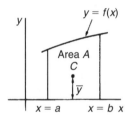

Figure 58.8

Problem 7. Determine the position of the centroid of a semicircle of radius r by using the theorem of Pappus. Check the answer by using integration (given that the equation of a circle, centre 0, radius r is $x^2+y^2=r^2$)

A semicircle is shown in Fig. 58.9 with its diameter lying on the x-axis and its centre at the origin. Area of semicircle $=\dfrac{\pi r^2}{2}$. When the area is rotated about the x-axis one revolution a sphere is generated of volume $\dfrac{4}{3}\pi r^3$.

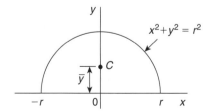

Figure 58.9

Let centroid C be at a distance $\bar{y}$ from the origin as shown in Fig. 58.9. From the theorem of Pappus,

volume generated = area × distance moved through by centroid i.e.

$$\frac{4}{3}\pi r^3=\left(\frac{\pi r^2}{2}\right)(2\pi\bar{y})$$

Hence $\bar{y}=\dfrac{\frac{4}{3}\pi r^3}{\pi^2 r^2}=\dfrac{4r}{3\pi}$

By integration,

$$\bar{y}=\frac{\frac{1}{2}\int_{-r}^{r}y^2\,dx}{\text{area}}$$

$$=\frac{\frac{1}{2}\int_{-r}^{r}(r^2-x^2)\,dx}{\frac{\pi r^2}{2}}=\frac{\frac{1}{2}\left[r^2x-\frac{x^3}{3}\right]_{-r}^{r}}{\frac{\pi r^2}{2}}$$

$$=\frac{\frac{1}{2}\left[\left(r^3-\frac{r^3}{3}\right)-\left(-r^3+\frac{r^3}{3}\right)\right]}{\frac{\pi r^2}{2}}=\frac{4r}{3\pi}$$

Hence the centroid of a semicircle lies on the axis of symmetry, distance $\dfrac{4r}{3\pi}$ (or $0.424r$) from its diameter.

Problem 8. Calculate the area bounded by the curve $y=2x^2$, the x-axis and ordinates $x=0$ and $x=3$. (b) If this area is revolved (i) about the x-axis and (ii) about the y-axis, find the volumes of the solids produced. (c) Locate the position of the centroid using (i) integration, and (ii) the theorem of Pappus

---

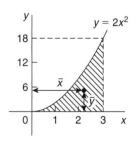

Figure 58.10

Section 9

502 Engineering Mathematics

(a) The required area is shown shaded in Fig. 58.10.

$$\text{Area} = \int_0^3 y\, dx = \int_0^3 2x^2\, dx = \left[\frac{2x^3}{3}\right]_0^3$$

$$= \textbf{18 square units}$$

(b) (i) When the shaded area of Fig. 58.10 is revolved 360° about the x-axis, the volume generated

$$= \int_0^3 \pi y^2\, dx = \int_0^3 \pi(2x^2)^2\, dx$$

$$= \int_0^3 4\pi x^4\, dx = 4\pi\left[\frac{x^5}{5}\right]_0^3 = 4\pi\left(\frac{243}{5}\right)$$

$$= \textbf{194.4}\pi \textbf{ cubic units}$$

(ii) When the shaded area of Fig. 58.10 is revolved 360° about the y-axis, the volume generated = (volume generated by x = 3) − (volume generated by y = 2x²)

$$= \int_0^{18} \pi(3)^2\, dy - \int_0^{18} \pi\left(\frac{y}{2}\right) dy$$

$$= \pi\int_0^{18}\left(9 - \frac{y}{2}\right) dy = \pi\left[9y - \frac{y^2}{4}\right]_0^{18}$$

$$= \textbf{81}\pi \textbf{ cubic units}$$

(c) If the co-ordinates of the centroid of the shaded area in Fig. 58.10 are $(\bar{x}, \bar{y})$ then:

(i) by integration,

$$\bar{x} = \frac{\displaystyle\int_0^3 xy\, dx}{\displaystyle\int_0^3 y\, dx} = \frac{\displaystyle\int_0^3 x(2x^2)\, dx}{18}$$

$$= \frac{\displaystyle\int_0^3 2x^3\, dx}{18} = \frac{\left[\dfrac{2x^4}{4}\right]_0^3}{18} = \frac{81}{36} = \textbf{2.25}$$

$$\bar{y} = \frac{\dfrac{1}{2}\displaystyle\int_0^3 y^2\, dx}{\displaystyle\int_0^3 y\, dx} = \frac{\dfrac{1}{2}\displaystyle\int_0^3 (2x^2)^2\, dx}{18}$$

$$= \frac{\dfrac{1}{2}\displaystyle\int_0^3 4x^4\, dx}{18} = \frac{\dfrac{1}{2}\left[\dfrac{4x^5}{5}\right]_0^3}{18} = \textbf{5.4}$$

(ii) using the theorem of Pappus:
Volume generated when shaded area is revolved about $Oy = (\text{area})(2\pi\bar{x})$

i.e. $\qquad 81\pi = (18)(2\pi\bar{x})$,

from which, $\qquad \bar{x} = \dfrac{81\pi}{36\pi} = \textbf{2.25}$

Volume generated when shaded area is revolved about $Ox = (\text{area})(2\pi\bar{y})$

i.e. $\qquad 194.4\pi = (18)(2\pi\bar{y})$,

from which, $\qquad \bar{y} = \dfrac{194.4\pi}{36\pi} = \textbf{5.4}$

Hence the centroid of the shaded area in Fig. 58.10 is at (2.25, 5.4)

Problem 9. A cylindrical pillar of diameter 400 mm has a groove cut round its circumference. The section of the groove is a semicircle of diameter 50 mm. Determine the volume of material removed, in cubic centimetres, correct to 4 significant figures

A part of the pillar showing the groove is shown in Fig. 58.11.
The distance of the centroid of the semicircle from its base is $\dfrac{4r}{3\pi}$ (see Problem 7) $= \dfrac{4(25)}{3\pi} = \dfrac{100}{3\pi}$ mm.
The distance of the centroid from the centre of the pillar $= \left(200 - \dfrac{100}{3\pi}\right)$ mm.

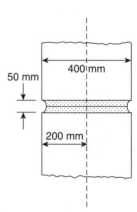

Figure 58.11

The distance moved by the centroid in one revolution

$$= 2\pi \left(200 - \frac{100}{3\pi}\right) = \left(400\pi - \frac{200}{3}\right) \text{mm}.$$

From the theorem of Pappus,
volume = area × distance moved by centroid

$$= \left(\frac{1}{2}\pi 25^2\right)\left(400\pi - \frac{200}{3}\right) = 1168250 \text{ mm}^3$$

Hence the volume of material removed is 1168 cm^3 correct to 4 significant figures.

Problem 10. A metal disc has a radius of 5.0 cm and is of thickness 2.0 cm. A semicircular groove of diameter 2.0 cm is machined centrally around the rim to form a pulley. Determine, using Pappus' theorem, the volume and mass of metal removed and the volume and mass of the pulley if the density of the metal is 8000 kg m^{-3}

A side view of the rim of the disc is shown in Fig. 58.12.

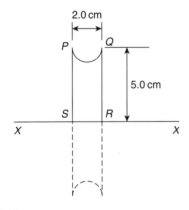

Figure 58.12

When area *PQRS* is rotated about axis *XX* the volume generated is that of the pulley. The centroid of the semicircular area removed is at a distance of $\frac{4r}{3\pi}$ from its diameter (see Problem 7), i.e. $\frac{4(1.0)}{3\pi}$, i.e. 0.424 cm from *PQ*. Thus the distance of the centroid from *XX* is (5.0 − 0.424), i.e. 4.576 cm. The distance moved through in one revolution by the centroid is $2\pi(4.576)$ cm.

Area of semicircle

$$= \frac{\pi r^2}{2} = \frac{\pi(1.0)^2}{2} = \frac{\pi}{2} \text{ cm}^2$$

By the theorem of Pappus, volume generated
= area × distance moved by centroid

$$= \left(\frac{\pi}{2}\right)(2\pi)(4.576)$$

i.e. **volume of metal removed = 45.16 cm^3**
Mass of metal removed = density × volume

$$= 8000 \text{ kg m}^{-3} \times \frac{45.16}{10^6} \text{ m}^3$$

$$= \textbf{0.3613 kg} \text{ or } \textbf{361.3 g}$$

Volume of pulley = volume of cylindrical disc
− volume of metal removed

$$= \pi(5.0)^2(2.0) - 45.16 = \textbf{111.9 cm}^3$$

Mass of pulley = density × volume

$$= 8000 \text{ kg m}^{-3} \times \frac{111.9}{10^6} \text{ m}^3$$

$$= \textbf{0.8952 kg or 895.2 g}$$

Now try the following exercise

Exercise 200   Further problems on the theorem of Pappus

1. A right angled isosceles triangle having a hypotenuse of 8 cm is revolved one revolution about one of its equal sides as axis. Determine the volume of the solid generated using Pappus' theorem. [189.6 cm^3]

2. A rectangle measuring 10.0 cm by 6.0 cm rotates one revolution about one of its longest sides as axis. Determine the volume of the resulting cylinder by using the theorem of Pappus. [1131 cm^2]

3. Using (a) the theorem of Pappus, and (b) integration, determine the position of the centroid of a metal template in the form of a quadrant of a circle of radius 4 cm. (The equation of a circle, centre 0, radius r is $x^2 + y^2 = r^2$).

$$\begin{bmatrix} \text{On the centre line, distance 2.40 cm} \\ \text{from the centre, i.e. at coordinates} \\ (1.70, 1.70) \end{bmatrix}$$

4. (a) Determine the area bounded by the curve $y = 5x^2$, the x-axis and the ordinates $x = 0$ and $x = 3$.

   (b) If this area is revolved 360° about (i) the x-axis, and (ii) the y-axis, find the volumes of the solids of revolution produced in each case.

   (c) Determine the co-ordinates of the centroid of the area using (i) integral calculus, and (ii) the theorem of Pappus

$$\begin{bmatrix} \text{(a) 45 square units (b) (i) } 1215\pi \\ \text{cubic units (ii) } 202.5\pi \text{ cubic units} \\ \text{(c) } (2.25, 13.5) \end{bmatrix}$$

5. A metal disc has a radius of 7.0 cm and is of thickness 2.5 cm. A semicircular groove of diameter 2.0 cm is machined centrally around the rim to form a pulley. Determine the volume of metal removed using Pappus' theorem and express this as a percentage of the original volume of the disc. Find also the mass of metal removed if the density of the metal is $7800 \, \text{kg m}^{-3}$.

$$[64.90 \, \text{cm}^3, 16.86\%, 506.2 \, \text{g}]$$

Chapter 59

Second moments of area

59.1 Second moments of area and radius of gyration

The **first moment of area** about a fixed axis of a lamina of area A, perpendicular distance y from the centroid of the lamina is defined as Ay cubic units. The **second moment of area** of the same lamina as above is given by Ay^2, i.e. the perpendicular distance from the centroid of the area to the fixed axis is squared. Second moments of areas are usually denoted by I and have limits of mm^4, cm^4, and so on.

Radius of gyration

Several areas, $a_1, a_2, a_3, \ldots$ at distances $y_1, y_2, y_3, \ldots$ from a fixed axis, may be replaced by a single area A, where $A = a_1 + a_2 + a_3 + \cdots$ at distance k from the axis, such that $Ak^2 = \sum ay^2$. k is called the **radius of gyration** of area A about the given axis. Since $Ak^2 = \sum ay^2 = I$ then the radius of gyration, $k = \sqrt{\dfrac{I}{A}}$

The second moment of area is a quantity much used in the theory of bending of beams, in the torsion of shafts, and in calculations involving water planes and centres of pressure.

59.2 Second moment of area of regular sections

The procedure to determine the second moment of area of regular sections about a given axis is (i) to find the second moment of area of a typical element and (ii) to sum all such second moments of area by integrating between appropriate limits.

For example, the second moment of area of the rectangle shown in Fig. 59.1 about axis PP is found by initially considering an elemental strip of width δx, parallel to and distance x from axis PP. Area of shaded strip $= b\delta x$. Second moment of area of the shaded strip about $PP = (x^2)(b\delta x)$.

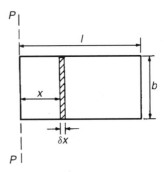

Figure 59.1

The second moment of area of the whole rectangle about PP is obtained by summing all such strips between $x = 0$ and $x = l$, i.e. $\sum_{x=0}^{x=l} x^2 b\delta x$. It is a fundamental theorem of integration that

$$\lim_{\delta x \to 0} \sum_{x=0}^{x=l} x^2 b\delta x = \int_0^l x^2 b \, dx$$

Thus the second moment of area of the rectangle about PP is

$$PP = b\int_0^l x^2 dx = b\left[\frac{x^3}{3}\right]_0^l = \frac{bl^3}{3}$$

Since the total area of the rectangle, $A = lb$, then

$$I_{pp} = (lb)\left(\frac{l^2}{3}\right) = \frac{Al^2}{3}$$

$$I_{pp} = Ak_{pp}^2 \text{ thus } k_{pp}^2 = \frac{l^2}{3}$$

i.e. the radius of gyration about axes PP,

$$k_{pp} = \sqrt{\frac{l^2}{3}} = \frac{l}{\sqrt{3}}$$

59.3 Parallel axis theorem

In Fig. 59.2, axis GG passes through the centroid C of area A. Axes DD and GG are in the same plane, are parallel to each other and distance d apart. The parallel axis theorem states:

$$I_{DD} = I_{GG} + Ad^2$$

Using the parallel axis theorem the second moment of area of a rectangle about an axis through the centroid may be determined. In the rectangle shown in Fig. 59.3, $I_{pp} = \dfrac{bl^3}{3}$ (from above). From the parallel axis theorem

$$I_{pp} = I_{GG} + (bl)\left(\frac{l}{2}\right)^2$$

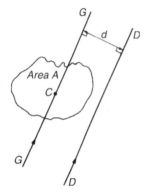

Figure 59.2

i.e. $\dfrac{bl^3}{3} = I_{GG} + \dfrac{bl^3}{4}$

from which, $I_{GG} = \dfrac{bl^3}{3} - \dfrac{bl^3}{4} = \dfrac{bl^3}{12}$

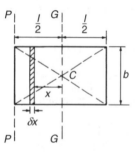

Figure 59.3

59.4 Perpendicular axis theorem

In Fig. 59.4, axes OX, OY and OZ are mutually perpendicular. If OX and OY lie in the plane of area A then the perpendicular axis theorem states:

$$I_{OZ} = I_{OX} + I_{OY}$$

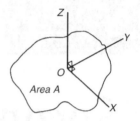

Figure 59.4

59.5 Summary of derived results

A summary of derive standard results for the second moment of area and radius of gyration of regular sections are listed in Table 59.1.

Table 59.1 Summary of standard results of the second moments of areas of regular sections

Shape	Position of axis	Second moment of area, I	Radius of gyration, k
Rectangle length l breadth b	(1) Coinciding with b	$\dfrac{bl^3}{3}$	$\dfrac{l}{\sqrt{3}}$
	(2) Coinciding with l	$\dfrac{lb^3}{3}$	$\dfrac{b}{\sqrt{3}}$
	(3) Through centroid, parallel to b	$\dfrac{bl^3}{12}$	$\dfrac{l}{\sqrt{12}}$
	(4) Through centroid, parallel to l	$\dfrac{lb^3}{12}$	$\dfrac{b}{\sqrt{12}}$
Triangle Perpendicular height h base b	(1) Coinciding with b	$\dfrac{bh^3}{12}$	$\dfrac{h}{\sqrt{6}}$
	(2) Through centroid, parallel to base	$\dfrac{bh^3}{36}$	$\dfrac{h}{\sqrt{18}}$
	(3) Through vertex, parallel to base	$\dfrac{bh^3}{4}$	$\dfrac{h}{\sqrt{2}}$
Circle radius r	(1) Through centre, perpendicular to plane (i.e. polar axis)	$\dfrac{\pi r^4}{2}$	$\dfrac{r}{\sqrt{2}}$
	(2) Coinciding with diameter	$\dfrac{\pi r^4}{4}$	$\dfrac{r}{2}$
	(3) About a tangent	$\dfrac{5\pi r^4}{4}$	$\dfrac{\sqrt{5}}{2}r$
Semicircle radius r	Coinciding with diameter	$\dfrac{\pi r^4}{8}$	$\dfrac{r}{2}$

59.6 Worked problems on second moments of area of regular sections

Problem 1. Determine the second moment of area and the radius of gyration about axes AA, BB and CC for the rectangle shown in Fig. 59.5.

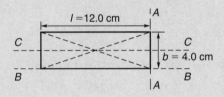

Figure 59.5

From Table 59.1, the second moment of area about axis AA, $I_{AA} = \dfrac{bl^3}{3} = \dfrac{(4.0)(12.0)^3}{3} = 2304 \text{ cm}^4$

Radius of gyration,

$$k_{AA} = \frac{l}{\sqrt{3}} = \frac{12.0}{\sqrt{3}} = 6.93 \text{ cm}$$

Similarly, $\quad I_{BB} = \dfrac{lb^3}{3} = \dfrac{(12.0)(4.0)^3}{3} = 256 \text{ cm}^4$

and $\quad k_{BB} = \dfrac{b}{\sqrt{3}} = \dfrac{4.0}{\sqrt{3}} = 2.31 \text{ cm}$

The second moment of area about the centroid of a rectangle is $\dfrac{bl^3}{12}$ when the axis through the centroid is parallel with the breadth, b. In this case, the axis CC is parallel with the length l.

Hence $\quad I_{CC} = \dfrac{lb^3}{12} = \dfrac{(12.0)(4.0)^3}{12} = 64 \text{ cm}^4$

and $\quad k_{CC} = \dfrac{b}{\sqrt{12}} = \dfrac{4.0}{\sqrt{12}} = 1.15 \text{ cm}$

Problem 2. Find the second moment of area and the radius of gyration about axis PP for the rectangle shown in Fig. 59.6

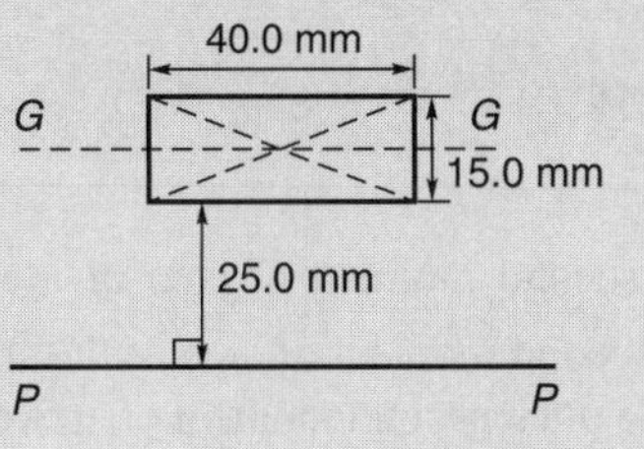

40.0 mm

15.0 mm

25.0 mm

Figure 59.6

$I_{GG} = \dfrac{lb^3}{12}$ where $l = 40.0 \text{ mm}$ and $b = 15.0 \text{ mm}$

Hence $I_{GG} = \dfrac{(40.0)(15.0)^3}{12} = 11\,250 \text{ mm}^4$

From the parallel axis theorem, $I_{PP} = I_{GG} + Ad^2$, where $A = 40.0 \times 15.0 = 600 \text{ mm}^2$ and $d = 25.0 + 7.5 = 32.5 \text{ mm}$, the perpendicular distance between GG and PP.

Hence, $\quad I_{PP} = 11\,250 + (600)(32.5)^2$

$$= 645\,000 \text{ mm}^4$$

$$I_{PP} = Ak_{PP}^2$$

from which, $\quad k_{PP} = \sqrt{\dfrac{I_{PP}}{\text{area}}}$

$$= \sqrt{\frac{645\,000}{600}} = 32.79 \text{ mm}$$

Problem 3. Determine the second moment of area and radius of gyration about axis QQ of the triangle BCD shown in Fig. 59.7

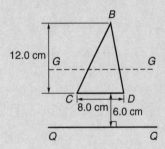

Figure 59.7

Using the parallel axis theorem: $I_{QQ} = I_{GG} + Ad^2$, where I_{GG} is the second moment of area about the centroid of the triangle,

i.e. $\dfrac{bh^3}{36} = \dfrac{(8.0)(12.0)^3}{36} = 384 \text{ cm}^4$, A is the area of the triangle $= \frac{1}{2}bh = \frac{1}{2}(8.0)(12.0) = 48 \text{ cm}^2$ and d is the distance between axes GG and $QQ = 6.0 + \frac{1}{3}(12.0) = 10 \text{ cm}$.

Hence the second moment of area about axis QQ,

$$I_{QQ} = 384 + (48)(10)^2 = 5184 \text{ cm}^4$$

Radius of gyration,

$$k_{QQ} = \sqrt{\frac{I_{QQ}}{\text{area}}} = \sqrt{\frac{5184}{48}} = 10.4 \text{ cm}$$

---

508 Engineering Mathematics

Problem 4. Determine the second moment of area and radius of gyration of the circle shown in Fig. 59.8 about axis *YY*

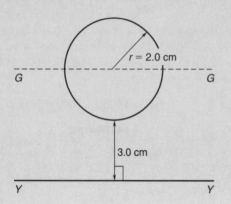

Figure 59.8

In Fig. 59.8, $I_{GG} = \dfrac{\pi r^4}{4} = \dfrac{\pi}{4}(2.0)^4 = 4\pi \text{ cm}^4$. Using the parallel axis theorem, $I_{YY} = I_{GG} + Ad^2$, where $d = 3.0 + 2.0 = 5.0 \text{ cm}$.

Hence $\quad I_{YY} = 4\pi + [\pi(2.0)^2](5.0)^2$

$$= 4\pi + 100\pi = 104\pi = \mathbf{327 \text{ cm}^4}$$

Radius of gyration,

$$k_{YY} = \sqrt{\dfrac{I_{YY}}{\text{area}}} = \sqrt{\dfrac{104\pi}{\pi(2.0)^2}} = \sqrt{26} = \mathbf{5.10 \text{ cm}}$$

Problem 5. Determine the second moment of area and radius of gyration for the semicircle shown in Fig. 59.9 about axis *XX*

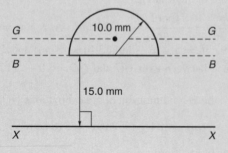

Figure 59.9

The centroid of a semicircle lies at $\dfrac{4r}{3\pi}$ from its diameter.

Using the parallel axis theorem: $I_{BB} = I_{GG} + Ad^2$,

where $\quad I_{BB} = \dfrac{\pi r^4}{8}$ (from Table 59.1)

$$= \dfrac{\pi(10.0)^4}{8} = 3927 \text{ mm}^4,$$

$$A = \dfrac{\pi r^2}{2} = \dfrac{\pi(10.0)^2}{2} = 157.1 \text{ mm}^2$$

and $\quad d = \dfrac{4r}{3\pi} = \dfrac{4(10.0)}{3\pi} = 4.244 \text{ mm}$

Hence $\quad 3927 = I_{GG} + (157.1)(4.244)^2$

i.e. $\quad 3927 = I_{GG} + 2830$,

from which, $\quad I_{GG} = 3927 - 2830 = 1097 \text{ mm}^4$

Using the parallel axis theorem again:

$I_{XX} = I_{GG} + A(15.0 + 4.244)^2$

i.e. $\boldsymbol{I_{XX}} = 1097 + (157.1)(19.244)^2$

$= 1097 + 58\,179 = 59\,276 \text{ mm}^4$ or $\mathbf{59\,280 \text{ mm}^4}$, correct to 4 significant figures.

Radius of gyration, $k_{XX} = \sqrt{\dfrac{I_{XX}}{\text{area}}} = \sqrt{\dfrac{59\,276}{157.1}}$

$$= \mathbf{19.42 \text{ mm}}$$

Problem 6. Determine the polar second moment of area of the propeller shaft cross-section shown in Fig. 59.10

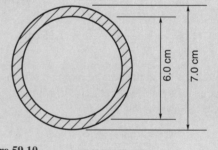

Figure 59.10

The polar second moment of area of a circle $= \dfrac{\pi r^4}{2}$. The polar second moment of area of the shaded area is given by the polar second moment of area of the 7.0 cm diameter circle minus the polar second moment of area of the 6.0 cm diameter circle. Hence the polar second moment of area of the

$$\text{cross-section shown} = \frac{\pi}{2}\left(\frac{7.0}{2}\right)^4 - \frac{\pi}{2}\left(\frac{6.0}{2}\right)^4$$

$$= 235.7 - 127.2 = \mathbf{108.5\,cm^4}$$

Problem 7. Determine the second moment of area and radius of gyration of a rectangular lamina of length 40 mm and width 15 mm about an axis through one corner, perpendicular to the plane of the lamina

The lamina is shown in Fig. 59.11.

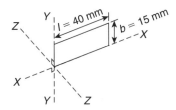

Figure 59.11

From the perpendicular axis theorem:

$$I_{ZZ} = I_{XX} + I_{YY}$$

$$I_{XX} = \frac{lb^3}{3} = \frac{(40)(15)^3}{3} = 45\,000\,\text{mm}^4$$

and $I_{YY} = \frac{bl^3}{3} = \frac{(15)(40)^3}{3} = 320\,000\,\text{mm}^4$

Hence $I_{ZZ} = 45\,000 + 320\,000$

$$= \mathbf{365\,000\,mm^4}\text{ or }\mathbf{36.5\,cm^4}$$

Radius of gyration,

$$k_{ZZ} = \sqrt{\frac{I_{ZZ}}{\text{area}}} = \sqrt{\frac{365\,000}{(40)(15)}}$$

$$= \mathbf{24.7\,mm}\text{ or }\mathbf{2.47\,cm}$$

Now try the following exercise

Exercise 201 Further problems on second moments of area of regular sections

1. Determine the second moment of area and radius of gyration for the rectangle shown in

Fig. 59.12 about (a) axis AA (b) axis BB, and (c) axis CC.

$$\begin{bmatrix}\text{(a) } 72\,\text{cm}^4,\ 1.73\,\text{cm}\\ \text{(b) } 128\,\text{cm}^4,\ 2.31\,\text{cm}\\ \text{(c) } 512\,\text{cm}^4,\ 4.62\,\text{cm}\end{bmatrix}$$

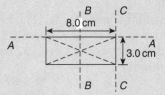

Figure 59.12

2. Determine the second moment of area and radius of gyration for the triangle shown in Fig. 59.13 about (a) axis DD (b) axis EE, and (c) an axis through the centroid of the triangle parallel to axis DD.

$$\begin{bmatrix}\text{(a) } 729\,\text{cm}^4,\ 3.67\,\text{cm}\\ \text{(b) } 2187\,\text{cm}^4,\ 6.36\,\text{cm}\\ \text{(c) } 243\,\text{cm}^4,\ 2.12\,\text{cm}\end{bmatrix}$$

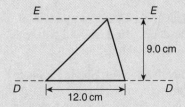

Figure 59.13

3. For the circle shown in Fig. 59.14, find the second moment of area and radius of gyration about (a) axis FF, and (b) axis HH.

$$\begin{bmatrix}\text{(a) } 201\,\text{cm}^4,\ 2.0\,\text{cm}\\ \text{(b) } 1005\,\text{cm}^4,\ 4.47\,\text{cm}\end{bmatrix}$$

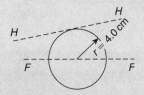

Figure 59.14

4. For the semicircle shown in Fig. 59.15, find the second moment of area and radius of gyration about axis JJ. [3927 mm^4, 5.0 mm]

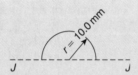

Figure 59.15

5. For each of the areas shown in Fig. 59.16 determine the second moment of area and radius of gyration about axis LL, by using the parallel axis theorem.

$$\left[\begin{array}{l} \text{(a) } 335\,\text{cm}^4,\ 4.73\,\text{cm} \\ \text{(b) } 22\,030\,\text{cm}^4,\ 14.3\,\text{cm} \\ \text{(c) } 628\,\text{cm}^4,\ 7.07\,\text{cm} \end{array}\right]$$

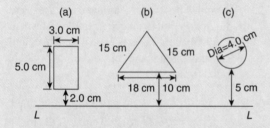

Figure 59.16

6. Calculate the radius of gyration of a rectangular door 2.0 m high by 1.5 m wide about a vertical axis through its hinge.   [0.866 m]

7. A circular door of a boiler is hinged so that it turns about a tangent. If its diameter is 1.0 m, determine its second moment of area and radius of gyration about the hinge.
    [0.245 m^4, 0.599 m]

8. A circular cover, centre 0, has a radius of 12.0 cm. A hole of radius 4.0 cm and centre X, where $OX = 6.0$ cm, is cut in the cover. Determine the second moment of area and the radius of gyration of the remainder about a diameter through 0 perpendicular to OX.
    [14 280 cm^4, 5.96 cm]

59.7   Worked problems on second moments of area of composite areas

Problem 8.   Determine correct to 3 significant figures, the second moment of area about XX for the composite area shown in Fig. 59.17

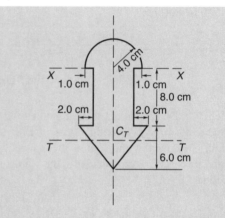

Figure 59.17

For the semicircle, $I_{XX} = \dfrac{\pi r^4}{8} = \dfrac{\pi (4.0)^4}{8}$

$$= 100.5\,\text{cm}^4$$

For the rectangle, $I_{XX} = \dfrac{bl^3}{3} = \dfrac{(6.0)(8.0)^3}{3}$

$$= 1024\,\text{cm}^4$$

For the triangle, about axis TT through centroid C_T,

$$I_{TT} = \frac{bh^3}{36} = \frac{(10)(6.0)^3}{36} = 60\,\text{cm}^4$$

By the parallel axis theorem, the second moment of area of the triangle about axis XX

$$= 60 + \left[\tfrac{1}{2}(10)(6.0)\right]\left[8.0 + \tfrac{1}{3}(6.0)\right]^2 = 3060\,\text{cm}^4.$$

Total second moment of area about XX.
$$= 100.5 + 1024 + 3060 = 4184.5 = \mathbf{4180\,cm^4}, \quad \text{correct}$$
to 3 significant figures

Problem 9.   Determine the second moment of area and the radius of gyration about axis XX for the I-section shown in Fig. 59.18

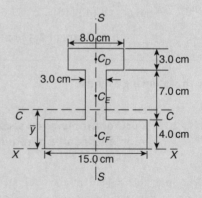

Figure 59.18

The *I*-section is divided into three rectangles, *D*, *E* and *F* and their centroids denoted by C_D, C_E and C_F respectively.

For rectangle D:
The second moment of area about C_D (an axis through C_D parallel to *XX*)

$$= \frac{bl^3}{12} = \frac{(8.0)(3.0)^3}{12} = 18\,\text{cm}^4$$

Using the parallel axis theorem: $I_{XX} = 18 + Ad^2$ where $A = (8.0)(3.0) = 24\,\text{cm}^2$ and $d = 12.5\,\text{cm}$

Hence $I_{XX} = 18 + 24(12.5)^2 = \mathbf{3768\,cm^4}$

For rectangle E:
The second moment of area about C_E (an axis through C_E parallel to *XX*)

$$= \frac{bl^3}{12} = \frac{(3.0)(7.0)^3}{12} = 85.75\,\text{cm}^4$$

Using the parallel axis theorem:
$I_{XX} = 85.75 + (7.0)(3.0)(7.5)^2 = \mathbf{1267\,cm^4}$

For rectangle F:

$$I_{XX} = \frac{bl^3}{3} = \frac{(15.0)(4.0)^3}{3} = \mathbf{320\,cm^4}$$

Total second moment of area for the *I*-section about axis *XX*,
$$I_{XX} = 3768 + 1267 + 320 = \mathbf{5355\,cm^4}$$
Total area of *I*-section
$$= (8.0)(3.0) + (3.0)(7.0) + (15.0)(4.0) = 105\,\text{cm}^2.$$
Radius of gyration,

$$k_{XX} = \sqrt{\frac{I_{XX}}{\text{area}}} = \sqrt{\frac{5355}{105}} = \mathbf{7.14\,cm}$$

Now try the following exercise

Exercise 202   Further problems on section moment of areas of composite areas

1. For the sections shown in Fig. 59.19, find the second moment of area and the radius of gyration about axis *XX*.

$$\left[\begin{array}{l}\text{(a) } 12\,190\,\text{mm}^4, \ 10.9\,\text{mm} \\ \text{(b) } 549.5\,\text{cm}^4, \ 4.18\,\text{cm}\end{array}\right]$$

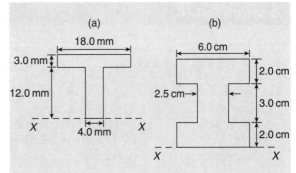

Figure 59.19

2. Determine the second moment of area about the given axes for the shapes shown in Fig. 59.20. (In Fig. 59.20(b), the circular area is removed.)

$$\left[\begin{array}{l} I_{AA} = 4224\,\text{cm}^4, \quad I_{BB} = 6718\,\text{cm}^4, \\ I_{CC} = 37\,300\,\text{cm}^4 \end{array}\right]$$

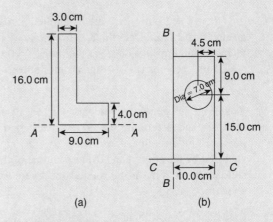

Figure 59.20

3. Find the second moment of area and radius of gyration about the axis *XX* for the beam section shown in Fig. 59.21.   [1350 cm⁴, 5.67 cm]

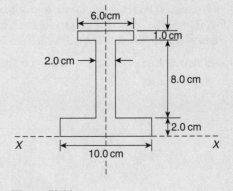

Figure 59.21

This Revision test covers the material contained in Chapters 55 to 59. *The marks for each question are shown in brackets at the end of each question.*

1. The force F newtons acting on a body at a distance x metres from a fixed point is given by: $F = 2x + 3x^2$. If work done $= \int_{x_1}^{x_2} F\,dx$, determine the work done when the body moves from the position when $x = 1$ m to that when $x = 4$ m.
(4)

2. Sketch and determine the area enclosed by the curve $y = 3\sin\dfrac{\theta}{2}$, the θ-axis and ordinates $\theta = 0$ and $\theta = \dfrac{2\pi}{3}$.
(4)

3. Calculate the area between the curve $y = x^3 - x^2 - 6x$ and the x-axis.
(10)

4. A voltage $v = 25\sin 50\pi t$ volts is applied across an electrical circuit. Determine, using integration, its mean and r.m.s. values over the range $t = 0$ to $t = 20$ ms, each correct to 4 significant figures.
(12)

5. Sketch on the same axes the curves $x^2 = 2y$ and $y^2 = 16x$ and determine the co-ordinates of the points of intersection. Determine (a) the area enclosed by the curves, and (b) the volume of the solid produced if the area is rotated one revolution about the x-axis.
(13)

6. Calculate the position of the centroid of the sheet of metal formed by the x-axis and the part of the curve $y = 5x - x^2$ which lies above the x-axis.
(9)

7. A cylindrical pillar of diameter 500 mm has a groove cut around its circumference as shown in Fig. R16.1. The section of the groove is a semicircle of diameter 40 mm. Given that the centroid of a semicircle from its base is $\dfrac{4r}{3\pi}$, use the theorem

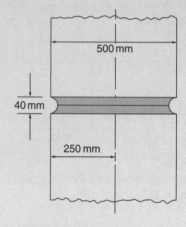

Figure R16.1

of Pappus to determine the volume of material removed, in cm^3, correct to 3 significant figures.
(8)

8. For each of the areas shown in Fig. R16.2 determine the second moment of area and radius of gyration about axis XX.
(15)

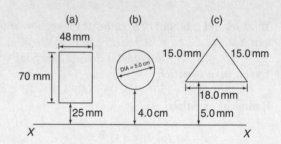

Figure R16.2

9. A circular door is hinged so that it turns about a tangent. If its diameter is 1.0 m find its second moment of area and radius of gyration about the hinge.
(5)

Further Number and Algebra

Chapter 60

Boolean algebra and logic circuits

60.1 Boolean algebra and switching circuits

A **two-state device** is one whose basic elements can only have one of two conditions. Thus, two-way switches, which can either be on or off, and the binary numbering system, having the digits 0 and 1 only, are two-state devices. In Boolean algebra, if A represents one state, then $\overline{A}$, called 'not-A', represents the second state.

The or-function

In Boolean algebra, the **or**-function for two elements A and B is written as $A + B$, and is defined as 'A, or B, or both A and B'. The equivalent electrical circuit for a two-input **or**-function is given by two switches connected in parallel. With reference to Fig. 60.1(a), the lamp will be on when A is on, when B is on, or when both A and B are on. In the table shown in Fig. 60.1(b), all the possible switch combinations are shown in columns 1 and 2, in which a 0 represents a switch being off and a 1 represents the switch being on, these columns being called the inputs. Column 3 is called the output and a 0 represents the lamp being off and a 1 represents the lamp being on. Such a table is called a **truth table**.

The and-function

In Boolean algebra, the **and**-function for two elements A and B is written as $A \cdot B$ and is defined as 'both A and B'. The equivalent electrical circuit for a two-input **and**-function is given by two switches connected in series. With reference to Fig. 60.2(a) the lamp will be on only

when both A and B are on. The truth table for a two-input **and**-function is shown in Fig. 60.2(b).

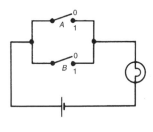

1	2	3
Input (switches)		Output (lamp)
A	B	$Z = A + B$
0	0	0
0	1	1
1	0	1
1	1	1

(a) Switching circuit for or - function   (b) Truth table for or - function

Figure 60.1

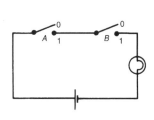

Input (switches)		Output (lamp)
A	B	$Z = A \cdot B$
0	0	0
0	1	0
1	0	0
1	1	1

(a) Switching circuit for and - function   (b) Truth table for and - function

Figure 60.2

The not-function

In Boolean algebra, the **not**-function for element A is written as $\overline{A}$, and is defined as 'the opposite to A'. Thus if A means switch A is on, $\overline{A}$ means that switch A is off. The truth table for the **not**-function is shown in Table 60.1.

Table 60.1

Input A	Output $Z = \bar{A}$
0	1
1	0

In the above, the Boolean expressions, equivalent switching circuits and truth tables for the three functions used in Boolean algebra are given for a two-input system. A system may have more than two inputs and the Boolean expression for a three-input **or**-function having elements A, B and C is $A + B + C$. Similarly, a three-input **and**-function is written as $A \cdot B \cdot C$. The equivalent electrical circuits and truth tables for three-input **or** and **and**-functions are shown in Figs. 60.3(a) and (b) respectively.

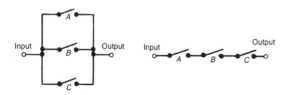

Input A B C	Output $Z = A + B + C$
0 0 0	0
0 0 1	1
0 1 0	1
0 1 1	1
1 0 0	1
1 0 1	1
1 1 0	1
1 1 1	1

Input A B C	Output $Z = A \cdot B \cdot C$
0 0 0	0
0 0 1	0
0 1 0	0
0 1 1	0
1 0 0	0
1 0 1	0
1 1 0	0
1 1 1	1

(a) The or - function electrical circuit and truth table

(b) The and - function electrical circuit and truth table

Figure 60.3

To achieve a given output, it is often necessary to use combinations of switches connected both in series and in parallel. If the output from a switching circuit is given by the Boolean expression $Z = A \cdot B + \bar{A} \cdot \bar{B}$, the truth table is as shown in Fig. 60.4(a). In this table, columns 1 and 2 give all the possible combinations of A and B. Column 3 corresponds to $A \cdot B$ and column 4 to $\bar{A} \cdot \bar{B}$, i.e. a 1 output is obtained when $A = 0$ and when $B = 0$. Column 5 is the

1 A	2 B	3 $A \cdot B$	4 $\bar{A} \cdot \bar{B}$	5 $Z = A \cdot B + \bar{A} \cdot \bar{B}$
0	0	0	1	1
0	1	0	0	0
1	0	0	0	0
1	1	1	0	1

(a) Truth table for $Z = A \cdot B + \bar{A} \cdot \bar{B}$

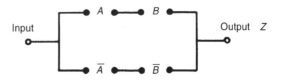

(b) Switching circuit for $Z = A \cdot B + \bar{A} \cdot \bar{B}$

Figure 60.4

or-function applied to columns 3 and 4 giving an output of $Z = A \cdot B + \bar{A} \cdot \bar{B}$. The corresponding switching circuit is shown in Fig. 60.4(b) in which A and B are connected in series to give $A \cdot B$, $\bar{A}$ and $\bar{B}$ are connected in series to give $\bar{A} \cdot \bar{B}$, and $A \cdot B$ and $\bar{A} \cdot \bar{B}$ are connected in parallel to give $A \cdot B + \bar{A} \cdot \bar{B}$. The circuit symbols used are such that A means the switch is on when A is 1, $\bar{A}$ means the switch is on when A is 0, and so on.

Problem 1. Derive the Boolean expression and construct a truth table for the switching circuit shown in Fig. 60.5.

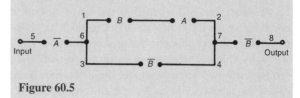

Figure 60.5

The switches between 1 and 2 in Fig. 60.5 are in series and have a Boolean expression of $B \cdot A$. The parallel circuit 1 to 2 and 3 to 4 have a Boolean expression of $(B \cdot A + \bar{B})$. The parallel circuit can be treated as a single switching unit, giving the equivalent of switches 5 to 6, 6 to 7 and 7 to 8 in series. Thus the output is given by:

$$Z = \bar{A} \cdot (B \cdot A + \bar{B}) \cdot \bar{B}$$

The truth table is as shown in Table 60.2. Columns 1 and 2 give all the possible combinations of switches A and B. Column 3 is the **and**-function applied to columns 1 and 2, giving $B \cdot A$. Column 4 is $\overline{B}$, i.e., the opposite to column 2. Column 5 is the **or**-function applied to columns 3 and 4. Column 6 is $\overline{A}$, i.e. the opposite to column 1. The output is column 7 and is obtained by applying the **and**-function to columns 4, 5 and 6.

Table 60.2

1	2	3	4	5	6	7
A	B	$B\cdot A$	$\overline{B}$	$B\cdot A+\overline{B}$	$\overline{A}$	$Z=\overline{A}\cdot(B\cdot A+\overline{B})\cdot\overline{B}$
0	0	0	1	1	1	1
0	1	0	0	0	1	0
1	0	0	1	1	0	0
1	1	1	0	1	0	0

Problem 2. Derive the Boolean expression and construct a truth table for the switching circuit shown in Fig. 60.6.

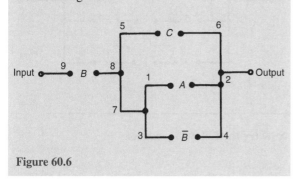

Figure 60.6

The parallel circuit 1 to 2 and 3 to 4 gives $(A+\overline{B})$ and this is equivalent to a single switching unit between 7 and 2. The parallel circuit 5 to 6 and 7 to 2 gives $C+(A+\overline{B})$ and this is equivalent to a single switching unit between 8 and 2. The series circuit 9 to 8 and 8 to 2 gives the output

$$Z = B \cdot [C + (A + B)]$$

The truth table is shown in Table 60.3. Columns 1, 2 and 3 give all the possible combinations of A, B and C. Column 4 is $\overline{B}$ and is the opposite to column 2. Column 5 is the **or**-function applied to columns 1 and 4, giving $(A+\overline{B})$. Column 6 is the **or**-function applied to columns 3 and 5 giving $C+(A+\overline{B})$. The output is given in column 7 and is obtained by applying the **and**-function to columns 2 and 6, giving $Z=B\cdot[C+(A+\overline{B})]$.

Table 60.3

1	2	3	4	5	6	7
A	B	C	$\overline{B}$	$A+\overline{B}$	$C+(A+\overline{B})$	$Z=B\cdot[C+(A+\overline{B})]$
0	0	0	1	1	1	0
0	0	1	1	1	1	0
0	1	0	0	0	0	0
0	1	1	0	0	1	1
1	0	0	1	1	1	0
1	0	1	1	1	1	0
1	1	0	0	1	1	1
1	1	1	0	1	1	1

Problem 3. Construct a switching circuit to meet the requirements of the Boolean expression:
$$Z = A \cdot \overline{C} + \overline{A} \cdot B + \overline{A} \cdot B \cdot \overline{C}$$
Construct the truth table for this circuit.

The three terms joined by **or**-functions, $(+)$, indicate three parallel branches.

having: branch 1 A **and** $\overline{C}$ in series

            branch 2 $\overline{A}$ **and** B in series

and   branch 3 $\overline{A}$ **and** B **and** $\overline{C}$ in series

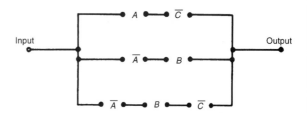

Figure 60.7

Hence the required switching circuit is as shown in Fig. 60.7. The corresponding truth table is shown in Table 60.4.

Column 4 is $\overline{C}$, i.e. the opposite to column 3

Column 5 is $A\cdot\overline{C}$, obtained by applying the **and**-function to columns 1 and 4

Column 6 is $\overline{A}$, the opposite to column 1

Column 7 is $\overline{A}\cdot B$, obtained by applying the **and**-function to columns 2 and 6

Table 60.4

1	2	3	4	5	6	7	8	9
A	B	C	$\overline{C}$	$A \cdot \overline{C}$	$\overline{A}$	$\overline{A} \cdot B$	$\overline{A} \cdot B \cdot \overline{C}$	$Z = A \cdot \overline{C} + \overline{A} \cdot B + \overline{A} \cdot B \cdot \overline{C}$
0	0	0	1	0	1	0	0	0
0	0	1	0	0	1	0	0	0
0	1	0	1	0	1	1	1	1
0	1	1	0	0	1	1	0	1
1	0	0	1	1	0	0	0	1
1	0	1	0	0	0	0	0	0
1	1	0	1	1	0	0	0	1
1	1	1	0	0	0	0	0	0

Column 8 is $\overline{A} \cdot B \cdot \overline{C}$, obtained by applying the **and**-function to columns 4 and 7

Column 9 is the output, obtained by applying the **or**-function to columns 5, 7 and 8

Problem 4. Derive the Boolean expression and construct the switching circuit for the truth table given in Table 60.5.

Table 60.5

	A	B	C	Z
1	0	0	0	1
2	0	0	1	0
3	0	1	0	1
4	0	1	1	1
5	1	0	0	0
6	1	0	1	1
7	1	1	0	0
8	1	1	1	0

Examination of the truth table shown in Table 60.5 shows that there is a 1 output in the Z-column in rows 1, 3, 4 and 6. Thus, the Boolean expression and switching circuit should be such that a 1 output is obtained for row 1 **or** row 3 **or** row 4 **or** row 6. In row 1, A is

0 **and** B is 0 **and** C is 0 and this corresponds to the Boolean expression $\overline{A} \cdot \overline{B} \cdot \overline{C}$. In row 3, A is 0 **and** B is 1 **and** C is 0, i.e. the Boolean expression in $\overline{A} \cdot B \cdot \overline{C}$. Similarly in rows 4 and 6, the Boolean expressions are $\overline{A} \cdot B \cdot C$ and $A \cdot \overline{B} \cdot C$ respectively. Hence the Boolean expression is:

$$Z = \overline{A} \cdot \overline{B} \cdot \overline{C} + \overline{A} \cdot B \cdot \overline{C} + \overline{A} \cdot B \cdot C + A \cdot \overline{B} \cdot C$$

The corresponding switching circuit is shown in Fig. 60.8. The four terms are joined by **or**-functions, (+), and are represented by four parallel circuits. Each term has three elements joined by an **and**-function, and is represented by three elements connected in series.

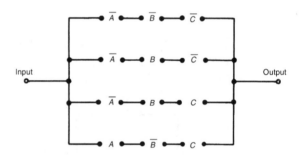

Figure 60.8

Now try the following exercise

Exercise 203   Further problems on Boolean algebra and switching circuits

In Problems 1 to 4, determine the Boolean expressions and construct truth tables for the switching circuits given.

1. The circuit shown in Fig. 60.9

$$\left[\begin{array}{l} C \cdot (A \cdot B + \overline{A} \cdot B); \\ \text{see Table 60.6, col. 4} \end{array} \right]$$

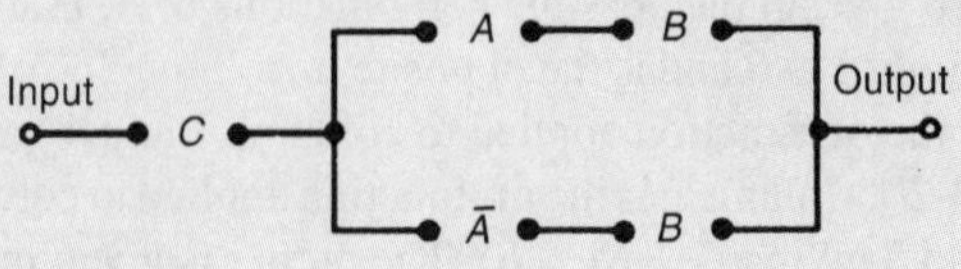

Figure 60.9

Table 60.6

1	2	3	4	5	6	7
A	B	C	$C \cdot (A \cdot B + \bar{A} \cdot B)$	$C \cdot (A \cdot \bar{B} + \bar{A})$	$A \cdot B \cdot (B \cdot \bar{C} + \bar{B} \cdot C + \bar{A} \cdot B)$	$C \cdot [B \cdot C \cdot \bar{A} + A \cdot (B + \bar{C})]$
0	0	0	0	0	0	0
0	0	1	0	1	0	0
0	1	0	0	0	0	0
0	1	1	1	1	0	1
1	0	0	0	0	0	0
1	0	1	0	1	0	0
1	1	0	0	0	1	0
1	1	1	1	0	0	1

2. The circuit shown in Fig. 60.10

$$\left[\begin{array}{c} C \cdot (A \cdot \bar{B} + \bar{A}); \\ \text{see Table 60.6, col. 5} \end{array} \right]$$

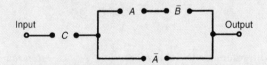

Figure 60.10

3. The circuit shown in Fig. 60.11

$$\left[\begin{array}{c} A \cdot B \cdot (B \cdot \bar{C} + \bar{B} \cdot C + \bar{A} \cdot B); \\ \text{see Table 60.6, col. 6} \end{array} \right]$$

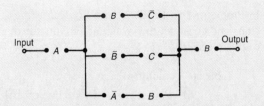

Figure 60.11

4. The circuit shown in Fig. 60.12

$$\left[\begin{array}{c} C \cdot [B \cdot C \cdot \bar{A} + A \cdot (B + \bar{C})], \\ \text{see Table 60.6, col. 7} \end{array} \right]$$

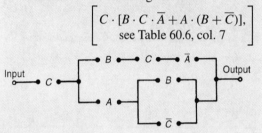

Figure 60.12

In Problems 5 to 7, construct switching circuits to meet the requirements of the Boolean expressions given.

5. $A \cdot C + A \cdot \bar{B} \cdot C + A \cdot B$ [See Fig. 60.13]

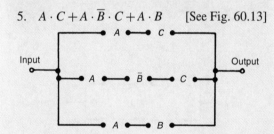

Figure 60.13

6. $A \cdot B \cdot C \cdot (A + B + C)$ [See Fig. 60.14]

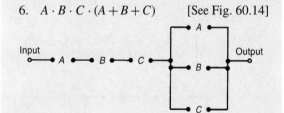

Figure 60.14

7. $A \cdot (A \cdot \bar{B} \cdot C + B \cdot (A + \bar{C}))$ [See Fig. 60.15]

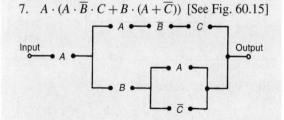

Figure 60.15

Section 10

In Problems 8 to 10, derive the Boolean expressions and construct the switching circuits for the truth table stated.

8. Table 60.7, column 4

$$[\overline{A} \cdot \overline{B} \cdot C + A \cdot B \cdot \overline{C}; \text{ see Fig. 60.16}]$$

Table 60.7

1	2	3	4	5	6
A	B	C			
0	0	0	0	1	1
0	0	1	1	0	0
0	1	0	0	0	1
0	1	1	0	1	0
1	0	0	0	1	1
1	0	1	0	0	1
1	1	0	1	0	0
1	1	1	0	0	0

Figure 60.16

9. Table 60.7, column 5

$$\left[\overline{A} \cdot \overline{B} \cdot \overline{C} + \overline{A} \cdot B \cdot C + A \cdot \overline{B} \cdot \overline{C}; \atop \text{see Fig. 60.17} \right]$$

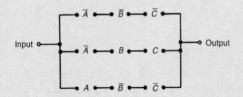

Figure 60.17

10. Table 60.7, column 6

$$\left[\overline{A} \cdot \overline{B} \cdot \overline{C} + \overline{A} \cdot B \cdot \overline{C} + A \cdot \overline{B} \cdot \overline{C} \atop + A \cdot \overline{B} \cdot C; \text{see Fig. 60.18} \right]$$

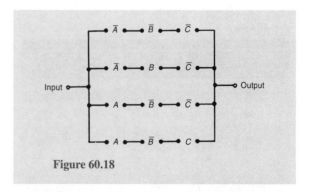

Figure 60.18

60.2 Simplifying Boolean expressions

A Boolean expression may be used to describe a complex switching circuit or logic system. If the Boolean expression can be simplified, then the number of switches or logic elements can be reduced resulting in a saving in cost. Three principal ways of simplifying Boolean expressions are:

(a) by using the laws and rules of Boolean algebra (see Section 60.3),

(b) by applying de Morgan's laws (see Section 60.4), and

(c) by using Karnaugh maps (see Section 60.5).

60.3 Laws and rules of Boolean algebra

A summary of the principal laws and rules of Boolean algebra are given in Table 60.8. The way in which these laws and rules may be used to simplify Boolean expressions is shown in Problems 5 to 10.

Problem 5. Simplify the Boolean expression:

$$\overline{P} \cdot \overline{Q} + \overline{P} \cdot Q + P \cdot \overline{Q}$$

With reference to Table 60.8:   *Reference*

$$\overline{P} \cdot \overline{Q} + \overline{P} \cdot Q + P \cdot \overline{Q}$$

$$= \overline{P} \cdot (\overline{Q} + Q) + P \cdot \overline{Q} \qquad 5$$

$$= \overline{P} \cdot 1 + P \cdot \overline{Q} \qquad 10$$

$$= \overline{P} + P \cdot \overline{Q} \qquad 12$$

Table 60.8

Ref.	Name	Rule or law
1	Commutative laws	$A+B=B+A$
2		$A \cdot B = B \cdot A$
3	Associative laws	$(A+B)+C=A+(B+C)$
4		$(A \cdot B) \cdot C = A \cdot (B \cdot C)$
5	Distributive laws	$A \cdot (B+C) = A \cdot B + A \cdot C$
6		$A + (B \cdot C)$ $= (A+B) \cdot (A+C)$
7	Sum rules	$A+0=A$
8		$A+1=1$
9		$A+A=A$
10		$A+\overline{A}=1$
11	Product rules	$A \cdot 0 = 0$
12		$A \cdot 1 = A$
13		$A \cdot A = A$
14		$A \cdot \overline{A} = 0$
15	Absorption rules	$A + A \cdot B = A$
16		$A \cdot (A+B) = A$
17		$A + \overline{A} \cdot B = A + B$

Problem 6. Simplify

$$(P + \overline{P} \cdot Q) \cdot (Q + \overline{Q} \cdot P)$$

With reference to Table 60.8:   *Reference*

$(P + \overline{P} \cdot Q) \cdot (Q + \overline{Q} \cdot P)$

$\quad = P \cdot (Q + \overline{Q} \cdot P)$

$\qquad + \overline{P} \cdot Q \cdot (Q + \overline{Q} \cdot P)$ 5

$\quad = P \cdot Q + P \cdot \overline{Q} \cdot P + \overline{P} \cdot Q \cdot Q$

$\qquad + \overline{P} \cdot Q \cdot \overline{Q} \cdot P$ 5

$\quad = P \cdot Q + P \cdot \overline{Q} + \overline{P} \cdot Q$

$\qquad + \overline{P} \cdot Q \cdot \overline{Q} \cdot P$ 13

$\quad = P \cdot Q + P \cdot \overline{Q} + \overline{P} \cdot Q + 0$ 14

$\quad = P \cdot Q + P \cdot \overline{Q} + \overline{P} \cdot Q$ 7

$\quad = P \cdot (Q + \overline{Q}) + \overline{P} \cdot Q$ 5

$\quad = P \cdot 1 + \overline{P} \cdot Q$ 10

$\quad = \boldsymbol{P + \overline{P} \cdot Q}$ 12

Problem 7. Simplify

$$F \cdot G \cdot \overline{H} + F \cdot G \cdot H + \overline{F} \cdot G \cdot H$$

With reference to Table 60.8   *Reference*

$F \cdot G \cdot \overline{H} + F \cdot G \cdot H + \overline{F} \cdot G \cdot H$

$\quad = F \cdot G \cdot (\overline{H} + H) + \overline{F} \cdot G \cdot H$ 5

$\quad = F \cdot G \cdot 1 + \overline{F} \cdot G \cdot H$ 10

$\quad = F \cdot G + \overline{F} \cdot G \cdot H$ 12

$\quad = \boldsymbol{G \cdot (F + \overline{F} \cdot H)}$ 5

Problem 8. Simplify
$$\overline{F} \cdot \overline{G} \cdot H + \overline{F} \cdot G \cdot H + F \cdot \overline{G} \cdot H + F \cdot G \cdot H$$

With reference to Table 60.8:   *Reference*

$\overline{F} \cdot \overline{G} \cdot H + \overline{F} \cdot G \cdot H + F \cdot \overline{G} \cdot H + F \cdot G \cdot H$

$\quad = \overline{G} \cdot H \cdot (\overline{F} + F) + G \cdot H \cdot (\overline{F} + F)$ 5

$\quad = \overline{G} \cdot H \cdot 1 + G \cdot H \cdot 1$ 10

$\quad = \overline{G} \cdot H + G \cdot H$ 12

$\quad = H \cdot (\overline{G} + G)$ 5

$\quad = H \cdot 1 = \boldsymbol{H}$ 10 and 12

Problem 9. Simplify
$$A \cdot \overline{C} + \overline{A} \cdot (B + C) + A \cdot B \cdot (C + \overline{B})$$
using the rules of Boolean algebra.

With reference to Table 60.8 *Reference*

$$A \cdot \overline{C} + \overline{A} \cdot (B + C) + A \cdot B \cdot (C + \overline{B})$$

$$= A \cdot \overline{C} + \overline{A} \cdot B + \overline{A} \cdot C + A \cdot B \cdot C$$

$$+ A \cdot B \cdot \overline{B} \qquad 5$$

$$= A \cdot \overline{C} + \overline{A} \cdot B + \overline{A} \cdot C + A \cdot B \cdot C$$

$$+ A \cdot 0 \qquad 14$$

$$= A \cdot \overline{C} + \overline{A} \cdot B + \overline{A} \cdot C + A \cdot B \cdot C \qquad 11$$

$$= A \cdot (\overline{C} + B \cdot C) + \overline{A} \cdot B + \overline{A} \cdot C \qquad 5$$

$$= A \cdot (\overline{C} + B) + \overline{A} \cdot B + \overline{A} \cdot C \qquad 17$$

$$= A \cdot \overline{C} + A \cdot B + \overline{A} \cdot B + \overline{A} \cdot C \qquad 5$$

$$= A \cdot \overline{C} + B \cdot (A + \overline{A}) + \overline{A} \cdot C \qquad 5$$

$$= A \cdot \overline{C} + B \cdot 1 + \overline{A} \cdot C \qquad 10$$

$$= \boldsymbol{A \cdot \overline{C} + B + \overline{A} \cdot C} \qquad 12$$

Problem 10. Simplify the expression
$P \cdot \overline{Q} \cdot R + P \cdot Q \cdot (\overline{P} + R) + Q \cdot R \cdot (\overline{Q} + P)$
using the rules of Boolean algebra.

With reference to Table 60.8: *Reference*

$$P \cdot \overline{Q} \cdot R + P \cdot Q \cdot (\overline{P} + R) + Q \cdot R \cdot (\overline{Q} + P)$$

$$= P \cdot \overline{Q} \cdot R + P \cdot Q \cdot \overline{P} + P \cdot Q \cdot R$$

$$+ Q \cdot R \cdot \overline{Q} + Q \cdot R \cdot P \qquad 5$$

$$= P \cdot \overline{Q} \cdot R + 0 \cdot Q + P \cdot Q \cdot R + 0 \cdot R$$

$$+ P \cdot Q \cdot R \qquad 14$$

$$= P \cdot \overline{Q} \cdot R + P \cdot Q \cdot R + P \cdot Q \cdot R \qquad 7 \text{ and } 11$$

$$= P \cdot \overline{Q} \cdot R + P \cdot Q \cdot R \qquad 9$$

$$= P \cdot R \cdot (Q + \overline{Q}) \qquad 5$$

$$= P \cdot R \cdot 1 \qquad 10$$

$$= \boldsymbol{P \cdot R} \qquad 12$$

Now try the following exercise

Exercise 204   Further problems on the laws the rules of Boolean algebra

Use the laws and rules of Boolean algebra given in Table 60.8 to simplify the following expressions:

1. $\overline{P} \cdot \overline{Q} + \overline{P} \cdot Q$ $[\overline{P}]$

2. $\overline{P} \cdot Q + P \cdot Q + \overline{P} \cdot \overline{Q}$ $[\overline{P} + P \cdot Q]$

3. $\overline{F} \cdot \overline{G} + F \cdot \overline{G} + G \cdot (F + \overline{F})$ $[\overline{G}]$

4. $F \cdot \overline{G} + F \cdot (G + \overline{G}) + F \cdot G$ $[F]$

5. $(P + P \cdot Q) \cdot (Q + Q \cdot P)$ $[P \cdot Q]$

6. $\overline{F} \cdot \overline{G} \cdot H + \overline{F} \cdot G \cdot H + F \cdot \overline{G} \cdot H$
      $[H \cdot (\overline{F} + F \cdot \overline{G}]$

7. $F \cdot \overline{G} \cdot \overline{H} + F \cdot G \cdot H + \overline{F} \cdot G \cdot H$
      $[F \cdot \overline{G} \cdot \overline{H} + G \cdot H]$

8. $\overline{P} \cdot \overline{Q} \cdot \overline{R} + \overline{P} \cdot Q \cdot R + P \cdot \overline{Q} \cdot \overline{R}$
      $[\overline{Q} \cdot \overline{R} + \overline{P} \cdot Q \cdot R]$

9. $\overline{F} \cdot \overline{G} \cdot \overline{H} + \overline{F} \cdot \overline{G} \cdot H + F \cdot \overline{G} \cdot \overline{H} + F \cdot \overline{G} \cdot H$
      $[\overline{G}]$

10. $F \cdot \overline{G} \cdot H + F \cdot G \cdot H + F \cdot G \cdot \overline{H} + \overline{F} \cdot G \cdot \overline{H}$
      $[F \cdot H + G \cdot \overline{H}]$

11. $R \cdot (P \cdot Q + P \cdot \overline{Q}) + \overline{R} \cdot (\overline{P} \cdot \overline{Q} + \overline{P} \cdot Q)$
      $[P \cdot R + \overline{P} \cdot \overline{R}]$

12. $\overline{R} \cdot (\overline{P} \cdot \overline{Q} + P \cdot Q + P \cdot \overline{Q}) + P \cdot (Q \cdot R + \overline{Q} \cdot R)$
      $[P + \overline{Q} \cdot \overline{R}]$

60.4   De Morgan's laws

De Morgan's laws may be used to simplify **not**-functions having two or more elements. The laws state that:

$$\overline{A + B} = \overline{A} \cdot \overline{B} \quad \text{and} \quad \overline{A \cdot B} = \overline{A} + \overline{B}$$

and may be verified by using a truth table (see Problem 11). The application of de Morgan's laws in simplifying Boolean expressions is shown in Problems 12 and 13.

Problem 11. Verify that $\overline{A + B} = \overline{A} \cdot \overline{B}$

A Boolean expression may be verified by using a truth table. In Table 60.9, columns 1 and 2 give all the possible arrangements of the inputs A and B. Column 3 is the **or**-function applied to columns 1 and 2 and column 4 is the **not**-function applied to column 3. Columns 5 and 6 are the **not**-function applied to columns 1 and 2 respectively and column 7 is the **and**-function applied to columns 5 and 6.

Table 60.9

1	2	3	4	5	6	7
A	B	$A+B$	$\overline{A+B}$	$\overline{A}$	$\overline{B}$	$\overline{A} \cdot \overline{B}$
0	0	0	1	1	1	1
0	1	1	0	1	0	0
1	0	1	0	0	1	0
1	1	1	0	0	0	0

Since columns 4 and 7 have the same pattern of 0's and 1's this verifies that $\overline{A+B}=\overline{A}\cdot\overline{B}$.

> **Problem 12.** Simplify the Boolean expression $(\overline{A\cdot B})+(\overline{A+B})$ by using de Morgan's laws and the rules of Boolean algebra.

Applying de Morgan's law to the first term gives:

$$\overline{A\cdot B}=\overline{\overline{A}}+\overline{B}=A+\overline{B} \quad \text{since } \overline{\overline{A}}=A$$

Applying de Morgan's law to the second term gives:

$$\overline{A+B}=\overline{\overline{A}}\cdot\overline{B}=A\cdot\overline{B}$$

Thus, $(\overline{A\cdot B})+(\overline{A+B})=(A+\overline{B})+A\cdot\overline{B}$

Removing the bracket and reordering gives:
$A+A\cdot\overline{B}+\overline{B}$
But, by rule 15, Table 60.8, $A+A\cdot B=A$. It follows that: $A+A\cdot\overline{B}=A$

Thus: $(\overline{A\cdot B})+(\overline{A+B})=A+\overline{B}$

> **Problem 13.** Simplify the Boolean expression $(A\cdot\overline{B}+C)\cdot(\overline{A+B\cdot\overline{C}})$ by using de Morgan's laws and the rules of Boolean algebra.

Applying de Morgan's laws to the first term gives:

$$\overline{A\cdot\overline{B}+C}=\overline{A\cdot\overline{B}}\cdot\overline{C}=(\overline{A}+\overline{\overline{B}})\cdot\overline{C}$$
$$=(\overline{A}+B)\cdot\overline{C}=\overline{A}\cdot\overline{C}+B\cdot\overline{C}$$

Applying de Morgan's law to the second term gives:

$$\overline{A+B\cdot\overline{C}}=\overline{A}+(\overline{B+\overline{\overline{C}}})=\overline{A}+(\overline{B}+C)$$

Thus $(A\cdot\overline{B}+C)\cdot(\overline{A+B\cdot\overline{C}})$
$$=(\overline{A}\cdot\overline{C}+B\cdot\overline{C})\cdot(\overline{A}+\overline{B}+C)$$
$$=\overline{A}\cdot\overline{A}\cdot\overline{C}+\overline{A}\cdot\overline{B}\cdot\overline{C}+\overline{A}\cdot\overline{C}\cdot C$$
$$\quad+\overline{A}\cdot B\cdot\overline{C}+B\cdot\overline{B}\cdot\overline{C}+B\cdot\overline{C}\cdot C$$

But from Table 60.8, $\overline{A}\cdot\overline{A}=\overline{A}$ and $\overline{C}\cdot C=B\cdot\overline{B}=0$ Hence the Boolean expression becomes:

$$\overline{A}\cdot\overline{C}+\overline{A}\cdot\overline{B}\cdot\overline{C}+\overline{A}\cdot B\cdot\overline{C}$$
$$=\overline{A}\cdot\overline{C}(1+\overline{B}+B)$$
$$=\overline{A}\cdot\overline{C}(1+B)$$
$$=\overline{A}\cdot\overline{C}$$

Thus: $(A\cdot\overline{B}+C)\cdot(\overline{A+B\cdot\overline{C}})=\overline{A}\cdot\overline{C}$

Now try the following exercise

> **Exercise 205 Further problems on simplifying Boolean expressions using de Morgan's laws**
>
> Use de Morgan's laws and the rules of Boolean algebra given in Table 60.8 to simplify the following expressions.
>
> 1. $(\overline{A}\cdot\overline{B})\cdot(\overline{A}\cdot B)$ $[\overline{A}\cdot\overline{B}]$
>
> 2. $(A+\overline{B\cdot C})+(\overline{A}\cdot\overline{B}+C)$ $[\overline{A}+\overline{B}+C]$
>
> 3. $(\overline{A\cdot B+B\cdot\overline{C}})\cdot\overline{A\cdot\overline{B}}$ $[\overline{A}\cdot\overline{B}+A\cdot B\cdot C]$
>
> 4. $(A\cdot\overline{B}+B\cdot\overline{C})+(\overline{A}\cdot B)$ $[1]$
>
> 5. $(\overline{P\cdot\overline{Q}+\overline{P}\cdot R})\cdot(\overline{P\cdot\overline{Q}\cdot R})$ $[P\cdot(\overline{Q}+\overline{R})]$

60.5 Karnaugh maps

(i) Two-variable Karnaugh maps

A truth table for a two-variable expression is shown in Table 60.10(a), the '1' in the third row output showing that $Z=A\cdot\overline{B}$. Each of the four possible Boolean expressions associated with a two-variable function can be depicted as shown in Table 60.10(b) in which one

Table 60.10

Inputs		Output	Boolean
A	B	Z	expression
0	0	0	$\overline{A} \cdot \overline{B}$
0	1	0	$\overline{A} \cdot B$
1	0	1	$A \cdot \overline{B}$
1	1	0	$A \cdot B$

(a)

B \ A	0 ($\overline{A}$)	1 (A)
0($\overline{B}$)	$\overline{A}\cdot\overline{B}$	$A\cdot\overline{B}$
1(B)	$\overline{A}\cdot B$	$A\cdot B$

B \ A	0	1
0	0	1
1	0	0

(b)   (c)

cell is allocated to each row of the truth table. A matrix similar to that shown in Table 60.10(b) can be used to depict $Z = A \cdot \overline{B}$, by putting a 1 in the cell corresponding to $A \cdot \overline{B}$ and 0's in the remaining cells. This method of depicting a Boolean expression is called a two-variable **Karnaugh map**, and is shown in Table 60.10(c).

To simplify a two-variable Boolean expression, the Boolean expression is depicted on a Karnaugh map, as outlined above. Any cells on the map having either a common vertical side or a common horizontal side are grouped together to form a **couple**. (This is a coupling together of cells, not just combining two together). The simplified Boolean expression for a couple is given by those variables common to all cells in the couple. See Problem 14.

(ii) Three-variable Karnaugh maps

A truth table for a three-variable expression is shown in Table 60.11(a), the 1's in the output column showing that:

$$Z = \overline{A} \cdot \overline{B} \cdot C + \overline{A} \cdot B \cdot C + A \cdot B \cdot \overline{C}$$

Each of the eight possible Boolean expressions associated with a three-variable function can be depicted as shown in Table 60.11(b) in which one cell is allocated to each row of the truth table. A matrix similar to that shown in Table 60.11(b) can be used to depict: $Z = \overline{A} \cdot \overline{B} \cdot C + \overline{A} \cdot B \cdot C + A \cdot B \cdot \overline{C}$, by putting 1's in the cells corresponding to the Boolean terms on the right of the Boolean equation and 0's in the remaining cells. This method of depicting a three-variable Boolean expression is called a three-variable Karnaugh map, and is shown in Table 60.11(c).

Table 60.11

Inputs			Output	Boolean
A	B	C	Z	expression
0	0	0	0	$\overline{A} \cdot \overline{B} \cdot \overline{C}$
0	0	1	1	$\overline{A} \cdot \overline{B} \cdot C$
0	1	0	0	$\overline{A} \cdot B \cdot \overline{C}$
0	1	1	1	$\overline{A} \cdot B \cdot C$
1	0	0	0	$A \cdot \overline{B} \cdot \overline{C}$
1	0	1	0	$A \cdot \overline{B} \cdot C$
1	1	0	1	$A \cdot B \cdot \overline{C}$
1	1	1	0	$A \cdot B \cdot C$

(a)

C \ A.B	00 ($\overline{A}\cdot\overline{B}$)	01 ($\overline{A}\cdot B$)	11 (A·B)	10 (A·$\overline{B}$)
0($\overline{C}$)	$\overline{A}\cdot\overline{B}\cdot\overline{C}$	$\overline{A}\cdot B\cdot\overline{C}$	$A\cdot B\cdot\overline{C}$	$A\cdot\overline{B}\cdot\overline{C}$
1(C)	$\overline{A}\cdot\overline{B}\cdot C$	$\overline{A}\cdot B\cdot C$	$A\cdot B\cdot C$	$A\cdot\overline{B}\cdot C$

(b)

C \ A.B	00	01	11	10
0	0	0	1	0
1	1	1	0	0

(c)

To simplify a three-variable Boolean expression, the Boolean expression is depicted on a Karnaugh map as outlined above. Any cells on the map having common edges either vertically or horizontally are grouped together to form couples of four cells or two cells. During coupling the horizontal lines at the top and bottom of the cells are taken as a common edge, as are the vertical lines on the left and right of the cells. The simplified Boolean expression for a couple is given by those variables common to all cells in the couple. See Problems 15 to 17.

(iii) Four-variable Karnaugh maps

A truth table for a four-variable expression is shown in Table 60.12(a), the 1's in the output column showing that:

$$Z = \overline{A} \cdot \overline{B} \cdot C \cdot \overline{D} + \overline{A} \cdot B \cdot C \cdot \overline{D}$$
$$+ A \cdot \overline{B} \cdot C \cdot \overline{D} + A \cdot B \cdot C \cdot \overline{D}$$

Each of the sixteen possible Boolean expressions associated with a four-variable function can be depicted as shown in Table 60.12(b), in which one cell is allocated to each row of the truth table. A matrix similar to that shown in Table 60.12(b) can be used to depict

$$Z = \overline{A} \cdot \overline{B} \cdot C \cdot \overline{D} + \overline{A} \cdot B \cdot C \cdot \overline{D}$$
$$+ A \cdot \overline{B} \cdot C \cdot \overline{D} + A \cdot B \cdot C \cdot \overline{D}$$

by putting 1's in the cells corresponding to the Boolean terms on the right of the Boolean equation and 0's in the remaining cells. This method of depicting a four-variable expression is called a four-variable Karnaugh map, and is shown in Table 60.12(c).

Table 60.12

Inputs				Output	Boolean
A	B	C	D	Z	expression
0	0	0	0	0	$\overline{A} \cdot \overline{B} \cdot \overline{C} \cdot \overline{D}$
0	0	0	1	0	$\overline{A} \cdot \overline{B} \cdot \overline{C} \cdot D$
0	0	1	0	1	$\overline{A} \cdot \overline{B} \cdot C \cdot \overline{D}$
0	0	1	1	0	$\overline{A} \cdot \overline{B} \cdot C \cdot D$
0	1	0	0	0	$\overline{A} \cdot B \cdot \overline{C} \cdot \overline{D}$
0	1	0	1	0	$\overline{A} \cdot B \cdot \overline{C} \cdot D$
0	1	1	0	1	$\overline{A} \cdot B \cdot C \cdot \overline{D}$
0	1	1	1	0	$\overline{A} \cdot B \cdot C \cdot D$
1	0	0	0	0	$A \cdot \overline{B} \cdot \overline{C} \cdot \overline{D}$
1	0	0	1	0	$A \cdot \overline{B} \cdot \overline{C} \cdot D$
1	0	1	0	1	$A \cdot \overline{B} \cdot C \cdot \overline{D}$
1	0	1	1	0	$A \cdot \overline{B} \cdot C \cdot D$
1	1	0	0	0	$A \cdot B \cdot \overline{C} \cdot \overline{D}$
1	1	0	1	0	$A \cdot B \cdot \overline{C} \cdot D$
1	1	1	0	1	$A \cdot B \cdot C \cdot \overline{D}$
1	1	1	1	0	$A \cdot B \cdot C \cdot D$

(a)

C·D \ A·B	00 ($\overline{A}\cdot\overline{B}$)	01 ($\overline{A}\cdot B$)	11 ($A\cdot B$)	10 ($A\cdot\overline{B}$)
00 ($\overline{C}\cdot\overline{D}$)	$\overline{A}\cdot\overline{B}\cdot\overline{C}\cdot\overline{D}$	$\overline{A}\cdot B\cdot\overline{C}\cdot\overline{D}$	$A\cdot B\cdot\overline{C}\cdot\overline{D}$	$A\cdot\overline{B}\cdot\overline{C}\cdot\overline{D}$
01 ($\overline{C}\cdot D$)	$\overline{A}\cdot\overline{B}\cdot\overline{C}\cdot D$	$\overline{A}\cdot B\cdot\overline{C}\cdot D$	$A\cdot B\cdot\overline{C}\cdot D$	$A\cdot\overline{B}\cdot\overline{C}\cdot D$
11 ($C\cdot D$)	$\overline{A}\cdot\overline{B}\cdot C\cdot D$	$\overline{A}\cdot B\cdot C\cdot D$	$A\cdot B\cdot C\cdot D$	$A\cdot\overline{B}\cdot C\cdot D$
10 ($C\cdot\overline{D}$)	$\overline{A}\cdot\overline{B}\cdot C\cdot\overline{D}$	$\overline{A}\cdot B\cdot C\cdot\overline{D}$	$A\cdot B\cdot C\cdot\overline{D}$	$A\cdot\overline{B}\cdot C\cdot\overline{D}$

(b)

C·D \ A·B	0·0	0·1	1·1	1·0
0·0	0	0	0	0
0·1	0	0	0	0
1·1	0	0	0	0
1·0	1	1	1	1

(c)

To simplify a four-variable Boolean expression, the Boolean expression is depicted on a Karnaugh map as outlined above. Any cells on the map having common edges either vertically or horizontally are grouped together to form couples of eight cells, four cells or two cells. During coupling, the horizontal lines at the top and bottom of the cells may be considered to be common edges, as are the vertical lines on the left and the right of the cells. The simplified Boolean expression for a couple is given by those variables common to all cells in the couple. See Problems 18 and 19.

Summary of procedure when simplifying a Boolean expression using a Karnaugh map

(a) Draw a four, eight or sixteen-cell matrix, depending on whether there are two, three or four variables.

(b) Mark in the Boolean expression by putting 1's in the appropriate cells.

(c) Form couples of 8, 4 or 2 cells having common edges, forming the largest groups of cells possible. (Note that a cell containing a 1 may be used more than once when forming a couple. Also note that each cell containing a 1 must be used at least once).

(d) The Boolean expression for the couple is given by the variables which are common to all cells in the couple.

Problem 14. Use Karnaugh map techniques to simplify the expression $\overline{P} \cdot \overline{Q} + \overline{P} \cdot Q$

Using the above procedure:

(a) The two-variable matrix is drawn and is shown in Table 60.13.

Table 60.13

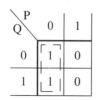

(b) The term $\overline{P} \cdot \overline{Q}$ is marked with a 1 in the top left-hand cell, corresponding to $P = 0$ and $Q = 0$; $\overline{P} \cdot Q$ is marked with a 1 in the bottom left-hand cell corresponding to $P = 0$ and $Q = 1$.

(c) The two cells containing 1's have a common horizontal edge and thus a vertical couple, can be formed.

(d) The variable common to both cells in the couple is $P = 0$, i.e. $\overline{P}$ thus

$$\overline{P} \cdot \overline{Q} + \overline{P} \cdot Q = \overline{P}$$

Problem 15. Simplify the expression $\overline{X} \cdot Y \cdot \overline{Z} + \overline{X} \cdot \overline{Y} \cdot Z + X \cdot Y \cdot \overline{Z} + X \cdot \overline{Y} \cdot Z$ by using Karnaugh map techniques.

Using the above procedure:

(a) A three-variable matrix is drawn and is shown in Table 60.14.

Table 60.14

Z \\ X·Y	0·0	0·1	1·1	1·0
0	0	1	1	0
1	1	0	0	1

(b) The 1's on the matrix correspond to the expression given, i.e. for $\overline{X} \cdot Y \cdot \overline{Z}$, $X = 0$, $Y = 1$ and $Z = 0$ and hence corresponds to the cell in the two row and second column, and so on.

(c) Two couples can be formed as shown. The couple in the bottom row may be formed since the vertical lines on the left and right of the cells are taken as a common edge.

(d) The variables common to the couple in the top row are $Y = 1$ and $Z = 0$, that is, $Y \cdot \overline{Z}$ and the variables common to the couple in the bottom row are $Y = 0$, $Z = 1$, that is, $\overline{Y} \cdot Z$. Hence:

$$\overline{X} \cdot Y \cdot \overline{Z} + \overline{X} \cdot \overline{Y} \cdot Z + X \cdot Y \cdot \overline{Z}$$
$$+ X \cdot \overline{Y} \cdot Z = Y \cdot \overline{Z} + \overline{Y} \cdot Z$$

Problem 16. Using a Karnaugh map technique to simplify the expression $\overline{(\overline{A} \cdot B)} \cdot \overline{(\overline{A} + B)}$.

Using the procedure, a two-variable matrix is drawn and is shown in Table 60.15.

Table 60.15

B \\ A	0	1
0	1	1 2
1		1

$\overline{A} \cdot B$ corresponds to the bottom left-hand cell and $\overline{(\overline{A} \cdot B)}$ must therefore be all cells except this one, marked with a 1 in Table 60.15. $(\overline{A} + B)$ corresponds to all the cells except the top right-hand cell marked with a 2 in Table 60.15. Hence $\overline{(\overline{A} + B)}$ must correspond to the cell marked with a 2. The expression $\overline{(\overline{A} \cdot B)} \cdot \overline{(\overline{A} + B)}$ corresponds to the cell having both 1 and 2 in it, i.e.

$$\overline{(\overline{A} \cdot B)} \cdot \overline{(\overline{A} + B)} = A \cdot \overline{B}$$

Problem 17. Simplify $\overline{(P + \overline{Q} \cdot R)} + \overline{(P \cdot Q + \overline{R})}$ using a Karnaugh map technique.

The term $(P + \overline{Q} \cdot R)$ corresponds to the cells marked 1 on the matrix in Table 60.16(a), hence $\overline{(P + \overline{Q} \cdot R)}$ corresponds to the cells marked 2. Similarly, $(P \cdot Q + \overline{R})$ corresponds to the cells marked 3 in Table 60.16(a), hence $\overline{(P \cdot Q + \overline{R})}$ corresponds to the cells marked 4. The expression $\overline{(P + \overline{Q} \cdot R)} + \overline{(P \cdot Q + \overline{R})}$ corresponds to cells marked with either a 2 or with a 4 and is shown in Table 60.16(b) by X's. These cells may be coupled as shown. The variables common to the group of four cells is $P = 0$, i.e. $\overline{P}$, and those common to the group of two cells are $Q = 0$, $R = 1$, i.e. $\overline{Q} \cdot R$.

Thus: $\overline{(P + \overline{Q} \cdot R)} + \overline{(P \cdot Q + \overline{R})} = \overline{P} + \overline{Q} \cdot R$.

Table 60.16

P·Q R	0·0	0·1	1·1	1·0
0	3 2	3 2	3 1	3 1
1	4 1	4 2	3 1	4 1

(a)

P·Q R	0·0	0·1	1·1	1·0
0	X	X		
1	X	X		X

(b)

Table 60.18

A·B C·D	0·0	0·1	1·1	1·0
0·0	1			1
0·1				
1·1			1	
1·0	1			1

Problem 18. Use Karnaugh map techniques to simplify the expression:

$$A \cdot B \cdot \overline{C} \cdot \overline{D} + A \cdot B \cdot C \cdot D + \overline{A} \cdot B \cdot C \cdot D$$
$$+ A \cdot B \cdot C \cdot \overline{D} + \overline{A} \cdot B \cdot C \cdot \overline{D}.$$

Using the procedure, a four-variable matrix is drawn and is shown in Table 60.17. The 1's marked on the matrix correspond to the expression given. Two couples can be formed as shown. The four-cell couple has $B = 1$, $C = 1$, i.e. $\mathbf{B} \cdot \mathbf{C}$ as the common variables to all four cells and the two-cell couple has $A \cdot B \cdot \overline{D}$ as the common variables to both cells. Hence, the expression simplifies to:

$$\mathbf{B} \cdot \mathbf{C} + \mathbf{A} \cdot \mathbf{B} \cdot \overline{\mathbf{D}} \quad \text{i.e.} \quad \mathbf{B} \cdot (\mathbf{C} + \mathbf{A} \cdot \overline{\mathbf{D}})$$

Table 60.17

A·B C·D	0·0	0·1	1·1	1·0
0·0			1	
0·1				
1·1		1	1	
1·0		1	1	

Problem 19. Simplify the expression $\overline{A} \cdot \overline{B} \cdot \overline{C} \cdot \overline{D} + A \cdot \overline{B} \cdot \overline{C} \cdot \overline{D} + \overline{A} \cdot \overline{B} \cdot C \cdot \overline{D} + A \cdot \overline{B} \cdot C \cdot \overline{D} + A \cdot B \cdot C \cdot D$ by using Karnaugh map techniques.

The Karnaugh map for the expression is shown in Table 60.18. Since the top and bottom horizontal lines are common edges and the vertical lines on the left and right of the cells are common, then the four corner cells form a couple, $\overline{B} \cdot \overline{D}$, (the cells can be considered as if they are stretched to completely cover a sphere, as far as common edges are concerned). The cell $A \cdot B \cdot C \cdot D$ cannot be coupled with any other. Hence the expression simplifies to

$$\overline{B} \cdot \overline{D} + A \cdot B \cdot C \cdot D$$

Now try the following exercise

Exercise 206   Further problems on simplifying Boolean expressions using Karnaugh maps

In Problems 1 to 11 use Karnaugh map techniques to simplify the expressions given.

1. $\overline{X} \cdot Y + X \cdot Y$ $[Y]$

2. $\overline{X} \cdot \overline{Y} + \overline{X} \cdot Y + X \cdot Y$ $[\overline{X} + Y]$

3. $(\overline{P} \cdot \overline{Q}) \cdot (\overline{\overline{P} \cdot Q})$ $[\overline{P} \cdot \overline{Q}]$

4. $A \cdot \overline{C} + \overline{A} \cdot (B + C) + A \cdot B \cdot (C + \overline{B})$
    $[B + A \cdot \overline{C} + \overline{A} \cdot C]$

5. $\overline{P} \cdot \overline{Q} \cdot \overline{R} + \overline{P} \cdot Q \cdot \overline{R} + P \cdot Q \cdot \overline{R}$ $[\overline{R} \cdot (\overline{P} + Q)]$

6. $\overline{P} \cdot \overline{Q} \cdot \overline{R} + P \cdot Q \cdot \overline{R} + P \cdot Q \cdot R + P \cdot \overline{Q} \cdot R$
    $[P \cdot (Q + R) + \overline{P} \cdot \overline{Q} \cdot \overline{R}]$

7. $\overline{A} \cdot \overline{B} \cdot \overline{C} \cdot \overline{D} + \overline{A} \cdot B \cdot \overline{C} \cdot \overline{D} + \overline{A} \cdot B \cdot \overline{C} \cdot D$
    $[\overline{A} \cdot \overline{C} \cdot (B + \overline{D})]$

8. $\overline{A} \cdot \overline{B} \cdot C \cdot D + \overline{A} \cdot \overline{B} \cdot C \cdot \overline{D} + A \cdot \overline{B} \cdot C \cdot \overline{D}$
    $[\overline{B} \cdot C \cdot (\overline{A} + \overline{D})]$

9. $\overline{A} \cdot B \cdot \overline{C} \cdot D + A \cdot B \cdot \overline{C} \cdot D + A \cdot B \cdot C \cdot D$
    $+ A \cdot \overline{B} \cdot \overline{C} \cdot D + A \cdot \overline{B} \cdot C \cdot D$
    $[D \cdot (A + B \cdot \overline{C})]$

10. $\overline{A} \cdot \overline{B} \cdot \overline{C} \cdot D + A \cdot B \cdot \overline{C} \cdot D + A \cdot \overline{B} \cdot \overline{C} \cdot D$
    $+ A \cdot B \cdot C \cdot \overline{D} + A \cdot \overline{B} \cdot C \cdot D$
    $[A \cdot \overline{D} + \overline{A} \cdot \overline{B} \cdot \overline{C} \cdot D]$

11. $A \cdot B \cdot \overline{C} \cdot \overline{D} + \overline{A} \cdot \overline{B} \cdot \overline{C} \cdot \overline{D} + \overline{A} \cdot B \cdot C \cdot D +$
    $\overline{A} \cdot \overline{B} \cdot C \cdot D + A \cdot \overline{B} \cdot \overline{C} \cdot \overline{D} + \overline{A} \cdot \overline{B} \cdot C \cdot \overline{D} +$
    $\overline{A} \cdot B \cdot C \cdot \overline{D}$
    $[\overline{A} \cdot C + A \cdot \overline{C} \cdot \overline{D} + \overline{B} \cdot \overline{D} \cdot (\overline{A} + \overline{C})]$

60.6 Logic circuits

In practice, logic gates are used to perform the **and, or** and **not**-functions introduced in Section 60.1. Logic gates can be made from switches, magnetic devices or fluidic devices, but most logic gates in use are electronic devices. Various logic gates are available. For example, the Boolean expression $(A \cdot B \cdot C)$ can be produced using a three-input, **and**-gate and $(C + D)$ by using a two-input **or**-gate. The principal gates in common use are introduced below. The term 'gate' is used in the same sense as a normal gate, the open state being indicated by a binary '1' and the closed state by a binary '0'. A gate will only open when the requirements of the gate are met and, for example, there will only be a '1' output on a two-input **and**-gate when both the inputs to the gate are at a '1' state.

The and-gate

The different symbols used for a three-input, **and**-gate are shown in Fig. 60.19(a) and the truth table is shown in Fig. 60.19(b). This shows that there will only be a '1' output when A is 1, or B is 1, or C is 1, written as:

$$Z = A \cdot B \cdot C$$

The or-gate

The different symbols used for a three-input **or**-gate are shown in Fig. 60.20(a) and the truth table is shown in Fig. 60.20(b). This shows that there will be a '1' output when A is 1, or B is 1, or C is 1, or any combination of A, B or C is 1, written as:

$$Z = A + B + C$$

The invert-gate or not-gate

The different symbols used for an **invert**-gate are shown in Fig. 60.21(a) and the truth table is shown in Fig. 60.21(b). This shows that a '0' input gives a '1' output and vice versa, i.e. it is an 'opposite to' function. The invert of A is written $\overline{A}$ and is called 'not-A'.

The nand-gate

The different symbols used for a **nand**-gate are shown in Fig. 60.22(a) and the truth table is shown in Fig. 60.22(b). This gate is equivalent to an **and**-gate and an **invert**-gate in series (not-and = nand) and the output is written as:

$$Z = \overline{A \cdot B \cdot C}$$

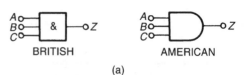

BRITISH   AMERICAN

(a)

INPUTS			OUTPUT
A	B	C	$Z = A \cdot B \cdot C$
0	0	0	0
0	0	1	0
0	1	0	0
0	1	1	0
1	0	0	0
1	0	1	0
1	1	0	0
1	1	1	1

(b)

Figure 60.19

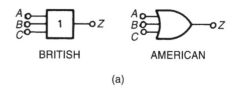

BRITISH   AMERICAN

(a)

INPUTS			OUTPUT
A	B	C	$Z = A + B + C$
0	0	0	0
0	0	1	1
0	1	0	1
0	1	1	1
1	0	0	1
1	0	1	1
1	1	0	1
1	1	1	1

(b)

Figure 60.20

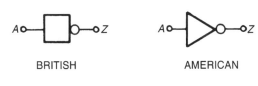

BRITISH   AMERICAN

(a)

INPUT A	OUTPUT $Z = \overline{A}$
0	1
1	0

(b)

Figure 60.21

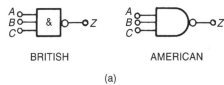

BRITISH   AMERICAN

(a)

INPUTS				OUTPUT
A	B	C	$A \cdot B \cdot C$	$Z = \overline{A \cdot B \cdot C}$
0	0	0	0	1
0	0	1	0	1
0	1	0	0	1
0	1	1	0	1
1	0	0	0	1
1	0	1	0	1
1	1	0	0	1
1	1	1	1	0

(b)

Figure 60.22

The nor-gate

The different symbols used for a **nor**-gate are shown in Fig. 60.23(a) and the truth table is shown in Fig. 60.23(b). This gate is equivalent to an **or**-gate and an **invert**-gate in series, (not-or = nor), and the output is written as:

$$Z = \overline{A + B + C}$$

Combinational logic networks

In most logic circuits, more than one gate is needed to give the required output. Except for the **invert**-gate,

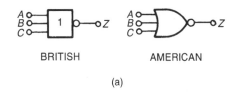

BRITISH   AMERICAN

(a)

INPUTS				OUTPUT
A	B	C	$A + B + C$	$Z = \overline{A + B + C}$
0	0	0	0	1
0	0	1	1	0
0	1	0	1	0
0	1	1	1	0
1	0	0	1	0
1	0	1	1	0
1	1	0	1	0
1	1	1	1	0

(b)

Figure 60.23

logic gates generally have two, three or four inputs and are confined to one function only. Thus, for example, a two-input, **or**-gate or a four-input **and**-gate can be used when designing a logic circuit. The way in which logic gates are used to generate a given output is shown in Problems 20 to 23.

Problem 20. Devise a logic system to meet the requirements of: $Z = A \cdot \overline{B} + C$

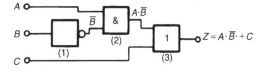

Figure 60.24

With reference to Fig. 60.24 an **invert**-gate, shown as (1), gives $\overline{B}$. The **and**-gate, shown as (2), has inputs of A and $\overline{B}$, giving $A \cdot \overline{B}$. The **or**-gate, shown as (3), has inputs of $A \cdot \overline{B}$ and C, giving:

$$Z = A \cdot \overline{B} + C$$

Problem 21. Devise a logic system to meet the requirements of $(P + \overline{Q}) \cdot (\overline{R} + S)$. `

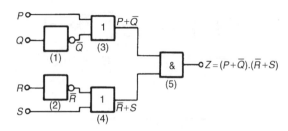

Figure 60.25

The logic system is shown in Fig. 60.25. The given expression shows that two **invert**-functions are needed to give $\overline{Q}$ and $\overline{R}$ and these are shown as gates (1) and (2). Two **or**-gates, shown as (3) and (4), give $(P + \overline{Q})$ and $(\overline{R} + S)$ respectively. Finally, an **and**-gate, shown as (5), gives the required output.

$$Z = (P + \overline{Q}) \cdot (\overline{R} + S)$$

Problem 22. Devise a logic circuit to meet the requirements of the output given in Table 60.19, using as few gates as possible.

Table 60.19

Inputs			Output
A	B	C	Z
0	0	0	0
0	0	1	0
0	1	0	0
0	1	1	0
1	0	0	0
1	0	1	1
1	1	0	1
1	1	1	1

The '1' outputs in rows 6, 7 and 8 of Table 60.19 show that the Boolean expression is:

$$Z = A \cdot \overline{B} \cdot C + A \cdot B \cdot \overline{C} + A \cdot B \cdot C$$

The logic circuit for this expression can be built using three, 3-input **and**-gates and one, 3-input **or**-gate, together with two **invert**-gates. However, the number of gates required can be reduced by using the techniques introduced in Sections 60.3 to 60.5, resulting in the cost

of the circuit being reduced. Any of the techniques can be used, and in this case, the rules of Boolean algebra (see Table 60.8) are used.

$$Z = A \cdot \overline{B} \cdot C + A \cdot B \cdot \overline{C} + A \cdot B \cdot C$$
$$= A \cdot [\overline{B} \cdot C + B \cdot \overline{C} + B \cdot C]$$
$$= A \cdot [\overline{B} \cdot C + B(\overline{C} + C)] = A \cdot [\overline{B} \cdot C + B]$$
$$= A \cdot [B + \overline{B} \cdot C] = A \cdot [B + C]$$

The logic circuit to give this simplified expression is shown in Fig. 60.26.

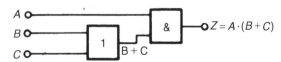

Figure 60.26

Problem 23. Simplify the expression:

$$Z = \overline{P} \cdot \overline{Q} \cdot \overline{R} \cdot \overline{S} + \overline{P} \cdot \overline{Q} \cdot \overline{R} \cdot S + \overline{P} \cdot Q \cdot \overline{R} \cdot \overline{S}$$
$$+ \overline{P} \cdot Q \cdot \overline{R} \cdot S + P \cdot \overline{Q} \cdot \overline{R} \cdot \overline{S}$$

and devise a logic circuit to give this output.

The given expression is simplified using the Karnaugh map techniques introduced in Section 60.5. Two couples are formed as shown in Fig. 60.27(a) and the simplified expression becomes:

$$Z = \overline{Q} \cdot \overline{R} \cdot \overline{S} + \overline{P} \cdot \overline{R}$$

i.e. $Z = \overline{R} \cdot (\overline{P} + \overline{Q} \cdot \overline{S})$

The logic circuit to produce this expression is shown in Fig. 60.27(b).

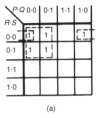

(a)

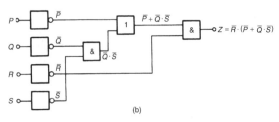

(b)

Figure 60.27

Now try the following exercise

Exercise 207   Further problems on logic circuits

In Problems 1 to 4, devise logic systems to meet the requirements of the Boolean expressions given.

1. $Z = \overline{A} + B \cdot C$ [See Fig. 60.28(a)]

2. $Z = A \cdot \overline{B} + B \cdot \overline{C}$ [See Fig. 60.28(b)]

3. $Z = A \cdot B \cdot \overline{C} + \overline{A} \cdot \overline{B} \cdot C$ [See Fig. 60.28(c)]

4. $Z = (\overline{A} + B) \cdot (\overline{C} + D)$ [See Fig. 60.28(d)]

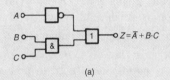

(a)

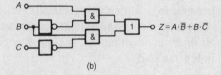

(b)

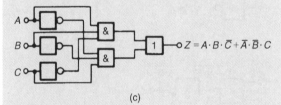

(c)

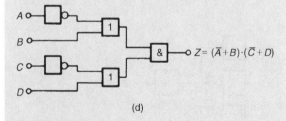

(d)

Figure 60.28

In Problems 5 to 7, simplify the expression given in the truth table and devise a logic circuit to meet the requirements stated.

5.   Column 4 of Table 60.20
$[Z_1 = A \cdot B + C$, see Fig. 60.29(a)]

6.   Column 5 of Table 60.20
$[Z_2 = A \cdot \overline{B} + B \cdot C$, see Fig. 60.29(b)]

Table 60.20

1	2	3	4	5	6
A	B	C	Z_1	Z_2	Z_3
0	0	0	0	0	0
0	0	1	1	0	0
0	1	0	0	0	1
0	1	1	1	1	1
1	0	0	0	1	0
1	0	1	1	1	1
1	1	0	1	0	1
1	1	1	1	1	1

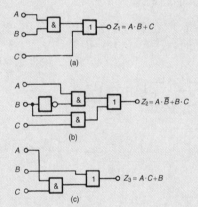

Figure 60.29

7.   Column 6 of Table 60.20
$[Z_3 = A \cdot C + B$, see Fig. 60.29(c)]

In Problems 8 to 12, simplify the Boolean expressions given and devise logic circuits to give the requirements of the simplified expressions.

8. $\overline{P} \cdot \overline{Q} + \overline{P} \cdot Q + P \cdot Q$
$[\overline{P} + Q$, see Fig. 60.30(a)]

9. $\overline{P} \cdot \overline{Q} \cdot \overline{R} + P \cdot Q \cdot \overline{R} + P \cdot \overline{Q} \cdot \overline{R}$
$[\overline{R} \cdot (P + \overline{Q})$, see Fig. 60.30(b)]

10. $P \cdot \overline{Q} \cdot R + P \cdot \overline{Q} \cdot \overline{R} + \overline{P} \cdot \overline{Q} \cdot \overline{R}$
$[\overline{Q} \cdot (P + \overline{R})$, see Fig. 60.30(c)]

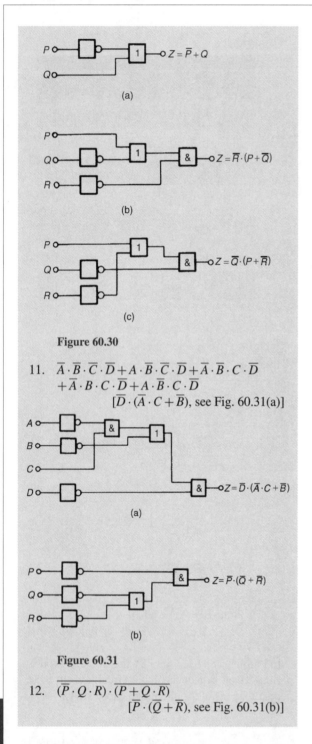

(a) $Z = \overline{P} + Q$

(b) $Z = \overline{R} \cdot (P + \overline{Q})$

(c) $Z = \overline{Q} \cdot (P + \overline{R})$

Figure 60.30

11. $\overline{A} \cdot \overline{B} \cdot \overline{C} \cdot \overline{D} + A \cdot \overline{B} \cdot \overline{C} \cdot \overline{D} + \overline{A} \cdot \overline{B} \cdot C \cdot \overline{D}$
$+ \overline{A} \cdot B \cdot C \cdot \overline{D} + A \cdot \overline{B} \cdot C \cdot \overline{D}$
$[\overline{D} \cdot (\overline{A} \cdot C + \overline{B}), \text{ see Fig. 60.31(a)}]$

(a) $Z = \overline{D} \cdot (\overline{A} \cdot C + \overline{B})$

(b) $Z = \overline{P} \cdot (\overline{Q} + \overline{R})$

Figure 60.31

12. $\overline{(\overline{P} \cdot Q \cdot R)} \cdot \overline{(P + Q \cdot R)}$
$[\overline{P} \cdot (\overline{Q} + \overline{R}), \text{ see Fig. 60.31(b)}]$

60.7 Universal logic gates

The function of any of the five logic gates in common use can be obtained by using either **nand**-gates or **nor**-gates and when used in this manner, the gate selected in called a **universal gate**. The way in which a universal **nand**-gate is used to produce the **invert, and, or** and **nor**-functions is shown in Problem 24. The way in which a universal **nor**-gate is used to produce the **invert, or, and** and **nand**-function is shown in Problem 25.

Problem 24. Show low **invert, and, or** and **nor**-functions can be produced using nand-gates only.

A single input to a **nand**-gate gives the **invert**-function, as shown in Fig. 60.32(a). When two **nand**-gates are connected, as shown in Fig. 60.32(b), the output from the first gate is $\overline{A \cdot B \cdot C}$ and this is inverted by the second gate, giving $Z = \overline{\overline{A \cdot B \cdot C}} = A \cdot B \cdot C$ i.e. the **and**-function is produced. When $\overline{A}$, $\overline{B}$ and $\overline{C}$ are the inputs to a **nand**-gate, the output is $\overline{\overline{A} \cdot \overline{B} \cdot \overline{C}}$.

By de Morgan's law, $\overline{\overline{A} \cdot \overline{B} \cdot \overline{C}} = \overline{\overline{A}} + \overline{\overline{B}} + \overline{\overline{C}} = A + B + C$, i.e. a **nand**-gate is used to produce **or**-function. The logic circuit shown in Fig. 60.32(c). If the output from the logic circuit in Fig. 60.32(c) is inverted by adding an additional **nand**-gate, the output becomes the invert of an **or**-function, i.e. the **nor**-function, as shown in Fig. 60.32(d).

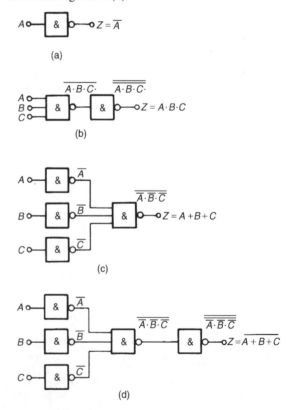

(a) $Z = \overline{A}$

(b) $Z = A \cdot B \cdot C$

(c) $Z = A + B + C$

(d) $Z = \overline{A + B + C}$

Figure 60.32

Problem 25. Show how **invert**, **or**, **and** and **nand**-functions can be produced by using **nor**-gates only.

A single input to a **nor**-gate gives the **invert**-function, as shown in Fig. 60.33(a). When two **nor**-gates are connected, as shown in Fig. 60.33(b), the output from the first gate is $\overline{A+B+C}$ and this is inverted by the second gate, giving $Z = \overline{\overline{A+B+C}} = A+B+C$, i.e. the **or**-function is produced. Inputs of $\overline{A}$, $\overline{B}$, and $\overline{C}$ to a **nor**-gate give an output of $\overline{\overline{A}+\overline{B}+\overline{C}}$.

By de Morgan's law, $\overline{\overline{A}+\overline{B}+\overline{C}} = \overline{\overline{A}} \cdot \overline{\overline{B}} \cdot \overline{\overline{C}} = A \cdot B \cdot C$, i.e. the **nor**-gate can be used to produce the **and**-function. The logic circuit is shown in Fig. 60.33(c). When the output of the logic circuit, shown in Fig. 60.33(c), is inverted by adding an additional **nor**-gate, the output then becomes the invert of an **or**-function, i.e. the **nor**-function as shown in Fig. 60.33(d).

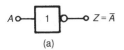

(a)

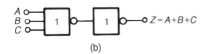

(b)

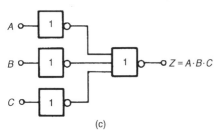

(c)

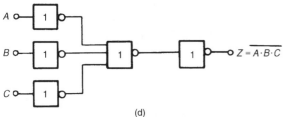

(d)

Figure 60.33

Problem 26. Design a logic circuit, using **nand**-gates having not more than three inputs, to meet the requirements of the Boolean

expression

$$Z = \overline{A} + \overline{B} + C + \overline{D}$$

When designing logic circuits, it is often easier to start at the output of the circuit. The given expression shows there are four variables joined by **or**-functions. From the principles introduced in Problem 24, if a four-input **nand**-gate is used to give the expression given, the input are $\overline{\overline{A}}$, $\overline{\overline{B}}$, $\overline{C}$ and $\overline{\overline{D}}$ that is A, B, $\overline{C}$ and D. However, the problem states that three-inputs are not to be exceeded so two of the variables are joined, i.e. the inputs to the three-input **nand**-gate, shown as gate (1) Fig. 60.34, is A, B, $\overline{C}$, and D. From Problem 24, the **and**-function is generated by using two **nand**-gates connected in series, as shown by gates (2) and (3) in Fig. 60.34. The logic circuit required to produce the given expression is as shown in Fig. 60.34.

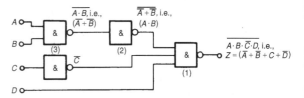

Figure 60.34

Problem 27. Use **nor**-gates only to design a logic circuit to meet the requirements of the expressions:

$$Z = \overline{D} \cdot (\overline{A} + B + \overline{C})$$

It is usual in logic circuit design to start the design at the output. From Problem 25, the **and**-function between $\overline{D}$ and the terms in the bracket can be produced by using inputs of $\overline{\overline{D}}$ and $\overline{\overline{A}+B+\overline{C}}$ to a **nor**-gate, i.e. by de Morgan's law, inputs of D and $A \cdot \overline{B} \cdot C$. Again, with reference to Problem 25, inputs of $\overline{A} \cdot B$ and $\overline{C}$ to a **nor**-gate give an output of $\overline{A}+B+\overline{C}$, which by de Morgan's law is $A \cdot \overline{B} \cdot C$. The logic circuit to produce the required expression is as shown in Fig. 60.35

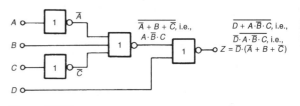

Figure 60.35

Problem 28. An alarm indicator in a grinding mill complex should be activated if (a) the power supply to all mills is off and (b) the hopper feeding the mills is less than 10% full, and (c) if less than two of the three grinding mills are in action. Devise a logic system to meet these requirements.

Let variable A represent the power supply on to all the mills, then $\overline{A}$ represents the power supply off. Let B represent the hopper feeding the mills being more than 10% full, then $\overline{B}$ represents the hopper being less than 10% full. Let C, D and E represent the three mills respectively being in action, then $\overline{C}$, $\overline{D}$ and $\overline{E}$ represent the three mills respectively not being in action. The required expression to activate the alarm is:

$$Z = \overline{A} \cdot \overline{B} \cdot (\overline{C} + \overline{D} + \overline{E})$$

There are three variables joined by **and**-functions in the output, indicating that a three-input **and**-gate is required, having inputs of $\overline{A}$, $\overline{B}$ and $(\overline{C} + \overline{D} + \overline{E})$.

The term $(\overline{C} + \overline{D} + \overline{E})$ is produced by a three-input **nand**-gate. When variables C, D and E are the inputs to a **nand**-gate, the output is $C \cdot D \cdot E$ which, by de Morgan's law is $\overline{C} + \overline{D} + \overline{E}$. Hence the required logic circuit is as shown in Fig. 60.36.

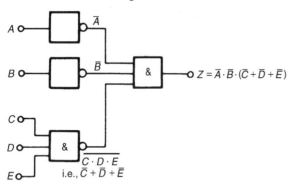

Figure 60.36

Now try the following exercise

Exercise 208   Further problems on universal logic gates

In Problems 1 to 3, use **nand**-gates only to devise the logic systems stated.

1. $Z = A + B \cdot C$ [See Fig. 60.37(a)]

2. $Z = A \cdot \overline{B} + B \cdot \overline{C}$ [See Fig. 60.37(b)]

3. $Z = A \cdot B \cdot \overline{C} + \overline{A} \cdot \overline{B} \cdot C$ [See Fig. 60.37(c)]

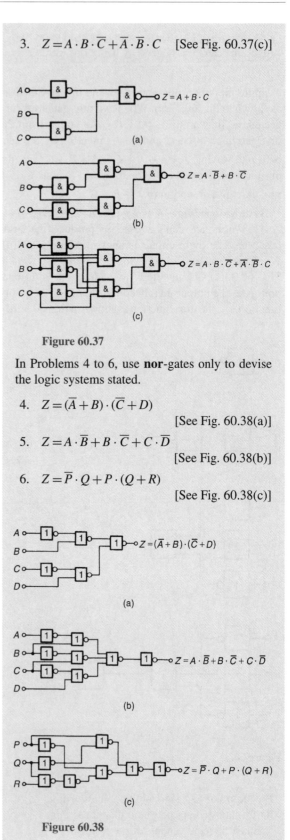

Figure 60.37

In Problems 4 to 6, use **nor**-gates only to devise the logic systems stated.

4. $Z = (\overline{A} + B) \cdot (\overline{C} + D)$

      [See Fig. 60.38(a)]

5. $Z = A \cdot \overline{B} + B \cdot \overline{C} + C \cdot \overline{D}$

      [See Fig. 60.38(b)]

6. $Z = \overline{P} \cdot Q + P \cdot (Q + R)$

      [See Fig. 60.38(c)]

Figure 60.38

7. In a chemical process, three of the transducers used are P, Q and R, giving output signals of either 0 or 1. Devise a logic system to give a 1 output when:

   (a) P and Q and R all have 0 outputs, or when:

   (b) P is 0 and (Q is 1 or R is 0)
   $$[\overline{P} \cdot (Q + \overline{R}), \text{ See Fig. 60.39(a)}]$$

8. Lift doors should close, (Z), if:

   (a) the master switch, (A), is on and either

   (b) a call, (B), is received from any other floor, or

   (c) the doors, (C), have been open for more than 10 seconds, or

   (d) the selector push within the lift (D), is pressed for another floor.

   Devise a logic circuit to meet these requirements.
   $$\left[\begin{array}{l} Z = A \cdot (B + C + D), \\ \text{see Fig. 60.39(b)} \end{array} \right]$$

9. A water tank feeds three separate processes. When any two of the processes are in operation at the same time, a signal is required to start a pump to maintain the head of water in the tank. Devise a logic circuit using **nor**-gates only to give the required signal.
   $$\left[\begin{array}{l} Z = A \cdot (B + C) + B \cdot C, \\ \text{see Fig. 60.39(c)} \end{array} \right]$$

10. A logic signal is required to give an indication when:

    (a) the supply to an oven is on, and

    (b) the temperature of the oven exceeds 210°C, or

    (c) the temperature of the oven is less than 190°C

    Devise a logic circuit using **nand**-gates only to meet these requirements.
    $$[Z = A \cdot (B + C), \text{ see Fig. 60.39(d)}]$$

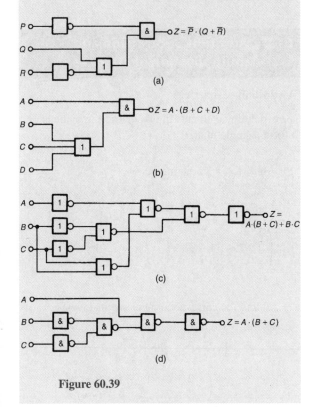

Figure 60.39

Chapter 61

The theory of matrices and determinants

61.1 Matrix notation

Matrices and determinants are mainly used for the solution of linear simultaneous equations. The theory of matrices and determinants is dealt with in this chapter and this theory is then used in Chapter 62 to solve simultaneous equations.

The coefficients of the variables for linear simultaneous equations may be shown in matrix form. The coefficients of x and y in the simultaneous equations

$$x + 2y = 3$$
$$4x - 5y = 6$$

become $\begin{pmatrix} 1 & 2 \\ 4 & -5 \end{pmatrix}$ in matrix notation.

Similarly, the coefficients of p, q and r in the equations

$$1.3p - 2.0q + r = 7$$
$$3.7p + 4.8q - 7r = 3$$
$$4.1p + 3.8q + 12r = -6$$

become $\begin{pmatrix} 1.3 & -2.0 & 1 \\ 3.7 & 4.8 & -7 \\ 4.1 & 3.8 & 12 \end{pmatrix}$ in matrix form.

The numbers within a matrix are called an **array** and the coefficients forming the array are called the **elements** of the matrix. The number of rows in a matrix is usually specified by m and the number of columns by n and a matrix referred to as an 'm by n' matrix. Thus, $\begin{pmatrix} 2 & 3 & 6 \\ 4 & 5 & 7 \end{pmatrix}$ is a '2 by 3' matrix. Matrices cannot be expressed as a single numerical value, but they can often

be simplified or combined, and unknown element values can be determined by comparison methods. Just as there are rules for addition, subtraction, multiplication and division of numbers in arithmetic, rules for these operations can be applied to matrices and the rules of matrices are such that they obey most of those governing the algebra of numbers.

61.2 Addition, subtraction and multiplication of matrices

(i) Addition of matrices

Corresponding elements in two matrices may be added to form a single matrix.

Problem 1. Add the matrices

(a) $\begin{pmatrix} 2 & -1 \\ -7 & 4 \end{pmatrix}$ and $\begin{pmatrix} -3 & 0 \\ 7 & -4 \end{pmatrix}$

(b) $\begin{pmatrix} 3 & 1 & -4 \\ 4 & 3 & 1 \\ 1 & 4 & -3 \end{pmatrix}$ and $\begin{pmatrix} 2 & 7 & -5 \\ -2 & 1 & 0 \\ 6 & 3 & 4 \end{pmatrix}$

(a) Adding the corresponding elements gives:

$$\begin{pmatrix} 2 & -1 \\ -7 & 4 \end{pmatrix} + \begin{pmatrix} -3 & 0 \\ 7 & -4 \end{pmatrix}$$
$$= \begin{pmatrix} 2+(-3) & -1+0 \\ -7+7 & 4+(-4) \end{pmatrix}$$
$$= \begin{pmatrix} -1 & -1 \\ 0 & 0 \end{pmatrix}$$

(b) Adding the corresponding elements gives:

$$\begin{pmatrix} 3 & 1 & -4 \\ 4 & 3 & 1 \\ 1 & 4 & -3 \end{pmatrix} + \begin{pmatrix} 2 & 7 & -5 \\ -2 & 1 & 0 \\ 6 & 3 & 4 \end{pmatrix}$$

$$= \begin{pmatrix} 3+2 & 1+7 & -4+(-5) \\ 4+(-2) & 3+1 & 1+0 \\ 1+6 & 4+3 & -3+4 \end{pmatrix}$$

$$= \begin{pmatrix} \mathbf{5} & \mathbf{8} & \mathbf{-9} \\ \mathbf{2} & \mathbf{4} & \mathbf{1} \\ \mathbf{7} & \mathbf{7} & \mathbf{1} \end{pmatrix}$$

(ii) Subtraction of matrices

If A is a matrix and B is another matrix, then $(A - B)$ is a single matrix formed by subtracting the elements of B from the corresponding elements of A.

Problem 2. Subtract

(a) $\begin{pmatrix} -3 & 0 \\ 7 & -4 \end{pmatrix}$ from $\begin{pmatrix} 2 & -1 \\ -7 & 4 \end{pmatrix}$

(b) $\begin{pmatrix} 2 & 7 & -5 \\ -2 & 1 & 0 \\ 6 & 3 & 4 \end{pmatrix}$ from $\begin{pmatrix} 3 & 1 & -4 \\ 4 & 3 & 1 \\ 1 & 4 & -3 \end{pmatrix}$

To find matrix A minus matrix B, the elements of B are taken from the corresponding elements of A. Thus:

(a) $\begin{pmatrix} 2 & -1 \\ -7 & 4 \end{pmatrix} - \begin{pmatrix} -3 & 0 \\ 7 & -4 \end{pmatrix}$

$$= \begin{pmatrix} 2-(-3) & -1-0 \\ -7-7 & 4-(-4) \end{pmatrix}$$

$$= \begin{pmatrix} \mathbf{5} & \mathbf{-1} \\ \mathbf{-14} & \mathbf{8} \end{pmatrix}$$

(b) $\begin{pmatrix} 3 & 1 & -4 \\ 4 & 3 & 1 \\ 1 & 4 & -3 \end{pmatrix} - \begin{pmatrix} 2 & 7 & -5 \\ -2 & 1 & 0 \\ 6 & 3 & 4 \end{pmatrix}$

$$= \begin{pmatrix} 3-2 & 1-7 & -4-(-5) \\ 4-(-2) & 3-1 & 1-0 \\ 1-6 & 4-3 & -3-4 \end{pmatrix}$$

$$= \begin{pmatrix} \mathbf{1} & \mathbf{-6} & \mathbf{1} \\ \mathbf{6} & \mathbf{2} & \mathbf{1} \\ \mathbf{-5} & \mathbf{1} & \mathbf{-7} \end{pmatrix}$$

Problem 3. If

$$A = \begin{pmatrix} -3 & 0 \\ 7 & -4 \end{pmatrix}, B = \begin{pmatrix} 2 & -1 \\ -7 & 4 \end{pmatrix} \text{ and}$$

$$C = \begin{pmatrix} 1 & 0 \\ -2 & -4 \end{pmatrix} \text{ find } A + B - C.$$

$$A + B = \begin{pmatrix} -1 & -1 \\ 0 & 0 \end{pmatrix}$$

(from Problem 1)

Hence, $A + B - C = \begin{pmatrix} -1 & -1 \\ 0 & 0 \end{pmatrix} - \begin{pmatrix} 1 & 0 \\ -2 & -4 \end{pmatrix}$

$$= \begin{pmatrix} -1-1 & -1-0 \\ 0-(-2) & 0-(-4) \end{pmatrix}$$

$$= \begin{pmatrix} \mathbf{-2} & \mathbf{-1} \\ \mathbf{2} & \mathbf{4} \end{pmatrix}$$

Alternatively $A + B - C$

$$= \begin{pmatrix} -3 & 0 \\ 7 & -4 \end{pmatrix} + \begin{pmatrix} 2 & -1 \\ -7 & 4 \end{pmatrix} - \begin{pmatrix} 1 & 0 \\ -2 & -4 \end{pmatrix}$$

$$= \begin{pmatrix} -3+2-1 & 0+(-1)-0 \\ 7+(-7)-(-2) & -4+4-(-4) \end{pmatrix}$$

$$= \begin{pmatrix} \mathbf{-2} & \mathbf{-1} \\ \mathbf{2} & \mathbf{4} \end{pmatrix} \text{ as obtained previously}$$

(iii) Multiplication

When a matrix is multiplied by a number, called **scalar multiplication**, a single matrix results in which each element of the original matrix has been multiplied by the number.

Problem 4. If $A = \begin{pmatrix} -3 & 0 \\ 7 & -4 \end{pmatrix}$,

$B = \begin{pmatrix} 2 & -1 \\ -7 & 4 \end{pmatrix}$ and $C = \begin{pmatrix} 1 & 0 \\ -2 & -4 \end{pmatrix}$ find $2A - 3B + 4C$.

For scalar multiplication, each element is multiplied by the scalar quantity, hence

$$2A = 2\begin{pmatrix} -3 & 0 \\ 7 & -4 \end{pmatrix} = \begin{pmatrix} -6 & 0 \\ 14 & -8 \end{pmatrix},$$

$$3B = 3\begin{pmatrix} 2 & -1 \\ -7 & 4 \end{pmatrix} = \begin{pmatrix} 6 & -3 \\ -21 & 12 \end{pmatrix},$$

Section 10

and $4C = 4\begin{pmatrix} 1 & 0 \\ -2 & -4 \end{pmatrix} = \begin{pmatrix} 4 & 0 \\ -8 & -16 \end{pmatrix}$

Hence $2A - 3B + 4C$

$$= \begin{pmatrix} -6 & 0 \\ 14 & -8 \end{pmatrix} - \begin{pmatrix} 6 & -3 \\ -21 & 12 \end{pmatrix} + \begin{pmatrix} 4 & 0 \\ -8 & -16 \end{pmatrix}$$

$$= \begin{pmatrix} -6 - 6 + 4 & 0 - (-3) + 0 \\ 14 - (-21) + (-8) & -8 - 12 + (-16) \end{pmatrix}$$

$$= \begin{pmatrix} -8 & 3 \\ 27 & -36 \end{pmatrix}$$

When a matrix A is multiplied by another matrix B, a single matrix results in which elements are obtained from the sum of the products of the corresponding rows of A and the corresponding columns of B.

Two matrices A and B may be multiplied together, provided the number of elements in the rows of matrix A are equal to the number of elements in the columns of matrix B. In general terms, when multiplying a matrix of dimensions (m by n), by a matrix of dimensions (n by r), the resulting matrix has dimensions (m by r). Thus a 2 by 3 matrix multiplied by a 3 by 1 matrix gives a matrix of dimensions 2 by 1.

Problem 5. If $A = \begin{pmatrix} 2 & 3 \\ 1 & -4 \end{pmatrix}$ and

$B = \begin{pmatrix} -5 & 7 \\ -3 & 4 \end{pmatrix}$ find $A \times B$.

Let $A \times B = C$ where $C = \begin{pmatrix} C_{11} & C_{12} \\ C_{21} & C_{22} \end{pmatrix}$

C_{11} is the sum of the products of the first row elements of A and the first column elements of B taken one at a time,

i.e. $C_{11} = (2 \times (-5)) + (3 \times (-3)) = -19$

C_{12} is the sum of the products of the first row elements of A and the second column elements of B, taken one at a time,

i.e. $C_{12} = (2 \times 7) + (3 \times 4) = 26$

C_{21} is the sum of the products of the second row elements of A and the first column elements of B, taken one at a time.

i.e. $C_{21} = (1 \times (-5)) + (-4) \times (-3)) = 7$

Finally, C_{22} is the sum of the products of the second row elements of A and the second column elements of B, taken one at a time

i.e. $C_{22} = (1 \times 7) + ((-4) \times 4) = -9$

Thus, $A \times B = \begin{pmatrix} -19 & 26 \\ 7 & -9 \end{pmatrix}$

Problem 6. Simplify

$$\begin{pmatrix} 3 & 4 & 0 \\ -2 & 6 & -3 \\ 7 & -4 & 1 \end{pmatrix} \times \begin{pmatrix} 2 \\ 5 \\ -1 \end{pmatrix}$$

The sum of the products of the elements of each row of the first matrix and the elements of the second matrix, (called a **column matrix**), are taken one at a time. Thus:

$$\begin{pmatrix} 3 & 4 & 0 \\ -2 & 6 & -3 \\ 7 & -4 & 1 \end{pmatrix} \times \begin{pmatrix} 2 \\ 5 \\ -1 \end{pmatrix}$$

$$= \begin{pmatrix} (3 \times 2) + (4 \times 5) + (0 \times (-1)) \\ (-2 \times 2) + (6 \times 5) + (-3 \times (-1)) \\ (7 \times 2) + (-4 \times 5) + (1 \times (-1)) \end{pmatrix}$$

$$= \begin{pmatrix} 26 \\ 29 \\ -7 \end{pmatrix}$$

Problem 7. If $A = \begin{pmatrix} 3 & 4 & 0 \\ -2 & 6 & -3 \\ 7 & -4 & 1 \end{pmatrix}$ and

$B = \begin{pmatrix} 2 & -5 \\ 5 & -6 \\ -1 & -7 \end{pmatrix}$ find $A \times B$.

The sum of the products of the elements of each row of the first matrix and the elements of each column of the second matrix are taken one at a time. Thus:

$$\begin{pmatrix} 3 & 4 & 0 \\ -2 & 6 & -3 \\ 7 & -4 & 1 \end{pmatrix} \times \begin{pmatrix} 2 & -5 \\ 5 & -6 \\ -1 & -7 \end{pmatrix}$$

$$= \begin{pmatrix} [(3 \times 2) + (4 \times 5) + (0 \times (-1))] & [(3 \times (-5)) + (4 \times (-6)) + (0 \times (-7))] \\ [(-2 \times 2) + (6 \times 5) + (-3 \times (-1))] & [(-2 \times (-5)) + (6 \times (-6)) + (-3 \times (-7))] \\ [(7 \times 2) + (-4 \times 5) + (1 \times (-1))] & [(7 \times (-5)) + (-4 \times (-6)) + (1 \times (-7))] \end{pmatrix}$$

$$= \begin{pmatrix} 26 & -39 \\ 29 & -5 \\ -7 & -18 \end{pmatrix}$$

Problem 8. Determine

$$\begin{pmatrix} 1 & 0 & 3 \\ 2 & 1 & 2 \\ 1 & 3 & 1 \end{pmatrix} \times \begin{pmatrix} 2 & 2 & 0 \\ 1 & 3 & 2 \\ 3 & 2 & 0 \end{pmatrix}$$

The sum of the products of the elements of each row of the first matrix and the elements of each column of the second matrix are taken one at a time. Thus:

$$\begin{pmatrix} 1 & 0 & 3 \\ 2 & 1 & 2 \\ 1 & 3 & 1 \end{pmatrix} \times \begin{pmatrix} 2 & 2 & 0 \\ 1 & 3 & 2 \\ 3 & 2 & 0 \end{pmatrix}$$

$$= \begin{pmatrix} \begin{aligned}&[(1 \times 2) \\ &+ (0 \times 1) \\ &+ (3 \times 3)]\end{aligned} & \begin{aligned}&[(1 \times 2) \\ &+ (0 \times 3) \\ &+ (3 \times 2)]\end{aligned} & \begin{aligned}&[(1 \times 0) \\ &+ (0 \times 2) \\ &+ (3 \times 0)]\end{aligned} \\ \begin{aligned}&[(2 \times 2) \\ &+ (1 \times 1) \\ &+ (2 \times 3)]\end{aligned} & \begin{aligned}&[(2 \times 2) \\ &+ (1 \times 3) \\ &+ (2 \times 2)]\end{aligned} & \begin{aligned}&[(2 \times 0) \\ &+ (1 \times 2) \\ &+ (2 \times 0)]\end{aligned} \\ \begin{aligned}&[(1 \times 2) \\ &+ (3 \times 1) \\ &+ (1 \times 3)]\end{aligned} & \begin{aligned}&[(1 \times 2) \\ &+ (3 \times 3) \\ &+ (1 \times 2)]\end{aligned} & \begin{aligned}&[(1 \times 0) \\ &+ (3 \times 2) \\ &+ (1 \times 0)]\end{aligned} \end{pmatrix}$$

$$= \begin{pmatrix} 11 & 8 & 0 \\ 11 & 11 & 2 \\ 8 & 13 & 6 \end{pmatrix}$$

In algebra, the commutative law of multiplication states that $a \times b = b \times a$. For matrices, this law is only true in a few special cases, and in general $A \times B$ is **not** equal to $B \times A$.

Problem 9. If $A = \begin{pmatrix} 2 & 3 \\ 1 & 0 \end{pmatrix}$ and

$B = \begin{pmatrix} 2 & 3 \\ 0 & 1 \end{pmatrix}$ show that $A \times B \neq B \times A$.

$$A \times B = \begin{pmatrix} 2 & 3 \\ 1 & 0 \end{pmatrix} \times \begin{pmatrix} 2 & 3 \\ 0 & 1 \end{pmatrix}$$

$$= \begin{pmatrix} [(2 \times 2) + (3 \times 0)] & [(2 \times 3) + (3 \times 1)] \\ [(1 \times 2) + (0 \times 0)] & [(1 \times 3) + (0 \times 1)] \end{pmatrix}$$

$$= \begin{pmatrix} 4 & 9 \\ 2 & 3 \end{pmatrix}$$

$$B \times A = \begin{pmatrix} 2 & 3 \\ 0 & 1 \end{pmatrix} \times \begin{pmatrix} 2 & 3 \\ 1 & 0 \end{pmatrix}$$

$$= \begin{pmatrix} [(2 \times 2) + (3 \times 1)] & [(2 \times 3) + (3 \times 0)] \\ [(0 \times 2) + (1 \times 1)] & [(0 \times 3) + (1 \times 0)] \end{pmatrix}$$

$$= \begin{pmatrix} 7 & 6 \\ 1 & 0 \end{pmatrix}$$

Since $\begin{pmatrix} 4 & 9 \\ 2 & 3 \end{pmatrix} \neq \begin{pmatrix} 7 & 6 \\ 1 & 0 \end{pmatrix}$ then $A \times B \neq B \times A$

Now try the following exercise

Exercise 209 Further problems on addition, subtraction and multiplication of matrices

In Problems 1 to 13, the matrices A to K are:

$$A = \begin{pmatrix} 3 & -1 \\ -4 & 7 \end{pmatrix} \quad B = \begin{pmatrix} \frac{1}{2} & \frac{2}{3} \\ -\frac{1}{3} & -\frac{3}{5} \end{pmatrix}$$

$$C = \begin{pmatrix} -1.3 & 7.4 \\ 2.5 & -3.9 \end{pmatrix}$$

$$D = \begin{pmatrix} 4 & -7 & 6 \\ -2 & 4 & 0 \\ 5 & 7 & -4 \end{pmatrix}$$

$$E = \begin{pmatrix} 3 & 6 & \frac{1}{2} \\ 5 & -\frac{2}{3} & 7 \\ -1 & 0 & \frac{5}{3} \end{pmatrix}$$

$$F = \begin{pmatrix} 3.1 & 2.4 & 6.4 \\ -1.6 & 3.8 & -1.9 \\ 5.3 & 3.4 & -4.8 \end{pmatrix} \quad G = \begin{pmatrix} \frac{3}{4} \\ \frac{2}{1\frac{2}{5}} \end{pmatrix}$$

$$H = \begin{pmatrix} -2 \\ 5 \end{pmatrix} \quad J = \begin{pmatrix} 4 \\ -11 \\ 7 \end{pmatrix} \quad K = \begin{pmatrix} 1 & 0 \\ 0 & 1 \\ 1 & 0 \end{pmatrix}$$

In Problems 1 to 12, perform the matrix operation stated.

1. $A + B$

$$\left[\begin{pmatrix} 3\frac{1}{2} & -\frac{1}{3} \\ -4\frac{1}{3} & 6\frac{2}{5} \end{pmatrix} \right]$$

2. $D + E$

$$\left[\begin{pmatrix} 7 & -1 & 6\frac{1}{2} \\ 3 & 3\frac{1}{3} & 7 \\ 4 & 7 & -3\frac{2}{5} \end{pmatrix} \right]$$

3. $A - B$

$$\left[\begin{pmatrix} 2\frac{1}{2} & -1\frac{2}{3} \\ -3\frac{2}{3} & 7\frac{3}{5} \end{pmatrix} \right]$$

Section 10

4. $A + B - C$ $\qquad \left[\begin{pmatrix} 4.8 & -7.7\dot{3} \\ -6.8\dot{3} & 10.3 \end{pmatrix} \right]$

5. $5A + 6B$ $\qquad \left[\begin{pmatrix} 18.0 & -1.0 \\ -22.0 & 31.4 \end{pmatrix} \right]$

6. $2D + 3E - 4F$

$$\left[\begin{pmatrix} 4.6 & -5.6 & -12.1 \\ 17.4 & -9.2 & 28.6 \\ -14.2 & 0.4 & 13.0 \end{pmatrix} \right]$$

7. $A \times H$ $\qquad \left[\begin{pmatrix} -11 \\ 43 \end{pmatrix} \right]$

8. $A \times B$ $\qquad \left[\begin{pmatrix} 1\frac{5}{6} & 2\frac{3}{5} \\ -4\frac{1}{3} & -6\frac{13}{15} \end{pmatrix} \right]$

9. $A \times C$ $\qquad \left[\begin{pmatrix} -6.4 & 26.1 \\ 22.7 & -56.9 \end{pmatrix} \right]$

10. $D \times J$ $\qquad \left[\begin{pmatrix} 135 \\ -52 \\ -85 \end{pmatrix} \right]$

11. $E \times K$ $\qquad \left[\begin{pmatrix} 3\frac{1}{2} & 6 \\ 12 & -\frac{2}{3} \\ -\frac{2}{5} & 0 \end{pmatrix} \right]$

12. $D \times F$ $\qquad \left[\begin{pmatrix} 55.4 & 3.4 & 10.1 \\ -12.6 & 10.4 & -20.4 \\ -16.9 & 25.0 & 37.9 \end{pmatrix} \right]$

13. Show that $A \times C \neq C \times A$

$$\left[\begin{array}{l} A \times C = \begin{pmatrix} -6.4 & 26.1 \\ 22.7 & -56.9 \end{pmatrix} \\ C \times A = \begin{pmatrix} -33.5 & -53.1 \\ 23.1 & -29.8 \end{pmatrix} \\ \text{Hence they are not equal} \end{array} \right]$$

61.3 The unit matrix

A **unit matrix, I,** is one in which all elements of the leading diagonal (\) have a value of 1 and all other elements have a value of 0. Multiplication of a matrix by I is the equivalent of multiplying by 1 in arithmetic.

61.4 The determinant of a 2 by 2 matrix

The determinant of a 2 by 2 matrix, $\begin{pmatrix} a & b \\ c & d \end{pmatrix}$ is defined as $(ad - bc)$

The elements of the determinant of a matrix are written between vertical lines. Thus, the determinant of $\begin{pmatrix} 3 & -4 \\ 1 & 6 \end{pmatrix}$ is written as $\begin{vmatrix} 3 & -4 \\ 1 & 6 \end{vmatrix}$ and is equal to $(3 \times 6) - (-4 \times 1)$, i.e. $18 - (-4)$ or 22. Hence the determinant of a matrix can be expressed as a single numerical value, i.e. $\begin{vmatrix} 3 & -4 \\ 1 & 6 \end{vmatrix} = 22$.

Problem 10. Determine the value of

$$\begin{vmatrix} 3 & -2 \\ 7 & 4 \end{vmatrix}$$

$$\begin{vmatrix} 3 & -2 \\ 7 & 4 \end{vmatrix} = (3 \times 4) - (-2 \times 7)$$

$$= 12 - (-14) = \mathbf{26}$$

Problem 11. Evaluate $\begin{vmatrix} (1+j) & j2 \\ -j3 & (1-j4) \end{vmatrix}$

$$\begin{vmatrix} (1+j) & j2 \\ -j3 & (1-j4) \end{vmatrix} = (1+j)(1-j4) - (j2)(-j3)$$

$$= 1 - j4 + j - j^2 4 + j^2 6$$

$$= 1 - j4 + j - (-4) + (-6)$$

since from Chapter 35, $j^2 = -1$

$$= 1 - j4 + j + 4 - 6$$

$$= \mathbf{-1 - j3}$$

Problem 12. Evaluate $\begin{vmatrix} 5\angle 30° & 2\angle -60° \\ 3\angle 60° & 4\angle -90° \end{vmatrix}$

$$\begin{vmatrix} 5\angle 30° & 2\angle -60° \\ 3\angle 60° & 4\angle -90° \end{vmatrix} = (5\angle 30°)(4\angle -90°)$$

$$- (2\angle -60°)(3\angle 60°)$$

$$= (20\angle -60°) - (6\angle 0°)$$

$$= (10 - j17.32) - (6 + j0)$$

$$= \mathbf{(4 - j17.32)} \text{ or } \mathbf{17.78\angle -77°}$$

Now try the following exercise

1. Calculate the determinant of $\begin{pmatrix} 3 & -1 \\ -4 & 7 \end{pmatrix}$
   [17]

2. Calculate the determinant of
   $$\begin{pmatrix} \frac{1}{2} & \frac{2}{3} \\ -\frac{1}{3} & -\frac{3}{5} \end{pmatrix}$$
   $\left[-\dfrac{7}{90}\right]$

3. Calculate the determinant of
   $\begin{pmatrix} -1.3 & 7.4 \\ 2.5 & -3.9 \end{pmatrix}$
   [-13.43]

4. Evaluate $\begin{vmatrix} j2 & -j3 \\ (1+j) & j \end{vmatrix}$
   [-5+j3]

5. Evaluate $\begin{vmatrix} 2\angle 40° & 5\angle -20° \\ 7\angle -32° & 4\angle -117° \end{vmatrix}$
   $\left[\begin{array}{c}(-19.75+j19.79)\\ \text{or } 27.95\angle 134.94°\end{array}\right]$

61.5 The inverse or reciprocal of a 2 by 2 matrix

The inverse of matrix A is A^{-1} such that $A \times A^{-1} = I$, the unit matrix.
Let matrix A be $\begin{pmatrix} 1 & 2 \\ 3 & 4 \end{pmatrix}$ and let the inverse matrix, A^{-1} be $\begin{pmatrix} a & b \\ c & d \end{pmatrix}$.

Then, since $A \times A^{-1} = I$,
$$\begin{pmatrix} 1 & 2 \\ 3 & 4 \end{pmatrix} \times \begin{pmatrix} a & b \\ c & d \end{pmatrix} = \begin{pmatrix} 1 & 0 \\ 0 & 1 \end{pmatrix}$$

Multiplying the matrices on the left hand side, gives
$$\begin{pmatrix} a+2c & b+2d \\ 3a+4c & 3b+4d \end{pmatrix} = \begin{pmatrix} 1 & 0 \\ 0 & 1 \end{pmatrix}$$

Equating corresponding elements gives:
$$b + 2d = 0, \quad \text{i.e.} \quad b = -2d$$
and $3a + 4c = 0$, i.e. $a = -\dfrac{4}{3}c$

Substituting for a and b gives:
$$\begin{pmatrix} -\frac{4}{3}c+2c & -2d+2d \\ 3\left(-\frac{4}{3}c\right)+4c & 3(-2d)+4d \end{pmatrix} = \begin{pmatrix} 1 & 0 \\ 0 & 1 \end{pmatrix}$$

i.e.
$$\begin{pmatrix} \frac{2}{3}c & 0 \\ 0 & -2d \end{pmatrix} = \begin{pmatrix} 1 & 0 \\ 0 & 1 \end{pmatrix}$$

showing that $\frac{2}{3}c = 1$, i.e. $c = \frac{3}{2}$ and $-2d = 1$, i.e. $d = -\dfrac{1}{2}$

Since $b = -2d$, $b = 1$ and since $a = -\dfrac{4}{3}c, a = -2.$

Thus the inverse of matrix $\begin{pmatrix} 1 & 2 \\ 3 & 4 \end{pmatrix}$ is $\begin{pmatrix} a & b \\ c & d \end{pmatrix}$ that is,
$$\begin{pmatrix} -2 & 1 \\ \frac{3}{2} & -\frac{1}{2} \end{pmatrix}$$

There is, however, **a quicker method of obtaining the inverse** of a 2 by 2 matrix.

For any matrix $\begin{pmatrix} p & q \\ r & s \end{pmatrix}$ the inverse may be obtained by:

(i) interchanging the positions of p and s,
(ii) changing the signs of q and r, and
(iii) multiplying this new matrix by the reciprocal of the determinant of $\begin{pmatrix} p & q \\ r & s \end{pmatrix}$.

Thus the inverse of matrix $\begin{pmatrix} 1 & 2 \\ 3 & 4 \end{pmatrix}$ is
$$\frac{1}{4-6}\begin{pmatrix} 4 & -2 \\ -3 & 1 \end{pmatrix} = \begin{pmatrix} -2 & 1 \\ \frac{3}{2} & -\frac{1}{2} \end{pmatrix}$$
as obtained previously.

The inverse of matrix $\begin{pmatrix} p & q \\ r & s \end{pmatrix}$ is obtained by interchanging the positions of p and s, changing the signs of q and r and multiplying by the reciprocal of the determinant $\begin{vmatrix} p & q \\ r & s \end{vmatrix}$. Thus, the inverse of
$$\begin{pmatrix} 3 & -2 \\ 7 & 4 \end{pmatrix} = \frac{1}{(3 \times 4)-(-2 \times 7)}\begin{pmatrix} 4 & 2 \\ -7 & 3 \end{pmatrix}$$

$$= \frac{1}{26}\begin{pmatrix} 4 & 2 \\ -7 & 3 \end{pmatrix} = \begin{pmatrix} \dfrac{2}{13} & \dfrac{1}{13} \\ \dfrac{-7}{26} & \dfrac{3}{26} \end{pmatrix}$$

Now try the following exercise

Exercise 211   Further problems on the inverse of 2 by 2 matrices

1. Determine the inverse of $\begin{pmatrix} 3 & -1 \\ -4 & 7 \end{pmatrix}$

$$\left[\begin{pmatrix} \dfrac{7}{17} & \dfrac{1}{17} \\ \dfrac{4}{17} & \dfrac{3}{17} \end{pmatrix} \right]$$

2. Determine the inverse of $\begin{pmatrix} \dfrac{1}{2} & \dfrac{2}{3} \\ -\dfrac{1}{3} & -\dfrac{3}{5} \end{pmatrix}$

$$\left[\begin{pmatrix} 7\dfrac{5}{17} & 8\dfrac{4}{7} \\ -4\dfrac{2}{7} & -6\dfrac{3}{7} \end{pmatrix} \right]$$

3. Determine the inverse of $\begin{pmatrix} -1.3 & 7.4 \\ 2.5 & -3.9 \end{pmatrix}$

$$\left[\begin{array}{l} \begin{pmatrix} 0.290 & 0.551 \\ 0.186 & 0.097 \end{pmatrix} \\ \text{correct to 3 dec. places} \end{array} \right]$$

61.6   The determinant of a 3 by 3 matrix

(i)   The **minor** of an element of a 3 by 3 matrix is the value of the 2 by 2 determinant obtained by covering up the row and column containing that element.

Thus for the matrix $\begin{pmatrix} 1 & 2 & 3 \\ 4 & 5 & 6 \\ 7 & 8 & 9 \end{pmatrix}$ the minor of element 4 is obtained by covering the row (4 5 6)

and the column $\begin{pmatrix} 1 \\ 4 \\ 7 \end{pmatrix}$, leaving the 2 by determinant $\begin{vmatrix} 2 & 3 \\ 8 & 9 \end{vmatrix}$, i.e. the minor of element 4 is $(2 \times 9) - (3 \times 8) = -6$.

(ii)   The sign of a minor depends on its position within the matrix, the sign pattern being $\begin{pmatrix} + & - & + \\ - & + & - \\ + & - & + \end{pmatrix}$. Thus the signed-minor of element 4 in the matrix $\begin{pmatrix} 1 & 2 & 3 \\ 4 & 5 & 6 \\ 7 & 8 & 9 \end{pmatrix}$ is $-\begin{vmatrix} 2 & 3 \\ 8 & 9 \end{vmatrix} = -(-6) = 6$.

The signed-minor of an element is called the **cofactor** of the element.

(iii)   **The value of a 3 by 3 determinant is the sum of the products of the elements and their cofactors of any row or any column of the corresponding 3 by 3 matrix.**

There are thus six different ways of evaluating a 3×3 determinant — and all should give the same value.

Problem 14.   Find the value of

$$\begin{vmatrix} 3 & 4 & -1 \\ 2 & 0 & 7 \\ 1 & -3 & -2 \end{vmatrix}$$

The value of this determinant is the sum of the products of the elements and their cofactors, of any row or of any column. If the second row or second column is selected, the element 0 will make the product of the element and its cofactor zero and reduce the amount of arithmetic to be done to a minimum.

Supposing a second row expansion is selected.

The minor of 2 is the value of the determinant remaining when the row and column containing the 2 (i.e. the second row and the first column), is covered up. Thus the cofactor of element 2 is $\begin{vmatrix} 4 & -1 \\ -3 & -2 \end{vmatrix}$ i.e. -11. The sign of element 2 is minus, (see (ii) above), hence the cofactor of element 2, (the signed-minor) is $+11$. Similarly the minor of element 7 is $\begin{vmatrix} 3 & 4 \\ 1 & -3 \end{vmatrix}$ i.e. -13, and its cofactor is $+13$. Hence the value of the sum of the products of the elements and their cofactors is $2 \times 11 + 7 \times 13$, i.e.,

$$\begin{vmatrix} 3 & 4 & -1 \\ 2 & 0 & 7 \\ 1 & -3 & -2 \end{vmatrix} = 2(11) + 0 + 7(13) = \mathbf{113}$$

The same result will be obtained whichever row or column is selected. For example, the third column expansion is

$$(-1)\begin{vmatrix} 2 & 0 \\ 1 & -3 \end{vmatrix} - 7\begin{vmatrix} 3 & 4 \\ 1 & -3 \end{vmatrix} + (-2)\begin{vmatrix} 3 & 4 \\ 2 & 0 \end{vmatrix}$$

$$= 6 + 91 + 16 = \mathbf{113}, \text{ as obtained previously.}$$

Problem 15. Evaluate $\begin{vmatrix} 1 & 4 & -3 \\ -5 & 2 & 6 \\ -1 & -4 & 2 \end{vmatrix}$

Using the first row: $\begin{vmatrix} 1 & 4 & -3 \\ -5 & 2 & 6 \\ -1 & -4 & 2 \end{vmatrix}$

$$= 1\begin{vmatrix} 2 & 6 \\ -4 & 2 \end{vmatrix} - 4\begin{vmatrix} -5 & 6 \\ -1 & 2 \end{vmatrix} + (-3)\begin{vmatrix} -5 & 2 \\ -1 & -4 \end{vmatrix}$$

$$= (4+24) - 4(-10+6) - 3(20+2)$$

$$= 28 + 16 - 66 = \mathbf{-22}$$

Using the second column: $\begin{vmatrix} 1 & 4 & -3 \\ -5 & 2 & 6 \\ -1 & -4 & 2 \end{vmatrix}$

$$= -4\begin{vmatrix} -5 & 6 \\ -1 & 2 \end{vmatrix} + 2\begin{vmatrix} 1 & -3 \\ -1 & 2 \end{vmatrix} - (-4)\begin{vmatrix} 1 & -3 \\ -5 & 6 \end{vmatrix}$$

$$= -4(-10+6) + 2(2-3) + 4(6-15)$$

$$= 16 - 2 - 36 = \mathbf{-22}$$

Problem 16. Determine the value of

$$\begin{vmatrix} j2 & (1+j) & 3 \\ (1-j) & 1 & j \\ 0 & j4 & 5 \end{vmatrix}$$

Using the first column, the value of the determinant is:

$$(j2)\begin{vmatrix} 1 & j \\ j4 & 5 \end{vmatrix} - (1-j)\begin{vmatrix} (1+j) & 3 \\ j4 & 5 \end{vmatrix}$$

$$+ (0)\begin{vmatrix} (1+j) & 3 \\ 1 & j \end{vmatrix}$$

$$= j2(5 - j^2 4) - (1-j)(5 + j5 - j12) + 0$$

$$= j2(9) - (1-j)(5 - j7)$$

$$= j18 - [5 - j7 - j5 + j^2 7]$$

$$= j18 - [-2 - j12]$$

$$= j18 + 2 + j12 = \mathbf{2 + j30} \text{ or } \mathbf{30.07\angle 86.19°}$$

Now try the following exercise

Exercise 212 Further problems on 3 by 3 determinants

1. Find the matrix of minors of
$$\begin{pmatrix} 4 & -7 & 6 \\ -2 & 4 & 0 \\ 5 & 7 & -4 \end{pmatrix}$$
$$\left[\begin{pmatrix} -16 & 8 & -34 \\ -14 & -46 & 63 \\ -24 & 12 & 2 \end{pmatrix}\right]$$

2. Find the matrix of cofactors of
$$\begin{pmatrix} 4 & -7 & 6 \\ -2 & 4 & 0 \\ 5 & 7 & -4 \end{pmatrix}$$
$$\left[\begin{pmatrix} -16 & -8 & -34 \\ 14 & -46 & -63 \\ -24 & -12 & 2 \end{pmatrix}\right]$$

3. Calculate the determinant of
$$\begin{pmatrix} 4 & -7 & 6 \\ -2 & 4 & 0 \\ 5 & 7 & -4 \end{pmatrix} \qquad [-212]$$

4. Evaluate $\begin{vmatrix} 8 & -2 & -10 \\ 2 & -3 & -2 \\ 6 & 3 & 8 \end{vmatrix}$ $[-328]$

5. Calculate the determinant of
$$\begin{pmatrix} 3.1 & 2.4 & 6.4 \\ -1.6 & 3.8 & -1.9 \\ 5.3 & 3.4 & -4.8 \end{pmatrix} \qquad [-242.83]$$

6. Evaluate $\begin{vmatrix} j2 & 2 & j \\ (1+j) & 1 & -3 \\ 5 & -j4 & 0 \end{vmatrix}$ $[-2-j]$

7. Evaluate $\begin{vmatrix} 3\angle 60° & j2 & 1 \\ 0 & (1+j) & 2\angle 30° \\ 0 & 2 & j5 \end{vmatrix}$
$$\left[\begin{array}{l} 29.94\angle -139.52° \text{ or} \\ (-20.49 - j17.49) \end{array}\right]$$

61.7 The inverse or reciprocal of a 3 by 3 matrix

The **adjoint** of a matrix A is obtained by:

(i) forming a matrix B of the cofactors of A, and
(ii) **transposing** matrix B to give B^T, where B^T is the matrix obtained by writing the rows of B as the columns of B^T. Then **adj** $A = B^T$

The **inverse of matrix A, A^{-1}** is given by

$$A^{-1} = \frac{\text{adj} A}{|A|}$$

where adj A is the adjoint of matrix A and $|A|$ is the determinant of matrix A.

Problem 17. Determine the inverse of the matrix $\begin{pmatrix} 3 & 4 & -1 \\ 2 & 0 & 7 \\ 1 & -3 & -2 \end{pmatrix}$

The inverse of matrix A, $A^{-1} = \dfrac{\text{adj} A}{|A|}$

The adjoint of A is found by:

(i) obtaining the matrix of the cofactors of the elements, and
(ii) transposing this matrix.

The cofactor of element 3 is $+\begin{vmatrix} 0 & 7 \\ -3 & -2 \end{vmatrix} = 21$.

The cofactor of element 4 is $-\begin{vmatrix} 2 & 7 \\ 1 & -2 \end{vmatrix} = 11$, and so on.

The matrix of cofactors is $\begin{pmatrix} 21 & 11 & -6 \\ 11 & -5 & 13 \\ 28 & -23 & -8 \end{pmatrix}$

The transpose of the matrix of cofactors, i.e. the adjoint of the matrix, is obtained by writing the rows as columns, and is $\begin{pmatrix} 21 & 11 & 28 \\ 11 & -5 & -23 \\ -6 & 13 & -8 \end{pmatrix}$

From Problem 14, the determinant of $\begin{vmatrix} 3 & 4 & -1 \\ 2 & 0 & 7 \\ 1 & -3 & -2 \end{vmatrix}$ is 113.

Hence the inverse of $\begin{pmatrix} 3 & 4 & -1 \\ 2 & 0 & 7 \\ 1 & -3 & -2 \end{pmatrix}$ is

$$\frac{\begin{pmatrix} 21 & 11 & 28 \\ 11 & -5 & -23 \\ -6 & 13 & -8 \end{pmatrix}}{113} \text{ or } \frac{1}{113}\begin{pmatrix} 21 & 11 & 28 \\ 11 & -5 & -23 \\ -6 & 13 & -8 \end{pmatrix}$$

Problem 18. Find the inverse of

$$\begin{pmatrix} 1 & 5 & -2 \\ 3 & -1 & 4 \\ -3 & 6 & -7 \end{pmatrix}$$

$$\text{Inverse} = \frac{\text{adjoint}}{\text{determinant}}$$

The matrix of cofactors is $\begin{pmatrix} -17 & 9 & 15 \\ 23 & -13 & -21 \\ 18 & -10 & -16 \end{pmatrix}$

The transpose of the matrix of cofactors (i.e. the adjoint) is $\begin{pmatrix} -17 & 23 & 18 \\ 9 & -13 & -10 \\ 15 & -21 & -16 \end{pmatrix}$

The determinant of $\begin{pmatrix} 1 & 5 & -2 \\ 3 & -1 & 4 \\ -3 & 6 & -7 \end{pmatrix}$

$$= 1(7 - 24) - 5(-21 + 12) - 2(18 - 3)$$

$$= -17 + 45 - 30 = -2$$

Hence the inverse of $\begin{pmatrix} 1 & 5 & -2 \\ 3 & -1 & 4 \\ -3 & 6 & -7 \end{pmatrix}$

$$= \frac{\begin{pmatrix} -17 & 23 & 18 \\ 9 & -13 & -10 \\ 15 & -21 & -16 \end{pmatrix}}{-2}$$

$$= \begin{pmatrix} 8.5 & -11.5 & -9 \\ -4.5 & 6.5 & 5 \\ -7.5 & 10.5 & 8 \end{pmatrix}$$

Now try the following exercise

Exercise 213 Further problems on the inverse of a 3 by 3 matrix

1. Write down the transpose of

$$\begin{pmatrix} 4 & -7 & 6 \\ -2 & 4 & 0 \\ 5 & 7 & -4 \end{pmatrix}$$

$$\left[\begin{pmatrix} 4 & -2 & 5 \\ -7 & 4 & 7 \\ 6 & 0 & -4 \end{pmatrix} \right]$$

2. Write down the transpose of

$$\begin{pmatrix} 3 & 6 & \frac{1}{2} \\ 5 & -\frac{2}{3} & 7 \\ -1 & 0 & \frac{3}{5} \end{pmatrix}$$

$$\left[\begin{pmatrix} 3 & 5 & -1 \\ 6 & -\frac{2}{3} & 0 \\ \frac{1}{2} & 7 & \frac{3}{5} \end{pmatrix} \right]$$

3. Determine the adjoint of $\begin{pmatrix} 4 & -7 & 6 \\ -2 & 4 & 0 \\ 5 & 7 & -4 \end{pmatrix}$

$$\left[\begin{pmatrix} -16 & 14 & -24 \\ -8 & -46 & -12 \\ -34 & -63 & 2 \end{pmatrix} \right]$$

4. Determine the adjoint of $\begin{pmatrix} 3 & 6 & \frac{1}{2} \\ 5 & -\frac{2}{3} & 7 \\ -1 & 0 & \frac{3}{5} \end{pmatrix}$

$$\left[\begin{pmatrix} -\frac{2}{5} & -3\frac{3}{5} & 42\frac{1}{3} \\ -10 & 2\frac{3}{10} & -18\frac{1}{2} \\ -\frac{2}{3} & -6 & -32 \end{pmatrix} \right]$$

5. Find the inverse of $\begin{pmatrix} 4 & -7 & 6 \\ -2 & 4 & 0 \\ 5 & 7 & -4 \end{pmatrix}$

$$\left[-\frac{1}{212} \begin{pmatrix} -16 & 14 & -24 \\ -8 & -46 & -12 \\ -34 & -63 & 2 \end{pmatrix} \right]$$

6. Find the inverse of $\begin{pmatrix} 3 & 6 & \frac{1}{2} \\ 5 & -\frac{2}{3} & 7 \\ -1 & 0 & \frac{3}{5} \end{pmatrix}$

$$\left[-\frac{15}{923} \begin{pmatrix} -\frac{2}{5} & -3\frac{3}{5} & 42\frac{1}{3} \\ -10 & 2\frac{3}{10} & -18\frac{1}{2} \\ -\frac{2}{3} & -6 & -32 \end{pmatrix} \right]$$

Chapter 62

The solution of simultaneous equations by matrices and determinants

62.1 Solution of simultaneous equations by matrices

(a) The procedure for solving linear simultaneous equations in **two unknowns using matrices** is:

(i) write the equations in the form
$$a_1x + b_1y = c_1$$
$$a_2x + b_2y = c_2$$

(ii) write the matrix equation corresponding to these equations,

i.e. $\begin{pmatrix} a_1 & b_1 \\ a_2 & b_2 \end{pmatrix} \times \begin{pmatrix} x \\ y \end{pmatrix} = \begin{pmatrix} c_1 \\ c_2 \end{pmatrix}$

(iii) determine the inverse matrix of $\begin{pmatrix} a_1 & b_1 \\ a_2 & b_2 \end{pmatrix}$

i.e. $\dfrac{1}{a_1b_2 - b_1a_2} \begin{pmatrix} b_2 & -b_1 \\ -a_2 & a_1 \end{pmatrix}$

(from Chapter 61)

(iv) multiply each side of (ii) by the inverse matrix, and

(v) solve for x and y by equating corresponding elements.

Problem 1. Use matrices to solve the simultaneous equations:
$$3x + 5y - 7 = 0 \qquad (1)$$
$$4x - 3y - 19 = 0 \qquad (2)$$

(i) Writing the equations in the $a_1x + b_1y = c$ form gives:
$$3x + 5y = 7$$
$$4x - 3y = 19$$

(ii) The matrix equation is
$$\begin{pmatrix} 3 & 5 \\ 4 & -3 \end{pmatrix} \times \begin{pmatrix} x \\ y \end{pmatrix} = \begin{pmatrix} 7 \\ 19 \end{pmatrix}$$

(iii) The inverse of matrix $\begin{pmatrix} 3 & 5 \\ 4 & -3 \end{pmatrix}$ is
$$\frac{1}{3 \times (-3) - 5 \times 4} \begin{pmatrix} -3 & -5 \\ -4 & 3 \end{pmatrix}$$

i.e. $\begin{pmatrix} \dfrac{3}{29} & \dfrac{5}{29} \\ \dfrac{4}{29} & \dfrac{-3}{29} \end{pmatrix}$

(iv) Multiplying each side of (ii) by (iii) and remembering that $A \times A^{-1} = I$, the unit matrix, gives:
$$\begin{pmatrix} 1 & 0 \\ 0 & 1 \end{pmatrix} \begin{pmatrix} x \\ y \end{pmatrix} = \begin{pmatrix} \dfrac{3}{29} & \dfrac{5}{29} \\ \dfrac{4}{29} & \dfrac{-3}{29} \end{pmatrix} \times \begin{pmatrix} 7 \\ 19 \end{pmatrix}$$

Thus
$$\begin{pmatrix} x \\ y \end{pmatrix} = \begin{pmatrix} \dfrac{21}{29} + \dfrac{95}{29} \\ \dfrac{28}{29} - \dfrac{57}{29} \end{pmatrix}$$

i.e.
$$\begin{pmatrix} x \\ y \end{pmatrix} = \begin{pmatrix} 4 \\ -1 \end{pmatrix}$$

(v) By comparing corresponding elements:

$$x = 4 \quad \text{and} \quad y = -1$$

Checking:
equation (1),

$$3 \times 4 + 5 \times (-1) - 7 = 0 = \text{RHS}$$

equation (2),

$$4 \times 4 - 3 \times (-1) - 19 = 0 = \text{RHS}$$

(b) The procedure for solving linear simultaneous equations in **three unknowns using matrices** is:

(i) write the equations in the form

$$a_1 x + b_1 y + c_1 z = d_1$$
$$a_2 x + b_2 y + c_2 z = d_2$$
$$a_3 x + b_3 y + c_3 z = d_3$$

(ii) write the matrix equation corresponding to these equations, i.e.

$$\begin{pmatrix} a_1 & b_1 & c_1 \\ a_2 & b_2 & c_2 \\ a_3 & b_3 & c_3 \end{pmatrix} \times \begin{pmatrix} x \\ y \\ z \end{pmatrix} = \begin{pmatrix} d_1 \\ d_2 \\ d_3 \end{pmatrix}$$

(iii) determine the inverse matrix of

$$\begin{pmatrix} a_1 & b_1 & c_1 \\ a_2 & b_2 & c_2 \\ a_3 & b_3 & c_3 \end{pmatrix} \text{ (see Chapter 61)}$$

(iv) multiply each side of (ii) by the inverse matrix, and

(v) solve for x, y and z by equating the corresponding elements.

Problem 2. Use matrices to solve the simultaneous equations:

$$x + y + z - 4 = 0 \quad (1)$$
$$2x - 3y + 4z - 33 = 0 \quad (2)$$
$$3x - 2y - 2z - 2 = 0 \quad (3)$$

(i) Writing the equations in the $a_1 x + b_1 y + c_1 z = d_1$ form gives:

$$x + y + z = 4$$
$$2x - 3y + 4z = 33$$
$$3x - 2y - 2z = 2$$

(ii) The matrix equation is

$$\begin{pmatrix} 1 & 1 & 1 \\ 2 & -3 & 4 \\ 3 & -2 & -2 \end{pmatrix} \times \begin{pmatrix} x \\ y \\ z \end{pmatrix} = \begin{pmatrix} 4 \\ 33 \\ 2 \end{pmatrix}$$

(iii) The inverse matrix of

$$A = \begin{pmatrix} 1 & 1 & 1 \\ 2 & -3 & 4 \\ 3 & -2 & -2 \end{pmatrix}$$

is given by

$$A^{-1} = \frac{\text{adj} A}{|A|}$$

The adjoint of A is the transpose of the matrix of the cofactors of the elements (see Chapter 61). The matrix of cofactors is

$$\begin{pmatrix} 14 & 16 & 5 \\ 0 & -5 & 5 \\ 7 & -2 & -5 \end{pmatrix}$$

and the transpose of this matrix gives

$$\text{adj} A = \begin{pmatrix} 14 & 0 & 7 \\ 16 & -5 & -2 \\ 5 & 5 & -5 \end{pmatrix}$$

The determinant of A, i.e. the sum of the products of elements and their cofactors, using a first row expansion is

$$1 \begin{vmatrix} -3 & 4 \\ -2 & -2 \end{vmatrix} - 1 \begin{vmatrix} 2 & 4 \\ 3 & -2 \end{vmatrix} + 1 \begin{vmatrix} 2 & -3 \\ 3 & -2 \end{vmatrix}$$

$$= (1 \times 14) - (1 \times (-16)) + (1 \times 5) = 35$$

Hence the inverse of A,

$$A^{-1} = \frac{1}{35} \begin{pmatrix} 14 & 0 & 7 \\ 16 & -5 & -2 \\ 5 & 5 & -5 \end{pmatrix}$$

(iv) Multiplying each side of (ii) by (iii), and remembering that $A \times A^{-1} = I$, the unit matrix, gives

$$\begin{pmatrix} 1 & 0 & 0 \\ 0 & 1 & 0 \\ 0 & 0 & 1 \end{pmatrix} \times \begin{pmatrix} x \\ y \\ z \end{pmatrix}$$

$$= \frac{1}{35} \begin{pmatrix} 14 & 0 & 7 \\ 16 & -5 & -2 \\ 5 & 5 & -5 \end{pmatrix} \times \begin{pmatrix} 4 \\ 33 \\ 2 \end{pmatrix}$$

Section 10

$$\begin{pmatrix} x \\ y \\ z \end{pmatrix} =$$

$$\frac{1}{35} \begin{pmatrix} (14 \times 4) + (0 \times 33) + (7 \times 2) \\ (16 \times 4) + ((-5) \times 33) + ((-2) \times 2) \\ (5 \times 4) + (5 \times 33) + ((-5) \times 2) \end{pmatrix}$$

$$= \frac{1}{35} \begin{pmatrix} 70 \\ -105 \\ 175 \end{pmatrix}$$

$$= \begin{pmatrix} 2 \\ -3 \\ 5 \end{pmatrix}$$

(v) By comparing corresponding elements, $x = 2$, $y = -3$, $z = 5$, which can be checked in the original equations.

Now try the following exercise

Exercise 214   Further problems on solving simultaneous equations using matrices

In Problems 1 to 5 use **matrices** to solve the simultaneous equations given.

1. $3x + 4y = 0$
   $2x + 5y + 7 = 0$ $[x = 4, y = -3]$

2. $2p + 5q + 14.6 = 0$
   $3.1p + 1.7q + 2.06 = 0$
                                        $[p = 1.2, q = -3.4]$

3. $x + 2y + 3z = 5$
   $2x - 3y - z = 3$
   $-3x + 4y + 5z = 3$
                                  $[x = 1, y = -1, z = 2]$

4. $3a + 4b - 3c = 2$
   $-2a + 2b + 2c = 15$
   $7a - 5b + 4c = 26$
                             $[a = 2.5, b = 3.5, c = 6.5]$

5. $p + 2q + 3r + 7.8 = 0$
   $2p + 5q - r - 1.4 = 0$
   $5p - q + 7r - 3.5 = 0$
                           $[p = 4.1, q = -1.9, r = -2.7]$

6. In two closed loops of an electrical circuit, the currents flowing are given by the simultaneous equations:

$$I_1 + 2I_2 + 4 = 0$$
$$5I_1 + 3I_2 - 1 = 0$$

Use matrices to solve for I_1 and I_2.
                                        $[I_1 = 2, I_2 = -3]$

7. The relationship between the displacement, s, velocity, v, and acceleration, a, of a piston is given by the equations:

$$s + 2v + 2a = 4$$
$$3s - v + 4a = 25$$
$$3s + 2v - a = -4$$

Use matrices to determine the values of s, v and a. $[s = 2, v = -3, a = 4]$

8. In a mechanical system, acceleration $\ddot{x}$, velocity $\dot{x}$ and distance x are related by the simultaneous equations:

$$3.4\ddot{x} + 7.0\dot{x} - 13.2x = -11.39$$
$$-6.0\ddot{x} + 4.0\dot{x} + 3.5x = 4.98$$
$$2.7\ddot{x} + 6.0\dot{x} + 7.1x = 15.91$$

Use matrices to find the values of $\ddot{x}$, $\dot{x}$ and x.
                        $[\ddot{x} = 0.5, \dot{x} = 0.77, x = 1.4]$

62.2   Solution of simultaneous equations by determinants

(a) When solving linear simultaneous equations in **two unknowns using determinants**:

(i) write the equations in the form

$$a_1x + b_1y + c_1 = 0$$
$$a_2x + b_2y + c_2 = 0$$

and then

(ii) the solution is given by

$$\frac{x}{D_x} = \frac{-y}{D_y} = \frac{1}{D}$$

where $\quad D_x = \begin{vmatrix} b_1 & c_1 \\ b_2 & c_2 \end{vmatrix}$

i.e. the determinant of the coefficients left when the x-column is covered up,

$$D_y = \begin{vmatrix} a_1 & c_1 \\ a_2 & c_2 \end{vmatrix}$$

i.e. the determinant of the coefficients left when the y-column is covered up,

and
$$D = \begin{vmatrix} a_1 & b_1 \\ a_2 & b_2 \end{vmatrix}$$

i.e. the determinant of the coefficients left when the constants-column is covered up.

> **Problem 3.** Solve the following simultaneous equations using determinants:
>
> $$3x - 4y = 12$$
> $$7x + 5y = 6.5$$

Following the above procedure:

(i) $3x - 4y - 12 = 0$
    $7x + 5y - 6.5 = 0$

(ii)
$$\frac{x}{\begin{vmatrix} -4 & -12 \\ 5 & -6.5 \end{vmatrix}} = \frac{-y}{\begin{vmatrix} 3 & -12 \\ 7 & -6.5 \end{vmatrix}} = \frac{1}{\begin{vmatrix} 3 & -4 \\ 7 & 5 \end{vmatrix}}$$

i.e.
$$\frac{x}{(-4)(-6.5) - (-12)(5)}$$

$$= \frac{-y}{(3)(-6.5) - (-12)(7)}$$

$$= \frac{1}{(3)(5) - (-4)(7)}$$

i.e. $\dfrac{x}{26 + 60} = \dfrac{-y}{-19.5 + 84} = \dfrac{1}{15 + 28}$

i.e. $\dfrac{x}{86} = \dfrac{-y}{64.5} = \dfrac{1}{43}$

Since $\dfrac{x}{86} = \dfrac{1}{43}$ then $x = \dfrac{86}{43} = \mathbf{2}$

and since

$$\frac{-y}{64.5} = \frac{1}{43} \text{ then}$$

$$y = -\frac{64.5}{43} = \mathbf{-1.5}$$

> **Problem 4.** The velocity of a car, accelerating at uniform acceleration a between two points, is given by $v = u + at$, where u is its velocity when passing the first point and t is the time taken to pass between the two points. If $v = 21$ m/s when $t = 3.5$ s and $v = 33$ m/s when $t = 6.1$ s, use determinants to find the values of u and a, each correct to 4 significant figures.

Substituting the given values in $v = u + at$ gives:

$$21 = u + 3.5a \tag{1}$$

$$33 = u + 6.1a \tag{2}$$

(i) The equations are written in the form
    $$a_1 x + b_1 y + c_1 = 0,$$
    i.e. $u + 3.5a - 21 = 0$
    and $u + 6.1a - 33 = 0$

(ii) The solution is given by

$$\frac{u}{D_u} = \frac{-a}{D_a} = \frac{1}{D}$$

where D_u is the determinant of coefficients left when the u column is covered up,

i.e. $D_u = \begin{vmatrix} 3.5 & -21 \\ 6.1 & -33 \end{vmatrix}$

$$= (3.5)(-33) - (-21)(6.1)$$

$$= 12.6$$

Similarly, $D_a = \begin{vmatrix} 1 & -21 \\ 1 & -33 \end{vmatrix}$

$$= (1)(-33) - (-21)(1)$$

$$= -12$$

and $D = \begin{vmatrix} 1 & 3.5 \\ 1 & 6.1 \end{vmatrix}$

$$= (1)(6.1) - (3.5)(1) = 2.6$$

Thus $\dfrac{u}{12.6} = \dfrac{-a}{-12} = \dfrac{1}{2.6}$

i.e. $u = \dfrac{12.6}{2.6} = \mathbf{4.846\,m/s}$

and $a = \dfrac{12}{2.6} = \mathbf{4.615\,m/s^2},$

each correct to 4 significant figures

> **Problem 5.** Applying Kirchhoff's laws to an electric circuit results in the following equations:
>
> $$(9 + j12)I_1 - (6 + j8)I_2 = 5$$
> $$-(6 + j8)I_1 + (8 + j3)I_2 = (2 + j4)$$
>
> Solve the equations for I_1 and I_2

Following the procedure:

(i) $(9 + j12)I_1 - (6 + j8)I_2 - 5 = 0$

    $-(6 + j8)I_1 + (8 + j3)I_2 - (2 + j4) = 0$

(ii)
$$\frac{I_1}{\begin{vmatrix} -(6+j8) & -5 \\ (8+j3) & -(2+j4) \end{vmatrix}}$$

$$= \frac{-I_2}{\begin{vmatrix} (9+j12) & -5 \\ -(6+j8) & -(2+j4) \end{vmatrix}}$$

$$= \frac{I}{\begin{vmatrix} (9+j12) & -(6+j8) \\ -(6+j8) & (8+j3) \end{vmatrix}}$$

$$\frac{I_1}{(-20+j40)+(40+j15)}$$

$$= \frac{-I_2}{(30-j60)-(30+j40)}$$

$$= \frac{1}{(36+j123)-(-28+j96)}$$

$$\frac{I_1}{20+j55} = \frac{-I_2}{-j100} = \frac{1}{64+j27}$$

Hence $I_1 = \dfrac{20+j55}{64+j27}$

$$= \frac{58.52\angle 70.02°}{69.46\angle 22.87°} = \mathbf{0.84\angle 47.15° A}$$

and $I_2 = \dfrac{100\angle 90°}{69.46\angle 22.87°} = \mathbf{1.44\angle 67.13° A}$

(b) When solving simultaneous equations in **three unknowns using determinants:**

(i) Write the equations in the form

$$a_1 x + b_1 y + c_1 z + d_1 = 0$$
$$a_2 x + b_2 y + c_2 z + d_2 = 0$$
$$a_3 x + b_3 y + c_3 z + d_3 = 0$$

and then
(ii) the solution is given by

$$\frac{x}{D_x} = \frac{-y}{D_y} = \frac{z}{D_z} = \frac{-1}{D}$$

where $D = x\begin{vmatrix} b_1 & c_1 & d_1 \\ b_2 & c_2 & d_2 \\ b_3 & c_3 & d_3 \end{vmatrix}$

i.e. the determinant of the coefficients obtained by covering up the x column.

$$D_y = \begin{vmatrix} a_1 & c_1 & d_1 \\ a_2 & c_2 & d_2 \\ a_3 & c_3 & d_3 \end{vmatrix}$$

i.e., the determinant of the coefficients obtained by covering up the y column.

$$D_z = \begin{vmatrix} a_1 & b_1 & d_1 \\ a_2 & b_2 & d_2 \\ a_3 & b_3 & d_3 \end{vmatrix}$$

i.e. the determinant of the coefficients obtained by covering up the z column.

and $\quad D = \begin{vmatrix} a_1 & b_1 & c_1 \\ a_2 & b_2 & c_2 \\ a_3 & b_3 & c_3 \end{vmatrix}$

i.e. the determinant of the coefficients obtained by covering up the constants column.

Problem 6. A d.c. circuit comprises three closed loops. Applying Kirchhoff's laws to the closed loops gives the following equations for current flow in milliamperes:

$$2I_1 + 3I_2 - 4I_3 = 26$$
$$I_1 - 5I_2 - 3I_3 = -87$$
$$-7I_1 + 2I_2 + 6I_3 = 12$$

Use determinants to solve for I_1, I_2 and I_3

(i) Writing the equations in the $a_1 x + b_1 y + c_1 z + d_1 = 0$ form gives:

$$2I_1 + 3I_2 - 4I_3 - 26 = 0$$
$$I_1 - 5I_2 - 3I_3 + 87 = 0$$
$$-7I_1 + 2I_2 + 6I_3 - 12 = 0$$

(ii) The solution is given by

$$\frac{I_1}{D_{I_1}} = \frac{-I_2}{D_{I_2}} = \frac{I_3}{D_{I_3}} = \frac{-1}{D}$$

where D_{I_1} is the determinant of coefficients obtained by covering up the I_1 column, i.e.

$$D_{I_1} = \begin{vmatrix} 3 & -4 & -26 \\ -5 & -3 & 87 \\ 2 & 6 & -12 \end{vmatrix}$$

$$= (3)\begin{vmatrix} -3 & 87 \\ 6 & -12 \end{vmatrix} - (-4)\begin{vmatrix} -5 & 87 \\ 2 & -12 \end{vmatrix}$$

$$+ (-26)\begin{vmatrix} -5 & -3 \\ 2 & 6 \end{vmatrix}$$

$$= 3(-486) + 4(-114) - 26(-24)$$

$$= \mathbf{-1290}$$

$$D_{I_2} = \begin{vmatrix} 2 & -4 & -26 \\ 1 & -3 & 87 \\ -7 & 6 & -12 \end{vmatrix}$$

$$= (2)(36 - 522) - (-4)(-12 + 609)$$
$$+ (-26)(6 - 21)$$
$$= -972 + 2388 + 390$$
$$= \mathbf{1806}$$

$$D_{I_3} = \begin{vmatrix} 2 & 3 & -26 \\ 1 & -5 & 87 \\ -7 & 2 & -12 \end{vmatrix}$$

$$= (2)(60 - 174) - (3)(-12 + 609)$$
$$+ (-26)(2 - 35)$$
$$= -228 - 1791 + 858 = \mathbf{-1161}$$

and $\quad D = \begin{vmatrix} 2 & 3 & -4 \\ 1 & -5 & -3 \\ -7 & 2 & 6 \end{vmatrix}$

$$= (2)(-30 + 6) - (3)(6 - 21)$$
$$+ (-4)(2 - 35)$$
$$= -48 + 45 + 132 = \mathbf{129}$$

Thus
$$\frac{I_1}{-1290} = \frac{-I_2}{1806} = \frac{I_3}{-1161} = \frac{-1}{129}$$
giving
$$I_1 = \frac{-1290}{-129} = \mathbf{10\,mA},$$

$$I_2 = \frac{1806}{129} = \mathbf{14\,mA}$$

and $\quad I_3 = \frac{1161}{129} = \mathbf{9\,mA}$

Now try the following exercise

Exercise 215   Further problems on solving simultaneous equations using determinants

In Problems 1 to 5 use **determinants** to solve the simultaneous equations given.

1. $3x - 5y = -17.6$
   $7y - 2x - 22 = 0$
   $$[x = -1.2, y = 2.8]$$

2. $2.3m - 4.4n = 6.84$
   $8.5n - 6.7m = 1.23$
   $$[m = -6.4, n = -4.9]$$

3. $3x + 4y + z = 10$
   $2x - 3y + 5z + 9 = 0$
   $x + 2y - z = 6$
   $$[x = 1, y = 2, z = -1]$$

4. $1.2p - 2.3q - 3.1r + 10.1 = 0$
   $4.7p + 3.8q - 5.3r - 21.5 = 0$
   $3.7p - 8.3q + 7.4r + 28.1 = 0$
   $$[p = 1.5, q = 4.5, r = 0.5]$$

5. $\dfrac{x}{2} - \dfrac{y}{3} + \dfrac{2z}{5} = -\dfrac{1}{20}$

   $\dfrac{x}{4} + \dfrac{2y}{3} - \dfrac{z}{2} = \dfrac{19}{40}$

   $x + y - z = \dfrac{59}{60}$

   $$\left[x = \frac{7}{20}, \; y = \frac{17}{40}, \; z = -\frac{5}{24} \right]$$

6. In a system of forces, the relationship between two forces F_1 and F_2 is given by:
   $$5F_1 + 3F_2 + 6 = 0$$
   $$3F_1 + 5F_2 + 18 = 0$$
   Use determinants to solve for F_1 and F_2
   $$[F_1 = 1.5, F_2 = -4.5]$$

7. Applying mesh-current analysis to an a.c. circuit results in the following equations:
   $$(5 - j4)I_1 - (-j4)I_2 = 100\angle 0°$$
   $$(4 + j3 - j4)I_2 - (-j4)I_1 = 0$$
   Solve the equations for I_1 and I_2.
   $$\left[\begin{array}{l} I_1 = 10.77\angle 19.23°\,A, \\ I_2 = 10.45\angle -56.73°\,A \end{array} \right]$$

8. Kirchhoff's laws are used to determine the current equations in an electrical network and show that
   $$i_1 + 8i_2 + 3i_3 = -31$$
   $$3i_1 - 2i_2 + i_3 = -5$$
   $$2i_1 - 3i_2 + 2i_3 = 6$$
   Use determinants to solve for i_1, i_2 and i_3.
   $$[i_1 = -5, i_2 = -4, i_3 = 2]$$

9. The forces in three members of a framework are F_1, F_2 and F_3. They are related by the simultaneous equations shown below.
   $$1.4F_1 + 2.8F_2 + 2.8F_3 = 5.6$$
   $$4.2F_1 - 1.4F_2 + 5.6F_3 = 35.0$$
   $$4.2F_1 + 2.8F_2 - 1.4F_3 = -5.6$$

Section 10

Find the values of F_1, F_2 and F_3 using determinants

$$[F_1 = 2, F_2 = -3, F_3 = 4]$$

10. Mesh-current analysis produces the following three equations:

$$20\angle 0° = (5 + 3 - j4)I_1 - (3 - j4)I_2$$
$$10\angle 90° = (3 - j4 + 2)I_2 - (3 - j4)I_1 - 2I_3$$
$$-15\angle 0° - 10\angle 90° = (12 + 2)I_3 - 2I_2$$

Solve the equations for the loop currents I_1, I_2 and I_3.

$$\begin{bmatrix} I_1 = 3.317\angle 22.57°\,\text{A}, \\ I_2 = 1.963\angle 40.97°\,\text{A and} \\ I_3 = 1.010\angle -148.32°\,\text{A} \end{bmatrix}$$

62.3 Solution of simultaneous equations using Cramers rule

Cramers rule states that if

$$a_{11}x + a_{12}y + a_{13}z = b_1$$
$$a_{21}x + a_{22}y + a_{23}z = b_2$$
$$a_{31}x + a_{32}y + a_{33}z = b_3$$

then $x = \dfrac{D_x}{D}$, $y = \dfrac{D_y}{D}$ and $z = \dfrac{D_z}{D}$

where

$$D = \begin{vmatrix} a_{11} & a_{12} & a_{13} \\ a_{21} & a_{22} & a_{23} \\ a_{31} & a_{32} & a_{33} \end{vmatrix}$$

$$D_x = \begin{vmatrix} b_1 & a_{12} & a_{13} \\ b_2 & a_{22} & a_{23} \\ b_3 & a_{32} & a_{33} \end{vmatrix}$$

i.e. the x-column has been replaced by the R.H.S. b column,

$$D_y = \begin{vmatrix} a_{11} & b_1 & a_{13} \\ a_{21} & b_2 & a_{23} \\ a_{31} & b_3 & a_{33} \end{vmatrix}$$

i.e. the y-column has been replaced by the R.H.S. b column,

$$D_z = \begin{vmatrix} a_{11} & a_{12} & b_1 \\ a_{21} & a_{22} & b_2 \\ a_{31} & a_{32} & b_3 \end{vmatrix}$$

i.e. the z-column has been replaced by the R.H.S. b column.

Problem 7. Solve the following simultaneous equations using Cramers rule

$$x + y + z = 4$$
$$2x - 3y + 4z = 33$$
$$3x - 2y - 2z = 2$$

(This is the same as Problem 2 and a comparison of methods may be made). Following the above method:

$$D = \begin{vmatrix} 1 & 1 & 1 \\ 2 & -3 & 4 \\ 3 & -2 & -2 \end{vmatrix}$$
$$= 1(6 - (-8)) - 1((-4) - 12)$$
$$\quad + 1((-4) - (-9)) = 14 + 16 + 5 = \mathbf{35}$$

$$D = \begin{vmatrix} 4 & 1 & 1 \\ 33 & -3 & 4 \\ 2 & -2 & -2 \end{vmatrix}$$
$$= 4(6 - (-8)) - 1((-66) - 8)$$
$$\quad + 1((-66) - (-6)) = 56 + 74 - 60 = \mathbf{70}$$

$$D_y = \begin{vmatrix} 1 & 4 & 1 \\ 2 & 33 & 4 \\ 3 & 2 & -2 \end{vmatrix}$$
$$= 1((-66) - 8) - 4((-4) - 12) + 1(4 - 99)$$
$$= -74 + 64 - 95 = \mathbf{-105}$$

$$D_z = \begin{vmatrix} 1 & 1 & 4 \\ 2 & -3 & 33 \\ 3 & -2 & 2 \end{vmatrix}$$
$$= 1((-6) - (-66)) - 1(4 - 99)$$
$$\quad + 4((-4) - (-9)) = 60 + 95 + 20 = \mathbf{175}$$

Hence

$$x = \frac{D_x}{D} = \frac{70}{35} = \mathbf{2},\ y = \frac{D_y}{D} = \frac{-105}{35} = \mathbf{-3}$$

and $z = \dfrac{D_z}{D} = \dfrac{175}{35} = \mathbf{5}$

Now try the following exercise

Exercise 216 Further problems on solving simultaneous equations using Cramers rule

1. Repeat problems 3, 4, 5, 7 and 8 of Exercise 214 on page 548, using Cramers rule.

2. Repeat problems 3, 4, 8 and 9 of Exercise 215 on page 551, using Cramers rule.

Revision Test 17

This Revision test covers the material contained in Chapters 60 to 62. *The marks for each question are shown in brackets at the end of each question.*

1. Use the laws and rules of Boolean algebra to simplify the following expressions:

    (a) $B \cdot (A + \overline{B}) + A \cdot \overline{B}$

    (b) $\overline{A} \cdot \overline{B} \cdot \overline{C} + \overline{A} \cdot B \cdot \overline{C} + \overline{A} \cdot B \cdot C + \overline{A} \cdot \overline{B} \cdot C$

    (9)

2. Simplify the Boolean expression:

    $\overline{A \cdot B} + \overline{A \cdot B \cdot \overline{C}}$ using de Morgan's laws. (5)

3. Use a Karnaugh map to simplify the Boolean expression:

    $$\overline{A} \cdot \overline{B} \cdot \overline{C} + \overline{A} \cdot B \cdot \overline{C} + \overline{A} \cdot B \cdot C + A \cdot \overline{B} \cdot C$$

    (6)

4. A clean room has two entrances, each having two doors, as shown in Fig. R17.1. A warning bell must sound if both doors A and B or doors C and D are open at the same time. Write down the Boolean expression depicting this occurrence, and devise a logic network to operate the bell using NAND-gates only. (8)

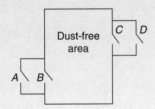

Figure R17.1

In questions 5 to 9, the matrices stated are:

$$A = \begin{pmatrix} -5 & 2 \\ 7 & -8 \end{pmatrix} \quad B = \begin{pmatrix} 1 & 6 \\ -3 & -4 \end{pmatrix}$$

$$C = \begin{pmatrix} j3 & (1+j2) \\ (-1-j4) & -j2 \end{pmatrix}$$

$$D = \begin{pmatrix} 2 & -1 & 3 \\ -5 & 1 & 0 \\ 4 & -6 & 2 \end{pmatrix} \quad E = \begin{pmatrix} -1 & 3 & 0 \\ 4 & -9 & 2 \\ -5 & 7 & 1 \end{pmatrix}$$

5. Determine $A \times B$ (4)

6. Calculate the determinant of matrix C (4)

7. Determine the inverse of matrix A (4)

8. Determine $E \times D$ (9)

9. Calculate the determinant of matrix D (5)

10. Use matrices to solve the following simultaneous equations:

    $$4x - 3y = 17$$
    $$x + y + 1 = 0 \quad (6)$$

11. Use determinants to solve the following simultaneous equations:

    $$4x + 9y + 2z = 21$$
    $$-8x + 6y - 3z = 41$$
    $$3x + y - 5z = -73 \quad (10)$$

12. The simultaneous equations representing the currents flowing in an unbalanced, three-phase, star-connected, electrical network are as follows:

    $$2.4I_1 + 3.6I_2 + 4.8I_3 = 1.2$$
    $$-3.9I_1 + 1.3I_2 - 6.5I_3 = 2.6$$
    $$1.7I_1 + 11.9I_2 + 8.5I_3 = 0$$

    Using matrices, solve equations for I_1, I_2 and I_3 (10)

Section 11

Differential Equations

Chapter 63

Introduction to differential equations

63.1 Family of curves

Integrating both sides of the derivative $\dfrac{dy}{dx} = 3$ with respect to x gives $y = \int 3\, dx$, i.e. $y = 3x + c$, where c is an arbitrary constant.

$y = 3x + c$ represents a **family of curves**, each of the curves in the family depending on the value of c. Examples include $y = 3x + 8$, $y = 3x + 3$, $y = 3x$ and $y = 3x - 10$ and these are shown in Fig. 63.1.

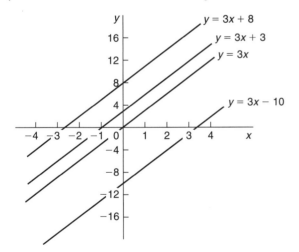

Figure 63.1

Each are straight lines of gradient 3. A particular curve of a family may be determined when a point on the curve is specified. Thus, if $y = 3x + c$ passes through the point (1, 2) then $2 = 3(1) + c$, from which, $c = -1$. The equation of the curve passing through (1, 2) is therefore $y = 3x - 1$.

Problem 1. Sketch the family of curves given by the equations $\dfrac{dy}{dx} = 4x$ and determine the equation of one of these curves which passes through the point (2, 3).

Integrating both sides of $\dfrac{dy}{dx} = 4x$ with respect to x gives:

$$\int \frac{dy}{dx}\, dx = \int 4x\, dx, \quad \text{i.e. } y = 2x^2 + c$$

Some members of the family of curves having an equation $y = 2x^2 + c$ include $y = 2x^2 + 15$, $y = 2x^2 + 8$, $y = 2x^2$ and $y = 2x^2 - 6$, and these are shown in Fig. 63.2. To determine the equation of the curve passing

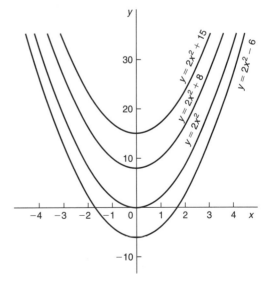

Figure 63.2

through the point (2, 3), $x = 2$ and $y = 3$ are substituted into the equation $y = 2x^2 + c$.

Thus $3 = 2(2)^2 + c$, from which $c = 3 - 8 = -5$.

Hence the equation of the curve passing through the point (2, 3) is $y = 2x^2 - 5$.

Now try the following exercise

Exercise 217 Further problems on families of curves

1. Sketch a family of curves represented by each of the following differential equations:

   (a) $\dfrac{dy}{dx} = 6$ (b) $\dfrac{dy}{dx} = 3x$ (c) $\dfrac{dy}{dx} = x + 2$

2. Sketch the family of curves given by the equation $\dfrac{dy}{dx} = 2x + 3$ and determine the equation of one of these curves which passes through the point (1, 3). $[y = x^2 + 3x - 1]$

63.2 Differential equations

A **differential equation** is one that contains differential coefficients.

Examples include

(i) $\dfrac{dy}{dx} = 7x$ and (ii) $\dfrac{d^2y}{dx^2} + 5\dfrac{dy}{dx} + 2y = 0$

Differential equations are classified according to the highest derivative which occurs in them. Thus example (i) above is a **first order differential equation**, and example (ii) is a **second order differential equation**.

The **degree** of a differential equation is that of the highest power of the highest differential which the equation contains after simplification.

Thus $\left(\dfrac{d^2x}{dt^2}\right)^3 + 2\left(\dfrac{dx}{dt}\right)^5 = 7$ is a second order differential equation of degree three.

Starting with a differential equation it is possible, by integration and by being given sufficient data to determine unknown constants, to obtain the original function. This process is called **'solving the differential equation'**. A solution to a differential equation which contains one or more arbitrary constants of integration is called the **general solution** of the differential equation.

When additional information is given so that constants may be calculated the **particular solution** of the

differential equation is obtained. The additional information is called **boundary conditions**. It was shown in Section 63.1 that $y = 3x + c$ is the general solution of the differential equation $\dfrac{dy}{dx} = 3$.

Given the boundary conditions $x = 1$ and $y = 2$, produces the particular solution of $y = 3x - 1$. Equations which can be written in the form

$$\frac{dy}{dx} = f(x), \frac{dy}{dx} = f(y) \text{ and } \frac{dy}{dx} = f(x) \cdot f(y)$$

can all be solved by integration. In each case it is possible to separate the y's to one side of the equation and the x's to the other. Solving such equations is therefore known as solution by **separation of variables**.

63.3 The solution of equations of the form $\dfrac{dy}{dx} = f(x)$

A differential equation of the form $\dfrac{dy}{dx} = f(x)$ is solved by direct integration,

i.e. $$y = \int f(x)\, dx$$

Problem 2. Determine the general solution of

$$x\frac{dy}{dx} = 2 - 4x^3$$

Rearranging $x\dfrac{dy}{dx} = 2 - 4x^3$ gives:

$$\frac{dy}{dx} = \frac{2 - 4x^3}{x} = \frac{2}{x} - \frac{4x^3}{x} = \frac{2}{x} - 4x^2$$

Integrating both sides gives:

$$y = \int \left(\frac{2}{x} - 4x^2\right) dx$$

i.e. $$y = 2\ln x - \frac{4}{3}x^3 + c,$$

which is the general solution.

Problem 3. Find the particular solution of the differential equation $5\dfrac{dy}{dx} + 2x = 3$, given the boundary conditions $y = 1\dfrac{2}{5}$ when $x = 2$.

Since $5\dfrac{dy}{dx} + 2x = 3$ then $\dfrac{dy}{dx} = \dfrac{3 - 2x}{5} = \dfrac{3}{5} - \dfrac{2x}{5}$

Hence $y = \displaystyle\int \left(\dfrac{3}{5} - \dfrac{2x}{5} \right) dx$

i.e. $\quad y = \dfrac{3x}{5} - \dfrac{x^2}{5} + c,$

which is the general solution.

Substituting the boundary conditions $y = 1\dfrac{2}{5}$ and $x = 2$ to evaluate c gives:

$1\dfrac{2}{5} = \dfrac{6}{5} - \dfrac{4}{5} + c$, from which, $c = 1$

Hence the particular solution is $y = \dfrac{3x}{5} - \dfrac{x^2}{5} + 1.$

Problem 4. Solve the equation

$2t\left(t - \dfrac{d\theta}{dt} \right) = 5$, given $\theta = 2$ when $t = 1$.

Rearranging gives:

$$t - \dfrac{d\theta}{dt} = \dfrac{5}{2t} \quad \text{and} \quad \dfrac{d\theta}{dt} = t - \dfrac{5}{2t}$$

Integrating gives:

$$\theta = \int \left(t - \dfrac{5}{2t} \right) dt$$

i.e. $\quad \theta = \dfrac{t^2}{2} - \dfrac{5}{2} \ln t + c,$

which is the general solution.

When $\theta = 2$, $t = 1$, thus $2 = \frac{1}{2} - \frac{5}{2} \ln 1 + c$ from which, $c = \frac{3}{2}$.

Hence the particular solution is:

$$\theta = \dfrac{t^2}{2} - \dfrac{5}{2} \ln t + \dfrac{3}{2}$$

i.e. $\quad \boldsymbol{\theta = \dfrac{1}{2}(t^2 - 5\ln t + 3)}$

Problem 5. The bending moment M of the beam is given by $\dfrac{dM}{dx} = -w(l - x)$, where w and x are constants. Determine M in terms of x given: $M = \frac{1}{2}wl^2$ when $x = 0$.

$$\dfrac{dM}{dx} = -w(l - x) = -wl + wx$$

Integrating with respect to x gives:

$$M = -wlx + \dfrac{wx^2}{2} + c$$

which is the general solution.

When $M = \frac{1}{2}wl^2$, $x = 0$.

Thus $\dfrac{1}{2}wl^2 = -wl(0) + \dfrac{w(0)^2}{2} + c$

from which, $c = \dfrac{1}{2}wl^2$.

Hence the particular solution is:

$$M = -wlx + \dfrac{wx^2}{2} + \dfrac{1}{2}wl^2$$

i.e. $\quad \boldsymbol{M = \dfrac{1}{2}w(l^2 - 2lx + x^2)}$

or $\quad \boldsymbol{M = \dfrac{1}{2}w(l - x)^2}$

Now try the following exercise

Exercise 218 Further problems on equations of the form $\dfrac{dy}{dx} = f(x).$

In Problems 1 to 5, solve the differential equations.

1. $\dfrac{dy}{dx} = \cos 4x - 2x \qquad \left[y = \dfrac{\sin 4x}{4} - x^2 + c \right]$

2. $2x\dfrac{dy}{dx} = 3 - x^3 \qquad \left[y = \dfrac{3}{2} \ln x - \dfrac{x^3}{6} + c \right]$

3. $\dfrac{dy}{dx} + x = 3$, given $y = 2$ when $x = 1$.

$$\left[y = 3x - \dfrac{x^2}{2} - \dfrac{1}{2} \right]$$

4. $3\dfrac{dy}{d\theta} + \sin \theta = 0$, given $y = \dfrac{2}{3}$ when $\theta = \dfrac{\pi}{3}$

$$\left[y = \dfrac{1}{3} \cos \theta + \dfrac{1}{2} \right]$$

5. $\dfrac{1}{e^x} + 2 = x - 3\dfrac{dy}{dx}$, given $y = 1$ when $x = 0$.

$$\left[y = \dfrac{1}{6}\left(x^2 - 4x + \dfrac{2}{e^x} + 4 \right) \right]$$

6. The gradient of a curve is given by:

$$\dfrac{dy}{dx} + \dfrac{x^2}{2} = 3x$$

Find the equation of the curve if it passes through the point $\left(1, \frac{1}{3}\right)$.

$$\left[y = \frac{3}{2}x^2 - \frac{x^3}{6} - 1\right]$$

7. The acceleration, a, of a body is equal to its rate of change of velocity, $\dfrac{dv}{dt}$. Find an equation for v in term of t, given that when $t = 0$, velocity $v = u$. $[v = u + at]$

8. An object is thrown vertically upwards with an initial velocity, u, of 20 m/s. The motion of the object follows the differential equation $\dfrac{ds}{dt} = u - gt$, where s is the height of the object in metres at time t seconds and $g = 9.8\,\text{m/s}^2$. Determine the height of the object after 3 seconds if $s = 0$ when $t = 0$. $[15.9\,\text{m}]$

63.4 The solution of equations of the form $\dfrac{dy}{dx} = f(y)$

A differential equation of the form $\dfrac{dy}{dx} = f(y)$ is initially rearranged to give $dx = \dfrac{dy}{f(y)}$ and then the solution is obtained by direct integration,

i.e. $$\int dx = \int \frac{dy}{f(y)}$$

Problem 6. Find the general solution of

$$\frac{dy}{dx} = 3 + 2y$$

Rearranging $\dfrac{dy}{dx} = 3 + 2y$ gives:

$$dx = \frac{dy}{3 + 2y}$$

Integrating both sides gives:

$$\int dx = \int \frac{dy}{3 + 2y}$$

Thus, by using the substitution $u = (3 + 2y)$ — see Chapter 49,

$$x = \tfrac{1}{2}\ln(3 + 2y) + c \qquad (1)$$

It is possible to give the general solution of a differential equation in a different form. For example, if $c = \ln k$, where k is a constant, then:

$$x = \tfrac{1}{2}\ln(3 + 2y) + \ln k,$$

i.e. $$x = \ln(3 + 2y)^{\frac{1}{2}} + \ln k$$

or $$x = \ln[k\sqrt{(3 + 2y)}] \qquad (2)$$

by the laws of logarithms, from which,

$$e^x = k\sqrt{(3 + 2y)} \qquad (3)$$

Equations (1), (2) and (3) are all acceptable general solutions of the differential equation

$$\frac{dy}{dx} = 3 + 2y$$

Problem 7. Determine the particular solution of $(y^2 - 1)\dfrac{dy}{dx} = 3y$ given that $y = 1$ when $x = 2\dfrac{1}{6}$

Rearranging gives:

$$dx = \left(\frac{y^2 - 1}{3y}\right) dy = \left(\frac{y}{3} - \frac{1}{3y}\right) dy$$

Integrating gives:

$$\int dx = \int \left(\frac{y}{3} - \frac{1}{3y}\right) dy$$

i.e. $$x = \frac{y^2}{6} - \frac{1}{3}\ln y + c,$$

which is the general solution.

When $y = 1$, $x = 2\dfrac{1}{6}$, thus $2\dfrac{1}{6} = \dfrac{1}{6} - \dfrac{1}{3}\ln 1 + c$, from which, $c = 2$.

Hence the particular solution is:

$$x = \frac{y^2}{6} - \frac{1}{3}\ln y + 2$$

Problem 8. (a) The variation of resistance R ohms, of an aluminium conductor with temperature $\theta°\text{C}$ is given by $\dfrac{dR}{d\theta} = \alpha R$, where α is the temperature coefficient of resistance of aluminum. If $R = R_0$ when $\theta = 0°\text{C}$, solve the equation for R.
(b) If $\alpha = 38 \times 10^{-4}/°\text{C}$, determine the resistance of an aluminum conductor at 50°C, correct to 3 significant figures, when its resistance at 0°C is 24.0 Ω.

(a) $\dfrac{dR}{d\theta} = \alpha R$ is the form $\dfrac{dy}{dx} = f(y)$

Rearranging givens: $d\theta = \dfrac{dR}{\alpha R}$

Integrating both sides gives:

$$\int d\theta = \int \dfrac{dR}{\alpha R}$$

i.e. $\theta = \dfrac{1}{\alpha} \ln R + c,$

which is the general solution.

Substituting the boundary conditions $R = R_0$ when $\theta = 0$ gives:

$$0 = \dfrac{1}{\alpha} \ln R_0 + c$$

from which $c = -\dfrac{1}{\alpha} \ln R_0$

Hence the particular solution is

$$\theta = \dfrac{1}{\alpha} \ln R - \dfrac{1}{\alpha} \ln R_0 = \dfrac{1}{\alpha}(\ln R - \ln R_0)$$

i.e. $\theta = \dfrac{1}{\alpha} \ln\left(\dfrac{R}{R_0}\right)$ or $\alpha\theta = \ln\left(\dfrac{R}{R_0}\right)$

Hence $e^{\alpha\theta} = \dfrac{R}{R_0}$ from which, $\boldsymbol{R = R_0 e^{\alpha\theta}}$

(b) Substituting $\alpha = 38 \times 10^{-4}$, $R_0 = 24.0$ and $\theta = 50$ into $R = R_0 e^{\alpha\theta}$ gives the resistance at 50°C, i.e. $\boldsymbol{R_{50}} = 24.0\, e^{(38 \times 10^{-4} \times 50)} = \boldsymbol{29.0\, ohms}.$

Now try the following exercise

Exercise 219   Further problems on equations of the form $\dfrac{dy}{dx} = f(y)$

In Problems 1 to 3, solve the differential equations.

1. $\dfrac{dy}{dx} = 2 + 3y$ $\left[x = \dfrac{1}{3} \ln(2 + 3y) + c\right]$

2. $\dfrac{dy}{dx} = 2 \cos^2 y$ $[\tan y = 2x + c]$

3. $(y^2 + 2)\dfrac{dy}{dx} = 5y$, given $y = 1$ when $x = \dfrac{1}{2}$ $\left[\dfrac{y^2}{2} + 2 \ln y = 5x - 2\right]$

4. The current in an electric circuit is given by the equation

$$Ri + L\dfrac{di}{dt} = 0$$

where L and R are constants. Shown that $i = Ie^{-\frac{Rt}{L}}$, given that $i = I$ when $t = 0$.

5. The velocity of a chemical reaction is given by $\dfrac{dx}{dt} = k(a - x)$. where x is the amount transferred in time t, k is a constant and a is the concentration at time $t = 0$ when $x = 0$. Solve the equation and determine x in terms of t. $[x = a(1 - e^{-kt})]$

6. (a) Charge Q coulombs at time t seconds is given by the differential equation $R\dfrac{dQ}{dt} + \dfrac{Q}{C} = 0$, where C is the capacitance in farads and R the resistance in ohms. Solve the equation for Q given that $Q = Q_0$ when $t = 0$.

    (b) A circuit possesses a resistance of $250 \times 10^3\,\Omega$ and a capacitance of $8.5 \times 10^{-6}\,\mathrm{F}$, and after 0.32 seconds the charge falls to 8.0 C. Determine the initial charge and the charge after 1 second, each correct to 3 significant figures.
    $[\text{(a) } Q = Q_0 e^{-\frac{t}{CR}} \quad \text{(b) } 9.30\,\mathrm{C}, 5.81\,\mathrm{C}]$

7. A differential equation relating the difference in tension T, pulley contact angle θ and coefficient of friction μ is $\dfrac{dT}{d\theta} = \mu T$. When $\theta = 0$, $T = 150\,\mathrm{N}$, and $\mu = 0.30$ as slipping starts. Determine the tension at the point of slipping when $\theta = 2$ radians. Determine also the value of θ when T is 300 N. $[273.3\,\mathrm{N}, 2.31\,\mathrm{rads}]$

8. The rate of cooling of a body is given by $\dfrac{d\theta}{dt} = k\theta$, where k is a constant. If $\theta = 60°\mathrm{C}$ when $t = 2$ minutes and $\theta = 50°\mathrm{C}$ when $t = 5$ minutes, determine the time taken for θ to fall to 40°C, correct to the nearest second. $[8\,\mathrm{m}\,40\,\mathrm{s}]$

63.5 The solution of equations of the form $\frac{dy}{dx} = f(x) \cdot f(y)$

A differential equation of the form $\frac{dy}{dx} = f(x) \cdot f(y)$, where $f(x)$ is a function of x only and $f(y)$ is a function of y only, may be rearranged as $\frac{dy}{f(y)} = f(x)dx$, and then the solution is obtained by direct integration, i.e.

$$\int \frac{dy}{f(y)} = \int f(x)\,dx$$

Problem 9. Solve the equation $4xy\frac{dy}{dx} = y^2 - 1$

Separating the variables gives:

$$\left(\frac{4y}{y^2 - 1}\right)dy = \frac{1}{x}dx$$

Integrating both sides gives:

$$\int \left(\frac{4y}{y^2 - 1}\right)dy = \int \left(\frac{1}{x}\right)dx$$

Using the substituting $u = y^2 - 1$, the general solution is:

$$2\ln(y^2 - 1) = \ln x + c \qquad (1)$$

or $\qquad \ln(y^2 - 1)^2 - \ln x = c$

from which, $\qquad \ln\left\{\frac{(y^2 - 1)^2}{x}\right\} = c$

and $\qquad \frac{(y^2 - 1)^2}{x} = e^c \qquad (2)$

If in equation (1), $c = \ln A$, where A is a different constant,

then $\quad \ln(y^2 - 1)^2 = \ln x + \ln A$

i.e. $\quad \ln(y^2 - 1)^2 = \ln Ax$

i.e. $\quad (y^2 - 1)^2 = Ax \qquad (3)$

Equations (1) to (3) are thus three valid solutions of the differential equations

$$4xy\frac{dy}{dx} = y^2 - 1$$

Problem 10. Determine the particular solution of $\frac{d\theta}{dt} = 2e^{3t-2\theta}$, given that $t = 0$ when $\theta = 0$.

$$\frac{d\theta}{dt} = 2e^{3t-2\theta} = 2(e^{3t})(e^{-2\theta}),$$

by the laws of indices.

Separating the variables gives:

$$\frac{d\theta}{e^{-2\theta}} = 2e^{2t}\,dt$$

i.e. $\qquad e^{2\theta}\,d\theta = 2e^{3t}\,dt$

Integrating both sides gives:

$$\int e^{2\theta}\,d\theta = \int 2e^{3t}\,dt$$

Thus the general solution is:

$$\frac{1}{2}e^{2\theta} = \frac{2}{3}e^{3t} + c$$

When $t = 0$, $\theta = 0$, thus:

$$\frac{1}{2}e^0 = \frac{2}{3}e^0 + c$$

from which, $c = \frac{1}{2} - \frac{2}{3} = -\frac{1}{6}$

Hence the particular solution is:

$$\frac{1}{2}e^{2\theta} = \frac{2}{3}e^{3t} - \frac{1}{6}$$

or $\qquad 3e^{2\theta} = 4e^{3t} - 1$

Problem 11. Find the curve which satisfies the equation $xy = (1 + x^2)\frac{dy}{dx}$ and passes through the point $(0, 1)$

Separating the variables gives:

$$\frac{x}{(1 + x^2)}dx = \frac{dy}{y}$$

Integrating both sides gives:

$$\frac{1}{2}\ln(1 + x^2) = \ln y + c$$

When $x = 0$, $y = 1$ thus $\frac{1}{2}\ln 1 = \ln 1 + c$, from which, $c = 0$.

Hence the particular solution is $\frac{1}{2}\ln(1+x^2) = \ln y$

i.e. $\ln(1+x^2)^{\frac{1}{2}} = \ln y$, from which, $(1+x^2)^{\frac{1}{2}} = y$.

Hence the equation of the curve is $y = \sqrt{(1+x^2)}$.

Problem 12. The current i in an electric circuit containing resistance R and inductance L in series with a constant voltage source E is given by the differential equation $E - L\left(\dfrac{di}{dt}\right) = Ri$. Solve the equation and find i in terms of time t given that when $t = 0$, $i = 0$.

In the $R - L$ series circuit shown in Fig. 63.3, the supply p.d., E, is given by

$$E = V_R + V_L$$

$$V_R = iR \quad \text{and} \quad V_L = L\frac{di}{dt}$$

Hence $\qquad E = iR + L\dfrac{di}{dt}$

from which $\quad E - L\dfrac{di}{dt} = Ri$

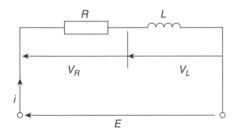

Figure 63.3

Most electrical circuits can be reduced to a differential equation.

Rearranging $E - L\dfrac{di}{dt} = Ri$ gives $\dfrac{di}{dt} = \dfrac{E - Ri}{L}$

and separating the variables gives:

$$\frac{di}{E - Ri} = \frac{dt}{L}$$

Integrating both sides gives:

$$\int \frac{di}{E - Ri} = \int \frac{dt}{L}$$

Hence the general solution is:

$$-\frac{1}{R}\ln(E - Ri) = \frac{t}{L} + c$$

(by making a substitution $u = E - Ri$, see Chapter 49).

When $t = 0$, $i = 0$, thus $-\dfrac{1}{R}\ln E = c$

Thus the particular solution is:

$$-\frac{1}{R}\ln(E - Ri) = \frac{t}{L} - \frac{1}{R}\ln E$$

Transposing gives:

$$-\frac{1}{R}\ln(E - Ri) + \frac{1}{R}\ln E = \frac{t}{L}$$

$$\frac{1}{R}[\ln E - \ln(E - Ri)] = \frac{t}{L}$$

$$\ln\left(\frac{E}{E - Ri}\right) = \frac{Rt}{L}$$

from which $\dfrac{E}{E - Ri} = e^{\frac{Rt}{L}}$

Hence $\dfrac{E - Ri}{E} = e^{-\frac{Rt}{L}}$ and $E - Ri = Ee^{-\frac{Rt}{L}}$ and

$$Ri = E - Ee^{-\frac{Rt}{L}}$$

Hence current,

$$i = \frac{E}{R}\left(1 - e^{-\frac{Rt}{L}}\right)$$

which represents the law of growth of current in an inductive circuit as shown in Fig. 63.4.

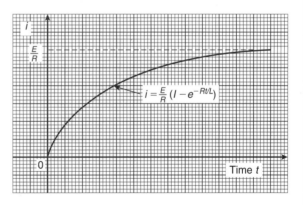

Figure 63.4

Problem 13. For an adiabatic expansion of a gas

$$C_v\frac{dp}{p} + C_p\frac{dV}{V} = 0$$

where C_p and C_v are constants. Given $n = \dfrac{C_p}{C_v}$ show that $pV^n = $ constant.

Separating the variables gives:

$$C_v \frac{dp}{p} = -C_p \frac{dV}{V}$$

Integrating both sides gives:

$$C_v \int \frac{dp}{p} = -C_p \int \frac{dV}{V}$$

i.e. $\qquad C_v \ln p = -C_p \ln V + k$

Dividing throughout by constant C_v gives:

$$\ln p = -\frac{C_p}{C_v} \ln V + \frac{k}{C_v}$$

Since $\dfrac{C_p}{C_v} = n$, then $\ln p + n \ln V = K$,

where $K = \dfrac{k}{C_v}$

i.e. $\ln p + \ln V^n = K$ or $\ln pV^n = K$, by the laws of logarithms.

Hence $pV^n = e^K$ i.e. $\mathbf{pV^n = constant}$.

Now try the following exercise

Exercise 220   Further problems on equations of the form $\dfrac{dy}{dx} = f(x) \cdot f(y)$

In Problems 1 to 4, solve the differential equations.

1. $\dfrac{dy}{dx} = 2y \cos x$ $\qquad$ [$\ln y = 2 \sin x + c$]

2. $(2y - 1)\dfrac{dy}{dx} = (3x^2 + 1)$, given $x = 1$ when $y = 2$.

$$[y^2 - y = x^3 + x]$$

3. $\dfrac{dy}{dx} = e^{2x-y}$, given $x = 0$ when $y = 0$.

$$\left[e^y = \frac{1}{2}e^{2x} + \frac{1}{2} \right]$$

4. $2y(1 - x) + x(1 + y)\dfrac{dy}{dx} = 0$, given $x = 1$ when $y = 1$.

$$[\ln(x^2 y) = 2x - y - 1]$$

5. Show that the solution of the equation $\dfrac{y^2 + 1}{x^2 + 1} = \dfrac{y}{x}\dfrac{dy}{dx}$ is of the form

$$\sqrt{\left(\frac{y^2 + 1}{x^2 + 1} \right)} = \text{constant}.$$

6. Solve $xy = (1 - x^2)\dfrac{dy}{dx}$ for y, given $x = 0$ when $y = 1$.

$$\left[y = \frac{1}{\sqrt{(1 - x^2)}} \right]$$

7. Determine the equation of the curve which satisfies the equation $xy\dfrac{dy}{dx} = x^2 - 1$, and which passes through the point $(1, 2)$.

$$[y^2 = x^2 - 2\ln x + 3]$$

8. The p.d., V, between the plates of a capacitor C charged by a steady voltage E through a resistor R is given by the equation $CR\dfrac{dV}{dt} + V = E$.

   (a) Solve the equation for V given that at $t = 0$, $V = 0$.
   (b) Calculate V, correct to 3 significant figures, when $E = 25$ V, $C = 20 \times 10^{-6}$ F, $R = 200 \times 10^3 \,\Omega$ and $t = 3.0$ s.

$$\left[\begin{array}{l} \text{(a)} \quad V = E\left(1 - e^{-\frac{t}{CR}}\right) \\ \text{(b)} \quad 13.2 \text{ V} \end{array} \right]$$

9. Determine the value of p, given that $x^3 \dfrac{dy}{dx} = p - x$, and that $y = 0$ when $x = 2$ and when $x = 6$. $\qquad$ [3]

Revision Test 18

This Revision test covers the material contained in Chapter 63. *The marks for each question are shown in brackets at the end of each question.*

1. Solve the differential equation: $x\dfrac{dy}{dx} + x^2 = 5$ given that $y = 2.5$ when $x = 1$. (5)

2. Determine the equation of the curve which satisfies the differential equation $2xy\dfrac{dy}{dx} = x^2 + 1$ and which passes through the point $(1, 2)$ (6)

3. A capacitor C is charged by applying a steady voltage E through a resistance R. The p.d. between the plates, V, is given by the differential equation:

$$CR\frac{dV}{dt} + V = E$$

(a) Solve the equation for E given that when time $t = 0$, $V = 0$.

(b) Evaluate voltage V when $E = 50\,V$, $C = 10\,\mu F$, $R = 200\,k\Omega$ and $t = 1.2\,s$. (14)

All questions have only one correct answer (answers on page 570).

1. Differentiating $y = 4x^5$ gives:

   (a) $\dfrac{dy}{dx} = \dfrac{2}{3}x^6$ (b) $\dfrac{dy}{dx} = 20x^4$

   (c) $\dfrac{dy}{dx} = 4x^6$ (d) $\dfrac{dy}{dx} = 5x^4$

2. $\int (5 - 3t^2)\,dt$ is equal to:

   (a) $5 - t^3 + c$ (b) $-3t^3 + c$

   (c) $-6t + c$ (d) $5t - t^3 + c$

3. The gradient of the curve $y = -2x^3 + 3x + 5$ at $x = 2$ is:

   (a) -21 (b) 27 (c) -16 (d) -5

4. $\int \left(\dfrac{5x - 1}{x}\right) dx$ is equal to:

   (a) $5x - \ln x + c$ (b) $\dfrac{5x^2 - x}{\frac{x^2}{2}}$

   (c) $\dfrac{5x^2}{2} + \dfrac{1}{x^2} + c$ (d) $5x + \dfrac{1}{x^2} + c$

5. For the curve shown in Figure M4.1, which of the following statements is incorrect?
   (a) P is a turning point
   (b) Q is a minimum point
   (c) R is a maximum value
   (d) Q is a stationary value

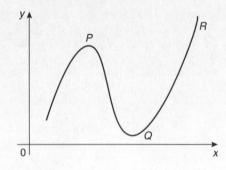

Figure M4.1

6. The value of $\int_0^1 (3 \sin 2\theta - 4 \cos \theta)\,d\theta$, correct to 4 significant figures, is:

   (a) -1.242 (b) -0.06890

   (c) -2.742 (d) -1.569

7. If $y = 5\sqrt{x^3} - 2$, $\dfrac{dy}{dx}$ is equal to:

   (a) $\dfrac{15}{2}\sqrt{x}$ (b) $2\sqrt{x^5} - 2x + c$

   (c) $\dfrac{5}{2}\sqrt{x} - 2$ (d) $5\sqrt{x} - 2x$

8. $\int xe^{2x}\,dx$ is:

   (a) $\dfrac{x^2}{4}e^{2x} + c$ (b) $2e^{2x} + c$

   (c) $\dfrac{e^{2x}}{4}(2x - 1) + c$ (d) $2e^{2x}(x - 2) + c$

9. An alternating current is given by $i = 4 \sin 150t$ amperes, where t is the time in seconds. The rate of change of current at $t = 0.025$s is:

   (a) 3.99 A/s   (b) -492.3 A/s

   (c) -3.28 A/s   (d) 598.7 A/s

10. A vehicle has a velocity $v = (2 + 3t)$ m/s after t seconds. The distance travelled is equal to the area under the v/t graph. In the first 3 seconds the vehicle has travelled:

    (a) 11 m   (b) 33 m   (c) 13.5 m   (d) 19.5 m

11. Differentiating $y = \dfrac{1}{\sqrt{x}} + 2$ with respect to x gives:

    (a) $\dfrac{1}{\sqrt{x^3}} + 2$ (b) $-\dfrac{1}{2\sqrt{x^3}}$

    (c) $2 - \dfrac{1}{2\sqrt{x^3}}$ (d) $\dfrac{2}{\sqrt{x^3}}$

12. The area, in square units, enclosed by the curve $y = 2x + 3$, the x-axis and ordinates $x = 1$ and $x = 4$ is:

    (a) 28 (b) 2 (c) 24 (d) 39

13. The resistance to motion F of a moving vehicle is given by $F = \dfrac{5}{x} + 100x$. The minimum value of resistance is:

    (a) -44.72 (b) 0.2236

    (c) 44.72 (d) -0.2236

14. Differentiating $i = 3\sin 2t - 2\cos 3t$ with respect to t gives:

    (a) $3\cos 2t + 2\sin 3t$

    (b) $6(\sin 2t - \cos 3t)$

    (c) $\frac{3}{2}\cos 2t + \frac{2}{3}\sin 3t$

    (d) $6(\cos 2t + \sin 3t)$

15. $\int \frac{2}{9}t^3\, dt$ is equal to:

    (a) $\frac{t^4}{18} + c$

    (b) $\frac{2}{3}t^2 + c$

    (c) $\frac{2}{9}t^4 + c$

    (d) $\frac{2}{9}t^3 + c$

16. Given $y = 3e^x + 2\ln 3x$, $\frac{dy}{dx}$ is equal to:

    (a) $6e^x + \frac{2}{3x}$

    (b) $3e^x + \frac{2}{x}$

    (c) $6e^x + \frac{2}{x}$

    (d) $3e^x + \frac{2}{3}$

17. $\int \left(\frac{t^3 - 3t}{2t}\right) dt$ is equal to:

    (a) $\frac{\frac{t^4}{4} - \frac{3t^2}{2}}{t^2} + c$

    (b) $\frac{t^3}{6} - \frac{3}{2}t + c$

    (c) $\frac{t^3}{3} - \frac{3t^2}{2} + c$

    (d) $\frac{1}{2}\left(\frac{t^4}{4} - 3t\right) + c$

18. The vertical displacement, s, of a prototype model in a tank is given by $s = 40\sin 0.1t$ mm, where t is the time in seconds. The vertical velocity of the model, in mm/s, is:

    (a) $-\cos 0.1t$

    (b) $400\cos 0.1t$

    (c) $-400\cos 0.1t$

    (d) $4\cos 0.1t$

19. Evaluating $\int_0^{\pi/3} 3\sin 3x\, dx$ gives:

    (a) 2   (b) 1.503   (c) -18 (d) 6

20. The equation of a curve is $y = 2x^3 - 6x + 1$. The maximum value of the curve is:

    (a) -3 (b) 1   (c) 5   (d) -6

21. The mean value of $y = 2x^2$ between $x = 1$ and $x = 3$ is:

    (a) 2   (b) 4   (c) $4\frac{1}{3}$ (d) $8\frac{2}{3}$

22. Given $f(t) = 3t^4 - 2$, $f'(t)$ is equal to:

    (a) $12t^3 - 2$

    (b) $\frac{3}{4}t^5 - 2t + c$

    (c) $12t^3$

    (d) $3t^5 - 2$

23. $\int \ln x\, dx$ is equal to:

    (a) $x(\ln x - 1) + c$

    (b) $\frac{1}{x} + c$

    (c) $x\ln x - 1 + c$

    (d) $\frac{1}{x} + \frac{1}{x^2} + c$

24. The current i in a circuit at time t seconds is given by $i = 0.20(1 - e^{-20t})A$. When time $t = 0.1$ s, the rate of change of current is:

    (a) -1.022 A/s

    (b) 0.541 A/s

    (c) 0.173 A/s

    (d) 0.373 A/s

25. $\int_2^3 \frac{3}{x^2 + x - 2}\, dx$ is equal to:

    (a) $3\ln 2.5$

    (b) $\frac{1}{3}\lg 1.6$

    (c) $\ln 40$

    (d) $\ln 1.6$

26. The gradient of the curve $y = 4x^2 - 7x + 3$ at the point $(1, 0)$ is

    (a) 1   (b) 3   (c) 0   (d) -7

27. $\int (5\sin 3t - 3\cos 5t)dt$ is equal to:

    (a) $-5\cos 3t + 3\sin 5t + c$

    (b) $15(\cos 3t + \sin 3t) + c$

    (c) $-\frac{5}{3}\cos 3t - \frac{3}{5}\sin 5t + c$

    (d) $\frac{3}{5}\cos 3t - \frac{5}{3}\sin 5t + c$

28. The derivative of $2\sqrt{x} - 2x$ is:

    (a) $\frac{4}{3}\sqrt{x^3} - x^2 + c$

    (b) $\frac{1}{\sqrt{x}} - 2$

    (c) $\sqrt{x} - 2$

    (d) $-\frac{1}{2\sqrt{x}} - 2$

29. The velocity of a car (in m/s) is related to time t seconds by the equation $v = 4.5 + 18t - 4.5t^2$. The maximum speed of the car, in km/h, is:

    (a) 81   (b) 6.25   (c) 22.5   (d) 77

30. $\int (\sqrt{x} - 3)dx$ is equal to:

    (a) $\frac{3}{2}\sqrt{x^3} - 3x + c$

    (b) $\frac{2}{3}\sqrt{x^3} + c$

    (c) $\frac{1}{2\sqrt{x}} + c$

    (d) $\frac{2}{3}\sqrt{x^3} - 3x + c$

31. An alternating voltage is given by $v = 10\sin 300t$ volts, where t is the time in seconds. The rate of change of voltage when $t = 0.01$ s is:

    (a) -2996 V/s

    (b) 157 V/s

    (c) -2970 V/s

    (d) 0.523 V/s

32. The r.m.s. value of $y = x^2$ between $x = 1$ and $x = 3$, correct to 2 decimal places, is:

    (a) 2.08   (b) 4.92   (c) 6.96   (d) 24.2

33. If $f(t) = 5t - \dfrac{1}{\sqrt{t}}$, $f'(t)$ is equal to:

    (a) $5 + \dfrac{1}{2\sqrt{t^3}}$ (b) $5 - 2\sqrt{t}$

    (c) $\dfrac{5t^2}{2} - 2\sqrt{t} + c$ (d) $5 + \dfrac{1}{\sqrt{t^3}}$

34. The value of $\int_0^{\pi/6} 2\sin\left(3t + \dfrac{\pi}{2}\right) dt$ is:

    (a) 6   (b) $-\dfrac{2}{3}$ (c) -6 (d) $\dfrac{2}{3}$

35. The equation of a curve is $y = 2x^3 - 6x + 1$. The minimum value of the curve is:

    (a) -6 (b) 1   (c) 5   (d) -3

36. The volume of the solid of revolution when the curve $y = 2x$ is rotated one revolution about the x-axis between the limits $x = 0$ and $x = 4$ cm is:

    (a) $85\dfrac{1}{3}\pi$ cm^3 (b) 8 cm^3

    (c) $85\dfrac{1}{3}$ cm^3 (d) 64π cm^3

37. The length l metres of a certain metal rod at temperature $t°C$ is given by $l = 1 + 4 \times 10^{-5}t + 4 \times 10^{-7}t^2$. The rate of change of length, in mm/°C, when the temperature is 400°C, is:

    (a) 3.6×10^{-4} (b) 1.00036

    (c) 0.36   (d) 3.2×10^{-4}

38. If $y = 3x^2 - \ln 5x$ then $\dfrac{d^2y}{dx^2}$ is equal to:

    (a) $6 + \dfrac{1}{5x^2}$ (b) $6x - \dfrac{1}{x}$

    (c) $6 - \dfrac{1}{5x}$ (d) $6 + \dfrac{1}{x^2}$

39. The area enclosed by the curve $y = 3\cos 2\theta$, the ordinates $\theta = 0$ and $\theta = \dfrac{\pi}{4}$ and the θ axis is:

    (a) -3 (b) 6   (c) 1.5   (d) 3

40. $\displaystyle\int \left(1 + \dfrac{4}{e^{2x}}\right) dx$ is equal to:

    (a) $\dfrac{8}{.e^{2x}} + c$ (b) $x - \dfrac{2}{e^{2x}} + c$

    (c) $x + \dfrac{4}{e^{2x}}$ (d) $x - \dfrac{8}{e^{2x}} + c$

41. The turning point on the curve $y = x^2 - 4x$ is at:

    (a) (2, 0)   (b) (0, 4)

    (c) $(-2, 12)$ (d) $(2, -4)$

42. Evaluating $\int_1^2 2e^{3t} dt$, correct to 4 significant figures, gives:

    (a) 2300   (b) 255.6

    (c) 766.7   (d) 282.3

43. An alternating current, i amperes, is given by $i = 100 \sin 2\pi f t$ amperes, where f is the frequency in hertz and t is the time in seconds. The rate of change of current when $t = 12$ ms and $f = 50$ Hz is:

    (a) 31 348 A/s   (b) -58.78 A/s

    (c) 627.0 A/s   (d) $-25 416$ A/s

44. A metal template is bounded by the curve $y = x^2$, the x-axis and ordinates $x = 0$ and $x = 2$. The x-co-ordinate of the centroid of the area is:

    (a) 1.0   (b) 2.0   (c) 1.5   (d) 2.5

45. If $f(t) = e^{2t} \ln 2t$, $f'(t)$ is equal to:

    (a) $\dfrac{2e^{2t}}{t}$ (b) $e^{2t}\left(\dfrac{1}{t} + 2\ln 2t\right)$

    (c) $\dfrac{e^{2t}}{2t}$ (d) $\dfrac{e^{2t}}{2t} + 2e^{2t}\ln 2t$

46. The area under a force/distance graph gives the work done. The shaded area shown between p and q in Figure M4.2 is:

    (a) $c(\ln p - \ln q)$ (b) $-\dfrac{c}{2}\left(\dfrac{1}{q^2} - \dfrac{1}{p^2}\right)$

    (c) $\dfrac{c}{2}(\ln q - \ln p)$ (d) $c\ln\dfrac{q}{p}$

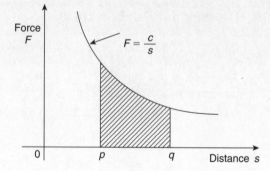

Figure M4.2

47. Evaluating $\int_0^1 \cos 2t \, dt$, correct to 3 decimal places, gives:

    (a) 0.455   (b) 0.070

    (c) 0.017   (d) 1.819

48. $\int_{-1}^{3}(3-x^2)dx$ has a value of:

(a) $3\frac{1}{3}$ (b) -8 (c) $2\frac{2}{3}$ (d) -16

49. The value of $\int_{0}^{\pi/3}16\cos^4\theta\sin\theta\,d\theta$ is:

(a) -0.1 (b) 3.1 (c) 0.1 (d) -3.1

50. $\int_{0}^{\pi/2}2\sin^3t\,dt$ is equal to:

(a) 1.33 (b) -0.25 (c) -1.33 (d) 0.25

51. The matrix product $\begin{pmatrix}2 & 3\\-1 & 4\end{pmatrix}\begin{pmatrix}1 & -5\\-2 & 6\end{pmatrix}$ is equal to:

(a) $\begin{pmatrix}-13\\26\end{pmatrix}$ (b) $\begin{pmatrix}3 & -2\\-3 & 10\end{pmatrix}$

(c) $\begin{pmatrix}-4 & 8\\-9 & 29\end{pmatrix}$ (d) $\begin{pmatrix}1 & -2\\-3 & -2\end{pmatrix}$

52. The Boolean expression $A+\overline{A}.B$ is equivalent to:

(a) A (b) B (c) $A+B$ (d) $A+\overline{A}$

53. The inverse of the matrix $\begin{pmatrix}5 & -3\\-2 & 1\end{pmatrix}$ is:

(a) $\begin{pmatrix}-5 & -3\\2 & -1\end{pmatrix}$ (b) $\begin{pmatrix}-1 & -3\\-2 & -5\end{pmatrix}$

(c) $\begin{pmatrix}-1 & 3\\2 & -5\end{pmatrix}$ (d) $\begin{pmatrix}1 & 3\\2 & 5\end{pmatrix}$

54. For the following simultaneous equations:

$$3x-4y+10=0$$
$$5y-2x=9$$

the value of x is:

(a) -2 (b) 1 (c) 2 (d) -1

55. The Boolean expression $\overline{P}.\overline{Q}+\overline{P}.Q$ is equivalent to:

(a) $\overline{P}$ (b) $\overline{Q}$ (c) P (d) Q

56. The value of $\begin{vmatrix}j2 & -(1+j)\\(1-j) & 1\end{vmatrix}$ is:

(a) $2(1+j)$ (b) 2 (c) $-j2$ (d) $2+j2$

57. The Boolean expression: $\overline{F}.\overline{G}.\overline{H}+\overline{F}.\overline{G}.H$ is equivalent to:

(a) $F.G$ (b) $F.\overline{G}$ (c) $\overline{F}.H$ (d) $\overline{F}.\overline{G}$

58. The value of the determinant $\begin{vmatrix}2 & -1 & 4\\0 & 1 & 5\\6 & 0 & -1\end{vmatrix}$ is:

(a) 4 (b) 52 (c) -56 (d) 8

59. Given $x=3t-1$ and $y=3t(t-1)$ then:

(a) $\frac{d^2y}{dx^2}=2$ (b) $\frac{dy}{dx}=3t-1$

(c) $\frac{d^2y}{dx^2}=\frac{1}{2}$ (d) $\frac{dy}{dx}=\frac{1}{2t-1}$

60. $\frac{d}{dx}(3x^2y^5)$ is equal to:

(a) $6xy^5$ (b) $3xy^4\left(2y\frac{dx}{dy}+5x\right)$

(c) $15x^2y^4+6xy^5$ (d) $3xy^4\left(5x\frac{dy}{dx}+2y\right)$

61. If $y=3x^{2x}$ then $\frac{dy}{dx}$ is equal to:

(a) $\ln 3+2x\ln x$ (b) $6x^{2x}(1+\ln x)$

(c) $2x(3x)^x$ (d) $3x^{2x}(2x\ln 3x)$

62. A solution of the differential equation $3x^2y\frac{dy}{dx}=y^2-3$ given that $x=1$ when $y=2$ is:

(a) $3\ln(y^2-3)=\ln x^2-\ln 4$

(b) $y=\dfrac{\frac{y^3}{3}-3}{x^3y}+\dfrac{13}{6}$

(c) $\frac{3}{2}\ln(y^2-3)=1-\frac{1}{x}$

(d) $2x^3=y^2-6\ln x-2$

Multiple choice questions on chapters 1–17 (page 134)

1. (b)	**14.** (d)	**27.** (a)	**39.** (b)	**51.** (a)
2. (b)	**15.** (a)	**28.** (d)	**40.** (d)	**52.** (d)
3. (c)	**16.** (a)	**29.** (b)	**41.** (c)	**53.** (d)
4. (b)	**17.** (a)	**30.** (d)	**42.** (d)	**54.** (d)
5. (a)	**18.** (d)	**31.** (a)	**43.** (b)	**55.** (d)
6. (a)	**19.** (c)	**32.** (c)	**44.** (c)	**56.** (b)
7. (a)	**20.** (d)	**33.** (d)	**45.** (d)	**57.** (d)
8. (c)	**21.** (c)	**34.** (c)	**46.** (b)	**58.** (d)
9. (c)	**22.** (a)	**35.** (a)	**47.** (c)	**59.** (b)
10. (c)	**23.** (c)	**36.** (b)	**48.** (b)	**60.** (c)
11. (a)	**24.** (b)	**37.** (c)	**49.** (c)	**61.** (b)
12. (a)	**25.** (a)	**38.** (a)	**50.** (a)	**62.** (c)
13. (a)	**26.** (c)			

Multiple choice questions on chapters 28–41 (page 375)

1. (d)	**12.** (a)	**23.** (a)	**34.** (d)	**45.** (b)
2. (b)	**13.** (d)	**24.** (c)	**35.** (c)	**46.** (d)
3. (a)	**14.** (d)	**25.** (b)	**36.** (c)	**47.** (a)
4. (d)	**15.** (a)	**26.** (c)	**37.** (a)	**48.** (d)
5. (c)	**16.** (a)	**27.** (b)	**38.** (d)	**49.** (a)
6. (d)	**17.** (c)	**28.** (d)	**39.** (b)	**50.** (c)
7. (c)	**18.** (c)	**29.** (b)	**40.** (b)	**51.** (b)
8. (d)	**19.** (b)	**30.** (b)	**41.** (c)	**52.** (c)
9. (b)	**20.** (a)	**31.** (a)	**42.** (a)	**53.** (a)
10. (b)	**21.** (b)	**32.** (a)	**43.** (c)	**54.** (b)
11. (d)	**22.** (c)	**33.** (d)	**44.** (d)	

Multiple choice questions on chapters 18–27 (page 242)

1. (d)	**13.** (c)	**25.** (a)	**37.** (d)	**49.** (c)
2. (a)	**14.** (c)	**26.** (b)	**38.** (c)	**50.** (c)
3. (b)	**15.** (d)	**27.** (a)	**39.** (b)	**51.** (c)
4. (a)	**16.** (d)	**28.** (b)	**40.** (d)	**52.** (d)
5. (b)	**17.** (d)	**29.** (d)	**41.** (d)	**53.** (b)
6. (a)	**18.** (b)	**30.** (a)	**42.** (a)	**54.** (a)
7. (c)	**19.** (d)	**31.** (b)	**43.** (d)	**55.** (b)
8. (c)	**20.** (d)	**32.** (d)	**44.** (b)	**56.** (d)
9. (c)	**21.** (b)	**33.** (d)	**45.** (b)	**57.** (b)
10. (a)	**22.** (c)	**34.** (a)	**46.** (c)	**58.** (a)
11. (b)	**23.** (a)	**35.** (b)	**47.** (b)	**59.** (a)
12. (d)	**24.** (c)	**36.** (a)	**48.** (c)	**60.** (d)

Multiple choice questions on chapters 42–63 (page 566)

1. (b)	**14.** (d)	**27.** (c)	**39.** (c)	**51.** (c)
2. (d)	**15.** (a)	**28.** (b)	**40.** (b)	**52.** (c)
3. (a)	**16.** (b)	**29.** (c)	**41.** (d)	**53.** (b)
4. (a)	**17.** (b)	**30.** (d)	**42.** (b)	**54.** (a)
5. (c)	**18.** (d)	**31.** (c)	**43.** (d)	**55.** (a)
6. (a)	**19.** (a)	**32.** (b)	**44.** (c)	**56.** (a)
7. (a)	**20.** (a)	**33.** (a)	**45.** (b)	**57.** (d)
8. (c)	**21.** (d)	**34.** (d)	**46.** (d)	**58.** (c)
9. (b)	**22.** (c)	**35.** (d)	**47.** (a)	**59.** (a)
10. (d)	**23.** (a)	**36.** (a)	**48.** (c)	**60.** (d)
11. (b)	**24.** (b)	**37.** (c)	**49.** (b)	**61.** (b)
12. (c)	**25.** (d)	**38.** (d)	**50.** (a)	**62.** (c)
13. (c)	**26.** (a)			

Index